BRITISH GEOLOGICAL SURVEY

BGS RESEARCH REPORT RR/11/01

JARÐFEINGI

JARÐFEINGI RESEARCH REPORT RR/11/01

Geology of the Faroe–Shetland Basin and adjacent areas

J D Ritchie[1], H Ziska[2], H Johnson[1], and D Evans[1] (editors)

1 British Geological Survey
2 Jarðfeingi (Faroese Earth and Energy Directorate)

Keyworth, Nottingham, UK: British Geological Survey 2011
Tórshavn, Faroe Islands: Jarðfeingi 2011

BRITISH GEOLOGICAL SURVEY

The full range of Survey publications is available from the BGS Sales Desks at Nottingham and Edinburgh; see contact details below or shop online at www.thebgs.co.uk

The London Information Office maintains a reference collection of BGS publications including maps for consultation.

The Survey publishes an annual catalogue of its maps and other publications; this catalogue is available online from www.bgs.ac.uk/catalogue.

The British Geological Survey carries out the geological survey of Great Britain and Northern Ireland (the latter as an agency service for the government of Northern Ireland), and of the surrounding continental shelf, as well as basic research projects. It also undertakes programmes of British technical aid in geology in developing countries as arranged by the Department for International Development and other agencies.

The British Geological Survey is a component body of the Natural Environment Research Council.

British Geological Survey Offices

Keyworth, Nottingham NG12 5GG
☎ 0115–936 3241 Fax 0115–936 3488
email enquiries@bgs.ac.uk

Murchison House, West Mains Road, Edinburgh EH9 3LA
☎ 0131–667 1000 Fax 0131–668 2683
e-mail: scotsales@bgs.ac.uk

BGS London, Natural History Museum, Cromwell Road, London SW7 5BD
☎ 020–7589 4090 Fax 020–7584 8270
☎ 020–7942 5344/45 e-mail: bgslondon@bgs.ac.uk

Columbus House, Greenmeadow Springs, Tongwynlais, Cardiff CF15 7NE
☎ 029–2052 1962 Fax 029–2052 1963

Geological Survey of Northern Ireland, Colby House, Stranmillis Court, Belfast BT9 5BF
☎ 028–9038 8462 Fax 028–9038 8461

Maclean Building, Crowmarsh Gifford, Wallingford Oxfordshire OX10 8BB
☎ 01491–838800 Fax 01491–692345

Parent Body

Natural Environment Research Council, Polaris House, North Star Avenue, Swindon SN2 1EU
☎ 01793–411500 Fax 01793–411501

Jarðfeingi (Faroese Earth and Energy Directorate)
Brekkutún 1, PO Box 3059,
FO 110 Tórshavn, Faroe Islands
☎ +298 357000 Fax +298 357001
e-mail: jardfeingi@jardfeingi.fo

Bibliographical reference

Ritchie, J D, Ziska, H, Johnson, H, and Evans, D (editors). 2011. Geology of the Faroe–Shetland Basin and adjacent areas. *British Geological Survey Research Report*, RR/11/01; *Jarðfeingi Research Report*, RR/11/01.

Example for individual chapters

Stoker, M, and Varming, T. 2011. Cenozoic (sedimentary). 151–208 *in* Geology of the Faroe–Shetland Basin and adjacent areas. Ritchie, J D, Ziska, H, Johnson, H, and Evans, D (editors). *British Geological Survey Research Report*, RR/11/01, *Jarðfeingi Research Report*, RR/11/01

Chapter authors:

ISBN 978 085272 643 3

Printed in the UK for the British Geological Survey by Hawthornes, Nottingham.

Contents

Figures

Tables

ACKNOWLEDGEMENTS

Oil and gas industry sponsors

We are very grateful to oil and gas industry sponsors and their representatives who not only kindly provided some financial support to help fund the compilation and production of this report, but also helped provide steerage, information and key datasets. In particular, thanks are due to:

BP	Jon Trueman, Dave Barr, Richard Corfield, Merv Davis, Susan Fowler, Trevor Ricketts, and Andrew Witt
ENI	Simon Beswetherick, Alessandro Barbera, Gianpiero Miglio and Sarah Poynter
Chevron	Dave Lewis, Dag Helland-Hansen and David Jones
ConocoPhillips	Fred Inch and Gus MacLeod
Hess	Dave Burnett
Statoil	Dave Ellis
Total	Rosa-Maria Campo, Tony Whitbread and Andy Gordon
Talisman	Colin Tannock and Mike Scotting

Data contributors

Apart from that provided from the aforementioned oil and gas industry sponsors, we are also indebted to a number of service companies, research establishments and consortia for the provision of data and images for use in the compilation of the report. In particular, we are grateful to:

Fugro Multi Client Services
Zyg Sarnowski and Tony Pedley

PGS Geophysical
Mark Martin

CGG Veritas
Steve Toothill and Richard Morgan

WesternGeco
Rupert Hoare and Phil Schearer

Western Frontiers Association (WFA)
Espen Andersen (Norsk Hydro), Dave Burnett (Hess), Dave Ellis (Statoil), Andrew Howard (ENI), Dave Lewis (Chevron), Christian-Henri Mathieu (Total), Gus MacLeod (ConocoPhillips), Oyvind Ruden (Shell), Jon Trueman (BP) and Sveinulf Vagene (ExxonMobil)

Faroes Oil Industry Group (FOIB) (formerly the Faroese Geotechnical Environmental and Meterological Group, GEM)
Ben Arabo (Hess) and Nils Sørensen (Faroe Petroleum)

Kaldbak Marine Biological Laboratory
Jan Sørenson

Department of Energy and Climate Change (DECC), formerly the Department of Business, Enterprise and Regulatory Reform (DBERR) and the UK Department of Trade and Industry (DTI).
Quentin Hugget (Geotek) on behalf of DBERR

Thanks are also due to The Geological Society of London, Blackwell, Springer Science and Business Media and AAPG publishing, IHS Energy (I Booker), The Geological Survey of Denmark and Greenland (GEUS), the Passive Margin Modelling Project (PmmP) sponsors, B Bell (University of Glasgow), J F Dewey (University of Oxford), N Grant (ConocoPhillips), R Herries (Mitra Energy), E Lamers (Shell), A Morton (HM Research Associates), T Nielsen (GEUS), A Kuijpers (GEUS), S Passey (Jarðfeingi), R W White (University of Cambridge), J Smallwood (Hess) and associated authors, who kindly allowed the use within the report of reproductions or modified reproductions from their published work (see appropriate figure caption acknowledgements in the body of the main text) and to the British Oceanographic Data Centre (BODC) who permitted the compilation and use of General Bathymetric Chart of the Oceans (GEBCO). Cambridge Paleomap Services Ltd graciously allowed the use of an image from their refit software. We are also indebted to F Gradstein et al. (2004), who permitted the use of their most recent timescale, with modifications. Finally thanks are due to R Gatliff (BGS), M Krabbendam (BGS), J Riding (BGS, palaeontology) and C Vane (BGS) in particular regarding editorial work. Diagrams were prepared by C Woodward, A Henderson, S Wild, K Henderson, S Piper, C Shearer, S Horsburgh, E Mann, K MacKenzie, L Wilson, J Wardlow and J Bow (BGS CartoGIS); design and page-setting was by A Hill and D Rayner (BGS); IPR advice was provided by G Dredge and T Gallagher (BGS). The report was edited by J Thomas (BGS) and indexed by A Jackson.

Foreword

John Ludden[1] and Petur Joensen[2]

[1] *Executive Director, British Geological Survey*

[2] *Director, Jarðfeingi (Faroese Earth and Energy Directorate)*

The production of this 'Faroe–Shetland Regional Geological Report' brings together results from a collaborative applied research project undertaken by the British Geological Survey (BGS) and Jarðfeingi (Faroese Earth and Energy Directorate) with the support of the oil and gas industry. Hopefully, it marks the beginning of future fruitful cooperative ventures. A key goal of this study was to produce a cogent geoscientific summary of the Faroe–Shetland Basin and surrounding area as a whole, rather than separate studies of the national sectors — a concept that the oil and gas industry enthusiastically supports.

Although the Faroe–Shetland region had its first hydrocarbon discovery 30 years ago, when the Clair Field was proved by BP approximately 75 km west of Shetland, it did not blossom in terms of exploration and development until the discovery of the Foinaven and Schiehallion fields by BP in 1992 and 1993, respectively. Since then, much new data has been collected and analysed by both industry and academia, the results of which are only gradually being made publicly accessible. The results of this study are, we believe, timely, as the oil and gas industry is currently undergoing a transitional phase, with a significant proportion of exploration effort now switching from the mature North Sea areas, to the north-east Atlantic margin, particularly focusing within the Faroe–Shetland Basin and adjacent areas.

This report, it is hoped, will provide an invaluable summary regarding the status of our geoscientific knowledge within the Faroe–Shetland area, particularly for the oil and gas industry in both UK and Faroese national sectors, and for academia too. Inevitably, a study of this type raises as many new geoscientific questions as it provides answers. We hope that some of these questions in this key area for future exploration will be addressed in a new era of cooperation and partnerships between national surveys, the oil industry and academics – a potent mix that should produce exciting new science and solutions that will be beneficial for all.

1 Introduction

Derek Ritchie[1] and Heri Ziska[2]
[1] *British Geological Survey*
[2] *Jarðfeingi (Faroese Earth and Energy Directorate)*

REPORT AREA DEFINITION

The Faroe–Shetland Regional Geological Report produced jointly by the BGS and Jarðfeingi provides an illustrated descriptive summary of diverse topics of geoscience organised into twelve separate chapters (see contents) within the offshore area to the north and north-west of Scotland, Orkney and Shetland Isles, and around the Faroe Islands (Figure 1). Although the UK/Faroese median line effectively separates the area into north-west and south-east portions, this national boundary has little relevance with regard to the geoscientific description of the Faroe–Shetland report area. In terms of geographical area and geoscientific content, this report partially supersedes a previous version entitled *'The geology of the Hebrides and West Shetland shelves, and adjacent deep-water area'* by Stoker et al. (1993). The eastern and southern margins of the report area utilise the same boundary as that of Stoker et al. (1993), whereas the south-west limit marked by the Wyville Thomson Ridge (Figure 1) is specifically adopted to align with the north-east boundary of a sister report prepared concurrently and focused on the Rockall Basin.

DATA

Since production of the previous Hebrides and west of Shetland regional report nearly 15 years ago (Stoker et al., 1993), vast new quantities of mainly oil and gas industry-related data have been acquired. Consequently, there have been significant advances in our understanding of the Faroe–Shetland area, particularly as a result of:

- The acquisition of high-quality 2D and 3D commercial seismic and academic deep seismic data, including surveys principally configured to focus on imaging beneath extensive Palaeogene lavas that cover a significant proportion of the report area
- The drilling and release from confidentiality of a number of new key exploration, appraisal and development wells (e.g. from Foinaven and Schiehallion fields)
- The formation of research consortia and academic study groups to investigate various geoscientific topics related to the report area, including deep seismic profiling below Palaeogene lavas, potential field modelling, geohazards, the sea-bed environment and various stratigraphical studies for example the BGS-led WFA and PmmP, GEM, SINDRI, FLARE (Faroese Large Aperture Experiment led by Hess Limited Partner Group), STRATAGEM (EC-funded Stratigraphical Development of the Glaciated European Margin) and iSIMM (integrated Seismic Imaging and Modelling of Margins)
- The publication of numerous research articles, especially within thematic conference proceedings that focus on the Faroe–Shetland area

For the purposes of this study, the following key data sources have been utilised.

Wells and boreholes

This report has made use the results from a significant number of released exploration and appraisal wells and BGS boreholes (Figure 2) and shallow sample sites from within and immediately adjacent to the Faroe–Shetland report area. A large proportion of the well data was kindly donated to the project by the oil and gas company sponsors. These data were augmented by released publicly available information from numerous sources including for example BGS publications and the DECC website (see www.og.decc.gov.uk/information/wells.htm).

Seismic

A number of 2D regional seismic profiles made available to this study courtesy of Fugro Multi Client Services, BP, WesternGeco, CCG Veritas and Chevron were used to illustrate the nature of the majority of the main structural elements and their geological evolution, particularly within the Faroe–Shetland Basin (see chapters 2, 8, 9 and 12). In shelf areas marginal to the south-east of the Faroe–Shetland Basin, where suitable seismic data were less readily available, geological cross-sections constructed from BGS shallow seismic profiles were mainly derived from Stoker et al. (1993) (see Chapter 2). Sea-bed P-wave returns gathered from a mosaic of 3D commercial seismic data were used in the compilation of a regional sea-bed image (see Figure 138) and more focused derivatives (see chapters 8 and 11) within the Faroe–Shetland Basin.

Potential field

Gravity data

The gravity image shown in Figure 3 is based on a combination of marine and land surveys, which were mostly acquired by BGS in the 1980s, and data derived from satellite altimetry (Smith and Sandwell, 1997). The gravity field includes effects due to variations in water depth and corresponding topography on the Moho, but these can be reduced by applying Bouguer and isostatic corrections (Figure 4). This improves the resolution of anomalies due to geological causes, such as the gravity lows over sedimentary basins and gravity highs over basic igneous centres and basement highs.

Magnetic data

The magnetic data compilation imaged in Figure 5 is based on airborne surveys flown by BGS, Huntings Geology and Geophysics and the US Navy (Project Mag-

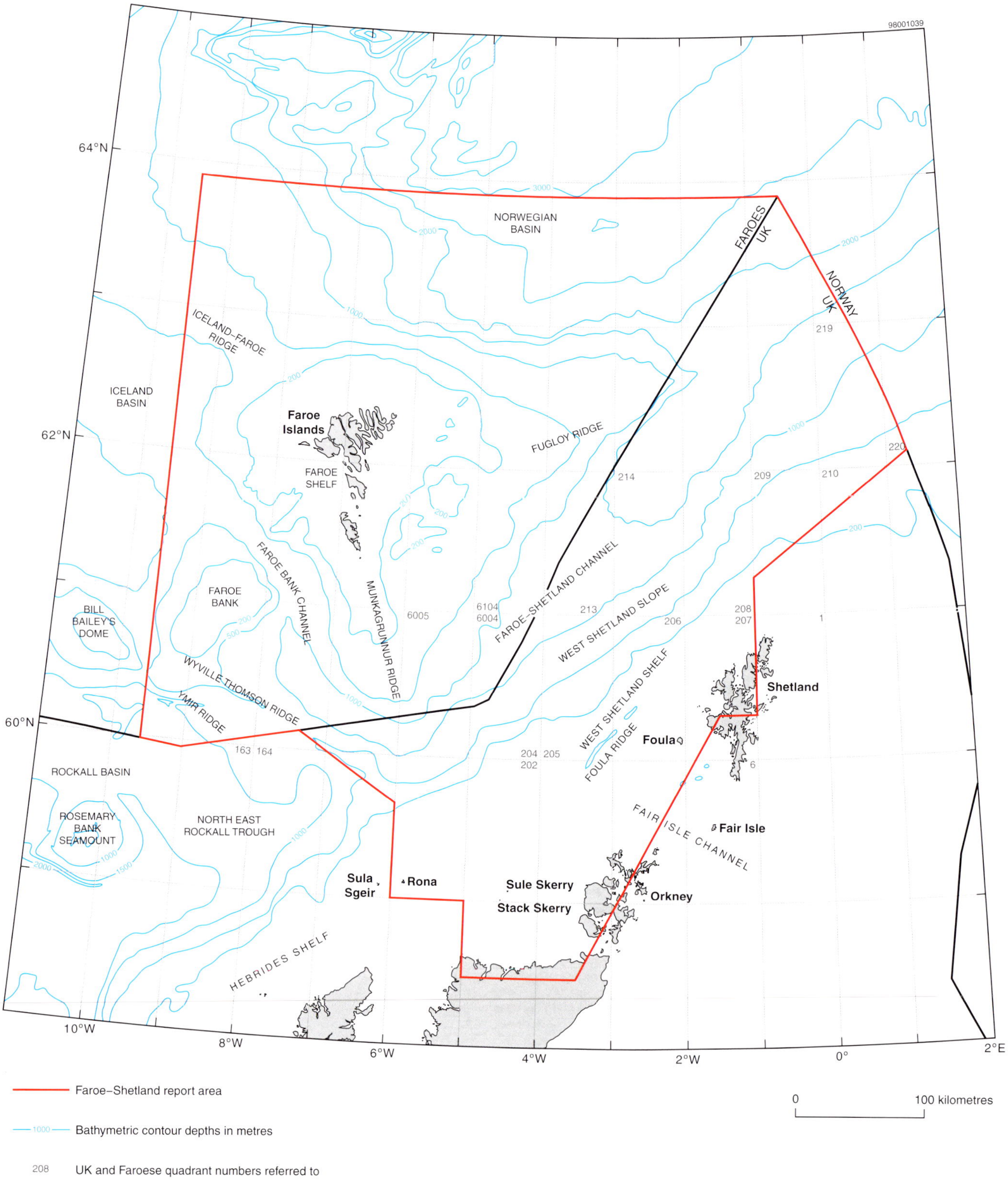

Figure 1 Outline map of Faroe–Shetland report area, including the main bathymetric features (GEBCO digital atlas bathymetric data courtesy of British Oceanographic Data Centre (BODC), NERC).

net) combined with an extract from a compilation by Verhoef et al. (1996) in the north. Strong magnetic anomalies are associated with the oceanic crust in the extreme north-west and Palaeogene igneous rocks elsewhere in the region. There is also clear evidence of magnetic anomalies associated with variations in the depth and magnetisation of the Precambrian crystalline basement.

Bathymetry

GEBCO bathymetric data were licensed from the BODC for the purposes of producing a low-resolution contour map of the Faroe–Shetland report area (Figure 1) and as a backdrop for other images (see chapters 8, 10 and 11).

Figure 2 Simplified structural elements map of the Faroe–Shetland report area showing the location and identifiers of commercial wells and BGS boreholes. For detailed structural elements map, see Figure 7.

Published literature

Published literature was extensively utilised in the compilation of this report, with approximately 700 references consulted. A selection of key publications that cover a diverse range of geoscientific topics within the Faroe–Shetland area includes, Duindam and van Hoorn (1987), Ziegler (1990), Boldreel and Andersen (1993),

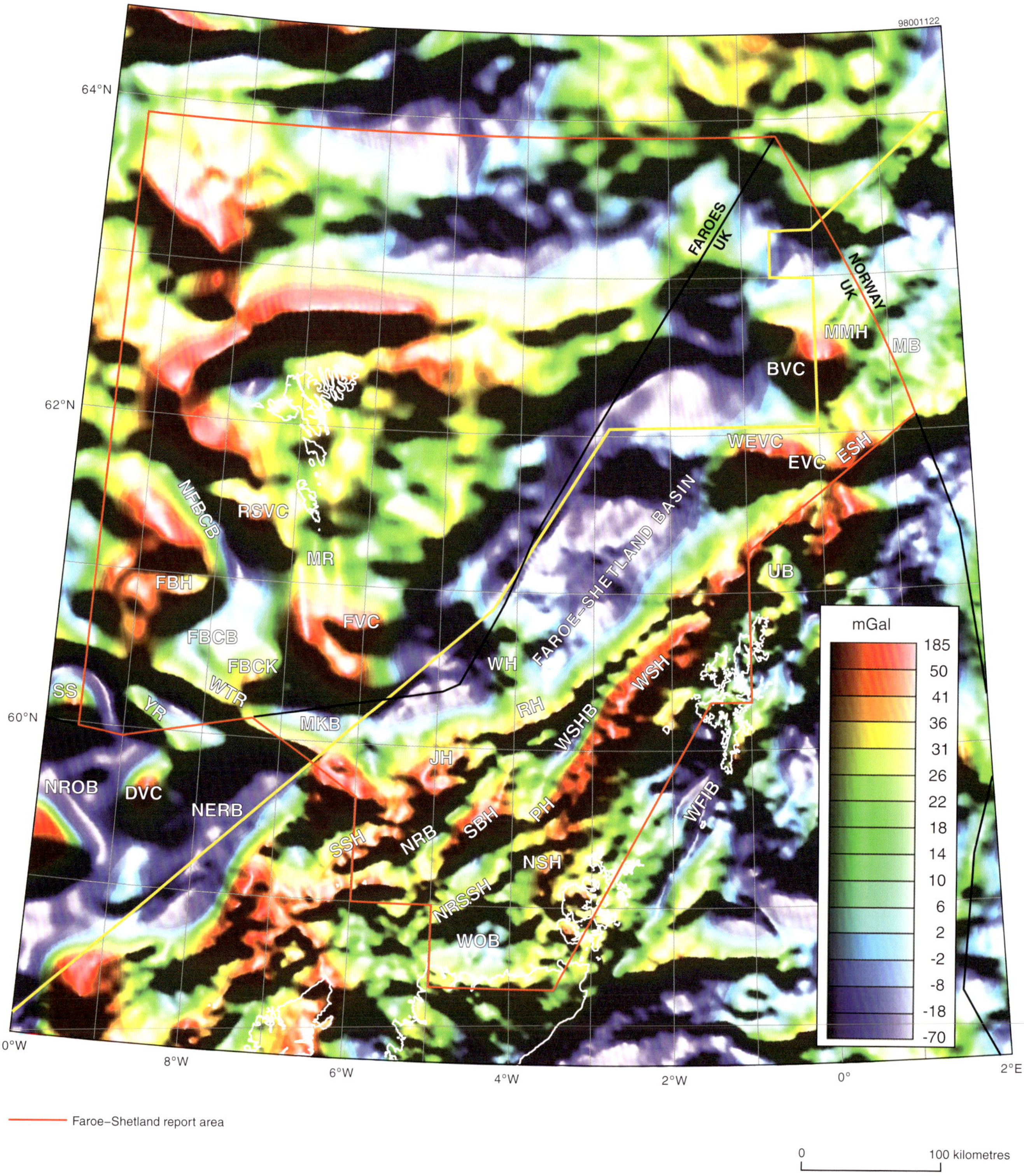

Figure 3 Image of gravity data over the Faroe–Shetland region, based on free-air gravity anomalies offshore and Bouguer anomalies onshore. Colour shaded-relief image with illumination from the north. The thick yellow line indicates the boundary between satellite-derived data to the north-west and marine data to the south-east. Abbreviations: see Figure 7.

Ebdon et al. (1995), Mitchell et al. (1993), Rumph et al. (1993), Stoker et al. (1993), Ritchie et al. (1996), Knox et al. (1997), Waagstein (1988), Cooper et al. (1999), Dean et al. (1999), Doré et al. (1999), Herries et al. (1999), Lamers and Carmichael (1999), Leach et al. (1999), Ritchie et al. (1999a), Roberts et al. (1999), Faroes GEM Network (2001a), Davies and Cartwright (2002), Ellis et al. (2002), Smallwood and Gill (2002), Coward et al. (2003), Keser Neish (2003), White et al. (2003), Long et al. (2004), England et al. (2005), Johnson et al. (2005b), Jolley et al. (2005), Keser Neish and Ziska (2005), Kimbell et al. (2005), Loizou (2005), Raum et al. (2005), Scotchman and Carr (2005), Smallwood and Kirk (2005), Stoker et al. (2005a, b and c), Passey and Jolley (2009) and the British Geological Survey (BGS) 1:250 000 map series.

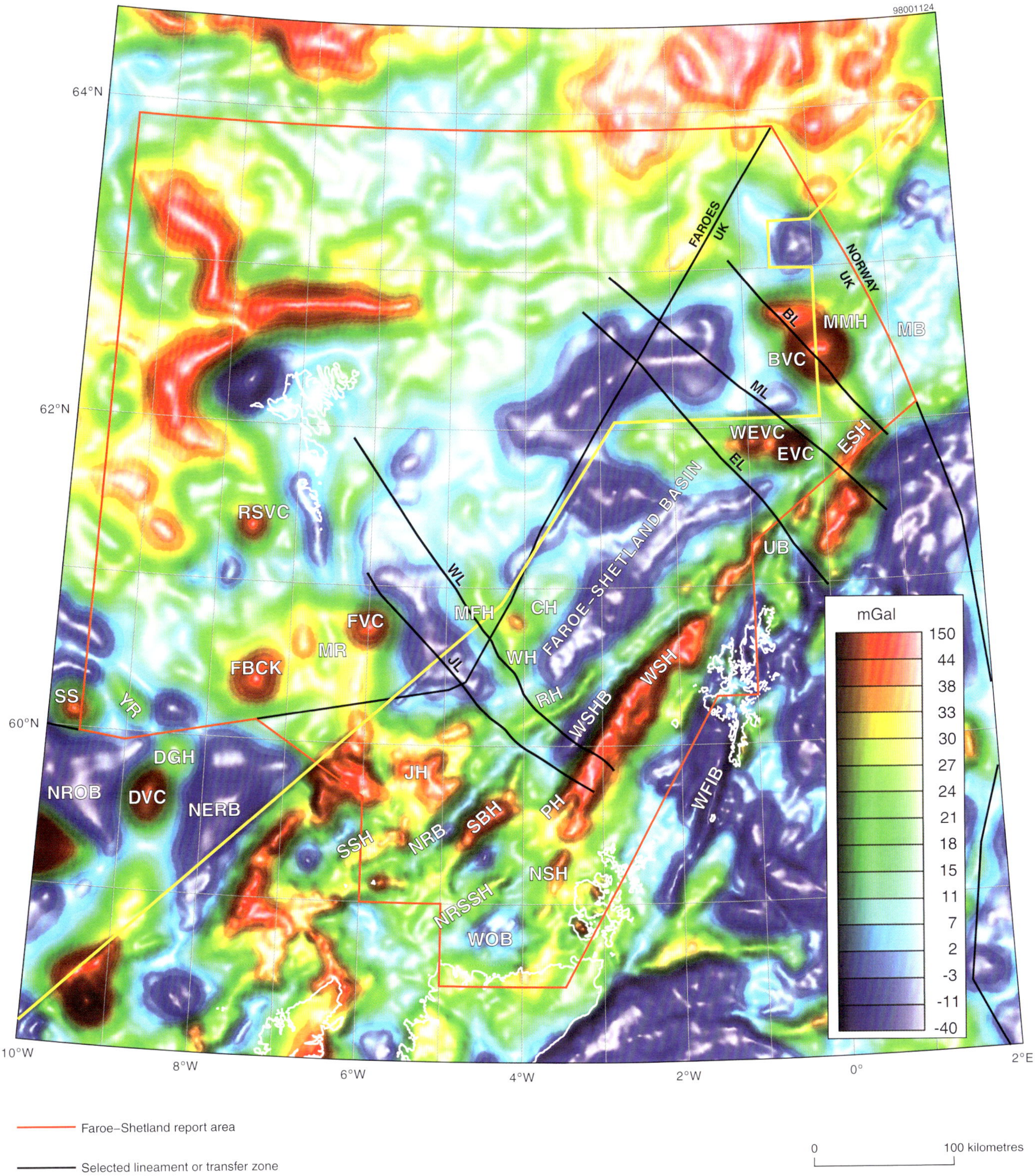

Figure 4 Isostatically corrected Bouguer gravity anomalies over the Faroe–Shetland region. Colour shaded-relief image with vertical illumination. The thick yellow line indicates the boundary between satellite-derived data to the north-west and marine data to the south-east. The bathymetry used in estimating the gravity corrections was based on the model of Smith and Sandwell, (1997), which employs a combination of ship soundings and satellite altimetry. Abbreviations: see Figure 7.

Timescale

Subdivision of the Phanerozoic timescale, including the nomenclature and the absolute ages that define the stage names, is constantly under review. Unless stated otherwise, the timescale and terminology adopted for this report are based on Gradstein et al. (2004), but with some minor amendments (Figure 6).

HISTORY OF FAROE–SHETLAND RESEARCH

Within the Faroe–Shetland report area, aspects of the onshore geology of the Shetland (e.g. Flinn, 1956), Orkney (e.g. Mykura, 1976) and the Faroe Islands (e.g. Rasmussen et al., 1956; Rasmussen and Noe-Nygaard, 1970) have been studied over many decades. In the offshore area, early pioneering work by academia

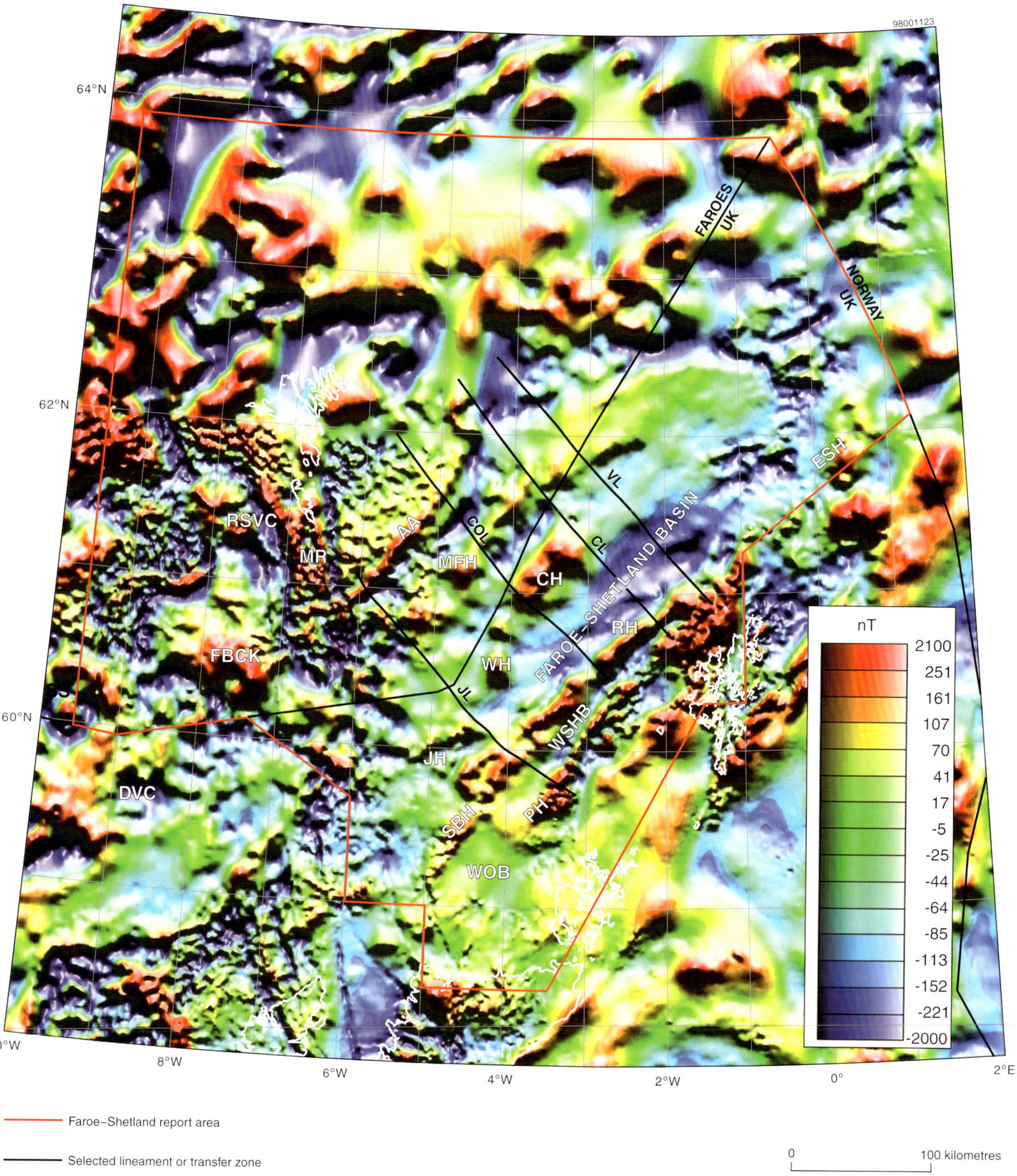

Figure 5 Image of magnetic data over the Faroe–Shetland region. Total-field magnetic anomalies displayed as colour shaded relief with illumination from the north. Abbreviations: see Figure 7.

focused on the general definition of basins and highs, igneous centres and nature of the underlying crust, principally from a combination of potential field data (e.g. Bott and Watts, 1971; Himsworth, 1973) and deep seismic refraction experimental data within the Faroe–Shetland and West Shetland basins (e.g. Bott and Smith, 1984), Faroe Platform (e.g. Bott et al., 1974), Iceland–Faroe Ridge (e.g. Bott et al., 1976) and the Faroe Bank Channel Basin (Bott and Stacey, 1967) (Figure 7). Subsequently, the deep structure of the Faroe–Shetland region has been investigated by a significant number of deep seismic experiments, sometimes coupled with potential field modelling. These studies have provided more, though sometimes conflicting information, regarding for example, depth to Moho, thickness of crystalline crust, Palaeogene lavas etc (e.g. see Hughes et al., 1997; Smallwood et al., 2001; White et al., 2003; England et al., 2005; Raum et al., 2005). Largely driven by the needs of the oil and gas industry, some geophysical studies have been

Era	Sub-Era	System/Period	Series/Epoch		Stage	Age (Ma)
Cenozoic	*Quaternary	Neogene	Holocene			0.01
			Pleistocene		Late	0.13
					Middle	0.78
					Early	1.8
	** *Tertiary*		Pliocene	L	Gelasian	2.6
					Piacenzian	3.6
				E	Zanclean	5.3
			Miocene	L	Messinian	7.2
					Tortonian	11.6
				M	Serravallian	13.6
					Langhian	16.0
				E	Burdigalian	20.4
					Aquitanian	23.0
		Palaeogene	Oligocene	L	Chattian	28.4
				E	Rupelian	33.9
			Eocene	L	Priabonian	37.2
				M	Bartonian	40.4
					Lutetian	48.6
				E	Ypresian	55.8
			Paleocene	L	Thanetian	58.7
				M	Selandian	61.7
				E	Danian	65.5
Mesozoic		Cretaceous	Late		Maastrichtian	70.6
					Campanian	83.5
					Santonian	85.8
					Coniacian	89.3
					Turonian	93.5
					Cenomanian	99.6
			Early		Albian	112.0
					Aptian	125.0
					Barremian	130.0
					Hauterivian	136.4
					Valanginian	140.2
					Berriasian	145.5
		Jurassic	Late		Tithonian	150.8
					Kimmeridgian	155.7
					Oxfordian	161.2
			Middle		Callovian	164.7
					Bathonian	167.7
					Bajocian	171.6
					Aalenian	175.6
			Early		Toarcian	183.0
					Pliensbachian	189.6
					Sinemurian	196.5
					Hettangian	199.6
		Triassic	Late		Rhaetian	203.6
					Norian	216.5
					Carnian	228.0
			Middle		Ladinian	237.0
					Anisian	245.0
			Early		Olenekian	249.7
					Induan	251.0

98001173

Era	System/Period	Series/Epoch		Stage	Regional	Sub-stage	Age (Ma)
							251.0
Palaeozoic	Permian	Lopingian		Changhsingian			253.8
				Wuchiapingian			260.4
		Guadalupian		Capitanian			265.0
				Wordian			268.0
				Roadian			270.6
		Cisuralian		Kungurian			275.6
				Artinskian			284.4
				Sakmarian			294.6
				Asselian			299.0
	Carboniferous	Pennsylvanian	Late	Gzhelian	*Stephan-ian		303.9
				Kasimovian			306.5
			Middle	Moscovian	*Westphalian	*Westphalian D*	
						Bolsovian	
						Duckmantian	311.7
			Early	Bashkirian		*Langsettian*	
					*Namurian	*Yeadonian*	
						Marsdenian	
						Kinderscoutian	
						Alportian	
						Chokierian	318.1
		Mississippian	Late	Serpuk-hovian		*Arnsbergian*	
						Pendleian	326.4
			Middle	Visean		*Brigantian*	
						Asbian	
						Holkerian	
						Arundian	
						Chadian	345.3
			Early	Tournaisian		*Courceyan*	359.2
	Devonian	Late		Famennian			374.5
				Frasnian			385.3
		Middle		Givetian			391.8
				Eifelian			397.5
		Early		Emsian			407.0
				Pragian			411.2
				Lochkovian			416.0
	Silurian	Pridoli					418.7
		Ludlow		Ludfordian			421.3
				Gorstian			422.9
		Wenlock		Homerian			426.2
				Sheinwoodian			428.2
		Llandovery		Telychian			436.0
				Aeronian			439.0
				Rhuddanian			443.7
	Ordovician	Late		Hirnantian		*Ashgill*	445.6
				Katian			455.8
				Sandbian		*Caradoc*	460.9
		Middle		Darriwilian		*Llanvirn*	468.1
				Stage 3		*Arenig*	471.8
		Early		Floian			478.6
				Tremadocian		*Tremadoc*	488.3
	Cambrian	Furongian		Stage 10			~492.0
				Stage 9			~496.0
				Paibian			501.0
		Series 3		Stage 7			~503.0
				Stage 6			~506.5
				Stage 5			~510.0
		Series 2		Stage 4			~517.0
				Stage 3			~521.0
		Series 1		Stage 2			~534.6
				Stage 1			542.0

NOTES:
E : Early; M : Middle; L : Late
Italicised terms e.g. *Westphalian, Ashgill* indicate NW European Regional Standards
*Quaternary: formal chronostratigraphical unit (ICS-INQUA task force, 2005 and ICS.
** *Tertiary*: informal chronostratigraphical unit sensu Aubry *et al* (2005, Episodes 28/2).
*Namurian, Westphalian and Stephanian are Regional Stages (Epochs)
Some Ordovician and Cambrian series/stages are to be ratified.

Figure 6 Timescale utilised in this report is based on 'A Geologic Time Scale 2004' by F M Gradstein, J G Ogg, A G Smith, et al. (2004) and the International Stratigraphical Chart, 2006 (ICS) with additions.

specifically targeted at imaging sub-Palaeogene lava structural levels.

Between the early 1970s and late 1980s, the offshore BGS mapping programme (funded by the former UK Department of Energy) produced a series of 1:250 000 scale maps that are based on single channel seismic data, calibrated by shallow drilling and sampling at the sea bed, mainly within the shallow water regions in the south-east part of the report area. In the quest for hydrocarbons, the first deep commercial well 206/12-1 was drilled in shallow water by Exxon approximately 75 km to the north-west of Shetland in 1972, proving Cenozoic and Upper Cretaceous strata resting on Precambrian crystalline basement on the Rona High (Figures 2 and 7). Since this time, the research history of the area has largely been led by the oil and gas industry. The acquisition of 2D commercial seismic surveys, potential field data and the drilling of new exploration wells provided the early impetus for publications principally focused on the elucidation of the structure, geology and hydrocarbon potential of the south-east flank of the Faroe–Shetland Basin (e.g. Ridd, 1983; Hitchen and Ritchie, 1987; Mudge and Rashid, 1987; Duindam and van Hoorn, 1987; Bailey et al., 1987; Rumph et al., 1993), and within the marginal basins to the south-east including the West Shetland Basin (e.g. Ridd, 1981; Ritchie and Darbyshire, 1984; Haszeldine et al., 1987; Blackbourn, 1987; Allen and Mange-Rajetsky, 1992), Unst Basin (Johns and Andrews, 1985), North Rona Basin (Kirton and Hitchen, 1987), West Orkney Basin (e.g. Enfield and Coward, 1987) and the Solan basins (Booth et al., 1993) (Figure 7). Early attempts at paleogeographical and plate tectonic reconstruction in the Faroe–Shetland area, set within the wider context of the north-west European margin were produced by Ziegler (1982 and 1990) and Knott et al. (1993). The results of this phase of research were summarised in the BGS regional report for the Hebrides and west of Shetland area (Stoker et al., 1993).

From the early 1980s onwards, significant interest also emerged in a small number of thematic topics within the Faroe–Shetland and adjacent areas that have continued until the present day. For example, its magmatic development during Palaeogene times, including its relationship with the Iceland Plume (e.g. White, 1988), the effects of underplating (e.g. Brodie and White 1995; Clift, 1999), the description of a number of offshore igneous complexes (e.g. Smythe, 1983; Gatliff et al., 1984), the distribution and radiometric age of sills and volcanic rocks recovered from core material (e.g. Hitchen and Ritchie, 1987; Gibb and Kanaris-Sotiriou, 1988; Hitchen and Ritchie, 1993; Ritchie and Hitchen, 1996; Ritchie et al., 1999a; Waagstein et al., 2002) and regional biostratigraphical correlations of volcanic intervals (including onshore areas) based on microfossil evidence (e.g. Jolley and Bell, 2002a; Jolley et al., 2005). There have also been many studies of the evolution of the Neogene to Recent succession (e.g. STRATAGEM partners, 2002 and 2003; Stoker et al., 2005a and c). Of increasing importance is research into the causes and effects of Cenozoic compression, regional uplift and denudation set within both local (e.g. Andersen et al., 2002; Ritchie et al., 2003; Davis et al., 2004) and regional contexts (e.g. Boldreel and Andersen, 1993 and 1998; Lundin and Doré, 2002; Stoker et al., 2005b; Johnson et al., 2005b).

Following discovery in the early 1990s of the Foinaven and Schiehallion oilfields (see Figure 150) in Mid Paleocene (T31-T35) sands (e.g. Leach et al., 1999; Cooper et al., 1999) within the Judd Sub-basin (Figure 7), the focus of exploration activity shifted firmly to the Paleocene sedimentary succession (e.g. Smallwood et al., 2004; Smallwood, 2005a), including the areas beneath the regional flood basalts observed on the Faroe Islands and mapped offshore within the north-east part of the Faroe–Shetland Basin area (see Figure 120). Increasing numbers of 3D seismic surveys and new exploration wells significantly improved the understanding of the Cenozoic interval as a whole. As predicted by Ritchie et al., 1999b, the acquisition of the 3D seismic datasets proved pivotal in allowing the study of magmatic-related mechanisms and processes such as sill-sediment complex development and propagation; a topic which had largely been in stasis since the early 1970s. In areas covered by thick Palaeogene lavas, new commercial seismic imaging techniques e.g. low-frequency source (e.g. see Ziolkowski et al., 2003) and oil and gas industry/government research-funded deep seismic experimental data focused on the resolution of the sub-basalt structure of the north-west Faroe–Shetland Basin and adjacent areas, for example FLARE (White et al., 1999 and 2003) and iSIMM (e.g. White et al., 2005).

In the late 1990s, publication of a series of articles within an Atlantic margin theme in the *'Proceedings of the 5th Conference on the Petroleum Geology of north-west Europe'* by Fleet and Boldy (1999) appeared to signal a 'high water mark' in terms of exploration interest and activity in the Faroe–Shetland area. In particular, there were key regional syntheses on the palaeogeographical and tectonostratigraphical evolution of the Faroe–Shetland area within the context of the north-east Atlantic margin as a whole (e.g. Roberts et al., 1999; Doré et al., 1999; Spencer et al., 1999; Dean et al., 1999) and on topics such as Palaeogene magmatic development (e.g. Ritchie et al., 1999a; Naylor et al., 1999; Kiørboe, 1999), aspects of the Paleocene hydrocarbon 'play' (Lamers and Carmichael, 1999; Sullivan et al., 1999; Jowitt et al., 1999), status of field development (e.g. Leach et al., 1999; Cooper et al., 1999; Goodchild et al., 1999; Herries et al., 1999) and hydrocarbon systems (Holmes et al., 1999; Iliffe et al., 1999; Jowitt et al., 1999; Parnell et al., 1999).

In the new millennium, exploration wells have been drilled to Paleocene levels in the Judd Sub-basin and Sjúrður Ridge within the Faroese national sector (Figure 7). The publication of articles within the *'Faroe Islands Exploration Conference: Proceedings of the 1st Conference'* by Ziska et al. (2005) has marked the beginning of a new phase of research focused principally, on sub-basalt imaging challenges mainly within the Faroese national sector. At the time of writing, a second conference has been held and the results are just published (Varming and Ziska, 2009). Within the UK sector, drilling within the axial part of the Faroe–Shetland Basin on the Corona High from 2000 onwards may have signalled a partial shift in exploration focus to Mesozoic and older exploration targets, though the Paleocene section is still the prime target. The discovery in 2002 of the Cambo Paleocene hydrocarbon accumulation in block 204/10-1 and in 2004 of the Rosebank/Lochnagar hydrocarbon accumulation in Block 213/27 within the Palaeogene and pre-Cretaceous succession (Loizou, 2005; Helland-Hansen, 2006; 2009) are potentially crucial for future hydrocarbon exploration within the Faroe–Shetland Basin. At the time of writing, the Rosebank/Lochnagar discovery is under active drilling appraisal but should it prove to be a significant find, it could act as a catalyst for a new and exiting phase of geoscientific development within the Faroe–Shetland region.

2 Structure

Derek Ritchie[1], Heri Ziska[2], Geoff Kimbell[1], Martyn Quinn[1] and Andy Chadwick[1]
[1] *British Geological Survey*
[2] *Jarðfeingi (Faroese Earth and Energy Directorate)*

PLATE SETTING AND TECTONIC EVOLUTION

Within the context of the north-east Atlantic margin, the general plate setting and tectonic evolution of the Faroe–Shetland report area (Figure 7) is fairly well documented (e.g. Ziegler, 1990; Knott et al., 1993, Doré et al., 1999; Roberts et al., 1999; Coward et al., 2003). Following break-up of the supercontinent Rodinia in late Precambrian times, basement terranes began assembling during Proterozoic times, culminating at the end of the Caledonian Orogeny during the Early Devonian (Figure 8; e.g. McKerrow et al., 2000; Oliver, 2002; Strachan et al., 2002 and references therein). Following the close of the Caledonian Orogeny, the north-east Atlantic margin was influenced by episodes of rifting during Late Palaeozoic to Mesozoic times (e.g. Doré et al., 1999; Roberts et al., 1999; Tate et al., 1999). Within the Faroe–Shetland region, this extensional tectonic activity occurred mainly during Mid to Late Devonian, Permo-Triassic, Jurassic, Cretaceous and locally, Early to Mid Paleocene times (Figure 8; e.g. Dean et al., 1999; Smallwood and Gill, 2002; Coward et al., 2003). Extension initially focused on the north-east-trending Caledonian structural grain prevalent throughout basins of the north-east Atlantic Margin. According to Doré et al. (1997 and 1999) and Spencer et al. (1999), the locus of rift activity migrated in a north-westerly direction through time, towards the present location of the north-east Atlantic Ocean. For example, in the Faroe–Shetland area, Permo-Triassic half-graben development is mainly concentrated in 'marginal' basins such as the West Orkney and south-west West Shetland basins that occur to the west of the Orkney and Shetland islands (Figure 7; e.g. Hitchen et al., 1995b; Earle et al., 1989). In contrast, the contiguous Rockall, Faroe–Shetland and Møre basins mainly formed further to the west in response to Cretaceous extension (e.g. Duindam and van Hoorn, 1987; Dean et al., 1999; Musgrove and Mitchener, 1996). Considerable uncertainty remains regarding the timing, duration and significance of many of the postulated rift events that influenced the development of the central and north-west parts of the Faroe–Shetland Basin in particular. This is due to a combination of a lack of adequate deep well control and the poor seismic definition of the deep structure of the basin, with the latter caused by the high acoustic impedance contrasts associated with the widespread Palaeogene volcanic and intrusive rocks (see Figure 120). During earliest Eocene times, renewed rifting, possibly facilitated by the presence of the Iceland Plume (e.g. White, 1988), resulted in continental break-up between north-west Europe and Greenland, leading to the formation of the present day north-east Atlantic Ocean. The break-up process is linked with widespread mainly continental volcanism and intrusive activity during latest Paleocene to earliest Eocene times (see Chapter 9). Following separation, post-rift subsidence was the dominant tectonic process influencing the development of the north-east Atlantic rift basins as a whole, though this was interrupted by significant phases of compressional deformation, uplift and shelf/basin margin tilting (Figure 8).

Precambrian basement

Prior to the Caledonian Orogeny, the Precambrian crystalline basement of the north-west European Atlantic margin lay within three main super continents, namely Laurentia, Baltica and Eastern Avalonia (Figure 9; e.g. Oliver, 2002). Approximately 430 to 410 Ma ago, collision between Laurentia and Baltica finally closed the northern Iapetus Ocean during the Scandian phase of the Caledonian Orogeny (e.g. Strachan et al., 2002). At approximately the same time, Eastern Avalonia, a magmatic arc that formed at the margin of the southern continent of Gondwana (Pharaoh, 1999), collided with Laurentia and Baltica following subduction of the southern Iapetus Ocean and the Tornquist Sea. The Faroe–Shetland region occurs within the eastern part of the Laurentia Terrane (Figure 9) and comprises mainly late Archaean metamorphic basement that ranges in age between approximately 2800 and 2700 Ma but also a younger Proterozoic component (see Chapter 3; Chambers et al., 2005). In the offshore area immediately to the south-west of the Faroe–Shetland region, the nature of the crystalline basement within the Laurentia Terrane is very different, with juvenile 1750 Ma Palaeoproterozoic crust present on the Rockall and Stanton highs that flank the western and eastern margins of the Rockall Basin, respectively (Chambers et al., 2005).

Devonian and Carboniferous

Following the end of the Caledonian Orogenic cycle about 410 to 400 Ma ago (Strachan et al., 2002; Oliver, 2002; Coward et al., 2003), Mid to Late Devonian terrestrial intermontane basins formed within the orogenic belt, possibly by a combination of gravitational collapse of the thickened crust (McClay et al., 1986) and strike-slip tectonics (e.g. Roberts et al., 1999; Soper and Woodcock, 2003). Within the north-east Atlantic area, Early to Mid Devonian basin development appears restricted to occurrences within the East Greenland margin, around the south-west Norwegian coast and in a belt of north-east-trending marginal basins running from the Porcupine Basin in the south-west to the Clair Basin in the north-east (see Figure 47; Ziegler, 1990). Within the Faroe–Shetland region, Middle to Upper Devonian continental red bed development in the Clair Basin (see Chapter 4) was terminated by a early Carboniferous (Visean) marine transgression (Meadows et al., 1987). This is cited by some as evidence that the Faroe–Shetland and Rockall basins had subsided by the later part of early Carboniferous times, and that there was a marine connection with the Rheic Ocean to the south (Haszeldine and Russell, 1987). Lower (Visean) to Upper (Namurian) Carboniferous sediments have

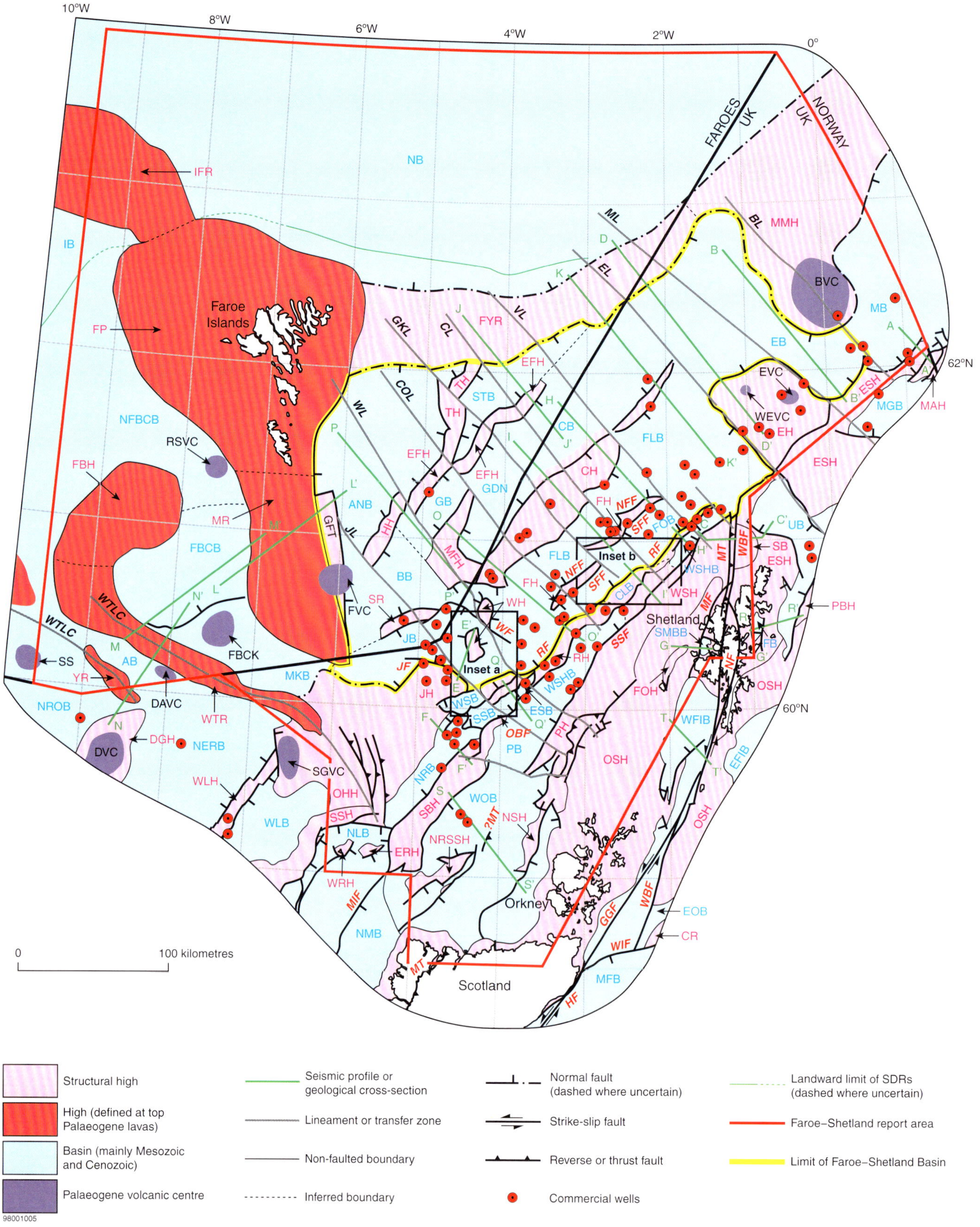

Figure 7 Structural elements of the Faroe–Shetland report area (mainly modified from Duindam and van Hoorn, 1987; Hitchen and Ritchie, 1987; Rumph et al., 1993; Stoker et al., 1993; Blystad et al., 1995; BGS, 1996; Dean et al., 1999; Herries et al., 1999, Lamers and Carmichael, 1999; Roberts et al., 1999; Ellis et al., 2002; Keser Neish, 2003; Kimbell et al., 2004; Smallwood et al., 2004; Kimbell et al., 2005; Smallwood and Kirk, 2005 and from commercial seismic data supplied by BP, Fugro Multi Client Services and CGG Veritas). For commercial well identifiers and location of BGS boreholes, see Figure 2.

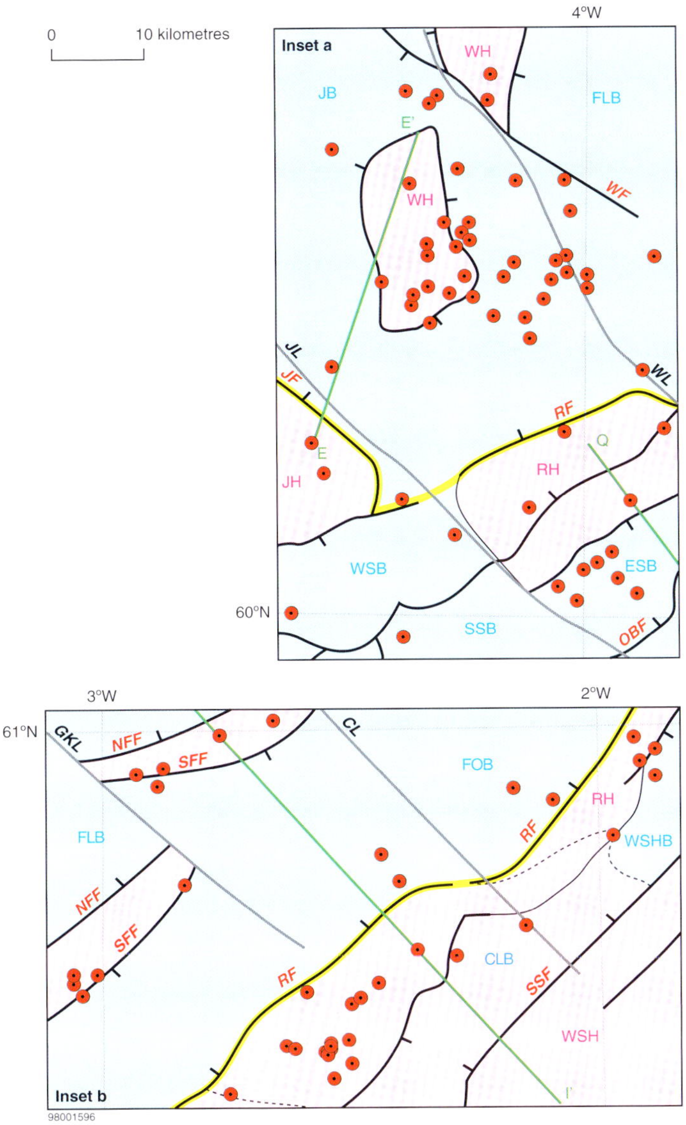

AB Auðhumla Basin
ANB Annika Sub-basin
BB Brynhild Sub-basin
BL Brendan Lineament
BVC Brendan Volcanic Centre
CB Corona Sub-basin
CH Corona High
CL Clair Lineament
CLB Clair Basin
COL Corona Lineament
CR Caithness Ridge
DAVC Drekaeygja Volcanic Centre
DGH Darwin–Geikie High
DVC Darwin Volcanic Centre
EB Erlend Sub-basin
EFH East Faroe High
EFIB East Fair Isle Basin
EH Erlend High
EL Erlend Lineament
EOB East Orkney Basin
ERH East Rona High
ESB East Solan Basin
ESH East Sheland High
EVC Erlend Volcanic Centre
FB Fetlar Basin
FBCB Faroe Bank Channel Basin
FBCK Faroe Bank Channel Knoll
FBH Faroe Bank High
FH Flett High
FLB Flett Sub-basin
FOB Foula Sub-basin
FOH Foula High
FP Faroe Platform
FVC Frænir Volcanic Centre
FYR Fugloy Ridge
GB Grimhild Sub-basin
GDN Guðrun Sub-basin
GFT Grani Fault Terrace
GGF Great Glen Fault
GKL Grímur Kamban Lineament
HF Helmsdale Fault
HH Heri High
IB Iceland Basin
IFR Iceland–Faroe Ridge
JB Judd Sub-basin
JF Judd Fault
JH Judd High
JL Judd Lineament
MAH Manet High
MB Møre Basin
MF Melby Fault
MFB Moray Forth Basin
MFH Mid Faroe High
MGB Magnus Basin
MIF Minch Fault
MKB Munkur Basin
ML Magnus Lineament
MMH Møre Marginal High
MR Munkagrunnur Ridge
MT Moine Thrust
NB Norwegian Basin
NERB NE Rockall Basin
NF Nesting Fault
NFBCB North Faroe Bank Channel Basin
NFF North Flett Fault
NLB North Lewis Basin
NMB North Minch Basin
NRB North Rona Basin
NROB North Rockall Basin
NRSSH Nun Rock–Sule Skerry High
NSH North Shoal High
OBF Otter Bank Fault
OHH Outer Hebrides High
OSH Orkney–Shetland High
PB Papa Basin
PBH Pobie High
PH Papa High
RF Rona Fault
RH Rona High
RSVC Regin Smiður Volcanic Centre
SB Sandwick Basin
SBH Solan Bank High
SFF South Flett Fault
SGVC Sula Sgeir Volcanic Centre
SMBB St Magnus Bay Basin
SR Sjúrður Ridge
SS Sigmundur Seamount
SSB South Solan Basin
SSF Shetland Spine Fault
SSH Sula Sgeir High
STB Steinvør Sub-basin
TH Tróndur High
UB Unst Basin
VL Victory Lineament
WBF Walls Boundary Fault
WEVC West Erlend Volcanic Centre
WF Westray Fault
WFIB West Fair Isle Basin
WH Westray High
WIF Wick Fault
WL Westray Lineament
WLB West Lewis Basin
WLH West Lewis High
WOB West Orkney Basin
WRH West Rona High
WSB West Solan Basin
WSH West Shetland High
WSHB West Shetland Basin
WTLC Wyville Thomson Lineament Complex
WTR Wyville Thomson Ridge
YR Ymir Ridge

Figure 7 *continued.*

been drilled on the Corona High (see Figure 52) within the central part of the Faroe–Shetland Basin, though according to Coward et al. (2003, figs. 2.8 and 2.9) and Ziegler (1990, enclosures 14 and 15), the high is located in an emergent area, far removed from other proven occurrences of Carboniferous strata. To the south-west of the report area, Upper Palaeozoic sediments have been drilled in the Porcupine, Slyne, Erris, Clair and Donegal basins, and have been predicted within the Rockall Basin (Naylor et al., 1999).

Permian and Triassic

In late Carboniferous to early Permian times, the Variscan Orogeny (also known as the Hercynian Orogeny) resulted from the collision of Laurussia and Gondwana and the formation of the Pangean supercontinent (see Chapter 5; Figure 10b). Glennie (2002) suggested that this mountain building episode was superseded by east–west tension across the north-west European margin in general. There was significant igneous activity in early Permian times, stretching in a generally north-west-trending belt from Poland–Denmark–Northern Permian Basin to the Faroe–Shetland region, where locally, volcanic rocks have been described from the Papa and Møre basins (Figures 7 and 10a; Hitchen et al., 1995b; Ritchie et al., 1996; Glennie, 2002, fig. 10.3). Towards the end of early Permian times, an incipient north-east-trending 'Arctic' rift system developed southwards between Greenland and Norway to western Ireland (Roberts et al., 1999; Coward et al., 2003), utilising the existing north-east-trending Caledonian structural grain (Figure 10a). This extensional activity was responsible for the formation of half-grabens or basins stretching from the Rockall Basin in the south-west to the Nordcapp Basin off the north-west coast of Norway (e.g. Ziegler, 1990; Doré et al., 1999, fig. 1; Coward et al., 2003, figs. 2.10 and 2.11). According to Torsvik et al. (1996), the main phase of extensional activity continued to at least Mid Triassic times (Figures 8 and 11), with nascent north-east-trending Early Triassic extensional faults particularly common throughout the north-west European margin (Coward et al., 2003). Within the Faroe–Shetland Basin however, Permo-Triassic rifting it is not thought to represent a major tectonic phase in the basin evolution (e.g. Duindam and van Hoorn, 1987; Glennie et al., 2003). Most extensional activity is restricted to the south-east within the

West Orkney, East Solan, south-west West Shetland and Unst basins (Figures 7 and 11). Here, vast thicknesses of mainly continental facies strata were deposited during late Permian to Mid Triassic times. During Late Triassic times, the influence of the Arctic rift system waned significantly (Roberts et al., 1999), with the northward propagation of the 'Atlantic' rift system becoming more dominant (Figure 11; Coward et al., 2003).

Jurassic

The Arctic rift system was relatively inactive in Early Jurassic times, although further to the south, latest Triassic to Early Jurassic rifting related to the opening of the Central North Atlantic is considered to have propagated northward through the basins off Nova Scotia and Newfoundland and into the Celtic Sea (Roberts et al., 1999, fig. 9). In the Faroe–Shetland region, isolated occurrences of Lower Jurassic strata have been drilled within the Faroe–Shetland and Solan basins (see Chapter 6; Ritchie et al., 1996). To the south-west, Jurassic rocks are widespread in the Sea of Hebrides–Little Minch, North Lewis, Donegal, North Porcupine, Slyne and Erris basins (Morton, 1989; Fyfe et al., 1993; Stoker et al., 1993; Dancer et al., 1999; Chapman et al., 1999; Naylor et al., 1999).

Sea-floor spreading commenced south of the Azores–Gibraltar Fracture Zone in the Central Atlantic Ocean during the Mid Jurassic (Aalenian) (Srivastava and Tapscott, 1986; Knott et al., 1993). The Arctic rift was rejuvenated, with extension documented from the North Sea, Norwegian Sea (Roberts et al., 1999, fig. 11) and even from within the Faroe–Shetland Basin too (Haszeldine et al., 1987). Others however, infer a period of uplift and erosion in the Faroe–Shetland region at this time (e.g. Doré et al., 1999), coincident with faunal separation of the Boreal and Tethyan environments and the Toarcian to Bathonian development of the North Sea hot spot (e.g. see Underhill and Partington, 1993). Drilled occurrences of Middle Jurassic strata are rare within the Faroe–Shetland Basin and adjacent area (see Chapter 6; Vestralen and Hurst, 1994) though to the south-west, they occur in the Porcupine, Slyne and Erris basins (e.g. Naylor et al., 1999; Dancer et al., 1999; Naylor and Shannon, 2005).

Active extension during latest Mid to Late Jurassic times resulted in the formation of a deep, fully marine connection between the Arctic and Atlantic rifts, and the creation of a continuous rift axis represented by the Vøring, Møre, Faroe–Shetland and Rockall basins (Figure 12; e.g. Doré et al., 1999; Roberts et al., 1999). The focus of this extensional activity centred on generally north to north-east-trending structures such as Halten Terrace, Slyne, Erris and Porcupine basins, and the main grabens within the North Sea (Figure 12; Roberts et al., 1999; Doré et al., 1999; Coward et al., 2003). Within the Faroe–Shetland region, the presence of a Late Jurassic seaway was proved by drilling (see Chapter 6), but according to Dean et al. (1999) the general lack of variation in the thickness of the Upper Jurassic strata in comparison with the adjacent West Shetland and East Solan basins suggests that the effects of any contemporaneous rift activity were negligible. To the south-west of the Faroe–Shetland report area, Upper Jurassic sediments have not been drilled in the Rockall Basin but interpretation of wide-angle deep seismic experimental data suggest that they may be present (e.g. Shannon et al., 1999).

Cretaceous

During Valanginian times, the Atlantic Ocean propagated north-eastwards into the region between the Azores Fracture Zone and the Bay of Biscay area. Further to the north, extension within the Arctic and North Sea rifts effectively ceased by earliest Cretaceous times and was replaced by a vigorous, north-east-trending, Early Cretaceous Atlantic rift system that extended from the Vøring Basin to the Bay of Biscay (Figure 13; Rattey and Hayward, 1993; Roberts et al., 1999; Doré et al., 1999; Coward et al., 2003). At the junction of the Viking Graben and Møre Basin, Hauterivian extension along the east-north-east-trending End of the World Fault system effectively truncated the north-trending Late Jurassic rift associated with the Viking Graben (Figure 13; Nelson and Lamy, 1987; Roberts et al., 1999). According to Coward et al. (2003), the two main Early Cretaceous phases of extensional fault activity within the Atlantic rift broadly occurred during Valanginian to Hauterivian and Aptian to Albian times (Figure 8). Within the Faroe–Shetland Basin however, there is significant variation in views regarding the timing and duration of rifting associated with development of proven Lower Cretaceous marine clastic rocks, with Berriasian to Valanginian (Dean et al., 1999), post-Berriasian (Lamers and Carmichael, 1999), Berriasian to Barremian (Turner and Scrutton, 1993), Late Jurassic to Valanginian (Roberts et al., 1999), Aptian to Albian (Stoker et al., 1993; Ritchie et al., 1996) and Hauterivian to Cenomanian (Grant et al., 1999) phases suggested. To the south-west of the Faroe–Shetland region, integrated seismic and well analysis on the eastern flank of the Rockall Basin by Musgrove and Mitchener (1996, fig. 5) indicated that the main phase of extension occurred during Early Cretaceous (Hauterivian to Albian) times. In contrast, Scrutton and Bentley (1988) proposed a Ryazanian to Barremian age for the main Rockall Basin rifting event.

According to Roberts et al. (1999), by Albian times, the Atlantic rift system had become relatively inactive and thermal subsidence was prevalent. Locally, the Faroe–Shetland Basin was far from tectonically quiescent, with thick, deep marine clastic successions developed during periods of extension in Hauterivian to Cenomanian (Grant et al., 1999), Turonian to early Maastrichtian (Turner and Scrutton, 1993), Cenomanian to Santonian and Campanian to Maastrichtian (Dean et al., 1999; Spencer et al., 1999) times. To the south-west of the Faroe–Shetland report area, Upper Cretaceous strata have been drilled within the Rockall Basin, though they are considered to have been deposited within a dominantly post-rift tectonic setting (Musgrove and Mitchener, 1996).

Cenozoic

Palaeogene

Within the Faroe–Shetland Basin, extensive Late Cretaceous rift activity was superseded by post-rift

Figure 8 *(opposite)* Generalised summary of the plate tectonic and regional tectonostratigraphical events that affected the north-east Atlantic margin including the Faroe–Shetland region. Timescale is based on 'A Geologic Time Scale 2004' by F M Gradstein, J G Ogg, A G Smith, et al. (2004) and the International Stratigraphical Chart, 2006 (ICS) with additions. Abbreviations: see Figure 7 except, FSB=Faroe–Shetland Basin and RB=Rockall Basin.

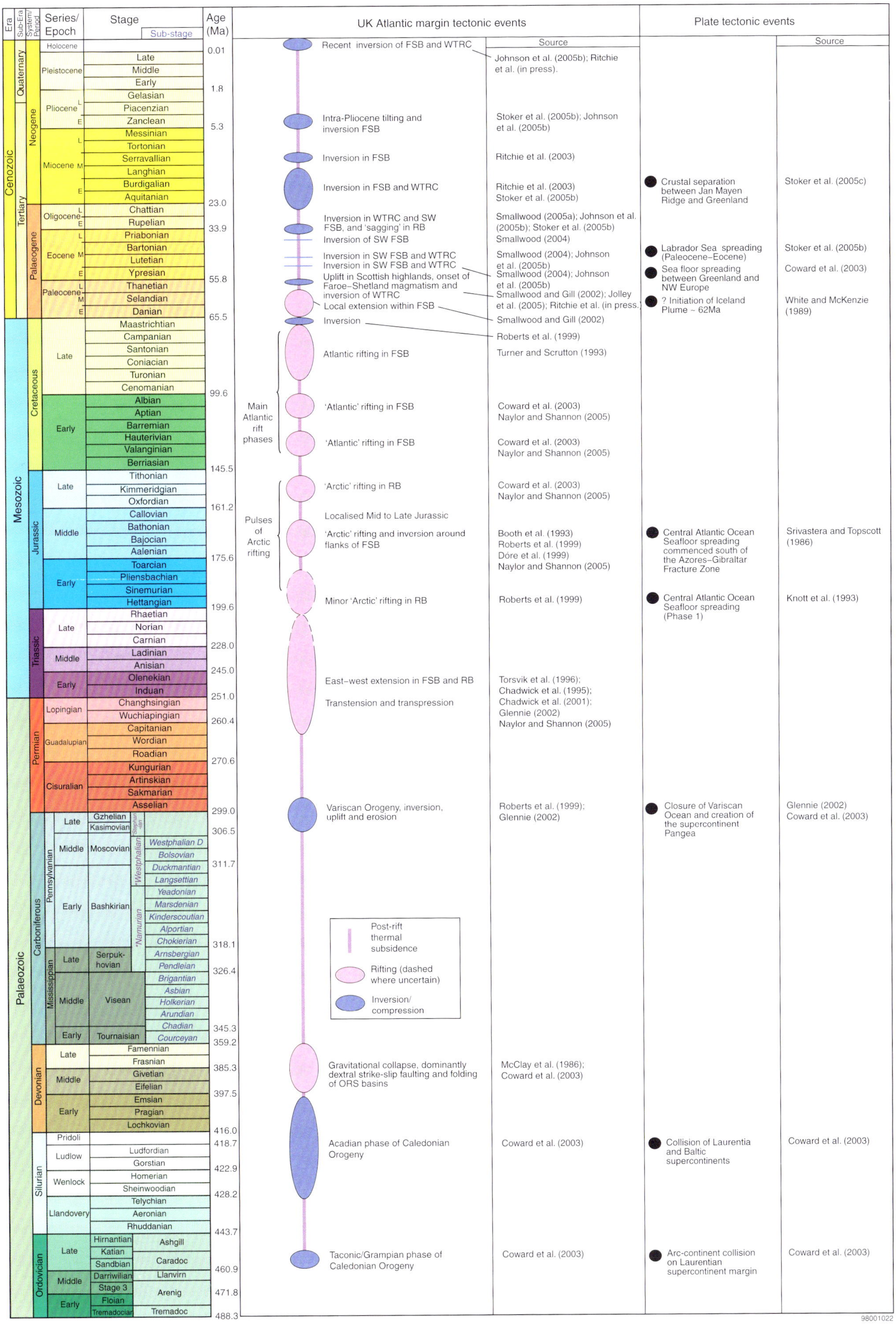

Era
Sub-Era
System/Period
Series/Epoch
Stage
Sub-stage
Age (Ma)
Cenozoic
Quaternary
Tertiary
Neogene
Palaeogene
Holocene
Pleistocene
Late
Middle
Early
Pliocene
Gelasian
Piacenzian
Zanclean
Miocene
Messinian
Tortonian
Serravallian
Langhian
Burdigalian
Aquitanian
Oligocene
Chattian
Rupelian
Eocene
Priabonian
Bartonian
Lutetian
Ypresian
Paleocene
Thanetian
Selandian
Danian
Mesozoic
Cretaceous
Maastrichtian
Campanian
Santonian
Coniacian
Turonian
Cenomanian
Albian
Aptian
Barremian
Hauterivian
Valanginian
Berriasian
Jurassic
Tithonian
Kimmeridgian
Oxfordian
Callovian
Bathonian
Bajocian
Aalenian
Toarcian
Pliensbachian
Sinemurian
Hettangian
Triassic
Rhaetian
Norian
Carnian
Ladinian
Anisian
Olenekian
Induan
Palaeozoic
Permian
Lopingian
Changhsingian
Wuchiapingian
Guadalupian
Capitanian
Wordian
Roadian
Cisuralian
Kungurian
Artinskian
Sakmarian
Asselian
Carboniferous
Pennsylvanian
Mississippian
Gzhelian
Kasimovian
Stephanian
Moscovian
Westphalian
Westphalian D
Bolsovian
Duckmantian
Langsettian
Bashkirian
Yeadonian
Marsdenian
Kinderscoutian
Alportian
Chokierian
Namurian
Serpuk-hovian
Arnsbergian
Pendleian
Visean
Brigantian
Asbian
Holkerian
Arundian
Chadian
Tournaisian
Courceyan
Devonian
Famennian
Frasnian
Givetian
Eifelian
Emsian
Pragian
Lochkovian
Silurian
Pridoli
Ludlow
Ludfordian
Gorstian
Wenlock
Homerian
Sheinwoodian
Llandovery
Telychian
Aeronian
Rhuddanian
Ordovician
Hirnantian
Katian
Sandbian
Ashgill
Caradoc
Darriwilian
Llanvirn
Stage 3
Floian
Tremadocian
Arenig
Tremadoc
0.01
1.8
5.3
23.0
33.9
55.8
65.5
99.6
145.5
161.2
175.6
199.6
228.0
245.0
251.0
260.4
270.6
299.0
306.5
311.7
318.1
326.4
345.3
359.2
385.3
397.5
416.0
418.7
422.9
428.2
443.7
460.9
471.8
488.3
UK Atlantic margin tectonic events
Source
Recent inversion of FSB and WTRC
Johnson et al. (2005b); Ritchie et al. (in press).
Intra-Pliocene tilting and inversion FSB
Stoker et al. (2005b); Johnson et al. (2005b)
Inversion in FSB
Ritchie et al. (2003)
Inversion in FSB and WTRC
Ritchie et al. (2003) Stoker et al. (2005b)
Inversion in WTRC and SW FSB, and 'sagging' in RB
Smallwood (2005a); Johnson et al. (2005b); Stoker et al. (2005b)
Inversion of SW FSB
Smallwood (2004)
Inversion in SW FSB and WTRC
Smallwood (2004); Johnson et al. (2005b)
Inversion in SW FSB and WTRC
Smallwood (2004); Johnson et al. (2005b)
Uplift in Scottish highlands, onset of Faroe–Shetland magmatism and inversion of WTRC
Smallwood and Gill (2002); Jolley et al. (2005); Ritchie et al. (in press.)
Local extension within FSB
Smallwood and Gill (2002)
Inversion
Roberts et al. (1999)
Atlantic rifting in FSB
Turner and Scrutton (1993)
Main Atlantic rift phases
'Atlantic' rifting in FSB
Coward et al. (2003) Naylor and Shannon (2005)
'Atlantic' rifting in FSB
Coward et al. (2003) Naylor and Shannon (2005)
Pulses of Arctic rifting
'Arctic' rifting in RB
Coward et al. (2003) Naylor and Shannon (2005)
Localised Mid to Late Jurassic
'Arctic' rifting and inversion around flanks of FSB
Booth et al. (1993) Roberts et al. (1999) Dòre et al. (1999) Naylor and Shannon (2005)
Minor 'Arctic' rifting in RB
Roberts et al. (1999)
East–west extension in FSB and RB
Transtension and transpression
Torsvik et al. (1996); Chadwick et al. (1995); Chadwick et al. (2001); Glennie (2002) Naylor and Shannon (2005)
Variscan Orogeny, inversion, uplift and erosion
Roberts et al. (1999); Glennie (2002)
Post-rift thermal subsidence
Rifting (dashed where uncertain)
Inversion/ compression
Gravitational collapse, dominantly dextral strike-slip faulting and folding of ORS basins
McClay et al. (1986); Coward et al. (2003)
Acadian phase of Caledonian Orogeny
Coward et al. (2003)
Taconic/Grampian phase of Caledonian Orogeny
Coward et al. (2003)
Plate tectonic events
Source
Crustal separation between Jan Mayen Ridge and Greenland
Stoker et al. (2005c)
Labrador Sea spreading (Paleocene–Eocene)
Stoker et al. (2005b)
Sea floor spreading between Greenland and NW Europe
Coward et al. (2003)
? Initiation of Iceland Plume ~ 62Ma
White and McKenzie (1989)
Central Atlantic Ocean Seafloor spreading commenced south of the Azores-Gibraltar Fracture Zone
Srivastera and Topscott (1986)
Central Atlantic Ocean Seafloor spreading (Phase 1)
Knott et al. (1993)
Closure of Variscan Ocean and creation of the supercontinent Pangea
Glennie (2002) Coward et al. (2003)
Collision of Laurentia and Baltic supercontinents
Coward et al. (2003)
Arc-continent collision on Laurentian supercontinent margin
Coward et al. (2003)
98001022

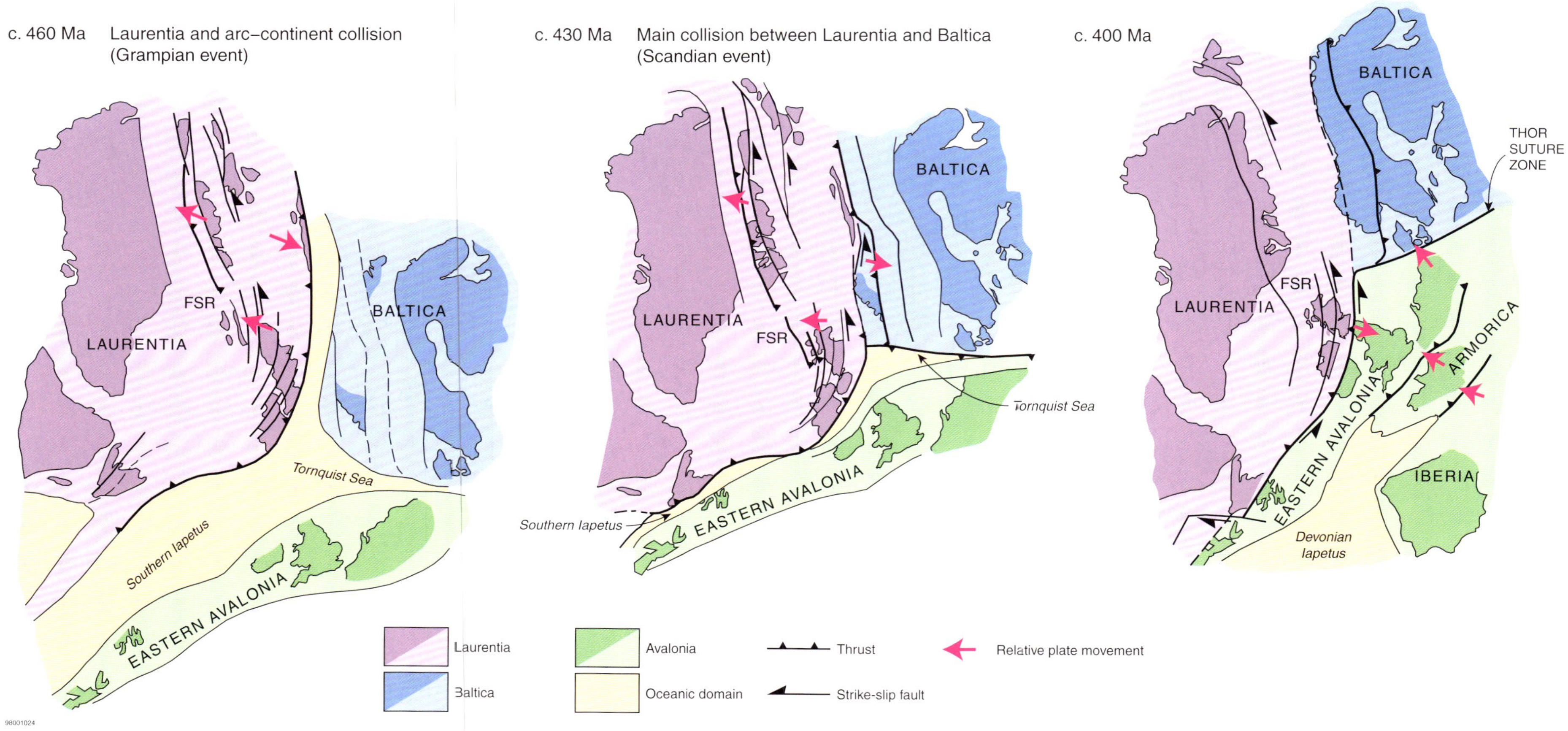

Figure 9 Evolution of the Caledonides with regard to the Faroe–Shetland region (slightly modified reproduction from Coward et al., (2003) by permission of The Geological Society, London). Abbreviations: FSR=Faroe–Shetland region.

Figure 10
a) Palaeogeographical and palaeofacies plate reconstruction of the Faroe–Shetland region and surrounding area for early Permian times; b) Global early Permian plate reconstruction (slightly modified reproductions from Coward et al., (2003) by permission of The Geological Society, London). Abbreviations: see Figures 7 and 8 except, MF=Moray Firth; NPB=Northern Permian Basin; SPB=Southern Permian Basin and VB=Vøring Basin.

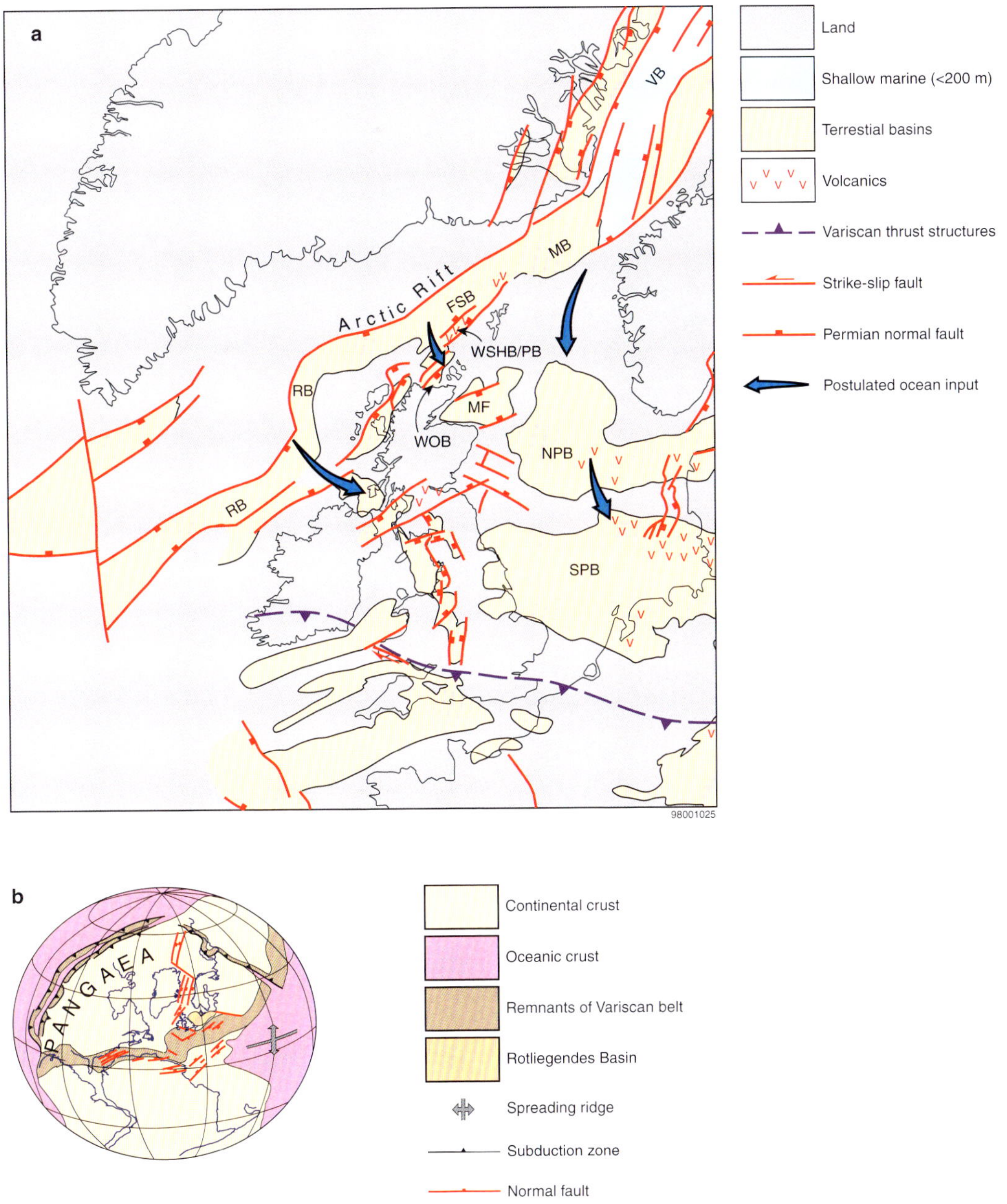

subsidence (e.g. Turner and Scrutton, 1993), with Early Paleocene deep marine clastic rocks infilling a drowned block-faulted topography. However, there were significant departures from this post-rift pattern of thermal subsidence, including episodes of extension, accelerated subsidence and pulses of compression and inversion (e.g. Doré et al., 1999, Japsen and Chalmers, 2000; Stoker et al., 2005b; Johnson et al., 2005b; Kimbell et al., 2005; Ceramicola et al., 2005).

The Labrador Sea opened during the Paleocene, with the formation of new oceanic crust between western Greenland and eastern Canada (see Figure 90; Doré et al., 1999; Coward et al., 2003). According to Roberts et al. (1999), this opening would have placed the Rockall–Hatton, Faroe–Shetland and Vøring areas in an oblique strike-slip tectonic setting, and may have been the cause for localised extension persisting within the Flett Sub-basin (Figure 7) until Mid Paleocene times (e.g. Smallwood and Gill, 2002), and the early growth of the Wyville Thomson Ridge (Johnson et al., 2005b). About 62 Ma ago, widespread thermal doming associated with the developing Iceland Plume affected an area some 2000 km in diameter, including Greenland, Vøring, Faroe–Shetland and Rockall–Hatton areas, and was responsible for the formation of the North Atlantic Igneous Province (NAIP) (e.g. White and McKenzie, 1989; Coffin and Eldholm, 1992). Mid Paleocene to earliest Eocene volcanic rocks of the NAIP cover an area of 106 km^3 (see Figure 90; White and McKenzie, 1989), and can be ascribed to pre- (62 to 56 Ma) and syn- (56 to 54 Ma) break-up phases of magmatism (White and Lovell, 1997; cf. Jolley et al., 2005). Northerly propagating oceanic spreading between eastern Greenland and north-west Europe commenced at approximately 55 to 54 Ma (Figure 8; Saunders et al., 1997; Ritchie et al., 1999a) along the Reykjanes, Aegir and Mohns spreading ridges. Break-up is interpreted to be marked by the widespread distribution of earliest Eocene Balder Formations tuffs, generated during the foundering of the incipient subaerial North Atlantic spreading axis (e.g. Ritchie et al., 1999a). The main mass of syn-break-up volcanic and intrusive rocks is mainly restricted to the East Greenland, Hatton–Rockall and Faroe–Shetland regions

Figure 11 Palaeogeographical and palaeofacies plate reconstruction of the Faroe–Shetland region and surrounding area for Mid Triassic times (slightly modified reproduction from Coward et al., (2003) by permission of The Geological Society, London). Abbreviations: see Figures 7, 8 and 10 except, CNS=Central North Sea; EB=Erris Basin; PB=Porcupine Basin; SB=Slyne Basin and VG=Viking Graben Basin.

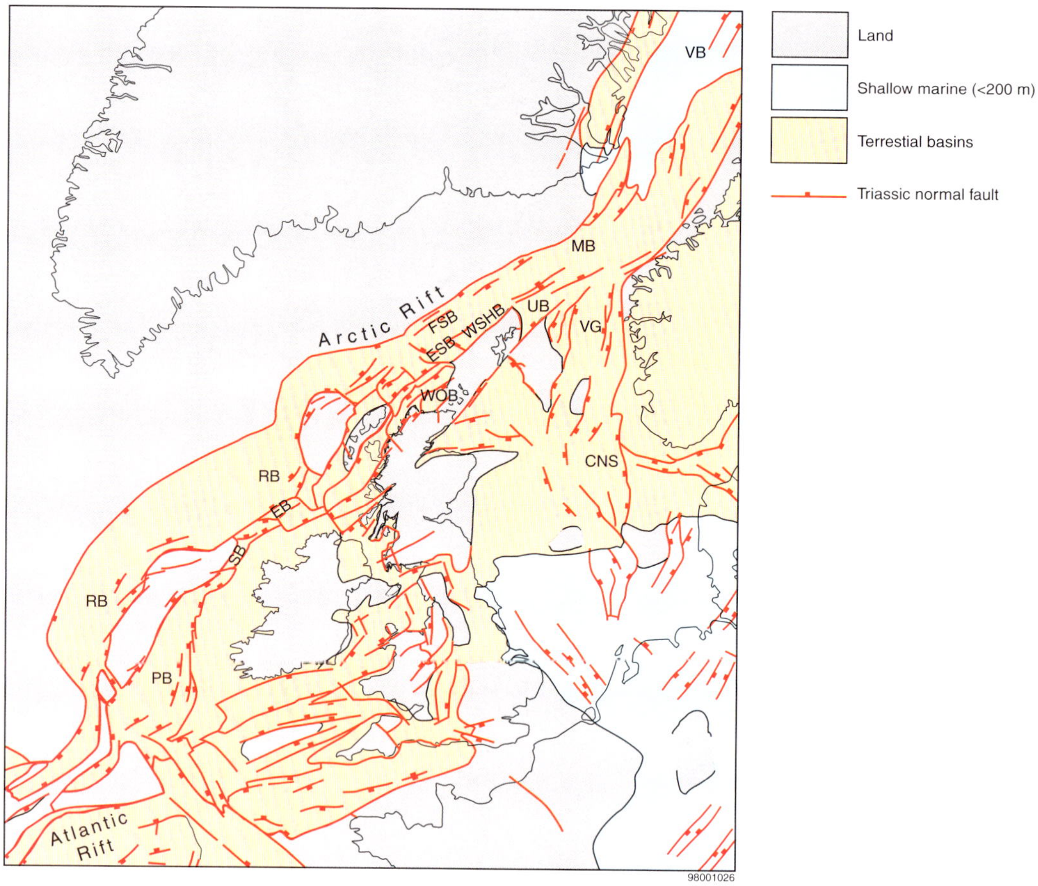

Figure 12 Palaeogeographical and palaeofacies plate reconstruction of the Faroe–Shetland region and surrounding area for Late Jurassic times (slightly modified reproduction from Coward et al., (2003) by permission of The Geological Society, London). Abbreviations: see Figures 7, 8, 10 and 11 except, HT=Halten Terrace and SHLMB=Sea of Hebrides–Little Minch Basin.

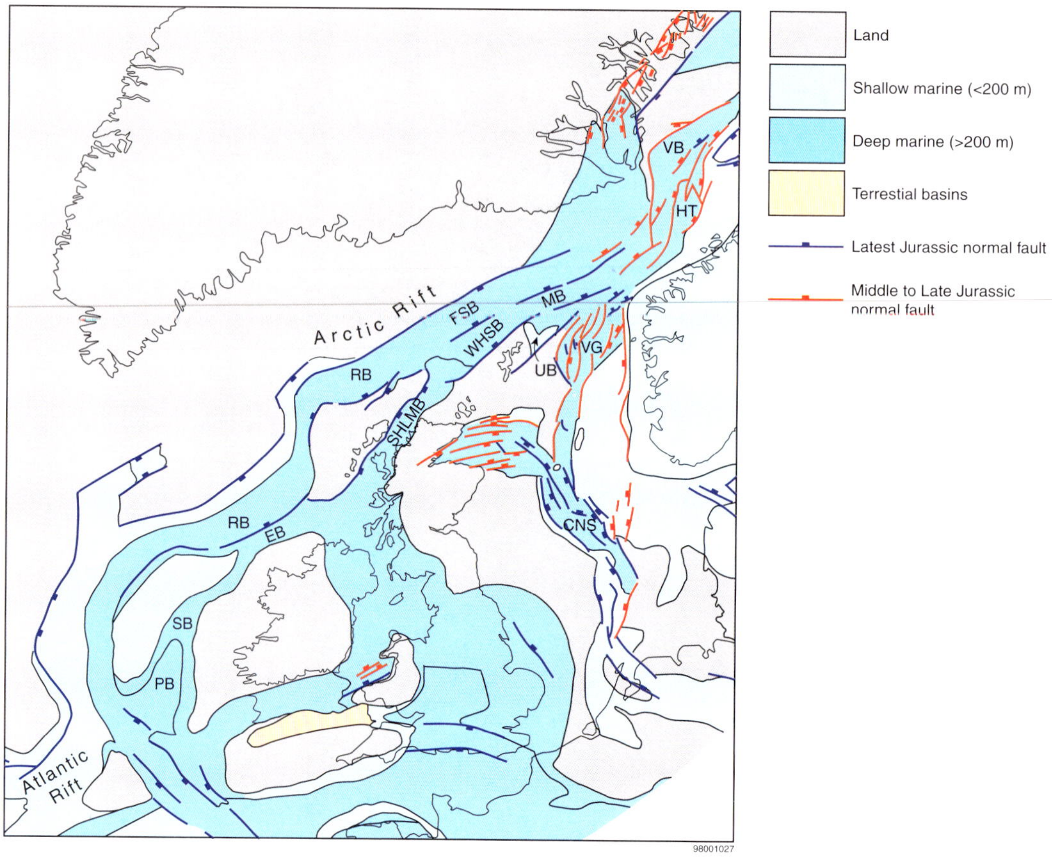

Table 1 Acronyms for deep seismic surveys in the Faroe–Shetland report area.

BIRPS	**B**ritish **I**nstitutions **R**eflection **P**rofiling **S**yndicate
DRUM	**D**eep **R**eflections from the **U**pper **M**antle
FAST	**F**aroe **S**hetland **T**raverse
FIRE	**F**aroe **I**celand **R**idge **E**xperiment
FLARE	**F**aroes **L**arge **A**perture **R**esearch **E**xperiment
GRID	Not an acronym (grid of profiles)
ISFA	Not known
LISPB	**Li**thospheric **S**eismic **P**rofile in **B**ritain
MOIST	**M**oine and **O**uter **I**sles **S**eismic **T**ransect
NSDP	**N**orth **S**ea **D**eep **P**rofiles
NASP	**N**orth **A**tlantic **S**eismic **P**roject
UNST	Not an acronym
WINCH	**W**estern **I**sles **N**orth **Ch**annel traverse

98001911

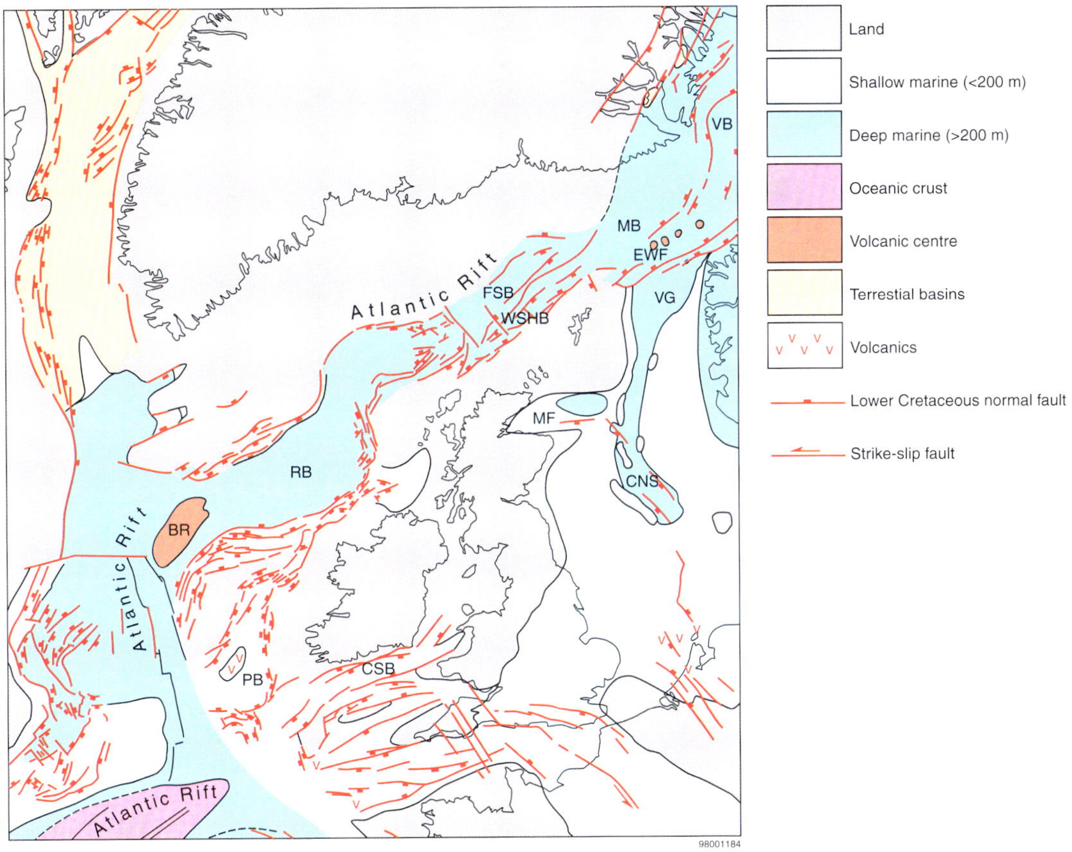

Figure 13 Palaeogeographical and palaeofacies plate reconstruction of the Faroe–Shetland region and surrounding area for Early Cretaceous times (slightly modified reproduction from Coward et al., (2003) by permission of The Geological Society, London). Abbreviations: see Figures 7, 8, 10 and 11 except, BR=Barra Volcanic Ridge Complex; CSB=Celtic Sea Basin and EWF=End Of The World Fault.

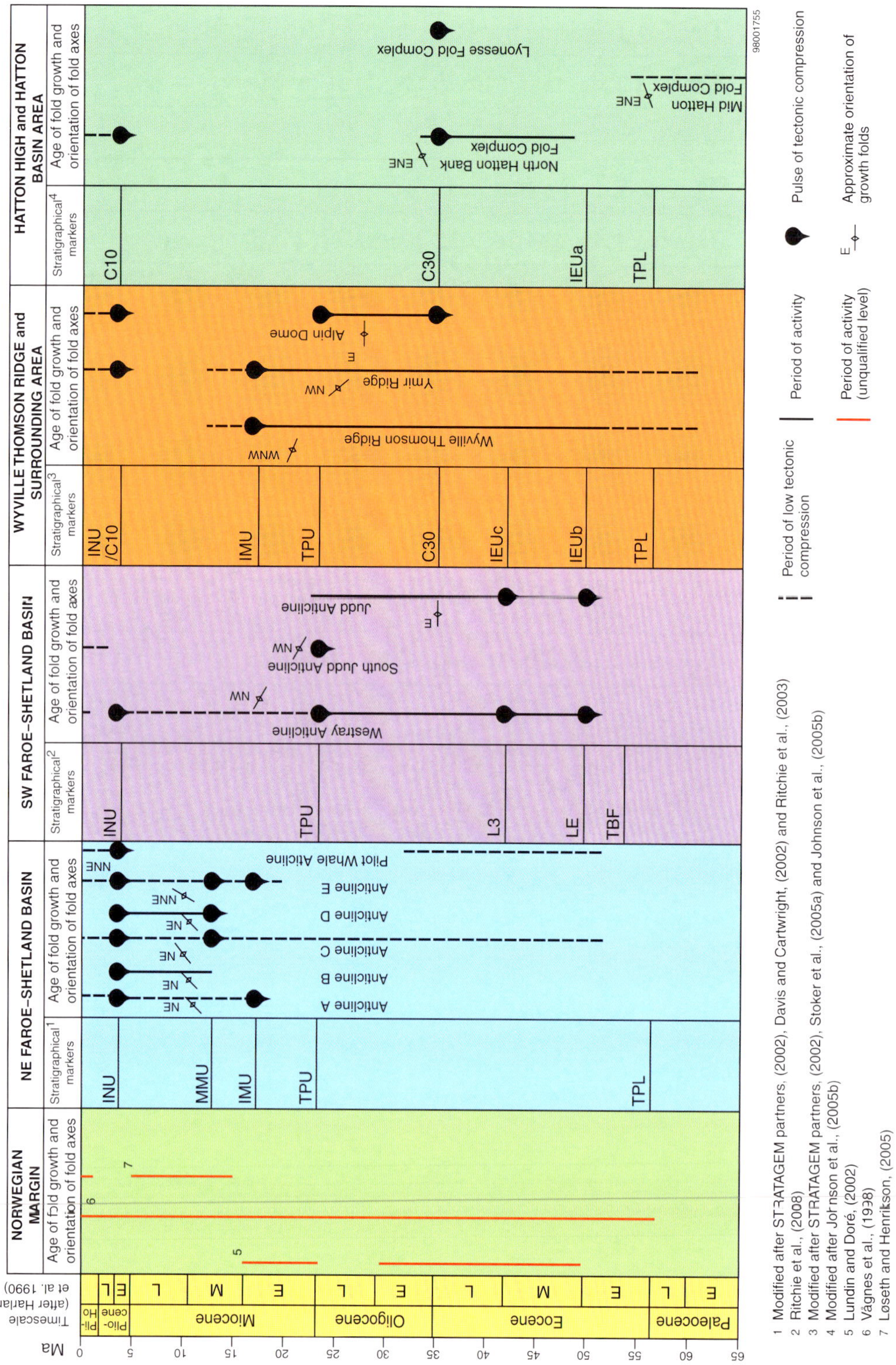

Figure 14 A comparison of the age of Cenozoic growth folds throughout the north-east Atlantic margin (modified from Ritchie et al., 2008). For location of folds, see Figure 15, Boldreel and Andersen, (1998), Lundin and Doré, (2002), Ritchie et al., (2003) and Johnson et al., (2005b). Abbreviations: INU=Intra-Neogene (Early Pliocene) Unconformity; C10=Early Pliocene Unconformity; MMU=Mid Miocene Unconformity; IMU=Intra-Miocene Unconformity; TPU=Top Palaeogene Unconformity; C30=Late Eocene Unconformity; IEUa, b and c=Intra-Eocene unconformities; L3=Mid Eocene (late Lutetian); LE= Early Eocene (latest Ypresian) Unconformity; TBF=Top Balder Formation and TPL=Top Palaeogene lavas.

(see Chapter 9; Saunders et al., 1997; Larsen et al., 1999; Ritchie et al. 1999a; Jolley and Bell, 2002a). A phase of poorly age-constrained Palaeogene uplift reported along the north-west European continent–ocean boundary (e.g. Kimbell et al., 2005) is related to the presence of underplated igneous material associated with the development of the NAIP (e.g. Morgan et al., 1989; Fowler et al., 1989; Richardson et al., 1998; Smallwood et al., 1999; White et al., 2005).

Following separation of eastern Greenland from north-west Europe, the north-east Atlantic margin is commonly regarded as 'passive' though it has been far from tectonically quiescent (Figure 8; e.g. Stoker et al., 2005a, b and c). During latest Paleocene to Oligocene times, regional compressive forces were responsible for the widespread growth of anticlines and dome-like structures in the Hatton Bank, Wyville Thomson Ridge (e.g. Johnson et al. 2005b; Ritchie et al. 2008), south-west Faroe–Shetland Basin e.g. Judd, Westray and South Judd anticlines (Smallwood, 2004; Ritchie et al., 2008) and Norwegian margin (e.g. Lundin and Doré, 2002) areas (Figure 14). Mechanisms to explain the age distribution and orientation of these (and Neogene) compressive structures observed throughout the north-east

Figure 15 The location of Cenozoic compressive anticlines and domes within the north-east Atlantic margin (modified from Ritchie et al., 2003; Kimbell et al., 2005 and Johnson et al., 2005b). Abbreviations: see Figure 7 except, ADL=Anton Dohrn Lineament; AR=Aegir Ridge; HA=Helland-Hansen Arch; JMFZ=Jan Mayen Fracture Zone; JML=Jan Mayen Lineament; MFL=Marflo Lineament; ND=Naglfar Dome; SHL=South Hatton Lineament and VD=Vema Dome.

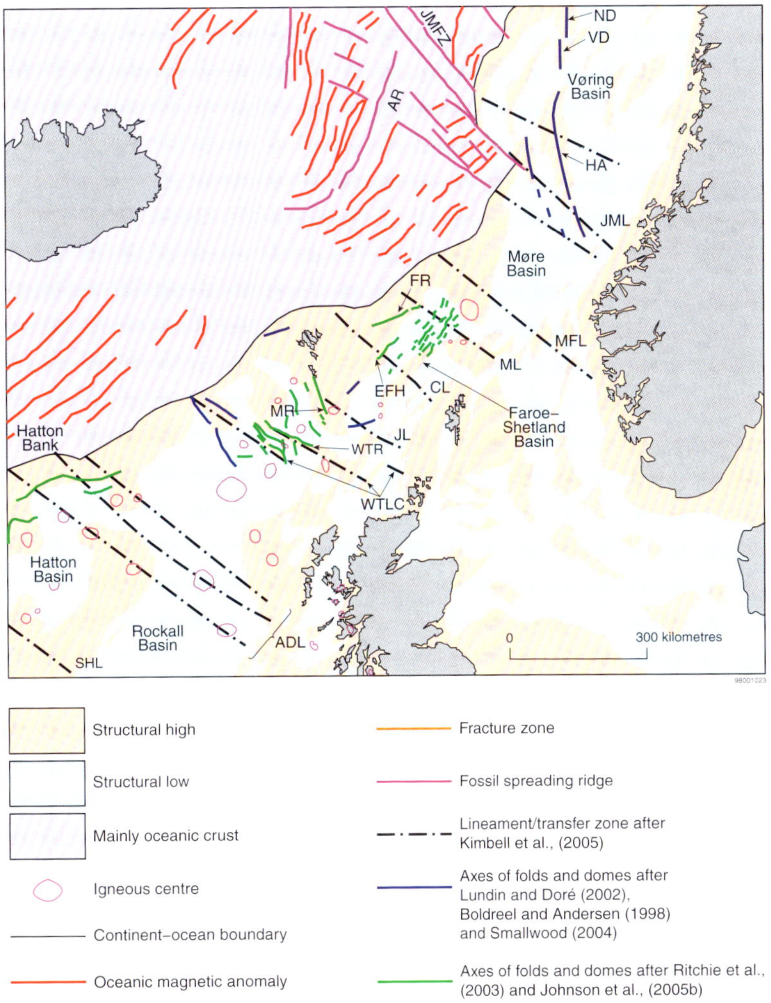

Atlantic margin (Figure 15) include ridge-push (e.g. Boldreel and Andersen, 1993), intra-plate stress (Cloetingh et al., 1990), the Alpine Orogeny (Boldreel and Andersen, 1998), plume-enhanced asthenospheric flow (pulsing plume) (Lundin and Doré, 2002) and supercharged asthenospheric flow associated with sea-floor spreading (Kusznir, 2005). Presently, no single mechanism offers a plausible explanation that can account for all identified domes and anticlines (Ritchie et al., 2008).

Neogene

In Early to Mid Miocene times, significant major phases of compressional tectonic activity affected the Vøring (e.g. Lundin and Dore, 2002; Løseth and Henriksen, 2005) and north-east Faroe–Shetland basins (Figures 8 and 14; e.g. Ritchie et al., 2003; Davis et al., 2004). Mid Miocene growth of the Wyville Thomson Ridge anticline and the complementary Faroe Bank Channel syncline (the Faroe Conduit) re-established a deep marine link between the Faroe–Shetland and Hatton–Rockall regions, facilitating the southerly transfer of cold North Atlantic Deep Water (see Chapter 8 and Figure 92; Stoker et al., 2005b). During Early Pliocene times, a further localised phase of compression was mainly focused within the north-east Faroe–Shetland Basin. Of perhaps greater significance, contemporaneous regional epeirogenic uplift and tilting of the north-east Atlantic continental margin was accompanied by accelerated offshore subsidence and the progradational development of sedimentary wedges (Andersen et al., 2000 and 2002; Stoker, 2002; Davies et al., 2004; Stoker et al., 2005b and c). There are currently no plausible mechanisms that explain this phenomenon.

DEEP STRUCTURE

The deep structure of the Faroe–Shetland region has been investigated by a number of seismic experiments (Figure 16). Wide-angle seismic surveys employing long offsets between sources and receivers have been used to investigate the velocity structure of the Earth's crust and upper mantle. Near-normal incidence seismic reflection surveys, with closely spaced sources and receivers, have detected detailed reflectivity patterns from similar depths but are less effective at resolving the velocity structure. This region has been a focus for the development of innovative technologies for deep seismic investigation, which integrate wide-angle and near-normal incidence approaches. These have been applied in the area south-east of the Faroe Islands, in order to image beneath the Palaeogene lavas that hinder exploration by conventional seismic methods, and in the area immediately north of the Scottish mainland, to investigate the sources of deep seismic reflections discovered there.

Figure 17 is a schematic map of variations in the depth to the Moho (the base of the crust) beneath the region. This was derived by integrating the results of the seismic experiments with those of regional three-dimensional gravity modelling (Kimbell et al., 2004 and

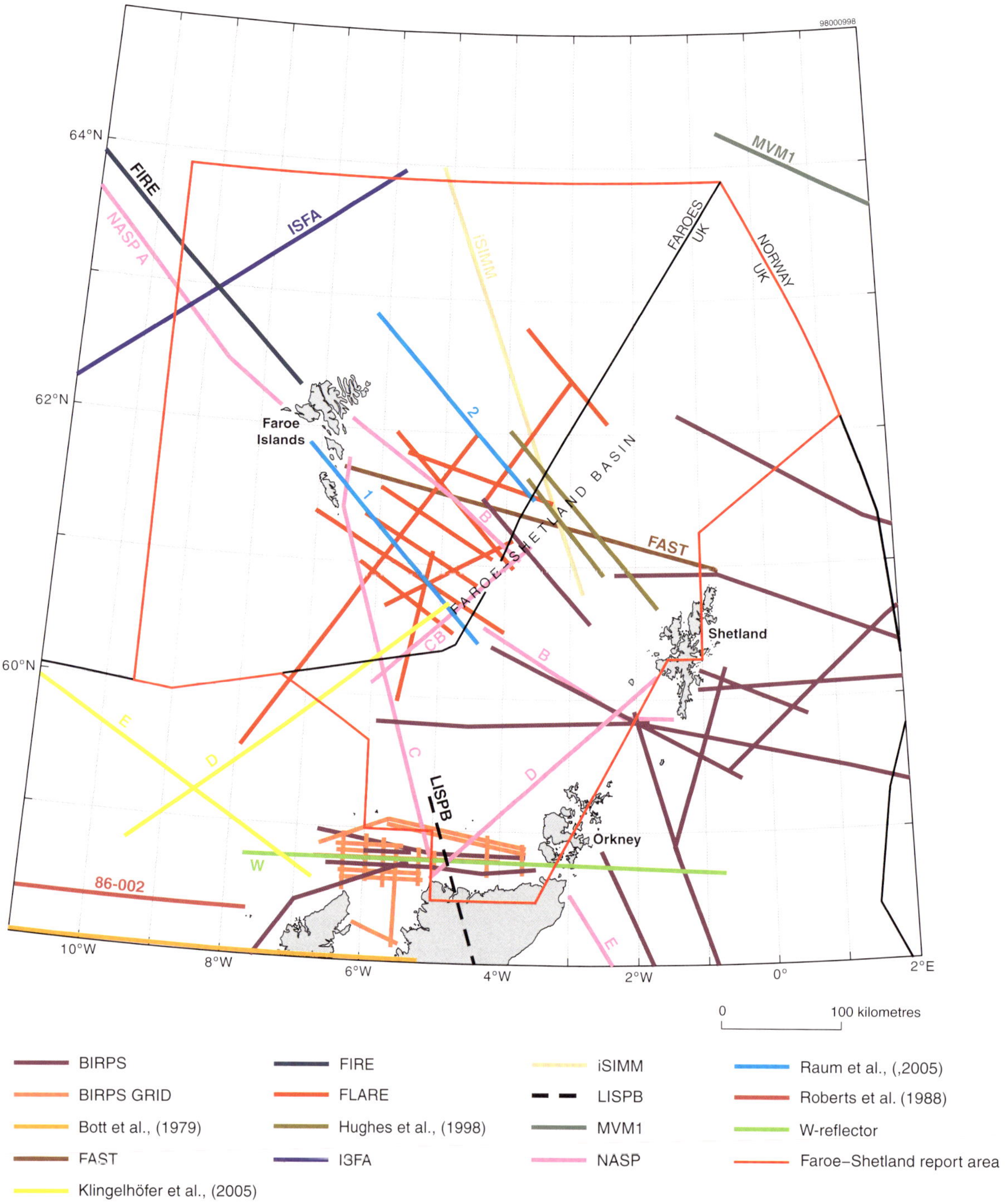

Figure 16 Location of deep seismic profiles in the Faroe–Shetland report area. The BIRPS surveys employed the near-normal incidence seismic reflection technique, while the other surveys incorporated wide-angle seismic acquisition with larger separations between sources and receivers. See Table 1 for acronyms for deep seismic surveys in the Faroe–Shetland report area.

2005). The Moho has only been reliably detected along a relatively small proportion of the seismic lines shown in Figure 16 and there are inconsistencies between the estimates from different surveys. Gravity modelling is less diagnostic, as a range of subsurface structures can give rise to the same gravity response, but does provide a regional overview which complements the local, two-dimensional 'snapshots' provided by seismic experiments.

The Moho is relatively shallow where the crust has been highly stretched to form the Faroe–Shetland and north-east Rockall basins, and lies at greater depth beneath the shelf areas to the north-west and south-east. Thin oceanic crust underlies the Norwegian Sea to the north of the report area and there is marked crustal thickening towards a north-west-trending axis that extends from the Faroe Islands to Iceland. The deep structure of these different areas is discussed in the following sections.

Iceland–Faroe Ridge

The Iceland–Faroe Ridge is a distinct bathymetric high (Figure 1) that trends north-westwards from the Faroe

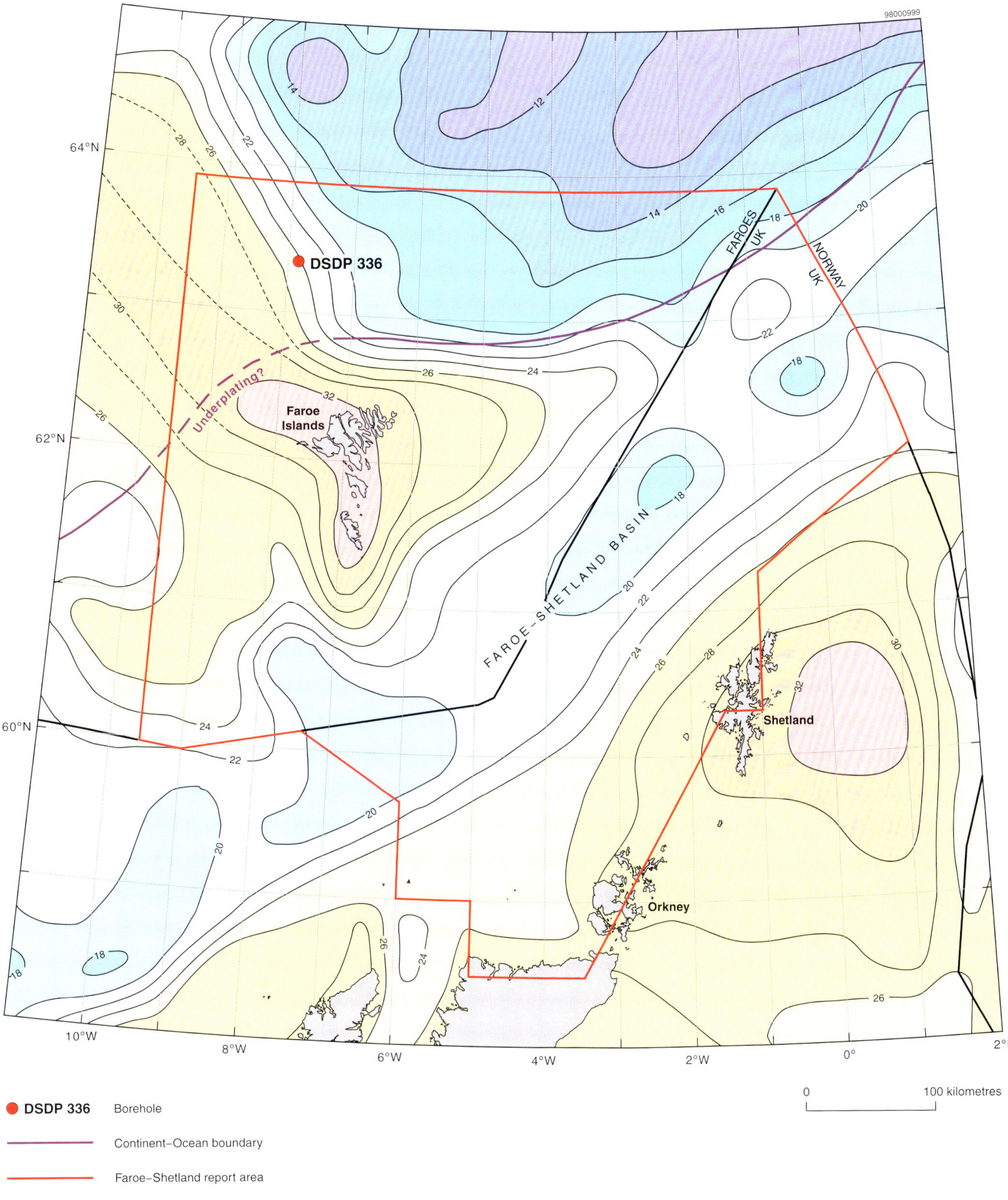

Figure 17 Schematic map of the depth to the Moho (in kilometres) beneath the Faroe–Shetland report area. Based on the results of wide-angle seismic experiments (see text for details and Figure 16), near-normal incidence reflection profiling (Chadwick and Pharaoh, 1998) and gravity modelling (Kimbell et al., 2004 and 2005).

Islands, and a similar feature can be traced along the same axis between Iceland and Greenland. Bott and Gunnarsson (1980) describe the results of a wide-angle seismic experiment along the axis of the ridge (NASP Line A; Figure 16 and Table 1) that revealed a substantial crustal thickness, with the Moho estimated to lie at depths of 30 to 35 km. The upper crust (to depths of 4 to 8 km) was interpreted to have a seismic velocity of 5.7 km/s and to be underlain by crust with velocities of 6.7 km/s or more. The FIRE experiment (Figure 16 and Table 1; Richardson et al., 1998; Smallwood et al., 1999) also suggested a depth of about 30 km to the

Moho beneath the Iceland–Faroe Ridge, and provided more information on its velocity structure. In the interpretation of Smallwood et al. (1999), an upper crust (down to about 10 km) characterised by a relatively high velocity gradient is underlain by crust in which the velocity increases more slowly with depth, from about 6.5 km/s at 10 km to 7.1–7.3 km/s immediately above a gradational Moho; the upper mantle velocity was relatively low at 7.7–7.9 km/s. This crustal velocity structure was interpreted to be characteristic of anomalously thick oceanic crust formed by sea-floor spreading above a thermal anomaly in the mantle (Bott and Gunnarsson, 1980; Richardson et al., 1998; Smallwood et al., 1999). The thermal anomaly has been ascribed to a mantle plume that developed in the Paleocene and persists to the present day (e.g. White and McKenzie, 1989). The resolution of the velocity structure of the deeper part of the crust beneath the FIRE line is, however, open to alternative interpretations because of limitations in data quality, and the preferred interpretation was guided by the results of gravity modelling (Richardson et al., 1998).

DSDP (Deep Sea Drilling Project) borehole 336 (Figure 17) intersected subaerially erupted basalt lavas on the northern flank on the Iceland–Faroe Ridge (Shipboard Scientific Party, 1976; see Figure 104). As with present day Iceland, the part of the ridge in the study area thus appears to have formed an emergent landmass for a period after it was created, and this was part of the Thulean land bridge linking Europe with North America (Nilsen, 1978a). The ridge subsequently subsided below sea level through a combination of thermal subsidence and erosion.

A different interpretation of the deep structure of the Iceland–Faroe Ridge is provided by Makris et al. (1995) and Bonhoff and Makris (2004) on the basis of their ISFA experiment, which crossed the ridge in a south-west–north-east direction about 110 km north-west of the Faroe Islands (Figure 16 and Table 1). They interpret the Moho to lie at a depth of 23 km beneath the axis of the ridge and 15 km beneath either end of the ISFA line, and consider the velocity structure of the ridge to be indicative of stretched continental, rather than oceanic, crust. They identified a low velocity zone beneath the subaerial basalts which they interpreted to be a sedimentary succession. The average velocity down to 23 km depth is similar to that revealed by the FIRE experiment, so the discrepancy in Moho depth hinges on the resolution of velocity structure and interfaces below this. In the ISFA interpretation the velocity increases sharply to 7.9 km/s at 23 km (Bonhoff and Makris, 2004). The two alternative interpretations for FIRE presented by Richardson et al. (1999) place the Moho at about 25 km and 30 km respectively where this line intersects with ISFA. A problem with the ISFA model is the symmetry it proposes for the Iceland–Faroe Ridge. Isostatic considerations (relating to the significant difference in water depth on either side of the ridge) and gravity modelling (Bott and Gunnarsson, 1980; Kimbell et al., 2005) strongly suggest that the Moho is significantly deeper beneath the south-west side of the ridge than beneath its north-east side. This discrepancy might arise because of seismic data coverage problems on the southern side of the ridge, which were acknowledged by Bonhoff and Makris (2004).

There are thus conflicting views regarding the nature (oceanic or continental) of the Iceland–Faroe Ridge in the north-western part of the report area, and further investigation is required to resolve this issue. As well as considering velocity structure, such investigation needs to address the feasibility of any proposed continental margin geometries in terms of North Atlantic plate reconstructions.

The continent–ocean boundary

An approximate position for the continent–ocean transition is shown in Figure 17. This has been identified on the iSIMM (Roberts et al., 2005) and MVM1 (Olafsson et al., 1992) seismic profiles (Figure 16), and by correlations with potential field data to the south-west of the present area of interest (Kimbell et al., 2005). Its position on the Iceland–Faroe Ridge is disputed, as discussed above. The magnetic anomaly pattern is less diagnostic because the anomalies over oceanic crust formed in a subaerial environment do not form the regular 'stripes' characteristic of oceanic crust formed in a submarine environment, and this makes it difficult to distinguish from the adjacent lava-covered continental areas. Data from the FIRE experiment suggest a zone 40 to 100 km north-west of the Faroe Islands where there is a north-westward increase in seismic velocity at mid-crustal levels, which Richardson et al. (1998) and Smallwood et al. (1999) have interpreted as representing the transition from continental to oceanic crust. Bonhoff and Makris (2004) suggested that this transition is located farther towards the north-west on the basis of the ISFA results. Richardson et al. (1998) and Smallwood et al. (1999) also identified a thickening of a high-velocity lower crustal zone beneath the continent–ocean transition, which they interpret as igneous underplated material formed during continental break-up. This may increase the crustal thickness to between 40 and 43 km beneath the Faroe Shelf (Richardson et al., 1998; Smallwood et al., 1999). Increased reflectivity in near-normal-incidence seismic reflection data has been correlated with the underplating/intrusion of continental crust in this area (McBride et al., 2004).

Area north-west of the Faroe–Shetland Basin

There is general consensus that the velocity structure of the crystalline crust beneath the Faroe Islands and surrounding structural high indicates a continental origin (Bott et al., 1974; Richardson et al., 1998; Raum et al., 2005) and this is compatible with geochemical evidence for continental contamination in the lavas that conceal this basement (Gariépy et al., 1983). The depth to the Moho in the immediate vicinity of the Faroe Islands is, however, poorly constrained. From the results of the early NASP surveys (Figure 16), Bott et al. (1974) estimated that it lay within the range 27 to 38 km. Wide-angle arrivals received at stations on the Faroe Islands from shots on the FAST profile (Figure 16) provided reflections from beneath the eastern side of the islands that were interpreted to indicate a depth to Moho of about 34 km (Richardson et al., 1999). In contrast, the interpretation by Raum et al. (2005) of a wide-angle seismic profile which crosses the same area (north-west part of their Profile 1, Figure 16), places the Moho at a depth of 22 to 23 km. Differences in the crustal velocity model and poor ray coverage of deeper interfaces towards the ends of the seismic profiles are contributing factors but cannot fully explain this large discrepancy.

Farther to the north, the Moho was detected at a depth of 28 to 29 km on Profile 2 of Raum et al. (2005). It also appears as a reflection on the seismic reflection profile along the iSIMM profile (Roberts et al., 2005). The latter demonstrates crustal thinning towards the continent–ocean boundary, which was not detected on the profile of Raum et al. (2005) because of the lack of ray coverage.

The Palaeogene lavas that extend across the Faroe Platform and north-west part of the Faroe–Shetland Basin (see Figure 120) impede the imaging of any underlying sedimentary rocks by conventional seismic reflection methods. The economic potential of these strata has encouraged the development of technologies for imaging beneath basalts. The FLARE surveys (Figure 16) used either single and multiple passes with two survey vessels to assemble synthetic large aperture profiles (White et al., 1999; Fliedner and White, 2003; White et al., 2003), and the iSIMM survey used a large-volume low-frequency source in combination with a very long streamer and ocean-bottom seismometers to record longer offsets (Roberts et al., 2005; Spitzer et al., 2005). Gallagher and Dromgoole (2005) demonstrated that careful processing of conventional seismic reflection data can yield good images of the sub-basalt succession. Figure 18 illustrates the sub-lava sedimentary thickness interpreted by White et al. (2003) from a combination of the FLARE wide-angle data and conventional seismic reflection data.

Faroe–Shetland Basin

Bott and Smith (1984) and Bott (1984) have reported interpretations of the NASP profile CB (Figure 16) along the axis of the southern part of the Faroe–Shetland Basin. Sedimentary and volcanic rocks were interpreted to extend down to a depth of about 9 km and be underlain by crystalline crust with a seismic velocity of 6.0–6.6 km/s. The Moho was estimated to lie at a depth of 15 to 17 km beneath the northern part of the profile, but at greater depth (>20 km) farther south, opposite the southward projection of the Munkagrunnur Ridge. These authors interpreted the basin to be underlain by pre-Cenozoic oceanic crust, but subsequent interpretations (e.g. Mudge and Rashid, 1987) have favoured a stretched continental origin. This has been corroborated by the wide-angle seismic surveys of Raum et al. (2005; Line 1) and Klingelhöfer et al. (2005) who detected a two-layer crystalline crust with velocities indicative of a continental origin. The south-east part of Line 1 of Raum et al. (2005) lies close to the north-east end of the Klingelhöfer et al. (2005) profile (Figure 16), but there are discrepancies between the respective interpretations. Raum et al. (2005) place the top basement at 9 km and Moho at 17 km in this area, whereas Klingelhöfer et al. (2005) interpret these interfaces to lie at 7 km and 21 km respectively. The model of Klingelhöfer et al. (2005) indicates that the Moho typically lies at depths of 19 to 20 km further south-west, beneath the Munkur Basin, Wyville Thomson Ridge and north-east Rockall Basin.

Further north, deep seismic surveys by Hughes et al. (1997 and 1998) suggest that the top of the basement lies at depths of 7 to 9 km beneath the central part of the Faroe–Shetland Basin, with the Moho identified at a depth of 18 ± 3 km on the basis of wide-angle reflections. The upper part of the presumed crystalline basement was interpreted to have a low velocity (5.0–5.5 km/s) by Hughes et al. (1998), and Richardson et al. (1999) identified similarly low velocities in crystalline basement beneath the FLARE and FAST lines. Raum et al. (2005; Line 2) suggested that the basement is somewhat deeper (approximately 10 km) and faster, and is overlain by a thicker layer of pre-Cretaceous sedimentary rocks. Once again, the velocity structure of the basement was considered indicative of stretched continental crust. Combined seismic and gravity modelling of the FAST profile by Smallwood et al. (2001) and England et al. (2005) indicated that the Moho lies at a depth of about 19 km where it is shallowest beneath the basin (Figure 19).

Although there are differences between the existing wide-angle interpretations described above, overall they suggest a crustal extension factor of about 3 along the axis of the Faroe–Shetland Basin. This is comparable with the amount of extension inferred for the northern part of the Rockall Basin but significantly less that that in the Møre and Vøring basins to the north and the southern Rockall Basin to the south (Kimbell et al., 2004; 2005). The central, less stretched region coincides with that part of the Atlantic margin where igneous activity, as expressed in the Iceland–Faroe Ridge, has been greatest (see, for example, Kimbell et al., 2005, fig. 5), but the reason for this is not clear as the main stretching events predate the development of the hot-spot. Subsequent (Palaeogene) thickening of the crust beneath the basins by igneous underplating would appear to offer a solution, but wide-angle seismic surveys have not detected such underplating (Raum et al., 2005; Klingelhöfer et al., 2005).

Area south-east of the Faroe–Shetland Basin

Wide-angle seismic reflection experiments across the highs and basins that fringe the southern and south-eastern margins of the Faroe–Shetland Basin have been undertaken on NASP lines B, C and D (Smith and Bott, 1975), the LISPB profile (Bamford et al., 1978; Barton, 1992), and the W-reflector profile (Warner et al., 1996; Morgan et al., 2000; Price and Morgan, 2000) (Figure 16). The W-reflector (Figure 20) and LISPB experiments suggested that the Moho lies at a depth of 26 to 27 km beneath a three-layer crystalline crust in the southern part of the study area, immediately north of the Scottish mainland. The results from the NASP lines also indicated a Moho depth of 26 ± 2 km beneath the shelf area farther north (Smith and Bott, 1975).

The Moho has been detected as either a discrete reflector or the base of a zone of lower-crustal reflectivity on the BIRPS (Table 1) near-normal incidence seismic reflection profiles in this area (Figures 21 and 22). It lies at a relatively uniform 9 to 10 seconds (s) two-way-travel-time (TWTT) over much of the area, despite the variations in the velocity structure of the overlying rocks (in particular the variable thickness of low-velocity sedimentary rocks). Warner (1987) has demonstrated how this can be explained by a combination of isostatic compensation and the velocity–density relationship of crustal rocks. The Moho is shallower beneath thick low-density sedimentary sequences because of isostasy, but this is approximately cancelled out in a time section by the 'push-down' of reflections beneath the basin due to the relatively low velocity of the sedimentary rocks. Chadwick and Pharaoh (1998) have mapped variations in Moho depth using depth-conversions of the BIRPS

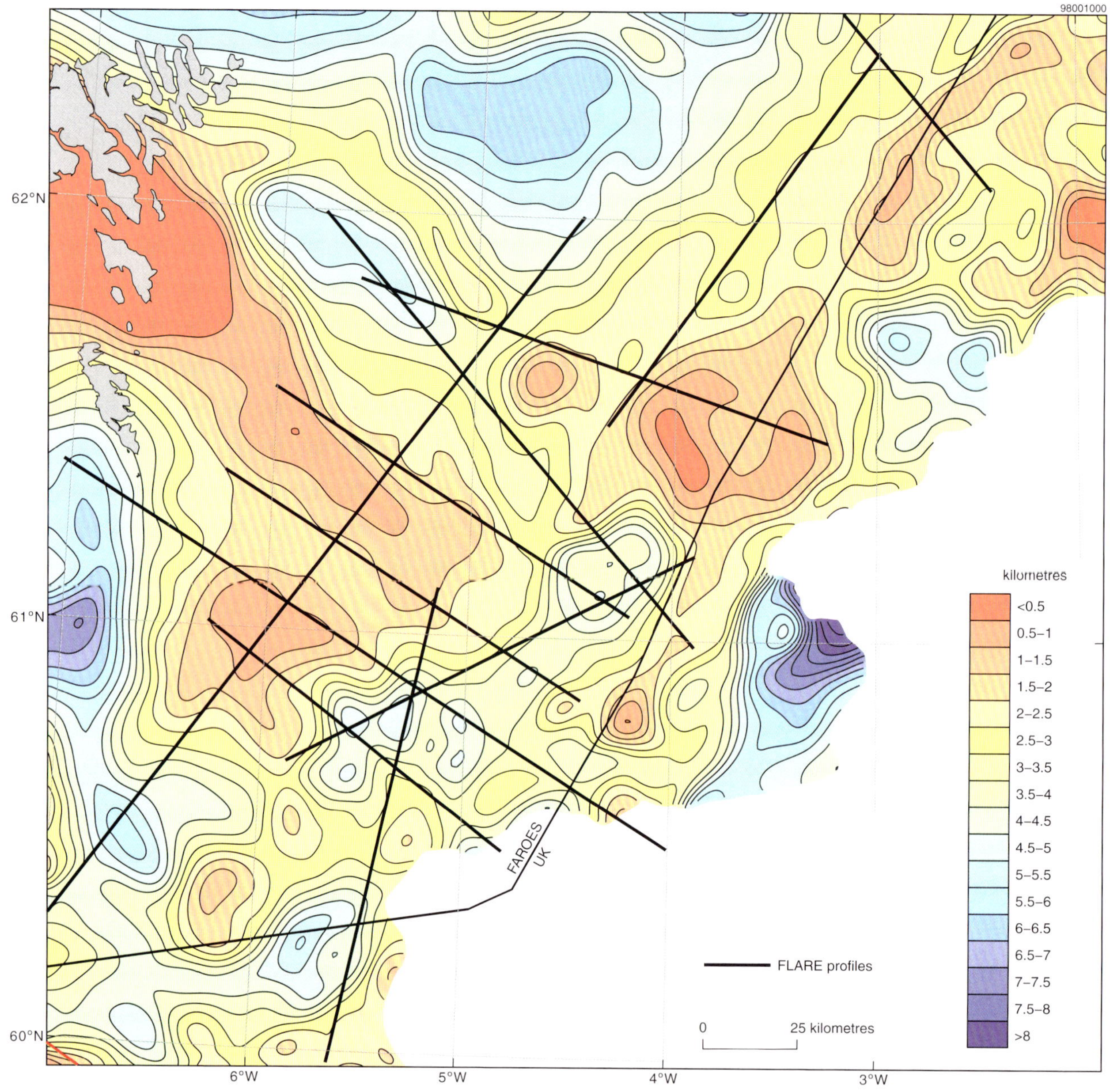

Figure 18 Interpreted thickness of the sedimentary rocks between the base of the basalts and the top of the pre-rift basement, interpreted from a combination of results from the FLARE profiles and conventional multichannel seismic survey data (slightly modified reproduction from White et al., (2003) by permission of Blackwell Publishing and J R Smallwood, personal communication).

seismic sections and their results are incorporated in the Moho map shown in Figure 17.

The BIRPS surveys also detected dipping structures in the mid-crust that appear to be associated with deformation in the vicinity of the Moine Thrust Zone (Caledonian Orogenic Front). A bright reflector with an eastward dip of 14 ± 2° (migrated apparent dip) at 4.1 to 5.7 s TWTT (11 to 16 km) beneath line NSDP85-8 separates an unreflective section above from a highly reflective section below (M-M' in Figure 21). McBride and England (1994) interpreted the reflector as the seismic signature of the Moine Thrust Zone, and correlated it with similar features seen on commercial seismic lines immediately south of Shetland (Andrews, 1985), which can in turn be correlated with the location of the Moine Thrust as inferred from nearby outcrop evidence at North Roe on Shetland. McBride and England (1994) suggested that the reflectivity underlying this interface could be due to Caledonian-age thrusts developed within the Lewisian basement. The NSDP84-3 (Figure 22) profile revealed some rather weak reflections with a south-east dip which provide some evidence of the continuity of the reflection pattern between NSDP85-8 and the profile of Andrews (1985), although there is an apparent dextral inflection or offset in the structure north of NSDP84-3 (McBride and England, 1994). The correlation of the reflector pattern with the Moine Thrust becomes more problematic farther south. For example, a very similar mid-crustal reflection pattern is seen beneath the GRID-15 seismic profile (Figure 21) but this underlies a north-east-trending ridge of autochthonous Lewisian basement and projects towards the surface well to the west of any feasible geological extrapolation of the Moine Thrust into this area (M-M' in Figure 21;

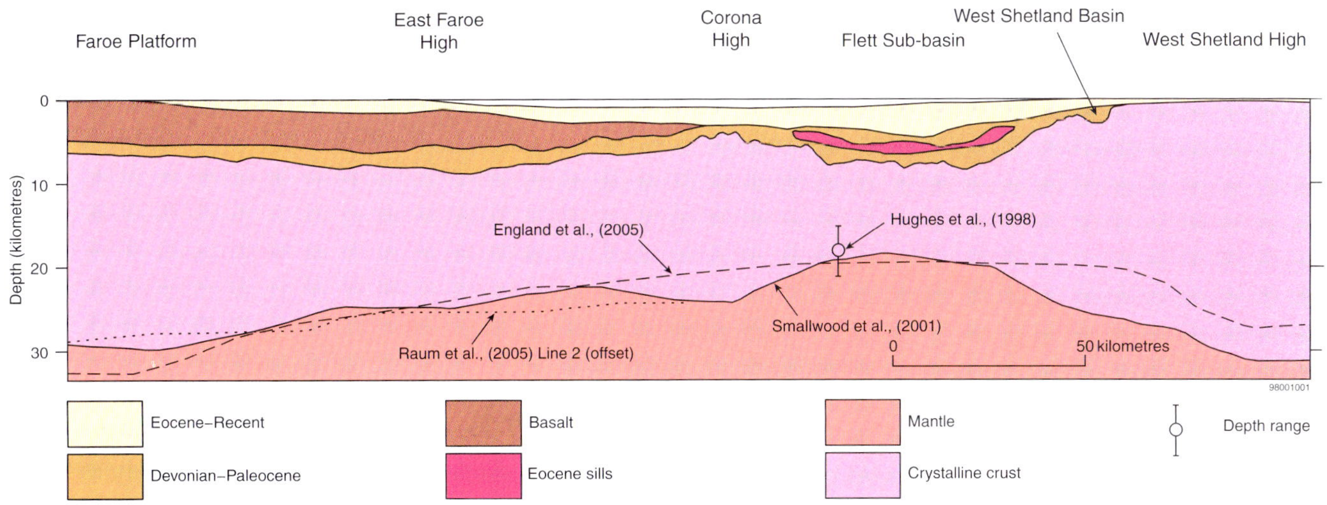

Figure 19 Interpretation of deep crustal structure beneath the FAST profile based on a combination of seismic interpretation and gravity modelling (modified from Smallwood et al., 2001). Alternative interpretations of Moho depth beneath this and neighbouring lines are shown. For location, see Figure 16.

McBride and England, 1995). One possible explanation for these observations is that the mid-crustal reflector (M-M') relates to a structure developed prior to the Caledonian Orogeny, for example as a Proterozoic normal fault; around Shetland this may have been reactivated or overprinted by the Moine Thrust, whereas closer to the mainland these two features diverge (McBride and England, 1995). Brewer and Smythe (1984) suggested two possible positions for the Moine Thrust on the MOIST seismic reflection profile, but this interpretation did not allow for the effects of the extension that formed the West Orkney Basin, which is interpreted to have involved westward translation of the footwall relative to a 'fixed' hanging wall to the east (Enfield and Coward, 1987; Stein and Blundell, 1990). Snyder (1990) incorporated these effects and correlated the features identified by Brewer and Smythe (1984) with the Naver and Swordly thrusts (north and south respectively in Figure 21). A more westerly location is thus predicted for the Moine Thrust, although this feature is not clearly imaged on the MOIST or DRUM profiles (Figure 21).

The patterns revealed by the offshore seismic reflection data suggests that the Caledonian Orogenic Front is a 'thick-skinned' feature, whereas structural interpretation in the onshore area favours a 'thin-skinned model (Butler and Coward, 1984). Such a change in structural style might be accommodated by a lateral ramp in the vicinity of the northern coast of Scotland.

The seismic reflection Moho lies at about 9 s TWTT beneath the UNST profile, to the north of the Shetland Islands and west of the Walls Boundary Fault

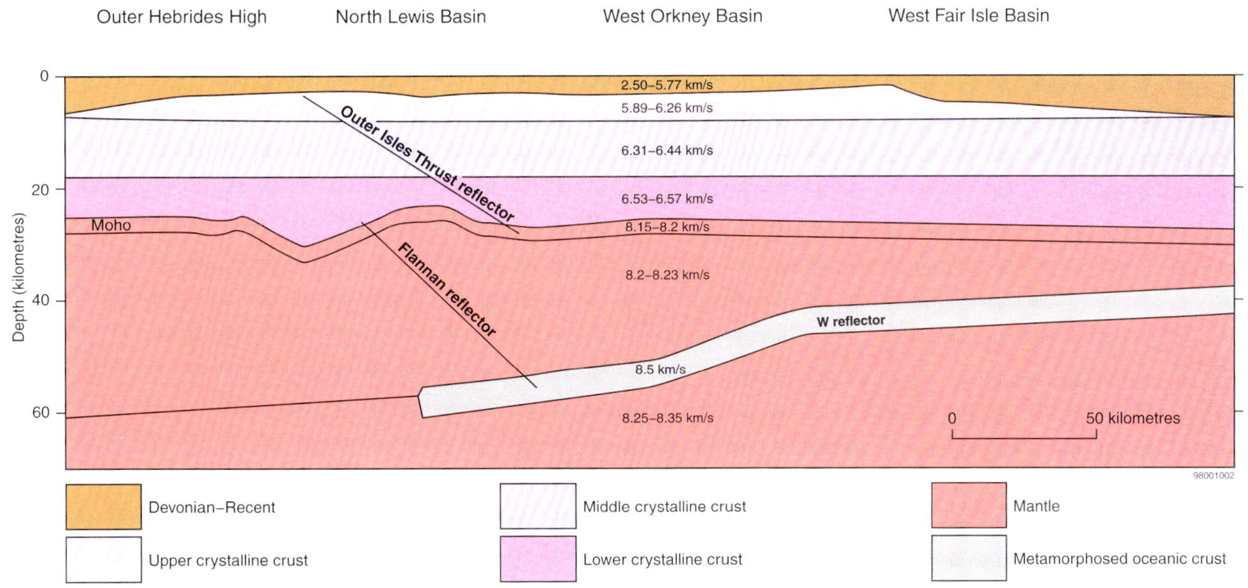

Figure 20 Deep lithospheric structure of the area to the north of Scotland from the results of the W-reflector integrated seismic experiment (modified from Morgan et al., 2000). For location, see Figure 16.

(Figure 22; McGeary, 1989; McBride and England, 1994). The Moho reflector is deflected downwards in the vicinity of the Walls Boundary Fault, and McGeary (1989) interpreted this, and associated diffractions, as evidence of a 2 to 3 km offset in the depth to Moho across the fault. McBride (1994), however, concluded that the Moho disruption took the form of a narrow (approximately 15 km) 'keel' which is bisected by the downward projection of the fault, and that the Moho lies at similar depth on either side. This interpretation is more compatible with the observed gravity field, which does not reveal the large gravity gradient that would characterise a Moho step of this amplitude. A zone of eastward-dipping reflections in the middle crust appears to correlate with features seen beneath the Moine Thrust farther south (Figure 22; McBride and England, 1994). These are cut off directly beneath the Walls Boundary Fault, supporting interpretations involving large-scale strike-slip displacements on the fault (McBride, 1994). Elsewhere, the Walls Boundary Fault is marked by reflector discontinuities and reduced reflectivity beneath the NSDP84-3 and FAIRISLE profiles (Figure 22).

The BIRPS experiments also resolved prominent reflectors within the upper mantle in this region (e.g. Smythe et al., 1982; Brewer et al., 1983; Brewer and Smythe, 1984; Flack and Warner, 1990; Snyder and Flack, 1990; McBride et al., 1995; Snyder and Hobbs, 1999; Morgan et al., 2000). The MOIST, WINCH, DRUM and GRID experiments mapped out two main reflective zones within the upper mantle beneath the North Lewis and West Orkney basins. The W-reflector occurs at a depth of 40 to 50 km, and appears to be truncated at its western end by the eastward-dipping (at about 30°) Flannan reflector (Figure 21, DRUM profile). Farther south, the mantle reflectors appear to have been offset by late Caledonian, sinistral, strike-slip displacements on the Great Glen Fault, indicating that they are of Caledonian or older age (Snyder and Flack, 1990; Snyder et al., 1997). The dipping mantle reflectors have been correlated with similar features farther north, to the west of Shetland, but appear to be disrupted in the intervening region (McBride et al., 1995). This offset can be seen in Figure 21, where the dipping mantle reflectors on MOIST and DRUM are not seen on GRID15 (they extend beneath and west of this line) but re-appear in a more easterly location beneath NSDP85-8. McBride et al. (1995) discussed alternative explanations for the offset in the mantle reflectors, including (i) differential extension relating to the formation of the West Orkney and Faroe–Shetland basins, and (ii) a broad shear zone that was largely confined to the mantle but influenced the formation of overlying basins. It is interesting to note the close correspondence between the mantle shear zone of McBride et al. (1995) and the Wyville Thomson Lineament Complex (Figure 7) identified by Kimbell et al. (2005), which itself assimilated components identified by previous authors i.e. the North Orkney Wyville Thomson Transfer Zone of Stoker et al. (1993) or the Orkney–Shetland Alignment of Earle et al. (1989).

A number of explanations have been put forward for the mantle reflections observed beneath north-west Scotland, including shear zones, igneous intrusions and relict subduction zones. Recent experiments indicate that there is a marked increase in acoustic impedance at the top of both the Flannan and W reflectors (Morgan et al., 1994; Warner et al., 1996; Morgan et al., 2000; Price and Morgan, 2000), suggesting that the reflections are not due to shearing or alteration, which would have the opposite acoustic impedance contrast. Warner et al. (1996) and Morgan et al. (2000) proposed that the reflections are caused by fragments of subducted ocean crust that have been metamorphosed to eclogite facies and preserved within the lithospheric mantle. Warner et al. (1996) suggested that the W-reflector might be a relict of low-angle subduction which was superseded by higher angle subduction recorded by the Flannan reflector. Such pre-Caledonian subduction may have been associated with the Proterozoic assembly of north-west Scottish terranes at about 1900 Ma (Dickin, 1992), and it is possible that the mantle shear zone postulated by McBride et al. (1995) originally formed at this time.

STRUCTURAL ELEMENTS

The structural nomenclature referred to within the Faroe–Shetland report area is summarised in Table 2. A description of individual structural elements is given below and broadly subdivided into platform areas, major Palaeogene basalt highs, major basinal areas and associated sub-basins and highs and other significant features such as faults and lineaments.

Platform areas

East Shetland High

The East Shetland High forms a generally north-north-east- to north-east-trending basement platform that encompasses the Unst, Magnus and Sandwick basins and extends for approximately 325 km, encroaching into the east and north-east part of the report area (Figure 7). The general morphology of the high is defined from commercial seismic (Figures 23 and 24) and BGS shallow seismic data (BGS, 1984; 1987; 1989a). The north-east part of the high is also well imaged from gravity data, corresponding with a positive free-air and isostatic gravity anomaly (Figures 3 and 4). The definition of the high incorporates parts of the former North Shetland Platform and Margarita Spur of Stoker et al. (1993, fig. 12) and the East Shetland High of Johnson et al. (1993, fig. 5). The East Shetland High is bounded to the north-west by a combination of the Flett Sub-basin, Erlend High, Erlend Sub-basin and the Møre Basin (Figure 7). The boundaries with the Møre Basin and Erlend sub-basins are defined by a series of steep, north-east-trending extensional faults with a downthrow of nearly 4s TWTT towards the north-west at top crystalline basement level (Figure 24). The high is interpreted to be cut by the north-west-trending Brendan, Erlend and Magnus lineaments. Towards the west and south-west, crystalline basement of the East Shetland High is separated from the West Shetland and Orkney–Shetland highs by the major north- to north-north-east-trending, strike-slip dominated Walls Boundary Fault (e.g. Figures 22 and 25). Outwith the report area to the east and south, the high is also flanked by a combination of the Orkney–Shetland High, Unst Basin, Magnus Basin and Manet High (see Blystad et al., 1995).

The part of the Shetland Isles to the east of the Walls Boundary Fault lies within the East Shetland High and comprises mainly Precambrian Moinian and Dalradian

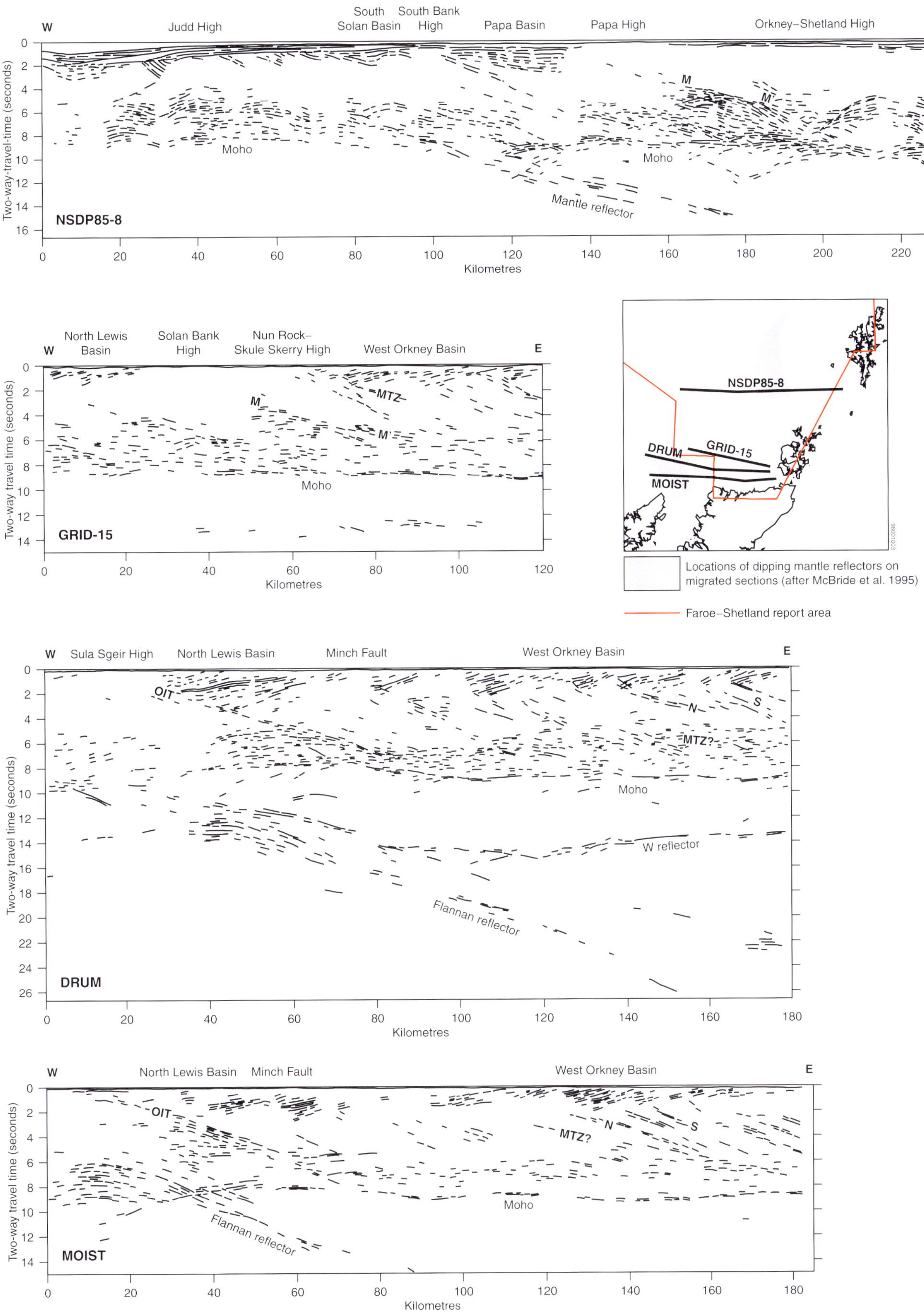

Figure 21 Line drawings based on BIRPS unmigrated deep seismic reflection profiles to the north of Scotland (see Klemperer and Hobbs, 1991). Abbreviations: M–M'=mid crustal reflector of McBride and England, (1994); MTZ=Moine Thrust Zone; N=Naver Thrust; OIT=Outer Isles Thrust Fault and S=Swordly Thrust (N and S after Snyder, 1990).

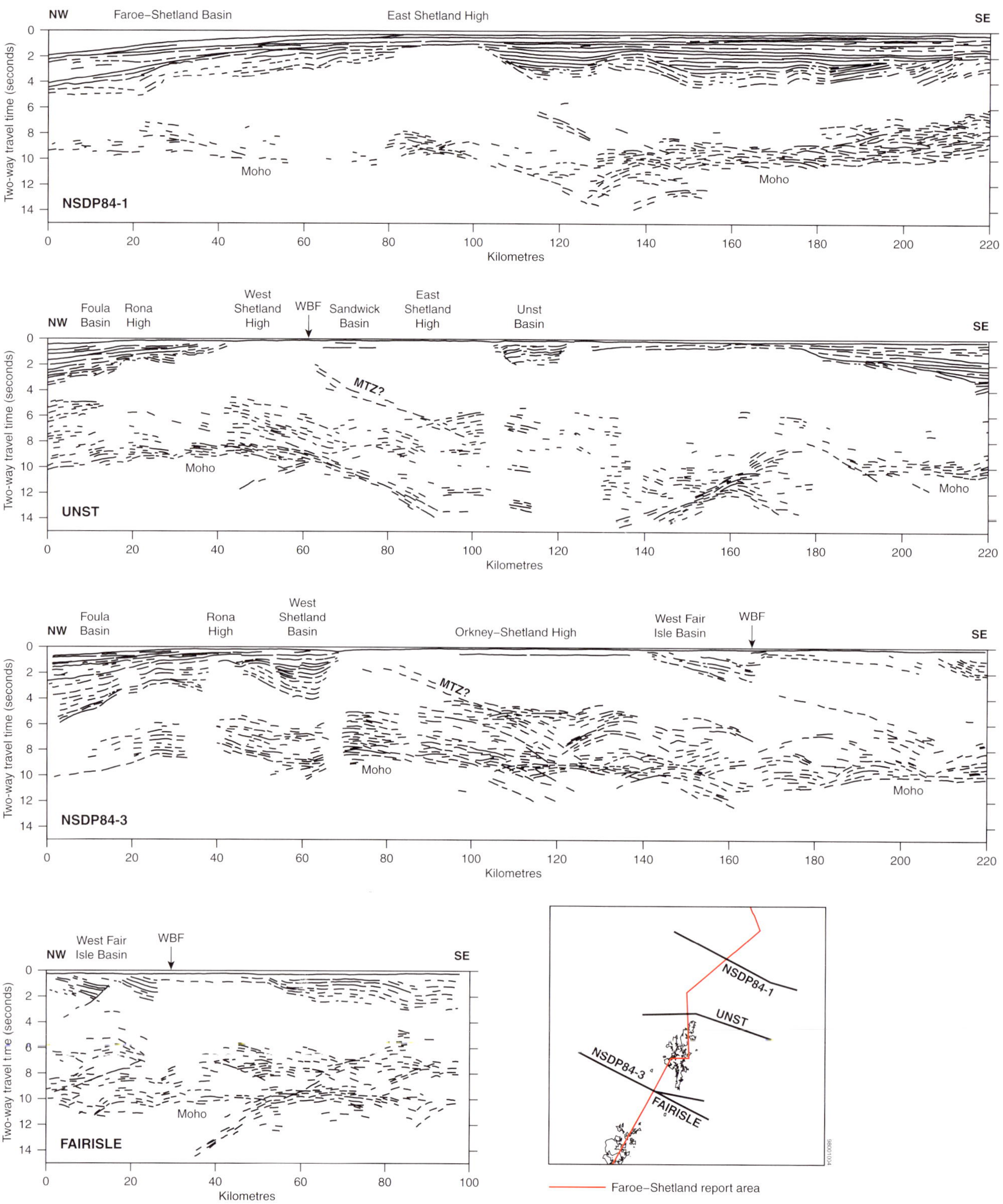

Figure 22 Line drawings based on BIRPS unmigrated deep seismic reflection profiles in the Shetland area (see Klemperer and Hobbs, 1991). Abbreviations: see Figures 7 and 21.

basement (e.g. see Chapter 3; Flinn, 1977a; BGS, 1984; Flinn, 1992;). In the offshore area, BGS borehole 81/17 (Figure 2) was drilled on the crest of the high, proving amphibolite just below the sea bed while close to the boundary with the Møre Basin, well 220/26-2 encountered approximately 1.7 km of Cenozoic to Recent, 1.2 km of Upper Cretaceous and 1 km of Permo-Triassic strata within which the well terminated.

Erlend High

The Erlend High forms a small, heavily intruded, fault-bounded basement terrace within the north-east part of report area that extends for a maximum of approximately 65 to 70 km in both north-easterly and north-westerly directions (Figure 7; Duindam and van Hoorn, 1987; Stoker et al., 1993). The high is well defined from seismic data, although its deep structure is mainly obscured by the presence of an overlying succession of Palaeogene lava and intrusive rocks associated with the Erlend and West Erlend Volcanic centres (see Chapter 9; Figure 26; Gatliff et al., 1984). The twin, circular positive free-air and isostatic gravity anomalies associated with the Erlend High appear mainly attributable to the presence of basic plutonic masses belonging to the Erlend and West Erlend Volcanic centres (Figures 3 and 4; Chalmers and Western, 1979; Gatliff et al., 1984). The Erlend High is flanked to the south-east by crystalline basement of the East Shetland High, and bounded to the north-east and south-west by the Magnus and Erlend lineaments (Figure 7), respectively. The north-west flank of the Erlend High is assumed to be fault-bounded (Duindam and van Hoorn, 1987) but is masked by the presence of a north-westerly thickening wedge of basic lavas that are associated with the Erlend Volcanic Escarpment (Figure 26). However, the presence of an elongate west-trending positive gravity anomaly that extends to the west of the West Erlend Volcanic Centre could indicate that the boundaries of the Erlend High should be migrated further in that direction (Figure 4).

The Erlend High has been drilled by released wells 208/15-1A, 209/03-1, 209/04-1A, 209/06-1 (Figure 26), 209/09-1A and 209/12-1 (Figure 2). These wells proved an approximate maximum thickness of 5.75 km of Mesozoic to Recent rocks, including 2.4 km of Cenozoic to Recent, 1.45 km of Palaeogene lavas, 1.8 km of heavily intruded Lower and Upper Cretaceous and 100 m of Upper Jurassic strata. Wells 209/09-1A and 209/12-1 reached terminal depth within crystalline basement approximately 2.7 to 3.5 km below the drillfloor.

Judd High

The Judd High occurs within the south-west part of the report area and is interpreted to form a large north-east-tapering platform area, with a maximum length and breadth of approximately 90 to 100 km (Figure 7; Stoker et al., 1993). The eastern parts of the high are generally well defined from both seismic data (Figures 27 and 28; Kirton and Hitchen, 1987, fig. 8) and gravity data, corresponding with a positive free-air and isostatic gravity anomaly (Figures 3 and 4). However, the nature and extent of the north-west and south-west margins of the high are largely inferred. The Judd High is separated from the Judd Sub-basin to the north-east by a combination of faults including the west-north-west- to west-trending Judd Fault and the north-west-trending Judd Lineament (Figure 7). Towards the south-east, it is bounded by a combination of the West Solan and North Rona basins, with the former contact represented by a normal fault dipping towards the south-east (Herries et al., 1999), and the latter by a south-east-dipping ramp that forms the western flank of a half-graben structure (Figure 28). The south-west margin of the high is interpreted to be separated from a combination of the Sula Sgeir and Outer Hebrides basement highs by a series of arcuate, generally north-north-west-trending inverted normal faults (Earle et al., 1989) and the Wyville Thomson Lineament Complex (Kimbell et al., 2005). The Munkur Basin flanks the Judd High to the north-west, though the nature of the contact is poorly understood.

The north-east margin of the Judd High has been drilled by a small number of released wells including 204/22-1, 204/26-1A, 204/27a-1, 204/28-1 (Figure 27) and 204/28-2 (Figure 2). These wells proved an approximate maximum thickness of 2.6 km of Mesozoic to Recent sediments, including 2.0 km of Cenozoic to Recent, 50 m of Lower Cretaceous, 350 m of Middle and Upper Jurassic and 200 m of Triassic strata. All the wells with the exception of 204/28-2 terminated within crystalline basement. Notably, Upper Cretaceous sediments have not been drilled on the Judd High, suggesting the possibility that it was either emergent at that time (see Chapter 7; Dean et al., 1999) or that subsequently, any Cretaceous sediments were removed by Early Paleocene erosion.

Møre Marginal High

The Møre Marginal High forms a large, elongate, north-east-trending basalt-capped basement platform approximately 350 km long and up to 160 km wide that occurs mainly within the Norwegian continental margin (e.g. Blystad et al., 1995; Brekke et al., 1999; Berndt et al., 2001; Planke et al., 2005), with only the south-west part extending into the report area (Figure 7; Duindam and van Hoorn, 1987, plate 4h; Doré et al., 1999, fig. 2f). The deep structure of the Møre Marginal High within the report area is not well resolved from commercial seismic data due to the presence of a variably thick succession of Palaeogene lavas (e.g. Hinz et al., 1984; Smythe, 1983; Smythe et al., 1983; Berndt et al., 2001, fig. 6), but close to the south-east margin of the high, a distinct sub-basalt dome-like structure has been identified (Hodges and Evans, 1999). A positive, circular free-air and isostatic gravity anomaly associated with the high appears mainly attributable to the presence of a basic plutonic mass belonging to the Brendan Volcanic Centre (see Chapter 9; Figures 3 and 4; Smythe, 1983; Smythe et al., 1983). Within the Norwegian margin, the north-east flank of the Møre Marginal High is offset in a sinistral sense from the Vøring Marginal High by the north-west-trending Jan Mayen Lineament/Fracture Zone (Blystad et al., 1995, fig. S9). Towards the north-west, the high is bounded by Lower Eocene oceanic crust of the Norwegian Basin whereas to the south-east, is separated from the Møre Basin by basaltic lavas of the Faroe–Shetland Escarpment. Within the report area, the south-west extension of the Møre Marginal High is poorly resolved (Figure 7), with conflicting ideas regarding the nature of the boundary between the high and the Møre Basin. For example, a largely non-faulted contact is preferred by Roberts et al. (1999) whereas south-east- and north-east-dipping fault planes are preferred by Duindam and van Hoorn (1987, plate 4h) and Doré et al. (1999, fig. 2f), respectively.

Within the Norwegian continental margin, the Møre Marginal High has not been tested by drilling but is interpreted to mainly comprise a thick Paleocene to Eocene volcanic succession (e.g. Hinz et al., 1982; Berndt et al., 2001), resting on undifferentiated Mesozoic and older rocks (Blystad et al., 1995, line M–M') The high is considered by Brekke et al. (1999) to have been emergent in Paleocene times, separating East Greenland and mid Norway, though from seismic volcanostatigraphical analysis Berndt et al. (2001) have described a mixture of marine and subaerial depositional environments for the lavas that are inferred to cap it. Within the report area, the south-east margin of the Møre Marginal High has been drilled by well

Table 2 Nomenclature of major structural elements within the Faroe–Shetland report area.

Nomenclature of major structural features	
Structural elements referred to in the report area	**Alternative terminology**
Annika Sub-basin (Keser Neish, 2003)	Part of the East Faroe Basin (Ellis et al., 2002)
Auðhumla Sub-basin (Keser Neish, 2003)	-
Brendan Lineament (this report)	Modified unnamed transfer zone (Rumph et al., 1993)
Brynhild Sub-basin (Keser Neish, 2003)	Part of the Sandoy Sub-basin or East Faroe Graben (Ellis et al., 2002), Corona or Faroe basins (see Keser Neish, 2003)
Clair Basin (e.g. Blackbourn, 1987; Grant et al.,1999; Nichols, 2005)	Part of the West Shetland Basin (e.g. Cashion, 1975; Roberts et al., 1999)
Clair Lineament (Kimbell et al., 2005)	Clair Transfer or Transfer Zone (e.g. Rumph et al., 1993; Grant et al., 1999)
Corona Sub-basin (e.g. Sorensen, 2003; Jolley et al., 2005)	Faroe–Shetland Channel Sub-basin (Roberts et al., 1999), Fugloy Basin (Andersen et al., 2002) or part of the Sissal Basin (Keser Neish, 2003)
Corona High (Sullivan et al., 1999)	Corona Ridge (e.g. Rumph et al., 1993; Lamers and Carmichael, 1999; Smallwood and Kirk, 2005) or Axial Opaque Zone (Ridd, 1983) or Central High–Axial Opaque Zone (Hitchen and Ritchie, 1987) or North Westray Ridge (Iliffe et al., 1999)
Corona Lineament (this report)	-
East Faroe High (e.g. Boldreel and Andersen, 1994; Keser Neish, 2003)	East Faroe Ridge (Lamers and Carmichael, 1999)
East Rona High (BGS, 1986a; Stoker et al., 1993)	-
East Shetland High (this report)	East Shetland Platform (e.g. Johnson et al., 1993); part of the North Shetland Platform (e.g. Hitchen and Ritchie, 1987; Stoker et al., 1993), Margarita's Spur (e.g. Duindam and van Hoorn, 1987; Stoker et al., 1993; Hodges and Evans, 1999), Magnus High (Coward et al., 2003) or Njordfjord High (Nelson and Lamy, 1987)
East Solan Basin (Booth et al., 1993; Herries et al., 1999)	Northern part of North Rona Basin (Stoker et al., 1993)
Erlend Sub-basin (Davis et al., 2004)	Part of the Fugloy Basin (Sorensen, 2003)
Erlend High (this report)	Erlend Platform (e.g. Duindam and van Hoorn, 1987; Stoker et al., 1993)
Erlend Lineament (this report)	Erlend Transfer (e.g. Doré et al., 1999) or Transfer Zone (e.g. Duindam and van Hoorn, 1987; Grant et al., 1999)
Faroe Platform (e.g. Rasmussen and Noe- Nygaard, 1970; Keser Neish, 2003)	Faroe Block, Faroe Plateau or Faroes Islands Shelf (see Keser Neish, 2003)
Faroe Bank Channel Basin (Keser Neish, 2003)	Faroe Bank Channel sub-Basin (Roberts et al., 1999) or part of the Faroe Bank Channel (e.g. Stride et al., 1967; Bott and Smith, 1984; Roberts et al., 1983)
Faroe Bank High (Keser Neish, 2003)	Faroe Bank (e.g. Bott, 1984; Keser Neish, 2003) or Faroe Bank Igneous Centre (Dobinson, 1970; Ritchie et al., 1999b)
Faroe–Shetland Basin (e.g. Hitchen and Ritchie, 1987; Stoker et al., 1993; Dean et al., 1999; Doré et al., 1999)	Faroe Basin (e.g. Ridd, 1981; Booth et al., 1993; Nichols, 2005), Faroe Trough (e.g. Duindam and Van Hoorn, 1987 and Ziegler, 1990); Faroe–Shetland Trough (e.g. Smythe, 1983; Ritchie and Darbyshire, 1984) or Faroe–Shetland Channel (Meadows et al., 1987)

98001893-1

219/21-1 (Figure 2). The well was located in the vicinity of a structure interpreted to represent the 'Ben Nevis' inversion dome, formed by re-activation of a Jurassic fault block during latest Paleocene times (Hodges and Evans, 1999, figs. 4 and 8).

Orkney–Shetland High

The Orkney–Shetland High represents a north-east-trending Devonian platform approximately 275 km long and at least 70 km wide that straddles the east and south-east boundary of the report area (Figure 7;

Table 2 *continued.*

Fetlar Basin (BGS, 1984; Stoker et al., 1993)	-
Flett Sub basin (Lamers and Carmichael, 1999; Davies et al., 2004)	Flett Basin (Hitchen and Ritchie, 1987) or part of the Faroe-Judd Basin (Herries et al., 1999), Erlend Basin (Davies et al., 2004) and Fugloy Basin (Sorensen, 2003)
Flett High (this report)	Flett Ridge (e.g. Mudge and Rashid, 1987; Hitchen and Ritchie, 1987; Smallwood and Kirk, 2005)
Foula Sub-basin (e.g. Lamers and Carmichael, 1999; Grant et al., 1999)	Rona Terrace (Mudge and Rashid, 1987) or Flett Basin (e.g. Goodchild et al., 1999)
Foula High (this report)	Foula Ridge (BGS, 1984)
Fugloy Ridge (e.g. Boldreel and Andersen, 1994; Keser Neish, 2003)	Fugloy Bank (Ziska and Andersen, 2005)
Grani Fault Terrace (Keser Neish, 2003)	-
Grimhild Sub-basin (Keser Neish, 2003)	Part of the East Faroe Graben (Ellis et al., 2002) or Fugloy Basin (Andersen et al., 2002)
Grimur Kamban Lineament (this report)	Grimur Kamban Transfer Zone (Keser Neish, 2003)
Guðrun Sub-basin (Keser Neish, 2003)	East Faroe Sub-basin, Corona, Faroe basins (see Keser Neish, 2003) or Fugloy Basin (Andersen et al., 2002)
Heri High (Keser Neish, 2003)	-
Iceland Basin	-
Iceland–Faroe Ridge (Bott and Watts, 1971)	-
Judd Sub-basin (Herries et al., 1999; Keser Neish, 2003; Smallwood and Kirk, 2005)	North Judd Basin (Goodchild et al., 1999; Jowitt et al., 1999), Faroe–Judd Basin (Herries et al., 1999) or Foinaven Sub-basin (e.g. Lamers and Carmichael, 1999; Ellis et al., 2002)
Judd Fault (e.g. Kirton and Hitchen, 1987; Herries et al., 1999)	Part of Judd Lineament (e.g. Kimbell et al., 2005)
Judd High (e.g. Dean et al., 1999; Smallwood and Kirk, 2005)	Judd Platform (Booth et al., 1993), North Lewis Platform (Mudge and Rashid, 1987), part of the North Rona High (Stoker et al., 1993) or including the Judd Terrace (Jowitt et al., 1999) and parts of the Sula Sgeir and Raasay highs (e.g. Mitchell et al., 1993; Herries et al., 1999)
Judd Lineament (Kimbell et al., 2005)	Faroe Transfer Zone (Duindam and van Hoorn, 1987), Judd Transfer (e.g. Doré et al.,1999) or Transfer Zone (e.g. Rumph et al., 1993; Keser Neish, 2003) or part of the Judd Fault (e.g. Kirton and Hitchen, 1987; Herries et al., 1999)
Magnus Lineament (Kimbell et al., 2005)	Magnus Transfer Zone (e.g. Duindam and van Hoorn, 1987; Ritchie et al., 2003)
Melby Fault (Flinn, 1977; BGS, 1984)	-
Mid Faroe High (this report)	-
Minch Fault (e.g. Jones, 1981)	-
Moine Thrust Zone (offshore) (e.g. Andrew, 1985; Ritchie et al., 1987)	-

98001893-2

Stoker et al., 1993). The general morphology of the Orkney–Shetland High is well defined from BGS shallow seismic data (IGS, 1982; BGS, 1984; 1985; 1988a; 1988b). The high is partially fault-bounded on its west and north-west flanks, juxtaposed with crystalline basement belonging to the Papa and West Shetland highs and Permo-Triassic red beds of the West Orkney Basin (Figure 7). The north-east and eastern margins of the high are defined by a combination of the West Shetland High (Figure 29) (including the onshore Walls Peninsula and Esha Ness areas of the Shetland Isles), St Magnus Bay Basin and West Fair Isle Basin. Towards the south, the high is interpreted to abut the Scottish mainland.

The Orkney–Shetland High has been drilled by a number of BGS shallow sample sites and boreholes e.g. 77/06, 80/10, 12, 82/06, 07, 08 and 09 (Figure 2). With the exception of borehole 80/02 (which encountered Permo-Triassic red beds) these proved Middle Devonian (Eifelian to Givetian) sediments close to the sea bed that are

Table 2 *continued.*

Møre Basin (Rønnevik et al., 1975; Brekke et al., 1999)	North Faroe Basin (e.g. Mitchell et al., 1993; Rumph *et al.*, 1993) or North Shetlands Basin (Duindam and van Hoorn, 1987)
Møre Marginal High (e.g. Blystad et al., 1995; Brekke et al., 1999)	Møre Platform (e.g. Hamar and Hjelle, 1984; Stoker et al., 1993), Outer Møre Platform (Coward et al., 2003) or including the Brendan and Ben Nevis domes (e.g. Hodges and Evans, 1999)
Munkagrunnur Ridge (e.g. Noe-Nygaard, 1962; Keser Neish, 2003)	Faroe Ridge (Ridd, 1981; Meadows et al., 1987)
Munkur Basin (Keser Neish, 2003)	Part of the Faroe Bank Channel Basin (e.g. Tate et al., 1999) or Munkun Basin (Waddams and Cordingley, 1999; Sørensen, 2003)
NE Rockall Basin (e.g. Waddams and Cordingley, 1999)	Part of Rockall Trough (e.g. Roberts, 1975; Smythe, 1989) or Rockall Basin (e.g. Naylor et al., 1999; Coward et al., 2003)
Nesting Fault (e.g. Miller and Flinn, 1966)	-
North Faroe Bank Channel Basin (Keser Neish, 2003)	Part of the Faroe Bank Channel (e.g. Bott and Smith, 1984; Roberts et al., 1983)
North Flett Fault (this report)	Flett Fault (Lamers and Carmichael, 1999)
North Lewis Basin (BGS, 1986a)	Outer Isles Basin (Coward and Enfield, 1987; Earle et al., 1989), Stack Skerry Basin (Duindam and van Hoorn, 1987; Earle et al., 1989) or part of the North Minch Basin (Naylor and Shannon, 1982)
North Rockall Basin (this report)	Part of the Rockall Trough (e.g. Roberts, 1975; Smythe 1989) or Rockall Basin (Naylor et al., 1999; Coward et al., 2003)
North Rona Basin (e.g. BGS, 1986a; Stoker et al., 1993)	Sula Sgeir Basin (Earle et al., 1989; Swiecicki et al., 1995), West Rona Basin (Earle et al., 1989) or Solan Basin (Bailey et al., 1987; Meadows et al., 1987)
North Shoal High (BGS, 1985)	-
Norwegian Basin	-
Nun Rock–Sule Skerry High (BGS, 1986a; 1989c; Stoker et al., 1993)	-
Orkney–Shetland High (this report)	Orkney–Shetland Platform (e.g. Stoker et al., 1993)
Otter Bank Fault (Duindam and van Hoorn, 1987; Herries et al., 1999)	Sula-Sgeir Fault (Booth et al., 1993)
Outer Hebrides High (this report)	Hebrides Platform (e.g. Ritchie et al., 1999b), Outer Hebrides Platform (e.g. BGS, 1992) or Outer Isles Platform (e.g. BGS, 1986a)
Papa Basin (e.g. BGS, 1985; Booth et al., 1993; Herries et al., 1999)	Foula Basin (Bailey et al., 1987; Meadows et al., 1987) or Back Basin (Roberts et al., 1999)
Papa High (this report)	Papa Bank High (BGS, 1985)
Rona Fault (e.g. Dean et al., 1999; Grant et al., 1999)	-
Rona High (this report)	Rona Ridge (e.g. Ridd, 1981; Dean et al., 1999; Nichols, 2005)
Sandwick Basin (Hitchen and Ritchie, 1987; BGS 1989a)	-
Shetland Spine Fault (e.g. Cashion, 1975; Nichols, 2005)	Spine Fault (Duindam and van Hoorn, 1987) or West Shetland Fault (Mudge and Rashid, 1987)

98001893-3

typical of the 'Orcadian Basin' (e.g. Marshall and Hewitt, 2003). From the results of a fission-track dating study of Devonian rocks of onshore north-east Scotland, the Orkney–Shetland High and surrounding platform area in general are considered to have had up to 3 km or so of Mesozoic and older rocks removed from them during phases of uplift, mainly in Mesozoic (2 km of denudation) and Palaeogene (1 km) times (Thomson et al., 1999).

Table 2 *continued.*

Sjúrður Ridge (Keser Neish, 2003)	Sandoy Ridge (Ellis et al., 2002) or partially overlaps the Central Ridge (Roberts et al., 1999; Cawley et al., 2005)
Solan Bank High (BGS, 1986a; Stoker et al., 1993)	Sula Sgeir High (Dean et al., 1999), Skerry Ridge (Brewer and Smythe, 1984), Stack Skerry Horst (Duindam and van Hoorn, 1987) or Rona Ridge (Earle et al., 1999)
South Flett Fault (this report)	Flett Fault (Grant et al., 1999)
South Solan Basin (Herries et al., 1999)	Southern part of East Solan Basin (Booth et al., 1993)
Steinvør Sub-basin (Keser Neish, 2003)	Part of the East Faroe Graben (Ellis et al., 2002)
St Magnus Bay Basin (BGS, 1984; Stoker et al., 1993)	-
Sula Sgeir High (Stoker et al., 1993)	-
Tróndur High (Keser Neish, 2003)	Part of the East Faroe High (Ellis et al., 2002)
Unst Basin (Johns and Andrews, 1985; Stoker et al., 1993)	-
Victory Lineament (Kimbell et al., 2005)	Victory Transfer Zone (Rumph et al., 1993; Goodchild et al., 1999)
Walls Boundary Fault (e.g. Flinn, 1961; BGS, 1984)	-
Westray Fault (Smallwood et al., 2004; Smallwood and Kirk, 2005)	-
Westray High (this report)	Westray Ridge (e.g. Duindam and van Hoorn, 1987; Lamers and Carmichael, 1999; Smallwood and Kirk, 2005)
Westray Lineament (this report)	Westray Transfer Zone (Rumph et al., 1993; Keser Neish, 2003)
West Fair Isle Basin (e.g. BGS, 1988a; Stoker et al., 1993)	Fair Isle Basin (Coward et al., 2003)
West Lewis Basin (e.g. Earle et al., 1989; Tate et al., 1999)	-
West Orkney Basin (Coward and Enfield 1987; Stoker et al., 1993)	Orkney Basin (Coward et al., 2003), West Orkney Basin Complex BGS, 1986a) or an amalgamation of the Skerry (BGS, 1985), Stormy Bank (BGS, 1985; 1989c) and Noup (BGS, 1985) basins
West Rona High (this report)	Rona High (BGS, 1986a; Stoker et al., 1993)
West Shetland High (this report)	West Shetland Platform (e.g. Ridd, 1981; Stoker et al., 1993; Nichols, 2005) or Shetland Margin (Robinson et al., 2004)
West Shetland Basin (e.g. Cashion, 1975; Stoker et al., 1993; Roberts et al., 1999)	-
West Solan Basin (Booth et al., 1993; Herries et al., 1999)	Northern part of North Rona Basin (Stoker et al., 1993)
Wyville Thomson Lineament Complex (Johnson et al., 2005; Kimbell et al., 2005)	Wyville Thomson Ridge Complex (Andersen et al., 2004), parts of the North Orkney/Wyville Thomson Transfer Zone (Stoker et al., 1993), the Wyville Thomson Transfer Zone (Waddams and Cordingley, 1999) and the Orkney–Faroe Alignment (Earle et al., 1989)
Wyville Thomson Ridge (e.g. Ellett and Roberts, 1973; Keser Neish, 2003)	-
Ymir Ridge (e.g. Boldreel and Andersen, 1993; Keser Neish, 2003)	-

98001893-4

Outer Hebrides High

The north-east-trending Outer Hebrides High (including the Outer Hebrides Isles) is a major basement platform approximately 400 km long and up to 190 km wide, the north-east extremity of which only just encroaches into south-west part of the report area (Figure 7). Outwith the study area, the high is clearly imaged on seismic and potential field data. The north and north-eastern margin of the Outer Hebrides High is inferred to be separated from a combination of the

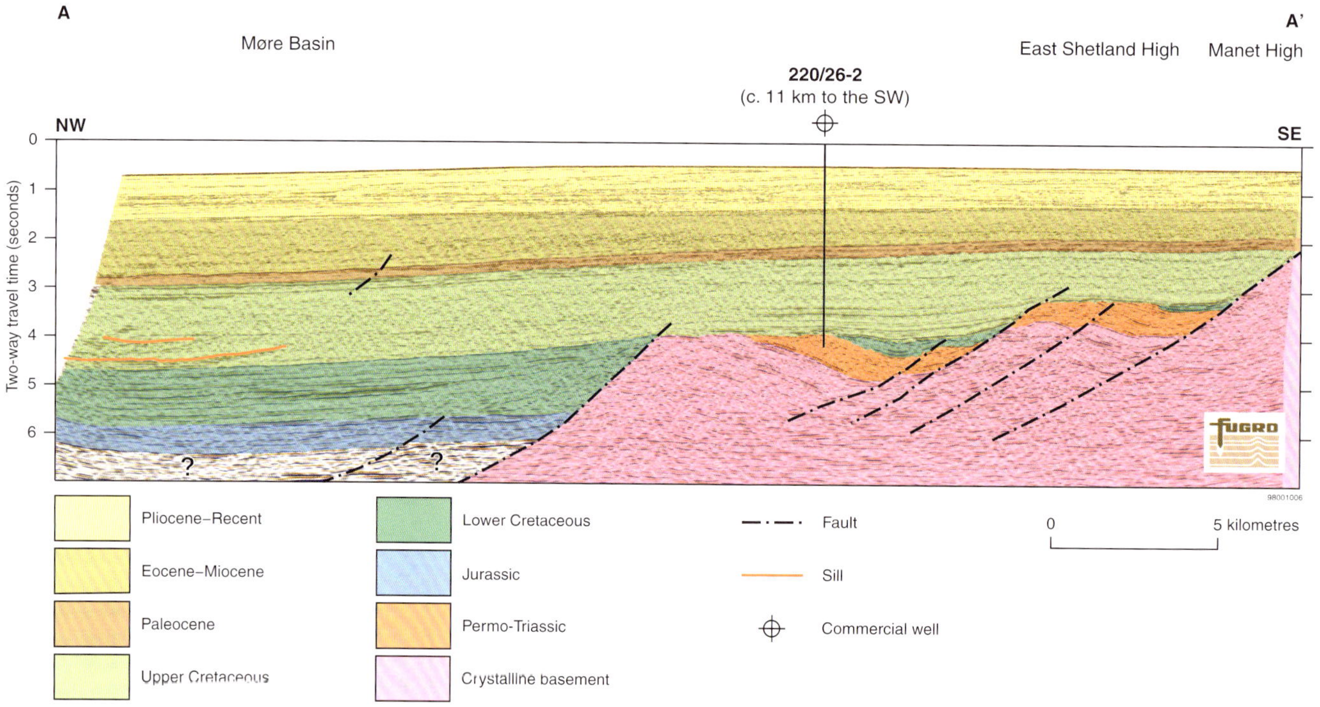

Figure 23 Seismic profile A–A' across the Møre Basin, East Shetland High and Manet High. For location of profile, see Figure 7. Seismic data courtesy of Fugro Multi Client Services.

Wyville Thomson Ridge and Judd High by the Wyville Thomson Lineament Complex and a north-north-east-trending belt of inverted normal faults.

The Outer Hebrides High has not been drilled within the report area, but further to the south-west the results from a number of BGS boreholes and shallow sample sites indicate that it comprises crystalline basement of Lewisian aspect, which in places, is unconformably overlain by thin Permo-Triassic and/or Palaeogene lavas and younger sedimentary rocks.

West Shetland High

The West Shetland High forms a north-east-trending basement platform approximately 175 km long and up to 30 km wide that occurs mainly to the west of the Shetland Islands (Figure 7; Stoker et al., 1993). The general morphology of the West Shetland High is well defined from a combination of both commercial seismic (e.g. Figures 30 and 31) and BGS shallow seismic data (BGS, 1984; 1985; 1988b; 1989a). It also corresponds with a large, elongate north-east-trending positive free-air and isostatic gravity anomaly (Figures 3 and 4). The West Shetland High is separated from the West Shetland Basin and Erlend Sub-basin to the north-west by the major extensional, north-east-trending Shetland Spine Fault (Figures 25, 30 and 31). Towards the south-west, the high is inferred to be terminated by the Westray Lineament, a transfer fault that is interpreted to coincide with a change from basement of Lewisian aspect in the north, to that of Moinian affinity on the Papa High to the south (see Chapter 3; BGS, 1985). Towards the south-east, the high has a partially fault-bounded contact with Devonian rocks of the Orkney–Shetland High, whereas to the north-east, it is separated from crystalline basement of the East Shetland High and Devonian rocks within the Sandwick Basin (Figure 25), by the Walls Boundary Fault.

The part of the Shetland Isles that comprises mainly gneisses of Lewisian affinities (see Chapter 3; BGS, 1984; Flinn, 1992) to the west of the Walls Boundary Fault is assigned to the West Shetland High. In the offshore area, the high has been drilled by BGS boreholes 80/07 and 09 (Figure 2), with the latter proving granitic gneiss close to the sea bed.

Major Palaeogene basalt-covered highs

Faroe Bank High

The Faroe Bank High is an elliptical, flat-topped, basalt-covered dome that occurs close to the south-west margin of the report area (Figure 7). The high has a marked positive bathymetric expression rising sharply to a depth of only 100 m (Figure 1). It is also well imaged from seismic data, with Palaeogene lavas outcropping at the sea bed (e.g. Keser Neish and Ziska, 2005, fig. 5, profiles C and D), though a definitive gravity signature is less clear. The Faroe Bank High is bounded to the north and east by a combination of the North Faroe Bank Channel and Faroe Bank Channel basins, respectively (Figure 7). Although these boundaries have been interpreted as steep normal faults at the level of top Palaeogene lavas (Keser Neish and Ziska, 2005, fig. 5, profiles C and D), little fault displacement is observed at this level. Towards the south, the transition with the Wyville Thomson Ridge remains poorly understood.

The Faroe Bank High has not been drilled, although dredge samples of basaltic lavas have been recovered (Waagstein, 1988, fig. 1). On the basis of potential field data, the high has been interpreted to be in part a ?Palaeogene volcanic centre (Dobinson, 1970).

Faroe Platform

With the Faroe Islands at its centre, the Faroe Platform is located towards the north-west of the report area and represents a major basalt-covered structural high (Figure 7). The general morphology of the platform is defined from seismic data (e.g. Keser Neish, 2003, enclosure 2, profiles E and G) although the presence of Palaeogene lavas mask its underlying structure. The boundaries that mark the transition between the Faroe Platform and other major juxtaposed basalt-covered structural highs, namely the Iceland–Faroe, Fugloy and Munkagrunnur ridges, are largely inferred (Figure 7). Similarly, the continent–ocean transition to the north, from the platform to the Norwegian and Iceland oceanic basins is poorly understood (see Deep Structure, The continent–ocean boundary). Towards the south-west and south-east, the high is flanked by the North Faroe Bank Channel Basin and Annika Sub-basin, respectively (Keser Neish, 2003, enclosure 2, profiles G and E), with the latter marked by the presence of a steep, generally north-trending, extensional fault (e.g. Ellis et al., 2002) defined at the level of top Palaeogene lavas.

On the Faroe Islands, the Palaeogene lavas have been drilled by a number of deep boreholes and the maximum stratigraphical thickness of the Faroe Island Basalt Group is estimated at 6.6 km, 1 km of which comprises intrusions and volcaniclastic and hyaloclastic rocks (see Chapter 9). According to Andersen et al. (2002) however, uplift due to effects of compression in Mid to Late Miocene and Late Pliocene times was responsible for the erosion of up to 2 km from the lava pile on the islands. In the near offshore area, the top of the lavas generally lies at or close to the sea bed (see Figure 121) but their thickness can only be derived from the results of deep geophysical experiments (Figure 16). For example, the integration of the FLARE seismic experiment with commercial seismic data (White et al., 2003, fig. 11) suggested that the bulk of eastern part of the Faroe Platform comprises between approximately 3 to 6 km of Palaeogene basaltic lavas and up to 3 km of sub-basalt strata resting on pre-rift (base Upper Jurassic and older) basement. A distinct local gravity anomaly low just to the north-west of the Faroe Islands (Figures 3 and 4) has been interpreted by Schrøder (1971) to represent a salic body within basalt. The deep crustal structure of the Faroe Platform is discussed above (see Deep Structure, Area north-west of the Faroe–Shetland Basin).

Fugloy Ridge

The Fugloy Ridge represents a slightly arcuate, east- to north-east-trending structural high that extends for approximately 240 km to the east of the Faroe Islands (Figure 7). On the basis of seismic data, it is interpreted as a deeply buried culmination at Mesozoic and older stratigraphical levels (Figure 32; Keser Neish, 2005, enclosure 2, profile A), that is slightly offset towards the south-east from a large, asymmetrical anticline defined at top

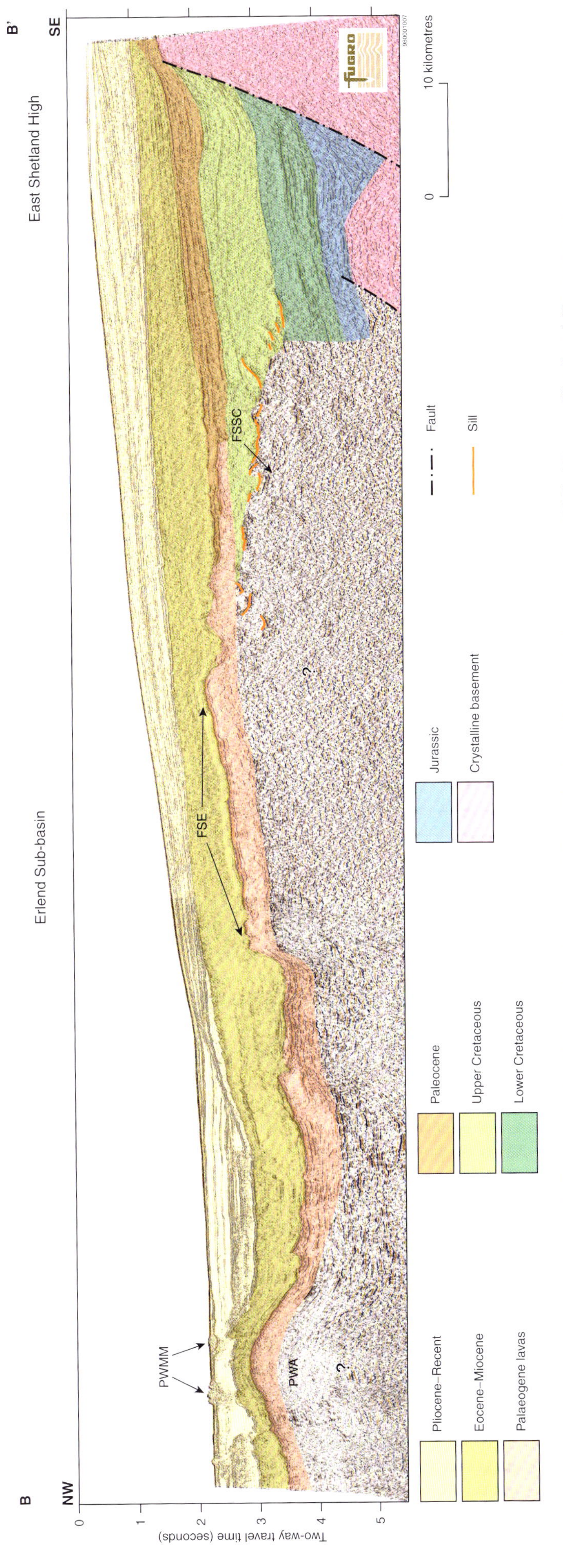

Figure 24 Merged seismic profile B–B' across the Erlend Sub-basin and East Shetland High. Abbreviations: FSE=Faroe–Shetland Escarpment; FSSC=Faroe–Shetland Sill Complex; PWA=Pilot Whale Anticline and PWMM=Pilot Whale mud mounds. For location of profile, see Figure 7. Seismic data courtesy of Fugro Multi Client Services.

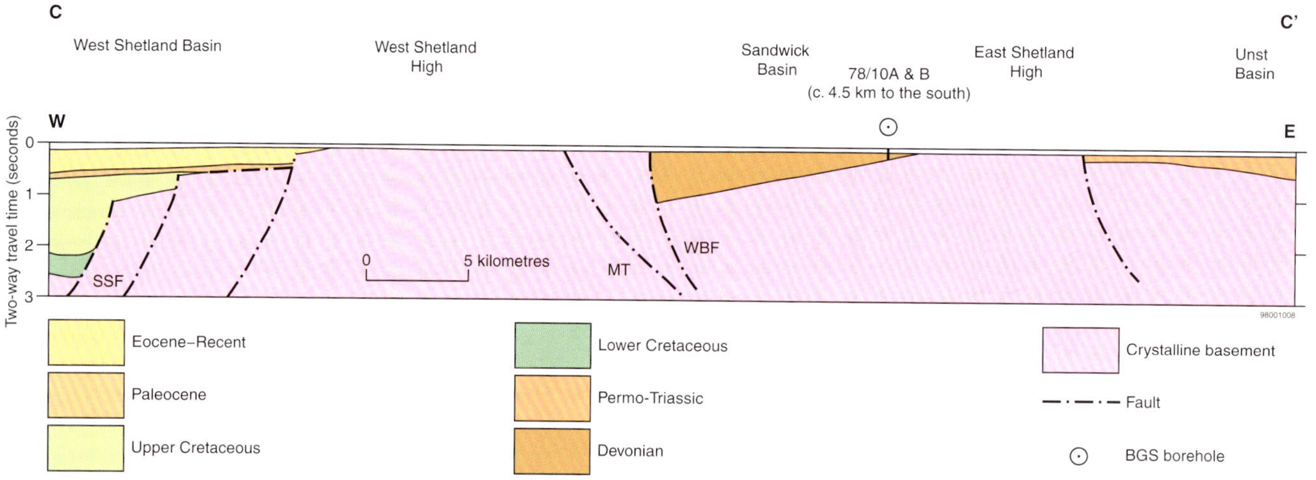

Figure 25 Generalised geological cross-section C–C' across the north-east West Shetland Basin, West Shetland High, Sandwick Basin, East Shetland High and Unst Basin (modified from BGS, 1989a). Abbreviations: see Figure 7. For location of profile, see Figure 7.

Palaeogene lavas and younger stratigraphical levels (e.g. Figures 26 and 33; White et al., 2005, fig. 10; Roberts et al., 2005, fig. 10; Keser Neish, 2005, fig. 9; Ziska and Andersen, 2005, fig. 3). The nature of the boundary between the Fugloy Ridge and Norwegian Basin is obscured by the presence of thick lavas and is largely conjectural (Figure 7; White et al., 2005, fig. 10). Towards the south-east, the ridge is separated from the Steinvør Sub-basin by a series of normal, north-east-trending, generally pre-Cenozoic faults, with small south-easterly downthrows at ?top crystalline basement level (Figure 32). The transition to the west and east with the Faroe Platform and Møre Marginal High, respectively, is poorly understood.

The Fugloy Ridge has not been drilled but results from conventional seismic mapping (Figure 32; Keser Neish, 2003, enclosure 2, profiles A, B and C) indicate that it may comprise up to 0.7 s TWTT of Eocene to Recent sediments, 1.2 s TWTT of Palaeogene lavas, 0.5 s TWTT of Paleocene sediments and 3 s TWTT of Mesozoic and older strata resting on ?crystalline basement. The results from a wide-angle seismic experiment towards the western margin of the Fugloy Ridge by Raum et al. (2005, fig. 9) (Figure 16) indicate the presence of approximately 1.5 km of Palaeogene lavas and 3 km of Cretaceous and older sediment resting on continental crystalline basement. According to White et al. (2003), the thickness of Palaeogene basaltic lavas decreases abruptly in an easterly direction along the ridge from approximately 6 km to 2 km. The thickness of the sub-basalt sedimentary strata varies, but a maximum of 4.5 km is interpreted to be present below the central part of the ridge. The Fugloy Ridge is thought by some to represent an Eocene to mainly Miocene compressional growth fold (e.g. Boldreel and Andersen, 1998; Ritchie et al., 2003). Although compression undoubtedly contributed significantly to its development, the ridge could represent the north-west margin of the Faroe–Shetland Basin as defined in Jurassic to Cretaceous times or simply a structural high that formed in response to the effects of post-Paleocene differential thermal subsidence.

Iceland–Faroe Ridge

The Iceland–Faroe Ridge forms a north-west-trending bathymetric high (Figure 1) that extends for approximately 300 km between the Faroe Islands and Iceland (Keser Neish, 2003). At its south-east margin, it is separated from the Faroe Platform by a series of short escarpments (Keser Neish, 2003). The ridge has been drilled by DSDP borehole 336, proving approximately 465 m of Middle Eocene to Recent sediments resting on basaltic lavas (Figure 17 and see Figure 104). These lavas yielded a K–Ar age of approximately 43 to 40 Ma (i.e. Mid Eocene). A full description of the origin and nature of the ridge is given above (see Deep Structure, Iceland–Faroe Ridge).

Munkagrunnur Ridge

The Munkagrunnur Ridge forms an elongate, slightly asymmetrical, north-north-west-trending, flat-topped anticline approximately 130 km long and up to 60 km wide that occurs within the western part of the report area (Figure 7). It is well defined from seismic data (Figures 34 and 35; Roberts et al., 1999, fig. 29; Smallwood et al., 2001, fig. 3; Keser Neish and Ziska, 2005, fig. 5f), and corresponds with a notable positive high frequency magnetic response (Figure 5) caused by the presence of lavas cropping out at the sea bed at relatively shallow water depths. The nature of the boundaries that define the Munkagrunnur Ridge is poorly understood, possibly due in part to the presence of Palaeogene lavas that mask the deep structure (Figure 34; Roberts et al., 1999, fig. 29). In particular, the transition between the Munkagrunnur Ridge and the Faroe Platform to the north is rather arbitrarily defined. According to Keser Neish (2003, enclosure 2, profile J) and Keser Neish and Ziska (2005, fig. 5, profile F), the flanks of the high are marked by series of near vertical extensional and reverse faults (e.g. Figures 34 and 35), though these have little or no throw at Palaeogene lavas and older stratigraphical levels in the basinward part of the feature. It should be noted, however, that poor seismic imaging probably hinders resolution of faults in key areas where the dip at the level of the Palaeogene lavas is greatest.

The Munkagrunnur Ridge has not been drilled but basaltic lava dredge samples have been recovered (Waagstein, 1988, fig. 1). The results of potential field modelling within the southern part of the ridge suggest that it represents a crystalline basement block, capped by 1 km of folded Palaeogene lavas and <2 km of pre-lava sedimentary rocks (Smallwood et al., 2001, fig. 3). The presence of an isostatic gravity anomaly along the axis of the Munkagrunnur Ridge (Figure 4) suggests that the lavas on the ridge are underlain by lower density (sedimentary or granitic?) rocks (G S Kimbell, pers comm.). Towards the north of the ridge however, results from the FLARE 1 seismic experiment indicate that up to 4.5 km of Palaeogene lavas and 4.5 km of weathered basement/high velocity sediments (5.5–5.60 km/s) rest on crystalline basement (White et al., 2005). According to Boldreel and Andersen (1993; 1998) and Tate et al. (1999, fig. 6), the ridge represents an inversion anticline, initially formed during Eocene times as part of a ramp anticline complex on a northward dipping major crustal fault associated with development of the incipient north-east Atlantic sea-floor spreading system. While Linnard and Nelson (2005) consider inversion an important component in the evolution of the ridge, they suggested that it mainly represents a 'Vaila (Mid Paleocene) plume trail'; a view at variance with the deep crustal analysis of Smallwood et al. (2001) and White et al. (2005).

Wyville Thomson Lineament Complex

The Wyville Thomson Lineament Complex is interpreted to comprise two main north-west to west-north-west-trending lineaments that potentially extend from the Iceland Basin in the north-west, to the intersection of the Outer Hebrides and Judd highs in the south-east (Figure 7; Kimbell et al., 2005, fig. 6). These lineaments were interpreted by Kimbell et al. (2005) to have acted as a transfer zone during Mesozoic times, separating basins with differing polarities (see Stoker et al., 1993). It may also represent the location of a long-lived mantle shear zone (see Deep structure, Area south-east of the Faroe–Shetland Basin). The individual lineaments or transfer faults that are speculated to underlie the Wyville Thomson and Ymir ridges could have been reactivated as part of a ramp anticline complex during Palaeogene times (see below). The interpretation of Kimbell et al. (2005) modified

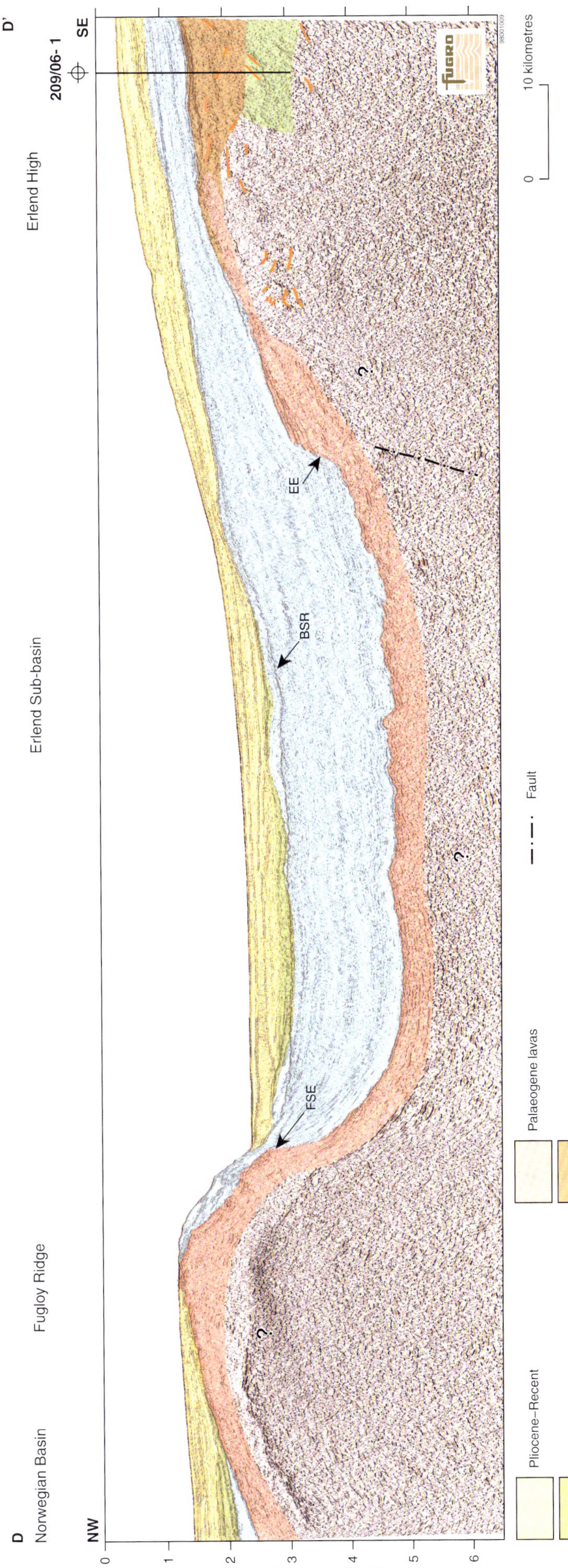

Figure 26 Merged seismic profile D–D' across the Fugloy Ridge, Erlend Sub-basin and Erlend High. Abbreviations: see Figure 24 except, BSR=Bottom Simulating Reflector and EE=Erlend Escarpment. For location of profile, see Figure 7. Seismic data courtesy of Fugro Multi Client Services.

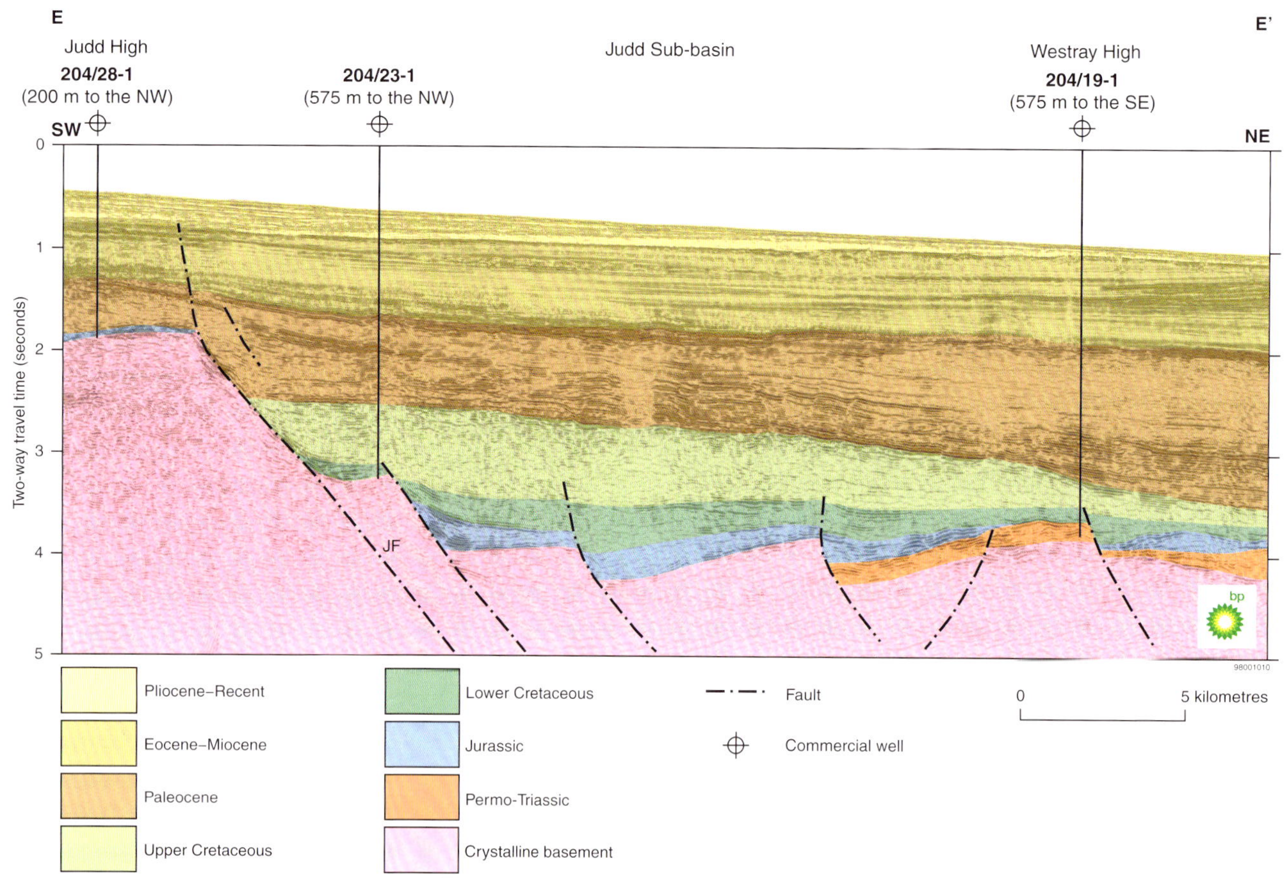

Figure 27 Seismic profile E–E' across the Judd High, Judd Sub-basin and Westray High. Abbreviations: see Figure 7. For location of profile, see Figure 7. Seismic data courtesy of BP.

and extended previously published versions of transfer zones in this area such as the Orkney–Faroe Alignment (Earle et al., 1989), the North Orkney/Wyville Thomson Transfer Zone (Stoker et al., 1993) and the Wyville Thomson Transfer Zone (Waddams and Cordingley, 1999). According to Johnson et al. (2005b), the strand of the lineament that underlies the Wyville Thomson Ridge is interpreted to have an element of sinistral strike-slip faulting associated with it, although the lack of a significant offset on the Outer Hebrides High (Figure 7) indicates that any strike-slip movement was either relatively small, or balanced by an earlier dextral offset. However, the results of potential field modelling by Kimbell et al. (2005) suggest that further north-west, there is a major transfer offset in the vicinity of the Wyville Thomson Ridge.

Wyville Thomson Ridge

The west-north-west-trending Wyville Thomson Ridge represents a symmetrical basalt anticline approximately 200 km long and 20 km wide that forms a marked bathymetric high (Figure 1) close to the south-west margin of the report area (Figure 7). It occupies a key structural location, occurring at the confluence of the Faroe Bank Channel, Munkur, Auðhumla, north-east Rockall and North Rockall basins. It is well defined from seismic data (e.g. Figures 35 and 36; Boldreel and Andersen, 1993; Johnson et al., 2005b). The nature of its boundaries with the surrounding basins are poorly understood, due in part to the presence Palaeogene lavas that mask the deep structure of the area.

The Wyville Thomson Ridge has not been drilled but basaltic lavas have been recovered from dredging operations at the sea bed (Waagstein 1988, fig. 1). Folded Palaeogene basaltic lavas are inferred to outcrop at the sea bed around the crest of the anticline, becoming progressively buried on both flanks by Eocene and younger sediments (Figures 35 and 36). The anticline was originally considered to comprise a 12 km thick pile of Palaeogene lavas that were deposited from a deeply buried fissure running along its axis (Roberts et al., 1983). However, the results of potential field modelling by Waddams and Cordingley (1999, fig. 5) has suggested the presence of as much as 10 km of Paleocene and older sedimentary strata below a relatively thin (600 m) cover of folded Palaeogene lavas. In contrast, Tate et al. (1999, fig. 3) preferred a model in which basement is no deeper than 4 to 6.5 km depth. Recent seismic and potential field modelling by Smith et al. (2009) along the central part of the ridge suggest a variation of between 500 m to 1.2 km and 4 to 6 km, in the respective thickness of the Palaeogene lavas and pre-volcanic sediment intervals. The results of a wide-angle seismic experiment by Klingelhöfer et al. (2005, fig. 9) across the south-east part of the ridge indicated the presence of 1 km of Palaeogene lavas and 5 km of older sediments resting on crystalline basement.

The Wyville Thomson Ridge is thought by many to represent a Cenozoic inversion structure (e.g. Boldreel and Andersen, 1993; Lundin and Doré, 2002; Johnson et al., 2005b; Smith et. al., 2009), though there are conflicting views regarding the mechanism of its

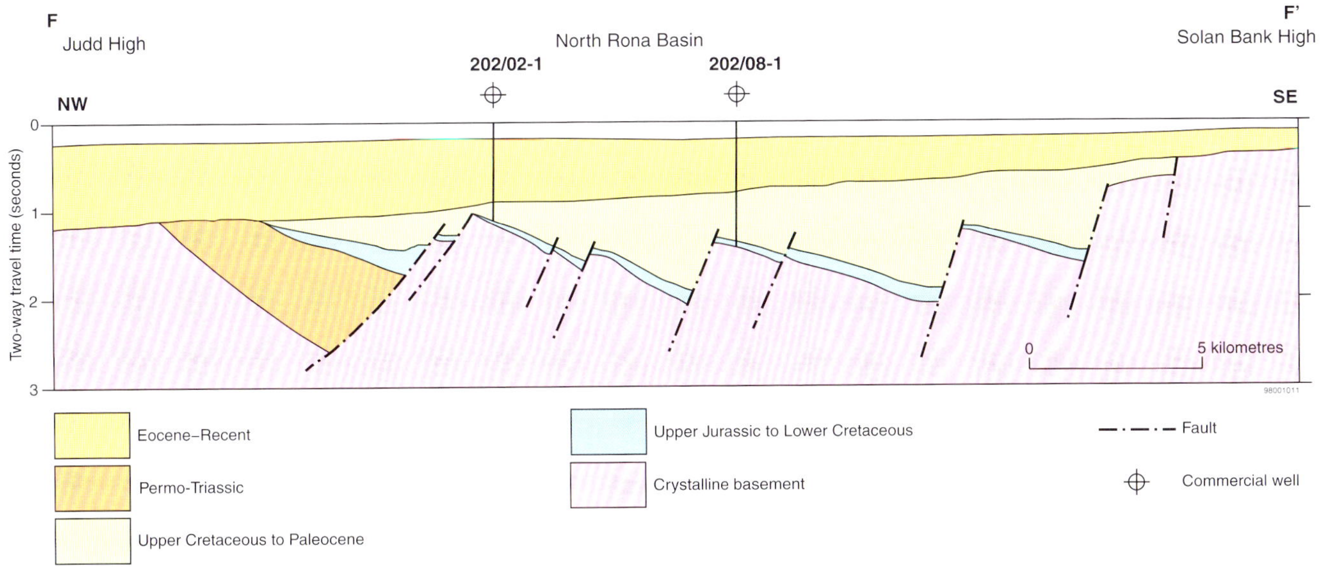

Figure 28 Generalised geological cross-section F–F' across the Judd High, North Rona Basin and Solan Bank High (modified from Stoker et al., 1993). For location of profile, see Figure 7.

origin. For example, it could form part of a system of ramp anticlines (that include the Ymir and Munkagrunnur ridges) caused by compressional reactivation of a northward-dipping extensional fault plane (Boldreel and Andersen, 1993; Doré and Lundin, 1996; Tate et al., 1999), a reactivated transfer zone (Duindam and van Hoorn, 1987; Stoker et al., 1993; Musgrove and Mitchener, 1996; Waddams and Cordingley 1999, Keser Neish, 2003) or even the location of a north-west-trending transient Paleocene (or older) rift system that was infilled with sediments and basaltic lavas (e.g. Lundin and Doré, 2005; Ziska and Varming, 2008) and then subsequently inverted. Growth of the Wyville Thomson Ridge anticline is thought to have started in late Paleocene times, with episodes of activity in Eocene and particularly Miocene times. (Figure 14; Johnson et al., 2005b; Stoker et al., 2005b and c).

Ymir Ridge

The north-west-trending Ymir Ridge is a faulted and asymmetrically folded and overturned basalt anticline at least 50 km long and up to 15 km wide that straddles the south-west boundary of the report area (Figure 7). According to Ziska and Varming (2008) it comprises northern, central and southern segments that are bounded to the south-west and north-east by the North Rockall and Auðhumla basins, respectively. The ridge is well defined on seismic data (Figure 36; Boldreel and Andersen, 1993; Johnson et al., 2005b; Stoker et al., 2005c).

The Ymir Ridge has not been drilled but eroded basaltic lavas are inferred to outcrop at the sea bed over its crest (Figure 36). There are different models regarding its genesis, including suggestions that it represents a frontal ramp anticline of a deeply buried system of Cenozoic fault-controlled ramp anticlines (Boldreel and Andersen, 1993), a segmented series of transpressional anticlines, possibly buttressed against the Darwin–Geikie High (Smith et al., 2009) or the location of a north-west-trending rift zone that has been infilled with lavas (Lundin and Doré, 2005) and shares a similar fold growth history to the more extensively studied Wyville Thomson Ridge. Ziska and

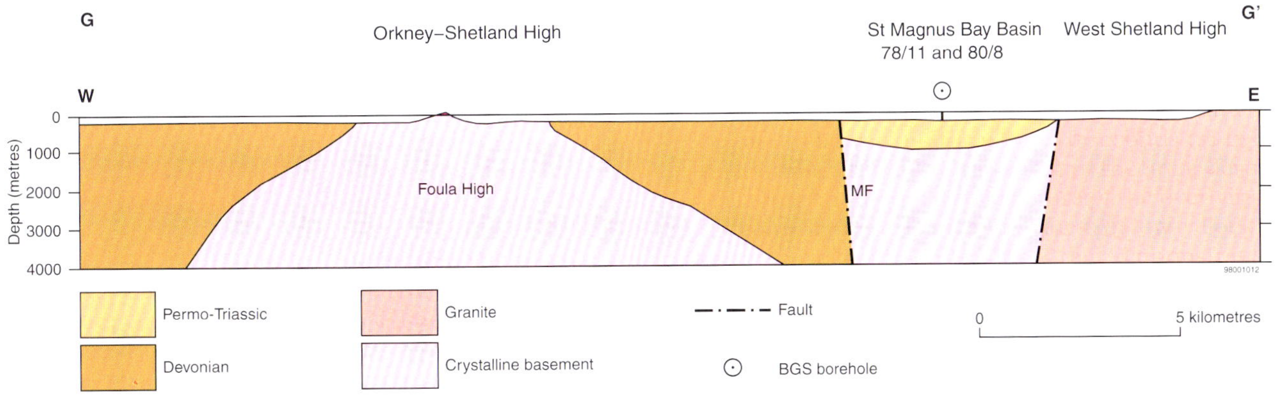

Figure 29 Generalised geological cross-section G–G' across the Orkney–Shetland High, Foula High, St Magnus Bay Basin and West Shetland High (modified from BGS, 1984). Abbreviations: see Figure 7. For location of profile, see Figure 7.

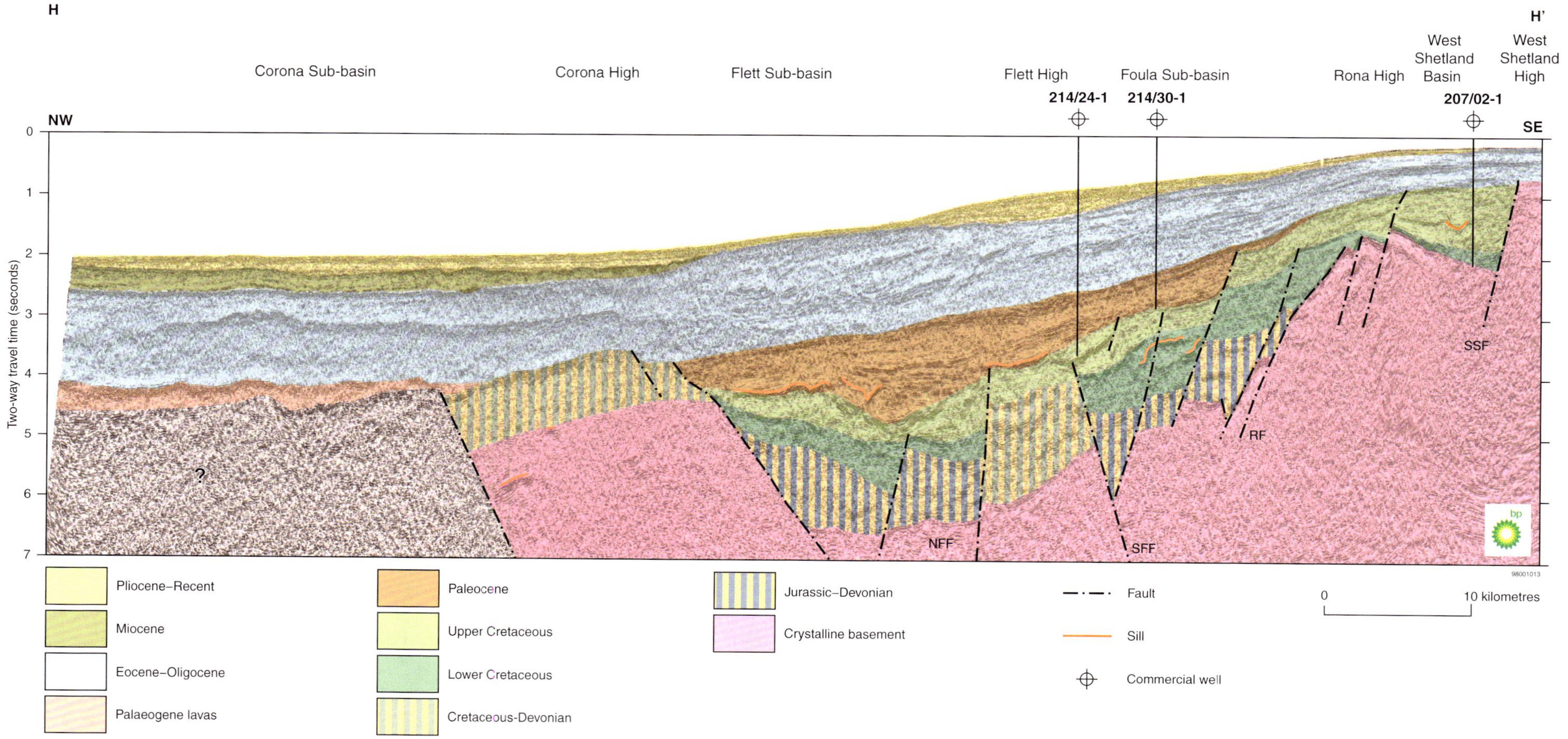

Figure 30 Seismic profile H–H' across the Corona Sub-basin, Corona High, Flett Sub-basin, Flett High, Foula Sub-basin, Rona High, West Shetland Basin and West Shetland High. Abbreviations: see Figure 7. For location of profile, see Figure 7. Seismic data courtesy of BP.

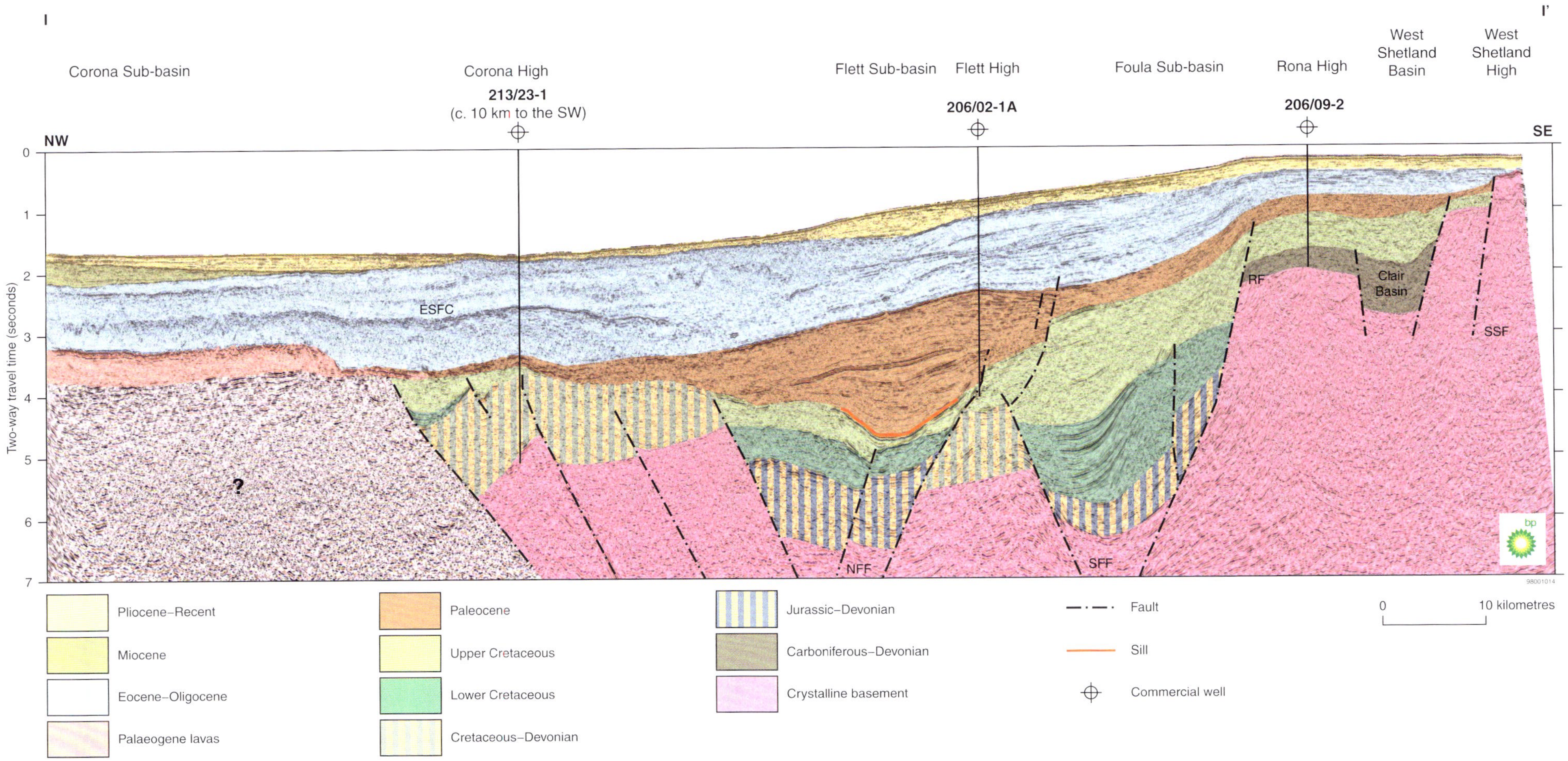

Figure 31 Seismic profile I–I' across the Corona Sub-basin, Corona High, Flett Sub-basin, Flett High, Foula Sub-basin, Rona High, Clair Basin, West Shetland Basin and West Shetland High. Abbreviations: see Figure 7, except, ESFC=Eocene submarine fan complex. For location of profile, see Figure 7. Seismic data courtesy of BP.

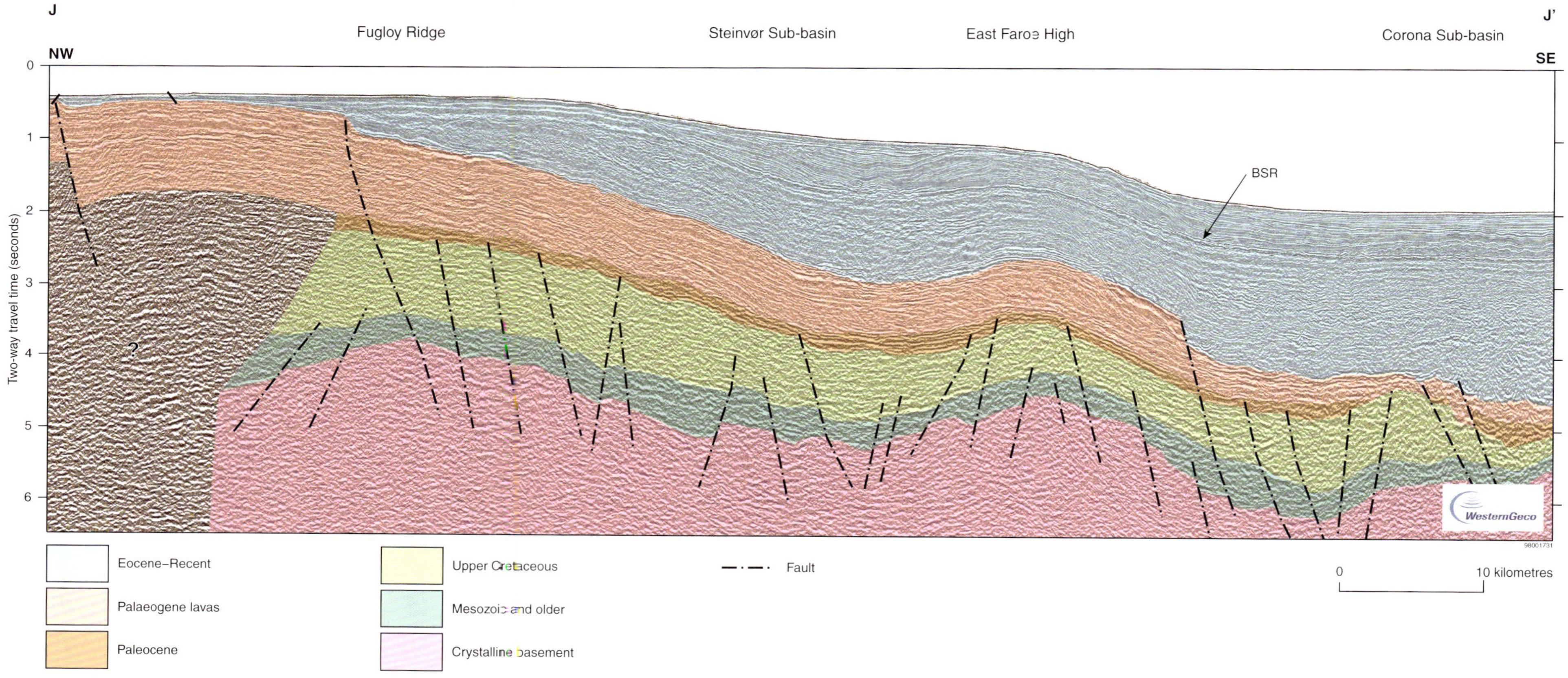

Figure 32 Seismic profile J–J' across the Fugloy Ridge, Steinvør Sub-basin, East Faroe High and Corona Sub-basin. Abbreviations: see Figure 26. For location of profile, see Figure 7. Seismic data courtesy of WesternGeco.

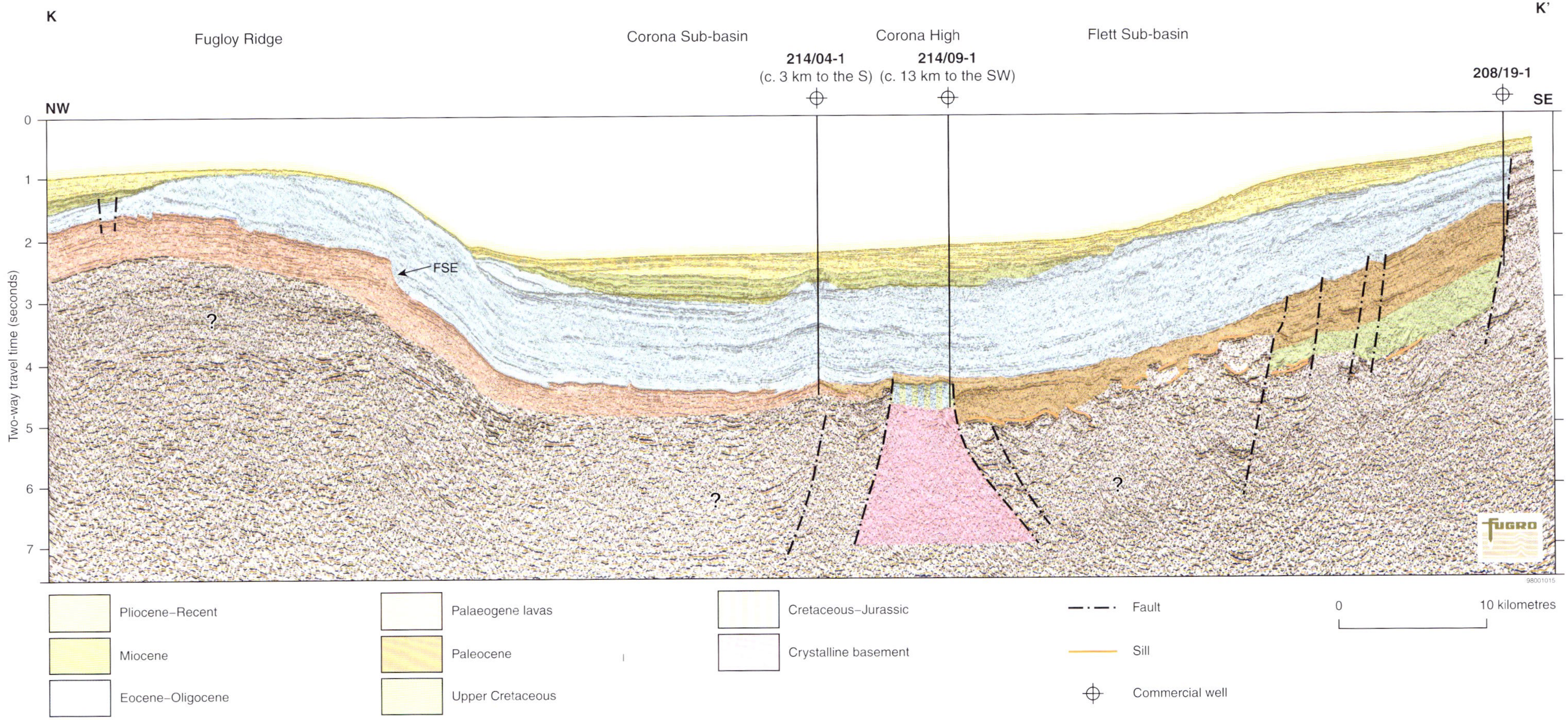

Figure 33 Merged seismic profile K–K' across the Fugloy Ridge, Corona Sub-basin, Corona High and Flett Sub-basin. Abbreviations: see Figure 24. For location of profile, see Figure 7. Seismic data courtesy of Fugro Multi Client Services.

Varming (2008) suggest that age of the Ymir Ridge varies, with an older central part and a younger southern margin.

Major basins, sub-basins and intrabasinal highs

Faroe–Shetland Basin

The Faroe–Shetland Basin represents a complex amalgam of mainly Late Palaeozoic to Recent sub-basins and intrabasinal highs up to 460 km long and 260 km wide that lie between the Faroe Islands and the Shetland/Orkney Islands in north-westerly and south-easterly directions, and the Munkagrunnur Ridge and the Møre Basin in south-westerly and north-easterly directions, respectively (Figure 7). Much is known about the general structure and tectonostratigraphical evolution of the Faroe–Shetland Basin (e.g. Ridd, 1981; Hitchen and Ritchie, 1987; Duindam and van Hoorn, 1987; Mudge and Rashid, 1987; Stoker et al., 1993; Rumph et al., 1993; Mitchell et al., 1993; Doré et al., 1999; Lamers and Carmichael, 1999; Dean et al., 1999; Roberts et al., 1999; Keser Neish, 2003 and Smallwood and Kirk, 2005), though the pre-Cenozoic development, particularly within the north-west part is less well understood. This is not only due to the obscuring presence of Palaeogene lavas and associated intrusions (e.g. Figures 24, 26 and 33), but also the relative lack of commercial wells (Figure 2). From Late Eocene to Early Oligocene times onwards, the structural expressions of all the individual sub-basins and intrabasinal highs that comprise the Faroe–Shetland Basin largely disappear (although locally parts are modified by the effects of Miocene inversion in particular), and it becomes one large, single basin (e.g. Figures 24, 30, 31 and 37).

The Faroe–Shetland Basin is floored by mainly Archaean metamorphic basement that has been drilled within the Corona High, Flett Sub-basin and Foula Sub-basin and also underlies the Rona High, Clair Basin and West Shetland Basin (see Chapter 3; e.g. Ritchie and Darbyshire, 1984; Ritchie et al., 1987) that flank its south-east margin.

Devonian continental red beds are the oldest known sedimentary rocks encountered within the basin, with up to approximately 300 m of isolated occurrences drilled on the Westray High, Corona High and south-east

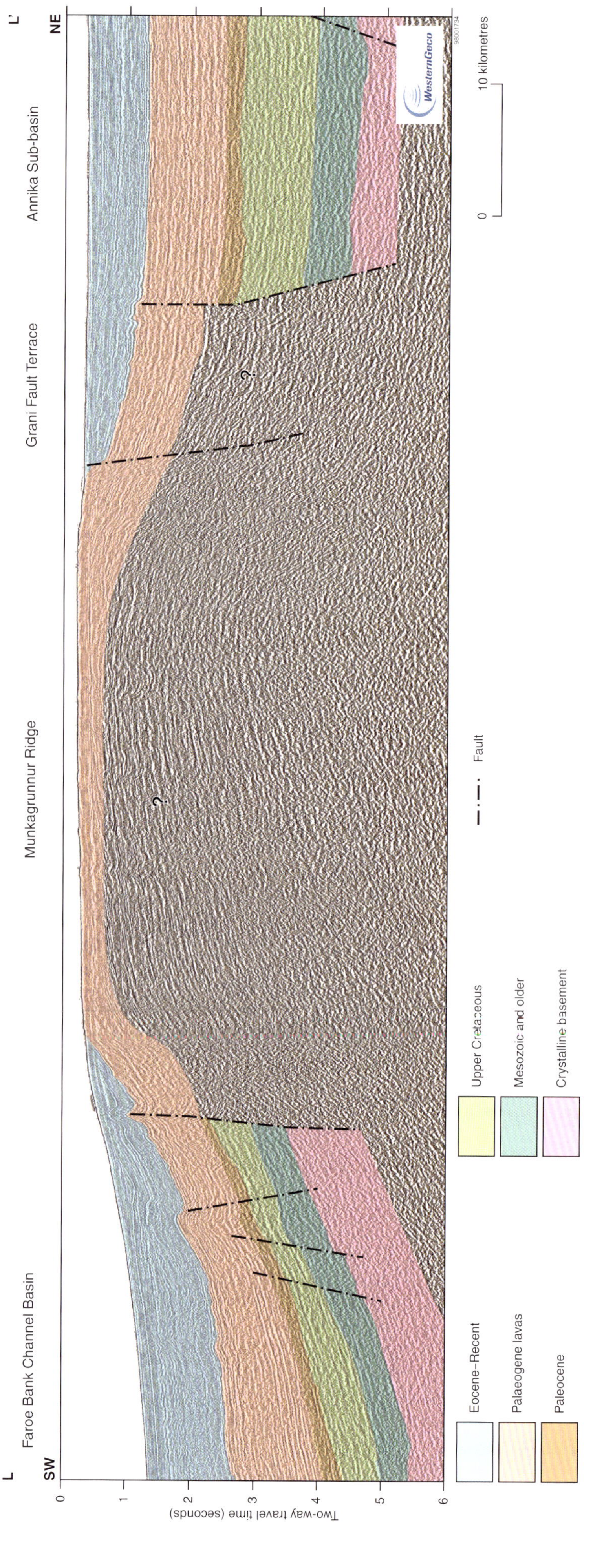

Figure 34 Seismic profile L–L' across the Faroe Bank Channel Basin, Munkagrunnur Ridge, Grani Fault Terrace and Annika Sub-basin. For location of profile, see Figure 7. Seismic data courtesy of WesternGeco.

flank of the Flett Sub-basin (see Chapter 4). Further to the south-east, rocks of similar nature and age have been described from the Clair Basin (e.g. Ridd, 1981; Blackbourn, 1987; Allen and Mange-Rajetzky, 1992; Stoker et al., 1993; Ritchie et al., 1996; Nichols, 2005). By analogy with the Clair Basin, Devonian strata within the Faroe–Shetland Basin are considered to have been deposited within a strike-slip or extensional intermontane basinal setting following collapse of the north-east-trending Caledonian orogenic thrust belts (Figure 8; e.g. Stoker et al., 1993; Roberts et al., 1999; Nichols, 2005). The detailed distribution and thickness of Devonian rocks within the Faroe–Shetland Basin is unknown but their presence on intrabasinal highs is suggestive that they are widespread, albeit probably thinly developed.

Within the Clair Basin, continental red bed facies sedimentation is known to have persisted into early Carboniferous times (e.g. Blackbourn, 1987), though in well 206/08-2 (Figure 2), marine sediments of Viséan age have been described (Meadows et al., 1987; Allen and Mange-Rajetzky, 1992). Elsewhere within the report area, Carboniferous (Viséan–Namurian) rocks have only been proved within well 213/23-1 on the Corona High (see Chapter 4 and Figure 52), which until recently, was considered to occur within an emergent Carboniferous hinterland (e.g. Ziegler, 1990, enclosure 15; Coward et al., 2003, figs. 2.8. and 2.9).

Following the late Carboniferous to early Permian Variscan Orogeny, Permo-Triassic extensional fault development within the Faroe–Shetland report area (Figure 8) appears to have been strongly influenced by the existing north-east-trending Caledonian basement structure (see Chapter 5; e.g. Ziegler, 1990, enclosures 17 to 23; Glennie, 2002). Extensional activity was concentrated to the south-east of the Faroe–Shetland Basin, with the north-east-trending West Orkney Basin, Papa and south-west West Shetland basins (Figure 7), forming major Permo-Triassic half-grabens, with a maximum of between 7.5 km and 10 km of syn-rift sediments interpreted to be present within the hanging wall of their basin-bounding faults (Enfield and Coward, 1987; Earle et al., 1989; Hitchen and Ritchie, 1987; Stoker et al., 1993; Hitchen et al., 1995b; Swiecicki et al., 1995). There is a notable switch in the polarity of the half-grabens located on either flank of the Judd Lineament (Figure 7), with a north-west-dipping Permo-Triassic sedimentary wedge within the West Orkney and south-east-dipping wedges within the Papa and south-west West Shetland basins (see Stoker et al., 1993, figs. 23 and 24). This structural

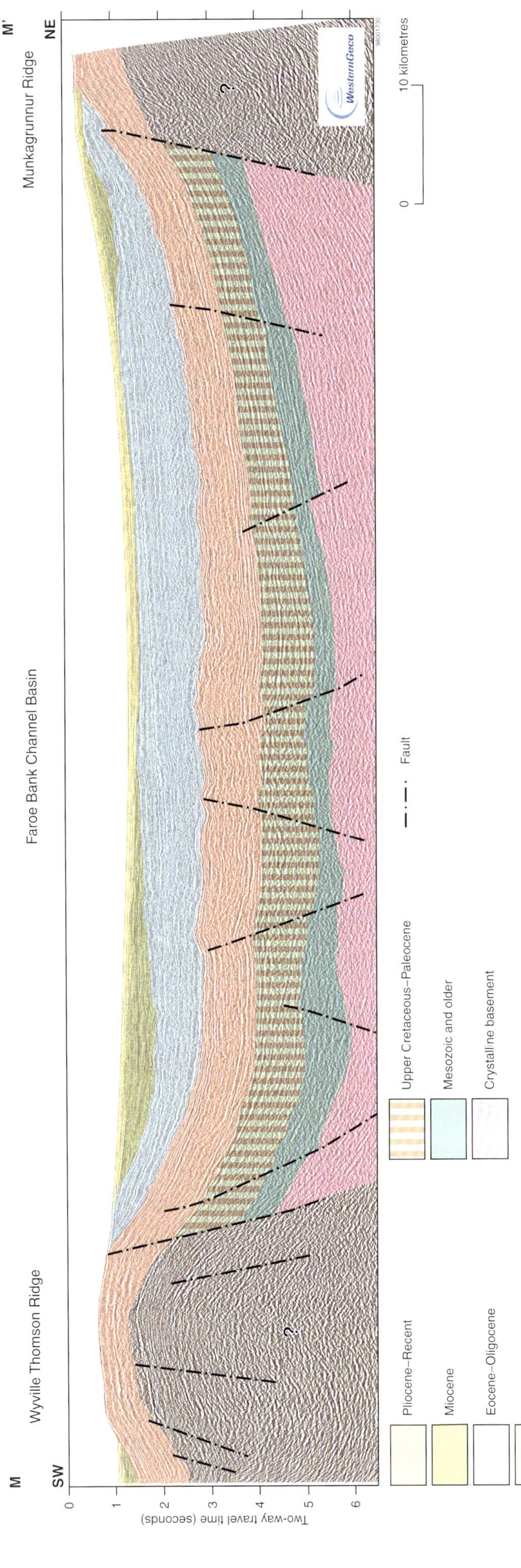

Figure 35 Seismic profile M–M' across the Wyville Thomson Ridge, Faroe Bank Channel Basin and Munkagrunnur Ridge. For location of profile, see Figure 7. Seismic data courtesy of WesternGeco.

compartmentalisation is cited by Earle et al. (1989) as supporting evidence for Permo-Triassic movement along the Judd Lineament. Evidence for Permo-Triassic extension within the adjacent Faroe–Shetland Basin is less clear (e.g. Doré et al., 1999; Dean et al., 1999), though many have inferred the presence of an active Arctic rift zone, particularly during Triassic times (Figure 8; e.g. Duindam and van Hoorn, 1987; Mudge and Rashid, 1987; Ziegler, 1990; Stoker et al., 1993; Roberts et al., 1999; Spencer et al., 1999; Glennie et al., 2003; Coward et al., 2003). However, a small number of wells have proved a maximum of only 200 m of mainly Triassic continental red bed facies sediments within the Foula Sub-basin, Corona High and the southern part of the Westray High (see Chapter 5; Figure 7), suggesting that any Triassic extension within the Faroe–Shetland Basin was of limited magnitude.

In Early Jurassic times, minor extensional tectonism affected the north-west European margin as a whole (Doré et al., 1999). Partly from the results of a study in the Sea of Hebrides–Little Minch area by Morton (1989), the presence of an Early (Roberts et al., 1999, fig. 9) or Early to Mid (Bajocian) Jurassic (Dean et al., 1999; Coward et al., 2003, fig. 2.12) rift system was inferred within the Faroe–Shetland region (Figure 8). Up to approximately 350 m of marginal marine Lower Jurassic sediments have been drilled within the Faroe–Shetland Basin and 750 m or so within the nearby West Solan Basin (Ritchie et al., 1996). Doré et al. (1999) postulated a major period of uplift and erosion during Mid Jurassic (Bajocian to Bathonian) times within the Faroe–Shetland region (Figure 8). This is supported by Booth et al. (1993), who invoked a deeply erosive Mid to earliest Late Jurassic unconformity caused by the removal of over 1.5 km of Lower Jurassic and older strata from the adjacent East Solan Basin. In contrast, from drilling results of well 206/05-1 located within

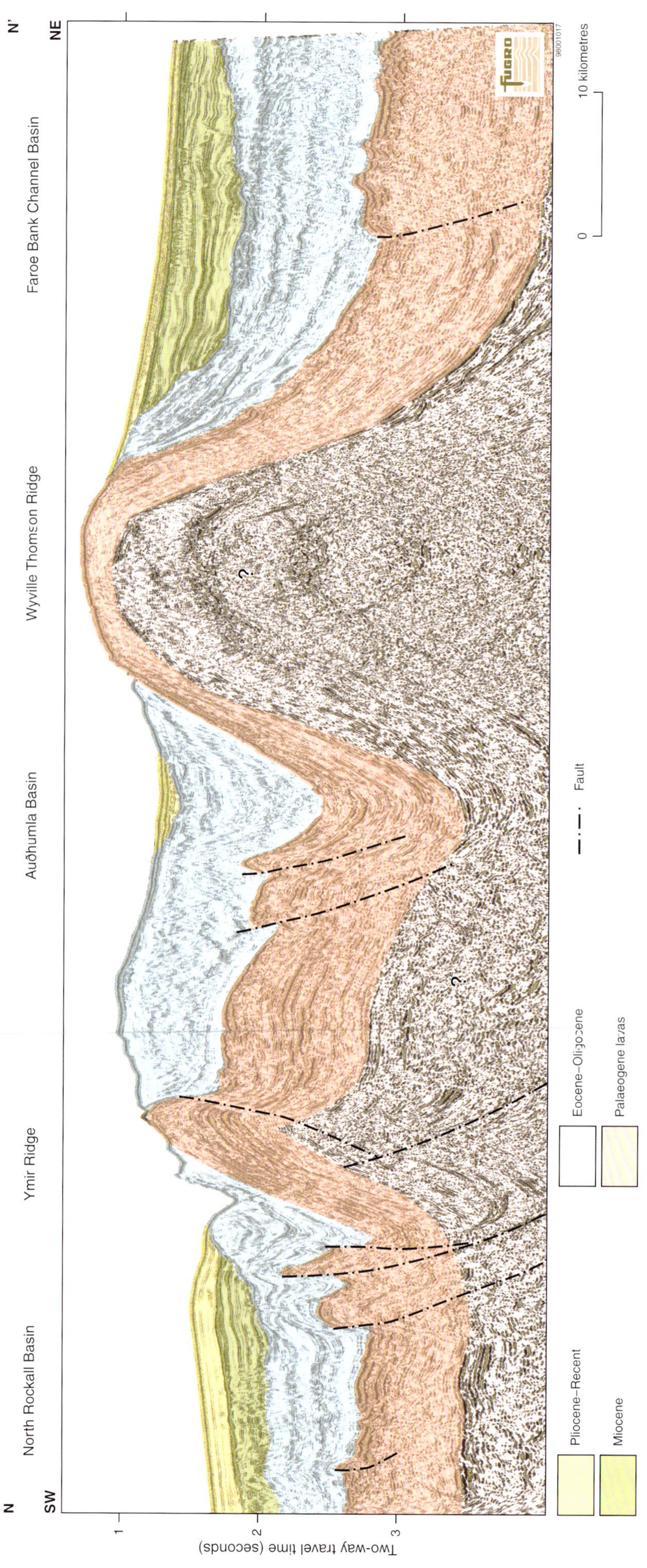

Figure 36 Seismic profile N–N' across the North Rockall Basin, Ymir Ridge, Auðhumla Basin, Wyville Thomson Ridge and Faroe Bank Channel Basin. For location of profile, see Figure 7. Seismic data courtesy of Fugro Multi Client Services.

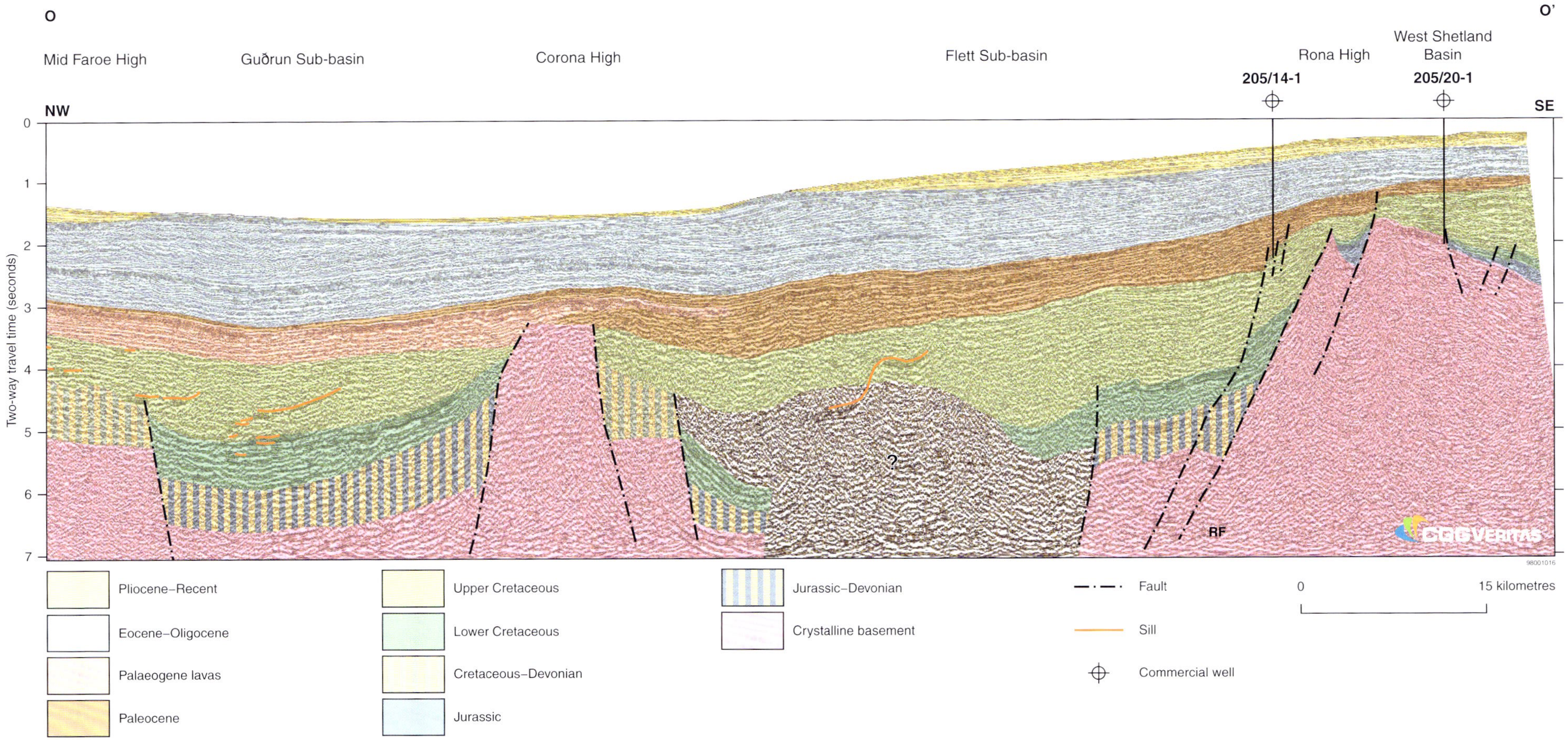

Figure 37 Merged seismic profile O–O' across the Mid Faroe High, Guðrun Sub-basin, Corona High, Flett Sub-basin, Rona High and West Shetland Basin. Abbreviations: see Figure 7. For location of profile, see Figure 7. Seismic data courtesy of CGG Veritas.

the Foula Sub-basin (Figures 2 and 7), Haszeldine et al. (1987) and Hitchen and Ritchie (1987) proposed a phase of Bajocian to Bathonian rifting. This is supported in part by Ziegler (1990), Knott et al. (1993) and Coward et al. (2003), who postulate the existence of a contemporaneous shallow marine seaway running through the Faroe–Shetland Basin. There are few proven Middle Jurassic occurrences within the Faroe–Shetland Basin, though a 600 m section was drilled in well 206/05-1 (see Chapter 6; Haszeldine et al., 1987). However, more widespread presence of Middle Jurassic source rocks is inferred from the results of oil analyses by Holmes et al. (1999) from other wells within the basin. According to Doré et al. (1997; 1999), significant latest Mid Jurassic to Late Jurassic rifting was focussed on north-trending structures within the north-west European margin. In the Faroe–Shetland report area, north-trending extensional faults that bound the Westray High (and in parts of the Corona High) are interpreted to be of Jurassic age and were truncated by north-east-trending reactivated Caledonian faults during Cretaceous times (Dean et al., 1999, fig. 3). The presence of a Late Jurassic seaway with or without the presence of contemporaneous rift systems within the Faroe–Shetland Basin is generally accepted (Figures 8 and 12; e.g. Ziegler, 1982; Haszeldine et al., 1987, Knott et al., 1993; Roberts et al., 1999) and is confirmed from the results of drilling by a number of wells located close to the south-east margins of the Foula (e.g. 400 m of Upper Jurassic marine clastic rocks in well 206/05-1), Flett and Judd sub-basins, and on the north-east Corona and Westray highs (see Chapter 6). Dean et al. (1999, fig. 2) suggested that the thickness of the strongly transgressive Upper Jurassic succession across the West Shetland Basin, East Solan Basin and south-east margin of the Faroe–Shetland Basin does not vary much, indicating that it is highly unlikely that there was contemporaneous rift activity of any significance.

The main phases of extension within the Faroe–Shetland Basin are generally thought to have occurred in Early to Late Cretaceous times (e.g. Hanisch, 1984; Price and Rattey, 1984; Duindam and van Hoorn, 1987; Hitchen and Ritchie, 1987; Dean et al., 1999; Doré et al., 1999; Grant et al., 1999; Lamers and Carmichael, 1999; Spencer et al., 1999; Roberts et al., 1999; Coward et al., 2003), though there is considerable disagreement regarding their detailed ages and intensity. For example, Coward et al. (2003) suggested two main Early Cretaceous phases of extensional fault activity within the Atlantic rift between Valanginian to Hauterivian and Aptian to Albian times (Figure 8). Dean et al. (1999) however, identified three main rift episodes within the Faroe–Shetland Basin, namely (1) Early Cretaceous (Valanginian to Berriasian) (2) Late Cretaceous (Cenomanian to Santonian) and (3) Late Cretaceous (Campanian to Maastrichtian). The age span of these three intervals broadly corresponds with two phases derived from the results of subsidence analyses using well data by Turner and Scrutton (1993). Lamers and Carmichael (1999) recognised significant Early Cretaceous (post-Berriasian) rifting, with the effects of differential subsidence forming the fault block topography associated with the Westray, Flett and Rona highs (e.g. Figures 30 and 31). This was superseded by Late Cretaceous normal fault movement along the Rona Fault the bounds the south-east margin of the Faroe–Shetland Basin (e.g. Figure 31), and the fault that defines the eastern margin of the Westray High. Dean et al. (1999, figs. 2 and 6) recognised significant lateral variation in the effects of extensional activity along the length of the Rona Fault and also noted offset in the locations of the Lower and Upper Cretaceous depocentres within the hanging wall block of the fault. Roberts et al. (1999) postulated the presence of a deep marine seaway throughout Cretaceous times, but restrict the main phase of rifting to the Late Jurassic to Early Cretaceous (Valanginian) interval. Following Duindam and van Hoorn (1987) and Earle et al. (1989) before them, Spencer et al. (1999) preferred two main phases of rifting during Early Cretaceous times (within the south-east part of the basin), followed by a Late Cretaceous (Campanian to Maastrichtian) episode. From a study of wells on the eastern flank of the Faroe–Shetland Basin, Grant et al. (1999, fig. 3) suggested that Cretaceous rift activity in the basin spanned the Hauterivian to Cenomanian interval. In contrast, Stoker et al. (1993, fig. 18) recognised Early Cretaceous (Albian to Aptian) extension, a view supported from a stratigraphical study of wells by Ritchie et al. (1996), with two further phases of Late Cretaceous rifting.

Within the Faroe–Shetland Basin, Lower and Upper Cretaceous strata were deposited in a progressively deepening marine environment (see Chapter 7; e.g. Stoker et al., 1993; Ritchie et al., 1996) and attain maximum drilled thicknesses of 1.5 km and 2.4 km, respectively. From seismic data, Lamers and Carmichael (1999) believed that the Upper Cretaceous succession could reach a thickness of 4.5 km or so on the eastern flank of the Westray High. The influence exerted on the distribution and thickness of Cretaceous sediment by north-west-trending lineaments that cross the Faroe–Shetland Basin was considerable (Rumph et al., 1993, fig. 12). For example, up to 4.9 km of combined Upper and Lower Cretaceous strata is developed close to the Victory Lineament within the north-east part of the Foula Sub-basin (Rumph et al., 1993).

Though post-rift subsidence is considered to be the dominant tectonic process that affected the Faroe–Shetland Basin during Cenozoic times (e.g. Turner and Scrutton, 1993), there is evidence for continuing minor extension from Late Cretaceous to Early to Mid Paleocene times (e.g. Smallwood and Gill, 2002), Late Paleocene regional uplift, numerous phases of compression (e.g. Boldreel and Andersen, 1993; Ritchie et al., 2003; Davis et al., 2004; Smallwood, 2004; Johnson et al., 2005b; Ritchie et al., 2008) and tilting and enhanced sagging (e.g. Stoker et al., 2005b and c) along the continental margin (Figures 8 and 14). For example, there is pronounced thickening of Paleocene deep-water marine sediments within the south-east part of the Flett Sub-basin (Figures 30 and 31; Lamers and Carmichael, 1999, fig. 4; Dean et al., 1999), extensional movement on the Westray Fault and the development of east-trending faults within the Judd Sub-basin (Smallwood and Kirk, 2005). However, faults controlling the deposition of the Paleocene succession are thought by Dean et al. (1999) to have decoupled from those associated with older rifts, soleing out within the vast thicknesses of ductile Upper Cretaceous mudstone within the basin. Locally, this activity, along with the effects of sediment loading, may have resulted in mobilisation of Cretaceous and Lower Paleocene mudstones and the formation of diapiric structures within parts of the Faroe–Shetland Basin (e.g. Lamers and Carmichael, 1999).

Extensional faulting is considered to have ceased during Mid Paleocene times with the deposition of the Kettla Tuff Member (Smallwood and Gill, 2002). In Mid to Late Paleocene to earliest Eocene times, regional uplift of the order of 200 m (Smallwood and Gill, 2002) to 900 m (Nadin et al., 1997) combined with a drop in sea level, resulted in the development of a dendritic drainage pattern on exhumed deep-marine Paleocene sediments (Smallwood and Gill, 2002). According to some, this uplift is attributable to dynamic support driven by ascending hot material associated with the inception of the Iceland Plume (e.g. White and McKenzie, 1989; Nadin et al., 1997; Smallwood and Gill, 2002). The Iceland Plume is also linked with the emplacement of the massive pile of basaltic lavas and associated intrusive rocks observed on the Faroe Islands and to the south-east within the Faroe–Shetland Basin (see Chapter 9). The Lopra, Beinisvørd, Malinstindur and Enni formations (formerly ascribed to the Lower, Middle and Upper Basalt formations) on the Faroe Islands are thought to have been emplaced during Late Paleocene to earliest Eocene times (i.e. 57.2 to 54.5 Ma; see Chapter 9), the beginning of which is coeval with the onset of regional plume-related uplift (Smallwood and Kirk, 2005). There is also some evidence for pulses of compression during Paleocene times, with growth observed on the Wyville Thomson Ridge (Boldreel and Andersen, 1993; Johnson et al., 2005b). If the age of the oldest lavas in the Faroe–Shetland area are approximately 57.2 Ma, then a Late Paleocene age is indicated for inception of the Wyville Thomson Ridge growth anticline. This is probably contemporaneous with the drop in sea level and development of the dendritic drainage pattern within the Judd Sub-basin. Following the onset of seafloor spreading within the Iceland and Norwegian basins during earliest Eocene times, post-rift subsidence and deepening marine conditions resumed within the Faroe–Shetland Basin.

Two main tectonic events influenced the Neogene evolution of the basin as a whole; Late Oligocene to Early/Mid Miocene compression, particularly within the north-east Faroe–Shetland Basin (Ritchie et al., 2003; Davis et al., 2004; Johnson et al., 2005b) and around the Wyville Thomson Ridge area (e.g. Boldreel and Andersen, 1993; Johnson et al., 2005b; Stoker et al., 2005b and c) and Early Pliocene uplift and tilting of the continental margin in general (e.g. Stoker et al., 2002; Davis et al., 2004; Stoker et al., 2005b and c). There is also some evidence for Early Pliocene and possibly even younger compressional activity throughout the Faroe–Shetland Basin (Ritchie et al., 2003; Johnson et al., 2005b; Ritchie et al., 2008; Figure 14).

The thickness of the Cenozoic succession varies considerably over the Faroe–Shetland Basin with a maximum drilled sedimentary thickness of approximately 4.3 km in well 214/28-1 over the Flett High, and an inferred stratigraphical thickness of 6.6 km for the basaltic lavas within the Faroe Islands (see Chapters 8 and 9).

The Faroe–Shetland Basin is cut by a number of north-west-trending lineaments (Figure 7) that are considered to have possibly originated as basement shear zones, similar to those described from the Lewisian Gneiss Complex on mainland Scotland (Park et al., 2002). These lineaments were subsequently reactivated as Permo-Triassic, Cretaceous and Cenozoic transfer zones (e.g. Earle et al., 1989; Rumph et al., 1993; Dean et al., 1999), partitioning extensional activity on mainly pre-existing north-east-trending Caledonian faults and exerting considerable influence over sedimentation patterns. Lineaments such as the Magnus, Erlend, Victory, Clair, Westray and Judd in particular have been extensively referenced (e.g. Duindam and van Hoorn, 1987; Rumph et al., 1993; Knox et al., 1997; Doré et al., 1999; Naylor et al., 1999; Iliffe et al., 1999; Lamers and Carmichael, 1999; Goodchild et al., 1999; Grant et al., 1999; Dean et al., 1999; Roberts et al., 1999; Ellis et al., 2002; Keser Neish, 2003; Jolley et al., 2005; Johnson et al., 2005b; Kimbell et al., 2005; Cawley et al., 2005) though it appears that their inferred locations vary in detail.

A more detailed description of the main structural elements that comprise the Faroe–Shetland Basin is given below in alphabetical order. It should be noted however, that due the presence of thick Palaeogene lavas within the north-west part of the Faroe–Shetland Basin, the definitions of some of the sub-basins and intrabasinal highs should be regarded as tentative.

Annika Sub-basin

The Annika Sub-basin forms a north-east-trending feature up to approximately 150 km long and 80 km wide that occurs towards the central part of the report area (Figure 7; Keser Neish, 2003). The sub-basin is defined from seismic data (e.g. Figures 34 and 38) at pre-Palaeogene lava stratigraphical levels (Keser Neish, 2003, enclosure 2, profiles D and E). The Annika Sub-basin is bounded to the south-east by a series of north-east- to north-north-east-trending, generally north-west-dipping, minor normal faults that mark the north-west flanks of the Heri, East Faroe and Tróndur highs (Figures 7 and 38; Keiser Neish, 2003, enclosure 2, profiles D and E; Keser Neish, 2005, fig. 7). The arcuate western and northern margins of the sub-basin are interpreted to be defined by east- and south-dipping normal faults that form the respective flanks of the Munkagrunnur Ridge/Faroe Platform and Fugloy Ridge. The fault that separates the Faroe Platform and Annika Sub-basin at Palaeogene lavas and older stratigraphical levels has an easterly downthrow of approximately 0.25 s TWTT at top Palaeogene lavas (Keser Neish, 2003, enclosure 2, profile E). The sub-basin is also interpreted to be traversed by the Grimur Kamban, Corona and Westray lineaments and at its south-west margin, bounded by a combination of the Grani Fault Terrace and Judd Lineament (Ellis et al., 2002).

The Annika Sub-basin has not been drilled, but is thought to represent a mainly Mesozoic basin that has been gently modified by the effects of post-Eocene compression. It is considered to comprise up to a maximum of 2.2 km of Eocene to Recent sediments (Keser Neish, 2003), 1.0 s TWTT of Palaeogene lavas, 0.2 s TWTT of Paleocene sediments and 2.0 s TWTT of Mesozoic and older strata that rest on ?crystalline basement (Figures 34 and 38; Keser Neish, 2003, enclosure 2, profiles D and E). Results from regional wide-angle seismic investigations of Raum et al. (2005) indicated that the sub-basin contains up to approximately 1 km of Eocene to Recent sediments, 1.5 km of Palaeogene lavas and 5 km of Cretaceous (?and older) strata resting on crystalline basement. According to White et al. (2003, fig. 11), the results from the FLARE deep seismic experiment suggested that the Palaeogene basalts vary in thickness between 2 and 4 km, whereas the sub-

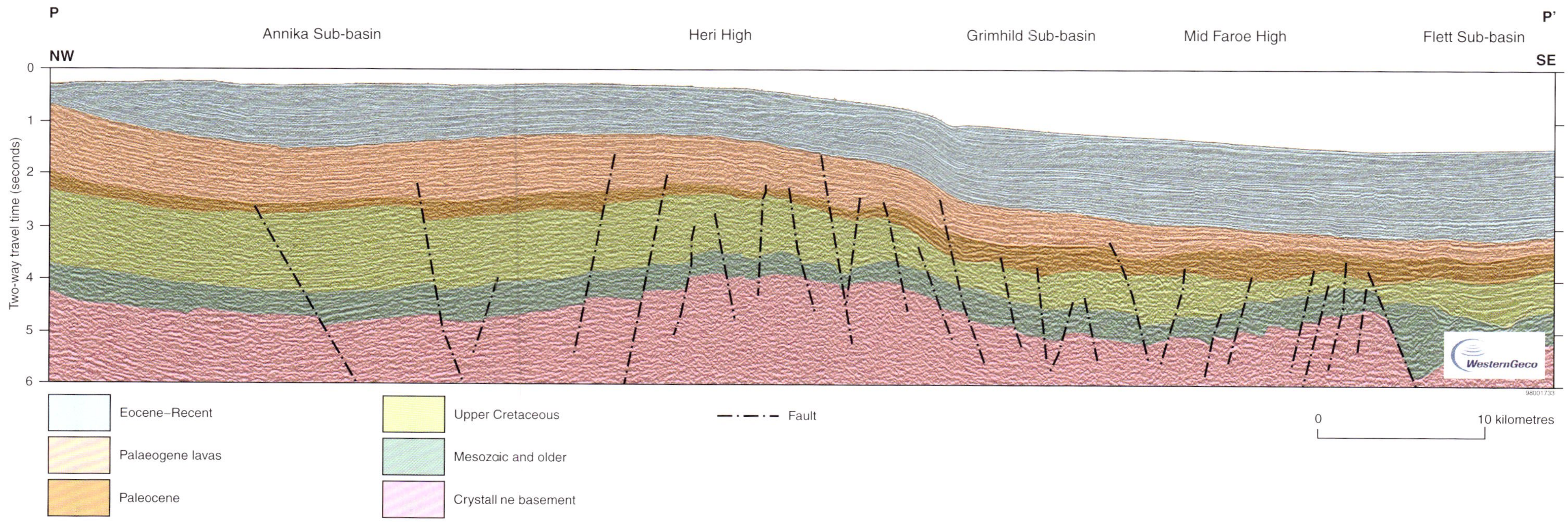

Figure 38 Seismic profile P–P' across the Annika Sub-basin, Heri High, Grimhild Sub-basin, Mid Faroe High and Flett Sub-basin. For location of profile, see Figure 7. Seismic data courtesy of WesternGeco.

basalt sedimentary succession thickness increases in a north-easterly direction from practically 0 to 4 km. It should be noted that the sub-basin is crossed by a conspicuous north-east to north-north-east-trending positive magnetic anomaly (Figure 5). The trace of the enigmatic 'Annika' anomaly does appear to correspond in part with a horst defined by Ellis et al. (2002), but could also represent a dyke (e.g. Sweetman, 1997; Roberts et al., 1999, fig. 22), a zone where a positively magnetised layer of Palaeogene lavas is truncated (Ziska and Morgan, 2005) or mark the south-east edge of the 'Lower Lava Formation' (Beinisvørð Formation) of Ellis et al. (2002).

Brynhild Sub-basin

The Brynhild Sub-basin forms a small, poorly defined structure with a maximum length and breath of approximately 70 km that occurs within the central part of the report area (Figure 7; Keser Neish, 2003). The sub-basin is defined from seismic data, with the Mesozoic and deeper stratigraphical levels better imaged towards the south-east where the Palaeogene lavas are thin to absent (cf Lamers and Carmichael 1999, fig. 4; Roberts et al., 1999, fig. 22; Keser Neish, 2003, enclosure 2, profile F). The Brynhild Sub-basin is bounded by a series of discontinuous structural culminations, including the Heri High, Mid Faroe High, Sjúrður Ridge and Grani Fault Terrace and by the Judd and Westray lineaments (Figure 7). Towards the north-east and southern margins of the sub-basin, the nature of its respective transitions with the Grimhild and Judd sub-basins is poorly understood.

The Brynhild Sub-basin has not been drilled but is considered to represent a substantial Mesozoic and older basin that has been folded by post-Eocene compressional deformation (Keser Neish, 2003). It is interpreted by Keser Neish (2003, enclosure 2, profile F) to comprise approximately 1.2 s TWTT of Eocene to Recent sediments, up to 0.8 s TWTT of Palaeogene lavas, 0.3 s TWTT of Paleocene sediments and 1.3 s TWTT of Mesozoic and older strata resting on ?crystalline basement. The results of a recent wide-angle seismic reflection experiment indicated that the southern part of the sub-basin potentially comprises 6 to 7 km of sediments, lavas and intrusive rocks resting on crystalline basement (Klingelhöfer et al., 2005). This is in very broad agreement with White et al. (2003, fig. 11), who suggested that the Palaeogene basalts thicken northward from 0 to 1.5 km, whereas the pre-basalt sedimentary succession is generally thickest (approximately 3.5 km) in the centre of the basin.

Corona Sub-basin

The Corona Sub-basin is interpreted to form a north-east-trending structure approximately 105 km long and up to 60 km wide that occurs within the central part of the report area (Figure 7). The pre-Cenozoic structure of the sub-basin is generally poorly imaged from seismic data due to the presence of Palaeogene lavas (e.g. Figures 30, 31, 32 and 33). The Corona Sub-basin is flanked to the north-west by a combination of the East Faroe High and Steinvør Sub-basin, though it should be noted that the nature of the boundary with the latter is largely inferred. The south-east margin of the sub-basin is interpreted to be marked by both south-east-dipping and north-west-dipping, planar normal faults, though the magnitude of throw on these cannot be established due to the poor resolution of the pre-Cenozoic succession within both footwall and hanging-wall blocks (e.g. Figures 30, 31 and 33). The north-east and south-west margins of the sub-basin are defined by the Erlend and Grimur Kamban lineaments, respectively.

Only the north-east part of the Corona Sub-basin has been drilled, where well 214/04-1 (Figure 33) proved 2.7 km of Eocene to Recent sediments and thin Palaeogene lavas. White et al. (2003) considers the thicknesses of Palaeogene basaltic lavas and pre-basalt sedimentary interval to be generally less than 1 km and 2 km, respectively. Although the pre-Cenozoic evolution of the Corona Sub-basin is unknown, it may share a similar history to that the nearby Flett Sub-basin.

Corona High

The Corona High forms an elongate, north-east-trending intrabasinal fault block approximately 200 km long and up to 30 km wide that partially straddles the UK–Faroese national boundary within the central part of the report area (Figure 7). The high is moderately well defined from seismic data (Figures 30, 31, 33 and 37; e.g. Dean et al., 1999, Roberts et al., 1999; Smallwood and Kirk, 2005), with the south-west part also corresponding with a positive isostatic gravity anomaly and a positive magnetic anomaly (Figures 4 and 5). The Corona High is interpreted to be separated from the Corona and Guðrun sub-basins and the Mid Faroe High to the north-west by largely planar, normal, north-east-trending faults that dip towards the north-west (Figures 33 and 37) and south-east (Figures 30 and 31). On Figure 37, a tentative throw of approximately 2.5 s TWTT towards the north-west has been estimated at the level of top crystalline basement. Changes in the direction of fault dip are inferred to be due to partitioning effects caused by the presence of the north-west-trending Corona, Grimur Kamban and Victory lineaments (Figure 7). Other authors, however, prefer the north-west margin of the high to be defined by a ramp (e.g. Dean et al., 1999; Lamers and Carmichael, 1999; Smallwood and Kirk, 2005). The south-east margin of the Corona High is separated from the Flett Sub-basin by a series of north-east-trending, south-east-dipping, planar normal faults (Figures 30, 31, 33 and 37). On Figure 37, a tentative downthrow of approximately 2 s TWTT towards the south-east has been estimated at top crystalline basement level. The north-east and south-west margins of the Corona High are defined by the Erlend Lineament and a large, unnamed east-trending fault, respectively (Smallwood and Kirk, 2005).

The Corona High has been drilled by released wells 204/10-1, 214/09-1, 214/17-1 and 213/23-1 (Figures 2, 31 and 33), proving an approximate maximum thickness of 3.85 km of Late Palaeozoic to Recent rocks, including 2.4 km of Cenozoic to Recent, 600 m of Lower and Upper Cretaceous, 50 m of Upper Jurassic, 200 m of Triassic and 600 m of Devono-Carboniferous strata. Wells 204/10-1, 213/23-1 and 214/09-1 terminated within crystalline basement. The presence of Upper Jurassic Kimmeridge Clay Formation within 214/09-1 confirms the hypothesis of a seaway within the central or axial part of the Faroe–Shetland Basin (e.g. Ziegler, 1990; Knott et al., 1993; Roberts et al., 1999), though whether there was significant extensional fault movement associated with the development of the seaway at this time is still a matter of some conjecture. Furthermore, the presence of a considerable thickness of Carboniferous (Viséan–Namurian) and

Devonian strata on the Corona High indicates significant basin development in this area at a time when published models suggest emergent land dominated this region (e.g. Ziegler, 1990; Friend et al., 2000; Coward et al., 2003).

East Faroe High

The East Faroe High forms a narrow, segmented, north-east-trending structural high approximately 125 km long and 10 km wide that occurs within the central part of the report area (Figure 7). The general morphology of the high is well defined from seismic data (Figure 32; Lamers and Carmichael, 1999, fig. 4; Keser Neish, 2003, enclosure 2, profiles B, C and D; Keser Neish, 2005, fig. 7; Ziska and Andersen, 2005, figs. 3 and 7). The East Faroe High comprises two distinct segments; the south-western part is flanked to the north-west and south-east by the Annika and Grimhild sub-basins respectively and is dextrally offset by approximately 10 km from the north-eastern part that lies subparallel to the Tróndur High (Figure 7; Keser Neish, 2005, fig. 7). The north-eastern part of the high is interpreted to represent a fault-bounded basement horst, bounded to the north-west and south-east by the Steinvør and Guðrun/Corona sub-basins, respectively (Figure 32; Keser Neish, 2005). The normal faults that define these high have small throws, reaching a maximum of approximately only 0.2 s TWTT at base Cenozoic and older stratigraphical levels.

The south-west segment of the East Faroe High has recently been drilled by well 6104/21-1 (Figure 2). Although the detailed results of drilling have not been released as yet, the well reached a depth of approximately 4.2 km within ?sub-basalt strata (see www. Jarðfeingi.fo, news item dated 17 October 2006). The results of predrilling seismic mapping suggested that the high comprises approximately 1.5 s TWTT of Eocene to Recent sediments, 0.8 to 1.0 s of Palaeogene lavas, 0.2 s of Paleocene sediments and 1.1 s of Mesozoic and older strata resting on ?crystalline basement (Figure 32; Ziska and Andersen, 2005, figs. 3 and 7; Keser Neish, 2005, fig. 7). According to Ziska and Andersen (2005), the East Faroe High is a Mesozoic horst block, with most of the faults that define the flanks of the high dying out upwards at the base of the Cenozoic (see also Keser Neish, 2005; Keser Neish and Ziska, 2005). The high may have acted as a source for sediments that contributed to the fill of adjacent sub basins prior to deposition of the Palaeogene lavas. During post-Eocene times, the effects of compressional deformation were responsible for its rejuvenation (Keser Neish, 2003).

Erlend Sub-basin

The Erlend Sub-basin is approximately 90 km long and 150 km wide and occurs within the central to north-east part of the report area (Figure 7). The pre-Cenozoic structure of the sub-basin is poorly defined from seismic data due to the presence of Palaeogene lavas and sills (Figure 26). Within the extreme south-east part of the sub-basin, these sills and lavas are thin to absent and consequently, the deeper structure is better imaged (Figure 24). The Erlend Sub-basin is interpreted to be bounded to the north and north-west by a combination of the Fugloy Ridge and Møre Marginal High (Figure 7), though the nature of these boundaries is poorly understood. The Brendan and Erlend lineaments define the north-east and south-west margins of the sub-basin, respectively. The East Shetland and Erlend highs flank the south-east margin of the Erlend Sub-basin, with the former boundary defined by a north-east-trending, north-west-dipping normal fault with a large downthrow of >3.5 s TWTT at the level of top crystalline basement (Figure 24), and the latter by a similarly trending fault inferred to be present below the Erlend Escarpment (Figure 26).

The Erlend Sub-basin has been drilled by wells 219/27-1 also 219/28-1 located close to the boundary with the Møre Basin (Figure 2). These wells proved a maximum thickness of approximately 2.05 km of Cenozoic to Recent and 1.2 km of Lower and Upper Cretaceous sediments, with well 219/28-1 terminating after penetrating 150 m of breccio-conglomerates of unknown age. Seismic evidence suggests that the sub-basin could contain up to approximately 2.0 s TWTT of Cenozoic to Recent sediments and Palaeogene lavas and 3.0 s TWTT of Jurassic to Cretaceous strata (Figure 24). In terms of its tectonostratigraphical development, the broad similarity in the style of fault-block topographies developed on the north-west flank of the East Shetland High in both the Erlend Sub-basin and Møre Basin (Figures 23 and 24) suggest these basins may share a similar history. Although the Erlend Sub-basin is generally considered to form part of an axial Cretaceous system of rift basins within the Faroe–Shetland Basin as a whole (e.g. Spencer et al., 1999; Dean et al., 1999; Doré et al. 1999), the suggestion from the speculative interpretation of the seismic data presented here is that the main rift event in this area may have occurred during Jurassic times, with a south-east-thickening wedge of syn-rift sediments developed against the flank of the East Shetland High (Figure 24). The mainly post-rift Cenozoic to Recent sedimentary succession within the sub-basin is interrupted by Palaeogene volcanism and intrusive activity (Figures 24 and 26) and by episodes of mainly Miocene and Early Pliocene compression or transpression that formed, for example, the north-north-east-trending Pilot Whale Anticline (Figure 24; Ritchie et al., 2003; Johnson et al., 2005b; Ritchie et al., 2008).

Flett Sub-basin

The Flett Sub-basin forms an elongate, north-east-trending half-graben approximately 220 km long and up to 90 km wide that occurs within the central part of the report area (Figure 7). The sub-basin is well defined from seismic data (Figures 30, 31 and 37; Lamers and Carmichael 1999, figs. 5 and 14). It is flanked to the north-west by the Corona High (Figures 30, 31 and 37), and is separated from the Flett, Rona and East Shetland highs to the south-east by the north-east-trending, north-west-dipping, planar normal North Flett, Rona and Shetland Spine faults, respectively (Figure 7). The Flett Sub-basin is traversed by the Corona, Grimur Kamban, Clair and Victory lineaments, and is interpreted to be bounded at its north-east and south-west extremities by the Erlend and Westray lineaments, respectively.

The Flett Sub-basin has been drilled by a number of released wells including 204/20-3, -6A, 205/08-1, 205/09-1, 205/12-1, 205/14-1, -2, 205/16-1, -2, 205/17a-1, 17b-2, 205/21-2, 21b-3, 205/22-1A, 208/15-2, 208/17-1, -2, 208/19-1, 208/21-1, 208/22-1, 208/24-1A, 214/19-1, 214/26-1, 214/27-1, -2 and 27a-4 (Figure 2). These wells proved an approximate maximum thickness of 6.8 km of Late Palaeozoic to Recent sediments and lavas, including up to 4.05 km of Cenozoic to

Recent, 2.5 km of Lower and Upper Cretaceous, 50 m of Upper Jurassic and 200 m of Devonian strata. Pre-Cretaceous rocks have only been drilled in wells 205/16-1, 205/21-2, 205/21b-3, 205/22-1A and 208/24-1A.

The central part of the Flett Sub-basin is interpreted to have formed by rifting mainly during Early to Late Cretaceous times, with significant south-east thickening of the Upper Cretaceous succession in particular within the hanging wall of the north-east-trending North Flett Fault (Figure 30). This broadly supports the interpretation of Dean et al. (1999), who recognised Cenomanian to Santonian and Campanian to Maastrictian phases of rifting within the Faroe–Shetland Basin as a whole. The Paleocene succession is observed to thicken markedly towards the centre of the Flett Sub-basin (Figures 30 and 31). This basin geometry may be due in part to post-rift subsidence following Cretaceous extension, it has also been linked with tectonic subsidence (Turner and Scrutton, 1993) associated with Early Paleocene extension along the North Flett Fault and other faults (Lamers and Carmichael, 1999, fig. 4; Dean et al., 1999). The strong concave-upward seismic event that characterises the boundary of the Paleocene and Upper Cretaceous successions in Figure 31 is interpreted to be caused by a Palaeogene sill that propagated along this plane of weakness using a mechanism similar to that proposed by Francis (1982). The deep structure of the south-west part of the Flett Sub-basin appears less clear, where the Flett High is absent and the Paleocene succession does not markedly thicken towards the basin centre (Figure 37). The north-east part of the Flett Sub-basin is also poorly imaged, with the pre-Paleocene succession in particular being obscured by a progressive increase in the intensity of Palaeogene igneous intrusive activity (Figure 33).

Flett High

The Flett High forms a narrow, segmented, north-east-trending, deeply buried horst block approximately 105 km long and up to 10 km wide that occurs within the central part of the report area (Figure 7). Both its north-easterly (Figures 30 and 31) and south-westerly segments (e.g. Smallwood et al., 2004, fig. 10; Robinson et al., 2004, fig. 3) are well defined from seismic data. The high is bounded to the north-west and south-east by the North Flett and South Flett faults, respectively, although both have been previously been referred to as the Flett Fault by Lamers and Carmichael (1999) and Grant et al. (1999). The Flett High is interpreted to comprise two north-east-trending elements (see Dean et al., 1999) that are inferred to be offset by 15 km in a sinistral sense by the Grimur Kamban Lineament.

The Flett High has been drilled by a number of released wells including, 205/10-1A, -2, -3, 205/14-3, 206/01-1, -2, -3, 206/02-1A, 214/27a-3 and 214/28-1 (Figure 2). These wells proved a maximum thickness of approximately 7.55 km of Mesozoic to Recent sediments and volcanic rocks, including 4.3 km and 3.25 km of Cenozoic to Recent and Cretaceous strata, respectively. The north-east segment of the Flett High is interpreted to form a basement horst block that is capped by a Devonian to Cretaceous succession (Figures 30 and 31), and to have developed in response to Early to Late Cretaceous extensional movement on the North and South Flett faults that flank it. This interpretation is supported by Lamers and Carmichael (1999), who suggested that the main phase of rifting occurred at end Berriasian times. The Flett High is also considered to have been rejuvenated during Early Paleocene times, with notable north-west thickening of the Lower Paleocene succession within the Flett Sub-basin (Figures 30 and 31).

Foula Sub-basin

The Foula Sub-basin forms a narrow, north-east-trending and north-west-dipping half-graben approximately 90 km long and up to 30 km wide that occurs towards the eastern part of the report area (Figure 7). It is well defined from seismic data (Figures 30 and 31; Grant et al., 1999, fig. 2; Roberts et al., 1999, fig. 23; Lamers and Carmichael, 1999, fig. 4). To the north-west, the sub-basin is separated from the Flett High by the north-east-trending, planar normal, south-east-dipping South Flett Fault. To the south-east, a combination of the Devonian Clair Basin and crystalline basement of the Rona High, bound the sub-basin (Figures 7, 30 and 31). The Foula Sub-basin is traversed by the north-west-trending Clair, Grimur Kamban, Victory and Corona lineaments, the latter two of which define its north-east and south-west margins, respectively.

A number of released wells have been drilled within the Foula Sub-basin including 205/10-4, -5A, 206/03-1, 206/04-1, 206/05-1, -2, 206/11-1, 208/26-1, 214/24-1, 214/29-1 and 214/30-1 (Figure 2). These wells proved a maximum thickness of approximately 7.35 km of Mesozoic to Recent sediments, including 2.6 km of Cenozoic to Recent, 3.45 km of Lower and Upper Cretaceous, 1.1 km of Lower, Middle and Upper Jurassic and 200 m of Triassic strata. Lamers and Carmichael (1999) estimate that up to 4.5 km of Cretaceous sediment was deposited within the Foula Sub-basin. Of all wells drilled within the sub-basin, only 206/04-1 and 208/26-1 penetrated crystalline basement.

The Foula Sub-basin is interpreted to form a north-west-dipping mainly Cretaceous half-graben (Figure 30). The pre-Cretaceous structure of the sub-basin is poorly resolved, though well 206/05-1 drilled more than 1 km of Upper, Middle and Lower Jurassic strata that Haszeldine et al. (1987) interpreted as being deposited during phases of extension in Bajocian to Bathonian and also Oxfordian to Volgian times. There is some doubt however, as to whether this succession is in situ as it may form part of a rotated fault block that was derived from the north-west flank of the Rona High (see Dean et al., 1999).

Grani Fault Terrace

The Grani Fault Terrace is a narrow, north-north-west-trending high approximately 90 km long and up to 10 km wide that occurs within the central part of the report area (Figure 7; Keser Neish, 2003). The high is defined from seismic data, forming a terraced area flanking the eastern margin of the Munkagrunnur Ridge (Figure 34). It is separated from the Munkagrunnur Ridge in the west and the Brynhild and Annika sub-basins in the east, by a combination of north-north-west-trending, east-dipping extensional faults (Figure 7). The more westerly fault is interpreted to have a maximum easterly downthrow of approximately 0.3 s TWTT at the level of top Palaeogene lavas (Figure 34). Poor seismic imaging prevents confident mapping of a possible extension of the Grani Fault Terrace further towards the north.

The terrace is interpreted to be intruded by the ?Palaeogene Fraenir Volcanic Centre (Keser Neish, 2003) which possibly utilised the fault that defines the boundary with the Brynhild Sub-basin as a conduit for upward migration (Figure 7). The Grani Fault Terrace is interpreted to comprise up to approximately 0.9 s TWTT of Eocene to Recent sediments and 1 s TWTT of Palaeogene lavas, though the deeper structure remains largely unresolved (Figure 34).

Grimhild Sub-basin

The Grimhild Sub-basin forms a small, poorly defined, north-east-trending structure approximately 50 km long and up to 25 km wide that occurs within the central part of the report area (Figure 7; Keser Neish, 2003). The sub-basin is poorly defined from seismic data (Figure 38), but is interpreted to be bounded to the north-west by a combination of the Heri and East Faroe highs, and to the south-east, by a major north-east-trending extensional fault that defines the north-west margin of the Mid Faroe High (Figure 7). The north-east flank of the sub-basin is marked by the Corona Lineament, but the transition with the Brynhild Sub-basin to the south-west is poorly resolved.

The Grimhild Sub-basin has not been drilled but is considered by Keser Neish (2003) to represent a Mesozoic basin that has been subsequently modified by the effects of mainly Neogene and younger compression. It is considered to contain up to approximately 1.6 s TWTT of Eocene to Recent sediments, 1.3 s TWTT of Palaeogene lavas, 0.7 s TWTT of Paleocene sediments and 1.4 s TWTT of Mesozoic and older strata resting on ?crystalline basement (Figure 38). White et al. (2003, fig. 11) suggested that the sub-basin typically contains approximately 1 to 2 km of Palaeogene lavas resting on a similar thickness of pre-basalt sedimentary strata.

Guðrun Sub-basin

The Guðrun Sub-basin represents a small, poorly defined, north-east-trending half-graben up to approximately 70 km long and 45 km wide that occurs within the central part of the report area (Figure 7; Keser Neish, 2003). The sub-basin is imaged from seismic data (Figure 37; Keser Neish, 2003, enclosure 2, profiles C and D), and is interpreted to be separated from the Corona High to the south-east by a north-east-trending extensional fault (Figures 7 and 37). Speculatively, this fault has a north-westerly downthrow of approximately 2.5 s TWTT at top ?crystalline basement (compare with Keser Neish, 2003, enclosure 2, profiles C and D). The north-west margin of the sub-basin is defined by north-east-trending, south-east-dipping extensional faults that mark the south-east boundaries of the Mid Faroe and East Faroe highs. The north-east and south-west extents of the sub-basin are marked by the Grimur Kamban Lineament and a combination of the Corona Lineament and a Mid Faroe High ramp, respectively.

The Guðrun Sub-basin has not been drilled but is interpreted as a mainly Mesozoic basin (Keser Neish, 2003; Figure 37) that has been modified slightly by the effects of post-Eocene compression. Interpretations as to the fill of sub-basin vary (Keser Neish, 2003, enclosure 2, profiles C and D; Figure 37), but it may comprise up to approximately 1.7 s TWTT of Eocene to Recent sediments, 0.5 s TWTT of Palaeogene lavas, 1.1 s TWTT of Paleocene sediments and 2.3 s TWTT of Mesozoic and older strata resting on ?crystalline basement. According to White et al. (2003, fig. 11), the Palaeogene lavas and sub-basalt sedimentary interval thickens from practically 0 to 2.5 km and 0 to 4 km in north-westerly and south-westerly directions, respectively. The inferred presence of Mesozoic strata within the Guðrun Sub-basin is partly based on their occurrence on the nearby Corona High, supported by the recognition of Palaeogene sills on seismic (Figure 37) that are commonly associated with Upper Cretaceous strata within the Faroe–Shetland region.

Heri High

The Heri High forms a small north-east-trending structural horst approximately 45 km long and up to 10 km wide that occurs within the central part of the report area (Figure 7; Keser Neish, 2003). The high is reasonably well defined on seismic data (Figure 38; Keser Neish, 2003, enclosure 3, profile E) but has no notable potential field response associated with it. The north-west and south-east flanks of the horst are separated from the Annika and Brynhild/Grimhild sub-basins, respectively, by extensional faults with small throws of approximately 0.1 s TWTT or less at base Cenozoic or older stratigraphical levels (Figure 38; Keser Neish, 2003, enclosure 3, profile E). Towards the north-east and south-west, the Heri High is bounded by the Westray and Judd lineaments, respectively.

The Heri High has not been drilled but is interpreted to comprise up to approximately 1.1 s TWTT of Eocene to Recent sediments, 0.9 s TWTT of Palaeogene lavas, 0.2 s TWTT of Paleocene sediments and 1.7 s TWTT of Mesozoic and older strata resting on ?crystalline basement (Figure 38; Keser Neish, 2003, enclosure 2, profile E). The high is dextrally offset by 10 km from the south-west segment of the East Faroe High (Figure 7), a similar feature with which the Heri High probably shares a common tectonostratigraphical history. The high probably originated as a Mesozoic horst block that was rejuvenated by post-Eocene compressional tectonic activity. According to Keser Neish (2003, enclosure 2, profile E) the Oligocene succession is absent over the crest of the structure at the sea bed, probably though erosion.

Judd Sub-basin

The Judd Sub-basin forms a small, generally north-east-trending structure up to approximately 80 km long and 60 km wide that occurs within the southern part of the report area (Figure 7). It is generally well defined from seismic data (Figure 27; Roberts et al., 1999, fig. 23; Smallwood and Kirk, 2005, fig. 4). The Sub-basin is separated to the north and east from the Brynhild Sub-basin, Sjúrður Ridge, northern part of the Westray High and the Flett Sub-basin, mainly by a combination of the arcuate south-west- to north-west-trending Westray Fault (Smallwood and Kirk, 2005) and the north-north-west-trending Westray Lineament (Figure 7). The Rona and Judd Highs bound the southern flank of the sub-basin, with the former margin defined by the north-east-trending Rona Fault and the latter by a combination of the generally west-north-west-trending, north-north-east-dipping Judd Fault, north-west-trending Judd Lineament and an unnamed north-east-trending normal fault. The extreme western margin of the Judd Sub-basin occurs at the confluence of the Munkur Basin, Munkagrunnur Ridge, Brynhild Sub-basin and north-west Judd High, although the nature of their structural relationships is poorly understood.

The Judd Sub-basin has been drilled by a number of released wells within the UK sector mainly around the Suilven, Loyal and Schiehallion fields including, 204/14-1, -2, 204/19-5, -6, -7, -8, 204/20-1, -2, -4, -5, 20a-7, 204/22-2, 204/23-1, 204/24a-5, -6, -8, 204/25a-

2, -3, 25b-4, -5, -6, -7 and 204/29-1 (Figure 2). Apart from wells 204/29-1 and 204/23-1, these wells bottomed in the Cenozoic or Upper Cretaceous intervals, with approximate maximum drilled thicknesses of 3.6 km and 900 m recorded, respectively. Wells 6004/12-1Z, 6004/16-1Z and 6004/17-1 have been drilled close to the north-west margin of the basin within Faroese waters (Figure 2) and all have terminated within the Paleocene interval (e.g. Varming, 2009; Woodfin et al., 2005; Smallwood and Kirk, 2005).

From seismic evidence, Lamers and Carmichael (1999, fig. 5) suggested that the Upper Cretaceous could be up to 4.5 km thick to the east of the Westray High. The only wells to test pre-Upper Cretaceous rocks within the Judd Sub-basin are 204/23-1 and 204/29-1, which proved a total of approximately 150 m of Lower Cretaceous, 25 m of Upper Jurassic and 400 m of Triassic strata, with the former well terminating within crystalline basement on the crest a tilted fault block close to the north-east margin of the Judd High (Figure 27). The pre-Upper Cretaceous structure of the northern part of the Judd Sub-basin is relatively poorly resolved from seismic data (Lamers and Carmichael, 1999, fig. 5), but in the area between the Judd and Westray highs, the results from wells 204/23-1 and 204/19-1 suggest the potential presence of Lower Cretaceous to Permo-Triassic and older rocks within half-grabens (Figure 27). According to Dean et al. (1999, fig. 7), the Judd Sub-basin is interpreted to have developed as part of a system of north-east-trending Late Cretaceous rift basins that occurred throughout the Faroe–Shetland Basin area at this time. Although they suggested the possibility that this rift activity extended into Paleocene times, particularly associated with east-trending extensional faults (e.g. Smallwood and Kirk, 2005), post-rift thermal subsidence was probably the dominant process influencing the development of the Judd Sub-basin as a major depocentre in the Cenozoic. During Mid to Late Paleocene times, the formation of amalgamated channel complexes associated within submarine fan systems formed the reservoir for the Marjun, Schiehallion, Foinaven, Loyal and Suilven hydrocarbon accumulations within the subsiding sub-basin. Smallwood and Kirk (2005) considered that pulsed uplift associated with igneous intrusion beneath the Orkney–Shetland hinterland could be responsible for the generation of the coarse clastic material shed into the Judd Sub-basin. During Eocene to Pliocene times, the effects of compression were widely felt in the Sub-basin, with the development of the Westray, South Judd and Judd anticlines (Figure 14; Smallwood, 2004; Ritchie et al., 2008).

Mid Faroe High

The Mid Faroe High is a small, poorly defined, generally north-east-trending structural high that occurs within the central part of report area (Figure 7). The high is recognised from seismic data (Figures 37 and 38), although its internal structure is poorly resolved (cf Ellis et al., 2002). The high broadly correlates with positive isostatic gravity and positive magnetic anomalies (Figures 4 and 5). The Mid Faroe High is bounded to the south-east by a combination of the Guðrun Sub-basin, Corona High, Flett Sub-basin and Brynhild Sub-basin (Figures 7, 37 and 38). The transition with the Flett Sub-basin is marked by a steep, east-trending planar normal fault with a large displacement of approximately 1.2 s TWTT towards the south at the level of top ?crystalline basement (Figure 38). Towards the north-west margin of the high, the boundary with the Grimhild Sub-basin is less clearly defined. The Corona Lineament truncates the north-east extremity of the high, whereas the nature of its transition with the Brynhild Sub-basin to the south-west is poorly resolved.

The Mid Faroe High has not been drilled, but is interpreted to comprise up to approximately 1.7 s TWTT of Eocene to Recent sediments, 0.5 s TWTT of Palaeogene lavas, 0.4 s TWTT of Paleocene sediments and 1.3 s TWTT of Mesozoic and older sediments resting on ?crystalline basement (Figures 37 and 38). By analogy with drilling results from nearby well 213/23-1 on the Corona High, the pre-Cenozoic stratigraphical interval may comprise Devono-Carboniferous to Cretaceous strata resting on crystalline basement (cf. Figures 31 and 37).

Rona High

The Rona High forms an elongate north-east-trending basement ridge approximately 200 km long and up to 15 km wide that occurs within the south-east part of the report area (Figure 7; e.g. Stoker et al., 1993). It is well defined from seismic data (Figures 30, 31 and 37; e.g. Haszeldine et al., 1987; Booth et al., 1993; Coney et al., 1993; Swiecicki et al., 1995; Lamers and Carmichael, 1999; Herries et al., 1999) and also potential field data, corresponding with strong, segmented positive gravity and magnetic anomalies (Figures 3, 4 and 5). It should be noted that the Devonian Clair Basin is considered to encroach onto the central part of the Rona High (Figures 7 and 31). The Rona High is separated from the Flett, Foula, and Judd sub-basins to the north-west by the north-east-trending, north-west-dipping, planar normal Rona Fault. Towards the south-east, the high is flanked by a combination of the West Shetland, and East Solan basins. The nature of this boundary varies along strike (e.g. Figures 31 and 37). These lateral variations are interpreted to be linked with the north-north-west- to north-west-trending lineaments or transfer zones that segment of the Rona High (Figure 7). Two of these features, the Judd Lineament and an unnamed lineament or transfer zone, truncate the south-west and north-east ends of the Rona High, respectively.

The Rona High has been drilled by a number of released commercial wells including 204/25-1, 204/30-1, 205/20-1 (Figure 37), 205/21-1A, 205/23-2, 205/26-1 and 206/12-1 in the south-west and 207/01-1, -3, 208/27-1 and -2 in the north-east part of the ridge (Figure 2). Wells 206/07-1, 07a-2, 206/08-1A, -2, -3A, -4, -5, -6A, -7Z, -8, -9Z, -10Z, -11A, -12A, 206/09-2, 206/12-2, 206/13a-2 and 207/01a-4Z are also located on the Rona High but where they contain Devono-Carboniferous strata, that part of the succession has been ascribed to the Clair Basin (Figure 7; see below). These drilling results indicate that the Rona High is more deeply buried towards the south-west and is capped by a maximum stratigraphical thickness of approximately 4.05 km of Mesozoic to Recent sediments, including 1.2 km of Cenozoic to Recent, 2.5 km of Lower and Upper Cretaceous and 100 m of Upper Jurassic and 150 m of Triassic and 100 m of undated basal conglomeratic strata resting on crystalline basement. The high has exerted a considerable influence in the development of the West Shetland Basin and the Judd, Flett and Foula sub-basins that flank it, particularly from Early Cretaceous times onwards (e.g. Ridd, 1981; Dean et al., 1999; Goodchild et al., 1999). For example, the Rona Fault that bounds the north-west margin of the high is thought to have been active at various times throughout the Cretaceous, though the focus of this activity is interpreted to have shifted along the strike of the fault through time (e.g. Dean et al., 1999). According to Haszeldine et al. (1987, fig. 5) and Dean et al. (1999), the south-west margin of the Rona High was emergent during earliest Cretaceous times, with the development of coarse clastic sediment wedges within the West Shetland Basin that appear to onlap and thin towards it. Early Paleocene uplift of the Rona High resulted in shallow marine Upper Paleocene to Eocene sediments resting unconformably on eroded deep marine Upper Cretaceous (Maastrichtian) strata in the Victory (Goodchild et al., 1999) and Clair fields (Ridd, 1981) (see Chapter 12).

Sjúrður Ridge

The Sjúrður Ridge is a small, narrow, east-north-east-trending basement high approximately 45 km long and 5 km wide that occurs towards the south-east corner of the Faroese national sector (Figure 7; Roberts et al., 1999, Keser Neish, 2003; Smallwood, 2005b; Smallwood et al., 2004, Smallwood and Kirk, 2005). Formerly known as the Central Ridge (Roberts et al., 1999, fig. 2), it is defined from seismic data (Keser Neish, 2003, enclosure 2, profiles E and F), and forms a horst block separated from the Brynhild and Judd sub-basins to the north-west and south-east respectively, by north-east- to east-north-east-trending extensional faults. The south-east margin of the ridge is defined in part by the arcuate, generally south-south-east-dipping early Paleocene Westray Fault (Figure 7; Smallwood and Kirk, 2005), demonstrating a throw of more than 0.4 s TWTT at the base Cenozoic horizon. Throws of similar magnitude have been recorded from north-north-west-dipping faults that define the north-west margin of the ridge, but these are mainly at pre-Cenozoic stratigraphical levels, for example top ?crystalline basement (Keser Neish, 2003, enclosure 2, profile E). The Sjúrður Ridge is traversed by the north-west-trending Judd Lineament, which according to Keser Neish (2003), dextrally offsets it by 1.5 km.

The Sjúrður Ridge has been drilled by well 6005/15-1 but the results have not been published as yet. Currently, the high is considered to comprise approximately 2 s (Keser Neish, 2003, enclosure 3, profiles E and F) to 2.5 s TWTT (Smallwood, 2005b, fig. 2; Roberts et al., 1999, fig. 22) of Paleocene to Recent sediments and lavas overlying 0.6 s (Roberts et al. 1999, fig. 22) to 1.0 s TWTT (Keser Neish (2003, enclosure 2, profiles E and F) of Mesozoic and older strata. The Sjúrður Ridge is considered to be a mainly post-Eocene inversion structure, with the Eocene succession forming a dome at the sea bed (Keser Neish, 2003, enclosure 2, profile F).

Steinvør Sub-basin

The Steinvør Sub-basin represents a north-east-trending, basinal feature approximately 130 km long and up to 35 km wide that occurs within the central part of the report area (Figure 7). The basin is defined from seismic data, forming a shallow col between the Tróndur High/Fugloy Ridge and the East Faroe High (Figure 32; Keser Neish, 2003, enclosure 2, profile C; Ziska and Andersen, 2005, fig. 3; Keser Neish, 2003, fig. 7). The Steinvør Sub-basin is flanked to the north-west by north-north-east- to east-north-east-trending, generally south-east-dipping extensional faults that form the boundaries with the Tróndur High and Fugloy Ridge, and to the south-east by a combination of north-east-trending, north-west-dipping extensional faults that form the flank of the East Faroe High, and by non-faulted transitions with the north-west part of the Corona Sub-basin (Figure 7). The interpreted throw on the faults on both flanks of the basin is generally small, with the largest being approximately 0.3 s TWTT at the level of top ?crystalline basement (Keser Neish 2003, enclosure 2, profile C). The basin is interpreted to be dissected by the Victory, Clair and Grimur Kamban lineaments, with the Corona Lineament lying at its south-western extremity.

According to Keser Neish (2003) the Steinvør Sub-basin represents a significant and extensive Mesozoic Basin, and is interpreted to comprise up to approximately 1.6 s TWTT of Eocene to Recent sediments, 1.0 s TWTT of Palaeogene lavas, 0.4 s TWTT of Paleocene sediments and 1.5 s of Mesozoic and older sediments resting on ?crystalline basement (Keser Neish, 2003, enclosure 2, profile C; Ziska and Andersen, 2005, fig. 3; Keser Neish, 2003, fig. 7). According to White et al. (2003, fig. 11), the sub-basin typically contains approximately 2 to 4 km of Palaeogene lavas resting on a north-westward thickening sedimentary succession up to 2 km thick.

Tróndur High

The Tróndur High forms a segmented north-north-east- to north-east-trending structural feature approximately 65 km long and up to 25 km wide that occurs within the central part of the report area (Figure 7). It is well imaged on seismic data (Keser Neish, 2003, enclosure 2, profile C and D; Keser Neish, 2005, fig. 7). The high is bounded to the north-east and south-west by the north-west-trending Clair and Corona lineaments, respectively, with a distinct change in trend from north-north-east to north-east where it is traversed by the Grimur Kamban Lineament (Figure 7). The north-west and south-east flanks of the horst are separated from the Annika and Steinvør sub-basins respectively by extensional faults with generally small throws of less that 0.1 s TWTT at the base Cenozoic stratigraphical level (Keser Neish, 2005, fig. 7). The Tróndur High appears structurally similar to the subparallel East Faroe High (Keser Neish, 2005).

The Tróndur High has not been drilled but is interpreted to comprise up to approximately 1.1 s TWTT of Eocene to Recent sediments, 0.7 s TWTT of Palaeogene lavas, 0.5 s TWTT of Paleocene sediments and 1.3 s TWTT of Mesozoic and older sediments resting on ?crystalline basement (Keser Neish, 2003, enclosure 2, profile C; Keser Neish, 2005, fig. 7). Similar to the adjacent East Faroe High, the Tróndur High is considered to have been rejuvenated in response to compression in post-Eocene times.

Westray High

The Westray High is interpreted to comprise two small, north- to north-north-west-trending intrabasinal highs (e.g. Smallwood et al., 2004; Smallwood and Kirk, 2005) that occur towards the central part of the report area (Figure 7). Previously, the high was considered to comprise a single continuous feature (e.g. Mudge and Rashid, 1987; Rumph et al., 1999; Dean et al., 1999; Lamers and Carmichael, 1999; Grant et al., 1999). Whereas the more southerly part of the Westray High is well defined from seismic data (Figure 27; Lamers and Carmichael, 1999, fig. 4), the more northerly of the highs is well imaged from potential field data, corresponding with positive gravity and magnetic anomalies (Figures 3, 4 and 5). The northerly and southerly elements of the Westray High occur close to the margin of the Flett Sub-basin and within the Judd Sub-basin, respectively (Figure 7). The southern segment of the Westray High is fault bounded, with its western and eastern margins delineated by north-north-west-trending normal faults with downthrows of approximately 1.5 and 1.0 s TWTT at the level of top crystalline basement towards the west and east, respectively (Lamers and Carmichael, 1999, fig. 4). In comparison, little has been published regarding the more northerly element of the Westray High, though its southern margin is interpreted to be bounded by a combination of the Westray Fault (Smallwood et al., 2004; Smallwood and Kirk, 2005) and the Westray Lineament.

The Westray High has been drilled by a number of released commercial wells including 204/15-1, -2, 204/19-1, -2, -3A, -4A, -9, 204/24-1A, 24a-2, -3, -4 and -7 (Figure 2), though only 204/19-1 and 9 penetrated a pre-Upper Cretaceous succession. Wells on the high proved a maximum thickness of approximately 5.8 km of Cenozoic to Upper Palaeozoic strata including 3.85 km of Cenozoic to Recent, 1.4 km of Lower and Upper Cretaceous, 150 m of Upper Jurassic, 200 m of Triassic and 200 m of Devonian sediments. Only well 204/15-2 within the northern part of the high has penetrated crystalline basement. The Westray High in general is considered to have exerted a considerable influence on sedimentation patterns, particularly during Cretaceous times, with up to 4.5 km deposited close to its eastern flank. Upper Cretaceous (upper Campanian to Maastrictian) sediments are mainly absent on the high (see Figure 77), possibly due to tectonic effects (Roberts et al., 1999).

Iceland Basin

The Iceland Basin is a deep oceanic basin that encroaches within the extreme western part of the report area, where it is bounded to the north and east by the Iceland–Faroe Ridge and Faroe Platform, respectively (Figure 7). Towards the south-east, its transition with the North Faroe Bank Channel Basin and the Faroe Bank High is poorly understood. The Iceland Basin is floored by oceanic crust, the oldest part of which initially formed at about 55 to 54 Ma (magnetic anomaly chron 24r) during earliest Eocene times as a result of sea-floor spreading (e.g. see Nunns, 1983; Srivastava and Tapscott, 1986; Smallwood and White, 2002) associated with the developing Reykjanes Ridge that is still active today.

Møre Basin

The Møre Basin forms a large north-east-trending basin approximately 330 km long and up to 230 km wide that occurs mainly within the Norwegian continental margin (e.g. Blystad et al., 1995), with only the extreme south-western part extending into the report area (Figure 7; e.g. Stoker et al., 1993). From the results of published seismic mapping, the basin morphology and structure within the Norwegian margin is generally well understood (e.g. Rønnevik, 1975; Gabrielsen et al., 1984; Blystad et al., 1995; Gabrielsen et al., 1999; Brekke et al., 1999), though rather less so within the UK part of the continental margin (e.g. Duindam and van Hoorn, 1987; Nelson and Lamy, 1987). Within the report area, the basin is characterised by a negative free-air and isostatic gravity anomaly (Figures 3 and 4). Further to the north-east, offshore Norway, the north-east flank of the Møre Basin is separated from the Vøring Basin by the north-west-trending Jan Mayen Lineament (Blystad et al., 1995). Towards the north-west and south-east, the basin is bounded by the Faroe–Shetland Escarpment/Møre Marginal High and the Manet High, respectively. Within the report area, the Møre Basin is flanked to the north-west and south-east by the Møre Marginal and East Shetland highs, respectively (Figures 7 and 23). The boundary between the Møre Marginal High and the Møre Basin is inferred to be a faulted contact, though conflicting models have been suggested regarding the dip direction of the fault plane (e.g. Duindam and van Hoorn, 1987; Doré et al., 1999). The boundary of the Møre Basin with the East Shetland High is represented by a steep, north-east-trending, normal fault with a throw of up to 3 s TWTT towards the north-west at top ?crystalline basement (Figure 23; Roberts et al., 1999). Towards the south-west, the nature of transition between the Møre Basin and the contiguous north-east-trending Faroe–Shetland Basin is less clear (see Price and Rattey, 1984). This lack of clarity may be due to the presence of Palaeogene lavas and intrusive rocks associated with the Brendan, Erlend or West Erlend Volcanic centres (Figure 7). On the basis of potential field data, however, the preferred location for the boundary between the basins is defined by the crest of an elongate, north-west-trending free-air and isostatic positive gravity anomaly (Figures 3 and 4). This feature broadly corresponds with a north-west-trending unnamed transfer zone of Rumph et al. (1993), that has been modified and named the Brendan Lineament (Figures 4 and 7). Kimbell et al. (2005) traced a possible extension of the Marflo Lineament (Smethurst, 2000) through this area but placed it to the north-east of the Brendan Lineament.

The succession within the part of the Møre Basin represented in the report area has been drilled by released wells 219/20-1, 219/28-2Z and 220/26-1 (Figure 2). These wells proved an approximate maximum thickness of 5.2 km of Mesozoic to Recent sediments and sills including 2.3 km of Cenozoic to Recent, 2.75 km of Lower and Upper Cretaceous and 150 m of basal conglomeratic strata. Wells 219/28-2Z and 220/26-1 terminated within crystalline basement at depths of approximately 4 and 5.3 km, respectively. Combined seismic and well evidence suggest the possibility that ?Palaeozoic to Recent sediments within the easterly tilted half-grabens on the south-east flank of the Møre Basin could attain a thickness of up to 10 km (Doré et al., 1999, fig. 2; Kimbell et al., 2005, fig. 6), up to 6 km of which is likely to be Cretaceous in age (e.g. Nelson and Lamy, 1987; Spencer et al., 1999; Brekke, 2000; Færseth and Lien, 2002). Stratigraphical thicknesses towards the central and north-west flanks of the basin within the report area are more difficult to estimate due to the increasing presence of Palaeogene intrusive and extrusive igneous rocks. According to Blystad et al. (1995) and Gabrielsen et al. (1999), the Møre Basin as a whole is considered to have formed mainly as a result of Late Jurassic to Early Cretaceous rifting. Towards the south-east margin of the basin within the report area, this rifting is characterised by the development of north-east-trending fault blocks with south-east-tilted half-grabens developed in their hanging walls (Figure 23; Doré et al., 1999, fig. 2; Spencer et al., 1999, fig. 7). The extensional phase of activity responsible for basin formation may have continued into ?earliest Late Cretaceous times, but was followed by the development of a thick Upper Cretaceous and Cenozoic post-rift succession. Post-rift subsidence was interrupted by phases of widespread Palaeogene volcanic and intrusive (Figure 23) activity (see well 219/20-1 in Chapter 9) associated with the Erlend and Brendan Volcanic centres and in places, by the effects of mainly Miocene compression (e.g. Løseth and Henriksen, 2005).

North-east Rockall Basin

The north-north-east-trending North-east Rockall Basin forms an elongate graben approximately 160 km long and 80 km wide that only just encroaches into the south-west margin of the report area (Figure 7). Outwith the report area, the Cenozoic sedimentary succession within the basin is well imaged on seismic data but its deeper structure is poorly resolved due to the presence of the thick Palaeogene lavas (e.g. Tate et al., 1999; Waddams and Cordingley, 1999; Archer et al., 2005). The North-east Rockall Basin also corresponds with an elongate, negative free-air and isostatic gravity anomaly (Figures 3 and 4). The basin is bounded on its northern flank by a combination of the Ymir Ridge, Auðhumla Basin, the Wyville Thomson Ridge and the Outer Hebrides High (Figure 7), although the nature of these boundaries is not well defined.

Outwith the study area, the basin has been drilled by a number of wells, with the most recent well, 164/07-1 (Figure 2), proving approximately 2.55 km of Cenozoic to Recent sediments and volcanic rocks and 1.8 km of Upper Cretaceous sediments and intrusive rocks before terminating in Lower Cretaceous sedimentary rocks — the oldest proven strata within the North-east Rockall Basin (Archer et al., 2005). Waddams and Cordingley (1999) consider that the basin originally formed an integral part of the north-east-trending Faroe–Shetland Basin during Mesozoic and older times, but has only become separated as a result of growth of the cross-cutting Wyville Thomson Ridge anticline during early Cenozoic times onwards. In terms of its tectonostratigraphical development, the North-east Rockall Basin is considered to represent a mainly early Cretaceous (Doré et al., 1999, fig. 3; Spencer et al., 1999, fig. 1) broadly symmetrical rift basin (Waddams and Cordingley, 1999, fig. 4), with the post-Lower Cretaceous interval representing a dominantly thermal-sag or post-rift succession.

Norwegian Basin

The Norwegian Basin is a deep oceanic basin that occurs within the northern part of the report area where it is bounded to the south by a combination of the Iceland–Faroe Ridge, Faroe Platform, Fugloy Ridge and the Møre Marginal High (Figure 7). The basin is floored by oceanic crust, initially formed approximately at 55 to 54 Ma (magnetic anomaly chron 24r) during earliest Eocene times as a result of sea-floor spreading (e.g. see Nunns, 1983; Srivastava and Tapscott, 1986) associated with the now extinct Aegir Ridge.

North Rockall Basin

The Rockall Basin as a whole forms an elongate, north-east- to north-north-east-trending, sediment starved, deep-water rift basin approximately 1100 km long and 350 km wide. The term north Rockall Basin is introduced informally that describes the northern part of the Rockall Basin. This basin only just encroaches within the extreme south-west part of the report area (Figures 7 and 36). Outwith the study area, the post-Palaeogene lavas sedimentary succession within the basin is well imaged from commercial and BGS shallow data seismic data. Within the report area, the north Rockall Basin is flanked on its northern margin by a combination of the Sigmundur Seamount, Ymir Ridge and Wyville Thomson Ridge (Figure 7), although the nature of these boundaries is poorly constrained.

Immediately adjacent to the report area, well 163/06-1 proved approximately 1.25 km of Eocene to Recent sediments resting on 1.05 km of Paleocene basic and acidic lavas on the north-west flank of the Darwin Volcanic Centre (see Chapter 9; Figure 2; Abraham and Ritchie, 1991). The deeper structure of the basin below the Palaeogene lavas is poorly resolved on conventional seismic data, although results from wide-angle seismic reflection/refraction experiments (e.g. Roberts et al., 1988) suggest that the combined sediment and volcanic rock thickness within the North Rockall Basin typically averages approximately 5 km.

Although it is clear that the Rockall Basin as a whole has undergone significant extension, there are several hypotheses regarding the timing and number of rift episodes i.e. single or polyphase rifting. The current consensus of opinion is that the Rockall Basin is a mainly Early Cretaceous rift basin (e.g. Musgrove and Mitchener, 1996) that experienced earlier rifting during Permo-Triassic and Jurassic times (e.g. Cole and Peachey, 1999; Nadin et al., 1999; Shannon et al., 1999). Following Cretaceous rifting, the Rockall Basin underwent thermal cooling and post-rift subsidence during Cenozoic times. However, the development of the sedimentary infill does not merely reflect passive infilling of Cretaceous block-faulted rift topography within the basin, as a number of tectonic events have interrupted the pattern of thermal subsidence. This includes Paleocene regional dynamic uplift caused by the developing Iceland Plume (White and McKenzie, 1989) and episodes of regional tilting, enhanced sagging and compression in Paleocene, Late Eocene to Oligocene, Late Oligocene to Mid Miocene and Early Pliocene times (e.g. Stoker et al., 2005b and c; Johnson et al., 2005b).

Marginal basinal areas and associated highs

Auðhumla Basin

The Auðhumla Basin is a small, narrow, north-west-trending and tapering syncline approximately 80 km long and 30 km wide that straddles the southern boundary of the report area towards its south-west corner (Figure 7; Keser Neish, 2003). The basin is well defined from seismic data, forming a complementary syncline to the west-north-west-trending Wyville Thomson and north-west-trending Ymir anticlines (Figure 36). Towards the north-west and south-east, the margins of the basin are poorly defined, but in the latter case, it is conceivable that the Auðhumla Basin represents a north-westerly extension of the North-east Rockall Basin.

The Auðhumla Basin has not been drilled but Keser Neish (2003, enclosure 2, profile J) speculated that it comprises up to approximately 3 s TWTT of earliest Neogene to Jurassic sediments and lavas that have been deformed by Eocene and younger compression. The Drekaeyga Volcanic Centre (see Chapter 9; e.g. Keser Neish, 2005; Keser Neish and Ziska, 2005) has been inferred to intrude the basin (Figure 7).

Clair Basin

The Clair Basin forms a small, north-east-trending Devonian to Carboniferous depocentre approximately 50 km long and up to 20 km wide that occurs within the eastern part of the report area (Figure 7). It is interpreted by Nichols (2005) to separate the north-east and south-west parts of the West Shetland Basin, although at Devonian to Carboniferous stratigraphical levels only (see Chapter 4). The general structure of the Devonian basin or 'rollover' of Roberts et al. (1999) is poorly resolved from seismic data (e.g. Coney et al., 1993, figs. 14 and 15), but is interpreted to straddle the central parts of the West Shetland Basin and Rona High (Figures 7 and 31). The north-west and south-east margins of the basin are separated from the Foula Sub-basin and West Shetland High by the north-east-trending, Rona and Shetland Spine faults, respectively.

The basin limits are rather arbitrarily defined from the presence or absence of Devono-Carboniferous rocks drilled within released commercial wells, including 206/07-1, 07a-2, 206/08-1A, -2, -3A, -4, -5, -6A, -7Z, -8, -9Z, -10Z, -11A, Z, -12A, 206/09-1, -2, 206/10a-1, 206/12-2, 206/13a-2 and 207/01a-4Z (Figure 2). Within the basin there is a maximum proven vertical thickness of approximately 1 km. Following the collapse of the Caledonian Orogen, the Clair Basin is interpreted to have developed within an extensional or strike-slip intermontane basin setting during Mid to Late Devonian to early Carboniferous times (e.g. Meadows et al., 1987; Stoker et al., 1993; Ritchie et al., 1996; Friend et al., 2000; Coward et al., 2003; Nichols, 2005). Whether the basin represents a discrete Devonian depocentre or an erosional remnant of a once much more extensive complex of basins remains a matter of some conjecture.

East Rona High

The East Rona High forms a small, partially fault-bounded intrabasinal basement high within the North Lewis Basin close to the southern margin of the report area (Figure 7; e.g. Stoker et al., 1993). The extent of the high is defined from shallow seismic (BGS, 1986a) and potential field data. It has not been drilled but is inferred to comprise basement of Lewisian aspect, similar to that described from the nearby island of Rona (Nisbet and Bowes, 1961).

East Solan Basin

The East Solan Basin is a small, north-east-trending half-graben approximately 35 km long and up to 10 to 15 km wide that occurs within the southern part of the report area (Figure 7). The basin is well defined from seismic data (Figure 39; Booth et al., 1993; Swiecicki et al., 1995; Herries et al., 1999) and is interpreted to be separated from the Rona High and Papa Basin to the north-west and south-east, by a minor north-east-trending, south-east-dipping unnamed planar normal fault and the north-east-trending, north-west-dipping planar normal Otter Bank Fault, respectively (Figures 7 and 39). The basin is interpreted to be bounded to the south-west by the Judd Lineament.

The East Solan Basin has been drilled by a number of released wells including 204/30a-2, -3, 205/26a-2, -3, -4, -5Z, -6 and 205/27-2 (Figure 2). The majority of these wells are associated with the Strathmore (Triassic) and Solan (Upper Jurassic) hydrocarbon accumulations (see Chapter 12, Herries et al., 1999), proving an approximate maximum thickness of 4.55 km of Mesozoic to Recent sediments, including 1.25 km of Cenozoic to Recent, 2.2 km of Lower and Upper Cretaceous, 100 m of Upper Jurassic and 1.0 km of Triassic strata. The East Solan Basin is considered by some to have formed part of a much wider grouping of Permo-Triassic basins (including the Papa Basin) that has been separated by deep erosion (with up to 1.5 km of sediments removed) during Mid to earliest Late Jurassic times (e.g. Booth et al., 1993; Swiecicki et al., 1995). This erosional event is deemed responsible for the major unconformity that separates a thin Upper Jurassic succession from a strongly truncated Triassic one. On balance, the formation of the East Solan Basin was probably initiated in Late Jurassic to Early Cretaceous times (e.g. Herries et al., 1999), but it is thought to represent a mainly Late Cretaceous extensional basin (e.g. Booth et al., 1993; Swiecicki et al., 1995). However, from Figure 39, there is little suggestion for south-eastwardly thickening of the Cretaceous succession within in the hanging wall of the Otter Bank Fault. Active Palaeogene sedimentation may have been disrupted by compression and the contemporaneous development of 'pop-up' structures may provide evidence of this (Booth et al., 1993). A significant unconformity has been recognised which formed during Miocene times, with Pliocene sediments inferred to rest on Eocene strata in places (Booth et al., 1993). Such postulated Paleocene and Miocene compression may be related to similar events recognised within the Judd and Flett sub-basins to the north-west and on the Wyville Thomson Ridge to the south-west (Figure 14; e.g. Ritchie et al., 2003; Smallwood, 2004; Johnson et al., 2005b; Ritchie et al., 2008).

Faroe Bank Channel Basin

The Faroe Bank Channel Basin is a generally north-north-west- to north-west-trending synclinal structure, approximately 70 km long and up to 95 km wide that occurs within the south-west part of the report area (Figure 7; Roberts et al., 1983; Keser Neish, 2003). It is well defined from seismic data, particularly at Palaeogene lavas and younger stratigraphical levels (Figures 34, 35 and 36; Keser Neish, 2003, enclosure 2, profile H; Keser Neish, 2005, fig. 4; Keser Neish and Ziska, 2005, fig. 5). The basin is also clearly expressed on potential field data too, corresponding with a negative free-air gravity anomaly (Figure 3). The basin is bounded to the east and west by the Munkagrunnur Ridge and a combination of the Wyville Thomson Ridge and Faroe Bank High, respectively (Figure 7), although nature of these boundaries is not well resolved at deep structural levels (Figures 35 and 36; Keser Neish, 2003, enclosure 2, profile H; Keser Neish, 2005, fig. 4; Keser Neish and Ziska, 2005, fig. 5). To the north, the basin passes laterally to the North Faroe Bank Channel Basin, although the location of this boundary is rather arbitrarily defined (Keser Neish and Ziska, 2005, fig. 5, profile E). The transition to the Munkur Basin and

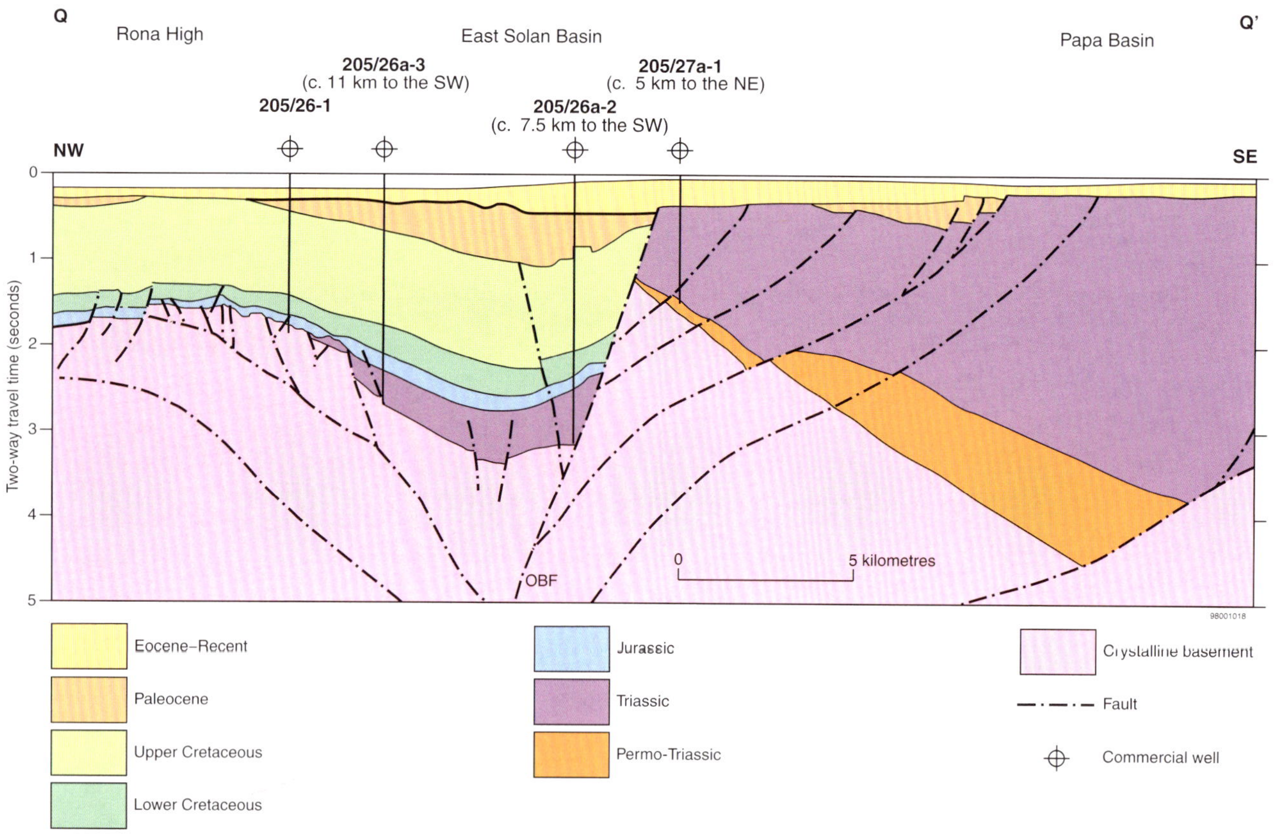

Figure 39 Generalised geological cross-section Q–Q' across the Rona High, East Solan Basin and Papa Basin (modified from Herries et al., 1999). Abbreviations: see Figure 7. For location of profile, see Figure 7.

Faroe Bank Channel Knoll to the south is defined by an east-north-east-trending, north-north-west-dipping normal fault, with an estimated downthrow of approximately 1 s TWTT at the level of base Palaeogene lavas (Keser Neish and Ziska, 2005, fig. 5, profile C).

The Faroe Bank Channel Basin has not been drilled, but it is thought to represent a mainly Mesozoic depocentre that is defined by series of east-north-east- to north-east-trending Caledonoid and north-trending faults (see Keser Neish and Ziska, 2005). From the interpretation on Figure 35, it comprises up to approximately 1.7 s TWTT of Eocene to Recent sediments, 1.4 s TWTT of Palaeogene lavas, 2.1 s TWTT of Paleocene sediments and Mesozoic and older strata that rest on ?crystalline basement. The basin has been significantly affected by Cenozoic compression and has been referred to as an 'inversion syncline', formed as part of a system of ramp anticlines and complimentary synclines on a northward dipping major crustal fault associated with changes in the north-east Atlantic sea-floor spreading geometry (e.g. Boldreel and Andersen, 1993; Tate et al., 1999). Most of the folding and thrusting of the Palaeogene succession in particular probably occurred during Early to Mid Miocene times (e.g. Johnson et al., 2005b; Stoker et al., 2005c). For example, within the central part of the basin there appears to have been significant inversion of the Eocene to Oligocene succession to form a high, with subsequent deposition of Middle to Upper Miocene and younger sediments progressively onlapping both flanks (Figure 35; see also Figures 110 and 114c).

Fetlar Basin

The Fetlar Basin forms a small, north-north-east-trending, eastward-tilted, partially fault-bounded half-graben approximately 20 km long and 8 km wide that only just encroaches on the eastern margin of the report area (Figure 7; Stoker et al., 1993). The general morphology of the basin is mainly defined from BGS shallow seismic data (BGS, 1984). The Fetlar Basin occurs within the East Shetland High and is bounded to the east by a north-north-east-trending extensional fault, with approximately 500 m of downthrow at the level of top crystalline basement (Figures 7 and 40).

The basin has been drilled by BGS borehole 80/02 (Figures 2 and 40), proving sediments of presumed Permo-Triassic age close to the sea bed. It probably represents a small remnant half-graben, developed as part of a more extensive series of basins on the East Shetland High and surrounding area during Triassic (Spencer et al., 1999) or Permo-Triassic (Stoker et al., 1993; Doré et al., 1999) times.

Foula High

The Foula High forms a small, arcuate, north-north-east- to north-east-trending ridge approximately 30 km long and 5 km wide close to the eastern margin of the report area (Figures 7; Stoker et al., 1993). The general morphology of the high is mainly defined from BGS shallow seismic data (BGS, 1984; 1988b) and is interpreted to represent a buried ridge composed of crystalline basement that is overstepped by Devonian sediments of the Orkney–Shetland High (Figure 29). It is terminat-

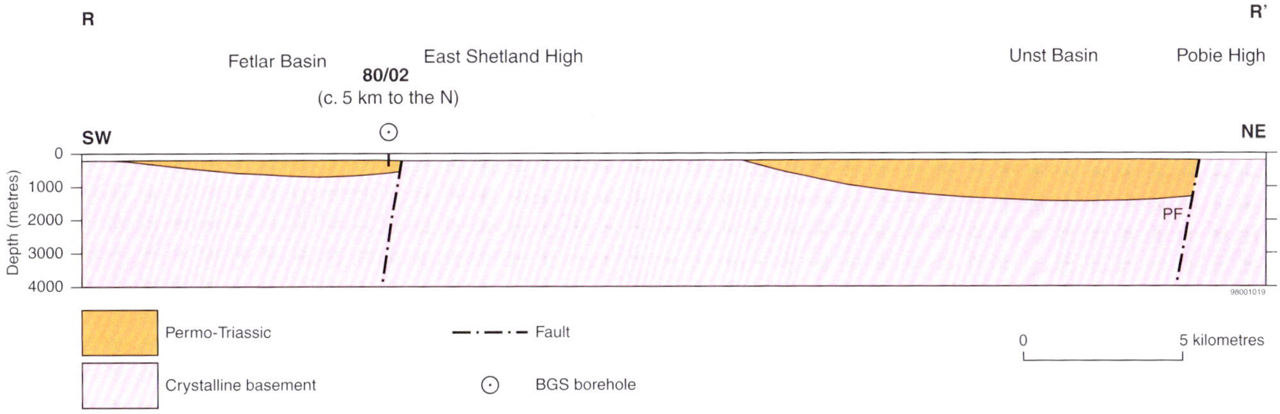

Figure 40 Generalised geological cross-section R–R' across the Fetlar Basin, East Shetland High, Unst Basin and Pobie High (modified from BGS, 1984). Abbreviations: PF=Pobie Fault. For location of profile, see Figure 7.

ed on its southern flank by a north-north-west-trending, west-south-west-dipping planar normal fault on Foula Island, with approximately 1.8 km of Middle Devonian sedimentary rocks present within its hanging wall (BGS, 1984). Both ends of the high have a subaerial expression, with basement rocks described from the Ve Skerries islets (Mykura, 1976) and Foula (BGS, 1984), respectively. Within the intervening offshore area, gneisses have been described from BGS shallow sample sites (BGS, 1984).

Munkur Basin

The Munkur Basin is a north-west-trending synclinal feature up to approximately 95 km long and 40 km wide that occurs close to the south-west edge of the report area (Figure 7; Waddams and Cordingley, 1999; Keser Neish, 2003). The uppermost stratigraphical levels within the basin are well defined from seismic data (Tate et al., 1999; fig. 3; Keser Neish, 2003, enclosure 2, profile H; Keser Neish, 2005; fig. 4; Keser Neish and Ziska, 2005, fig. 5, profile C). The Munkur Basin is bounded to the south-west and north-east by the Wyville Thomson and Munkagrunnur ridges, respectively (Figure 7), with the latter defined by a reverse fault with a small south-westerly downthrow of approximately 0.2 s TWTT at top Palaeogene lavas (Tate et al., 1999, fig. 3). Towards the south-east, the basin is interpreted to be bounded by a combination of the Judd High and Judd Sub-basin though the nature of these contacts is poorly understood. The north-west margin of the basin is defined by a combination of the Faroe Bank Channel Knoll Volcanic Centre and an east-north-east-trending, north-north-west-dipping normal fault that marks the transition to the Faroe Bank Channel Basin.

The Munkur Basin has not been drilled but according to Keser Neish (2003, enclosure 2, profile H), it forms a prominent Mesozoic basin that comprises up to approximately 1 s TWTT of Eocene sediments, 1.25 to 3.5 km of south-easterly thinning Palaeogene lavas, 0.7 s TWTT of Paleocene sediments and 1.75 s TWTT of Mesozoic and older strata resting on ?crystalline basement. The results of a recent wide-angle seismic reflection profile indicate that the south-east part of the Munkur Basin contains approximately 4 to 5 km of sediments, lavas and intrusive rocks resting on crystalline basement (Klingelhöfer et al., 2005).

North Faroe Bank Channel Basin

The North Faroe Bank Channel Basin forms a poorly defined, generally north-west-trending basinal feature that occurs close to the western margin of the report area (Figure 7; Keser Neish, 2003). The basin is defined from seismic data (Keser Neish, 2003, enclosure 2, profile G; Keser Neish and Ziska, 2005, fig. 2). The North Faroe Bank Channel Basin is bounded to the north-east and south-west by the Munkagrunnur Ridge and Faroe Bank High, respectively (Figure 7). Towards the south-east, the basin is considered to pass laterally to the Faroe Bank Channel Basin although there is no obvious structural break (Keser Neish, 2003; Keser Neish and Ziska, 2005). The relationship with the oceanic Iceland Basin towards the north-west is poorly understood.

The North Faroe Bank Channel Basin has not been drilled, but from the interpretation of Keser Neish (2003, enclosure 2, profile G), it is inferred to represent a mainly Palaeogene and younger basin that comprises up to approximately 1.2 s TWTT Eocene to Recent sediments, 1.5 s TWTT of Palaeogene lavas and 1.1. TWTT of Paleocene sediments resting on ?crystalline basement. A northward thickening of the volcanic succession is considered to be associated with the presence of the Regin Smiður Volcanic Centre (see Chapter 9; Keser Neish and Ziska, 2005, fig. 2; Ziska and Varming, 2008).

North Lewis Basin

The north-east-trending North Lewis Basin comprises two main north-west-tilted half-grabens, approximately 90 km long and 50 km wide that straddle the south-west margin of the report area (Figure 7). The general morphology of the basin is defined from both BGS shallow seismic (BGS, 1986a; 1988c and 1989b) and commercial seismic data (Stein, 1988). Within the report area, the North Lewis Basin is separated to the north from the Sula Sgeir High, Outer Hebrides High and North Rona Basin by a southward-dipping normal fault (Figure 7; BGS, 1986a). The eastern flank of the basin is defined by a combination of a north-trending, west-dipping normal fault that delimits the south-west margin of the Solan Bank High and the north-east-trending, south-east-dipping, Minch Fault that separates the North Lewis and North Minch basins (BGS, 1989b; Stein and Blundell, 1990).

The North Lewis Basin has been drilled by BGS boreholes 72/35 and 36, with the latter proving Upper Triassic to Lower Jurassic strata (Figures 2 and 7). The basin also includes the island of Rona, an intrabasinal horst block comprising Lewisian basement ascribed to the West Rona High. Outwith the study area, evidence from seismic data suggests that the North Lewis Basin may contain up to 4 km of north-west-dipping Permo-Triassic and Lower Jurassic strata that developed within the hanging wall of the Outer Isles Fault. The North Lewis Basin developed as a pair of westward-tilted Permo-Triassic (e.g. Kirton and Hitchen, 1987; Stoker et al., 1993; Doré et al., 1999) or Triassic to Jurassic (Spencer et al., 1999) half-grabens, although a pre-Devonian precursor for the basin cannot be discounted (Klemperer and Hobbs, 1991; Stoker et al., 1993).

North Rona Basin

The North Rona Basin forms an elongate, north-east-trending, south-east-dipping half-graben complex, approximately 90 km long and up to 25 km wide that occurs within the south-west part of the report area (Figure 7; Stoker et al., 1993). The structure of the basin is defined from a combination of BGS shallow seismic and commercial data (Figure 28). The basin is also well defined from gravity data, corresponding with a negative free-air and isostatic gravity anomaly (Figures 3 and 4). The North Rona Basin passes laterally to the Judd High towards the north-west, at the up-dip truncation of a Mesozoic, tilted fault block (Figures 7 and 28). The north-east margin of the basin is interpreted to be separated from the West and South Solan basins by stepped, generally east-trending, northward-dipping normal faults (Herries et al., 1999). Towards the south-east, it is separated from crystalline basement of the Solan Bank High by a combination of the north-east-trending, north-west-dipping extensional Otter Bank Fault (Herries et al., 1999) and an *en echelon* unnamed equivalent fault, with approximately 0.75 s TWTT of demonstrable Cretaceous downthrow towards the north-west. The North Rona Basin is inferred to be separated to the south from the North Lewis Basin, by an east-trending, south-dipping, normal fault.

The north east and central parts of the North Rona Basin have been drilled by five released wells i.e. 202/02-1, 202/03-1A, -2, 202/08-1 and 202/12-1 (Figure 2). These wells proved a maximum thickness of approximately 2.2 km of Cenozoic to Mesozoic rocks including 950 m of Cenozoic to Recent, 900 m of Lower and Upper Cretaceous, 200 m of Upper Jurassic and 150 m of Triassic strata. Apart from 202/12-1, all the other wells penetrated crystalline basement at relatively shallow depths varying between approximately 1.2 and 1.8 km. The North Rona Basin is considered to have formed as a suite of half-grabens during Permo-Triassic (Doré et al., 1999) or Triassic (Spencer et al., 1999) times. While this is undoubtedly true within the extreme western part of the basin (Figure 28), the central and eastern parts appear to have developed in response to an extensional phase of activity during Jurassic, Early Cretaceous and particularly, Late Cretaceous times.

North Shoal High

The North Shoal High forms a small, elongate, north-north-east-trending intrabasinal basement high approximately 50 km long and up to 10 km wide that occurs within the south-east part of the report area (Figure 7; Stoker et al., 1993). The high is partly defined from BGS shallow seismic data (BGS, 1985) and also corresponds to a positive, elongate free-air and isostatic gravity anomaly (Figures 3 and 4). The North Shoal High occurs close to the eastern margin of the West Orkney Basin and is bounded on its southern and eastern flanks by normal faults (Figure 7), with the latter having an estimated downthrow of approximately 2 km towards the east at top crystalline basement level (BGS 1985, section 2). The western flank of the high is interpreted to be a non-faulted contact, with onlapping Permo-Triassic strata.

The North Shoal High has been tested by a number of shallow BGS sample sites and comprises pelite, psammite and amphibolite of Moinian affinity (see Figure 43; BGS, 1985).

Nun Rock–Sule Skerry High

The Nun Rock–Sule Skerry High forms a small, elongate, north-east-trending, intrabasinal basement high approximately 50 km long and up to 15 km wide that occurs within the West Orkney Basin close to the southern boundary of the report area (Figure 7; Stoker et al., 1993). The high includes the islets of Sule and Stack Skerry. Its offshore extent is defined from BGS shallow seismic (BGS, 1986a; 1989b) and gravity (Figures 3 and 4) data. The Nun Rock–Sule Skerry High is partially fault-bounded, with the north-east-trending, south-east-dipping normal fault that defines its south-east margin displacing crystalline basement by approximately 1.5 km (BGS 1986a, section 1).

The islets of Sule and Stack Skerry comprise Lewisian basement and within the offshore area around these skerries, samples of amphibolite and gneiss of Lewisian affinity have been described from two BGS shallow sample sites (BGS, 1986a; Stoker et al., 1993).

Papa Basin

The Papa Basin is a small, generally north-north-east-trending Late Palaeozoic to early Mesozoic basin approximately 55 km long and up to 35 km wide that occurs towards the south-east part of the report area (Figure 7; Stoker et al., 1993). The morphology of the basin is well defined from seismic data (Figure 39; Herries et al., 1999, fig. 8; Booth et al., 1993, fig. 2). The Papa Basin is separated from the Papa High to the south-east by the north-north-east-trending Shetland Spine Fault (Figure 7). Towards the north-west, the basin is flanked at the Otter Bank Fault by a combination of the East and South Solan basins. The north-east and south-west margins of the Papa Basin are separated from the West Shetland and West Orkney basins by the north-west-trending Westray and an unnamed west-north-west-trending lineament, respectively. The Westray Lineament is considered to have a significant component of dip slip associated with it, as the Jurassic and Cretaceous successions present within the south-west part of the West Shetland Basin (Haszeldine et al., 1987, fig. 5) are thin to absent in the Papa Basin (Figure 39).

The western margin of the Papa Basin has been drilled by well 205/27a-1 (Figure 2), proving approximately 100 m of Cenozoic to Recent sediments resting on 2.45 km of Permo-Triassic sediments and a thin volcanic interval within which the well terminated. The basin has also been drilled by BGS boreholes 77/09, 82/01, 02 and 12. Although these boreholes proved a mixture of Cenozoic, Cretaceous and Jurassic strata close to the sea bed, post-Permo-Triassic sediments are mainly considered to be thin or absent within the basin (Figure 39; Booth et al., 1993; Herries et al., 1999). Although the Papa Basin has been interpreted to contain up to 7.6 km of Permo-Triassic sediments within a south-east-dipping half-graben (Swiecicki et al., 1995), there is little evidence from seismic data for dramatic thickening towards the south-east and the Shetland Spine Fault (Figure 39). Consequently, Swiecicki et al. (1995) and Herries et al. (1999) considered that it represents a down-faulted eroded remnant of a previously more extensive Permo-Triassic basin complex.

Papa High

The Papa High forms a small, generally north-north-east-trending basement platform approximately 50 km long and up to 20 km wide that occurs within the south-east part of the report area (Figure 7; Stoker et al., 1993). The general morphology of the high is well defined from BGS shallow seismic (BGS, 1985; 1988b) and potential field data (Figures 3, 4 and 5). The Papa High is separated from the Papa Basin to the north-west by the north-east-trending Shetland Spine Fault (Figure 7). Towards the south-east, its transition with the Orkney–Shetland High is partially fault-bounded, with a downthrow of approximately 1.5 km towards the south-east at top crystalline basement level along a north-east-trending fault (BGS, 1985). The Papa High is flanked to the south by a combination of the West Orkney Basin and the North Shoal High. The transition between the south-west extremity of the high and the West Orkney Basin is defined by an unnamed west-north-west-trending structure that has been termed the 'Wyville Thomson Transfer' (Roberts et al., 1999, fig. 22).

From the results of BGS shallow drilling, the western and eastern flanks of the Papa High are considered to comprise psammite, mylonite and granodiorite of Moinian affinity and gneiss, amphibolite and granodiorite of Lewisian aspect, respectively (see Figure 43; BGS, 1985).

Sandwick Basin

The Sandwick Basin forms a small, north-trending, westward-tilted, partially fault-bounded half-graben approximately 30 km long and up to 15 km wide that lies at the eastern margin of the report area (Figure 7; Stoker et al., 1993). The morphology of the basin is mainly defined from BGS shallow seismic data (Figure 25; BGS 1984; 1989a; Hitchen and Ritchie, 1987) and BIRPS deep experimental seismic data (Figure 22; McGeary, 1989; England et al., 2005). The Sandwick Basin occurs within the East Shetland High and is separated from the West Shetland High to the west by the north-trending Walls Boundary Fault (Figures 7 and 25).

The Sandwick Basin has been drilled by BGS boreholes 78/10A and B (Figure 2), proving Devonian strata close to the sea bed. The results from a deep seismic experiment were tentatively interpreted to suggest that the basin contains approximately 2 km (McGeary, 1989) of Devonian sediments resting unconformably on crystalline basement. The basin probably represents a small, isolated remnant half-graben, possibly similar in style to the Clair Basin that occurs approximately 50 km to the south-west (Figure 7; Nichols, 2005). These locally developed fluvial-continental intermontane depocentres most probably formed in response to the orogenic collapse and/or strike-slip fault movement of the areas marginal to the foreland of the Caledonian Orogeny.

Solan Bank High

The Solan Bank High forms an elongate, north-east-trending basement high approximately 130 km long and up to 25 km wide that occurs within the southern part of the report area (Figure 7; Stoker et al., 1993). The general morphology of this north-east-tapering high is defined from both BGS shallow seismic (BGS, 1986a) and commercial seismic data (e.g. Kirton and Hitchen, 1987) and is associated with well defined positive gravity and magnetic anomalies (Figures 3, 4 and 5). The Solan Bank High is flanked on its south-east margin by a combination of the Papa and West Orkney basins (Figure 7), with the latter defined by a significant north-east-trending normal fault with a minimum downthrow of 7.5 km towards the south-east at the level of top crystalline basement (Figure 41). The north-west margin of the high is separated from the North Rona and South Solan basins by a combination of the north-east-trending, north-west-dipping Otter Bank Fault and an unnamed similarly trending and dipping fault.

The crest of the Solan Bank High has been drilled by BGS borehole 77/07 (Figure 2) and recovered amphibolite basement of Lewisian aspect close to the sea bed (BGS, 1986a). A fault terrace on the north-west margin of the high has been drilled by well 202/09-1 proving approximately 450 m of Cenozoic to Recent, 200 m of Cretaceous, 100 m of Upper Jurassic and 500 m of Triassic strata resting on crystalline basement.

South Solan Basin

The South Solan Basin forms a small, north-east-trending, south-east-dipping half-graben up to 25 km long and 15 km wide that occurs within the southern part of the report area (Figure 7). Formerly, the basin was thought to form part of the East Solan Basin as defined by Booth et al. (1993) and Swiecicki et al. (1995). The South Solan Basin is separated from a combination of the Papa Basin and the Solan Bank High to the south-east by the north-east-trending Otter Bank Fault, and from the East Solan Basin to the north-east by the Judd Lineament (Figure 7). Towards the north-west and south-west, the basin is separated from the West Solan and North Rona basins by a north-east-trending normal fault and a combination of north-north-east- and east-trending normal faults, respectively.

The South Solan Basin has only been drilled by well 202/04-1 (Figure 2) which proved approximately 900 m of Cenozoic to Recent, 750 m of Upper and Lower Cretaceous and 150 m of Upper and Lower Jurassic strata before terminating within basal conglomerates and crystalline basement. Although little is known regarding the tectonostratigraphical evolution of the South Solan Basin, its juxtaposition with the better understood East Solan Basin could suggest that they share a similar developmental history (Figure 7). However, the presence of Lower Jurassic rocks only within the South Solan Basin does suggest that the deep erosive effects associated with the prominent Mid to Late Jurassic unconformity observed within the East Solan Basin (e.g. Booth et al., 1993), had a more lim-

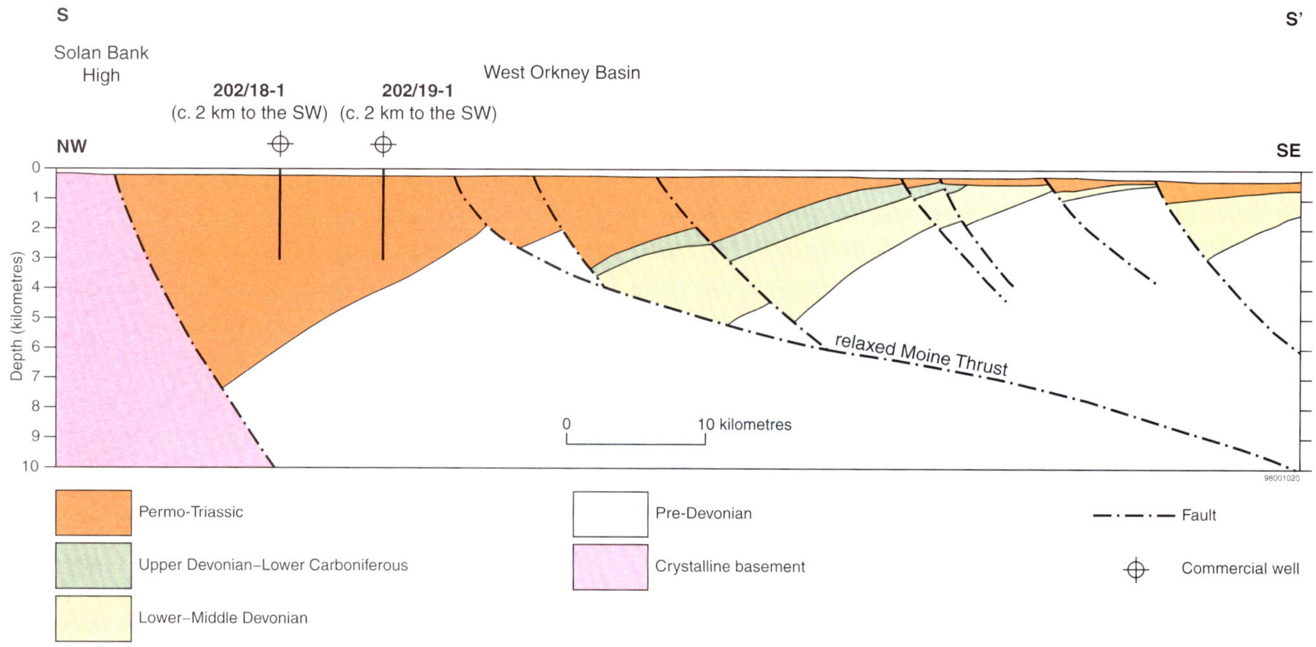

Figure 41 Generalised geological cross-section S–S' across the Solan Bank High and West Orkney Basin (modified from Earle et al., 1989 and Stoker et al., 1993). For location of profile, see Figure 7.

ited impact here. Both the East and South Solan basins are bounded by the Judd Lineament and it seems likely that this feature played a significant role in terms of shaping their evolution.

St Magnus Bay Basin

The St Magnus Bay Basin forms a very small, north-north-east-trending, westward-tilted, partially fault-bounded half-graben that is located just west of the Shetland Islands (Figure 7; Stoker et al., 1993). The general morphology of the basin is defined from BGS shallow seismic data (BGS, 1984). The basin is separated from the Orkney-Shetland High to the west by the main basin-bounding Melby Fault (Figure 29).

The St Magnus Bay Basin has been drilled by BGS boreholes 78/11 and 80/08 (Figures 2 and 29), proving Permo-Triassic sediments close to the sea bed. It is interpreted to contain up to 1 km of Permo-Triassic resting on crystalline basement (Figure 29) and probably represents an eroded remnant of one of a more extensive series of rift basins that were formerly developed over the West and East Shetland highs and surrounding area during Triassic (Spencer et al., 1999) or Permo-Triassic (e.g. Stoker et al., 1993; Doré et al., 1999) times.

Sula Sgeir High

The Sula Sgeir High forms a segmented north-east- to east-north-east-trending basement block at least 90 km long and up to 30 km wide that only just extends into the south-west part of the report area (Figure 7; Stoker et al., 1993). The high is well defined from BGS shallow seismic (BGS, 1986a and 1988c) and gravity (Figures 3 and 4) data. The Sula Sgeir High is flanked to the west and north by a combination of the West Lewis Basin and Outer Hebrides High, though the nature of these boundaries are poorly understood (Figure 7). Towards the east and south, it is interpreted to be separated from the Judd High and North Lewis Basin by a series of inverted normal faults associated with the Wyville Thomson Lineament Complex, and a mainly southward-dipping normal fault, respectively.

Within the report area, the eastern margin of the Sula Sgeir High has been drilled by BGS borehole 78/07 (Figure 2) which proved Permo-Triassic sediments close to the sea bed.

Unst Basin

The Unst Basin forms a 'trilete' complex of rifted half-grabens, the north-west arm of which only just encroaches into the report area to the north of Shetland (Figure 7; Johns and Andrews, 1985). At its maximum, the dimensions of the basin are approximately 80 km by 130 km. The morphology of the basin is well defined from both BGS shallow seismic (BGS, 1984; 1987 and 1989a) and commercial seismic data (Johns and Andrews, 1985), and also corresponds with a negative free-air and isostatic gravity anomaly (Figures 3 and 4).

The Unst Basin has not been drilled within the report area, but immediately to the south-east, wells 1/04-1 and 2 (Figure 2) are located at the confluence of the rifted half-graben complex. Well 1/04-1 proved thin Cenozoic to Recent and Lower Cretaceous sediments resting on approximately 650 m of Jurassic and 1.7 km of Permo-Triassic strata. In terms of its tectonostratigraphical development, the shape of the Unst Basin is considered by Johns and Andrews (1985) to be largely due to the interaction of north-east- and north-west-trending Caledonian faults, with later north-north-west-trending faults. They suggest that the Unst Basin system of half-grabens was largely formed during Permo-Triassic times (though the north-west arm may have a mainly post-Jurassic age), and that Palaeogene uplift was responsible for removing a thick Upper Cretaceous and Early Paleocene succession from the basin as a whole.

West Fair Isle Basin

The West Fair Isle Basin forms an elongate, north-north-east-trending, eastward-tilted, partially fault-bounded complex of half-graben approximately 200 km long and up to 30 km wide that only just straddles the south-east boundary of the report area (Figure 7; Stoker et al., 1993). The general morphology of the basin is mainly defined from BGS shallow seismic data (Figure 42; BGS, 1982; 1984; 1985 and 1988a) but also from BIRPS experimental data (Figure 22; McGeary, 1989; McBride, 1994). It is also well imaged by a negative free-air and isostatic gravity anomaly which coincides with a zone where magnetic field variations are subdued (Figures 3, 4 and 5). The West Fair Isle Basin occurs mainly within the Orkney–Shetland High (Figures 7 and 42). Towards the south, the basin margin is defined by the east-north-east-trending Wick Fault whereas in the west, its transition with the Orkney–Shetland High is marked by an eastward-dipping ramp.

The West Fair Isle Basin has been drilled by a small number of BGS boreholes with, for example, 80/11 proving Permo-Triassic sediments close to the sea bed (Figures 2 and 42). The Permo-Triassic succession generally thickens towards the east and the basin-bounding Walls Boundary Fault, with at least 2.5 km interpreted to be preserved locally in the hanging-wall block of the fault in parts of the basin (Figure 42). This is supported by the results BIRPS experimental data, which suggested that the basin could contain up to 1.75 s TWTT of sediments (McGeary, 1989) i.e. approximately 2.5 to 3.0 km. The West Fair Isle Basin is considered to have evolved as a complex of half-graben during mainly Triassic (e.g. Spencer et al., 1999) or Permo-Triassic (e.g. Doré et al., 1999) times. It is likely to have undergone significant subsequent uplift and erosion, probably during Early Palaeogene times onwards.

West Lewis Basin

The north-east-trending West Lewis Basin forms an elongate, westward-tilted half-graben approximately 85 km long and 50 km wide that only just encroaches into the south-west margin of the report area (Figure 7; Stoker et al., 1993). Outwith the study area, the basin is well imaged by seismic and gravity data. Within the report area, the West Lewis Basin is flanked to the north-east by the Outer Hebrides High (Figure 7) though the nature of the contact is poorly understood (Earle et al., 1989).

The West Lewis Basin has not been drilled within the report area, but further to the south-west, the results from a number of BGS boreholes and well 164/25-1 proved that it contains at least 2.7 km of Cenozoic to Recent and 900 m of Cretaceous to Permo-Triassic sediments. The results of gravity modelling indicate that the basin may comprise a maximum of between 8 and 11 km of strata (Waddams and Cordingley, 1999). The presence of thick Permo-Triassic sediments within the West Lewis Basin strongly suggests that it may have been initiated as a mainly Triassic (Spencer et al. 1999) or Permo-Triassic (Hitchen et al., 1995b; Doré et al., 1999, fig. 1) extensional half-graben.

West Orkney Basin

The West Orkney Basin forms a generally north-east-trending, north-west-dipping complex of half-grabens approximately up to 135 km long and 90 km wide that occur within the southern part of the report area (Figure 7; Stoker et al., 1993). This basin complex is well defined from both commercial and deep seismic experimental data (Figure 21; Brewer and Smythe, 1984; Coward et al., 1989; Enfield and Coward, 1989; Earle et al., 1989; Stoker et al., 1993), and also from potential field data, with separate gravity minima in the north-west and south-east parts of the basin and subdued magnetic field variations (Figures 3, 4 and 5). The West Orkney Basin is separated from the Solan Bank High to the north-west by a north-east-trending, south-east-dipping normal fault, with thick Permo-Triassic sediments present within its hanging-wall block (Figures 7 and 41). Towards the south-east, it is separated from the Orkney–Shetland High by a north-east-trending, north-west-dipping normal fault and a north-west-dipping ramp. The basin is bounded to the north-east by an unnamed lineament and to the south-west by a combination of the Scottish mainland and the North Minch Basin. The internal structure of the basin has been simplified in Figure 7 but includes the Nun Rock–Sule Skerry and North Shoal intrabasinal basement horsts and significant south-east-dipping faults, including a possible candidate for the offshore continuation of the Moine Thrust (Figures 7 and 41).

The West Orkney Basin has been drilled by wells 202/18-1 and 202/19-1 (Figure 2), proving approximately 150 m of Cenozoic to Recent sediments resting unconformably on 2.95 km of Permo-Triassic sediments. The basin has also been drilled by a number of BGS boreholes including 82/05 in the northern, 72/37, 82/03, 04 and 82/14 in the central and 72/25, 26, 27, 28, 34, 73/29, 31 and 82/17 in the southern areas (Figure 2). With the exception of 72/34 that recovered Lower Jurassic, all these boreholes proved Permo-Triassic at or close to the sea bed. Maximum preserved thicknesses of 7.5 km (Earle et al., 1999), 8 km (Stoker et al., 1993) or even 9 to 10 km (Enfield and Coward. 1987; Coward et al., 1989) of strata have been suggested to occur within the most north-westerly of the complex of half-graben. The West Orkney Basin was initially considered to have developed during Devonian times (Coward and Enfield, 1987; Enfield and Coward, 1987; Coward et al., 1989), with extension along reactivated (in a normal sense) low angle, south-east-dipping Caledonian thrusts. Although a few of the half-graben within the eastern part of the basin complex are interpreted to have wedges of Devonian to Carboniferous sediments present within the hanging-wall blocks of largely low-angle extensional faults (Figure 41; Earle et al., 1989, fig. 4; Stoker 1993 et al., fig. 33), the north-west-thickening wedge of Permo-Triassic sediments proved in wells 202/18-1 and 19-1 within the most westerly of the West Orkney Basin complex of half-graben indicates formation mainly during Triassic (e.g. Spencer et al., 1999), Permo-Triassic (e.g. Stoker et al., 1993; Hitchen et al., 1995b; Doré et al., 1999) or late Permian to Mid Triassic (Torsvik et al., 1996) phases of rifting. Analysis of velocity and density data from the Permo-Triassic sediments drilled within well 202/19-1 suggested that approximately 1.6 km of post-Triassic rocks have subsequently been removed from the basin (Evans, 1997). According to Evans (1997), this agrees with a figure of 2 km or so proposed by Brodie and White (1995) for Mesozoic basins north of Scotland in general. The timing of erosion is not clear, but an end Cretaceous to early Cenozoic age is most likely.

West Rona High

The West Rona High forms a small intrabasinal basement high that occurs within the North Lewis Basin,

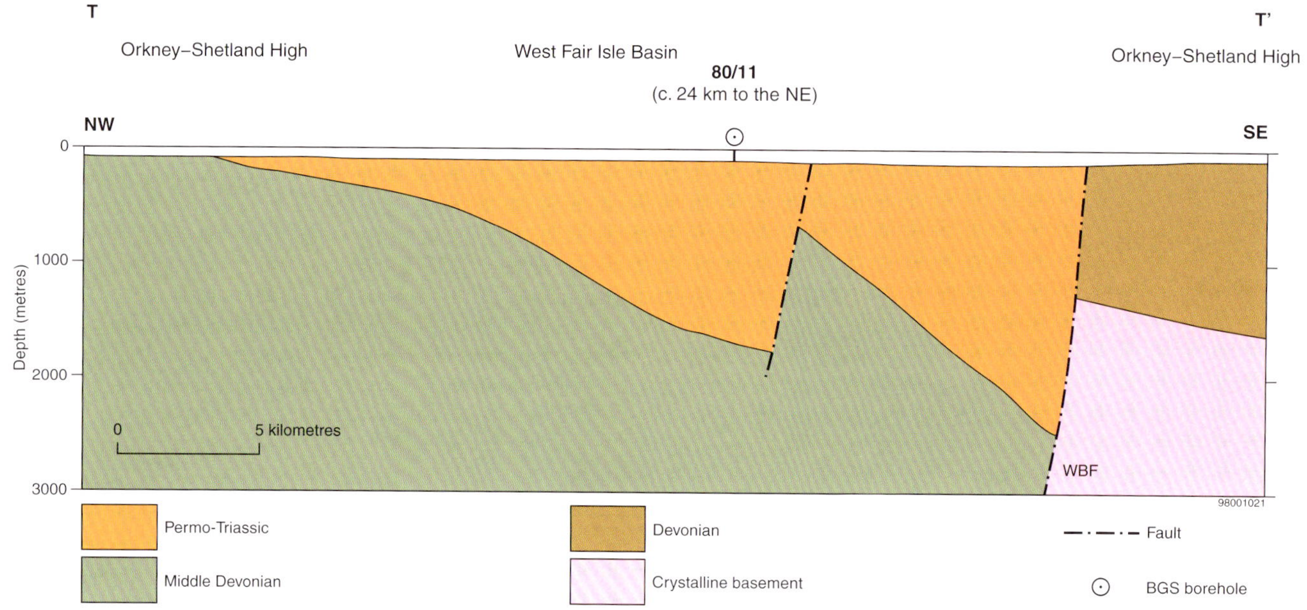

Figure 42 Generalised geological cross-section T–T' across the Orkney–Shetland High and West Fair Isle Basin (modified from BGS, 1988a). Abbreviations: see Figure 7. For location of profile, see Figure 7.

close to the south-west margin of the report area (Figure 7). The high includes the island of Rona, and its offshore extent is defined by a combination of shallow seismic (BGS, 1986a) and gravity data. The West Rona High is partially fault bounded, with normal faults defining the south-west and south-east margins of the high (Figure 7). The Island of Rona comprises mainly gneiss and amphibolite of Lewisian affinity (Stoker et al., 1993) and in the offshore area, a sample of granitic gneiss was recovered from a BGS shallow sample site.

West Shetland Basin

The West Shetland Basin forms an elongate, north-east-trending and tapering, south-east-tilted half-graben (cf. Knott et al., 1993) approximately 170 km long and up 25 km wide that occurs within the south-east part of the report area (Figure 7; Stoker et al., 1993). It is subdivided into two segments, separated at Devonian level only by the Clair Basin. The West Shetland Basin is well defined from seismic data (Figures 30, 31 and 37; Cashion, 1975, fig. 6; Haszeldine et al., 1987, fig. 2; Lamers and Carmichael, 1999, fig. 4; Goodchild et al., 1999, fig. 4; Grant et al., 1999 fig. 2) and potential field data, corresponding with negative gravity and magnetic anomalies (Figures 3, 4 and 5). The West Shetland Basin is separated to the south-east from crystalline basement of the West Shetland High by the planar, normal, north-east-trending, north-west-dipping Shetland Spine Fault (Figure 7). Towards the north-west, the basin is bounded by the Rona High though there is considerable lateral variation in the configuration of this contact (e.g. Figures 30, 31 and 37). The basin is interpreted to be traversed by a number of north-west-trending lineaments (Figure 7), with its south-west and north-east extents defined by the Westray and an unnamed lineament, respectively.

The West Shetland Basin has been drilled by a number of released commercial wells including 205/20-2, 205/23-1, 205/25-1, 205/30-1, 206/13-1, 206/16-1, 207/01-2, 01a-5, 207/02-1, and 208/23-1 (Figure 2). Wells 206/09-1 and 206/10a-1 also occur within the West Shetland Basin, though the Devono-Carboniferous successions they encountered have been ascribed to the Clair Basin (Figure 7 and 31). The wells indicate a maximum thickness of approximately 5.85 km of Mesozoic to Recent sediments, including up to 1.1 km of Cenozoic to Recent, 3.05 km of Lower and Upper Cretaceous, 200 m of Upper Jurassic and 1.5 km of Triassic strata. Crystalline basement was reached in wells 207/02-1, 207/01-2, 01a-5 and 208/23-1 within the north-east part of the basin only. The thickness of the sedimentary infill within the West Shetland Basin increases significantly towards the south-west (Goodchild et al., 1999; Stoker et al., 1993, fig. 24), where up to 8 km of sediments including 4 km of Permo-Triassic (Hitchen and Ritchie, 1987; Haszeldine et al., 1987) are inferred to be present. There is also considerable variation in the distribution of Mesozoic rocks, with Jurassic and Triassic rocks absent within the north-east part of the basin where in general, Lower Cretaceous (Albian) sediments generally rest on crystalline basement (Figure 30; Ridd, 1981; Goodchild et al., 1999; Stoker et al., 1993). Speculatively, the north-west-trending Westray, Corona, Clair or Victory lineaments could have acted as important controls on sedimentation patterns within the basin, though Dean et al. (1999) suggest that the different levels of structural/stratigraphical preservation are more likely associated with a Mid to earliest Late Jurassic unconformity that is thought to have affected the west of Shetland area in general (e.g. Booth et al., 1993; Stoker et al., 1993; Swiecicki et al., 1995).

The south-west part of the West Shetland Basin in particular is considered to represent a mainly Triassic or Permo-Triassic half-graben (e.g. Cashion, 1975; Hitchen and Ritchie, 1987; Duindam and van Hoorn, 1987; Booth et al., 1993; Stoker et al., 1993; Knott et al., 1993; Doré et al., 1999; Spencer et al., 1999). This rifting phase is interpreted to have been terminated by severe erosion and peneplanation around Mid to earliest Late Jurassic times. During Cretaceous times, spatially and temporally disparate phases of extension may have affected the basin as a whole (see Dean et al., 1999, fig. 5), with for example, late Albian (Goodchild et al., 1999) and Campanian (Dean et al., 1999) rifting causing the development of syn-rift wedges within the hanging wall of the Shetland Spine Fault. In contrast, within the central part of the basin around the Clair Field area, Dean et al. (1999) suggested that the rifting occurred during Late Cretaceous (Turonian to Coniacian) times. Towards the south-west part of the basin, Albian to Cenomanian and late Maastrichtian extension followed a period of subdued Late Jurassic activity. Uplift and erosion during early Paleocene times led to marginal marine Upper Paleocene and younger sediments resting on deep marine Maastrichtian sediments in the Victory Field area (Goodchild et al., 1999).

West Solan Basin

The West Solan Basin forms a small north-east-trending basin approximately 25 km long and 15 km wide within the southern part of the report area (Figure 7; Herries et al., 1999) that is well defined from seismic data (e.g. Booth et al., 1993, fig. 4). The basin is separated from the Judd High and a combination of the North Rona and South Solan basins by an east-north-east-trending normal fault, possibly the equivalent of the Solan Fault of Booth et al. (1993), and an east- to east-north-east-trending pair of normal faults, respectively (Figure 7). The north-east margin of the basin is delimited from the Rona High by the Judd Lineament.

The West Solan Basin has been drilled by released wells 202/03a-3 and 204/29-2 (Figure 2). These wells proved a maximum thickness of approximately 3.8 km of Mesozoic to Recent sediments including 1.55 km of Cenozoic to Recent, 800 m of Lower and Upper Cretaceous, 800 m of Lower and Upper Jurassic and 650 m of Triassic strata. The occurrence of thick Lower Jurassic sediments within the basin could suggest that it originally formed during Lower Jurassic times and may share a history similar to that of the adjacent South Solan Basin.

Other major structural features (faults, lineaments, lineament complexes etc.)

Brendan Lineament

The north-west-trending Brendan Lineament extends for approximately 165 km across the north-east part of the report area (Figure 7), and corresponds to an unnamed transfer zone originally described by Rumph et al. (1993). The lineament is interpreted to mark the boundary between the Møre and Erlend basins and corresponds with a narrow, north-west-trending Isostatic gravity high that links the East Shetland High and the Brendan Volcanic Centre (Figures 4 and 7).

Clair Lineament

The north-west-trending Clair Lineament is at least 230 km long, extending across the Faroe–Shetland Basin within the central part of the report area (Figure 7; Rumph et al., 1993). The lineament is interpreted to occur where there is a marked dextral offset in the trend of the Rona High and also a north-easterly truncation of a distinct positive magnetic anomaly coincident with the central part of the Corona High (Figures 5 and 7). Indirect support for the presence of the lineament is derived from closely associated 'pop-up' (Dean et al., 1999, fig. 13) and 'arch-like' (Grant et al., 1999, fig. 6) structures identified from seismic data in the area between the Flett and Foula sub-basins. According to Dean et al. (1999), the transpressional pop-up is related to oblique movement along the Clair Lineament during an episode of Paleocene compression. Additional support for its presence is derived from notable variations in Cretaceous (Rumph et al., 1993; fig. 12) and total gross sediment (Kimbell et al., 2005, fig. 6) thicknesses across the projected path of the lineament. Further to the north-west, the lineament is interpreted to define the north-east margin of the most northerly segment of the Tróndur High. Kimbell et al. (2005, fig. 7) consider the possibility that the Clair Lineament could have been a precursor to the Faroe Fracture Zone described within oceanic crust of the Norwegian Basin.

Corona Lineament

The north-west-trending Corona Lineament extends for approximately 230 km, crossing the Faroe–Shetland Basin within the central part of the report area (Figure 7). The south-east part of the Corona Lineament generally correlates with an unnamed transfer zone originally described by Rumph et al. (1993). The lineament is interpreted to correspond to a sinistral offset of the trend the Rona High, and also the south-westerly truncation of a distinct positive magnetic anomaly coincident with the central part of the Corona High (Figures 5 and 7). Further to the north-west, it possibly marks the location of the north-east margin of the Mid Faroe High and a marked bifurcation of the East Faroe High (Figure 7).

Erlend Lineament

The north-west-trending Erlend Lineament is approximately 250 km long and occurs within the north-east part of the report area (Figure 7). It was originally defined by Duindam and van Hoorn (1987) as a north-west-trending transfer zone with a significant component of dip slip associated with it that marks the boundary between the Erlend High and the Faroe–Shetland Basin. Rumph et al. (1993) redefined its location, placing it at the south-west margin of the West Erlend Volcanic Centre (Figure 7). The south-east part of the Erlend Lineament is interpreted to correspond with a gravity gradient associated with the south-western margin of the East Shetland High (Figure 4). Towards the north-west however, the location of the lineament is poorly defined, a view supported by Kimbell et al. (2005).

Grimur Kamban Lineament

The north-west-trending Grimur Kamban Lineament is approximately 235 km long and extends across the Faroe–Shetland Basin within the central part of the report area (Figure 7). It was initially described by Keser Neish (2003) within the Faroese area, but has been extended to the south-east to correlate with an unnamed transfer zone of Rumph et al. (1993). It is interpreted to be associated with the apparent segmentation of the Tróndur and East Faroe highs, a shift in polarity of the fault associated with the north-west margin of the Corona High and a 10 to 15 km sinistral offset of the Flett High (Figure 7).

Judd Fault

The Judd Fault forms a slightly arcuate, west- to north-west-trending, north- to north-east-dipping, planar normal fault that extends for approximately 45 km within the south of the report area (Figure 7). The fault was originally defined by Kirton and Hitchen (1987) as north-west-trending, but was modified by Smallwood and Kirk (2005) and appears now to form a splay from the inferred location of the north-west-trending Judd Lineament. The fault separates the Judd High from the Judd Sub-basin but where it merges with the Judd Lineament, it is interpreted to have a downthrow towards the north-east of between approximately 1.0 and 1.5 s TWTT at top crystalline basement (Figure 27; Lamers and Carmichael, 1999, fig. 5).

Judd Lineament

The generally north-west-trending Judd Lineament is approximately 235 km long and is interpreted to run between the Papa High and Grani Fault Terrace (Figure 7) within the south of the report area. The lineament was originally defined as the Faroe Transfer Zone by Duindam and van Hoorn (1987), and subsequently modified by Rumph et al. (1993) and Keser Neish (2003). It is interpreted to merge with the Judd Fault (Figure 27; Lamers and Carmichael, 1999, fig. 5) and corresponds to a sinistral offset in the positive magnetic anomaly associated with the Papa High and similarly offset positive isostatic gravity anomalies attributed to the Rona and Solan Bank highs (Figures 4 and 5). Towards the north-west, the lineament is interpreted to define the south-west margin of the Heri High (Ellis et al., 2002) and intersect with the Grani Fault Terrace (Figure 7).

Magnus Lineament

The north-west-trending Magnus Lineament is at least 240 km long and occurs within the north-east part of the report area (Figure 7). It was first described by Duindam and van Hoorn (1987) as forming an arcuate, mainly north-west-trending transfer zone that marks the boundary between the Erlend High/East Shetland High and the Erlend Basin. Rumph et al. (1993) partly redefined its location, interpreting it to coincide with a marked sinistral offset of a north-east trending positive gravity feature associated with the East Shetland High (Figure 4). Ritchie et al. (2003) considered that the lineament may have been active in Pliocene times, and implicated in the development of the Pilot Whale Anticline and associated mud diapirs and mounds within the Erlend Sub-basin (Figure 24).

Melby Fault

The slightly arcuate north-north-east- to north-east-trending Melby Fault is an approximately 70 km long, mainly east-dipping planar fault that occurs close to the eastern margin of the report area (Figure 7). On the Shetland Islands, it includes the Melby, St Magnus Bay and Eshaness fault components (Flinn, 1992). Here, the fault separates generally undifferentiated Devonian sediments and extrusive rocks from tightly folded Lower to Middle Devonian sediments and granites in its hanging wall and footwall, respectively. Mykura (1991) suggested a post-Devonian component of strike-slip movement associated with the fault although according to Flinn (1977a), the presence of slickensides indicates that it is also a reverse or inverted extensional fault. In the offshore area, the Melby Fault is interpreted to separate Devonian sediments on the Orkney–Shetland High, from crystalline basement and Permo-Triassic sediments belonging to the East Shetland High and St Magnus Bay Basin, respectively, with an estimated minimum of 500 m of syn- to post-Permo-Triassic extensional movement down to the south-east associated with it (Figure 29). Towards its northerly extremity, the Melby Fault merges with the Walls Boundary Fault (Figure 7).

Minch Fault

The north-north-east-trending Minch Fault forms an approximately 340 km long, east-south-east-dipping normal fault that only just encroaches within the south of the report area (Figure 7). Outwith the report area, the fault is imaged from BGS shallow seismic (BGS, 1989b; Chesher et al., 1983), commercial seismic (Stein, 1988) and deep seismic experimental data, including MOIST (Brewer and Smythe, 1984) and DRUM (Figure 21). The Minch Fault is interpreted as a mainly Permo-Triassic extensional fault, but with significant post-Lower Jurassic movement too (Stein, 1988), that separates the North Lewis Basin from the North Minch Basin. It has a net downthrow of about 3 km towards the east-south-east at top Permo-Triassic level.

Moine Thrust

The structure and geology of the Moine Thrust within the north-west Scotland highlands (Figure 7) has been the focus of intensive study for more than a century (see Strachan et al., 2002). Briefly, the Moine Thrust onshore forms an approximately 500 km long, east-south-east-dipping complex of thrusts with Proterozoic Moinian rocks generally resting on older Archaean foreland rocks. This thrust zone was probably emplaced approximately 437 to 430 Ma ago (e.g. Goodenough et al., 2006), and is thought by some to mark the north-west deformation front of the Caledonian Orogeny (e.g. Strachan et al., 2002). To the north of the Scottish mainland, the trace of the Moine Thrust cannot be defined with any certainty, although its presence has been inferred from deep seismic profiles (e.g. Figures 21 and 41; Brewer and Smythe, 1984, Snyder, 1990). The thrust zone is considered to pass to the west of Moinian rocks recovered form the North Shoal and Papa highs (see Chapter 3; Figure 7) and has also tentatively been identified from seismic data to the south-west of Foula (Andrews, 1985; Figure 21). On the north-west part of the Shetland Isles, a potential candidate for the Moine Thrust has been recognised (see Chapter 3; e.g. Pringle, 1970; Flinn et al., 1979). In the offshore area to the north of the Shetland, Ritchie et al. (1987) speculate that the Moine Thrust might run parallel with the Walls Boundary Fault before intersecting the north-west margin of the West Shetland High, though other interpretations are also possible (Figures 7 and 22).

Nesting Fault

The arcuate, generally north-north-east- to north-trending Nesting Fault occurs mostly on mainland Shetland close to the eastern margin of the report area (Figure 7). It is interpreted as a dextral strike-slip fault with an estimated displacement of 16 km (Miller and Flinn, 1966).

North Flett Fault

The North Flett Fault forms a generally north-east-trending, north-west-dipping, sinistrally-offset, planar normal fault approximately 100 km long that occurs within the central part of the report area (Figure 7). It is clearly imaged from seismic data (Figures 30 and 31). Formerly referred to as the Flett Fault (Lamers and Carmichael, 1999), the North Flett Fault forms the north-west margin of the Flett High where it is interpreted to comprise segments that are sinistrally offset by approximately 10 km across the Grimur Kamban Lineament (Figure 7). The North Flett Fault is considered to have been initiated in Cretaceous times, with the main phase of movement at end Berriasian times (Lamers and Carmichael, 1999). It is also is interpreted to have undergone renewed extension during the Early to Mid Paleocene, with notable north-west thickening of the Lower Paleocene (pre-Kettla Member) succession into the Flett Sub-basin (Figure 31).

Otter Bank Fault

The Otter Bank Fault forms a sinuous, north-east- to north-north-east trending, north-west- to west-north-west-dipping, planar normal fault that is approximately 75 km long and occurs within the south of the report area (Figure 7). The fault was first described by Duindam and van Hoorn (1987) and initially named the Sula-Sgeir Fault by Booth et al. (1993). It is interpreted to separate in part the north-west Solan Bank High and Papa Basin on the footwall block, from the North Rona, South Solan and East Solan basins within the hanging-wall block. The Otter Bank Fault was considered to be a mainly Late Cretaceous extensional fault by Booth et al. (1993). However, although there is an apparent downthrow of approximately 2.0 s TWTT towards the north-west at the level of top crystalline basement between the Papa and East Solan Basins, there is little indication of growth during Mesozoic times (Figure 39).

Rona Fault

The north-east-trending Rona Fault forms a slightly sinuous, north-west-dipping extensional fault approximately 200 km long that bounds the south-east margin of the Faroe–Shetland Basin (Figure 7). It is well imaged from seismic data (Figures 30, 31 and 37; e.g. Haszeldine et al., 1987; Booth et al., 1993; Lamers and Carmichael, 1999; Grant et al., 1999; Goodchild et al., 1999). The Rona Fault separates the Judd, Flett and Foula sub-basins to the north-west, from the Rona High and Clair Basin to the south-east. The Judd Lineament and an unnamed lineament bound the fault at its south-west and north-east extremities, respectively. The Rona Fault is interpreted to be a mainly Cretaceous extensional fault, though the temporal focus of this activity varied significantly along its length (Dean et al., 1999, fig. 5). Extension and the creation of a block-faulted topography within the hanging wall of the Rona Fault began in the Early Cretaceous, possibly during Berriasian (Lamers and Carmichael, 1999) or Albian to Aptian (Ritchie et al., 1996; Dean et al., 1999) times, with significant phases of activity during Late Cretaceous Cenomanian–Santonian and Campanian–Maastrichtian times (particularly towards the south-west end of the fault). The amount of downthrow towards the north-west on the Rona Fault varies significantly along its length, with between 2.5 and 6.0 s TWTT indicated at the level of top crystalline basement (Figures 30, 31 and 37; e.g. Lamers and Carmichael, 1999, fig. 4;).

Shetland Spine Fault

The north-east-trending Shetland Spine Fault forms a north-west-dipping extensional fault (or more accurately, a series of en echelon faults) approximately 260 km long that separates the East Shetland, West Shetland, and Papa highs in the east from the Flett Sub-basin, West Shetland and Papa basins in the west (Figure 7). It has been defined from BGS shallow seismic (BGS, 1984; 1985; 1988b; 1989a) and commercial seismic data (Figures 25, 30 and 31; Cashion, 1975; Booth et al., 1993; Dean et al., 1999; Herries et al., 1999). At its north-east and south-west limits, the fault is considered to terminate against the Erlend High and the west-north-west-trending 'Wyville–Thomson Transfer' lineament of Roberts et al. (1999), respectively. The extensional history of the Shetland Spine Fault varies significantly along its length. For example, towards the north-east, it is considered to be of mainly Early Cretaceous to pre-Campanian age (Goodchild et al., 1999, fig. 3), with a downthrow towards the north-west of approximately 2 s TWTT at the level of top crystalline basement. In contrast, towards the south-west the fault is mainly Permo-Triassic in age (but possibly Jurassic to Cretaceous too), controlling the development of 4 to 6 km of mainly syn-rift continental red beds within the West Shetland and Papa basins (e.g. Hitchen et al., 1995b; Swiecicki et al., 1995). According to Hitchen and Ritchie (1987), the south-west segment of the Shetland Spine Fault underwent renewed extension during Paleocene times, with the formation of the small, narrow, north-east-trending 'Scarvister Sub-basin' within its hanging-wall block.

South Flett Fault

The South Flett Fault is a generally north-east-trending, south-east-dipping, sinistrally-offset, planar normal fault approximately 100 km long that occurs within the central part of the report area (Figure 7). Alternatively referred to as the Flett Fault (Grant et al., 1999), the South Flett Fault forms the south-east margin of the Flett High (Figures 30 and 31) where it is interpreted to comprise strands that are sinistrally offset by 10 to 15 km or so by the Grimur Kamban Lineament. According to Grant et al. (1999, fig. 2) and Lamers and Carmichael (1999, fig. 4), the fault downthrows approximately 1.0 and 0.25 s TWTT to the south-east at top Upper Cretaceous (Cenomanian) and top Jurassic levels, respectively.

Victory Lineament

The north-west-trending Victory Lineament is at least 230 km long, extending across the Faroe–Shetland Basin and beyond within the central part of the report area (Figure 7). The lineament was first described by Rumph et al. (1993) and is interpreted to truncate the north-east margin of the main area of the Corona High and also correspond with an offset of the positive magnetic anomaly associated with the Rona High (Figures 5 and 7). According to Mitchell et al. (1993, figs. 2 and 9), a significant variation in the lateral thickness of the Paleocene succession across the Victory Lineament suggests that it was an active control on sedimentation at that time.

Walls Boundary Fault

First described on the Shetland Isles, the arcuate north-east- to north-trending Walls Boundary Fault is a large, 290 km long, reactivated mainly transcurrent fault system (e.g. Flinn, 1961; 1992; Watts et al., 2007) that runs close to the south-east margin of the report area (Figure 7). In the offshore area, the Walls Boundary Fault is imaged from seismic data immediately adjacent to the north and south of the Shetland Islands (Figure 22; e.g. McQuillin et al., 1982; Ritchie et al., 1987; Ritchie and Hitchen, 1993). To the south, the fault may terminate to the east of Orkney at the confluence of the West Fair Isle, East Orkney and Moray Firth basins (Figure 7; IGS, 1982; Zanella et al., 2003). Alternatively, it may comprise a component of a much larger suite of faults including the Helmsdale and Great Glen faults to the south-west (e.g. McQuillin et al., 1982; Flinn, 1992; Underhill, 1993).

The Walls Boundary Fault is considered to represent a steep, crustal penetrating feature that offsets the Moho by 2 to 3 km (see Deep Structure above; McGeary, 1989; McBride, 1994). The fault is interpreted to have a significant component of lateral movement associated with it, mainly due to the great differences in the ages of the successions on either side of the fault on the Shetland Islands (e.g. Flinn, 1961; 1992). From potential field and palaeogeographical evidence, Flinn (1992) considers that a net sinistral displacement of 100 km occurred in Carboniferous times (see also Zanella and Coward, 2003), followed by 65 km of dextral movement in Jurassic times. Underhill (1993) however, argues that even the latter figure appears too large. Largely in agreement with Flinn (1992), Watts et al. (2007) postulate that the Walls Boundary Fault began as a Caledonian strike-slip fault with a sinistral displacement of 100 km or so, followed by some reactivation in a dextral sense during latest Carboniferous times. He considers that there may also have been about 15 km of sinistral strike-slip movement during Cenozoic times (see also Ritchie and Hitchen, 1993). There is also a notable component of dip slip on both flanks of the fault, with estimated throws at top crystalline basement level of up to 2 km towards the east associated with the Sandwick Basin, and at least 1.5 km towards the west associated with the formation of the West Fair Isle Basin (Figure 42).

Westray Fault

The Westray Fault forms an arcuate, north-west- to south-west-trending, predominantly planar normal fault approximately 75 km long that defines the south-west margin of the northern part of the Westray High and the eastern boundary of the Sjúrður Ridge (Figure 7; Smallwood et al., 2004; Smallwood and Kirk, 2005; Smallwood, 2005a and b). The fault may be connected at depth with the Westray Lineament, and has a generally southerly downthrow of approximately 0.4 s TWTT at base Cenozoic (e.g. Smallwood and Kirk, 2005, fig. 4) and greater than 1.0 s TWTT at top crystalline basement (e.g. Lamers and Carmichael, 1999, fig. 5).

Westray Lineament

The Westray Lineament is a generally north-west-trending feature approximately 275 km long that extends across the Faroe–Shetland Basin and adjacent areas within the central part of the report area (Figure 7). Originally described by Rumph et al. (1993) (and the equivalent of the Schiehallion Transfer Zone of Herries et al. (1999)), the south-east part of the Westray Lineament separates the Flett Sub-basin, Westray High, West Shetland Basin and West Shetland High from the Brynhild Sub-basin, Judd Sub-basin, East Solan Basin, Papa Basin and Papa High on its north-east and south-west flanks, respectively. To the north-west of the Westray High, the location of the lineament is largely conjectural, although similar to that proposed by Ellis et al. (2002, fig. 4), it is thought to correspond with a distinct dextral displacement of the Heri High.

3 Pre-Devonian

Derek Ritchie, Stephen Noble, Fiona Darbyshire, Ian Millar and Lynne Chambers
British Geological Survey

Pre-Devonian basement rocks underlie most of the Faroe–Shetland report area, except towards the north-west, where Eocene and younger oceanic crust occurs within the Iceland and Norwegian basins. The nature, age and affinity of the pre-Devonian basement rocks, although relatively well known from exposed parts of the Shetland Islands and north-west Scotland, and from the Foula, Sule Kerry, Stack Skerry and Rona islands (Figures 1 and 43), is poorly resolved within the offshore area. This is mainly due to their burial beneath substantial thicknesses of Late Palaeozoic, Mesozoic and Cenozoic strata. As a consequence, information regarding their age and composition over large parts of the offshore area is mainly restricted to that derived from commercial wells and BGS boreholes (Figure 43).

DISTRIBUTION, NATURE, AGE AND AFFINITY

As far as is known, pre-Devonian basement rocks within the report area are of Archaean to Ordovician age. This basement is considered to form part of the Laurentia superterrane, which collided with the Baltica superterrane approximately 430 to 400 Ma ago during the Scandian phase of the Caledonian Orogeny (e.g. Oliver, 2002; Coward et al., 2003; Figure 9). For the purposes of discussion, the distribution, nature, age and affinity of the pre-Devonian basement rocks are described within the following five main subdivisions: Lewisian, Torridonian, Moine, Dalradian and Cambro-Ordovician.

Lewisian

The term *'Lewisian Gneiss Complex of mainland north-west Scotland and the Outer Hebrides'* is used to describe Archaean and Proterozoic lower crustal rocks that have suffered high-grade metamorphism and deformation before approximately 1600 Ma (Kinny et al., 2005). The complex is mainly comprised of grey quartzofeldspathic gneiss of tonalitic–trondhjemtic–dioritic (TTG) affinity (formerly igneous plutonic rocks) together with minor subsidiary metasedimentary and mafic to ultramafic bodies, and cut by later basic dykes and acidic bodies. The traditional interpretation is that the Lewisian Gneiss Complex represents a block of Archaean crust that was affected by two orogenic cycles, namely a late Archaean Badcallian event (characterised by granulite facies metamorphism at c.2500 Ma) and a Proterozoic Laxfordian event (characterised by amphibolite facies metamorphism at c.1700 Ma and reworking associated within shear zones), that was separated by intrusion of basic 'Scourie' dykes intruded at c.2400 Ma (e.g. Sutton and Watson, 1951; Park et al., 1994, 2002). This view has been challenged, primarily on the basis of recently acquired geochronology data (e.g. Friend and Kinny, 2001; Kinny et al., 2005; Park, 2005). According to Kinny et al. (2005), the traditional subdivision of the Lewisian Gneiss Complex into northern, central and southern regions (e.g. Moorbath and Park, 1971; Park and Tarney, 1987; Park et al., 2002, fig. 3.3) was based on the assumption that all the gneiss is approximately of the same age and affinity and that the rocks shared a common tectonothermal history; a premise that has not been confirmed by the results of modern U–Pb age dating (e.g. see Friend and Kinny, 2001; Love et al., 2004; Kinny et al., 2005). According to Friend and Kinny (2001) and Kinny et al. (2005), the Lewisian Gneiss Complex comprises nine separate terranes and an undifferentiated Uist 'block' (Figure 44), all with distinct geochronological histories prior to their amalgamation during Proterozoic times. They suggested that the terranes observed on mainland Scotland and the Outer Hebrides Isles are non-correlatable, with the later having potentially more in common with the Archaean and Proterozoic blocks of East Greenland. Park (2005) has partially refined their scheme on the basis of Proterozoic structural, metamorphic and igneous characteristics, suggesting that the Lewisian Gneiss Complex can be subdivided into 14 separate terranes and blocks (or suspect terranes).

Mainland of north Scotland and adjacent offshore area

Lewisian rocks within the extreme southern part of the report area have been described to the west of the Moine Thrust on the Cape Wrath Peninsula (Figure 43; BGS, 1989b). Formerly part of the northern region of the Lewisian Gneiss Complex, they have been informally ascribed to the Rhiconich Terrane (Friend and Kinny, 2001; Kinny et al., 2005) of the Lewisian Gneiss Complex, which is bounded to the south by the west-north-west-trending Laxford Shear Zone (Figure 44). The Rhiconich Terrane is described by Kinny et al. (2005) as comprising mainly TTG gneiss with protolith ages of 2840 to 2680 Ma, with minor mafic and metasedimentary rocks that are intruded by Scourie dykes at c.2400 Ma and the Rubha Ruadh granites at c.1855 Ma. All of these rocks have been affected by amphibolite facies metamorphism and deformation associated with the Laxfordian event at c.1740 Ma (e.g. Corfu et al., 1994). Park et al. (2002, fig. 3.7) describe the basement on the Cape Wrath Peninsula area as mainly comprising migmatites of Laxfordian age. Lewisian rocks of the Rhiconich Terrane are considered to extend into the offshore area immediately to the north of Cape Wrath (Figure 43; BGS, 1989b). Between the Moine Thrust and Navar Thrust, Lewisian gneiss occurs as shear-zone and anticlinal cores, mainly close to the base of the Moinian Morar Group (Figure 43; Strachan et al., 2002, fig. 4.3; BGS, 1989b). These tectonic slices probably extend northwards into the offshore area in the immediate vicinity of the coast. Interpretation of the MOIST deep seismic profile (Figure 21) suggested that they could also be present below thick Devonian and Permo-Triassic strata of the southern West Orkney Basin (Brewer and Smythe, 1984).

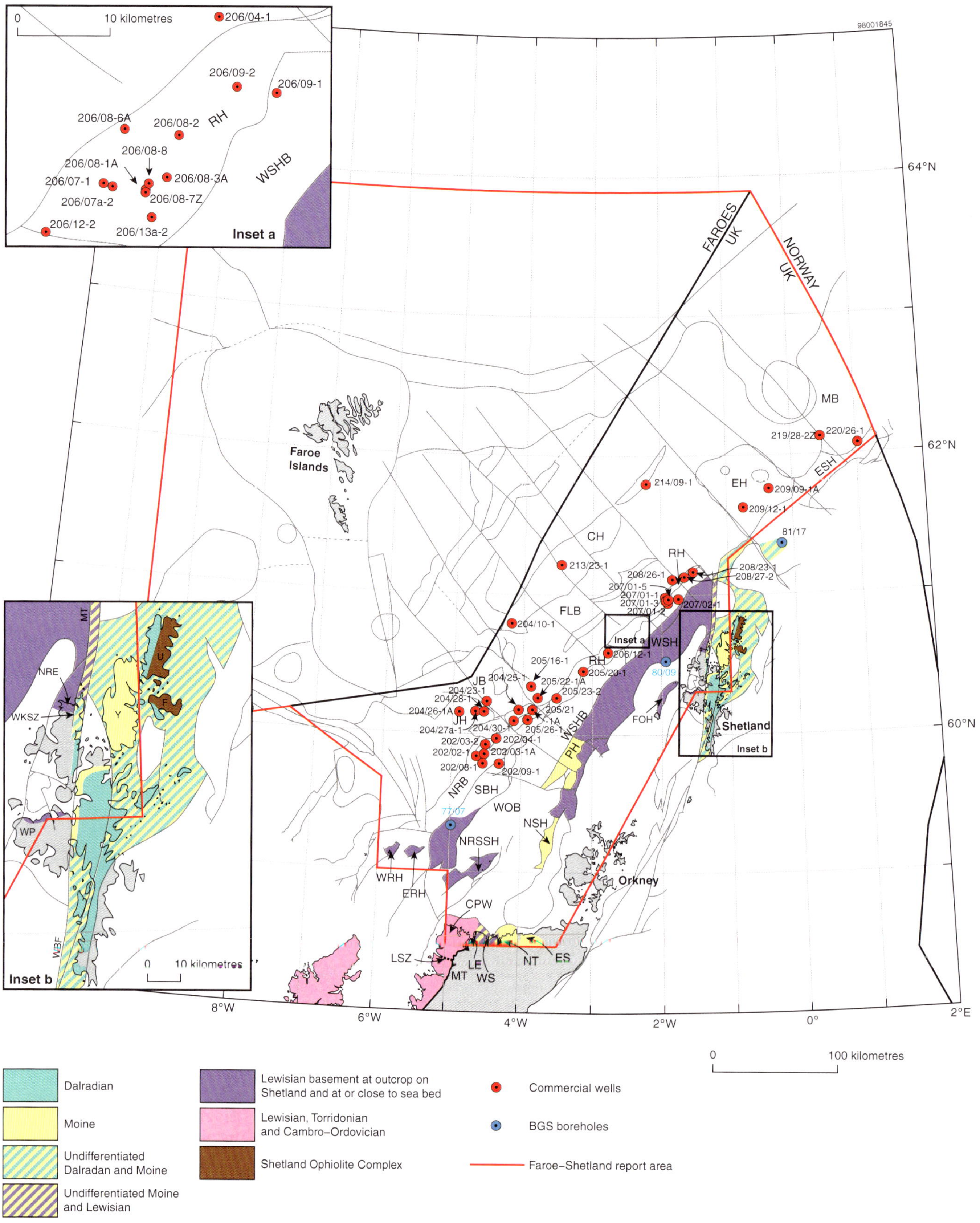

Figure 43 Drilled occurrences of crystalline basement within the Faroe–Shetland report area (partly modified after BGS 1984, 1986a, 1988b, 1989c; Flinn, 1992 and Strachan et al., 2002). Present day structural elements (grey) derived from Figure 7. Abbreviations: see Figure 7 except, CPW=Cape Wrath Peninsula; ES=East Sutherland; F=Fetlar; LE=Loch Eriboll; LSZ=Laxford Shear Zone; NRE=North Roe; NT=Navar Thrust; U=Unst; WKSZ=Western Keolka Shear Zone; WP=Walls Peninsula; WS=West Sutherland and Y=Yell.

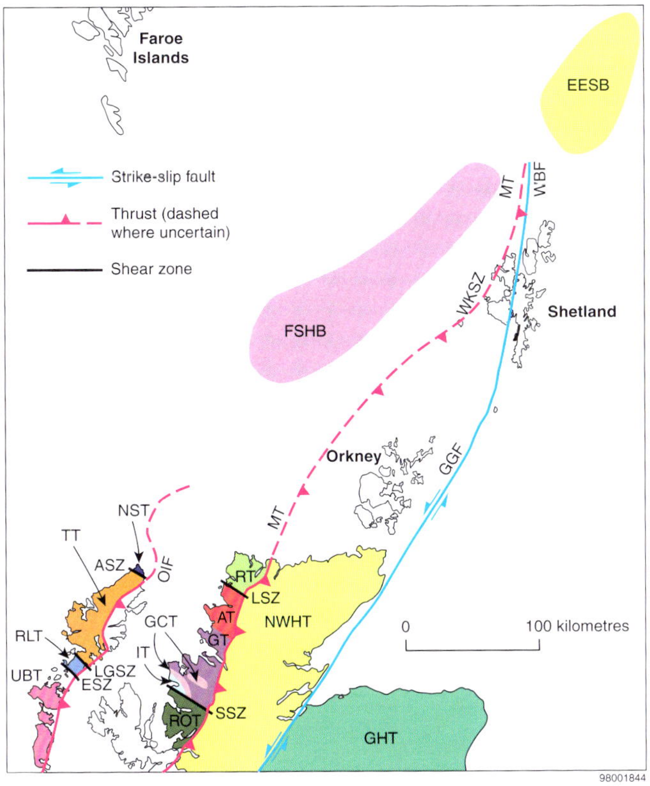

Figure 44 Schematic distribution of terranes and blocks within the north-west UK and surrounding offshore area (partly modified after Kinny et al., 2005 and Trewin and Rollin, 2002). Abbreviations: see Figures 7 and 43 except, ASZ=Alasdair Shear Zone; AT=Assynt Terrane; EESB=Erlend–East Shetland Block; ESZ=Ensay Shear Zone; FSHB=Faroe–Shetland Block; GCT=Gairloch Terrane; GHT=Grampian Highland Terrane; GT=Gruinard Terrane; IT=Ialltaig Terrane; LGSZ=Langavat Shear Zone; LSZ=Laxford Shear Zone; NWHT=North-west Highland Terrane; NST=Nis Terrane; OIF=Outer Isle Fault; RLT=Roineabhal Terrane; ROT=Rona Terrane; RT=Rhiconich Terrane; SSZ=Shieldaig Shear Zone; TT=Tarbet Terrane and UBT=Uist/Barra Block.

Rocks with similar characteristics (e.g. mineralogy, chemistry and age) to those of basement described from the Lewisian Gneiss Complex on north-west Scotland and the Outer Hebrides Isles have been described from Shetland and other small islets, and from a significant number of commercial well and BGS boreholes in the offshore area, particularly along the south-east flank of the Faroe–Shetland Basin (Figure 43 and Table 3). Their presence, at or close to, the sea bed has also been inferred from parts of the platform area to the west of Orkney and Shetland.

Shetland, Foula, Rona and other islets

On Shetland, Lewisian quartzofeldspathic gneiss of the Western Series is present close to the north-west limit of the mainland area at North Roe (Figure 43; Flinn et al., 1979), and separated from Moine rocks to the east by the steeply dipping Western Keolka Shear Zone (Pringle, 1970; Flinn et al., 1979). The gneiss has yielded K–Ar ages of between approximately 2900 and 2650 Ma, and is generally correlated with the Lewisian Gneiss Complex of mainland Scotland (e.g. Flinn et al., 1979). A depleted mantle Nd isotope-model age (T_{DM}) of 2900 Ma was also recorded from gneiss of the Western Series (Wilgi Goes Group) (Knudsen, 2000). Gneiss is also present as inliers within Moine rocks immediately to the east of the Western Keolka Shear Zone, and to the south along the north coast of the Walls peninsula (Figure 43; Strachan et al., 2002).

In the offshore area to the west of Shetland, Lewisian rocks outcrop at the south-west and north-east ends of the Foula High, on Foula Island (Figures 29 and 43; BGS, 1984; 1988b) and the Ve Skerries islets (Mykura, 1976), respectively. Within the area between these islands, mainly banded gneiss and granulite has been described from BGS shallow sample sites. In the south of the report area, Lewisian rocks also crop out on the Nun Rock–Sule Skerry High (Stack Skerry and Sule Skerry islets) and on the West Rona High (Rona Island) (Figure 43; BGS, 1986a; Nisbet and Bowes, 1961).

Faroe–Shetland Basin and marginal basins and highs

In a study into the nature, age and affinity of basement rocks within the offshore part of the report area, Ritchie and Darbyshire (1984) recorded an Rb–Sr (whole-rock) late Archaean 'Scourian' age of 2527±73 Ma from widely spaced samples in and around the West Shetland Basin (Figure 43). Samples for a few other scattered wells yielded a wide variety of K–Ar whole rock and mineral ages (e.g. Cashion, 1975; Ziegler, 1982; Hitchen and Ritchie, 1987; Ritchie et al., 1987). Chambers et al. (2005) have recently completed a study into the nature, age and affinity of the basement along a north-east-trending, 800 km-long transect between the Rockall High in the south-west, and the East Shetland High in the north-east (Figure 7). A summary of the lithologies and ages of the well/borehole samples analysed from the Faroe–Shetland area is presented in Table 3.

Petrography and geochemistry

A wide variety of lithologies typical of the Lewisian Gneiss Complex have been reported from basement rocks within the Faroe–Shetland area, including quartzofeldspathic, granitic, dioritic, tonalitic and augen gneiss, diorite, monzonite/granodiorite, amphibolite, protomylonite, mylonite and cataclasite (Table 3 and Figure 45). These rock types are considered to have undergone high-grade metamorphism, with most at amphibolite facies level (Chambers et al., 2005). Assessment of protolith lithologies is difficult, mainly due to a lack of context regarding overall rock associations and uncertainty regarding structural relationships. Most of the granitic rocks are considered to be of TTG or related affinity, although only samples from wells 206/07a-2, 206/08-2 and -7Z (Table 3 and Figure 43) meet with the strict geochemical criteria indicative of a TTG protolith (Defant and Drummond, 1990). Geochemically, the TTG samples are characterised by high alumina and low niobium contents, variable degrees of heavy rare earth element depletion and are variably depleted and/or enriched in uranium and thorium (Chambers et al., 2005).

Geochronology

Lewisian dioritic, quartzofeldspathic and augen gneiss recovered from wells drilled within the North Rona Basin (202/02-1), Flett Sub-basin (205/22-1A), Rona

Table 3 Summary of lithology and age data of selected commercial wells and BGS boreholes that penetrate crystalline basement within the Faroe–Shetland report area (partly compiled from Chambers et al., 2005). Abbreviations: Qf=quartzofeldspathic.

98001848

Well/BGS borehole	Depth (m)	Location	Description	Sm-Nd model age (Ma) (T_{DM})	U-Pb age (Ma)
			Faroe–Shetland Block		
202/02-1	1219.8 m	North Rona Basin	Qf gneiss	2940**	2829±46
202/03-1A	1774.5 m	North Rona Basin	Cataclasite	2830**	-
204/23-1	3847 m	Westray High	Qf gneiss	2920**	-
204/25-1	2870 m	Rona High	Tonalitic gneiss	2900**	-
205/16-1	4172 m	Flett Sub-basin	Pseudotachylite and diorite	2920**	-
205/22-1A	3225.5 m	Flett Sub-basin	Dioritic gneiss	2960**	2700±13
206/07a-2	2142.8–2600 m	Clair Basin	Monzonite to granodiorite and mylonitic amphibolite	2950 and 3050**	-
206/08-1A	2310.9 m	Clair Basin	Dioritic gneiss	3020**	2801.7+5.1/-4.6
206/08-2	1864.4 m	Clair Basin	Protomylonite	3040*	-
206/08-7Z	2338 m	Clair Basin	Amphibolite	3010**	-
206/08-8	2502 m	Clair Basin	Granitic gneiss	3070**	-
208/27-2	1380.3 m	Rona High	Granitic gneiss and cataclasite	3300**	-
			Erlend–North Shetland Block		
209/09-1A	2697.5 m	Erlend High	Augen gneiss	2980**	2738.3±4.1
209/12-1	3511.5 m	Erlend High	Biotite schist	1440*	-
220/26-1	5280 m	More Basin	Qf mylonite	2320*	-
81/17	137.6 m	East Shetland High	Amphibolite	1820*	-

* Single stage T_{DM} calculated according to Borg *et al.*, (1990)

* * Two stage T_{DM} calculated using U-Pb zircon age or assumed age of crystallisation, following DePaolo *et al.*, (1991)

High (206/08-1A) and the Erlend High (209/09-1A), yielded U-Pb zircon ages of 2829±46 Ma, 2700±13 Ma, 2801.7+5.1/-4.6 Ma and 2738.3±4.1 Ma, respectively (Table 3 and Figure 43; Chambers et al., 2005). These wells are spread over a 400 km-long, north-east-trending transect, indicating that the age of crystallisation of a substantial area of basement with the south-east part of the report area is likely to be of Archaean age. This finding is supported by results from Sm–Nd analysis of these wells, and a significant number of additional well samples along the length of the transect (see below). On the grounds of zircon inheritance characteristics, Chambers et al. (2005) suggested that these basement gneisses fall into older (c.2800 Ma) and slightly younger age categories (c.2740 to 2700 Ma), with the former representing new continental crust, while the latter was involved in some degree of crustal recycling.

Sm–Nd model ages of basement samples within the report area occur in two main categories. In the first category, samples from wells 202/02-1, 202/03-1A, 204/23-1, 204/25-1, 205/16-1, 205/22-1A, 206/07a-2, 206/08-1A, -2, -7Z and -8 and 208/27-2 located between the North Rona Basin in the south-west and the Rona High in the north-east yielded a relatively narrow range of Archaean T_{DM} model ages that fall between 3300 and 2830 Ma (Table 3 and Figure 43). In the second category, samples from wells 220/26-1, 209/09-1A, 209/12-1 and BGS borehole 81/17 located to the north-east of the study area yielded more variable Sm–Nd ages (Table 3 and Figure 43). For example, the mylonite from 220/26-1 on the East Shetland High had a T_{DM} crustal formation age of 2320 Ma, whereas the augen gneiss in 209/09-1A on the Erlend High yielded a model age 2980 Ma (similar to its U–Pb age). A biotite schist from well 209/12-1 on the Erlend High recorded a Proterozoic T_{DM} model age of 1440 Ma, possibly indicating a significant contribution from juvenile magmatic material, although its mineralogy could suggest feldspathic sandstone as the protolith (Chambers et al., 2005). BGS borehole 81/17 drilled on the East Shetland High also yielded a young T_{DM} model age of 1820 Ma from an amphibolite. Although Chambers et al. (2005) suggest that the age derived from 81/17, and also to a lesser extent from 220/26-1, should be treated with caution, it may be that these samples have Moinian or Dalradian rather than Lewisian affinity.

Terrane affinity and comparisons with the Lewisian Gneiss Complex

From the results of isotope geochemistry on a limited dataset, the Lewisian and undifferentiated basement drilled within the offshore area to the north-west and north of Orkney and Shetland have been crudely ascribed to two main crustal domains, namely a late Archaean 'Faroe–Shetland' block and an undifferentiated late Archaean to Proterozoic 'Erlend–North Shetland' block. Note that due to the relatively limited sampling and geochemical assessment of basement in the offshore area when compared with that on land, the use of the informal term *'block'* is preferred to *'terrane'*. The boundary between the blocks is tentatively defined by the north-trending Walls Boundary Fault and the north-west-trending Erlend Lineament, that marks the south-west margin of the Erlend High (Figure 43 and 44). The major Walls Boundary strike-slip fault may represent a particularly significant feature, as on Shetland it is considered to juxtapose Moine rocks of very different affinities (see below).

Faroe–Shetland Block: This block comprises an area of Archaean basement defined by 12 well penetrations within a narrow, 275 km-long transect between the North Rona High and the Erlend High (Figures 43, 44 and Table 3). The Faroe–Shetland block occurs within an area dominated by a north-east-trending Caledonian

Figure 45 Photomicrographs of Lewisian rock thin sections in crossed polarised light from a) an augen gneiss from well 202/09-1A (at 1640.1 m below drillfloor) containing mainly quartz, K-feldspar, plagioclase and biotite; b) an amphibolite from BGS borehole 81/17 (at a sub-sea depth of 137.6 to 38.2 m) containing mainly amphibole, plagioclase and quartz (as a vein); c) a quartzofeldspathic gneiss from well 202/02-1 (at 1219.8 m below drillfloor) containing mainly quartz, plagioclase, biotite and garnet. Field of view =2 mm (from Chambers et al., 2005). For location of wells and boreholes, see Figure 43.

a)

b)

c)

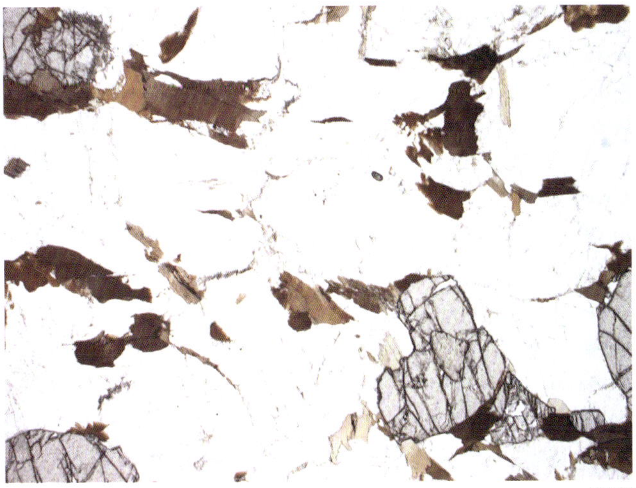

98001834

structural grain and is also cut by numerous north-west-trending transfer zones or lineaments (Figures 7 and 43; e.g. Rumph et al., 1993), which may represent former Laxfordian-style shear zones (e.g. see Park et al., 2002), similar to those that bound terranes within the Lewisian Gneiss Complex of Scotland (Figure 44).

Within the Faroe–Shetland block, U–Pb analysis of samples from three wells yielded Archaean crystallisation ages of between c.2800 and 2700 Ma i.e. 202/02-1, 205/22-1A, and 206/08-1A (Figure 43). The Archaean age is strongly supported by T_{DM} model ages (3300 to 2830 Ma) derived from all twelve wells within the block. These T_{DM} model ages are broadly similar to those of the Lewisian Gneiss Complex (3000 to 2700 Ma) (e.g. Whitehouse, 1989; Corfu et al., 1998). According to Chambers et al. (2005), these model ages are consistent with an interpretation suggesting widespread crust formation at 2800 to 2700 Ma, derived from slightly older basaltic precursors.

The Faroe–Shetland block lies 125 km to the north of the Lewisian Gneiss Complex, with the Rhiconich Terrane the closest for comparative purposes (Figure 44). The U–Pb ages of the TTG gneisses in both areas are fairly similar, with Inchard gneisses at 2840 to 2680 Ma and Faroe–Shetland gneisses at c.2830 to 2700 Ma. Within the Rhiconich Terrane, Laxfordian amphibolite facies metamorphism and deformation at about c.1740 Ma is pervasive, but although the majority of the rocks within the Faroe–Shetland block are at amphibolite facies, there are currently no indications of the age of the metamorphism. There are also no indications of the presence of the c.1855 Ma Laxfordian Rubha Ruadh granites observed within the Rhiconich Terrane. Nonetheless, there are clearly some similarities in the characteristics of both basement units, but new material and further work are required before more definitive correlations can be made. Speculatively, however, it may be that Proterozoic subduction-related events that affect the Lewisian Gneiss Complex on mainland Scotland are not represented within the Faroe–Shetland Block.

The Western Series of mainly quartzofeldspathic gneisses on Shetland represents the closest outcrop of crystalline basement to the Faroe–Shetland block. These gneisses recorded K–Ar ages of between c.2900 and 2650 Ma and a T_{DM} model age of 2900 Ma, broadly similar to those of the Faroe–Shetland Block.

Erlend–East Shetland Block: This block comprises an undifferentiated area of late Archaean and Proterozoic basement of unknown affinity and is delineated on the basis of only four wells that are approximately 100 km apart and located between the Erlend High and the Møre Basin (Figures 43 and 44). An Archaean U–Pb crystallisation age of 2738 Ma and a supporting 2980 Ma T_{DM} model age were obtained from 209/09-1A and are similar to that described from the Faroe–Shetland block. However, three other samples from well 209/12-1, BGS borehole 81/17 and well 220/26-1 yielded T_{DM} model ages of 1440, 1820 and 2320 Ma, respectively, that are at variance with the range of values associated with the Faroe–Shetland block (see above). The origin and affinity of these rocks are unknown but they may prove to have more in common with the metamorphic rocks observed on Shetland. For example, a Moinian psammite from Yell and quartzite from the Dalradian Whiteness and Cliff Hills divisions yielded model ages of 1800, 2000 and 2500 Ma, respectively (see below).

Torridonian

Torridonian strata are only definitely present on the Cape Wrath Peninsula in the extreme south of the report area, where they rest unconformably on Lewisian gneiss (Figure 43; BGS, 1989b). Here, they form the northerly onshore limit of a north-east-trending zone of exposed Torridonian rocks that extend for approximately 200 km to the island of Rum in the south (see Stewart, 2002). In ascending order of age, the Torridonian succession is subdivided into the Stoer, Sleat and Torridon groups, that are mainly comprised of red to grey fluviatile sandstone up to 2 km, 3.5 km and greater than 5 km thick, respectively (see Park et al., 2002). The Torridon Group is considered to have been deposited as an orogen-parallel foreland basin to the Grenville Orogeny (Krabbendam et al., 2008). Results from radiometric age dating suggests that the Stoer and Torridon groups were deposited at c.1200 and 1000 Ma, respectively (Turnbull et al., 1996; Rainbird et al., 2001). Between mainland Scotland and the Hebrides, Torridonian strata have been penetrated within a commercial well and the possibility exists that a succession up to 6 km thick may be present within the Sea of Hebrides–Little Minch Basin (Stein, 1992).

Within the report area onshore, only the Applecross Formation of the Torridon Group is preserved on the Cape Wrath Peninsula (BGS, 1989b; Williams, 2001). Here, the Applecross Formation is approximately 500 m thick and rests unconformably on Lewisian gneiss (Park et al., 2002). It mainly comprises fine to coarse-grained, planar and trough cross-bedded fluviatile red sandstone that commonly displays soft sediment deformation features and coarse sheet conglomerates. Torridonian strata are inferred to extend into the offshore area immediately north of Cape Wrath though they have not been proved by drilling.

Moine

Within the report area, Moine rocks have been described on Shetland and along the north of mainland Scotland (Figure 43; BGS, 1984; 1989b). On mainland Scotland, Moine rocks of the North-west Highlands Terrane (Figure 44) comprise a Neoproterozoic arenaceous and argillaceous sedimentary succession. These sediments were subsequently subjected to high-grade metamorphism and variable deformation during the Knoydartian tectonothermal event at approximately 800 Ma and during the later Grampian and Scandian orogenic cycles (Strachan et al., 2002). In ascending order of age, rocks of the Moine Supergroup on mainland Scotland are subdivided into three main tectonostratigraphically-defined groups, namely the Morar, Glenfinnan and Loch Eil, which are considered to be 5 km, 1 to 4 km and 5 km thick, respectively (Johnson et al., 1969; Strachan, 1985; Holdsworth et al., 1987; Holdsworth et al., 1994; Soper et al., 1998). The Morar Group is considered by Krabbendam et al. (2008) to have been deposited in a foreland basin to the Grenville Orogeny.

Shetland, Orkney and adjacent offshore area

Moine metamorphic rocks on Shetland occur within both the West Shetland Zone or East Mainland successions that are separated by the Walls Boundary Fault (Figure 43; Flinn et al., 1979). The West Shetland Zone and East Mainland Succession have very different characteristics, possibly due to postdepositional juxtaposition as a result of 100 to 200 km of strike-slip fault movement on the Walls Boundary Fault (e.g. Flinn, 1992; 1988). Within the East Mainland Succession, Moine rocks occur mainly on the island of Yell (but also along a narrow, north-trending tract adjacent to the Walls Boundary Fault) (Figure 43). These rocks are ascribed to the Yell Sound Division, and are described as comprising psammite, schist and quartzite that on Yell are interleaved with Lewisian gneisses (Flinn et al., 1979; Flinn, 1988). On lithological grounds, they have been correlated with the Glenfinnan Group of mainland Scotland (Flinn, 1988). A T_{DM} model age of 1800 Ma was recorded from a psammite analysed from the Yell Sound Division (Knudsen, 2000). Within the West Shetland Zone, a narrow strip of Moine rocks has been described adjacent to the Walls Boundary Fault in the extreme north of the mainland. These rocks comprise mainly psammite that is interleaved with Lewisian basement (Flinn, 1988). On lithological grounds and the lack of sedimentary structures, they have been correlated with the Morar Group described from mainland Scotland (Flinn, 1988). They are separated from Lewisian basement to the west, by the Western Keolka Shear Zone, a possible correlative for the Moine Thrust on mainland Scotland (e.g. Watson, 1984).

Within the offshore area around Shetland, basement rocks are categorised as undifferentiated Dalradian, Moinian and Lewisian, with the most likely occurrences of Moine rocks probably best developed within the vicinity of Yell Island (Figure 43). On Orkney, a small outcrop of Moine-like rocks occurs on the south coast of the mainland area at Scapa Flow (BGS, 1985; Strachan et al., 2002).

Mainland of north Scotland and adjacent offshore area

Within the report area, the Naver Thrust (Figure 43) is interpreted to separate Moine rocks of the Morar Group in the west from those of possible affinity to the Glenfinnan Group of East Sutherland to the east. The Morar Group comprises mainly psammitic and pelitic gneiss, interleaved with fault-bounded slices of Lewisian gneiss (Strachan et al., 2002). In places, Moine basal conglomerate is demonstrated to unconformably rest on Lewisian basement (e.g. Strachan and Holdsworth, 1988). The depositional thickness of psammite-dominated Morar Group of West Sutherland is greater than 3 km and perhaps even 7.5 km (Krabbendam et al., 2008).

Undifferentiated Moine and Lewisian rocks between the Naver Thrust and Moine Thrust are inferred to extend into the offshore area immediately north of mainland Scotland (Figure 43; BGS, 1989b). Interpretation of the MOIST deep seismic profile

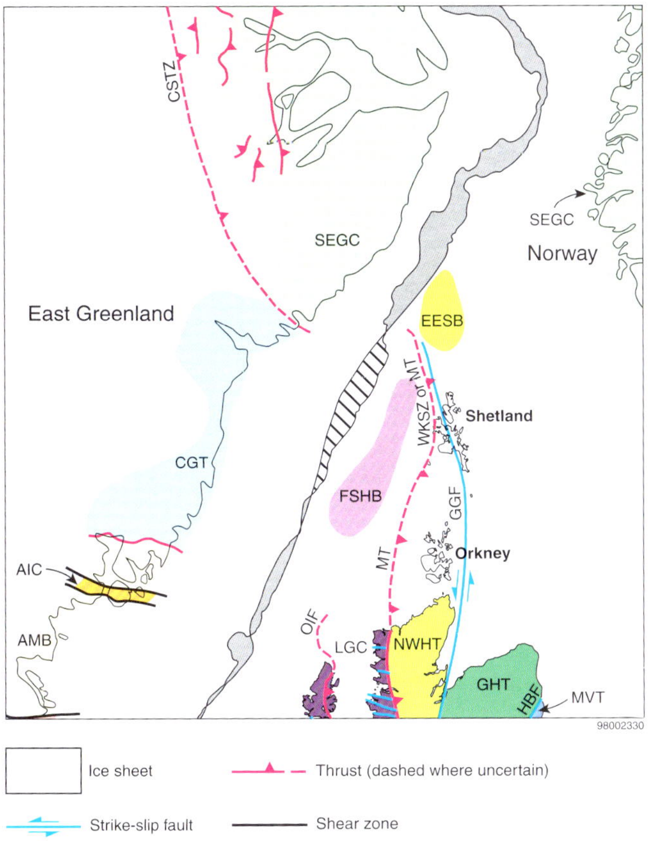

Figure 46 North-east Atlantic margin mid Palaeozoic reconstruction showing terrane juxtapositions between offshore north-west UK and East Greenland. Plate reconstructions were produced using Atlas plate reconstruction software, developed by Cambridge Paleomap Services Limited (CPSL). The reconstruction is based on a best fit using the 1000 m bathymetric contour, with areas of poor fit highlighted by zones shaded grey (areas of overlap) and hachured (areas where there are gaps). For terrane subdivision of Lewisian Gneiss Complex, see Figure 44. Abbreviations: see Figures 7, 43 and 44 except, AMB=Ammassalik Mobile Belt; AIC=Ammassalik Intrusive Complex; CGT=Central Greenland Terrane; CSTZ=Caledonian Sole Thrust Zone; HBF=Highland Boundary Fault; LGC=Lewisian Gneiss Complex; MVT=Midland Valley Terrane and SEGC=Scandian and East Greenland Complex.

suggested the possibility that Moine rocks are present below Devonian and Permo-Triassic strata within the southern part of the West Orkney Basin (Brewer and Smythe, 1984).

North Shoal and Papa High

The presence of Moine-like rocks at the sea bed on the North Shoal and Papa highs has been inferred from the recovery of core material from BGS shallow drilling operations with sample descriptions such as psammite and pelite etc (Figure 43; BGS, 1985; 1988b). These occurrences have been used by Ritchie et al. (1987) to suggest that the northward propagation of the Moine Thrust should lie to west of this. With the exception of the North Shoal and Papa highs, Moine rocks have not been drilled elsewhere within the report area.

Dalradian

Within the report area, Dalradian rocks have only been described from Shetland (Figure 43; BGS, 1984). On mainland Scotland, Dalradian rocks of the Grampian Highland Terrane (Figure 44) comprise a Neoproterozoic sandstone, siltstone, mudstone and limestone-dominated marine succession with a stratigraphical thickness of 25 km that was deposited between approximately 870 Ma and early Cambrian times during the break-up of the Rodinian super continent (Strachan et al., 2002). Following deposition, it has subsequently been affected by low to medium-grade metamorphism, mainly during the Grampian Orogeny (Strachan et al., 2002). In ascending order of age, the Dalradian Supergroup can be subdivided into the Grampian, Appin, Argyll and Southern Highland groups (e.g. Harris et al., 1978).

Shetland and adjacent offshore area

On Shetland, Dalradian rocks almost exclusively occur to the east of the Walls Boundary Fault as part of the East Mainland Succession (Figure 43). Currently, the Dalradian succession is subdivided from west to east into Scatsta (oldest), Whiteness and Cliff Hills (youngest) divisions, which have been correlated with the Dalradian succession of mainland Scotland (e.g. Flinn et al., 1972). Sm–Nd analysis of impure quartzite samples from both the Whiteness and Cliff Hills divisions has yielded T_{DM} model ages of 2000 and 2500 Ma, respectively (Knudsen, 2000). Within the West Shetland Zone, Dalradian rocks are restricted to a narrow north-trending tract in the most northerly part of mainland Shetland (e.g. Flinn 1988; 1992, Strachan et al., 2002, fig. 4.3). Here, they are interpreted to be thrust in a westerly direction over Moine rocks along the Virdibreck Shear Zone (Flinn et al., 1979; Flinn, 1988).

Within the offshore area around Shetland, basement rocks are categorised as undifferentiated Dalradian, Moinian and Lewisian, with the most likely occurrences of Dalradian strata inferred to be found around the majority of the mainland area and to the west and south of Unst (Figure 43).

Cambro-Ordovician

Rocks of Cambro-Ordovician age are present only in the extreme southern part of the report area where they have been described to the west of the Moine Thrust around Loch Eriboll and on the Cape Wrath Peninsula close to the north coast of mainland Scotland (Figure 43; BGS, 1989b). Here, they form the northerly onshore limit of a narrow, 175 km-long, north-east-trending belt of Cambro-Ordovician strata that extend immediately to the west of the Moine Thrust as far south as Skye (Park et al., 2002). The succession ranges in age from Lower Cambrian (Ardvrech Group) to Mid Ordovician (Durness Group), with both groups present within the report area (BGS, 1989b). The clastic-dominated Ardvrech Group, which can be subdivided into two formations, has a maximum thickness of only 175 m and comprises mainly quartz arenite deposited in tidal or storm-dominated environments with some minor conglomerate and dolomitic siltstone at the top. The younger carbonate-dominated Durness Group can be subdivided into seven formations, and comprises approximately 750 m of thick dolostone with subordinate limestone and chert. These deposits are likely to have been developed in a low-energy, sub-tidal carbonate shelf environment.

Cambro-Ordovician rocks are considered to extend into the offshore area immediately north of mainland Scotland (Figure 43; BGS, 1989b). Interpretation of the MOIST deep seismic profile suggested the possibility that they also occur at depth beneath Devonian and Permo-Triassic strata within the southern part of the West Orkney Basin (Brewer and Smythe, 1984). No rocks of this age have been proved by drilling within the report area.

Correlations with East Greenland

Over the past few decades, there have been numerous correlations produced of Archaean and Proterozoic basement terranes observed within the Canadian maritime provinces, Greenland, Ireland, UK, Norway and Finland (e.g. Winchester, 1988; Dickin, 1992; Park et al., 1994; Park, 1995; Baba, 1999; Buchan et al., 2000; Friend and Kinny, 2001; Park, 2005). Though the onshore basement terranes in these areas have been extensively studied, their correlation across the continental areas of the offshore north-east Atlantic shelves has been largely inferred. Using a pre-Atlantic Ocean opening refit set at Early Devonian times (410 Ma), results from the report area provisionally indicate that the Archaean Faroe–Shetland block may generally correlate in part with the Archaean to Proterozoic Central Greenland Craton of East Greenland (Figure 46). The mainly Proterozoic ages model ages (1440, 1820 and 2320 Ma) derived from the Erlend–North Shetland Block are more enigmatic and could suggest a correlation with either the Central Greenland Craton or possibly even the Caledonian Scandian East Greenland Terrane. Although the Scandian East Greenland Terrane has a strong Caledonian overprint, both Archaean and Proterozoic ages have been recorded from basement rocks within its southern part, though in the latter case, none are younger than 2000 Ma (e.g. Kalsbeek et al., 1993; Thrane, 2002).

4 Devonian and Carboniferous

Kevin Smith[1] and Heri Ziska[2]
[1] *British Geological Survey*
[2] *Jarðfeingi (Faroese Earth and Energy Directorate)*

Devonian and Carboniferous sediments and volcanic rocks in the British Isles accumulated in the interval between the Caledonian and Variscan orogenies. During the Early Devonian, following the closure of the Iapetus Ocean, phases of sinistral strike-slip tectonism continued to influence basin formation and magmatism (Soper, 1988; Soper et al., 1987; 1992; Soper and Woodcock, 2003; Dewey and Strachan, 2003). To the north of Scotland, the Orcadian, Clair and Faroe–Shetland basins developed as a transtensional or extensional regime persisted into Mid Devonian times (Figure 47). Evidence of Late Devonian and Carboniferous sedimentation is scanty in the Faroe–Shetland report area, but regional studies from elsewhere in the Variscan foreland suggest that a period of dextral transtension possibly controlled early Carboniferous basin formation by reactivating pre-existing Caledonian structures. Basins formed at this time were eventually uplifted and deformed during Variscan orogenesis, as the Rheic Ocean closed to the south of the British Isles to create the supercontinent of Pangaea by the end of the Carboniferous (Figure 10b).

DEVONIAN

Thick successions of continental red beds accumulated on the deeply eroded roots of the Caledonian orogenic belt, as Scotland lay between 30° and 20° south of the Equator throughout the Devonian (Mykura, 1991; Trewin and Thirlwall, 2002). Sedimentation was predominantly fluviatile, but long term variations in the amount of rainfall meant that ephemeral lakes expanded and coalesced during wetter intervals, while the area covered by sabkha and aeolian deposition increased in more arid times. For historical reasons, these Devonian sediments are commonly described as the Old Red Sandstone (ORS) to distinguish them from similar (New Red Sandstone) facies of Permo-Triassic age (see Chapter 5). The difficulty of differentiating between these two largely unfossiliferous intervals has sometimes led to stratigraphical misidentification, both onshore in Scotland and in the offshore area (Marshall and Hewett, 2003). Further problems of correlation and nomenclature have arisen because the Old Red

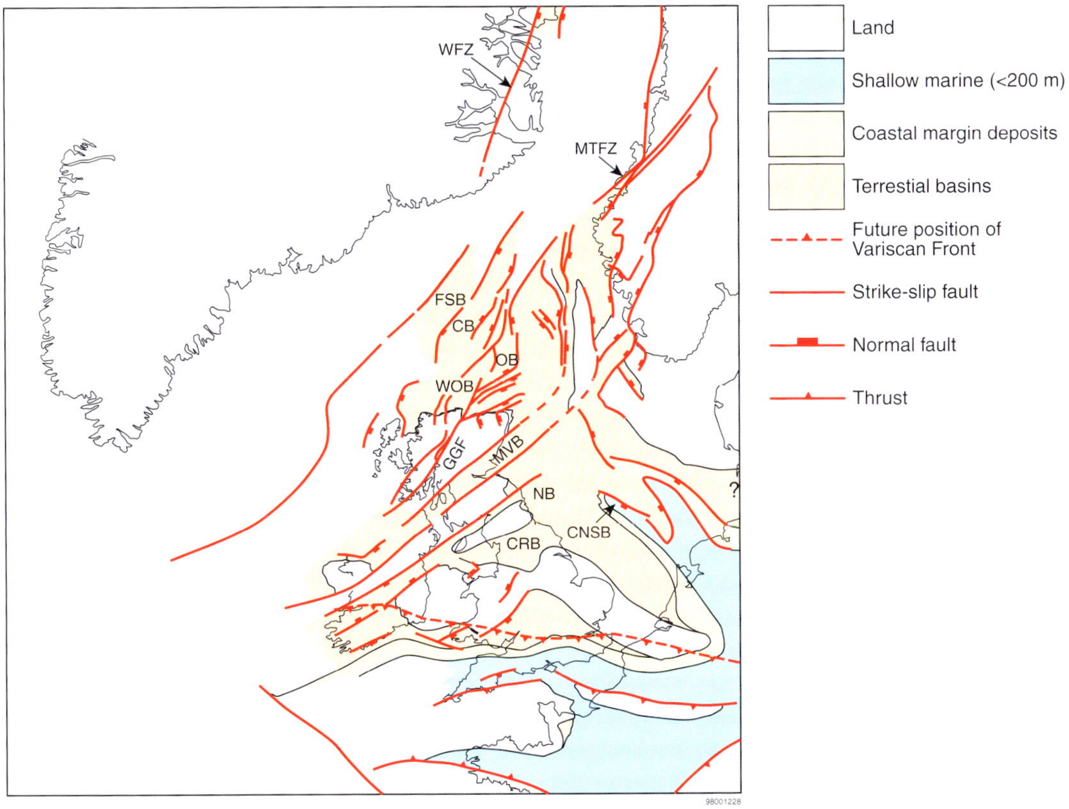

Figure 47 Palaeogeographical and palaeofacies plate reconstruction of the Faroe–Shetland region and surrounding area for the Mid Devonian times (slightly modified reproduction from Coward et al., (2003) by permission of The Geological Society, London). Abbreviations: see Figures 7 and 8 except, CNSB=Central North Sea Basin; CRB= Craven Basin; MTFZ=Møre–Trøndelag Fault Zone; MVB=Midland Valley Basin; NB=Northumberland Basin; OB=Orcadian Basin and WFZ=Western Fracture Zone.

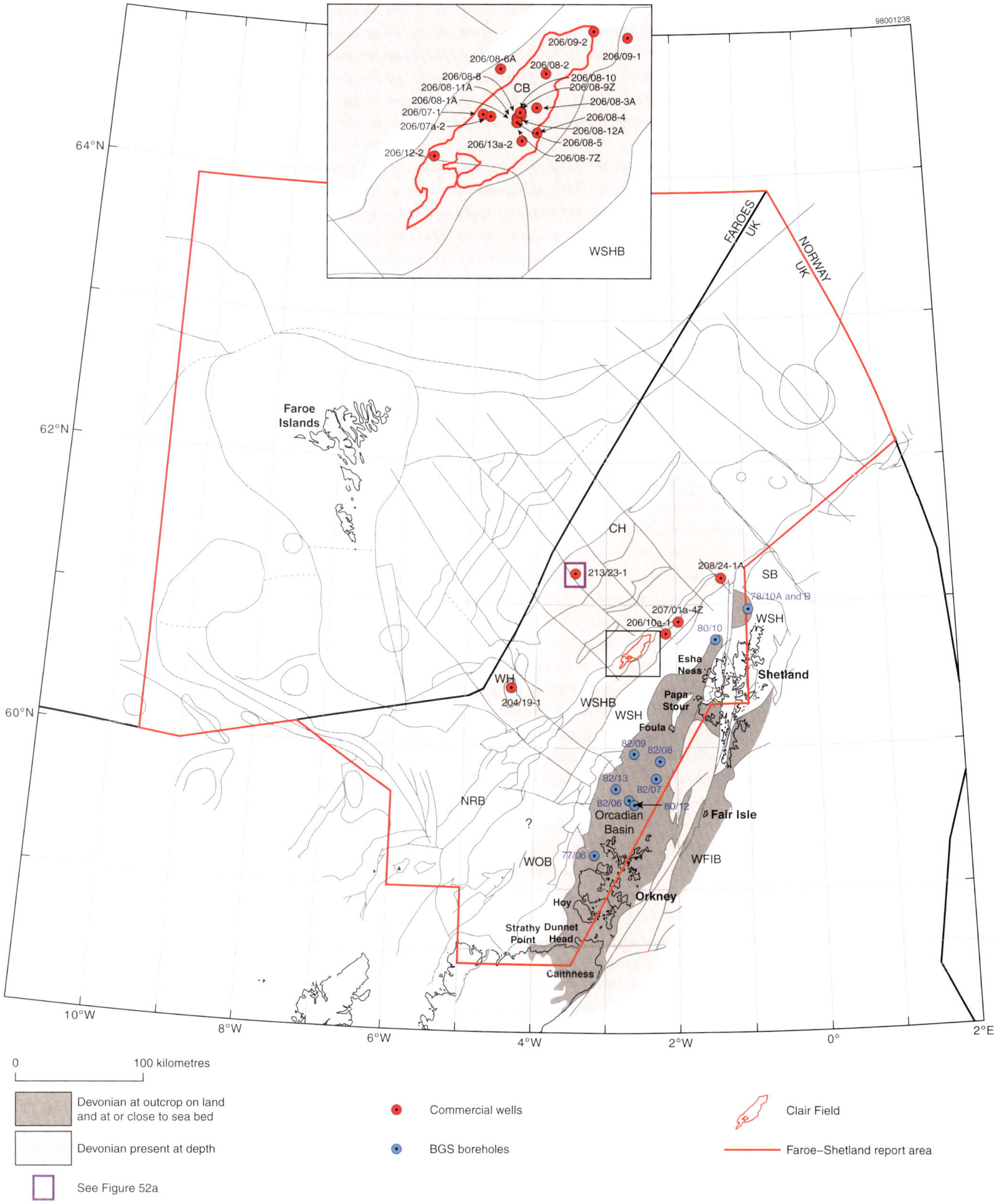

Figure 48 Distribution of Devonian strata within the Faroe–Shetland report area (partly modified from Stoker et al., 1993). Present day structural elements (grey) derived from Figure 7. Abbreviations: see Figure 7.

Sandstone is a diachronous facies, which is locally not restricted to the Devonian. In some parts of the Midland Valley of Scotland, Palaeozoic red beds were first laid down during the Silurian (Marshall, 1991), while elsewhere, at the top of the succession, similar deposition continued uninterrupted across the Devonian–Carboniferous boundary.

Onshore, at the southern and eastern margins of the report area, rocks of the predominantly Mid Devonian age Orcadian Basin crop out in Caithness, Orkney and Shetland (Figure 48; Mykura, 1991; Trewin and Thirlwall, 2002). Scattered offshore boreholes drilled by BGS have confirmed that similar strata occur at the sea bed on the Orkney–Shetland High, a shallow platform area that lies between Shetland and the mainland (Figure 48). Elsewhere offshore, Devonian rocks may be concealed by Permo-Triassic sediments within the West Orkney Basin (Figure 41; e.g. Earle et al., 1989; Evans, 1997), but drilling has indicated that they cannot be as thick or as extensive as formerly proposed by Enfield and Coward (1987) and Coward et al. (1989).

The discovery and development of the Clair Field (see Chapter 12) has shown that Devonian sediments overlie the Lewisian basement of the Rona High and extend beneath the West Shetland Basin (Figure 48) to form the Late Palaeozoic Clair Basin (Ridd, 1981; Blackbourn, 1987; Meadows et al., 1987 Allen and Mange-Rajetzky, 1992; Coney et al., 1993; McKie and Garden, 1996; Nicholls, 2005; Barr et al., 2007). Further to the west within the Faroe–Shetland Basin, other commercial wells have proven isolated occurrences of Devonian sediments on the Westray and Corona highs.

While the occurrence of thick Devonian to Permian successions in eastern Greenland (Larsen and Bengaard, 1991; Larsen and Marcussen, 1992) raises the possibility that Upper Palaeozoic sediments may be widespread within the Faroe–Shetland report area, the distribution of Devonian strata remains largely unknown (Figure 48). However, the interpretation of wide-angle seismic and gravity data has supplied some evidence for a sub-Palaeogene basalt sedimentary interval in the Faroes area (Hughes et al., 1997; Richardson et al., 1998; 1999; White et al., 1999; Spitzer et al., 2005; Raum et al., 2005). In a study of the FLARE wide-angle seismic profile, Richardson et al. (1999) invoked a layer of Devonian or Torridonian strata capping the basement to account for interval velocities of 5.7 km/s, which are uncharacteristically low for crystalline metamorphic rocks. Others have proposed that layers with similar velocities in the 5 km/s range, from elsewhere on the Atlantic margin, consist of sediments pervasively intruded by basic igneous rocks (Mjelde et al., 1998). A detailed velocity analysis of the Corona High carried out before drilling concluded that seismic velocities of 4.0 to 4.5 km/s indicated the presence of pre-Cretaceous sedimentary lithologies at 4.5 to 8.0 km depth (Hughes et al., 1997). When well 213/23-1 subsequently penetrated metamorphic basement at a depth of 4308.7 m below sea level in this area (Figure 48), it confirmed the observation of Richardson et al. (1999) that simple velocity analyses are incapable of resolving small-scale basement structures in structurally complex areas. In this case, a layer with a uniform seismic velocity of 4.0 to 4.5 km/s was shown to correspond to an unresolved tilted block composed of metamorphic basement (with a likely velocity in the range 5.9 to 6.4 km/s) overlain by Upper Palaeozoic strata (3.7 km/s) and a thickening wedge of Mesozoic sediments (2.7 km/s). Only a small part of the 3.5 km thick interval of pre-Cretaceous strata inferred from the original velocity analysis actually consists of Devono-Carboniferous rocks (576 m). Until further data are acquired by drilling, any regional interpretations of Devonian distribution are likely to remain conjectural.

Orcadian Basin

Devonian rocks of the Orcadian Basin crop out along the southern and eastern margins of the report area in Caithness, Orkney, Fair Isle, Shetland and Foula (Figure 48).

Caithness

The extent of post-Caledonian erosion is apparent in Caithness, where Lower Devonian sediments rest directly upon Moinian basement and the Caledonian Helmsdale Granite. This largely conglomeratic basal succession, which in places includes siltstone and mudstone, forms the western fringe of the Caithness outcrop and extends into the report area on the northern Scottish coast near Strathy Point (Collins and Donovan, 1977; Mykura, 1991). Most of the Caithness outcrop is composed of younger sediments of Mid Devonian age, which onlap the margins of the Lower Devonian basin and consist largely of flaggy sandstones. In southern Caithness, some of the sandstones are fluvial in origin, but these pass northwards into an area of mostly lacustrine facies, with repeated cycles of interbedded sandstone, limestone and siltstone rich in fish remains (Stephenson et al., 2006). One of these units, the Achanarras Fish Bed (Figure 49), forms a distinctive, fossil-rich marker horizon of Eifelian age (Marshall et al., 1996). Mid Devonian sedimentation continued in the Givetian, with deposition of the fluvial sandstones of the John O'Groats Sandstone Formation, which possibly persisted into Frasnian times (Figure 49).

Orkney

Lower Devonian sediments have only been tentatively identified on Orkney, where most of the exposed Devonian succession is of Mid Devonian age. Here, as in northern Scotland, widespread flaggy sandstones (the Upper and Lower Stromness Flagstone formations) are diversified by finely laminated siltstones, which include the Sandwick Fish Bed, a correlative of the Achanarras Fish Bed in Caithness (Figure 49). The Eday succession, which overlies the lacustrine sediments, is dominated by sandstone deposited in a braided, fluviatile environment, but also includes laminite and marl, as well as some locally developed aeolian deposits. Scattered outcrops of thin basalt flows, tuff and small intrusions also occur within this interval.

When marine Devonian rocks with crinoids, corals, bryozoans and other shell debris were discovered in the Argyll Field in the Central North Sea in 1972 (in well 30/24-3), it indicated that the sea had transgressed the Old Red Sandstone continent from the south and east, beyond the limit envisaged from the onshore outcrop, where marine rocks do not extend north of central England. Cameron (1993a) assigned all the marine rocks of Devonian age in the Central North Sea to the Kyle Group. Since then, evidence of a more northerly marine influence at the former continental margin has been detected in the Orcadian Basin, both at outcrop in Orkney (where rare scolecodonts and widespread pseudomorphs after halite occur in the Eday Marl) (Figure 49), and offshore on the adjoining Orkney–Shetland High, beyond the eastern margin of the report area (where sidewall cores in well 14/6-1 show that acritarchs and scolecodonts are present in mud rocks of late Givetian to early Frasnian age (Marshall et al., 1996)). These rocks indicate deposition in basins close to sea level and not in an elevated intermontane environment.

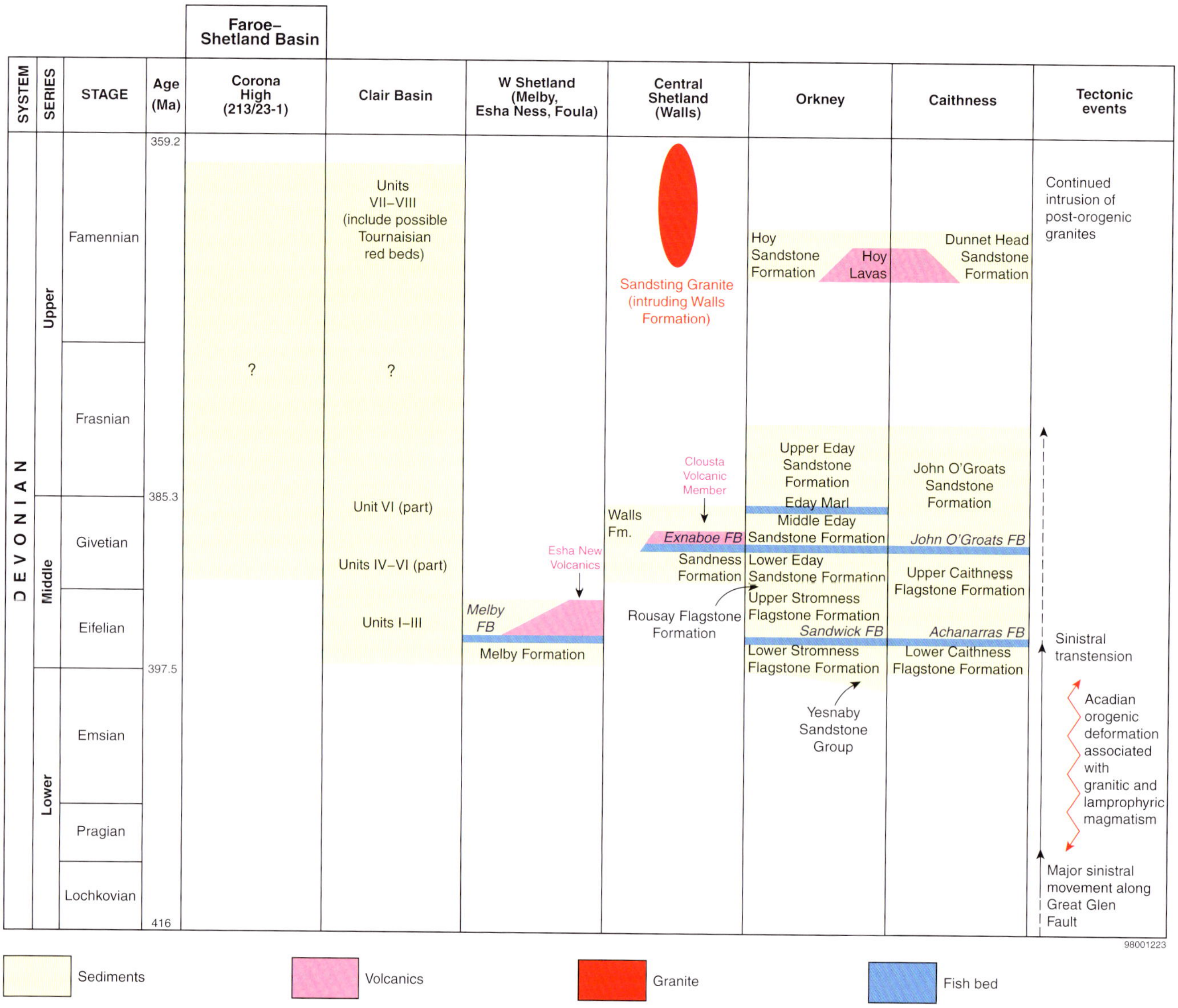

Figure 49 Devonian stratigraphy of the Faroe–Shetland region and surrounding area (modified from Marshall and Hewett, 2003; Trewin and Thirlwall, 2002). Timescale is based on 'A Geologic Time Scale 2004' by F M Gradstein, J G Ogg, A G Smith, et al. (2004) and the International Stratigraphical Chart, 2006 (ICS) with additions. Current stratigraphical nomenclature varies between authors, with notable differences in the allocation of group and formation status.

Shetland

The Devonian outcrop in Shetland is divided into separate structural domains by three regionally significant faults; the Melby Fault (here taken as equivalent to the St Magnus Bay Fault and its related onshore components, as used by Flinn et al., 1968; Flinn, 1992 and Marshall, 2000), the Walls Boundary Fault (Flinn, 1992; Ritchie and Hitchen, 1993; Underhill, 1993) and the Nesting Fault (Miller and Flinn, 1966; Figure 7). Stratigraphical changes across these faults have been used as supporting evidence for post-Caledonian strike-slip displacement in the Orcadian Basin.

The Melby Formation, which crops out to the west of the Melby Fault on the island of Foula and on the north-west headlands of Shetland (Figure 49), consists largely of sandstone. Sparse finer-grained sediments interbedded within the sequence include a bed of lacustrine origin rich in fossil fish, which is probably equivalent to the Achanarras Fish Bed (Marshall, 1988). Overlying these sediments, a succession of varied volcanic rocks, including basalt and rhyolite lava and tuff forms the island of Papa Stour and the headland at Esha Ness (Figure 48). Thirlwall (1979) has shown that each of the main outcrops of these volcanic rocks is geochemically distinct.

Sediments and igneous rocks of the Walls Group crop out between the Melby Fault and the Walls Boundary Fault. The Walls Group consists of the Sandness and Walls formations (Figure 49), which were formerly thought to be separated by the Sulma Water Fault, an east-north-east-trending structure extending between the Melby and Walls Boundary faults (Mykura and Phemister, 1976). The group also includes intrusive and extrusive rocks of the Clousta Volcanic Member. Intrusion of Upper Devonian Sandsting Granite immediately east of the report area (Figure 49) was probably responsible for generating the high vitrinite reflectance values of the Walls Group, which have been observed to range from 5.0 to 8.9%. Marshall (2000) suggests that the Walls Group is largely restricted to the early Givetian, although, with poor preservation of palynomorphs, the presence of younger Givetian sediments cannot be entirely ruled out. The microflora also confirm that the Sandness and Walls formations are partly age equivalent and probably represent contemporaneous facies variation with the Walls Basin, as originally proposed by Astin (1982). Previous proposals of an Early Devonian age for the Sandness Formation and a faulted contact with the Walls Formation (Mykura and Phemister, 1976) are not supported by the new miospore data. Correlating the Walls Group with the Lower and Middle Eday Sandstone formations of Orkney and the lower part of the John O'Groats Sandstone Formation of Caithness (Mid to Late Devonian age) means that all these sediments postdate the main Eifelian episode of lacustrine sedimentation in the Orcadian Basin. The partial equivalence of the Walls and Sandness formations, together with their more restricted time range, suggests that previous interpretations of up to 12 km of basin fill, based on cumulative stratigraphical thicknesses (Mykura and Phemister, 1976), are certain to be overestimates.

East of the Walls Boundary and Nesting faults on Shetland, beyond the eastern margin of the report area, a varied Middle Devonian succession includes diachronous basal conglomerate, flaggy sandstone, aeolian sandstone and finer-grained deposits including the distinctive Exnaboe Fish Bed (Figure 49), which possibly extend into the Upper Devonian (Mykura, 1976, 1991; Marshall and Hewett, 2003). On Fair Isle, the largely conglomeratic basal Devonian succession passes upwards to include some cross-bedded fluvial sandstones, interbedded calcareous mudstones, and siltstones, which contain a late Givetian flora. Acid and basic dykes of late Caledonian affinity are sparsely developed. Gross lithological comparisons between the Walls Group and the Fair Isle succession have been used to infer various amounts of post-Devonian dextral transcurrent movement along the Walls Boundary Fault (Rogers et al., 1989; Flinn, 1992).

Orkney–Shetland High

BGS boreholes drilled on the Orkney–Shetland High (but formerly the site of a Devonian basin) have confirmed that sandstone and finer-grained sediment of Mid Devonian age crop out on the sea bed between the northern islands and the Scottish mainland (Figure 48). Sparsely recovered biostratigraphical data from the cores have indicated late Eifelian to early Givetian ages.

Sandwick Basin

The Sandwick Basin forms a Devonian outlier on the East Shetland High to the north of Shetland (Figures 25 and 48). BGS borehole 78/10A and B proved red sandstone and arkose cropping out at the sea bed and seismic reflection data show a succession of undivided Devonian sediments up to 2 km thick dipping westwards towards the inferred northern extension of the Walls Boundary Fault (Hitchen and Ritchie, 1987; Ritchie and Hitchen, 1993).

Clair Basin

Exploration and appraisal wells within the Clair Field prove the Devonian (and Carboniferous) succession of the Clair Basin (Figures 48 and 50). Although the Devonian facies drilled in these wells have much in common with those of the Orcadian Basin, there is currently no evidence that Devonian marine sediments extended to the west of Shetland. Here, deposition probably took place in a separate tectonically controlled continental basin dominated by interbedded sediments of fluviatile, aeolian and lacustrine origin. Similar local depocentres, with internal drainage of Mid to Late Devonian age, are also found in the Canadian Maritimes, Ireland, Spitsbergen, Greenland and Norway (Bradley, 1982; Leeder, 1988). During the prolonged development of the Clair Field, a wide range of stratigraphical techniques have been used to study lengthy cores from the low-yielding Upper Palaeozoic reservoir in an effort to improve hydrocarbon recovery (Figures 50 and 51).

Blackbourn (1987), who analysed 900 m of core from five of the Clair Field wells, recognised four dominant sedimentary environments in the Clair Basin and informally divided the Upper Palaeozoic succession into four units, largely on the basis of grain size:

1 Basal Clair Unit (30 to 120 m); screes and alluvial fans with local interfan lakes.

2 Lower Clair Unit (>325 m); broad sandy plain with minor aeolian dunes.

3 Upper Clair Unit (>334 m); low-grade meandering rivers and floodplains.

4 Clair Carbonaceous Unit (<100 m); floodplain lakes, reflecting change to more humid conditions.

Subsequently, Allen and Mange-Rajetzky (1992) used 1.7 km of core from eight wells to define a new lithostratigraphical scheme, supported by heavy mineral analysis. They divided the Clair sediments into 10 units (I to X), of which units I to VI comprise the Lower Clair Group of Devonian age (Figure 50). The remaining units (VII-X), which have possible Carboniferous affinities, are included in the Upper Clair Group.

Unit I: *Fluviatile with braid plain lakes*

Conglomerate and pebbly sandstone form the two main facies in this unit. They were deposited on a coarse-grained fan delta as it prograded into a lake where laminated sandstone and siltstone were being laid down. Unit I forms the rift lake stage of Allen and Mange-Rajetzky (1992) and thickens north-eastwards and towards the main bounding faults in the west.

Unit II: *Sand-rich braid system*

This unit was deposited in a high-energy fluvial system. As a terminal fan built out into the Clair Basin, braidplain sands became increasingly common (forming the lower fluviatile stage of Allen and Mange-Rajetzky (1992)). A decrease in the sediment supply possibly associated with diminishing relief in the hinterland, produced a fining-up succession.

Unit III: *Aeolian facies association*

Sabkhas formed as aridity increased following Unit II. The area covered by sand flats expanded, and was crossed by widespread longitudinal dunes and ephemeral rivers. The presence of well-sorted aeolian sandstone within this unit considerably improves reservoir quality in the Clair Field. In well 206/08-4 (Figure 50), Middle Devonian spores occur in Unit III at 2973 m depth (Ritchie et al., 1996).

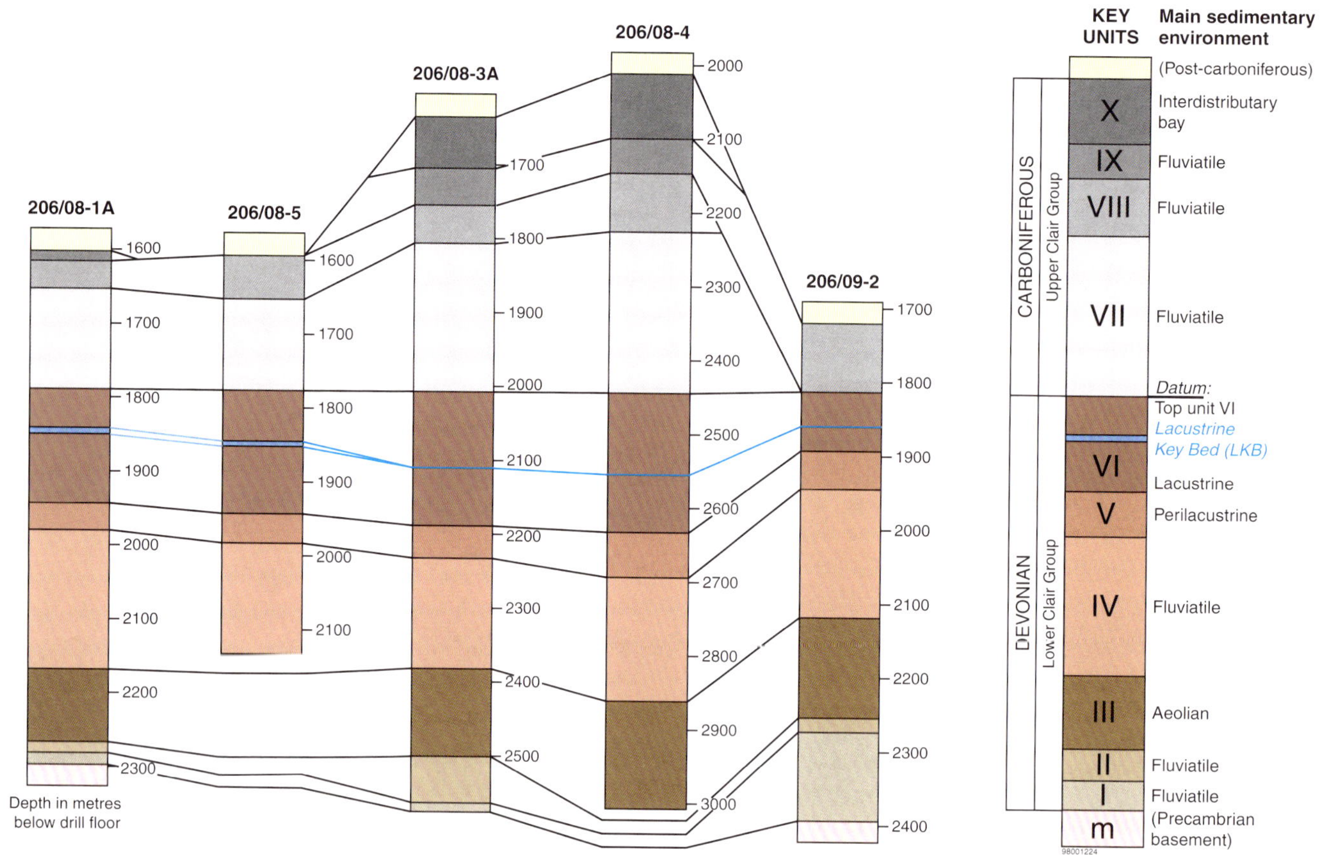

Figure 50 Devono-Carboniferous correlations from selected wells in the Clair Basin, showing stratigraphical subdivisions of Allen and Mange-Rajetzky, (1992) (modified from Ritchie et al., 1996). For location of wells, see Figure 48.

Unit IV: *Predominantly sand-rich fluvial environment with minor aeolian reworking*

A return to a wetter climate restored high-energy fluvial conditions in the area. The development of an unconformity at the base of the unit at the margins of the Clair Basin is consistent with the possible tectonic rejuvenation of the hinterland following Unit III.

Unit V: *Sandy sabkha, with sheetfloods and some aeolian reworking*

This unit forms the lower part of the perilacustrine stage of Allen and Mange-Rajetzky (1992). Although it consists predominantly of fluvial deposits, the presence of some aeolian sediment has improved reservoir quality. The top of Unit V was identified as a key marker horizon in a heavy mineral analysis carried out on well 206/08-8 (see later discussion) and is also commonly associated with a prominent seismic reflector (McKie and Garden, 1996). *Holoptychius* fish scales identified in core from Unit V in well 206/08-1A (Ridd, 1981) (Figure 50) are possibly related to the Late Devonian to Tournaisian age fish, *H. nobilissimus* (Ritchie et al., 1996).

Unit VI: *Lacustrine facies association*

The sediments associated with this unit are generally of low-energy fluvial or lacustrine type and form the upper part of perilacustrine stage of Allen and Mange-Rajetzky (1992). They include the fine-grained lake deposits of the Lacustrine Key Bed (LKB) (Figure 50), an important high gamma-ray and low interval velocity marker horizon on geophysical well logs. As lake levels rose and fell, some coarse-grained marginal sediments were interbedded with the open lake deposits.

Coney et al. (1993) focused upon Units II to VII of Allen and Mange-Rajetzky (1992), and proposed a sedimentary model based upon the following eight lithofacies types (with inferred sedimentary environment in *italics*):

1 Mudstone and siltstone with minor sandstone; *floodplains, including lakes or playas.*

2 Fine to very fine-grained sandstone with parallel and ripple cross-laminations; *peri-lacustrine sheet floods, or bar tops.*

3 Fine to very fine-grained sandstone, with interwoven ripple cross laminations and climbing ripples; *lacustrine shoreface and beach zones.*

4 Medium to fine-grained sandstone, bearing mud intraclasts, and exhibiting larger cross-sets than lithofacies 2; *sheet flows, crevasse splays and channel fills.*

5 Medium to very coarse-grained sandstone, with thick co-sets; *channel fills and bars.*

6 Conglomerates and pebbly sandstone; *channel lags and bar cores.*

7 Fine to medium-grained sandstone with wedge-shaped cross-sets; *aeolian dunes.*

8 Fine to very fine-grained sandstone with 'wispy' laminae of siltstone and mudstone; *interdune sabkha, aeolian and reworked lacustrine and fluvial deposits.*

Subsequently, McKie and Garden (1996) used the reservoir scheme of Allen and Mange-Rajetzky (1992) to identify stratigraphical cycles based on the changing proportion of fluviatile and non-fluviatile sediments within the Lower Clair Group (Units I-VI) (Figure 50). These cyclic variations were related to changes in accommodation space and sediment supply in the Clair Basin, and this in turn attributed to climatic variation, and to tectonic factors such as the uplift of the source area. The principles of high-resolution sequence stratigraphy formed the basis of some of this analysis. The Lacustrine Key Bed (in Unit VI) was interpreted as a lacustrine maximum flooding surface, while interfluvial palaeosols were associated with lowstand episodes.

In a recent summary and reappraisal of the Devonian succession, Nichols (2005) recognised seven principal facies in the Lower Clair Group (Units I-VI). In Unit I, thick beds of pebble conglomerate and coarse sandstone were interpreted as mass-flow deposits associated with a lacustrine fan delta. Better-stratified, upward-fining beds of conglomerates and sandstones are common in Units II, IV, V and VI. With their mudstone clasts and erosive bases, these were probably laid down in braided river channels. Interbedded within the same units are sandstones, up to 1 m thick, which were deposited on adjoining flood plains. These also contain signs of local redeposition in the form of ripped-up nodules and clasts. Thin siltstone and mudstone beds, with bedding disrupted by burrows, roots and calcrete nodules, provide evidence of palaeosol development on floodplains and lake margins. More finely laminated fine-grained sandstones, siltstones and mudstones accumulated from suspension in lakes. These pass laterally into slightly coarser beds of sandstone with common soft sediment deformation features, which were laid down as mouth bars associated with the lacustrine deltaic deposits.

In the Clair Basin, the Devonian sediments form part of a distributary fluvial system and are largely derived from a fan delta with a northern or north-western source (Nichols, 2005; Barr et al., 2007) (unlike the Orcadian Basin, where according to Mykura (1991) some of the basin fill has a southerly provenance). Clasts in the south of the basin are dominated by gneiss, schist and granitic igneous rocks, while in the north (Block 206/09) (Figure 48, see inset), quartz and quartzite predominate (Nichols, 2005).

In their lithostratigraphical analysis of the Clair Basin, Allen and Mange-Rajetzky (1992) incorporated data from a study of the heavy mineral assemblages of wells 206/08-2 (5 core samples), 206/08-4 (37 ditch cuttings, 12 core samples) and 206/08-7 (36 core samples) (Figure 48, see inset). They identified a group of chemically stable and ultrastable minerals, and used garnet:apatite, zircon:tourmaline, and apatite:tourmaline ratios, in conjunction with the morphology of apatite crystals (rounded, subrounded, prismatic or angular) as provenance and depositional process indicators. Although all the minerals are most likely to have been originally derived from high grade Lewisian and Moinian rocks, Allen and Mange-Rajetzky (1992) recognised both first cycle and polycyclic heavy mineral suites within the Clair Group. First cycle minerals were represented by hornblende, epidote, sphene, chlorotoid, micas, euhedral and subhedral zircon, euhedral tourmaline, rutile, apatite, angular non-etched garnet and staurolite. Polycyclic minerals were probably derived from pre-existing sediments, and consisted of rounded zircons, tourmaline, rutile, apatite, monazite, and the rest of the staurolite and garnet grains. In general, the Devonian part of the succession (the Lower Clair Group) was dominated by polycyclic minerals with good sorting and well-developed roundness.

Subsequent heavy mineral studies have focused upon the use of the technique as an aid to stratigraphical correlation in repetitive red-bed successions. An increase in the proportion of rounded apatite in Units III and V, for example, reflects a higher degree of aeolian influence and corresponding improved reservoir quality in those parts of the sequence. Recognition of significant facies and provenance variation between the heavy mineral contents of the stratigraphical units of Allen and Mange-Rajetzky (1992) has enabled the technique to be applied during the drilling of appraisal wells in the Clair Field (Morton et al., 2002b; 2003) (Figure 51c). This has proved useful in helping to locate the drill within the reservoir during the drilling of extended reach horizontal wells in the Devonian succession.

Faroe–Shetland Basin

Two isolated drilled occurrences of Devonian strata to the west of the Clair Field may form part of the northwest margin of the Clair Basin, but stratigraphical links between the basin flanks, across the deepest part of the Faroe–Shetland Basin, have yet to be confirmed by well or seismic data.

Westray High

Devonian sediments on the Westray High (Figure 48) are known only from well 204/19-1, which terminated in an approximately 190 m thick, sandstone-rich, red bed succession containing a range of Devonian miospores, including *Emphanisporites* sp., *Convolutispora* sp., and *Corbulispora* sp., together with tentatively identified *Ancyrospora* sp., *and Grandispora* sp. (Ritchie et al., 1996).

Corona High

Along strike from the Westray High, well 213/23-1 (Figure 48) proved a 314 m thick succession of undifferentiated Devonian sediments resting unconformably upon a westerly dipping tilted block composed of metamorphic basement forming the Corona High (Figure 52). The sediments consist of orange-stained, coarse to fine-grained sandstones interbedded with red-brown mudstone and siltstone (Figure 52c). Overlying these red beds are sediments of similar facies that contain some fossil evidence of an early Carboniferous age. As yet, there is no published correlation of the basal part of the succession with the long established stratigraphy of the Orcadian and Clair basins, where the Middle Devonian is bound by unconformities. In the Clair Basin, the basal units onlap the Rona High, while an unconformity at the top of Unit VI has been conjecturally proposed as the top of the Devonian (Allen and Mange-Rajetzky, 1992). The potential absence of both the top and the base of the equivalent interval in well 213/23-1, by non-deposition or erosion, hinders correlation of the repetitious Corona High succession with that of the Clair Basin. If the basal conglomeratic beds of well 213/23-1 are equivalent to the coarse-grained deposits of Unit I

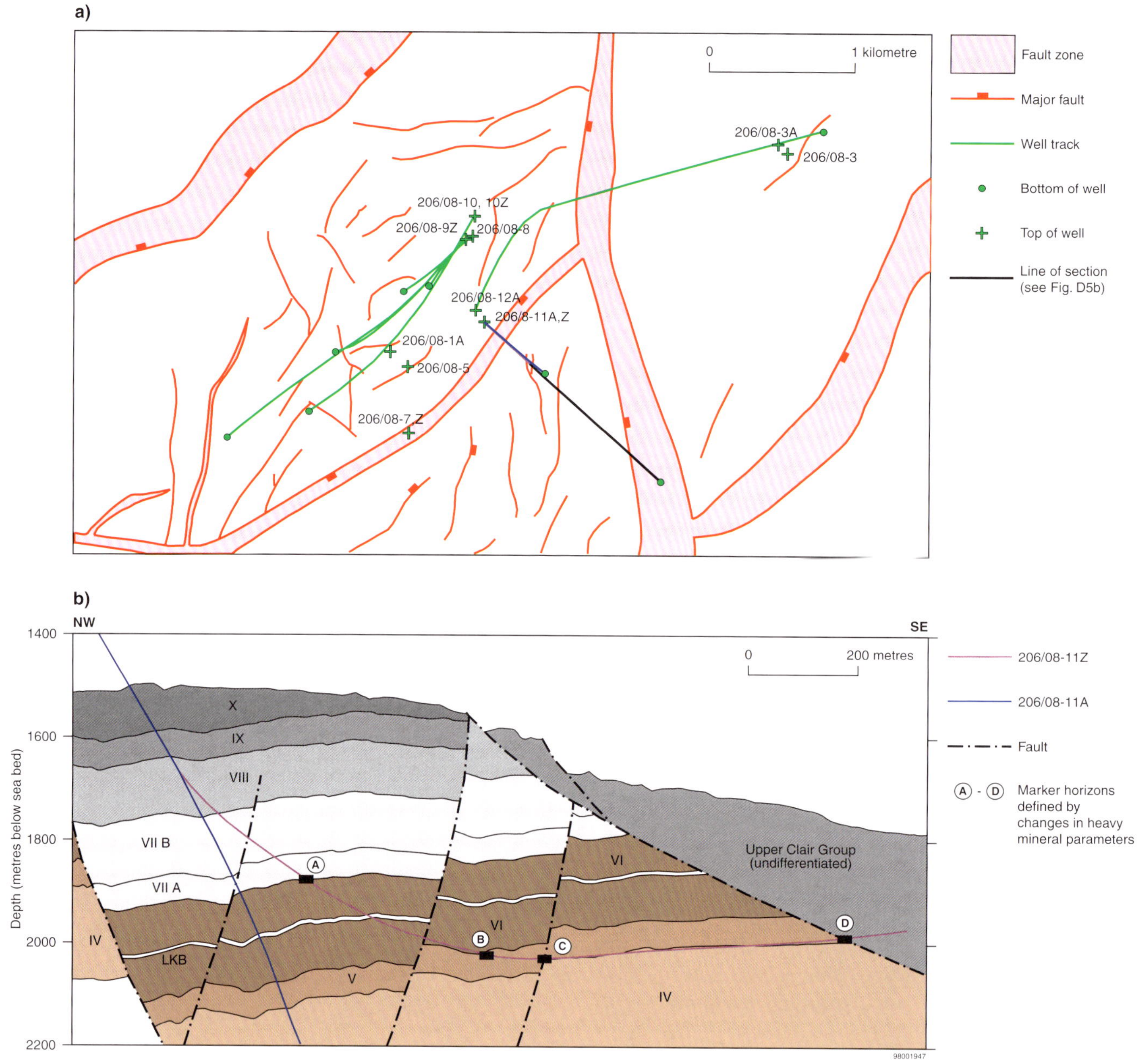

Figure 51a, b Heavy mineral analysis of the Devono-Carboniferous section in well 206/08-11Z, Clair Field; a) Location map of part of Clair Field area showing the track of 206/8-11Z and other deviated wells (modified from Morton et al., 2003, AAPG©2003, reprinted by permission of the AAPG whose approval is required for further use); b) Geological cross-section across part of the Clair Field, showing stratigraphical subdivisions of Allen and Mange-Rajetzky, (1992) (see Figure 50) and the track of deviated well 206/08-11Z (slightly modified reproduction from Morton et al., 2002 by permission of The Geological Society, London).

in the Clair Basin (Allen and Mange-Rajetzky, 1992) then the overlying Devonian succession, which coarsens upwards and is dominated by sandstones, might be equivalent to the mixed fluvial and aeolian deposits of the Lower Clair Group in the Clair Basin. Alternatively, the basal part of the Devonian in the Clair Basin could be absent from the Corona High, with the younger Devonian succession passing more conformably into the basal Carboniferous.

Devonian tectonics and basin development

Recent reviews of the post-orogenic structural setting of Devonian basins (Friend et al., 2000; Dewey and Strachan, 2003) have distinguished between tectonic models that emphasise the role of lithospheric extension linked to gravitational collapse of orogenically thickened crust (McClay et al., 1986; Norton et al., 1987), and those that consider that strike-slip tectonism to be one of the main controlling factors in Devonian basin formation (Larsen and Bengaard, 1991; Seranne, 1992; Coward, 1993; Dewey and Strachan, 2003). Simple models of lithospheric stretching were formerly supported by seismic interpretation of major half-graben features in the West Orkney Basin off the northern coast of Scotland (Figure 41). This basin was inferred to have originated by the extensional reactivation of Caledonian thrust faults during Devonian times (Enfield and Coward,

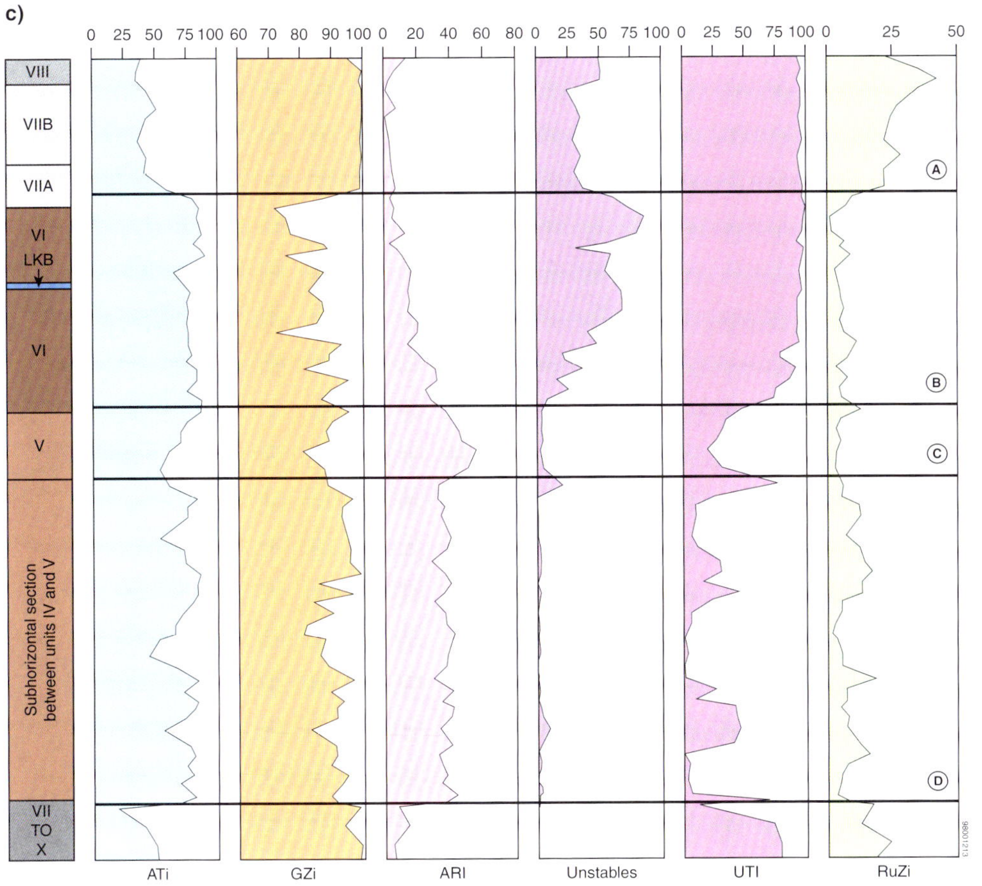

Figure 51c Graphs displaying the downhole variation in heavy mineral ratios in the Devono-Carboniferous section of well 206/08-11Z. A to D are marker horizons, which can be recognised during drilling (see Figure 51b), characterised by significant changes in heavy mineral parameters. Stratigraphical subdivision of drilled succession into units is based on subsequent well log interpretation. Key to heavy mineral ratios: ATi = apatite-tourmaline index (% apatite in total apatite + tourmaline); GZi = garnet-zircon index (% garnet in total garnet + zircon); ARI = apatite roundness index (% rounded apatite in apatite population); Unstables = % epidote + titanite, UTi = unstables-tourmaline index (% epidote + titanite in total epidote + titanite + tourmaline); RuZi = Rutile-zircon index (% rutile in total rutile + zircon) (modified from Morton et al., 2003, AAPG©2003, reprinted by permission of the AAPG whose permission is required for further use).

1987; Coward et al., 1989). However, subsequent drilling has confirmed that the West Orkney Basin is filled largely by Permo-Triassic sediments and formed as a result of latest Palaeozoic to Mesozoic extension.

Dewey and Strachan (2003) have argued that widespread sinistral transtension during the early Devonian followed the tectonic regime of sinistral transpression that dominated the end of the Caledonian Orogeny in late Silurian times. Following intense (Scandian) deformation in Scotland, wrench movement along faults and terrane boundaries parallel to the Great Glen Fault led to the formation of pull-apart basins at suitably oriented fault bends and offsets. Some areas of localised extension became sites of Devonian granitic magmatism (Jacques and Reavy, 1994). This period of basin formation preceded the Acadian tectonic event, which mostly affected large areas of the British Isles to the south of the Iapetus Suture (Soper, 1988; Soper et al., 1987; 1992; Soper and Woodcock, 2005).

Coward (1993) argued that the plate tectonic setting during Silurian and Devonian times can be envisaged as one in which a rigid indentor, consisting of the eastern Avalonian plate, caused sinistral movement along its western marginal contact with Laurentia, and dextral movement along its eastern margin with Baltica. Seranne (1992) suggested that the orientation of fold structures in the Walls Peninsula on Shetland was consistent with synsedimentary sinistral strike-slip faulting of the Walls Formation. Although evidence for the component of dextral transcurrent movement remains scanty, Coward's (1993) structural model is said to demonstrate the regional effects of escape tectonics, with an eastern extensional quadrant (mostly underlying the North Sea at the present day) forming an expelled fragment between the interacting plates. The Orcadian Basin, which lies in a strike-slip setting straddling the Great Glen and Walls Boundary faults, may have developed as a result of similar tectonic events persisting locally into Mid Devonian times.

Flinn (1961) first proposed significant transcurrent movement along the Walls Boundary Fault based on structural and stratigraphical discrepancies between the two sides of the fault, and the development of widespread cataclasis, onshore in Shetland. Since then, the offshore trace of the fault, and its history of transcurrent movement, has remained controversial (e.g. Flinn,

a)

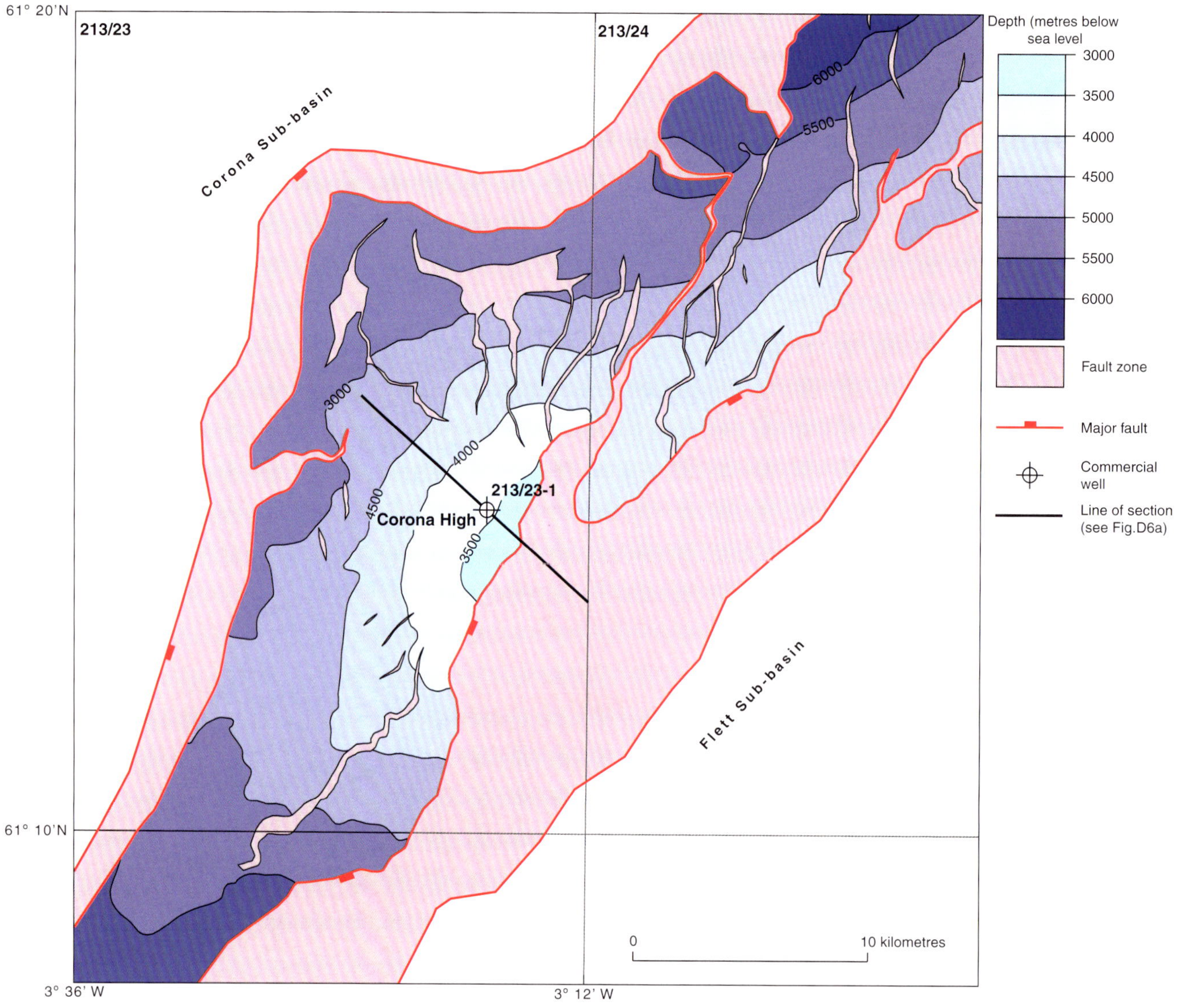
61° 20'N
213/23
213/24
Corona Sub-basin
6000
5500
3000
4000
4500
3500
213/23-1
Corona High
Flett Sub-basin
61° 10'N
0
10 kilometres
3° 36' W
3° 12' W
Depth (metres below sea level
3000
3500
4000
4500
5000
5500
6000
Fault zone
Major fault
Commercial well
Line of section (see Fig.D6a)

b)

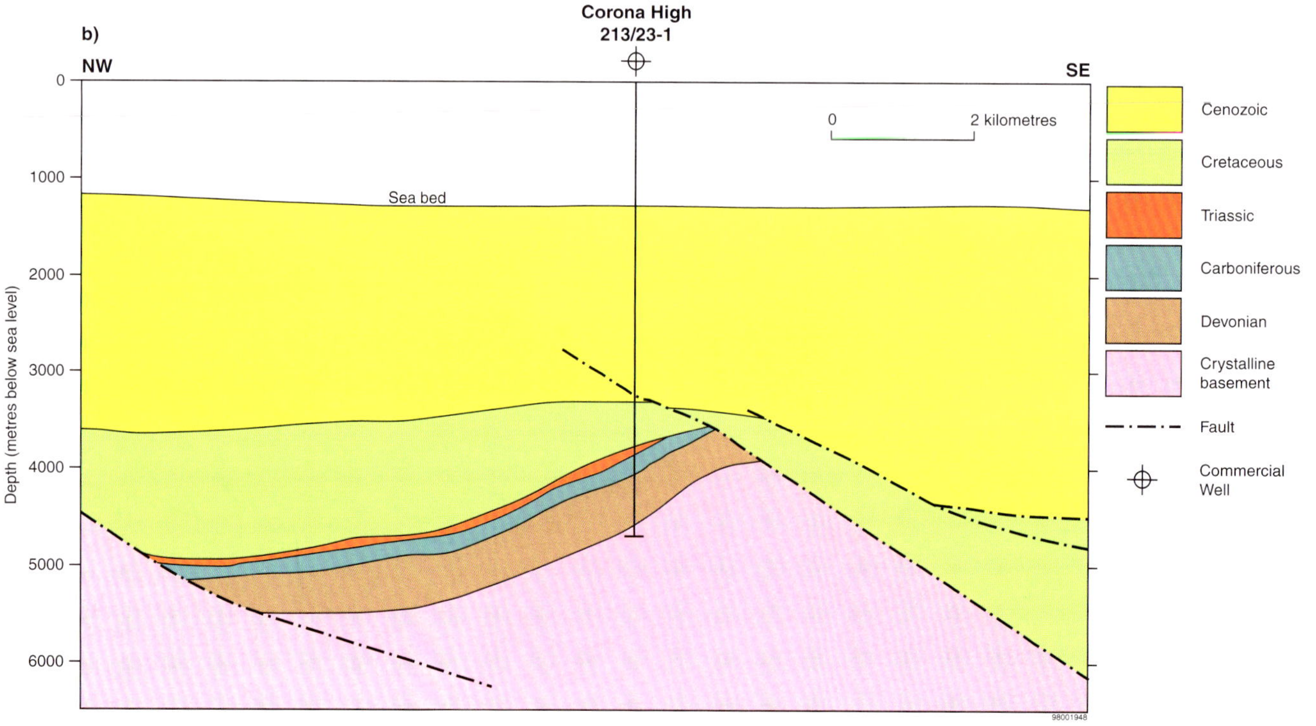
Corona High
213/23-1
NW
SE
0
1000
2000
3000
4000
5000
6000
Depth (metres below sea level)
Sea bed
0
2 kilometres
Cenozoic
Cretaceous
Triassic
Carboniferous
Devonian
Crystalline basement
Fault
Commercial Well

Figure 52 Devono-Carboniferous sediments on the Corona High (modified from IHS Energy well data release);
a) *(opposite)* Location of well 213/23-1 on the Corona High, with structural contours at Base Cretaceous level;
b) *(opposite)* Geoseismic section across the Corona High. For location, see Figure 52a;
c) Composite log of the Devono-Carboniferous section in well 213/23-1.

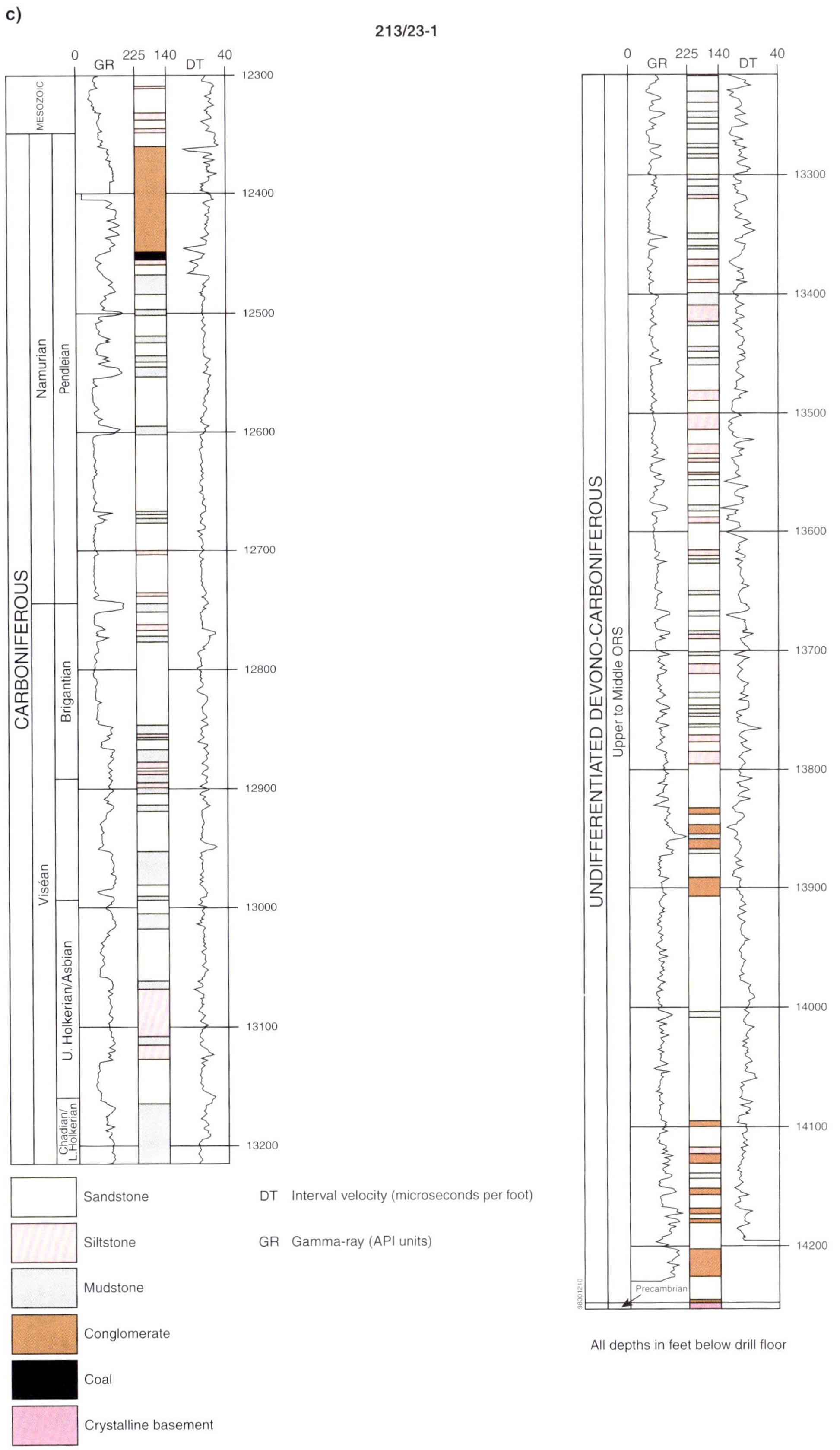

1992; Ritchie and Hitchen, 1993; Underhill, 1993, Watts et al., 2007).

Hydrocarbon exploration companies involved in the appraisal of the Clair Field carried out detailed studies of fault and fracture relationships in the Devonian red beds and Moinian basement of the north coast of Scotland and the Orkney Islands (Coney et al., 1993). Initial analysis revealed basement fractures with an average spacing of 1–2 /m, increasing to 10–30 /m in fracture 'corridors' and fault zones. Three scales of fracture corridor were recognised: regional fault zones (1 to1.5 km spacing), intermediate fault zones (100 to 200 m spacing) and small-scale faults (30–35 m spacing). Core studies in the Clair Field area show that the denser zones of basement fracturing extend into the overlying Devonian, but display a decreased fracture frequency. These structures have two main orientations; north-north-east-trending oblique-slip faults, which are parallel to the Rona High and lie close to the orientation of the Great Glen Fault, and north-east- to east-north-east-trending normal faults. Core inspection showed that while some associated veins were capable of bleeding oil, related microfaults and granulation seams were generally tight. In 1991, two appraisal wells (206/07a-2 and 206/08-8) established the presence of an open fracture network in the Clair Field, with fractures concentrated between the upper part of Unit IV and the basal part of Unit VII (Figure 53). Initial studies indicated that the open fractures show a different orientation to the other fractures, with two principal trends, north-north-west and west-north-west. The north-north-west-trending faults were thought to provide the main flow interval in the hydrocarbon reservoir and may have been kept open by the emplacement of hydrocarbons at the end of the Cretaceous (Coney et al., 1993). Subsequent core studies have shown that this fracture type is closely related to the lithology of the sedimentary matrix. Fractures are concentrated in the more competent units and 3D seismic data reveal they are closely aligned with Late Cretaceous faults, which vary in strike across the reservoir (Barr et al., 2007).

CARBONIFEROUS

Outside of the Midland Valley of Scotland, which marks the site of a former rift basin, sediments and volcanic rocks of Carboniferous age are only patchily preserved in Scotland and its adjoining offshore area (Fyfe et al., 1993; Stoker et al., 1993). During the Carboniferous, Scotland lay to the north of the closing Rheic Ocean and formed part of the foreland of the Variscan Orogeny. The effects of intra-Carboniferous folding, uplift and basin inversion are widely distributed in the Scottish part of the foreland, even though the most intense Variscan orogenic deformation in the British Isles is confined to the south of the Variscan Front, a major structural boundary transecting southern England, Wales and Ireland (Figure 47).

Orcadian Basin

Deposition of coarse-grained sediments of Old Red Sandstone facies in the Midland Valley of Scotland persisted across the Devono-Carboniferous boundary into Tournaisian times. Although similar successions have not been firmly identified within the onshore part of the report area, it is possible that small areas of red beds on Hoy, south-west Orkney, and nearby at Dunnet Head in Caithness, are also partly of Carboniferous age (Figure 48).

Clair Basin

Currently, the only proven Carboniferous sediments in the Faroe–Shetland report area occur offshore. These sediments are best known from exploration wells in the Clair Field, where Devonian strata are locally overlain by a more argillaceous, partly marine succession, which includes flora and fauna with Tournaisian and Viséan affinities (Meadows et al., 1987). Allen and Mange-Rajetzky (1992) placed a conjectural Devonian–Carboniferous boundary at the minor unconformity between Units VI and VII, which separates the Lower and the Upper Clair Group in the Clair Basin (Figure 50). The Upper Clair Group is everywhere truncated by the base Cretaceous unconformity (e.g. see Figure 84), but reaches up to 439 m in thickness in well 206/08-4. Although the Upper Clair Group is potentially all of early Carboniferous age, supporting biostratigraphical data have only been recovered from Units IX and X. This leaves open the possibility that Units VII and VIII may include sediments of Frasnian or Famennian age.

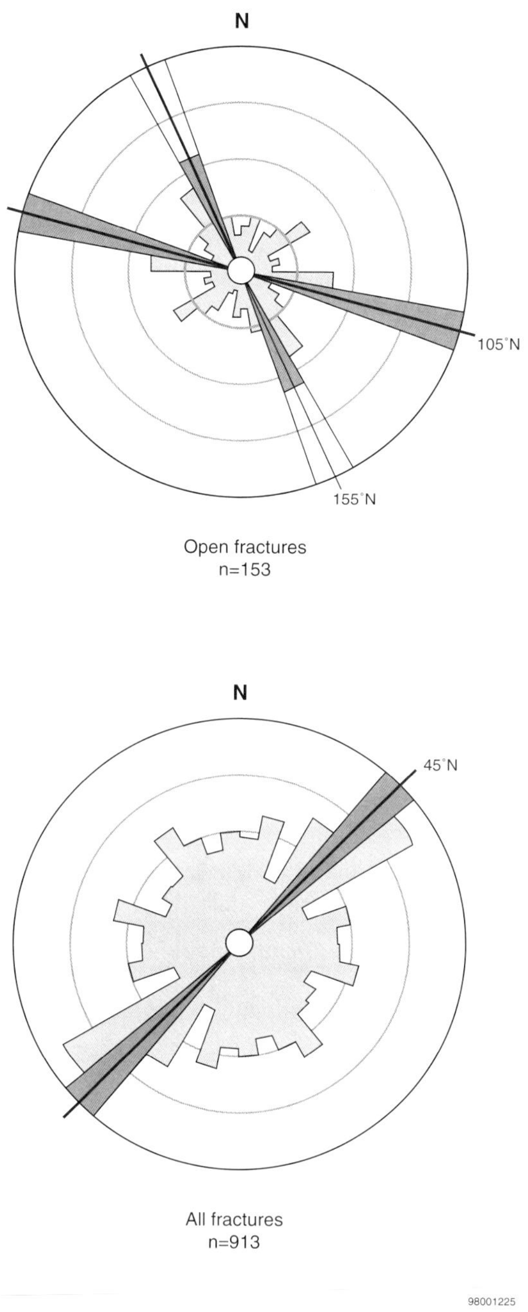

Figure 53 Fracture orientations measured in oriented core from the Devonian reservoir interval in well 206/08-8, Clair Field (modified from Coney et al., 1993). For location, see Figure 48.

UNIT VII: *Fluviatile*

Unit VII was deposited in a stable fluvial system with channel sandstone and basal lag sediments, and some finer-grained beds that show evidence of soil profile development. Allen and Mange-Rajetzky (1992) also distinguished between sub-units (VIIA and VIIB), with sub-unit VIIB forming a finer-grained upper interval.

UNIT VIII: *Fluviatile*

Fluvial deposition largely in sand bars continued in Unit VIII, which is mostly composed of cross-bedded sandstone.

UNIT IX: *Fluviatile braidplain*

Gravels and coarse-grained to pebbly fluvial sandstone characterise Unit IX, which onlaps the basin margin in places. A lack of calcrete, and the presence of coaly debris in these sandstones probably points to the development of a rising water table during this interval. Fauna in core at 1608.2 m measured depth below the drillfloor in well 206/08-1A (Figure 50) indicates that the z coral–brachiopod zone of late Tournaisian age occurs within Unit IX (Ritchie et al., 1996). This is supported by biostratigraphical evidence from well 206/08-4, where spores indicate a latest Tournaisian age at 2120.2 m measured depth below the drillfloor (Ritchie et al., 1996).

UNIT X: *Interdistributary bay*

Unit X forms the marginal marine stage of Allen and Mange-Rajetzky (1992). The presence of acanthomorph acritarchs in Unit X (from 1338 to1365 m depth) in well 206/08-2 (Meadows et al., 1987) supports a marine interpretation, and contemporaneous miospores indicate an early Visean age (c2s1-s2 coral–brachiopod zone).

In the heavy mineral analysis of the Clair reservoir succession carried out by Allen and Mange-Rajetzky (1992), the Upper Clair Group is characterised by assemblages containing euhedral forms of the stable minerals together with abundant unstable species. This implies a first-cycle source for the Carboniferous strata, in contrast to the older Devonian part of the succession.

In terms of chronostratigraphy of the Carboniferous, all these sediments are assigned to the early Visean Stage (Figure 6). The occurrence of Carboniferous strata with marine affinities in the Clair Basin clearly poses a problem for any palaeogeographical reconstructions that place Scotland in the middle of a huge continental landmass throughout the Late Palaeozoic (e.g. Coward et al., 2003). Even though remnants of cover on adjoining upland areas such as the Southern Uplands suggest that Carboniferous sediments were formerly more widespread on the uplifted flanks of the Midland Valley, simplified palaeogeographical maps of the British Isles commonly depict a fully emergent, eroding hinterland across the north of Scotland and adjoining shelves throughout the Carboniferous.

Faroe–Shetland Basin

Well 213/23-1 drilled on the Corona High established for the first time that sediments of early Namurian age are preserved in Faroe–Shetland area (Figure 52c). These sediments belong to the Pendleian, a local substage of the Namurian (Figure 6). On the Corona High, 120 m of Namurian sandstone, conglomerate and mudstone, with traces of coal, overlie a 142 m thick succession of older Carboniferous (Viséan) strata, which may correspond with that in the Clair Field area. An equivalent Namurian interval has not been recognised in the Clair Basin. The nearest successions with which the Namurian sediments can be correlated occur in the Midland Valley of Scotland, onshore in Ireland, and on parts of the Irish continental shelf.

Carboniferous tectonics and basin development

An early synthesis of Late Palaeozoic basin development in the British Isles proposed that east–west extension initiated large scale north-trending fractures related to the distribution of Carboniferous metalliferous mineralisation (Russell, 1968; 1972). Subsequently, this tectonic proposal was linked to an evolving hypothesis about the development of a proto-Atlantic rift in the Rockall area during the Carboniferous (Russell and Smythe, 1978; 1983; Russell, 1987; Smythe et al., 1995). Independently, Haszeldine (1984) collated regional stratigraphical evidence for early rifting of the Pangaean continent west of the British Isles, before reaching a similar conclusion that widespread dyke intrusion preceded the formation of late Carboniferous oceanic crust between Spain and the Faroes. As part of the tectonic scheme, it was envisaged that the proto-North Atlantic Ocean was linked northwards, through the conjectural rift system, to the Boreal Ocean.

In contrast, Leeder (1988) accounted for the limited evidence that requires Carboniferous marine deposition to the west of Shetland by postulating a southern marine connection along the north-east-trending structures that transect the continental shelf to the west of Britain. These structures formed parallel to the Great Glen Fault at the end of the Caledonian Orogeny, and possibly link with similar faults in Maritime Canada (Bradley, 1982), which have a similar tectonic history during the Palaeozoic. In this alternative model, basin formation and deformation during the Carboniferous, in both Canada and the British Isles, are linked to the effects of dextral shear, which have been recognised elsewhere within the Variscan foreland (Badham, 1982; Read, 1988; 1990; Serranne, 1992; Rippon et al., 1996). Recently, De Paola et al. (2005) have shown that the effects of partitioned dextral transtension can explain many of the features of folding and basin inversion in southern Scotland and northern England that were formerly attributed to the distant compressional effects of Variscan orogenesis. In their account, the distribution of this deformation can be commonly linked to structures inherited from previous orogenies.

5 Permian and Triassic

Martyn Quinn[1] and Heri Ziska[2]
[1] *British Geological Survey*
[2] *Jarðfeingi (Faroese Earth and Energy Directorate)*

The suturing of the northward migrating megacontinents of Gondwana and Laurasia, in late Carboniferous to early Permian times, resulted in the consolidation of the Variscan and Appalachian fold belts and formation of the supercontinent of Pangaea (Figure 10b). Pangaea was inherently unstable (Ziegler, 1990), and throughout Permo-Triassic times was subjected to repeated rifting that mainly focused, in the Faroe–Shetland region, on the existing north-east Caledonoid structural grain (Glennie, 2002; Figure 7). This rifting, together with eustatic rises in sea level, resulted in southward and northward transgressions of the Arctic and proto-Atlantic seas, respectively. Permian and Triassic sedimentation patterns and igneous activity within the Faroe–Shetland region reflect the break-up of the Pangaean supercontinent. Equatorial climatic conditions prevalent during the Carboniferous gave way to a hot and arid climate in Permo-Triassic times, caused by the northward migration of the Laurussian and Gondwana megacontinents, with aridity in particular a key influence in the determination of the environments of deposition and lithologies preserved in Permo-Triassic successions (Herries et al., 1999).

STRATIGRAPHY

Within the Faroe–Shetland report area, the age of much of the Permo-Triassic succession is very poorly constrained due to the absence of, or nondiagnostic nature of any contained fauna and flora (Figures 54 and 55). Consequently, meaningful subdivision and correlation of the sedimentary successions across the region has proved challenging. The first lithostratigraphical nomenclature scheme for the Permo-Triassic succession in the Faroe–Shetland area was established by Ritchie et al., 1996, though the scheme adopted for this report also partly reflects a subsequent study of Herries et al. (1999) (Figure 56). However, due to the paucity of wells that proved Permo-Triassic rocks within the area, these stratigraphical correlations should be treated with some caution.

Permian

With regard to the Permian succession, the terms Rotliegend Group and Zechstein Group are utilised throughout the Faroe–Shetland report area (Figure 56). Within these groups however, there is substantial lithological variation between the different formations that have been defined in the north-east part of the report area around the East Shetland High and Møre Basin (Cameron, 1993b), when compared with those much further to the south-west within the Papa and West Orkney basins (Figures 54, 56 and 57; e.g. Ritchie et al., 1996; Herries et al., 1999).

Triassic

Within the north-east part of the report area, the Triassic succession around the East Shetland High and Møre Basin is assigned to the Heron Group (Cameron, 1993b) whereas to the south-west in the West Orkney, Papa, West Shetland, East Solan and West Solan basins, and on the Solan Bank and Westray Highs, it is assigned to the Papa Group (Figures 54 and 56; e.g. Ritchie et al., 1996; Herries et al., 1999). In the East Solan Basin, the scheme of Ritchie et al. (1996) has been slightly modified, with for example, the Shoal High and Otter Bank formations being subdivided and renamed the Shoal High, Otter Bank Sandstone and Otter Bank Shale formations (Figure 56). It should also be noted that the informal term Westray High Limestone Formation has been introduced here to represent a distinctive unit originally included at the top of the F2 subdivision of the Foula Formation. For example, wells 204/19-1 and 204/19-9 located on the Westray High penetrated an Upper Triassic succession that included limestone intervals, suggesting deposition in a marine environment at this time.

DISTRIBUTION

Permo-Triassic strata are inferred to have a widespread distribution within the Faroe–Shetland report area, though they are best defined within east and south-eastern parts where they have been proved by a number of commercial wells and BGS boreholes (Figure 54). In the West Orkney, North Lewis, Papa, St Magnus Bay, Fetlar and West Fair Isle basins, and parts of the Unst Basin, Permo-Triassic rocks subcrop the sea bed though elsewhere, they lie beneath thick Mesozoic and/or Cenozoic strata. The considerable variation in the proven thickness of the Permo-Triassic succession throughout the report area (Figure 55) probably reflects to some extent, the effects of localised and regional uplift and erosion during Mid to earliest Late Jurassic (e.g. Booth et al., 1993) and Cenozoic (Smallwood and Gill, 2002) times. The present day distribution is considered to represent a remnant of a once very extensive Permo-Triassic sedimentary cover (Booth et al., 1993; Swiecicki et al., 1995; Glennie, 2002). A more detailed summary of the distribution and fill of the Permo-Triassic in basins and on highs within the Faroe–Shetland region is given below.

Møre Basin and East Shetland High

The north-east-trending Møre Basin straddles the UK/Norwegian median line in the far north-east of the report area (Figure 54) and is mainly defined by its Cretaceous fill (e.g. Brekke et al., 1999). Although Permo-Triassic strata have not been proven within the UK part of this basin, it is likely that they are present at depth. For example, a Permo-Triassic succession has been drilled within a south-east-dipping half-graben close to the south-east flank of the Møre Basin (Figure 23). Here, well 220/26-2 proved a 43.5 m thick succession of Lower Permian fine-grained, generally nonporphyritic

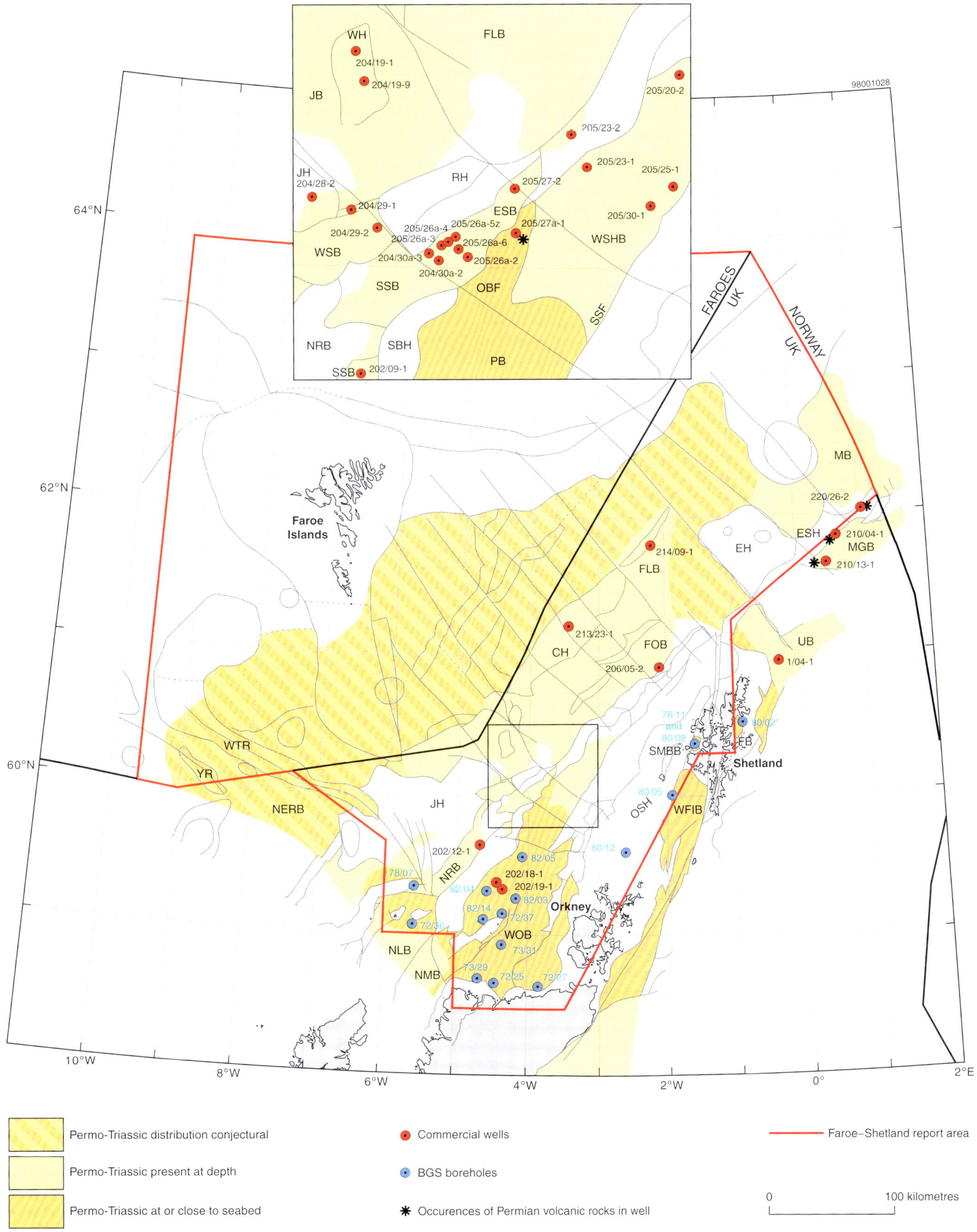

Figure 54 Distribution of Permo-Triassic strata within the Faroe–Shetland report area. Present-day structural elements (grey) derived from Figure 7. Abbreviations: see Figure 7.

trachyandesitic lava (e.g. Hitchen et al., 1995b), volcaniclastic sandstone and siltstone of the Margarita Volcanics Formation, overlain by 381 m of upper Permian anhydrite, limestone, dolomite and siltstone of the Turbot Anhydrite and Bosies Bank formations (Figures 55a, 56 and 57). The upper Permian succession is succeeded by a 25.5 m thickness of Lower Triassic mudstone and 546 m of Middle to Upper Triassic fine to granular sandstone with siltstone, which are assigned to the Otter Bank Shale and Cormorant formations, respectively. Well 210/04-1, drilled close to the boundary of the Magnus Basin and East Shetland High (Figure 54), proved 305 m of probable Lower Permian sandstone and subordinate highly altered alkaline basalt-trachyte (Stoker et al., 1993), 206 m of upper Permian carbonate, mudstone and anhydrite and 407 m of undivided Permo-Triassic sandstone unconformably overlain by Upper Cretaceous shale. Well 210/13-1, also drilled on the East Shetland High, penetrated 324 m of Triassic sandstone, conglomeratic sandstone, mudstone and siltstone containing fragments of lava of likely Permian age. The lava fragments may represent fairly acid volcanic rocks of dacitic or trachyandesitic affinity (Hitchen et al., 1995b).

From comparisons with Permo-Triassic successions described from East Greenland and in the northern North Sea, Hamar and Hjelle (1984) postulated that up to 3.5 km of Permo-Triassic rocks might be present within the part of the Møre Basin located in the Norwegian sector. Support for this inference is derived from the results of Norwegian well 6305/12-1 located in the north-east part of the Møre Basin in the Slørebotn Sub-basin, which penetrated more than 200 m of Triassic sediments comprising conglomerate, pebbly sandstone, siltstone and mudstone within which the well reached terminal depth (Jongepier et al., 1996). From the results of seismic interpretation within the sub-basin, Jongepier et al. (1996) estimate the Triassic may attain a thickness of at least 600 m.

Faroe–Shetland Basin

The presence of Permo-Triassic rocks within the Faroe–Shetland Basin is largely inferred from seismic profiles (e.g. Mudge and Rashid, 1987; Lamers and Carmichael, 1999; Roberts et al., 1999; Keser Neish, 2003). However, wells drilled on the Westray High, Corona High and the south-east part of the Foula Sub-basin have proved relatively thin successions of Triassic strata (Figures 54 and 55a). On the Westray High, well 204/19-1 penetrated a 205 m thick succession of sandstone, mudstone and thin limestone before terminating within Devonian sandstone, whereas 204/19-9 proved an unbottomed 128 m thick succession of sandstone and mudstone that included a 15 m thick unit of limestone of possible late Rhaetian to Sinemurian age (Figures 54 and 55a). Well 213/23-1 drilled on the Corona High recorded 177 m of fine to very coarse-grained, poorly sorted sandstone and minor noncalcareous mudstone of possible Triassic age before reaching terminal depth within Devonian sandstone (Figure 52b). Also on the Corona High, well 214/09-1 proved 73 m of mudstone, siltstone, sandstone and gneissic conglomerate of possible Permo-Triassic age resting unconformably on metamorphic basement (Figure 33). Within the Foula Sub-basin, well 206/05-2 terminated within 224 m of fine to very coarse-grained, poorly sorted sandstone interbedded with mudstone of possible Triassic age. Towards the north-west part of the Faroe–Shetland Basin, Palaeogene lavas and intrusive rocks partially mask the deep structure of the basin and the presence of Permo-Triassic rocks is largely conjectural (Figure 54; Keser Neish, 2003).

Marginal basins

West Orkney Basin

Evidence from seismic reflection data is interpreted by some to suggest that the West Orkney Basin comprises a series of half-grabens, possibly formed by relaxation of former Caledonian thrusts (Figures 41 and 54; e.g. Stoker *et al.* 1993, Hitchen *et al.* 1995b). These half-grabens are bounded by major north-east-trending, south-east-dipping faults, with north-west-thickening wedges of Permo-Triassic developed within their hanging-wall blocks. Triassic sediments occur at or near sea bed across the basin and lithologies from wells and boreholes indicate palaeo-environments were nonmarine (Figure 55b). Towards the east of the basin, the Permo-Triassic thins, with Devonian strata closer to sea bed (Figure 41; e.g. Earle et al., 1989). Wells 202/18-1 and 202/19-1 both drilled almost 3 km of upper Permian and Triassic strata that subcrops the Quaternary in the deepest and most north-westerly of the half-graben (Figures 41, 54 and 55b). However, between 7.5 km (Earle et al., 1989) and 10 km (Enfield and Coward, 1987) of Permo-Triassic (and ?Devonian) strata may be present in places within the hanging wall of the main basin-bounding fault with the Solan Bank High. BGS boreholes within the basin generally record red sandstone and occasional mudstone (Figure 55b). Well 202/19-1 terminated within the upper Permian West Orkney Evaporite Formation, which consists of 1432 m of anhydritic mudstone and mudstone interbedded with varying proportions of halite, anhydrite and sandstone (Figure 57). This succession is overlain by the Lower Triassic Shoal High Formation, and comprises 598 m of interbedded sandstone, mudstone and siltstone that also includes some thick individual beds of these lithologies (for example a 79 m thick bed of siltstone and 38 m thick bed of sandstone). The sandstone is fine to very fine-grained, and moderately to well-sorted. The mudstone is occasionally silty or dolomitic, whereas the siltstone is very micaceous and slightly calcareous. The Shoal High Formation is succeeded by 902 m of the Otter Bank Sandstone and Foula formations that range in age from Early to Late Triassic. These formations mainly consist of sandstone, grading in places to conglomerate and breccia-conglomerate with sporadic thin

Figure 55 *(opposite)* Stratigraphical range chart showing generalised lithology and thickness of Permo-Triassic rocks within a) the Judd, Rona, Solan Bank, Orkney–Shetland and East Shetland highs and the Faroe–Shetland, Fetlar, Foula, North Rona, St Magnus Bay, West Fair Isle, West Solan and West Shetland basins; b) *(overleaf)* the Outer Hebrides High and West Orkney, Papa, East Solan and North Lewis basins. Timescale is based on 'A Geologic Time Scale 2004' by F M Gradstein, J G Ogg, A G Smith, et al. (2004) and the International Stratigraphical Chart, 2006 (ICS) with additions. For location of wells and BGS boreholes, see Figure 54.

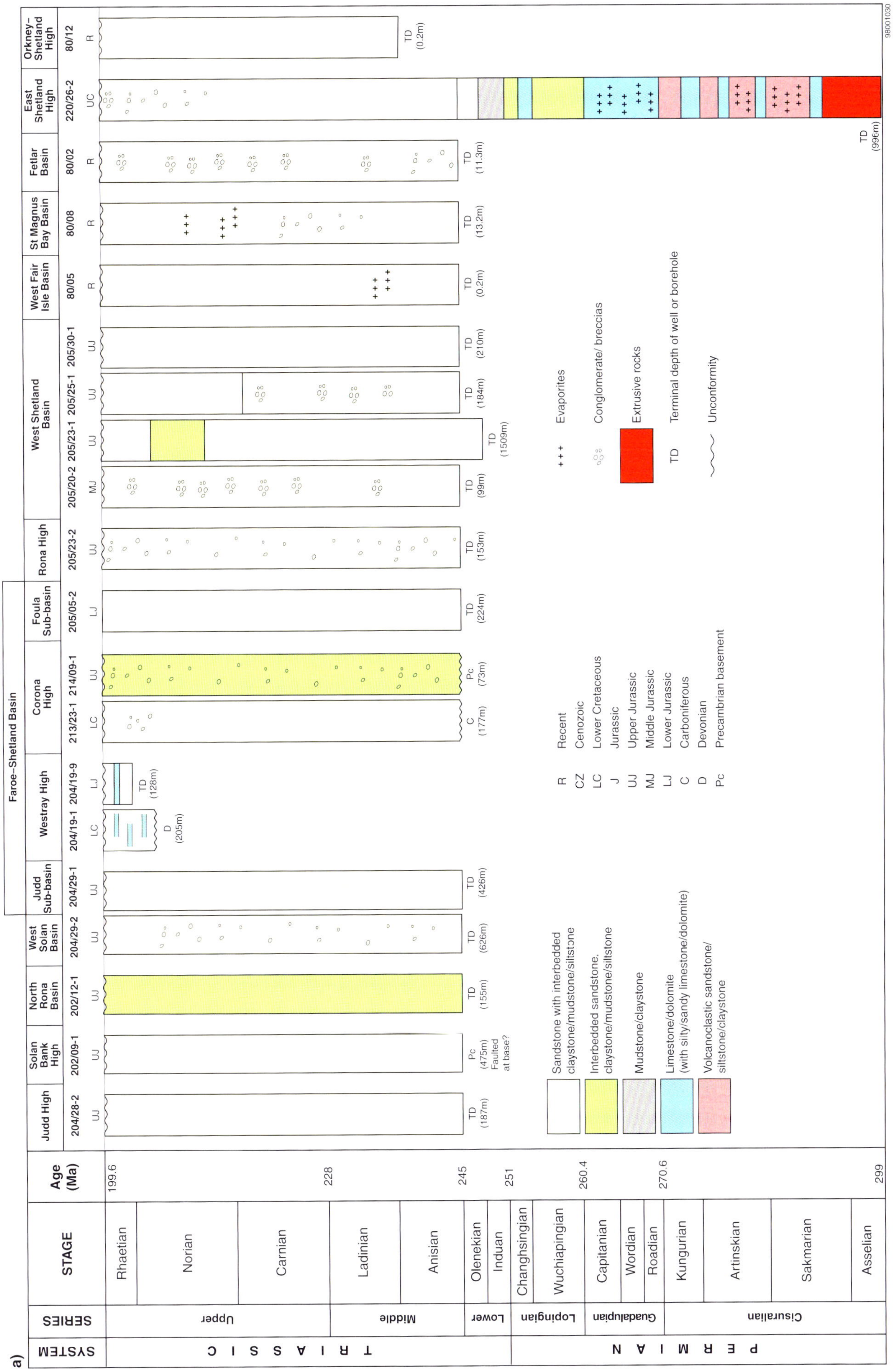

a)
SYSTEM
SERIES
STAGE
Age (Ma)
TRIASSIC
PERMIAN
Upper
Middle
Lower
Lopingian
Guadalupian
Cisuralian
Rhaetian
Norian
Carnian
Ladinian
Anisian
Olenekian
Induan
Changhsingian
Wuchiapingian
Capitanian
Wordian
Roadian
Kungurian
Artinskian
Sakmarian
Asselian
199.6
228
245
251
260.4
270.6
299
Faroe–Shetland Basin
Judd High
204/28-2
Solan Bank High
202/09-1
North Rona Basin
202/12-1
West Solan Basin
204/29-2
Judd Sub-basin
204/29-1
Westray High
204/19-1
204/19-9
Corona High
213/23-1
214/09-1
Foula Sub-basin
205/05-2
Rona High
205/23-2
West Shetland Basin
205/20-2
205/23-1
205/25-1
205/30-1
West Fair Isle Basin
80/05
St Magnus Bay Basin
80/08
Fetlar Basin
80/02
East Shetland High
220/26-2
Orkney–Shetland High
80/12
TD (187m)
Pc (475m) Faulted at base?
TD (155m)
TD (626m)
TD (426m)
D (205m)
TD (128m)
C (177m)
Pc (73m)
TD (224m)
TD (153m)
TD (99m)
TD (1509m)
TD (184m)
TD (210m)
TD (0.2m)
TD (13.2m)
TD (11.3m)
TD (996m)
TD (0.2m)
Sandstone with interbedded claystone/mudstone/siltstone
Interbedded sandstone, claystone/mudstone/siltstone
Mudstone/claystone
Limestone/dolomite (with silty/sandy limestone/dolomite)
Volcanoclastic sandstone/ siltstone/claystone
R Recent
CZ Cenozoic
LC Lower Cretaceous
J Jurassic
UJ Upper Jurassic
MJ Middle Jurassic
LJ Lower Jurassic
C Carboniferous
D Devonian
Pc Precambrian basement
+++ Evaporites
Conglomerate/ breccias
Extrusive rocks
TD Terminal depth of well or borehole
Unconformity
98001030

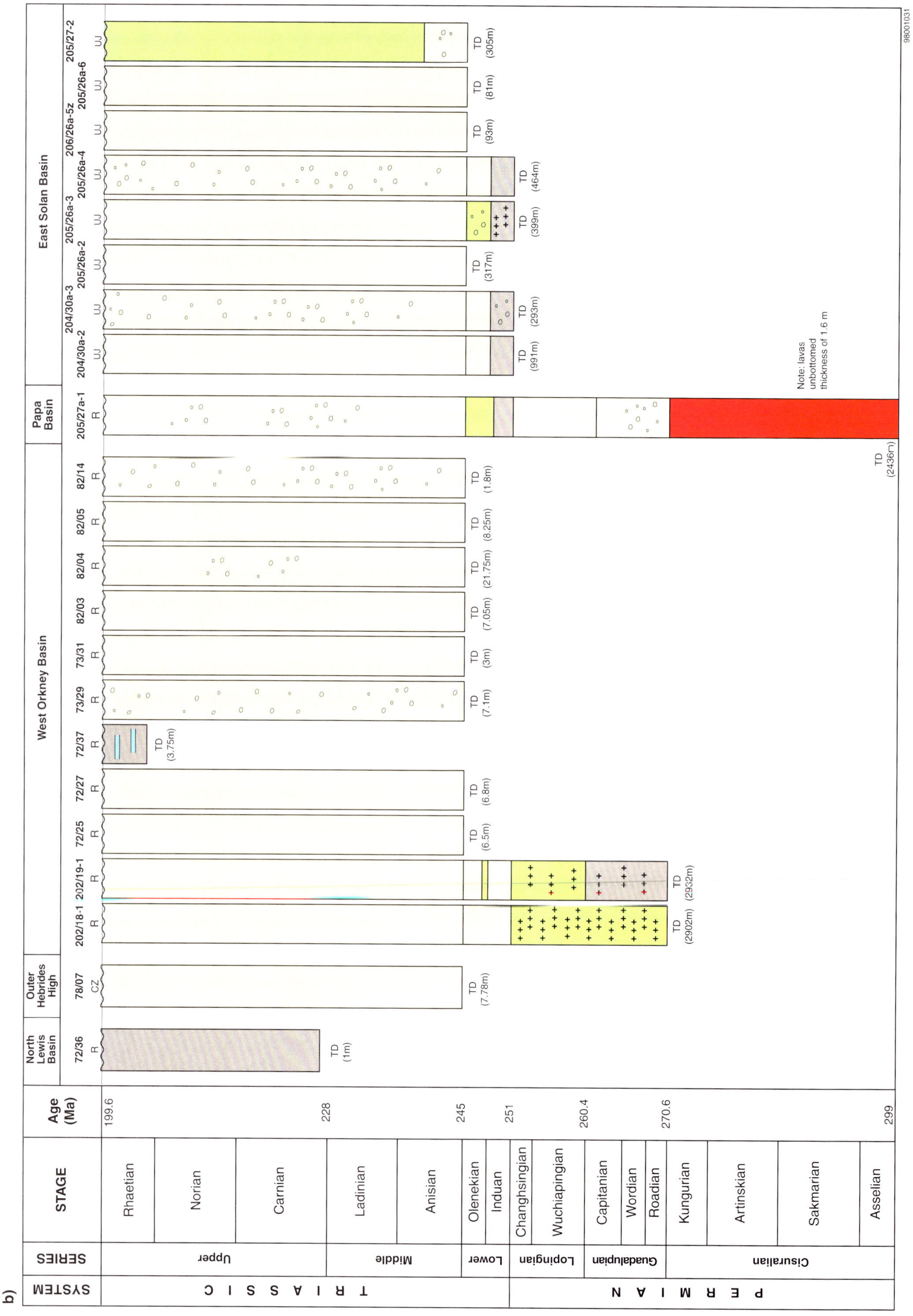
b)
SYSTEM
SERIES
STAGE
Age (Ma)
TRIASSIC
PERMIAN
Upper
Middle
Lower
Lopingian
Guadalupian
Cisuralian
Rhaetian
Norian
Carnian
Ladinian
Anisian
Olenekian
Induan
Changhsingian
Wuchiapingian
Capitanian
Wordian
Roadian
Kungurian
Artinskian
Sakmarian
Asselian
199.6
228
245
251
260.4
270.6
299
North Lewis Basin
Outer Hebrides High
West Orkney Basin
Papa Basin
East Solan Basin
72/36
78/07
202/18-1
202/19-1
72/25
72/27
72/37
73/29
73/31
82/03
82/04
82/05
82/14
205/27a-1
204/30a-2
204/30a-3
205/26a-2
205/26a-3
205/26a-4
206/26a-5z
205/26a-6
205/27-2
TD (1m)
TD (7.78m)
TD (2902m)
TD (2932m)
TD (6.5m)
TD (6.8m)
TD (3.75m)
TD (7.1m)
TD (3m)
TD (7.05m)
TD (21.75m)
TD (8.25m)
TD (1.8m)
TD (2436m)
TD (991m)
TD (293m)
TD (317m)
TD (399m)
TD (464m)
TD (93m)
TD (81m)
TD (305m)
Note: lavas unbottomed thickness of 1.6 m
98001031

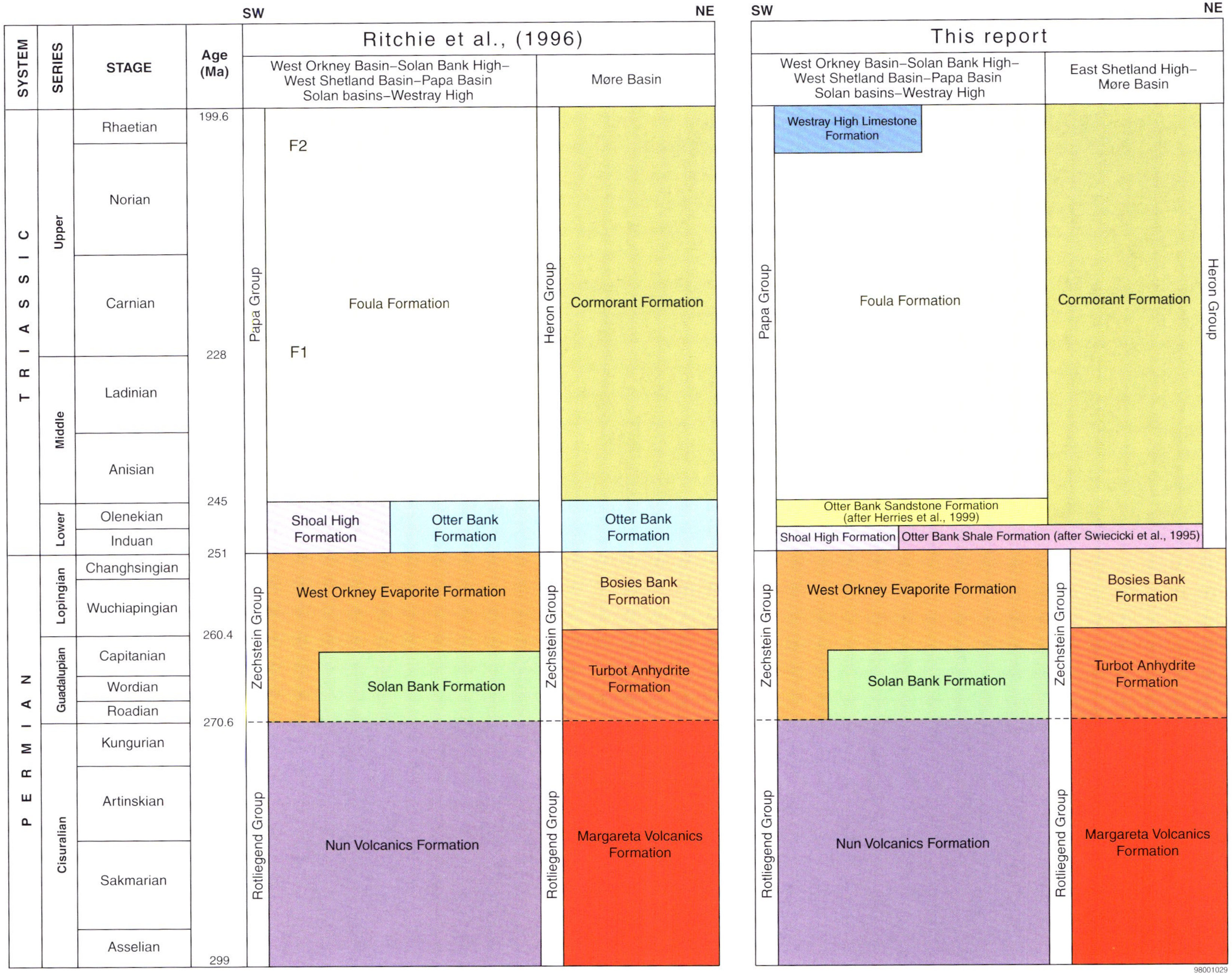

Figure 56 Permo-Triassic lithostratigraphical nomenclature for the Faroe–Shetland report area. Timescale is based on 'A Geologic Time Scale 2004' by F M Gradstein, J G Ogg, A G Smith, et al. (2004) and the International Stratigraphical Chart, 2006 (ICS) with additions.

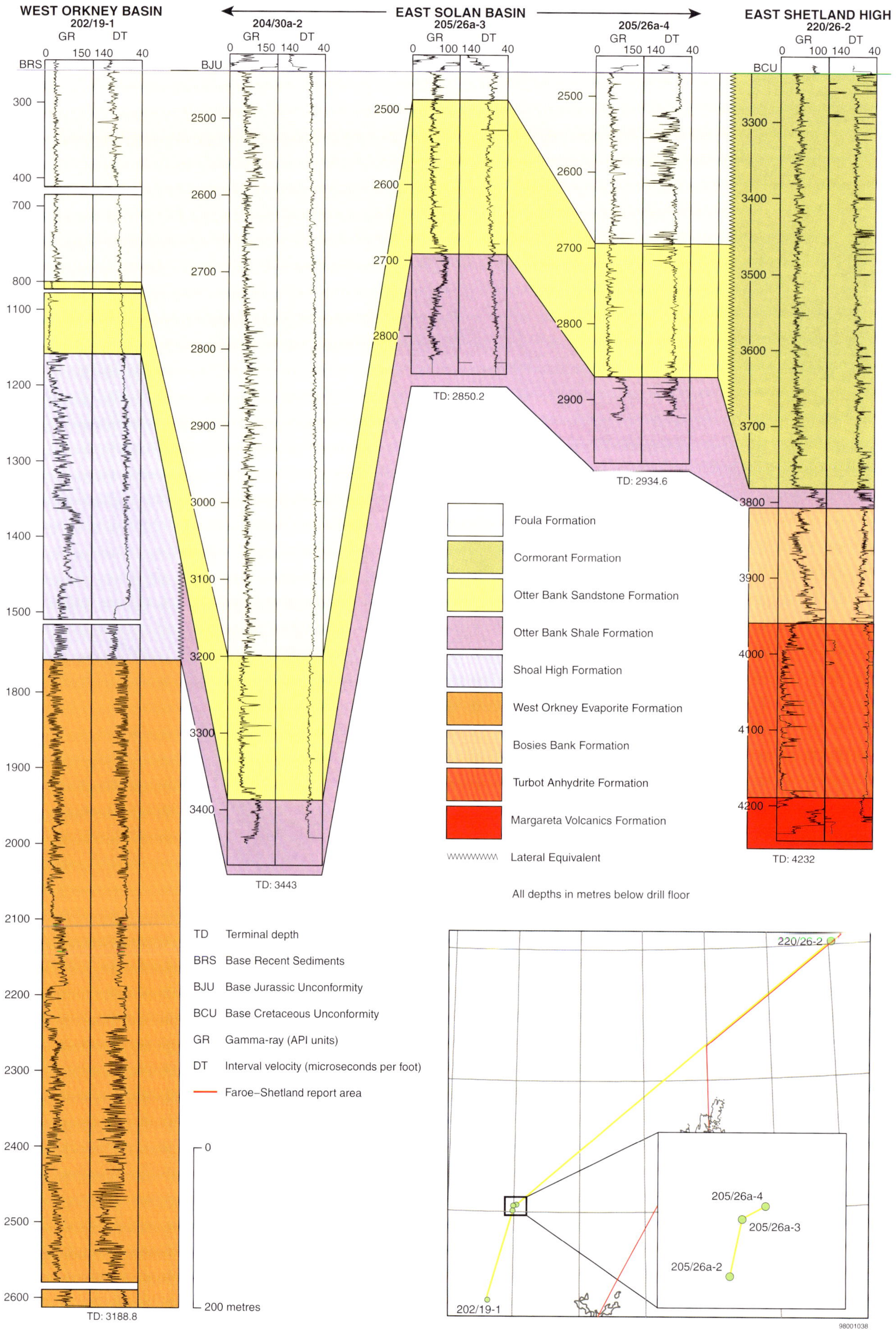

Figure 57 Well log correlation of the Permo-Triassic in wells 202/19-1, 204/30a-2, 205/26a-3, -4 and 220/26-2 within the West Orkney Basin, East Solan Basin and East Shetland High.

silty mudstone and calcareous siltstone. The transition between the Otter Bank Sandstone and Foula formations is taken at a significant increase in the gamma-ray log response, reflecting the more mature lithology of the former unit (Figure 57; Ritchie et al., 1996).

East, South And West Solan and North Rona Basins

The East Solan Basin lies to the south-east of the Faroe–Shetland Basin (Figure 54) and is economically important as it contains hydrocarbon accumulations in the Upper Jurassic Solan Sandstone Member and Lower Triassic Otter Bank Sandstone Formation within the Solan and Strathmore discoveries respectively (see Chapter 12; Herries et al., 1999). Due to exploration interest, the East Solan Basin has the highest concentration of wells that have penetrated the Triassic in the report area (Figures 39 and 55b). A maximum thickness of nearly 1 km of Triassic mudstone, sandstone and siltstone correlated with the Foula, Otter Bank Sandstone and Otter Bank Shale formations has been drilled in well 204/30a-2 (Figure 57). The Otter Bank Sandstone has a mature detrital mineralogy and this, coupled with heavy mineral analyses, suggests Lewisian affinities and a possible Permian or Devonian source (Herries et al., 1999). In contrast, the overlying Foula sandstone is comprised of immature, first-cycle detritus with a heavy mineral assemblage not seen in older rocks of the British Isles. This suggests that the Foula Formation was derived from a high-grade metamorphic terrane. Palaeocurrent analysis indicates that this source would have lain to the north, between the Rona Ridge and East Greenland (Herries et al., 1999).

Drilling within the West Solan Basin has proved a maximum of 626 m of possible Middle and Upper Triassic Foula Formation sandstone, mudstone and conglomerate in well 204/29-2 (Figures 54, 55a and 56). On the Solan Bank High and in the North Rona Basin, wells 202/09-1 and 202/12-1 proved 475 m and 155 m of Middle to Upper Triassic Foula Sandstone Formation, respectively, with the former succession resting on a possible faulted contact with Lewisian basement. It is likely that the wedge-shaped geometry of the Permo-Triassic succession observed on seismic profiles within this complex of basins (e.g. Figure 28) represents the remains of a larger and more extensive basin that has subsequently been geographically separated by post-Triassic local and regional uplift and erosion (Booth et al., 1993; Herries et al., 1999). For example, to the west of the Otter Bank Fault (Figures 54), Triassic sediments subcrop younger Mesozoic in the East, West and South Solan basins (e.g. Figure 39) whereas to the east, in the Papa Basin, the Triassic succession subcrops Quaternary sediments. In addition, the stratigraphical similarity of the Permo-Triassic successions within the East Solan and Papa basins suggests that they were probably originally contiguous (Swiecicki et al., 1995).

Papa Basin

Within the Papa Basin, thick Permo-Triassic strata within a south-east-dipping half-graben subcrops thin Quaternary sediments close to the sea bed (Figures 39 and 54). Well 205/27a-1 drilled within the basin adjacent to the Rona High proved a combined thickness of 2436 m of Permian and Triassic strata that includes 1180 m of Middle to Upper Triassic Foula Formation sandstone and conglomerate, 237 m of Lower Triassic sandstone and shale of the Otter Bank Sandstone Formation and 56 m of the Otter Bank Shale Formation (Figures 55b and 56). The Triassic succession rests on 465 m of anhydritic sandstone, mudstone and siltstone of the upper Permian West Orkney Evaporite Formation and 497 m of sandstone and conglomerate of the upper Permian Solan Bank Formation, before terminating in 1.5 m of fine-grained, strongly vesicular and highly altered trachyandesitic lava of the Lower Permian Nun Volcanics Formation. A Permo-Triassic succession of up to 7.6 km (Swiecicki et al., 1995) is estimated to be present against the basin-controlling Shetland Spine Fault (Figure 54).

West Shetland Basin

The north-east-trending and south-east-dipping West Shetland Basin forms an elongate 'perched' half-graben adjacent to the south-east flank of the Faroe–Shetland Basin (Figures 30 and 54). It is separated from the Papa Basin to the south-west by the Westray Lineament (Figure 7) that may be equivalent in part to the so-called NOWT (North Orkney/Wyville Thomson transfer zone) of Stoker et al. (1993). Calibrated seismic profiles within the south-west part of the basin show a Triassic and probable Permian succession thickening towards the Shetland Spine Fault (e.g. Haszeldine et al., 1987), indicating that this fault was active throughout Permo-Triassic times. Within the basin, wells 205/20-2, 205/23-1, 205/25-1, and 205/30-1 only penetrate the Triassic succession (Figures 54 and 55a). In all four wells, the lithology is typically continental in aspect, consisting of sandstone interbedded with mudstone. Well 205/23-1 probably penetrates the upper part of the Otter Bank Sandstone Formation whereas wells 205/20-2, 205/25-1 and 205/30-1 all reach terminal depth within the Foula Formation (Figure 56). Conglomerate and breccia occur throughout the succession, particularly in wells 205/20-2 and 205/25-1. In the extreme south-west part of the basin, seismic evidence suggests the possibility that up to 4 km of Permo-Triassic strata may be present (Hitchen and Ritchie, 1987; Haszeldine et al., 1987). Towards the north-east of the West Shetland Basin, the Permo-Triassic succession is absent and in well 207/02-1 (Figure 30), Cretaceous strata rest on metamorphic basement.

Remnant basins on the East, West Shetland and Orkney–Shetland highs

On the platform area straddling the south-east margin of the report area, the Unst, West Fair Isle, Fetlar and St Magnus Bay basins probably represent the remnants of a more extensive Permo-Triassic basin (Figure 54). For instance, an outlier of fine-grained dark red sandstone of possible Permo-Triassic age drilled by BGS borehole 80/12, indicates the presence of isolated occurrences of Permo-Triassic rocks on the Orkney–Shetland High.

West Fair Isle Basin

The West Fair Isle Basin only just encroaches within the south-east part of the report area where it has been drilled by BGS borehole 80/05 (Figure 54). The borehole penetrated only 0.2 m of a poorly sorted, fine to medium-grained sandstone with possible anhydrite of Permo-Triassic age (Figure 55a). Within other parts of the basin outwith the report area, BGS boreholes 80/11 (Figure 42), 77/10, 77/04 and 78/08 also drilled

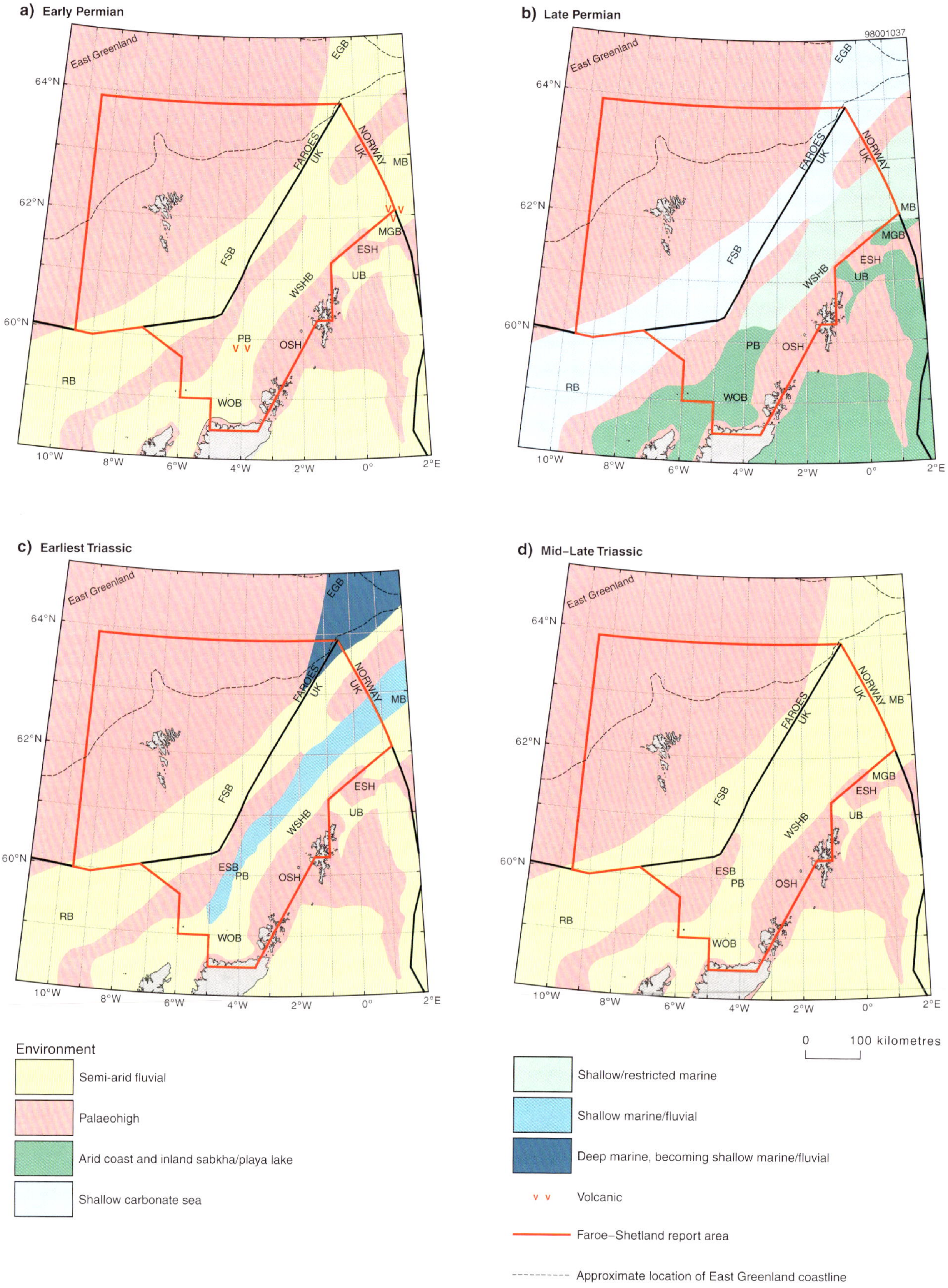

Figure 58 Generalised palaeogeography and depositional environments within the Faroe–Shetland report area during; a) early Permian; b) late Permian; c) earliest Triassic and; d) Mid to Late Triassic times. Abbreviations: see Figures 7 and 8 except, EGB=East Greenland Basin.

probable Permo-Triassic strata. A geological cross-section across the West Fair Isle Basin illustrates that the Permo-Triassic succession thickens south-eastwards towards major north-north-east-trending intra and basin-bounding faults (Figure 42). The thickness of this succession may reach more than 2.5 km adjacent to the Walls Boundary Fault. Towards the north-west, the interval thins and on the West Shetland High, is largely absent.

St Magnus Bay Basin

Off the western coast of Shetland, the St Magnus Bay Basin forms a very small Permo-Triassic basin that has been drilled by BGS borehole 80/08 (Figure 54). The borehole recorded 13.2 m of sandstone overlain by conglomerate, siltstone containing small lenses of anhydrite, and sandstone of possible Permo-Triassic age (Figure 55a). A geological cross-section across the basin indicates that approximately 1 km of Permo-Triassic sediments may be preserved close to the basin centre (Figure 29).

Fetlar Basin

The Fetlar Basin only just encroaches within the eastern part of the report area where it has been drilled by BGS borehole 80/02 (Figure 54). The borehole penetrated 11.3 m of sandstone and subordinate conglomerate of possible Permo-Triassic age (Figure 55a). A geological cross-section across the basin shows the Permo-Triassic succession attains a maximum thickness of approximately 500 m adjacent to the north-north-east-trending fault that bounds the basin to the east (Figure 40).

Unst Basin

To the north-east of Shetland, only the extreme north-west edge of the Unst Basin occurs within the Faroe–Shetland report area (Figure 54). The basin comprises three fault-controlled depocentres that contain between 1.8 and 3.6 km of Permo-Triassic red beds (Johns and Andrews, 1985). The Permo-Triassic succession subcrops Quaternary sediments in the southern and also part of the north-west depocentre, though elsewhere it subcrops younger Mesozoic and Cenozoic strata. Johns and Andrews (1985) subdivided the red bed succession drilled in well 1/04-1 (Figure 54) into a possible Lower Permian unit consisting of 575 m of conglomerate, overlain by 127 m of sandstone and mudstone, and a biostratigraphically well defined upper Permian unit comprising 398 m of red-brown mudstone. These upper Permian strata are unconformably overlain by 735 m of Triassic conglomerate, sandstone and limestone, with marl and nonmarine clastic rocks at higher stratigraphical levels. A geological cross-section across the southern depocentre shows the Permo-Triassic succession thickens towards the axis of the basin.

DEPOSITIONAL ENVIRONMENTS AND BASIN DEVELOPMENT

Consolidation of the Variscan fold belt in late Carboniferous times resulted in a change in the convergence direction between Gondwana and Laurussia and a shifting of orogenic activity westwards, manifested in the Alleghenian Orogeny (Appalachian–Mauretanides/Marathon–Ouachita fold belts) (Ziegler, 1990). As the Alleghenian Orogeny also drew to a close in Stephanian to early Permian times, major dextral translation between North Africa and Europe occurred, causing the development of a complex system of conjugate sinistral and dextral shears that extended far into the Variscan foreland. This resulted in transtensional and transpressional styles of tectonic activity and associated widespread magmatism (Ziegler, 1990; Wilson et al., 2004; De Paola et al., 2005).

Within the north-east Atlantic region, the Arctic rift system initially developed during early Carboniferous (Namurian) times (Coward et al., 2003). It was reactivated during the late Carboniferous to early Permian, propagating south-westwards into the Møre, Faroe–Shetland and Rockall Basin basins (Figure 10; Ziegler, 1990; Hitchen et al., 1995b; Coward et al., 2003). On East Greenland, the Lower Permian succession was described by Surlyk (1990) as being developed during a phase of intense extensional block faulting. More recently however, Coffield (1992) and Roberts et al. (1999) suggest that a dextral transtensional tectonic model is more apposite in terms of early Permian basin formation on the East Greenland margin.

Within the Faroe–Shetland report area, semi-arid, fluvial-continental conditions are interpreted to have predominated during deposition of Lower Permian syn-rift sediments that are inferred to be present within the deepest parts of half-grabens such as the West Orkney and West Shetland basins, and also to the north-west within the Faroe–Shetland and Møre basins (Figure 58a). Lower Permian volcanic rocks drilled on the East Shetland High and within the Papa Basin are considered to have been deposited in association with changes to the regional stress pattern at the end of the Variscan Orogeny during late Carboniferous to early Permian times (Ziegler, 1990; Read et al., 2002; Upton et al., 2004). In particular, existing north-west- to north-north-west-trending faults, some with a strike-slip component, are thought to have become important controls on the distribution of early Permian volcanism (McLean, 1978; Andersen, et al., 1995; Upton, et al., 2004). Within the Faroe–Shetland region, the Brendan and Judd lineaments (Figure 7) may have acted as conduits for migration of magma within the East Shetland High and Papa Basin areas. This early Permian magmatism is associated with the intrusion of tholeiitic dykes and sills across northern England and much of Scotland and the North Sea, with subdued alkaline igneous activity continuing until early Permian times (Upton et al., 2004). On East Greenland and within the offshore Norway and the Barents Sea areas, the recognition of a mid Permian unconformity is interpreted to mark a cessation of latest Carboniferous to early Permian rifting (Surlyk, 1990; Roberts et al., 1999). However, this unconformity has not been recognised within the Faroe–Shetland report area.

During late Permian times, the Arctic Sea transgressed rapidly south-westwards due to a combination of rifting within the Vøring Basin area (Roberts et al., 1999) and a glacioeustatic rise in sea level linked to the deglaciation of Gondwana (Ziegler, 1990; Coward et al., 2003). This transgression of a warm, shallow carbonate sea is interpreted to have connected East Greenland, Faroe–Shetland, and Rockall (Figure 58b) and culminated in the development of the Bakevellia Sea which is interpreted to have covered part of Northern Ireland, the East Irish Sea Basin and the Cheshire Basin. Within the Cheshire and Worcester basins, evidence for a late Permian phase of crustal extension has been described by Chadwick and Evans (1995), who recognised syndepositional thickening of upper Permian strata against basin-bounding faults. Within the Faroe–Shetland area, around the north-west flank of the East Shetland High and Møre, Magnus and Unst basins, the occurrence of upper Permian (Zechstein Group) anhydrite and limestone, indicative of a shallow or restricted marine environment (e.g. well 220/26-2) (Figure 58b), are interpreted to be succeeded by a fluvial/ lacustrine facies (Cameron, 1993b; Ritchie et al., 1996). In contrast, further to the south-west in the West Orkney and Papa basins, alluvial sediments with occasional debris-flow deposits are succeeded both laterally and vertically, by mudstone and evaporitic units that were probably deposited in a sabkha/playa lake environment (e.g. wells 202/18-1, 202/19-1 and 205/27a-1).

The transition from latest Permian to earliest Triassic times is marked by rifting and regional regression of the Arctic Sea (Ziegler, 1990; Coward et al., 2003). During earliest Triassic times, deposition of the Otter Bank Shale Formation in the East Solan Basin is interpreted to have occurred within a shallow marine/ fluvial setting (Swiecicki et al., 1995), contemporaneous with the southerly advancing marine transgression recorded in East Greenland (Figure 58c; Seidler et al., 2004). The progradational and upward coarsening braided fluvial environment recognised by Swiecicki et al. (1995) within the lower part of the overlying Otter Bank Sandstone Formation in the East Solan Basin is interpreted as indicative of Early Triassic rift activity. Alternatively, Herries et al. (1999) prefer deposition of the Otter Bank Sandstone Formation within a post-rift setting. The upper part of the Otter Bank Sandstone Formation consists of a repeated facies association comprising interbedded dry and damp sand flat, playa lake and small-scale sandy braided fluvial channels (Swiecicki et al., 1995). Contemporaneous Early Triassic strata within wells 202/19-1 and 202/18-1 in the northern part of the West Orkney Basin are considered to have developed in broadly similar depositional settings to those described by Swiecicki et al. (1995) in the East Solan Basin.

Outwith the Faroe–Shetland report area, support for Early Triassic rifting is derived from East Greenland, offshore mid Norway (Seidler et al., 2004) and in the East Irish Sea Basin. In the East Irish Sea Basin, thickening of the Sherwood Sandstone Group (the time equivalent of the Otter Bank Sandstone Formation) is recognised by Chadwick et al. (2001) within the hanging-wall blocks of major growth faults. In contrast, Ruffell and Shelton (1999), suggested that the Sherwood Sandstone Group was more likely developed during a period of post-rift subsidence.

Vigorous rift activity continued within the Faroe–Shetland area during Mid Triassic times (Figure 11; e.g. Torsvik et al., 1996; Coward et al., 2003) and also to the south in the Irish Sea Basin (Ruffell and Shelton, 1999), though with an increasing component of regional thermal subsidence (Chadwick et al., 2001). In contrast with the Early and Mid Triassic, the Late Triassic was a time of relative tectonic quiescence (e.g. Torsvik et al., 1996; Coward et al., 2003). Figure 58d summarises the prevalent depositional environments in Mid to Late Triassic times during deposition of the Foula and Cormorant formations (Figure 56) within the Faroe–Shetland area. In the East Solan Basin, a facies association similar to that described for the Otter Bank Sandstone Formation is interpreted to prevail during deposition of the Foula Formation, with interbedded fluvial and sabkha deposits superseded by wholly fluvial deposits (Herries et al., 1999). These Middle and Upper Triassic strata have probably been penetrated in many basins within the report area namely, the West Orkney, North Rona, West, East and South Solan, West Shetland, Papa, Faroe–Shetland, St Magnus Bay, West Fair Isle and Fetlar basins. In the West Orkney and Papa basins, nonmarine Middle and Upper Triassic sediments described from wells 202/19-1 and 205/27a-1 were probably deposited in braided or anastomising fluviatile or alluvial braid-plain with minor sheet flood environments (Ritchie et al., 1996). Within the East Solan Basin an aeolian sabkha environment has been inferred from the lower part of the Foula Formation in well 205/26a-4, succeeded by fluvial deposition in the upper part (Herries et al., 1999). Further to the west within the Faroe–Shetland Basin, fluvial depositional systems have been described from well 213/23-1 on the Corona High.

The Arctic and proto-Atlantic seas remained separate throughout most of the Triassic with communication only likely during a regional transgression through the Rockall, Faroe–Shetland and Møre areas in Rhaetian times. Within the Faroe–Shetland report area, the presence 15 m of Rhaetian limestone within well 204/19-9 on the Westray High could provide evidence of an open marine connection, a precursor to the development of a more a more established seaway during Lower Jurassic times.

6 Jurassic

Derek Ritchie[1] and Thomas Varming[2]
[1] *British Geological Survey*
[2] *Jarðfeingi (Faroese Earth and Energy Directorate)*

Following Mid Triassic continental rifting throughout the Arctic rift system within the north-east Atlantic region (Figure 11), sedimentary basins generally underwent Early Jurassic post-rift thermal subsidence, with Lower Jurassic shallow marine sediments infilling the rift topography (Coward et al., 2003). In places however, there is evidence for renewed rifting during Sinemurian to Toarcian times, for example within the Sea of Hebrides Basin–Little Minch Basin (Morton, 1989) and in parts of offshore Norway (Blystad et al., 1995). Lower Jurassic strata have been drilled in the northern part of the Porcupine Basin and in the Slyne and Erris basins (e.g. Naylor and Shannon, 2005). Within the Faroe–Shetland region, the existence of a seaway is confirmed by the presence of a significant thickness of Lower Jurassic shallow marine strata within the West Solan Basin (Figures 59 and 60a), and it is conceivable that the deposition of these sediments also occurred within an extensional tectonic regime (e.g. Dean et al., 1999).

Sea-floor spreading within the Central North Atlantic region began in Middle Jurassic times (e.g. Knott et al., 1993). Further to the north, however, Atlantic rift basins such as the Celtic Sea, Jeane D'Arc, North Porcupine, Slyne and Erris were still undergoing thermal subsidence (Roberts et al., 1999; Coward et al., 2003). Within the Faroe–Shetland area, proven occurrences of Middle Jurassic strata are extremely rare, although a significant thickness of mainly Bajocian to Callovian marine sandstone drilled within the Foula Sub-basin was interpreted by Haszeldine et al. (1987) to have been deposited in an extensional tectonic regime. In contrast, a phase of Mid to earliest Late Jurassic uplift and deep erosion is considered by some to have affected the Faroe–Shetland region (e.g. Doré et al., 1999), with for example, over 1.5 km of Lower Jurassic and older strata removed from the East Solan Basin (Booth et al., 1993). Although this phase of exhumation influenced the development of the marginal basins and highs that fringe the south-east margin of the Faroe–Shetland Basin, the processes that were active within the main basin remain unclear.

The presence of a significant deep marine seaway through the Faroe–Shetland Basin during latest Mid to Late Jurassic times is confirmed from the drilling results of a number of wells. This open marine connection linked the offshore Norwegian margin with the Rockall Basin and the Atlantic rift (Figure 12). The development of this seaway may at least be partly related to limited crustal extension (e.g. Haszeldine et al., 1987; Dean et al., 1999), possibly along mainly north- to north-east-trending structures (Doré et al., 1999) such as the Westray High.

Most of the proven occurrences of Jurassic strata within the Faroe–Shetland report area are mainly confined to the peripheral basins and highs that occur immediately adjacent to the south-west and south-east margins of Faroe–Shetland Basin, including the North Lewis, North Rona, West Shetland and West, South, and East Solan basins and the Judd and south-west Rona highs (Figure 59). Approximate maximum drilled thicknesses of 200, 250 and 770 m have been recorded for the Upper Jurassic (well 202/12-1 in the North Rona Basin), Middle Jurassic (well 204/22-1 on the Judd High) and Lower Jurassic (well 202/03a-3 in the West Solan Basin), respectively (Figure 60a). Within these marginal areas, the general level of preservation of Jurassic strata is patchy (Figure 59), though the presence of scrappy exposures of Jurassic rocks at or close to the sea bed within the West Orkney and Papa basins is highly suggestive that they may have been geographically more extensive in the past, but subsequently removed as a result of uplift and erosion in mainly Mid Jurassic and Cenozoic times. According to Hudson and Trewin (2002, fig. 1), it is unlikely that Jurassic strata were ever deposited on the Orkney–Shetland High (Figure 59).

Within the Faroe–Shetland Basin, Jurassic strata have only been drilled in eight wells along the south-east margin of the basin or on intrabasinal horsts including the Westray and Corona highs (Figures 59 and 60b). Maximum drilled thicknesses of approximately 1.05 km of Upper and Middle Jurassic (well 206/05-1 in the Foula Sub-basin) and 340 m of Lower Jurassic (well 206/05-2 in the Foula Basin) have been penetrated. However, the Fionaven and Schiehallion fields, and discoveries such as Tobermory and Laggan (see Chapter 12), contain hydrocarbons generated from Jurassic source rocks and this strongly suggests that these strata are widespread throughout the central and south-east parts of the basin at least (see also Roberts et al. 1999; Coward et al. 2003). Towards the north-west part of the Faroe–Shetland Basin, imaging of the Mesozoic interval beneath the Palaeogene lavas and intrusive rocks becomes increasingly difficult although the presence of Jurassic rocks has been predicted (e.g. Roberts et al., 1999; Tate et al., 1999; Keser Neish, 2003; Ziska and Andersen, 2005; Figure 59).

STRATIGRAPHY

Largely building upon the work of Haszeldine et al. (1987), Vesteralen and Hurst (1994), Vesteralen et al. (1995) and the results of well correlation and limited biostratigraphical information, the first published lithostratigraphical review of the Jurassic was completed by Ritchie et al. (1996). However, further insights by Herries et al. (1999), particularly with regard to the Upper Jurassic in the East Solan Basin and adjacent areas has led to the revised scheme presented here (Figure 61).

It should be noted that in the following descriptions of lithologies recorded from the various wells within the report area, the term mudstone is used to describe argillaceous rocks that include claystone, marl or shale.

Lower Jurassic

Lower Jurassic strata have only been encountered in five deep wells and a single BGS borehole in the Faroe–

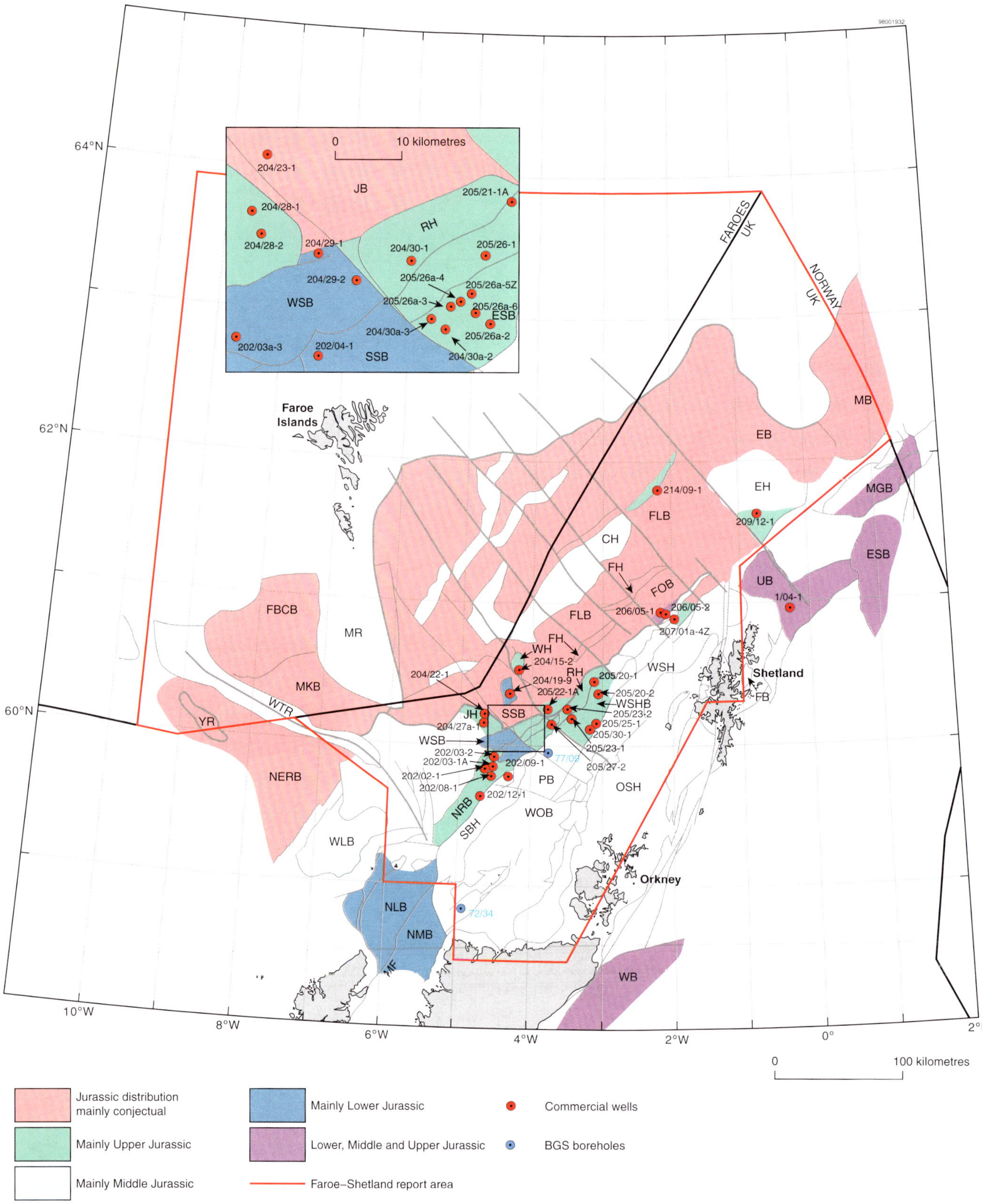

Figure 59 Distribution of Jurassic strata within the Faroe–Shetland report area. Present day structural elements (grey) derived from Figure 7. Abbreviations: see Figure 7.

Shetland report area (202/03a-3, 202/04-1, 204/19-9, 206/05-1, -2, and 72/34) (Figures 59 and 60). Following the scheme presented by Ritchie et al. (1996), the successions in the wells, with the exception of 206/05-1, are ascribed to the Skerry Group, which is comprised of the Sule Skerry and Stack Skerry formations (Figure 61). The Skerry Group is the probable lateral equivalent of the Pabba Shale Formation of the Inner Hebrides or the Dunlin Group of the northern North Sea (e.g. Morton, 1987; Richards et al., 1993). The 28.5 m thick, sand-rich ?Sinemurian to Pliensbachian succession at the base of well 206/05-1 is currently undivided

(Figure 60b), although Ritchie et al. (1996) suggested that it could conceivably even belong to the Devonian Clair Group.

STACK SKERRY FORMATION

In the type section within well 202/03a-3 (Ritchie et al., 1996), the Stack Skerry Formation mainly comprises 513 m of white to grey, firm to hard, coarse-grained, well-sorted, well-cemented micaceous and slightly carbonaceous sandstone (Figures 60a, 61 and 62) with subordinate interbedded grey, firm, blocky to subfissile, calcareous, slightly silty mudstone and minor grey to black, calcareous, slightly sandy siltstone which are both micromicaceous and microcarbonaceous. These sediments are interpreted as shallow marine inner shelf deposits and are biostratigraphically characterised by a rich palynomorph association including *Cerebropollenites macroverrucosus* (see Ritchie et al., 1996). This species, together with the dinoflagellate cyst *Liasidium variable* and the foraminifera *Glomospira* sp., *Marginulina prima* and *Nodosaria fontinensis* indicate a late Sinemurian to early Pliensbachian age. Drilling results from wells 202/04-1 in the South Solan Basin, 206/05-2 in the Foula Sub-basin and 204/19-9 on the Westray High could suggest that the formation spans the Hettangian to late Pliensbachian (Figures 60 and 61). The Stack Skerry Formation is well defined from wireline well logs, with the gamma-ray log in particular displaying a notable 'blocky' or 'box car' signature (Figure 62).

SULE SKERRY FORMATION

The Sule Skerry Formation as defined in the type section within well 202/03a-3 comprises 257 m of dark grey to black, hard, brittle, subfissile, calcareous, carbonaceous, pyritic mudstone (Figures 60a, 61 and 62) with traces of grey to brown, soft to hard, calcareous, micaceous siltstone and brown to grey, hard to friable, argillaceous carbonaceous limestone. This unit is interpreted as a shallow marine deposit, with the presence of the foraminifera *Lingulina tenera tenera*, *Lingulina pupa* and the agglutinating form *Verneuilinoides mauritti* being consistent with an early Pliensbachian age. The most significant wireline well log characteristic of the Sule Skerry Formation is its consistent low gamma-ray response (Figure 62).

Middle to Upper Jurassic

Although Middle Jurassic rocks were previously considered to be widespread (e.g. Stoker et al., 1993), they have only really been proved in three wells (204/22-1, 205/20-2 and 206/05-1) within the report area (Figures 59). However, only the succession in the well 206/05-1 is currently assigned to a lithostratigraphical formation (Figure 60b). Based on the schemes of Haszeldine et al. (1987) and Ritchie et al. (1996), the Bathonian to Oxfordian succession within well 206/05-1 is ascribed to the Heather Formation of the Humber Group (Figures 61 and 63). This assignation is in agreement with the age range of the Heather Formation where originally defined within the northern North Sea (Richards et al., 1993). Although the Oxfordian part of the Heather Formation within well 206/05-1 technically lies within the Upper Jurassic (Figure 61), the formation in its totality is described below.

Drilled occurrences of Upper Jurassic strata are relatively widespread and currently, are mainly assigned to the Kimmeridge Clay Formation (Figures 59, 60 and 61). However, it should be noted that the Kimmeridge Clay Formation does extend up into the Lower Cretaceous (Berriasian) succession in places (e.g. see Figures 70, 73 and 77).

HEATHER FORMATION

In the type section for the Faroe–Shetland region, as defined by Ritchie et al. (1996) within well 206/05-1, the Heather Formation is interpreted to be at least 520 m thick, spanning both the Upper and Middle Jurassic intervals (Oxfordian to Bathonian) (Figures 60b and 63). The lower part of the succession is described as dominantly comprising approximately 300 m of pale grey to white, fine to coarse-grained, angular to rounded, poorly sorted, pyritic, calcareous to noncalcareous, kaolinitic, micaceous, carbonaceous, locally quartz and dolomite cemented sandstone (including the Fair Sandstone Member), with thin interbedded olive to black, noncalcareous, pyritic, carbonaceous mudstone and an overlying 220 m thick dominantly argillaceous succession consisting of dark grey to black or rarely brown, noncalcareous to slightly calcareous, fissile, micaceous, carbonaceous and pyritic mudstone with sporadic beds of olive to grey, argillaceous laminated noncalcareous, micaceous, carbonaceous sandstone (Figure 63). The Heather Formation is considered to have been deposited in a shelf setting, where bottom waters were mainly oxygenated. It contains a number of common palynomorphs, but the dinoflagellate cysts *Rigaudella aemula* and *Wanaea* spp. are key biomarkers, particularly within the upper part of the formation. Like laterally equivalent deposits of the northern North Sea, the Heather Formation generally ranges in age from Bathonian/Callovian to Oxfordian i.e. Mid to Late Jurassic (e.g. Richards et al., 1993; Ritchie et al., 1996). The distinctive two-fold subdivision of the mudstone dominated Heather Formation recognised within the East Shetland Basin (see Richards et al., 1993) is also observed on wireline well logs from well 206/05-1. There is a downward decrease in gamma-ray log values and an increase in velocity, which corresponds to a downward transition from a mudstone to a more mixed facies that includes the Fair Sandstone Member at its base (Figure 63).

Fair Sandstone Member

The Fair Sandstone Member in well 206/05-1 consists of 216 m of mainly grey to white, fine to coarse-grained, subangular to subrounded, poorly sorted, pyritic, locally quartz and dolomite cemented sandstone with thin interbedded olive to black, noncalcareous, pyritic, carbonaceous mudstone (Figures 60b and 63). According to Haszeldine et al. (1987), this succession is considered to have been deposited within a submarine fan setting, possibly associated with extensional fault movement. The biostratigraphical age of the member is poorly constrained, as the Fair Sandstone Member appears to be barren (Haszeldine et al., 1987). The transition from the base of the argillaceous-dominated part of the Heather Formation to the top of the Fair Sandstone Member is marked by a sharp decrease in gamma-ray and an increase in velocity well log responses (Figure 63). The transition from the base of the member to the top of the underlying succession is defined by a sudden increase in gamma-ray and a decrease in velocity log response.

The Upper Jurassic interval has been penetrated by 39 wells and boreholes within the Faroe–Shetland

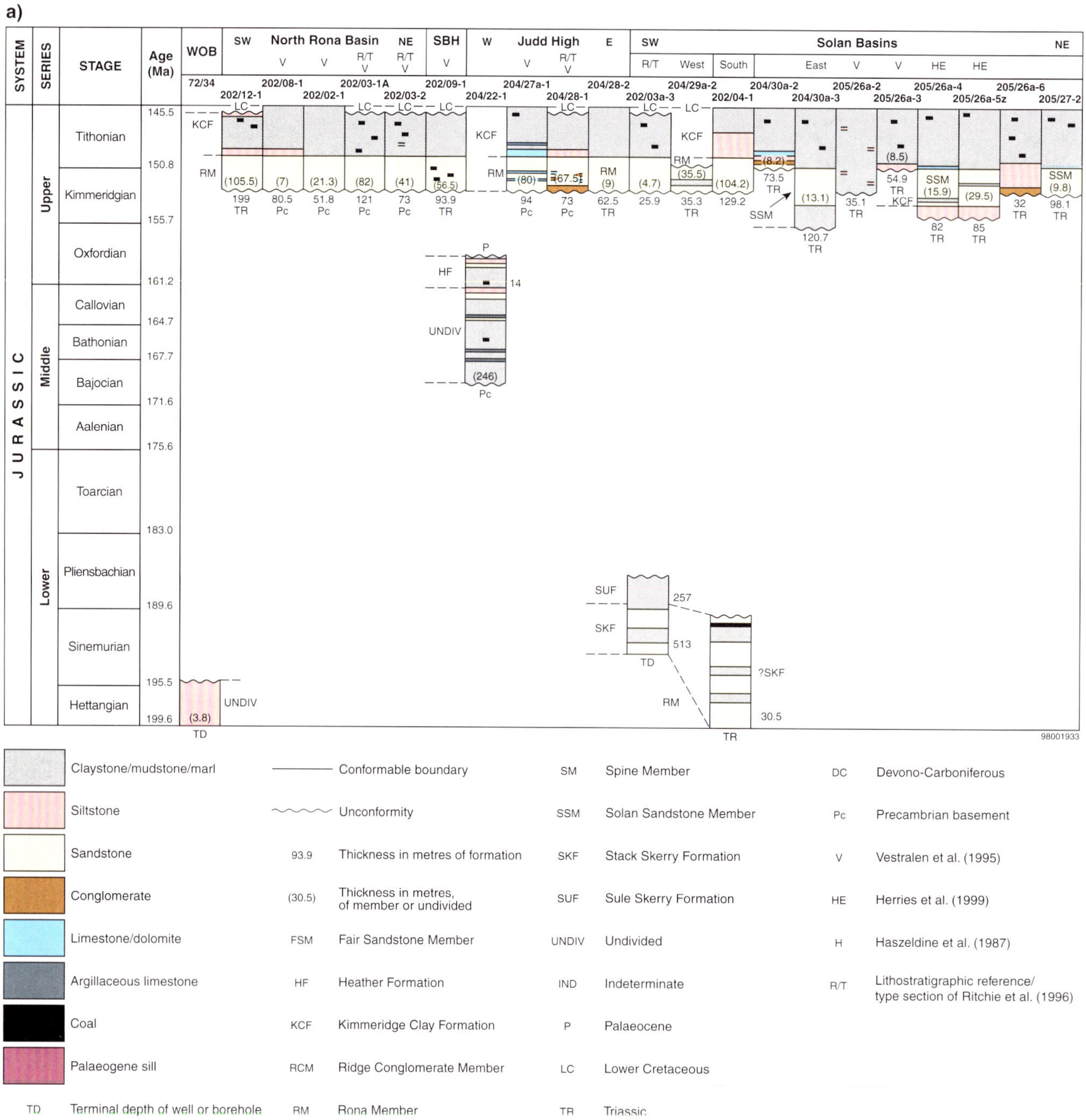

Figure 60 Stratigraphical range chart showing generalised lithology and thickness (m) of Jurassic rocks within a) the North Rona Basin, Solan Bank High, Judd High and Solan basins b) West Shetland Basin, Papa Basin, Rona High, Erlend High and Faroe–Shetland Basin (modified from Stoker et al., 1993; Vestralen and Hurst, 1994; Vestralen et al., 1995; Ritchie et al., 1996; Herries et al., 1999). Timescale is based on 'A Geologic Time Scale 2004' by F M Gradstein, J G Ogg, A G Smith, et al. (2004) and the International Stratigraphical Chart, 2006 (ICS) with additions. Abbreviations: see Figure 7. For location of wells and BGS boreholes, and Jurassic lithostratigraphical nomenclature, see Figures 59 and 61, respectively.

report area (Figures 59 and 60). Apart from the Heather Formation, the Kimmeridge Clay Formation is the only other formally recognised Upper Jurassic formation in the Faroe–Shetland region (Ritchie et al., 1996), although it does include the Ridge Conglomerate, Rona, Solan Sandstone and Spine members (Figure 61). It should be noted that Ritchie et al. (1996) defined a Rona Formation (which included the Spine Member) but this has been reduced in status following revelations by Herries et al. (1999), particularly regarding the Solan Sandstone Member and its stratigraphical relationship with the Rona and Spine members (see below).

b)

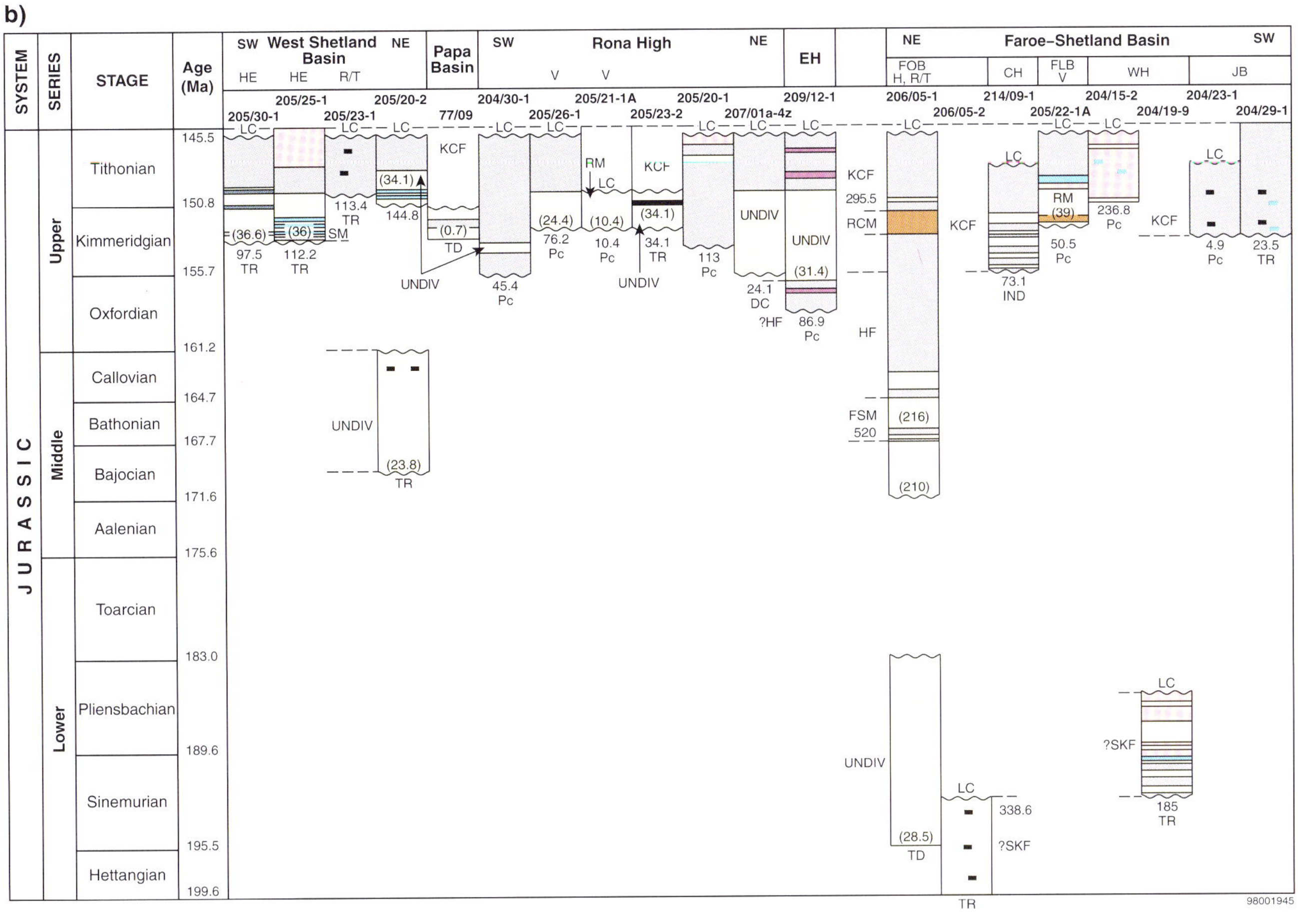

Figure 60 *continued.*

Kimmeridge Clay Formation

The Kimmeridge Clay Formation, as defined in the Faroe–Shetland type section within well 206/05-1 (Ritchie et al., 1996), has a maximum drilled thickness of 295.5 m and generally comprises dark grey to black or locally brown, slightly to noncalcareous, soft to firm, partly fissile, moderately to highly organic-rich, occasionally carbonaceous, micaceous and pyritic mudstone and minor siltstone (Figures 60b and 63). The mudstone is sometimes silty, displays slump structures and shows signs of bioturbation. These sediments grade to thin, white to grey, fine-grained sandstone and rudaceous rocks of the Ridge Conglomerate Member (see below). Towards the base of the Kimmeridge Clay Formation around the marginal areas to the south and east of the Faroe–Shetland Basin, the succession commonly becomes increasingly sand-prone, and can locally be subdivided into the Rona, Spine and Solan Sandstone members (see below).

The argillaceous upper part of the Kimmeridge Clay Formation represents marine hemipelagic deposition where the bottom waters are oxygen-starved. These strata contain a rich assemblage of palynomorphs and other organically amorphous material. The presence of key dinoflagellate cyst biomarkers including *Rotosphaeropsis thula, Egmontodinium expiratum, Glossodinium dimorphum, Muderongia* sp. A, *Cribroperidinium longicorne* and *Endoscrinium luridum* and the miospores *Aequitriradites spinulosus, Cicatricosisporites australiensis, Cicatricosisporites purbeckensis* and *Classopolis echinatus* is consistent with a Kimmeridgian to Berriasian age, and indicating that the Kimmeridge Clay Formation extends up into the Lower Cretaceous (e.g. see Figure 73). According to Herries et al. (1999), the age of the basal Kimmeridge Clay Formation varies laterally. The argillaceous part of the Kimmeridge Clay Formation is typically characterised on wireline well logs by a moderately high gamma-ray and low velocity log signature, with local 'hot shale' gamma-ray spikes attaining values of up 225 API (American Petroleum Institute defined units) (Figure 64).

Ridge Conglomerate Member

The term Ridge Conglomerate Member was first used by Ritchie et al. (1996) to describe a thick mainly rudaceous sedimentary unit of Kimmeridgian age that only occurs within well 206/05-1 located close to the southeast margin of the Foula Sub-basin (Figures 59, 60b, 63 and 65). The member principally consists of pebble to boulder-sized clasts of Devono-Carboniferous sandstone and mudstone and Lewisian gneiss within a matrix of dark grey, fine-grained, calcareous, micaceous sandstone. The Ridge Conglomerate Member is interpreted to represent submarine scree deposits that were derived from a nearby active fault scarp (Haszeldine et al., 1987).

The downward transition from mudstone typical of the Kimmeridge Clay Formation to the Ridge Conglomerate Member is recognised on wireline well logs by a decrease in gamma-ray and an increase in velocity responses whereas the boundary that defines the base of the conglomeratic unit with mudstones of the

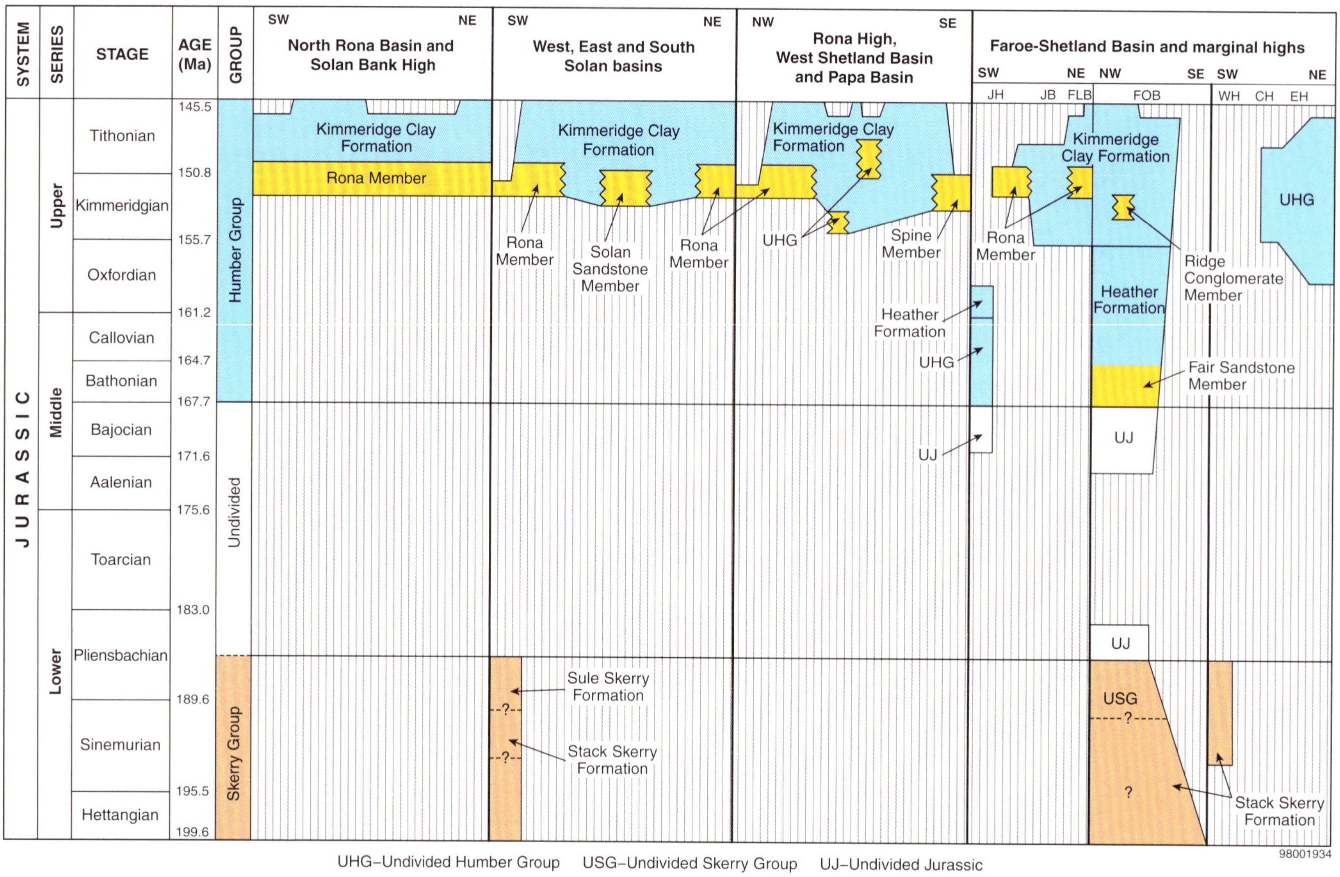

Figure 61 Generalised Jurassic lithostratigraphical nomenclature for the Faroe–Shetland report area (modified from Vestralen and Hurst, 1994; Vestralen et al., 1995; Ritchie et al., 1996; Herries et al., 1999). Timescale is based on 'A Geologic Time Scale 2004' by F M Gradstein, J G Ogg, A G Smith, et al. (2004) and the International Stratigraphical Chart, 2006 (ICS) with additions. Sand rich intervals within the Kimmeridge Clay and Heather formations are shaded yellow. Structural element abbreviations: see Figure 7. It should be noted that that the scheme is poorly constrained in detail.

Kimmeridge Clay Formation is defined by a downward increase in gamma-ray and a decrease in velocity values (Figure 63).

Spine Member

The term Spine Member was introduced by Vestralen et al. (1995) for a unit that has only been recognised in wells 205/25-1 and 205/30-1 within the south-west corner of the West Shetland Basin (Figures 59, 60b and 65). It is described as being comprised of white to grey, dark brown to black, friable, fine to coarse-grained, occasionally subfissile, poorly sorted, subangular to subrounded, moderately calcite-cemented sandstone including occasional coal fragments with subordinate white, grey, brown, grey-green, noncalcareous and micaceous mudstone and white to off-white, hard crystalline limestone. The sandstone is interpreted to have been deposited in a shallow marine environment (Herries et al., 1999). According to Vestralen and Hurst (1994) and Vestralen et al. (1995, fig. 4), the sandstone deposits of the Spine and Rona members, which occur in fairly close juxtaposition (Figure 65), share a similar late Kimmeridgian to Early Tithonian age. This age is based on palynomorphs that occur within the ?Eudoxus and Wheatleyensis standard (ammonite) zones, respectively (Table 4). According to Herries et al. (1999) however, the Spine Member sandstone is slightly older, although it is said to correspond with the late Kimmeridgian pre-Autissiodorensis standard (ammonite) zones.

Comparison of wireline well log signatures between the Rona and Spine members reveals that they are broadly similar (Vestralen et al., 1995, fig. 8), although the latter has a slightly lower gamma-ray and higher velocity response (cf. Figures 64 and 66) which may support the hypothesis of it representing a cleaner more mature sediment source. It is suggested that the source may have been Devono-Carboniferous sediment originally present on the West Shetland High, which was eroded in response to footwall uplift along the Shetland Spine Fault in Late Jurassic times (Figure 65).

Rona Member

First described by Vestralen et al. (1995), the Rona Member has a maximum drilled thickness of 105.5 m in well 202/12-1 and is interpreted to be present within the North Rona Basin, Solan Bank High, Judd High, West Solan Basin, south-west Flett Sub-basin and the Rona High (Figures 59, 60 and 65). On the basis of a study of 10 cored Jurassic successions in wells, the member can be subdivided into seven facies associations, with a generalised summary of, A (the oldest) to G (from Vestralen et al., 1995, figs. 5 and 6), provided below.

Figure 62 Well log summary of the Lower Jurassic Sule Skerry and Stack Skerry formations in well 202/03a-3 within the West Solan Basin. For location, see Figure 59.

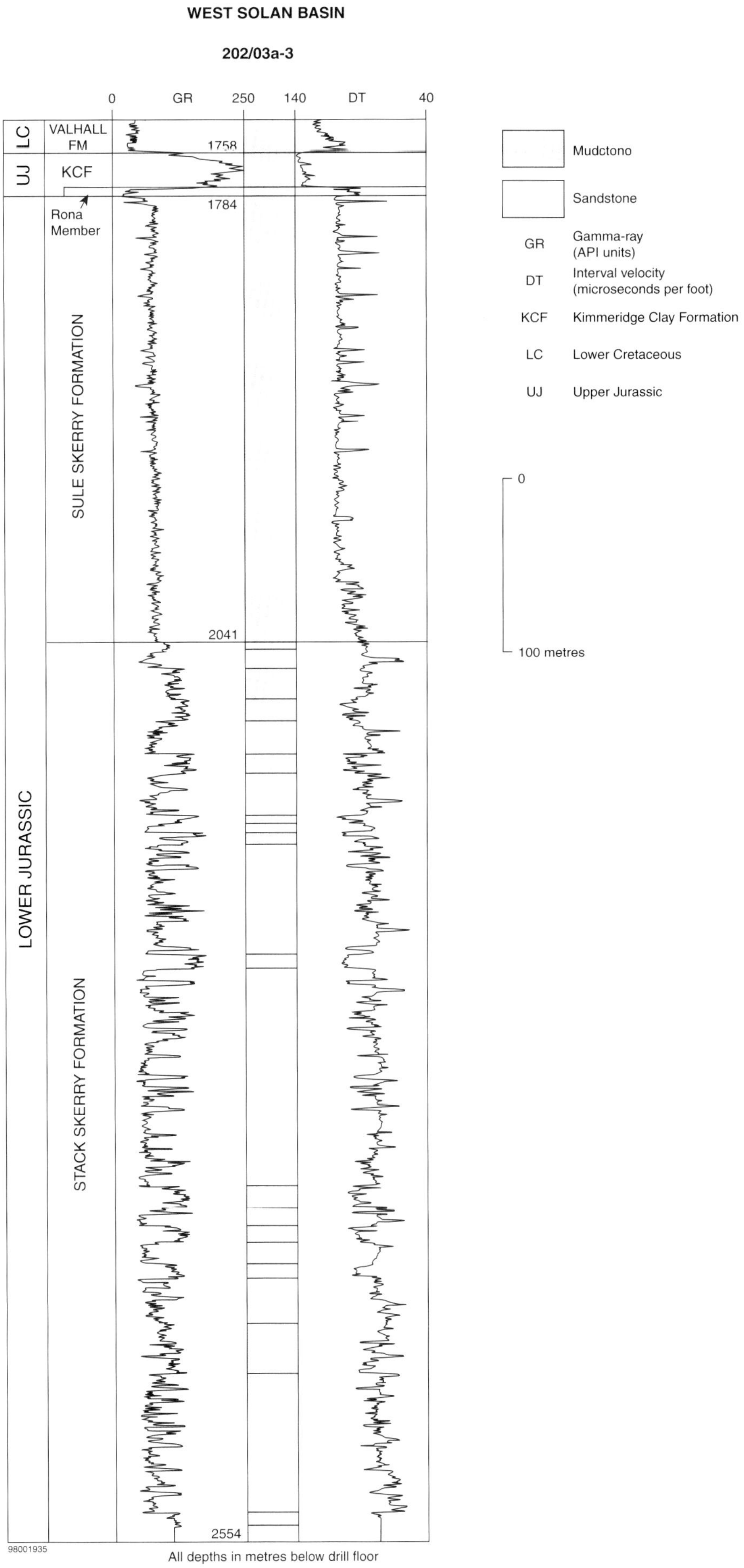

Facies association A — Matrix-supported conglomerate with randomly orientated clasts up to cobble size in a poorly sorted, fine to coarse-grained sandstone. Bed thicknesses are up to 3 m and separated by thin sandstone, mudstone or siltstone layers. Coal and pyrite are common and in well 204/27a-1 (Figure 59), a palaeosol has been described. Facies A is interpreted as representing cohesionless ?subaerial debris-flow deposits, with clasts derived mainly from Lewisian basement.

Facies association B1 — Upward-fining graded and massive pebbly sandstone units with pebble and granule-sized clasts occurring within a very fine to very coarse-grained sandstone/arkosic sandstone matrix. Beds thicknesses are up to 1 m or more. The massive beds are characterised by upward-increasing concentrations of bivalve debris. Facies B1 is interpreted as high-density turbidite deposits.

Facies association B2 — Poorly sorted pebbly sandstone that comprises angular to subrounded granules to small pebble-sized clasts within a matrix of coarse to very coarse-grained sandstone. This facies is interleaved with Facies A and has only been described in well 204/27a-1. These sediments are interpreted to represent poorly confined sheet-flow deposits.

Facies association C — A combination of a basal sandstone that includes cobble-sized clasts, followed by pebbly sandstone and coarse to very coarse-grained sandstone, the lower part of which is poorly sorted, structureless and contains rip-up clasts. Bivalve debris is common. This facies association has only been described from well 204/27a-1 but the presence of the shell debris and the well-rounded nature of the constituent sand grains suggest deposition within a high-energy shoreface environment.

Facies association D — Chaotic, structureless, poor to well-sorted, fine to very coarse-grained sandstone with subangular to subrounded grains and the occasional presence of bivalve debris, black carbonaceous films, water-escape structures and other soft sediment deformation features. This facies is interpreted as turbiditic.

Facies association E — Fine to very fine-grained, well-sorted and massively bioturbated mottled sandstone. Numerous burrow structures have been recognised along with bivalve and rare belemnite fragments. There is evidence of small-scale faulting and soft sediment deformational structures. The presence of an ichnofauna is interpreted to suggest a shallow marine, below wave-base environment of deposition.

Facies association F — Interbedded fine to very fine-grained laminated sandstone and very coarse-grained structureless clean sandstone, with the latter less common at higher stratigraphical levels. The effects of burrowing are commonly observed within both the fine-grained laminated and coarse-grained structureless units along with limited occurrences of coal streaks and shell debris. The laminated and clean sandstone units are interpreted to have been deposited as turbidites, in a shallow marine shelf environment.

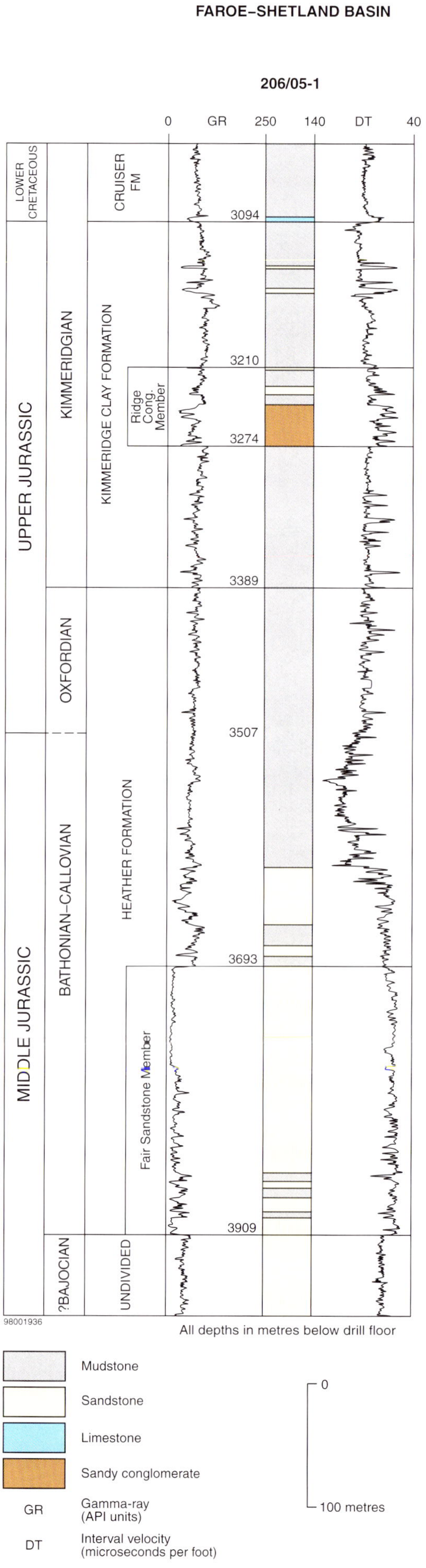

Figure 63 Well log summary of the Upper and Middle Jurassic Kimmeridge Clay and Heather formations and older strata in well 206/05-1 within the Foula Sub-basin (modified from Haszeldine et al., 1987 and Ritchie et al., 1996). For location, see Figure 59.

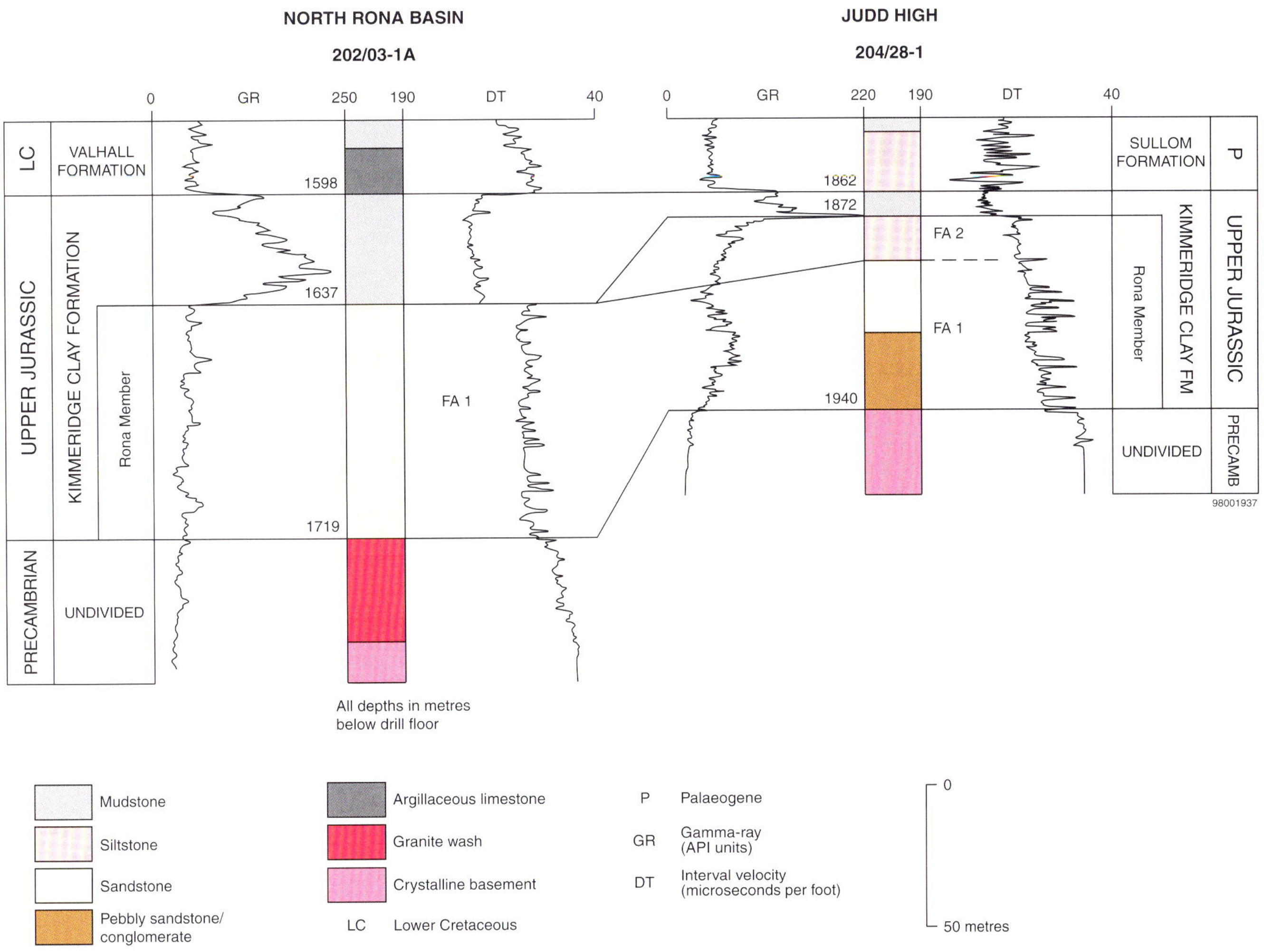

Figure 64 Well log correlation of the Upper Jurassic Kimmeridge Clay Formation, including the Rona Member in wells 202/03-1A and 204/28-1 within the North Rona Basin and on the Judd High, respectively. Abbreviations: FA1=Unit FA1 (including facies associations A, B1, B2, C and D) and FA2= Unit FA2 (including facies associations E, F and G) (see text for details). For location of wells, see Figure 59.

Facies association G — Very fine-grained, well-sorted sandstone and siltstone interbedded with very thin fine-grained sandstone laminae. Bioturbation is common in the lower part of the facies association. The sandstone and siltstone-dominated part of the succession is interpreted to have been deposited below storm wave-base within a marine environment by turbidity currents.

For the purposes of including wells that are interpreted to contain Rona Member strata but have not been cored, these facies associations were simplified by Vestralen et al. (1995) into a lower unit termed FA1 (comprising the coarser-grained A, B1, B2, C and D facies described above) and an upper unit termed FA2 (comprising the finer-grained E, F and G facies). The general fining-up facies transition from FA1 (oldest) through FA2 to argillaceous rocks of the Kimmeridge Clay Formation is interpreted to represent a deepening of the environment of deposition, with for example, nonmarine palaeosols recognised at the base of FA1 and relatively deep-water marine hemipelagic mudstone within the upper part of FA2 (Vestralen et al., 1995). An example correlation of these units in the North Rona Basin and Judd High is given in Figure 64. A study by Morton (1995, in Herries et al. (1999)) apparently indicated that the coarser-grained facies of the Rona Member have been locally transported from a source area with a highly variable mineral content. According to Vestralen et al. (1995), unit FA1 always rests on Lewisian basement and its variable thickness results from an infilling of existing topography by proximal and distal fan delta systems (see Herries et al., 1999). When FA1 is absent, FA2 is interpreted to rest on either Lewisian basement, Triassic or older Jurassic and is associated with a more widespread marine transgression. Vestralen and Hurst (1994) and Vestralen et al. (1995, fig. 4) suggested that the Rona Member has a late Kimmeridgian to Early Tithonian age, based on palynomorphs that occur within the ?Eudoxus and Wheatleyensis standard (ammonite) zones (Table 4). In contrast, Herries et al. (1999) prefer a pre-Hudlestoni age for the coarser-grained 'Lower Rona' succession that probably corresponds with facies association FA1 (although the oldest palynomorphs are of Oxfordian age but these considered to have been reworked) and a post-Hudlestoni age for the finer-grained 'Upper Rona 'or FA2 unit.

On wireline well logs, the signature of the Rona Member facies associations reflects stratigraphical preservation. For example, in well 202/03-1A, only the coarser facies (unit FA1) is interpreted to be present whereas in 204/28-1 both coarse- and fine-grained (FA2) units are present (Figure 64). The downward transition from

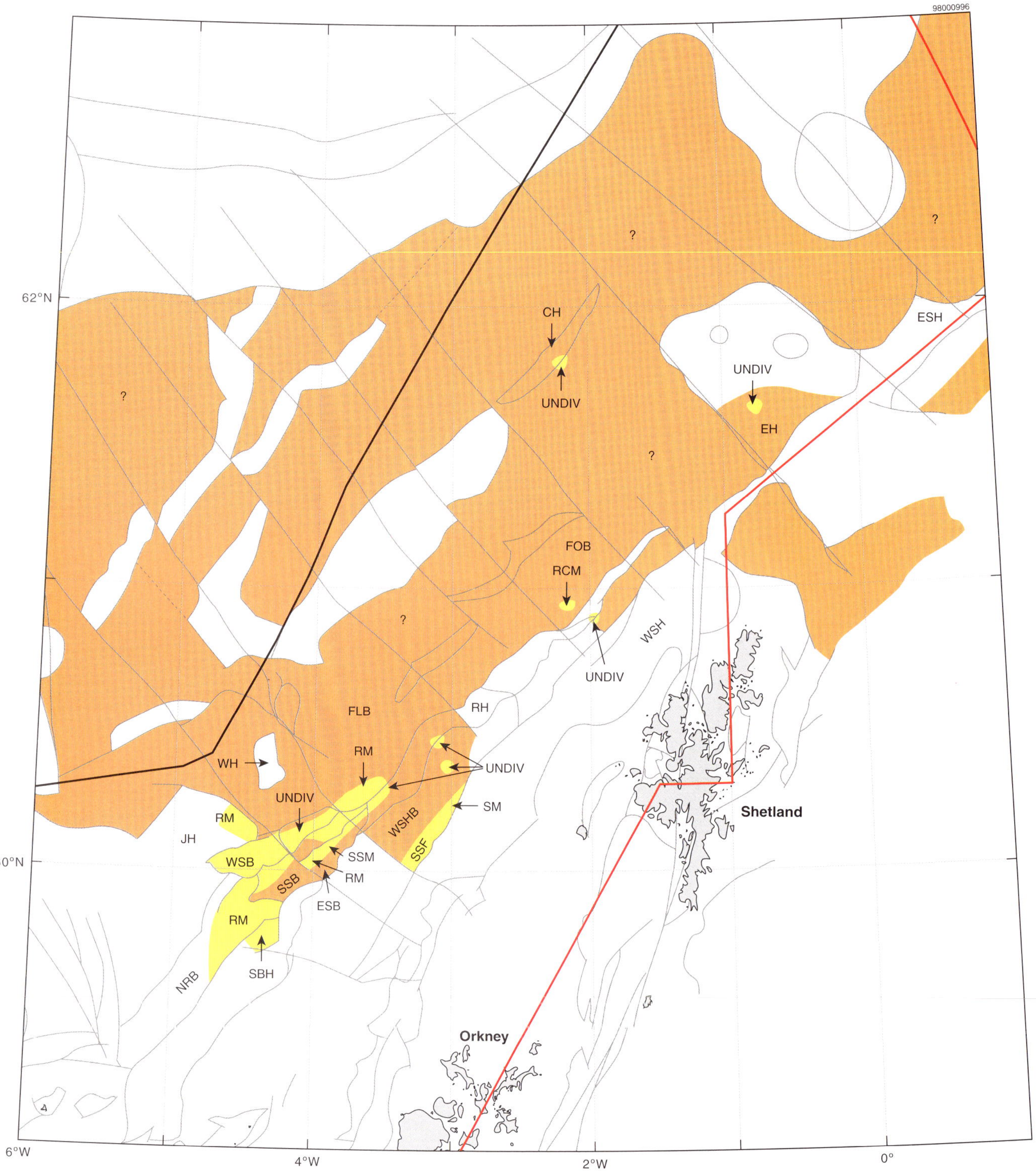

	Shelf to basinal marine mudstone
	Proven marginal marine sandstone
	Faroe-Shetland report area

KEY TO LITHOSTRATIGRAPHIC SYMBOLS

RCM	Ridge Conglomerate Member
RM	Rona Member
SM	Spine Member
SSM	Solan Sandstone Member
UNDIV	Undivided

0 100 kilometres

Figure 65 Generalised facies distribution map for the Upper Jurassic Kimmeridge Clay Formation within the Faroe–Shetland report area. Note that the facies distribution to the north-west of the Judd (JH), Rona, (RH), West Shetland (WSH) and East Shetland highs (ESH) within the Faroe–Shetland Basin is largely conjectural. Present day structural elements (grey) are derived from Figure 7. Structural element abbreviations: see Figure 7.

typical Kimmeridge Clay Formation mudstone to Rona Member sandstone or siltstone is generally marked by a sudden decrease in gamma ray and increase in velocity log response. In the vast majority of cases, the base of the Rona Member is marked by an unconformity with Lewisian basement which on wireline well logs, is defined by a slight to significant downward decrease in gamma-ray and an increase in velocity values.

Solan Sandstone Member

The Solan Sandstone Member, first described by Herries et al. (1999), is best preserved in wells 205/26a-4 and -5Z within the East Solan Basin (Figures 59, 60a and 67). It is described as being comprised of colourless, grey to brown, very fine to medium-grained, loose to friable, angular to rounded, moderately well-sorted, slightly glauconitic, pyritic and cemented sandstone with subordinate interbedded thin mudstone and grey to brown, soft to hard, micaceous, slightly calcareous and pyritic basal siltstone. The sandstone is generally massive in appearance with signs of de-watering and postdepositional slumping (Herries et al., 1999). The member has been subdivided into 'Upper' and 'Lower' Solan sands (Figures 67 and 68) and both of these subunits are interpreted to represent turbidites, deposited on slopes within a marine (confirmed by the presence of glauconite) shelf or basinal environment (Herries et al., 1999). This contrasts with the view of Vestralen (1996, in Herries et al. (1999)) who regarded them as gully-fill deposits, similar to that described from the Hareelv Formation on East Greenland. According to Herries et al. (1999), palynomorph assemblages indicate that the lower part of the Solan Sandstone Member occurs within the Eudoxus to Autissiodorensis standard (ammonite) zones i.e. late Kimmeridgian, whereas the upper part occurs within the Hudlestoni to Pectinatus standard (ammonite) zones i.e. Early Tithonian (Table 4 and Figure 68). Although these ages seem broadly similar to those of the Rona Member (see above), Herries et al. (1999) believe the Solan Sandstone Member to be slightly older. The Solan Sandstone Member has an easily recognisable wireline well log signature with the gamma-ray log in particular displaying a classic blocky or box car response that picks out the Upper and Lower Solan sandstone units (Figures 67 and 68).

DISTRIBUTION, NATURE AND AGE

The distribution, nature and age of the Jurassic strata within the Faroe–Shetland area is generally poorly understood, mainly because only 42 wells and boreholes (Figure 59 and 60) have reached this stratigraphical level. Furthermore, within the Faroe–Shetland Basin, the limited seismic resolution of pre-Cretaceous strata within the central and south-east part of the basin steadily worsens towards the north-west, where there is masking by Palaeogene lava and intrusive rocks (e.g. Figures 30 and 31). Although Jurassic rocks have not been proved within the north-west part of the basin, their presence has been predicted, for example, by Spencer et al. (1999), Keser Neish (2003) and Ziska and Andersen (2005).

Lower Jurassic

Lower Jurassic rocks have been drilled only by five wells and one BGS borehole within the report area

Table 4 Correlation of age verses ammonite zonation for the Late Jurassic to Early Cretaceous interval (modified from Riding and Fensome, 2002).

PERIOD	EPOCH	AGE		STANDARD (AMMONITE) ZONE
CRETACEOUS	EARLY	BERRIASIAN	Late	Albidum
				Stenomphalus
				Icenii
			E.	Kochi
				Runctoni
JURASSIC	LATE	TITHONIAN	Late	Lamplughi
				Preplicomphalus
				Primitivus
			Mid	Oppressus
				Anguiformis
				Kerberus
				Okusensis
				Glaucolithus
				Albani
				Fittoni
				Rotunda
				Pallasioides
			Early	Pectinatus
				Hudlestoni
				Wheatleyensis
				Scitulus
				Elegans
		KIMMERIDGIAN		Autissiodorensis
				Eudoxus
				Mutabilis
				Cymodoce
				Baylei

98001938

(Figures 59 and 60). In contrast, they are known to be extremely well developed in the marginal basins that lie to the south-west of the Faroe–Shetland area. For example, up to 800 m of Lower Jurassic is present on the footwall block of the Minch Fault within the North Lewis Basin (Figure 59; BGS 1989b). Within its hanging-wall block, even greater thicknesses of 1.5 km and 2.5 km of Lower Jurassic strata are present within Sea of Hebrides–Little Minch Basin and North Minch Basin, respectively (Fyfe et al., 1993). On South Skye and Raasay, a stratigraphical thickness of approximately 600 m of Lower Jurassic rocks is considered to be present (Hudson and Trewin, 2002), whereas within the Erris and Slyne basins on the eastern flank of the Rockall Basin, a considerable, though non-specified thicknesses of Lower Jurassic rocks are reported to have been drilled (e.g. Scotchman and Thomas, 1995).

Faroe–Shetland Basin

Lower Jurassic rocks have been proven in only three wells on the Westray High and within the Foula Subbasin (Figure 59 and 60b). Drilling results indicate that they are absent from parts of the Corona and Westray highs in the central part of the basin, suggesting that they may not be widely developed.

Westray High

Well 204/19-9 drilled on the southern part of the Westray High penetrated approximately 185 m of interbedded white, pale brown to grey, very fine to fine-grained, friable to well cemented, subangular to subspherical, poor to well-sorted sandstone, pale grey to dark brown, dull, earthy, firm, blocky to very fissile, slightly to noncalcareous, microcarbonaceous mudstone/siltstone and a minor band of white, clean, hard limestone of late

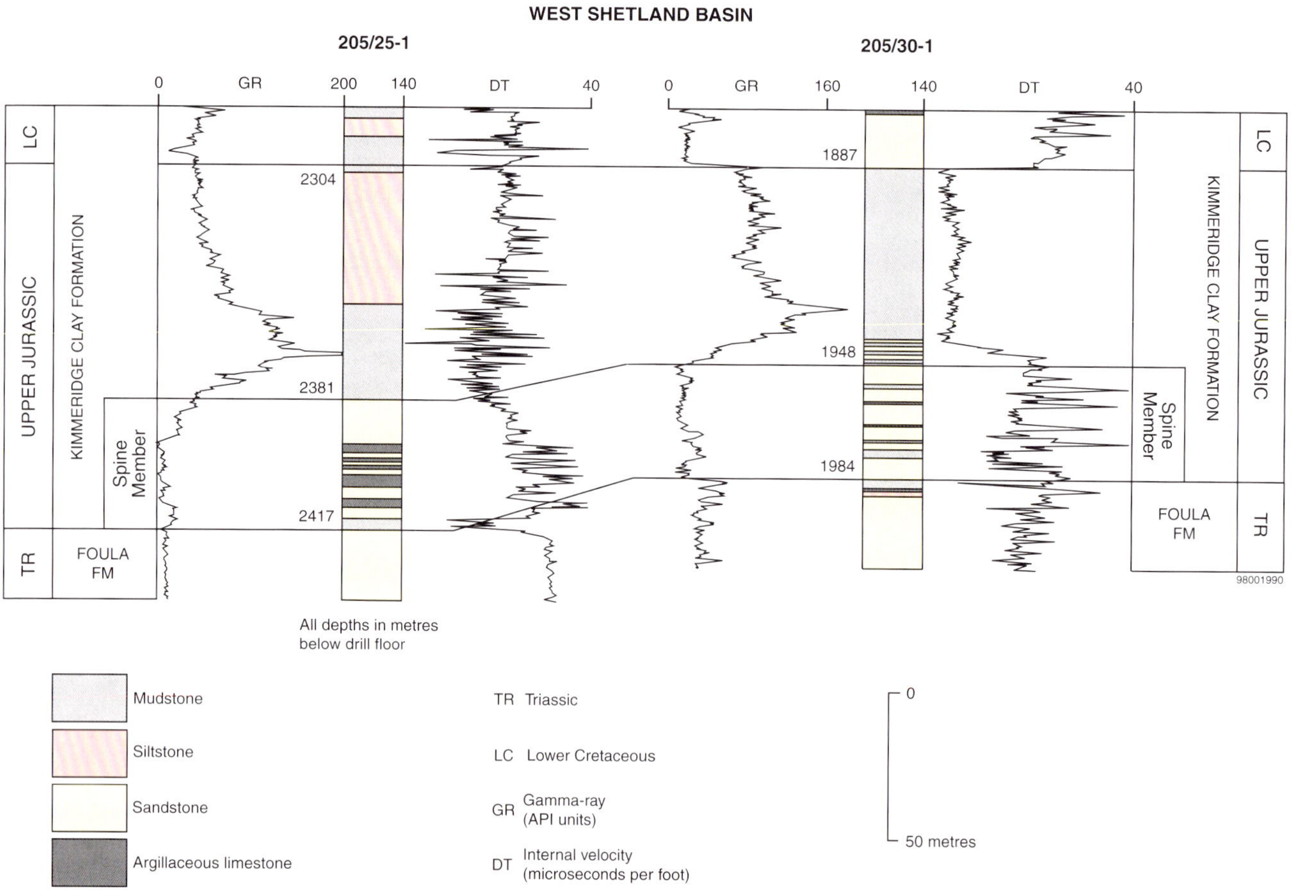

Figure 66 Well log correlation of the Upper Jurassic Kimmeridge Clay Formation, including the Spine Member in wells 205/25-1 and 205/30-1 within the south-west West Shetland Basin. For location of wells, see Figure 59.

Sinemurian to late Pliensbachian age (Figures 59 and Figure 60b). On the basis of age and a blocky wireline gamma-ray well log response, the succession has been tentatively ascribed to the Stack Skerry Formation as defined in type section described in well 202/03a-3. The Jurassic in well 204/19-9 rests unconformably on Triassic strata (Figure 55a) and is overlain by Lower Cretaceous Cruiser Formation (see Figure 77).

Foula Sub-basin

Lower Jurassic rocks have been drilled by adjacent wells 206/05-1 and -2 close to the south-east margin of the Foula Sub-basin (Figure 59). Well 206/05-1 terminated after encountering 28.5 m of pale grey to grey, moderately hard, micaceous, pyritic and noncalcareous sandstone (Figure 60b). Due to the presence of the dinoflagellate cyst ?*Mancodinium semitabulatum* (see Haszeldine et al., 1987), a Sinemurian to Pliensbachian age has been tentatively ascribed to this lithostratigraphically undivided unit. According to Haszeldine et al. (1987), the wireline well log response is typical of that expected from sediments deposited within a submarine fan environment. The Lower Jurassic in well 206/05-1 is interpreted to be unconformably overlain by undivided Bajocian strata, and younger Heather Formation rocks (Figure 60b). Nearby well 205/05-2 proved 338.6 m of mainly pale grey to off white, locally medium grey to brown, very fine to fine-grained, locally medium to very coarse-grained, angular to subspherical, poor to moderately well-sorted, partly calcite cemented sandstone, with traces of coal, gneissose and schistose lithic fragments, pyrite, kaolin and glauconite. There are also reported minor occurrences of pale to dark grey, soft to hard, waxy mudstone that is blocky to fissile, calcareous to noncalcareous, micaceous, carbonaceous, chloritic and containing traces of glauconite, pyrite and limestone. A Hettangian to early Sinemurian age has been ascribed to the unit and is tentatively assigned to the Skerry Group (possibly the Stack Skerry Formation) of Ritchie et al. (1996) (Figure 60b). This Jurassic succession rests on Triassic strata (Figure 60b) and is unconformably overlain by Lower Cretaceous Cruiser Formation (see Figure 79).

Marginal basins and highs

Immediately to the east and south of the Faroe–Shetland Basin, the preservation potential of Lower Jurassic strata is limited, probably mainly due to the local effects of Mid to Late Jurassic and regional Cenozoic episodes of uplift and erosion. Within the report area limits, drilled Lower Jurassic rocks have only been encountered in two wells from the marginal basins and structural highs and in a single BGS borehole (Figures 59 and 60a).

South Solan Basin

Well 202/04-1 drilled within the South Solan Basin encountered 30.5 m of pale grey, loose, fine to coarse-

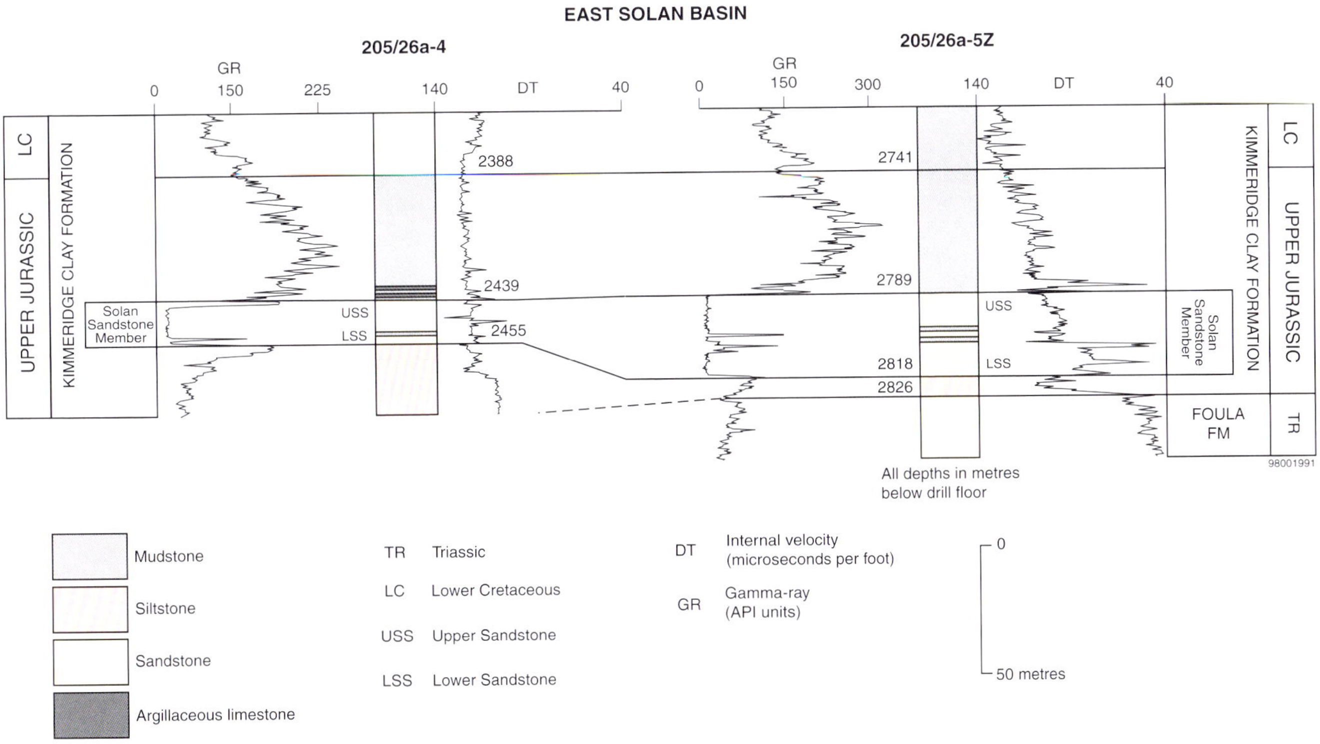

Figure 67 Well log correlation of the Upper Jurassic Kimmeridge Clay Formation, including the Solan Sandstone Member in wells 205/26a-4 and 205/26a-5Z within the East Solan Basin. For location of wells, see Figure 59.

grained, subangular to subspherical sandstone including traces of coal and pyrite interbedded with minor pale grey, brown to brown-black, dull, firm and non-calcareous, siliceous, silty mudstone and a single bed of coal (Figures 59 and 60a). A Hettangian to Sinemurian age has been ascribed to this unit and it is tentatively correlated with rocks of similar lithology and age from the Stack Skerry Formation defined in well 202/03a-3 within the adjacent West Solan Basin. Upper Jurassic Kimmeridge Clay Formation sandstone unconformably overlies the succession although the nature of the transition with underlying indeterminate rocks is unclear.

West Solan Basin

Well 202/03a-3 is located in the West Solan Basin and drilled 770 m of Lower Jurassic strata of the Skerry Group within which the well terminated (Figures 59 and 60a). The succession can be divided into an older Stack Skerry Formation and a younger Sule Skerry Formation. The Stack Skerry Formation comprises approximately 513 m of sandstone with subordinate interbedded silty mudstone and sandy siltstone (see description of Stack Skerry Formation above) and is late Sinemurian to early Pliensbachian in age (Ritchie et al., 1996). The overlying Sule Skerry Formation comprises 257 m of mudstone with minor very thin interbedded siltstone and argillaceous limestone (see description of Sule Skerry Formation above) of Early Pliensbachian age (Ritchie et al., 1996). The Sule Skerry Formation is unconformably overlain by the Upper Jurassic Rona Member. The Lower Jurassic succession within well 202/03a-3 is the thickest drilled within the Faroe–Shetland report area but is not proven in the adjacent North Rona Basin or East Solan basins (Figures 59 and 60a). A possible explanation for this could be the influence of nearby north-west-trending transfer zones on Early Jurassic sedimentation and uplift patterns.

West Orkney Basin

Towards the south-west margin of the West Orkney Basin, BGS borehole 72/34 terminated after drilling 3.8 m of brick-red bioturbated siltstone with minor sandy lenses of undivided ?Lower Jurassic that is thought to be of Hettangian to Sinemurian age (Figures 59 and 60a; Stoker et al., 1993).

Middle Jurassic

Strata of Middle Jurassic age have only been drilled by three wells within West Shetland Basin, Foula Sub-basin and on the Judd High (Figures 59 and 60) so their overall distribution is largely a matter of conjecture. However, to the south-west of the Faroe–Shetland region, approximately 500 m of Middle Jurassic is preserved on northern Skye (Hudson and Trewin, 2002). On the eastern flank of the Rockall Basin, considerable thicknesses of Middle Jurassic have been inferred to occur with the Erris and Slyne basins (e.g. Dancer et al., 1999; Chapman et al., 1999) whereas further to the north-east, a thin Middle Jurassic succession has been drilled in BGS borehole 88/01on the south-east flank of the West Lewis Basin (Isaksen et al., 2000). To the north-east of the report area, Middle Jurassic rocks are considered to be present in the Magnus, Unst and East Shetland basins, with less than 150 m of strata present in the latter (Johns and Andrews, 1985; Johnson et al., 2005a).

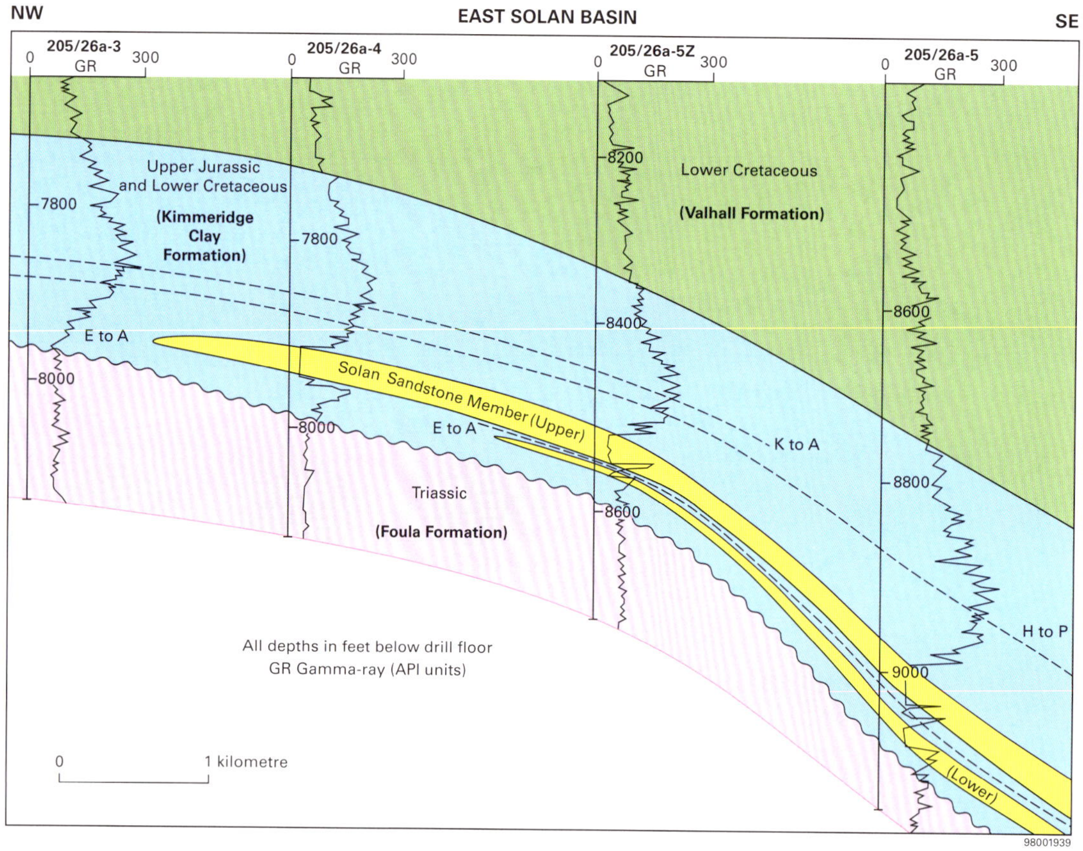

Figure 68 Geological cross-section showing the occurrence, relative age and well log response of the Solan Sandstone Member from wells 205/26a-3, -4, -5Z and 5 within the East Solan Basin (modified from Herries et al., 1999). Abbreviations of standard ammonite zones: K to A=kerberus to anguiformis; H to P=hudlestoni to pectinatus; E to A=eudoxus to autissiodorensis. For location of wells and relative age of ammonite zones, see Figure 59 and Table 4, respectively.

Faroe–Shetland Basin

Middle Jurassic strata have been drilled in well 206/05-1 within the Faroe–Shetland Basin (Figure 59). Previously, a sandstone-rich succession present within well 205/22-1A on the south-east flank of the Flett Sub-basin was thought to be of Middle Jurassic age (e.g. Stoker et al., 1993), but has subsequently been re-interpreted as Upper Jurassic by Vestralen and Hurst (1994) and Vestralen et al. (1995). Although absent on parts of the Corona and Westray highs, the Middle Jurassic is predicted by some to have widespread development throughout the basin (e.g. Holmes et al., 1999, fig. 5). In the south-west part of the Faroe–Shetland Basin, this inference is supported by Scotchman et al. (1998), who revealed that the geochemical composition of Middle Jurassic source rocks within well 204/22-1 on the Judd High (Figure 59) matched a component of the hydrocarbons recovered from the Foinaven Field within the Judd Sub-basin. This finding was confirmed by Cawley et al. (2005) who also suggested the presence of a Middle Jurassic source within wells 204/19-2, -6, 204/28-1 and 6004/16-1Z (Figure 2).

Foula Sub-basin

Middle Jurassic strata have been drilled within well 206/05-1 towards the south-east margin of the Foula Sub-basin (Figure 59). Here, approximately 600 m of Middle Jurassic clastic sediments occur, the upper part of which belongs to the oldest part of the Heather Formation whereas the lower part is lithostratigraphically undivided (Figures 60b and 63). The Middle Jurassic Heather Formation (including the Fair Sandstone Member) is described as comprising approximately 400 m of sandstone and mudstone (see description of Heather Formation above).

Marginal basins and highs

Middle Jurassic strata are thought to be largely absent in the area to the east and south of the Faroe–Shetland Basin, with only single wells in the West Shetland Basin and the Judd High proving their presence (Figures 59 and 60). Re-interpretation of a number of wells previously thought to contain Middle Jurassic within the North Rona Basin (202/02-1, 202/03-1A) West Shetland Basin (205/25-1, 205/30-1), Rona High (205/21-1A, 205/26-1) (e.g. see Stoker et al., 1993) are now believed to be of Late Jurassic age (Vestralen and Hurst, 1994; Vestralen et al., 1995). The absence of the Middle (and Lower) Jurassic is most likely as a result of local Mid to Late Jurassic and more regional Cenozoic inversion and denudation.

West Shetland Basin

Well 205/20-2 was drilled within the southern part of the basin and penetrated 23.8 m of buff to pale brown, very fine to medium-grained, well to subrounded, moderately well-sorted and consolidated, slightly to noncalcareous, ?kaolinite cemented muddy Middle Jurassic sandstone with intercalations of carbonaceous material (Figure 59 and 60b). The succession is apparently of Bajocian to Callovian in age but is lithostratigraphically undivided. The unit rests unconformably on Triassic sandstone (Figure 55a) and is overlain by an undivided Upper Jurassic sandstone-rich succession (Figure 60b).

Judd High

Well 204/22-1 occurs close to the boundary of the Judd High and Judd Sub-basin and drilled approximately 246 m of mainly pale grey to locally dark grey, medium brown to pale/medium grey-brown, dull, early, soft to firm, blocky to subfissile, homogenous to heterogeneous, moderately to noncalcareous, microcarbonaceous, pyritic mudstone of Middle Jurassic age (Figure 59 and 60a). This unit also includes occurrences of thin white to opaque, firm, blocky cryptocrystalline limestone, fine-grained subrounded and well- sorted glauconitic sandstone and siltstone. The lithological character of this succession, together with its wireline well log signature, could suggest a correlation with the Sule Skerry Formation as defined in the type section within well 202/03a-3 in the nearby West Solan Basin. Although the succession drilled within 204/22-1 is thought to be much younger i.e. Bajocian to Callovian, this does not necessarily preclude the age of the Sule Skerry Formation extending beyond the Pliensbachian age as defined in the type well. The unit rests on Precambrian crystalline basement and is conformably overlain by Heather Formation mudstone of Oxfordian age (Figure 60a).

Upper Jurassic

In contrast with the Lower and Middle Jurassic successions, Upper Jurassic rocks have been drilled by 39 wells within the Faroe–Shetland area (Figure 59 and 60). To the south-west of the report area, the Upper Jurassic is absent in the marginal basins but to the north-east, it is considered to be up to 250, 750 and 500 m thick in the Magnus, Unst, and north East Shetland basins, respectively (Johnson et al., 1993; Johnson et al., 2005a). Within the adjacent Møre Basin however, its distribution and thickness is poorly understood (e.g. Blystad et al., 1995; Brekke et al., 1999).

Faroe–Shetland Basin

Upper Jurassic strata have been drilled in only six wells within the Faroe–Shetland Basin i.e. 204/15-2, 204/23-1, 204/29-1, 205/22-1A, 206/05-1 and 214/09-1 (Figures 59 and 60b), although their presence has been predicted over large parts of the basin (e.g. Ziegler, 1990; Holmes et al., 1999; Spencer et al., 1999; Tate et al., 1999; Hudson and Trewin, 2002; Coward et al., 2003; Cawley et al., 2005; Ziska and Andersen, 2005; Figures 30 and 31). However, within the centre of the basin, Upper Jurassic rocks are known to be locally absent on the Corona and Westray highs (Figures 59 and 65).

Foula Sub-basin

A thick Upper Jurassic succession was penetrated within well 206/05-1 located on the south-east margin of the Foula Sub-basin (Figure 59). Approximately 411.5 m of Upper Jurassic sediments assigned to the Kimmeridge Clay (295.5 m) and Heather formations (118 m) have been drilled, although the latter extends down into the Middle Jurassic (Figures 60b and 63). The Upper Jurassic part of the Heather Formation is described as dominantly comprising 118 m of grey to black, micaceous, carbonaceous, noncalcareous, silty mudstone with fish remains and including minor sandstone beds. The overlying Kimmeridge Clay Formation is 295.5 m thick and generally comprises mudstone and minor siltstone (see Kimmeridge Clay Formation described above). These sediments grade to thin sandstone and rudaceous rocks of the Ridge Conglomerate Member (Figure 63). The Upper Jurassic Kimmeridge Clay and Heather formation is Oxfordian to Tithonian in age (Figure 63; Haszeldine et al., 1987). The Kimmeridge Clay is unconformably overlain by Lower Cretaceous Cruiser Formation sediments (see Figure 79). It should be noted that the Upper Jurassic succession is locally absent in adjacent well 206/05-2, although the reason for this remains unclear.

Corona High

Approximately 73.1 m of Upper Jurassic Kimmeridge Clay Formation were drilled in well 214/09-1, located towards the north-east end of the Corona High (Figures 59 and 60b). The succession mainly consists of colourless, clear, yellow-white to orange-brown, angular, very fine to medium-grained sandstone with grains coated in dark red haematite and fragments of jasper and medium to dark grey to brown-grey, firm to slightly hard, blocky to subfissile, slightly carbonaceous, pyritic mudstone with sporadic silty and sandy streaks. Minor pale to dark grey, very finely sandy, micaceous and very calcareous micro-laminated siltstone also occurs. According to biostratigraphical analyses from released well information, this unit has an early Kimmeridgian to Early Tithonian age. It is unconformably bounded at top and base by undivided Lower Cretaceous Cromer Knoll Group (see Figure 81) and strata of ?Triassic age (Figure 55a), respectively. Towards the central south-west part of the Corona High, Jurassic strata are absent within wells 213/23-1 and 204/10-1 (Figure 2).

Flett Sub-basin

Towards the south-west corner of the Flett Sub-basin, well 205/22-1A drilled 50.5 m of Upper Jurassic Kimmeridge Clay Formation, the basal 39 m of which is interpreted to belong to the Rona Member (Figures 59 and 60b). This thin Jurassic succession occurs within the hanging-wall block of the mainly Late Cretaceous Rona Fault (Haszeldine et al., 1987, fig. 5). Here, the Rona Member comprises mainly sandstone with intervals of conglomerate and breccia towards the base and capped by a pale grey limestone bed. The clastic rocks are ascribed to facies associations A (?subaerial debris-flow) and B1 (turbidite) of Vestralen et al. (1995, fig. 5). Although originally considered to be of Mid Jurassic age (e.g. Stoker et al. 1993), Vestralen et al. (1995, fig. 4) suggested that the Rona Member is actually late Kimmeridgian to Early Tithonian in age. The Rona Member is overlain by approximately 11.5 m of black mudstone, typical of the Kimmeridge Clay Formation, with limestone and siltstone. The Kimmeridge Clay Formation rests on Precambrian crystalline basement and is unconformably overlain by Lower Cretaceous Valhall Formation (see Figure 81).

Westray High

Well 204/15-2 was drilled within the northern part of the Westray High and proved 236.8 m of Tithonian Kimmeridge Clay Formation (Figures 59 and 60b). The succession is described as mainly comprising brown-grey to brown-black, soft, blocky, carbonaceous and micaceous siltstone. The basal part of the succession contains some thin sandstone intervals up to approximately 3 m thick, described as white to pale grey or dark red-brown, soft to locally firm, fine to coarse-grained, subrounded to subangular, moderately to well-sorted and cemented, and containing glauconite and carbonaceous material. The upper part contains sporadic thin white to pale brown and green speckled, soft to hard, locally pyritic limestone and a notable 1 m thick band of white to pale grey and yellow, soft, moderately calcite cemented, carbonaceous sandstone with subrounded to subspherical, well-sorted grains. The succession as a whole rests on Precambrian crystalline basement and is unconformably overlain by lithostratigraphically undivided Lower Cretaceous (see Figure 77).

Judd Sub-basin

Located close the south-west margin of the Judd Sub-basin, wells 204/23-1 and 204/29-1 proved 4.9 m and approximately 23.5 m of argillaceous Upper Jurassic Kimmeridge Clay Formation, respectively (Figures 59 and 60b). Well 204/23-1 was drilled on the crest of a hanging-wall block that forms one of a series of Jurassic to Cretaceous tilted-fault blocks (Figure 27) and proved a thin late Kimmeridgian to Tithonian succession of grey to black-brown, homogeneous, hard, subfissile, slightly silty, very calcareous, carbonaceous mudstone. This unit rests on Precambrian crystalline basement and is unconformably overlain by Lower Cretaceous Neptune Formation (see Figure 77). The Kimmeridge Clay Formation within well 204/29-1 comprises black, dark brown to grey, firm to moderately hard, subfissile to blocky, carbonaceous, slightly to noncalcareous mudstone with occasional traces of buff to tan, hard to firm microcrystalline dolomite. This succession rests on lithostratigraphically undifferentiated Triassic (Figure 55a) and passes vertically up into Kimmeridge Clay Formation of earliest Cretaceous (Berriasian) age (see Figure 77).

MARGINAL BASINS AND HIGHS

Upper Jurassic rocks are present within many of the basins and highs that fringe the eastern margin of the Faroe–Shetland Basin i.e. North Rona Basin, Solan Bank High, Solan basins, Judd High, south-west Rona High, West Shetland Basin, Papa Basin and Erlend High (Figures 59 and 60). Further to the east around Shetland and Orkney, Upper Jurassic strata are absent, possibly due to nondeposition (Hudson and Trewin, 2002).

North Rona Basin

Upper Jurassic Kimmeridge Clay Formation has been penetrated by five wells within the north-east part of the North Rona Basin i.e. 202/02-1, 202/03-1A, 202/03-2, 202/08-1 and 202/12-1 (Figures 59 and 60a). Here, the Jurassic succession is interpreted to occur as thin, south-easterly-dipping wedges developed within the hanging-wall blocks of a series of half-grabens.

Well 202/02-1 was drilled on the crest of a tilted fault block (Figure 28) and proved 51.8 m of Upper Jurassic Kimmeridge Clay Formation that includes 21.3 m of the Rona Member (Figures 59 and 60a). The Rona Member is described as comprising white to pale grey, unconsolidated coarse-grained, well-sorted, subrounded, calcareous, micaceous sandstone. Although this, and all other Jurassic sand-prone successions encountered in wells within the North Rona Basin (see below) were originally thought to be of Mid Jurassic age (e.g. Stoker et al., 1993), Vestralen and Hurst (1994) and Vestralen et al. (1995, fig. 4) suggested a late Kimmeridgian to Early Tithonian age to be more appropriate. The Rona Member is overlain by 30.5 m of dark grey-brown, soft, noncalcareous mudstone and minor siltstone. Here, the mainly Upper Jurassic Kimmeridge Clay Formation extends into the earliest Cretaceous (Berriasian) (see Figure 73) and rests on Precambrian crystalline basement. Well 202/08-1 is also drilled close to the crest of a tilted fault block (Figure 28) and contains approximately 80.5 m of Upper Jurassic Kimmeridge Clay Formation, the basal 7 m of which comprises pale grey, friable, bioturbated sandstone and siltstone of the Rona Member. The Rona Member is overlain by approximately 73.5 m of dark grey to black, soft, silty, carbonaceous mudstone and minor siltstone. Here too, the mainly Upper Jurassic Kimmeridge Clay Formation extends up to the earliest Cretaceous (Berriasian) (see Figure 73) and rests on Precambrian crystalline basement. Well 202/12-1 contains 199 m of Kimmeridge Clay Formation, the basal 105.5 m of which is assigned to the Rona Member. This unit comprises mainly buff, clear to white, loose to firm, friable, fine to medium and occasionally coarse-grained, locally pebbly sandstone with subrounded to spherical clasts and a very fine to silty matrix including kaolin. The sandstone is slightly carbonaceous, micaceous and pyritic with minor red to red-brown, dull, soft to moderately hard, subfissile, calcareous, slightly micromicaceous mudstone grading to siltstone. The Rona Member is overlain by 93.5 m of predominately grey, dull, hard, subfissile carbonaceous mudstone typical of the Kimmeridge Clay Formation, with fish remains, worm burrows and a part of an ammonite, together with traces mica and pyrite. The unit includes subordinate amounts of dark brown to pale grey, dull, soft to firm, blocky, amorphous, friable, slightly to noncalcareous, micromicaceous siltstone with carbonaceous traces and grey-white, firm, blocky, slightly sandy argillaceous limestone. This unconformity-bounded Kimmeridge Clay Formation rests on undifferentiated Triassic strata (Figure 55a) and is overlain by Lower Cretaceous Valhall Formation (see Figure 73). Well 202/03-1A encountered 121 m of Kimmeridge Clay Formation, the lower part of which comprises 82 m of pale grey, fine to medium-grained, moderately well-sorted, subrounded, poorly consolidated and friable, variably carbonaceous, glauconitic, occasionally calcareous sandstone with traces of pyrite that has been assigned to the Rona Member. This sandy succession is interpreted to form part of unit FA1 (subfacies D), deposited as a result of high-concentration turbidity currents (Vestralen et al., 1995, figs. 5 and 7; Figure 64). The Rona Member is overlain by 39 m of dark brown to black, carbonaceous, silty, slightly calcareous mudstone. The Kimmeridge Clay Formation rests on Precambrian crystalline basement and is unconformably overlain by Lower Cretaceous Valhall Formation (see Figure 73). Well 202/03-2 contains 73 m of Kimmeridge Clay Formation, the lower 41 m of which is assigned to the Rona Member. This unit comprises pale grey, very fine to fine-grained, silty, micaceous, carbonaceous sandstone with subrounded grains, traces of pyrite and tourmaline, which is interbedded with varicoloured kaolinitic mudstone with occasional quartz grains. The Rona Member can be subdivided into the FA1 and FA2 units of Vestralen et al. (1995, fig. 7) and is overlain by 32 m of dark brown, grey to black, very carbonaceous, silty, slightly calcareous mudstone with fish remains. Here, the Kimmeridge Clay Formation is separated by unconformities from the underlying Precambrian crystalline basement and the overlying Lower Cretaceous Valhall Formation (see Figure 73).

Solan Bank High

Well 202/09-1 drilled within the north-east part of the Solan Bank High proved 93.9 m of Kimmeridge Clay Formation (Figures 59 and 60a). The basal 56.5 m of the succession is described as comprising grey, firm to hard, fine to medium-grained, occasionally conglomeratic, calcareous to noncalcareous, carbonaceous, pyritic sandstone provisionally assigned to the Rona Member. The Rona Member is overlain by 37.4 m of dark grey to grey-black, firm, non-fissile and micromicaceous mudstone. The Kimmeridge Clay Formation rests unconformably on Triassic strata (Figure 55a) and is overlain by Lower Cretaceous Valhall Formation (see Figure 73).

West Solan Basin

The West Solan Basin has been drilled by wells 202/03a-3 and 204/29-2 that proved Upper Jurassic Kimmeridge Clay Formation (Figures 59 and 60a). Well 202/03a-3 encountered 25.9 m of Upper Jurassic Kimmeridge Clay Formation, the basal 4.7 m of which comprises greyish blue-green, crumbly to hard, fine to medium and locally coarse-grained, subangular to spherical, moderate to well-sorted sandstone with a silica cement and trace mica, carbonaceous fragments and glauconite, that is assigned to Rona Member (Figure 62). This unit is overlain by 21.2 m of dark brown to black, hard, blocky, greasy, slightly silty, slightly to non-calcareous, very carbonaceous mudstone. The Kimmeridge Clay Formation is unconformity-bounded, resting on Lower Jurassic Sule Skerry Formation and overlain by Lower Cretaceous Valhall Formation (Figure 62). Well 204/29-2 drilled in the north-east part of the basin, encountered 35.3 m of grey, red to brown, orange and yellow, fine to coarse-grained, subangular to spherical, poor to moderately well-sorted sandstone of the Rona Member, that is locally pebbly to conglomeratic, and cemented with kaolinite, feldspar, calcite and quartz. Subordinate mudstone is described as red-brown, grey-red, grey, and green to brown, firm to moderately hard, mottled, slightly carbonaceous, silty, noncalcareous with occasional fine grains of quartz. The Kimmeridge Clay Formation is bounded at its base and top by undivided Triassic (Figure 55a) and Lower Cretaceous Rødby Formation, respectively (see Figure 73).

South Solan Basin

Well 202/04-1 is the only well drilled within the basin and proved 129.2 m of Upper Jurassic Kimmeridge Clay Formation (Figures 59 and 60a). The basal 104.2 m of the succession is assigned to the Rona Member and is described as comprising white to pale grey, occasionally orange to pale brown, loose to friable, fine to occasionally coarse-grained, subangular to spherical, poor to well-sorted sandstone with patchy calcite and quartz cement and trace pyrite, kaolinite, glauconite and coal fragments. This sand-rich succession also includes occurrences of grey, dull, moderately hard, blocky to subfissile, noncalcareous indurated mudstone with traces of shell impressions, quartz grains and pyritised lignite fragments. The Rona Member is overlain by 25 m of dark brown, dull, firm, friable to compact, subfissile, slightly to noncalcareous, carbonaceous siltstone and mudstone with trace disseminated pyrite and fish remains. The Upper Jurassic Kimmeridge Clay Formation extends into the earliest Lower Cretaceous (Berriasian) (see Figure 73) and is unconformable on Lower Jurassic Stack Skerry Formation (Figure 60a).

Rona High

Upper Jurassic strata are present in six wells drilled on the Rona High i.e. 204/30-1, 205/20-1, 205/21-1A, 205/23-2, 205/26-1 and 207/01a-4Z (Figures 59 and 60b).

Towards the south-west end of the Rona High, well 204/30-1 (Herries et al., 1999, fig. 8) penetrated 45.4 m of Kimmeridgian to Tithonian Kimmeridge Clay Formation that mainly comprises dark brown to black, soft to firm, silty, noncalcareous mudstone with a minor bed of sandstone towards the base of the succession (Figures 59 and 60b). The sand-prone basal part of the succession is not considered to represent the Rona Member on the grounds of age and environment of deposition (see Vestralen and Hurst, 1994). The Upper Jurassic succession as a whole rests on Precambrian crystalline basement and is unconformably overlain by Lower Cretaceous Valhall Formation strata (see Figure 76). Well 205/21-1A drilled 10.4 m of Upper Jurassic Kimmeridge Clay Formation strata comprising mainly pebbly sandstone with pyrite and bioclastic material that becomes coarser towards the base. It was drilled in 1974 and was the first well to prove the late Kimmeridgian to Early Tithonian Rona Member within the Faroe–Shetland area (Vestralen and Hurst, 1994). In this well the Rona Member can be subdivided into the FA1 and FA2 facies association units of Vestralen et al. (1995, fig. 7). The Rona Member rests on Precambrian crystalline basement and is unconformably overlain by Lower Cretaceous Valhall Formation (see Figure 76). Well 205/26-1 proved 76.2 m of Upper Jurassic Kimmeridge Clay Formation (Figure 39). This succession comprises 24.4 m of white, grey to green, fine to coarse-grained, angular, hard, calcareous to noncalcareous sandstone with minor amounts of reddish brown, hard, slightly calcareous mudstone belonging to the Rona Member, overlain by 51.8 m of medium grey to dark brown, soft, earthy, carbonaceous, pyritic and partially calcareous mudstone typical of Kimmeridge Clay Formation. The Kimmeridge Clay Formation is unconformity-bounded, resting on Precambrian crystalline basement and overlain by Lower Cretaceous Valhall Formation (see Figure 76). Towards the north-east, Well 205/20-1 was drilled on the south-east flank of the Rona High and penetrated 113 m of Upper Jurassic Kimmeridge Clay Formation (Figure 37). This succession is considered to be mainly late Kimmeridgian to Tithonian in age, the majority of which comprises mainly grey-brown, very soft, very silty slightly calcareous mudstone with trace glauconite, fine-grained sandstone and lignite. There is also a thin sandstone unit, which is not thought to be representative of the Rona Member (Vestralen and Hurst, 1994). The uppermost part of the unit comprises grey to brown, very fine calcareous siltstone with grey to brown, very soft, very silty and slightly calcareous mudstone. The succession as a whole rests on Precambrian crystalline basement and is unconformably overlain by Lower Cretaceous Victory Formation (see Figure 76). Well 205/23-2 drilled 34.1 m of undivided Upper Jurassic strata comprising interbedded pale grey to grey, soft to moderately hard, blocky, noncalcareous mudstone with traces of pyrite and glauconite and poorly sorted sandstone with colourless and clear to cloudy, mainly coarse, angular to subrounded grains, and traces of pyrite and lignite and distinct layers of black to dark brown, shiny to dull, moderately hard, blocky, angular lignite with traces of pyrite. Towards the base, the succession is more arenaceous and described as comprising colourless to clear or cloudy, mainly coarse-grained to pebbly, locally argillaceous, sandstone with pyrite and lignite. This lithostratigraphically unassigned Upper Jurassic succession rests unconformably on Triassic strata (Figure 55a) and is unconformably overlain Upper Cretaceous Shetland Group (see Figure 76).

Well 207/01a-4Z drilled 24.1 m of Upper Jurassic strata and is the only well with the north-east part of the Rona High to prove rocks of this age (Figures 59 and 60b). The lower part of the succession comprises 19.8 m of clear to translucent, white to grey, fine to coarse-grained, friable, poorly sorted noncalcareous sandstone with subangular to rounded grains, traces of pyrite and sporadic pale grey to blue-grey, firm, noncalcareous argillaceous streaks and laminae. The stratigraphical relationship of this sandy unit to the Rona and Spine member described from wells in the Rona High and West Shetland Basin, respectively remains uncertain (Figure 65). The upper 4.3 m consists of firm, dark grey, noncalcareous, silty, micromicaceous mudstone and is typical of the Kimmeridge Clay Formation. The Jurassic succession as a whole is unconformity-bounded, resting on Devono-Carboniferous Clair Group and overlain by Lower Cretaceous Victory Formation (see Figure 76).

West Shetland Basin

Upper Jurassic strata have only been proved in four wells drilled within the south-west part of the West Shetland Basin i.e. 205/20-2, 205/23-1, 205/25-1 and 205/30-1 (Figures 59 and 60b). Jurassic strata are absent from the central part of the basin, probably due to uplift and erosion in Cretaceous times but are present within the north-east part of this basin.

Towards the eastern margin of the basin, wells 205/25-1 and 205/30-1 proved 112.2 m and 97.5 m of Upper Jurassic, respectively (Figures 59 and 60b). Both wells contain approximately 36 m thick basal units of sandstone and subordinate mudstone with limestone horizons (see Spine Member described above) (Figure 66). Although originally thought to be of Mid Jurassic age (e.g. Stoker et al., 1993), these successions are now assigned to the Spine Member of the Upper Jurassic Kimmeridge Clay Formation (Vestralen and Hurst, 1994; Vestralen et al., 1995). However, there are conflicting views as to the detailed age of the member with late Kimmeridgian to Early Tithonian (Vestralen and Hurst, 1994; Vestralen et al., 1995) or late Kimmeridgian ages advocated (Herries et al., 1999). Within 205/25-1 and 205/30-1, the 76.2 and 60.9 m thick upper parts of the Upper Jurassic successions comprise very dark brown to dark grey, firm, hard, slightly fissile, micaceous, moderately calcareous, lignitic mudstone with traces of pyrite and rare layers of calciferous sandstone of Kimmeridgian to Late Tithonian age. Within the former well, the argillaceous succession coarsens upwards to dark to medium grey, firm, slightly to non-calcareous and micromicaceous siltstone with traces of glauconite and heavy minerals. In places, the siltstone grades to mudstone and very fine-grained sandstone. The Jurassic successions in both wells rest unconformably on Triassic strata (Figure 55a) but pass upwards into Lower Cretaceous (Berriasian) Kimmeridge Clay and Victory formations (see Figure 73).

Well 205/23-1, located close the western margin of the West Shetland Basin, drilled 113.4 m of dark brown to black, soft to firm, sticky, subfissile, pyritic silty, carbonaceous mudstone with minor loose calcareous sand assigned to the Kimmeridge Clay Formation (Figures 59 and 60b). This Kimmeridgian to Tithonian (although probably mainly of the latter age) succession rests unconformably on Triassic rocks (Figure 55a) and is overlain by the Lower Cretaceous Victory Formation (see Figure 73). Well 205/20-2 occurs towards the central part of the basin and penetrated 144.8 m of Upper Jurassic Kimmeridge Clay Formation. The uppermost 110.7 m comprises dark or very dark grey to brown or black, soft to hard, blocky or platy to subfissile, micaceous, slightly calcareous and carbonaceous, silty, earthy, waxy Tithonian mudstone with trace pyrite and glauconite, and minor pale brown, very fine, friable sandstone and rare white, crumbly argillaceous limestone. This rests on 34.1 m of lithostratigraphically unassigned mainly clear to white, translucent to transparent commonly very fine to medium-grained but occasionally coarse to very coarse-grained, poor to medium-sorted, friable, weakly calcareous cemented sandstone with subangular to rounded grains and a chalky white matrix and traces of pyrite and glauconite. The unit also contains beds of limestone and grey to dark grey, carbonaceous, micaceous mudstone with bivalve fragments. The sandy section is apparently of Tithonian age but its relationship with the broadly coeval Spine and Rona members present in surrounding wells within the south-west West Shetland Basin and the Rona High is currently unresolved.

Papa Basin

BGS borehole 77/09 encountered 0.7 m of black, poorly consolidated muddy sandstone of presumed Upper Jurassic age close to the sea bed within the Papa Basin (Figures 59 and 60b). However, it is likely that the Jurassic is thin or absent over the majority of the basin with Permo-Triassic strata occurring close to, or at the sea bed.

Erlend High

Well 209/12-1 drilled towards the southern margin of the Erlend High proved 86.9 m of Upper Jurassic sandstone and mudstone assigned to the Kimmeridge Clay Formation (Figures 59 and 60b). The lower part of the succession comprises 31.4 m of mainly pale to dark grey, angular to subangular, very hard, mottled sandstone with an argillaceous matrix and dolomitic cement, and a basal unit composed of dark grey, moderately hard, blocky, silty, micromicaceous, carbonaceous mudstone with local concentrations of pyrite and intruded by a Palaeogene dyke. The age of the sand-prone part of this succession is Oxfordian to Early Tithonian and appears to be slightly older than the Rona Member described elsewhere to the south-west of the Erlend High, and consequently is not assigned to that member. The basal mudstone-rich unit has an apparent Mid Oxfordian age, possibly suggesting that it could form part of the Heather Formation. The uppermost 55.5 m of Jurassic section encountered within the well dominantly comprises dark grey to very dark grey, firm to moderately hard, blocky, micromicaceous, slightly silty, noncalcareous thinly bedded mudstone with traces of pyrite that is fairly typical of the Kimmeridge Clay Formation. This unit is also cut by thin Palaeogene sills. The Jurassic succession as a whole rests on Precambrian crystalline basement and is unconformably overlain by undivided Cromer Knoll Group (see Figure 82).

East Solan Basin

Eight wells have penetrated Upper Jurassic Kimmeridge Clay Formation strata within the East Solan Basin (Figures 39, 59 and 60a). Well 204/30a-2 drilled 73.5 m of Upper Jurassic strata, with the stratigraphically oldest 8.2 m comprising siltstone, buff to pale grey silty limestone and medium to pale grey coarsening upwards sandstone and conglomerate with gneissic clasts. According to the well log, this succession forms part of the Heather Formation but probably should be assigned to the late Kimmeridgian to Early Tithonian Rona Member (Vestralen and Hurst, 1994; fig. 7). The overlying 65.3 m is comprised of medium to dark grey-brown, firm to hard, blocky to subfissile, earthy, silty, micromicaceous, calcareous, carbonaceous, pyritic mudstone typical of the Kimmeridge Clay Formation. As is the case for all wells that penetrated the Jurassic within the East Solan Basin, the succession rests on Triassic strata (Figure 55b) and is conformably overlain by Lower Cretaceous (Berriasian) Kimmeridge Clay Formation (see Figure 73). Well 205/26a-3 is drilled on the north-west slope of the basin (Figure 39) and contains approximately 54.9 m of Upper Jurassic Kimmeridge Clay Formation, the basal 8.5 m of which comprises mainly siltstone (but grading from silty mudstone to very fine-grained sandstone) formerly assigned to the Heather Formation but now interpreted as the Rona Member. The overlying 46.4 m is described as medium grey-brown to dark olive-brown, soft, locally subfissile, slightly calcareous, micaceous, carbonaceous mudstone that is sporadically pyritic, glauconitic with very fine-grained sandstone and coal fragments. Well 205/26a-2 occurs within the hanging-wall block of the Otter Bank Fault (Figure 39) and proved 35.1 m of Upper Jurassic Kimmeridge Clay Formation. This succession comprises dark brown to dark grey, firm to hard, subfissile, noncalcareous mudstone of late Kimmeridgian to Late Tithonian age. Well 205/26a-6 drilled 32 m of mainly medium brown-grey to olive-brown or dark olive-grey, firm to hard, blocky to subfissile, slightly calcareous, carbonaceous, sometimes micromicaceous Upper Jurassic Kimmeridge Clay Formation mudstone and siltstone with traces of pyrite and glauconite and rare sand grains. At the base of the succession is less than 1 m of dark grey-brown conglomerate that include quartzite, shale or lithic clasts.

Well 205/26a-4 proved 82 m of Upper Jurassic Kimmeridge Clay Formation strata, the majority of which comprises dark grey to grey-brown, pale brown to brown, hard, occasionally soft to firm, blocky to subfissile, earthy, slightly to noncalcareous mudstone with subordinate very dark grey-brown, hard micaceous, pyritic siltstone at the base of the succession (Figures 59 and 60a). Approximately 15.9 m of grey to brown, very fine to medium-grained, moderately well-sorted, calcareous cemented sandstone with rounded to subangular grains and minor mudstone also occur near the base of the Upper Jurassic section. According to Herries et al. (1999), this sandstone is considered to be form part of the Solan Sandstone Member that is probably restricted to the Solan basins area (Figure 65). From biostratigraphical evidence, they suggest that it is broadly similar in age (late Kimmeridgian to Early Tithonian) but slightly older than the Rona Member, and slightly younger than the Spine Member present within the West Shetland Basin and surrounding area (see above). The Solan Sandstone Member is easily distinguishable from the Rona Member by its striking box-car gamma-ray wireline log response (Figures 67 and 68), reflecting a cleaner and more mature nature. Similarly, 85 m of Upper Jurassic Kimmeridge Clay Formation was drilled in nearby well 205/26a-5Z, comprising mainly medium to dark grey to brown, olive-brown to dark olive-brown, firm to occasionally soft or hard, blocky to subfissile, silty, earthy, carbonaceous, slightly to moderately calcareous, micromicaceous mudstone and siltstone with disseminated pyrite. However, the lower part of the succession comprises pale to medium grey to brown, soft to firm but locally hard, friable to blocky, micromicaceous, earthy, slightly calcareous siltstone, with 29.5 m of colourless to pale grey, loose to friable, very fine to medium-grained sandstone with fair to moderately well-rounded, angular to subrounded grains and rare calcareous cement and traces of glauconite and pyrite and minor mudstone. This sandstone unit is assigned to the Solan Sandstone Member (Figures 67 and 68). Well 204/30a-3 drilled 120.7 m of Upper Jurassic Kimmeridge Clay Formation, the majority of which is described as comprising mainly Tithonian dusky yellow-brown to locally brown-black or medium grey, soft to hard, blocky, fissile to subfissile, sporadically waxy to subresinous and silty, carbonaceous, pyritic, micromicaceous and noncalcareous mudstone. Towards the base of this succession is a 13.1 m unit of medium grey, very hard, moderately well-sorted, very fine to medium-grained sandstone with angular to subspherical grains and a mixture of local pyrite, calcite and quartz cements with abundant glauconite, carbonaceous and lithic fragments. This sandstone unit is interpreted to belong to the Solan Sandstone Member of Herries et al. (1999). Well 205/27-2 is located in the extreme north-east corner of the basin and drilled 98.1 m of Upper Jurassic Kimmeridge Clay Formation of mainly Tithonian age. The basal 9.8 m of section comprises buff, pale grey to colourless, fine to very-fine grained, moderately well-sorted carbonaceous sandstone with subangular to subrounded grains, which has a calciferous matrix and traces of pyrite, glauconite and chlorite. The sandstone unit has a blocky gamma-ray well log response, possibly suggesting that it forms part of the Solan Sandstone Member. The upper part of the succession comprises dark grey-brown, soft to moderately hard, amorphous, blocky to occasionally subfissile, slightly to noncalciferous and silty micromicaceous mudstone with traces of pyrite and carbonaceous material.

Judd High

Three wells have drilled Upper Jurassic Kimmeridge Clay Formation strata on the Judd High (Figures 59 and 60a). Well 204/28-1 was drilled close to the northeast margin of the high (Figure 27) and proved 73 m of Upper Jurassic Kimmeridge Clay Formation. The uppermost 5.5 m of the succession comprises pale grey to occasionally buff, soft, sticky, slightly silty, slightly to moderately calcareous mudstone considered typical of the Kimmeridge Clay Formation. The underlying 67.5 m comprises predominately pale to medium grey, mainly fine-grained, calcite cemented arkosic sandstone with minor glauconite and medium brown, occasionally buff, sandy siltstone. The succession grades downwards into pebbly sandstone and fine to coarse-grained, poorly sorted, calcite cemented conglomerate with rounded quartz pebbles and mixtures of pale grey, fine to very coarse-grained, poor to well-sorted, massively bedded, fining upwards sandstone/conglomerate horizons including boulders of crystalline basement. According to Stoker et al. (1993), the arenaceous part of the succession was thought to be of Mid Jurassic age but is now assigned to the Rona Member (Vestralen and Hurst, 1994; Vestralen et al., 1995). This unit can be divided into a lower coarse-grained unit termed FA1 and a fine-grained FA2 unit (Figure 64). The Upper Jurassic Kimmeridge Clay Formation rests on Precambrian crystalline basement and is interpreted to be overlain by Kimmeridge Clay Formation of early Cretaceous (Berriasian) age (see Figure 73). Similarly, nearby well 204/27a-1 drilled 94 m of Upper Jurassic Kimmeridge Clay Formation. The uppermost 14 m or so is described as olive-black, moderately hard, laminated, slightly silty, carbonaceous, pyritic mudstone, resting on 80 m of mainly sandstone, conglomerate and breccia of the Rona Member. In the facies association model for the Rona Member (Vestralen et al., 1995), the succession in this well is unique in being the only example described within the Faroe–Shetland area that is interpreted to contain all 7 (A-G) facies. The Upper Jurassic Kimmeridge Clay Formation rests on Precambrian crystalline basement and is interpreted to be overlain by Kimmeridge Clay Formation of Early Cretaceous (Berriasian) age (see Figure 73). Well 204/28-2 drilled 62.5 m of Upper Jurassic Kimmeridge Clay Formation. The basal 9 m is described as translucent, off white to very pale grey, soft to loose, amorphous to calcite cemented, poor to moderately well-sorted, fine to very fine-grained and locally medium to coarse-grained sandstone with subrounded to subspherical grains that has affinities with the Rona Member (Figure 60a), although it could be as young as Mid Tithonian in age. This unit is overlain by 53.5 m of medium to dark grey, locally pale grey and red-brown, moderately soft to firm, blocky, slightly to noncalcareous mudstone typical of the Kimmeridge Clay Formation. The succession rests on undivided Triassic strata (Figure 55a) and is conformably overlain by Kimmeridge Clay Formation of Early Cretaceous (Berriasian) age (see Figure 73).

7 Cretaceous

Martyn Stoker[1] and Heri Ziska[2]
[1] *British Geological Survey*
[2] *Jarðfeingi (Faroese Earth and Energy Directorate)*

Cretaceous strata have been proved in the central and north-eastern parts of the report area, where they occur in or marginal to the major downwarped areas of the Faroe–Shetland Basin, Møre Basin and surrounding marginal basins (Figures 13 and 69). Seismic evidence suggests that the Cretaceous succession may reach approximately 5 km in thickness along the eastern margin of the Faroe–Shetland Basin in the Foula Sub-basin (Figures 30 and 31), and up to 10 km of Cretaceous rocks have been postulated to occur in the Møre Basin (Doré et al., 1999). The axis of the Faroe–Shetland Basin–Møre Basin depocentre has been interpreted to continue to the south-west, beneath the Cenozoic inversion structure of the Wyville Thomson Ridge, to link with the North-east Rockall Basin (Waddams and Cordingley, 1999). Keser Neish and Ziska (2005) have suggested that Cretaceous rocks may also be present within the Faroe Bank Channel Basin (Figure 69), though the subcrop of Cretaceous strata in the Faroese national sector is largely obscured beneath a cover of Palaeogene basalts (e.g. Figures 30 and 31). Northwest of the study area, up to 1 km of Lower Cretaceous (Aptian) to Paleocene strata is exposed in the Kangerlussuaq Basin, onshore East Greenland (Larsen et al., 1999; Larsen and Whitham, 2005). In Scotland, the nearest onshore outcrops include Lower Cretaceous shallow-marine sandstone near Peterhead, and Upper Cretaceous carbonate and clastic sedimentary rocks in the Inner Hebrides and on the Argyll mainland, near Morvern (Harker, 2002). Lower and Upper Cretaceous rocks are widespread throughout the central and northern North Sea (Copestake et al., 2003; Surlyk et al., 2003; Figure 13).

The Jurassic–Cretaceous boundary in the Faroe–Shetland region, where conformable, occurs within the upper part of the Kimmeridge Clay Formation (Figure 70), which consists of dark grey to black, organic-rich mudstone with high-gamma and low velocity well log responses, deposited in an anaerobic environment (Ritchie et al., 1996; see Chapter 6). However, a significant change of depositional environment is recorded within upper Berriasian sediments that marks the end of anaerobic conditions, and the onset of deposition of open marine, less organic-rich, variably calcareous mudstone with sporadic thick sandstone. These sediments are all included within the Cromer Knoll Group, which extends to the top of the Lower Cretaceous (Figure 71).

The anoxic shales of the Kimmeridge Clay Formation are interpreted as a condensed post-rift sequence following Late Jurassic extension that created a deep-water marine connection between the Norwegian, Faroe–Rockall and North Sea rifts (Doré et al., 1999; Grant et al., 1999; Roberts et al., 1999; Figure 12). Deposition of the Cromer Knoll Group was instigated in response to the rejuvenation of the Faroe–Shetland region by a major episode of structuration and basin enlargement during the Early Cretaceous (e.g. Dean et al., 1999). This structural development was controlled by a north-west–south-east-directed extensional regime (replacing the east–west Jurassic extension) that was initiated as the Atlantic rift system began to propagate into the proto-north-east Atlantic region between Greenland and Eurasia (Ziegler, 1988; Knott et al., 1993; Rattey and Hayward 1993; Roberts et al., 1999; Doré et al., 1999; Lundin, 2002; Figure 13). The resulting broad rift system included a chain of basins that extended from the Bay of Biscay in the south to the Barents Sea margin in the north, and included the Faroe–Shetland and Møre basins, and adjacent marginal basins, e.g. West Shetland Basin, North Rona Basin and the Solan basins (Figure 69). Rawson and Riley (1982) have suggested that a major 'basin-flushing' event within this tectonically active environment, in combination with a rising eustatic sea level (Haq et al., 1987), are important factors responsible for overturning the anoxic water mass in which the Kimmeridge Clay Formation was deposited.

The onset of rifting in the Faroe–Shetland region is generally envisaged to have occurred in the Valanginian to Hauterivian interval (Dean et al., 1999; Coward et al., 2003; Figure 8), which is consistent with regional evidence from the north-east Atlantic margin between the Rockall Basin and mid Norway (Musgrove and Mitchener, 1996; Lundin and Doré, 1997; Doré et al., 1999). Further episodes of rifting occurred intermittently throughout the Early Cretaceous, though the timing and duration of these events remains ambiguous (see Chapter 2), not least due to the problematic dating of some of the sand-rich Lower Cretaceous rocks (Goodchild et al., 1999). From a regional perspective, a second, Early Cretaceous 'main Atlantic rift phase' is envisaged by Coward et al. (2003) to have occurred during the Aptian to Albian interval, though Doré et al. (1999) suggest that this event may extend into the Cenomanian.

During the Late Cretaceous, the Atlantic rift system as a whole had become relatively inactive and the region largely underwent passive thermal subsidence (Roberts et al., 1999; Coward et al., 2003). However, the Faroe–Shetland area was far from tectonically quiescent and episodic extension, rifting and inversion are postulated throughout this interval (e.g. Booth et al., 1993; Turner and Scrutton, 1993; Dean et al., 1999; Grant et al., 1999; Spencer et al., 1999). According to Roberts et al. (1999), Late Cretaceous extension in the area west of Shetland was directed north-east–south-west, subparallel to the developing ocean margin between Labrador and Iberia, and most probably linked to the opening of the Bay of Biscay. Although thick Upper Cretaceous sequences have accumulated in many of the basins and sub-basins of the Faroe–Shetland region, oblique to strike-slip reactivation of earlier north-east-trending faults resulted in localised uplift of the Westray, Rona and Judd highs. Campanian to Maastrichtian rocks are absent from the northern part of the Westray High, whereas Cenomanian to Campanian and upper Maastrichtian rocks are commonly absent from the Rona High. The complete absence of Upper Cretaceous strata on the Judd High (Figure 69) and possibly the Outer

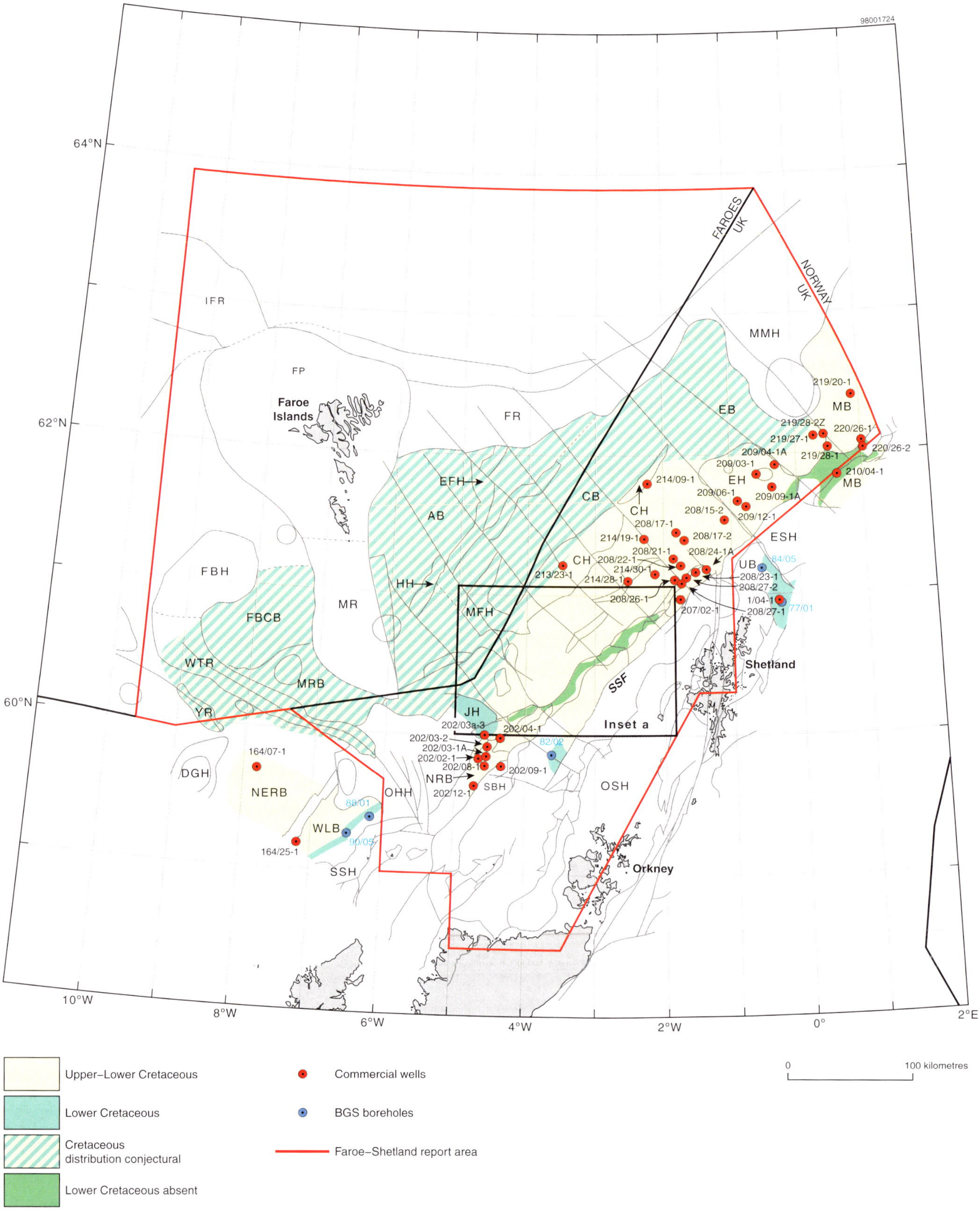

Figure 69 Distribution of Cretaceous strata within the Faroe–Shetland report area. Present day structural elements (grey) derived from Figure 7. Abbreviations: see Figure 7.

Hebrides High suggests that it could have been emergent throughout the Late Cretaceous interval. This pattern of sedimentation may reflect a rift model in which uplift and subsidence were variable in both time and space across the study area (Dean et al., 1999).

STRATIGRAPHY

The lithostratigraphical terminology used here is that of Ritchie et al. (1996) (Figure 70). This scheme was mainly constructed on the basis of wireline well log responses,

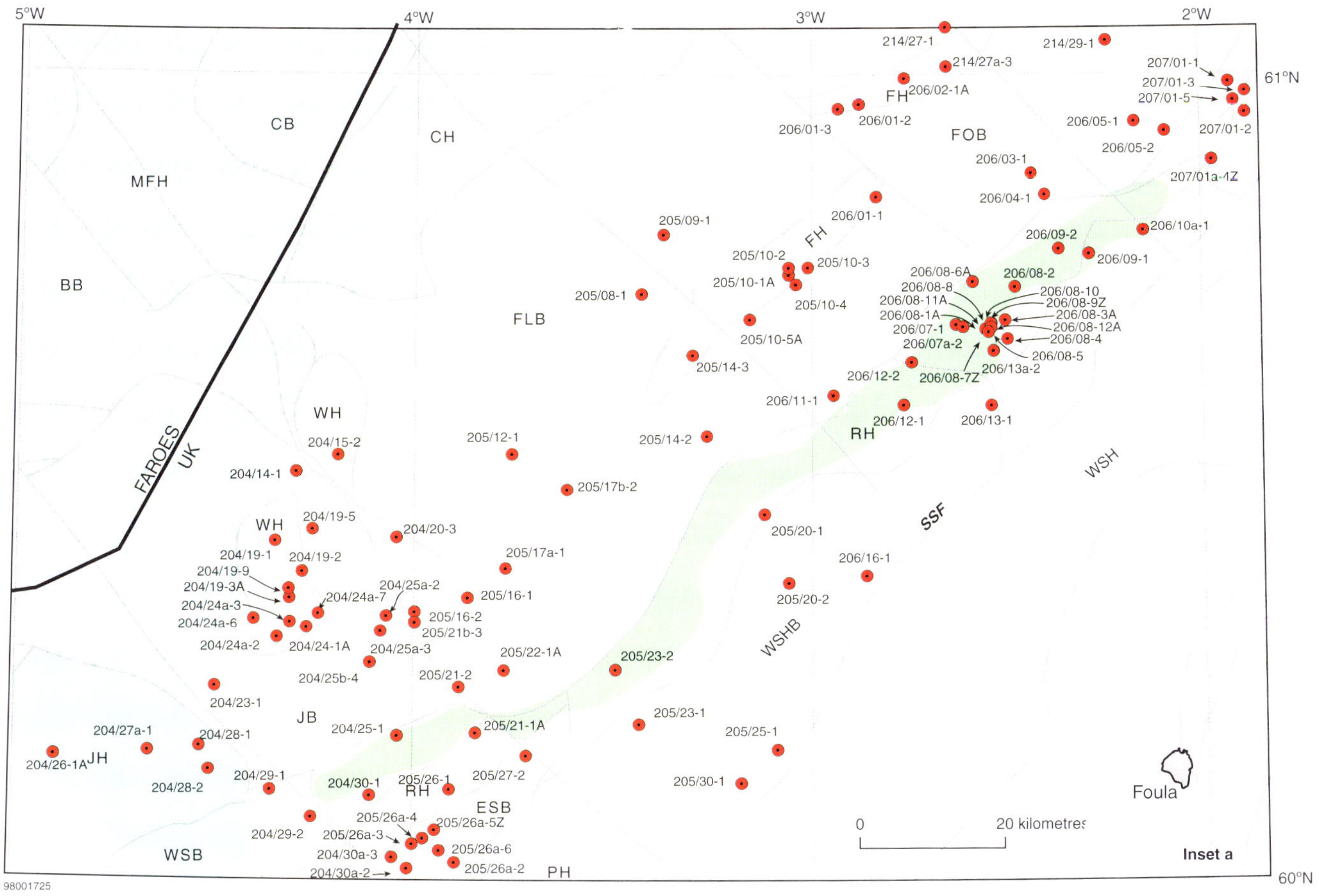

Figure 69 *continued.*

together with biostratigraphical data. The Cromer Knoll Group represents the post-Kimmeridgian Clay Formation Lower Cretaceous sequence, whereas the overlying Chalk and Shetland groups represent, respectively, limestone-dominated and mudstone-dominated successions within the Upper Cretaceous. The nature of the boundary between the Lower and Upper Cretaceous is variable. An unconformity predominates within the West Shetland, North Rona and Solan basins, and on the adjacent Rona and Solan Bank highs, though a conformable transition occurs locally in the southern part of the West Shetland Basin and Rona High (Figure 69). A conformable boundary is more commonly developed in the Faroe–Shetland Basin, including the intrabasinal highs. The relationship of the formations that comprise the Cromer Knoll, Chalk and Shetland groups are illustrated in Figure 70, and a summary of their lithology and environment of deposition is presented in Figure 71. In the report area, the Chalk Group is restricted to the Cenomanian to Turonian interval.

The subdivision of the Cromer Knoll Group is currently applicable only to the south-east part of the report area (Figures 70). In the West Shetland Basin, the Lower Cretaceous Victory Formation represents the whole of the Cromer Knoll Group and consists of a complex shallow marine to paralic succession of sandstone, limestone, conglomerate and coal. To the west, deep-water mass-flow sandstone wedges of the Royal Sovereign and Neptune formations fringe the Faroe–Shetland Basin (Figures 70 and 72a). These clastic wedges were sourced from the Rona, Judd and Westray highs, and are commonly cited as evidence of the Early Cretaceous tectonic rejuvenation of the Faroe–Shetland region (Stoker et al., 1993; Ritchie et al., 1996; Dean et al., 1999; Grant et al., 1999). At the same time within the basinal areas of the south-west Faroe–Shetland, North Rona and the Solan basins, mudstone and argillaceous limestone of the Valhall Formation accumulated in a well-oxygenated shelf to basinal environment (Figure 72a). In contrast, the mid to late Aptian to Albian Cruiser, Carrack and Rødby formations include organic-rich pyritic mudstone units indicative of a fluctuating aerobic-anaerobic basinal setting. A final Early Cretaceous pulse of mass-flow sedimentation is indicated by the deep-water, sandstone-dominated Commodore Formation that accumulated along the eastern margin of the Faroe–Shetland Basin until Cenomanian time (Figure 72b).

During the Late Cretaceous, a high eustatic sea level (Haq et al., 1987) resulted in much of the Faroe–Shetland area being drowned (Harker, 2002). In contrast to the warm tropical Chalk seas that existed over much of Britain and the adjacent shelf, colder boreal waters that were less favourable for coccolith productivity largely covered the Faroe–Shetland region. Consequently, the Chalk Group does not occur north of about 60° 20' N, with rocks characteristic of this group restricted to the North Rona Basin, the East Solan Basin and the southern part of the West Shetland Basin (Figure 69). More generally, the rocks of the Faroe–Shetland region are less calcareous and more mud-dominated, and belong to the Shetland Group (Figure 70). Subdivision of

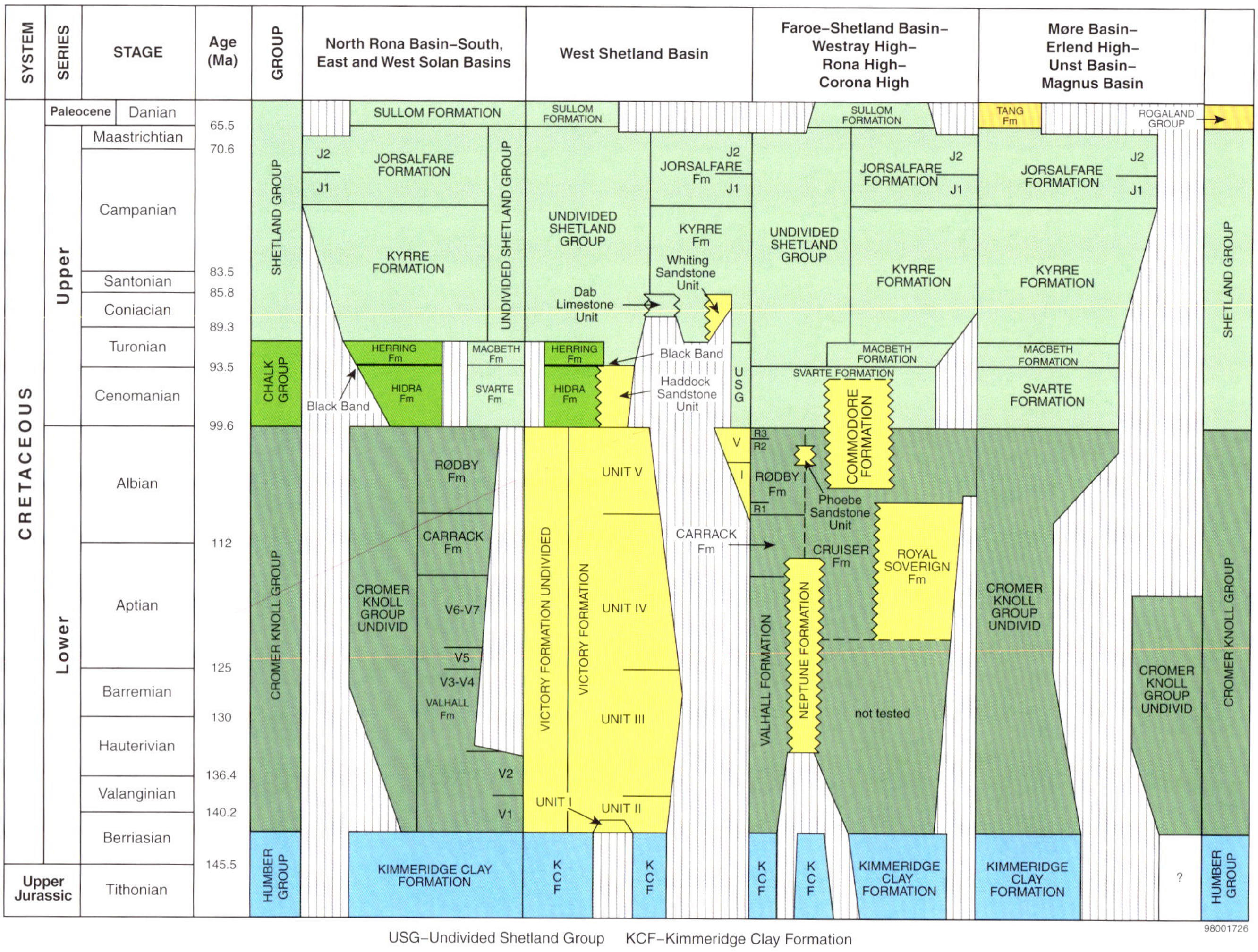

Figure 70 Cretaceous lithostratigraphical nomenclature for the Faroe–Shetland report area (modified from Ritchie et al., 1996). Timescale is based on 'A Geologic Time Scale 2004' by F M Gradstein, J G Ogg, A G Smith, et al. (2004) and the International Stratigraphical Chart, 2006 (ICS) with additions.

the Shetland Group has been undertaken throughout the report area on the basis of lithology and wireline well log response; though in all basins the succession remains locally undivided.

In the Cenomanian to Turonian interval, Chalk Group limestone of the Hidra and Herring formations was deposited on a well-oxygenated carbonate shelf in the southern part of the report area (Figures 70 and 72b). To the north, the limestone is replaced by calcareous, shelf-marine mudstone of the Svarte and Macbeth formations. The mass-flow sandstone units of the Commodore Formation imply continuing tectonic activity. The base of the Herring Formation is marked by the organic-rich Black Band (Figure 70), which indicates a time when anaerobic bottom waters were dominant. During the Coniacian to Maastrichtian interval, the Shetland Group facies extended farther south as represented by marine shelf-margin mudstones of the Kyrre and Jorsalfare formations (Figure 72c). Over large parts of the report area, deposition of the Shetland Group continued into the Paleocene where it is preserved as the Sullom and Tang formations.

As the lithostratigraphical subdivision and nomenclature of both the Lower and Upper Cretaceous successions varies between basins, groups of basins or highs (Figure 70), the distribution of Cretaceous rocks is described separately below for each basin or high. Generalised stratigraphical and lithological information is summarised in Figures 73 to 86, which include stratigraphical-range charts for the various basins and highs. These charts primarily display the reported chronostratigraphical range of the Cretaceous strata drilled in wells and boreholes. In addition, the charts also show the lithostratigraphical framework defined by Ritchie et al. (1996). As this framework has been formally defined in only 25% of the wells utilised in this report, its inclusion on the range charts is intended to act as a guide only to interpretation and lithostratigraphical subdivision in the remaining wells. Those wells in which formal lithostratigraphical definition has been undertaken for the Cretaceous succession are highlighted in red in Figures 73, 76, 77, 79, 81 and 82.

DISTRIBUTION OF LOWER CRETACEOUS

The distribution of Lower Cretaceous rocks in the various basins is described firstly from the marginal West Shetland, North Rona and the Solan basins, and the adjacent interbasinal highs, such as the Judd and Solan Bank highs (Figure 69). The Rona High separates the West Shetland and Faroe–Shetland basins (Figure 31),

Figure 71 Summary of lithology and depositional environment of the Cretaceous lithostratigraphical formations (see Figure 70). Information derived from Ritchie et al., (1996) and Harker, (2002).

Series	Lithostratigraphy		Lithology	Depositional environment
UPPER CRETACEOUS	Shetland Group	Jorsalfare Formation	Mudstone with sporadic interbedded argillaceous limestone and rare sandstone	Aerobic siliciclastic marine shelf to upper bathyal slope
		Kyrre Formation	Mudstone with sporadic limestone, dolomite, sandstone and siltstone. In West Shetland Basin, mudstone grades to basal sandstone and limestone-rich facies, named the Whiting Sandstone and Dab Limestone units	Outer sublittoral to bathyal, though Dab Limestone deposited on a marine inner shelf
		Macbeth Formation	Mudstone with interbedded limestone, minor dolomite and sporadic sandstone and siltstone	Aerobic siliciclastic marine shelf
		Svarte Formation	Mudstone with interbedded limestone, argillaceous limestone and sporadic siltstone	Aerobic siliciclastic marine shelf to bathyal slope
	Chalk Group	Herring Formation	Cryptocrystalline limestone with interbedded argillaceous limestone and mudstone, and high gamma-ray pyritic mudstone (Black Band) at base	Mostly aerobic carbonate shelf, though Black Band represents minor phase of anaerobic conditions
		Hidra Formation	Fine-grained limestone and argillaceous limestone with interbedded mudstone, which, in West Shetland Basin, pass laterally to the sandstone-rich Haddock Sandstone Unit	Aerobic carbonate shelf fringed by shallow-marine sands derived from West Shetland Platform
LOWER CRETACEOUS	Cromer Knoll Group	Commodore Formation	Fine- to medium-grained sandstone with interbedded thin mudstone and limestone	Deepwater mass-flow sandstones on eastern margin of Faroe-Shetland Basin sourced from Rona High. Correlative Phoebe Sandstone Unit sourced from North Rona or Westray highs
		Rodby Formation	Mudstone interbedded with thin limestone, siltstone and sandstone	Aerobic marine shelf, though base of formation deposited in an anaerobic environment (based on agglutinated foraminifera)
		Carrack Formation	Non-calcareous, carbonaceous and pyritic mudstone and siltstone	Basinal marine with anaerobic bottom waters and sporadic incursions of oxygenated surface waters (indicated by planktonic foraminifera)
		Cruiser Formation	Non-calcareous, carbonaceous and pyritic mudstone with sporadic thin siltstone, fine-grained sandstone and limestone	Basinal marine with fluctuating anaerobic–aerobic bottom waters
		Royal Sovereign Formation	Conglomerate and fine- to coarse-grained sandstone with interbedded mudstone	Deepwater mass-flow deposits on eastern margin of Faroe-Shetland Basin sourced from Rona High
		Neptune Formation	Fine- to medium-grained sandstone with interbedded thin mudstone	Deepwater mass-flow sandstones in SW Faroe-Shetland basin possibly sourced from the Westray or Judd highs
		Valhall Formation	Mudstone grading into thin argillaceous limestone	Aerobic marine basin
		Victory Formation	Fine- to medium-grained sandstone, locally conglomeratic at base, with sporadic mudstone and thin coal	Shallow marine to paralic setting in the West Shetland Basin. Coal beds indicate episodic exposure of the delta plain

98001727

and is described separately. This is followed by the Faroe–Shetland Basin, which is divided into a number of sub-basins and intrabasinal highs. Peripheral basins and highs, such as the Møre, Magnus and Unst basins, the Erlend High in the north-east and the West Lewis and north-east Rockall basins in the south-west are described at the end of the section.

West Shetland Basin

Lower Cretaceous rocks have been proved in numerous wells throughout the West Shetland Basin (Figure 73), but are absent over the structurally highest parts of the Rona High (Figures 31 and 69). In the north-east and south-west parts of the basin, the basal part of the Lower Cretaceous Cromer Knoll Group mainly rests unconformably on Upper Jurassic Kimmeridge Clay Formation or Precambrian crystalline basement. The top of the Lower Cretaceous succession is marked by a mainly conformable contact with Upper Cretaceous strata in the south-west part of the basin, whereas an unconformity between Lower and Upper Cretaceous rocks is generally observed in the north-east.

The most complete section of the Cromer Knoll Group, essentially spanning the entire Early Cretaceous age range, is preserved in the south-west part of the West Shetland Basin, with a maximum thickness of 1054.6 m proved in well 205/25-1 close to the Shetland Spine Fault (Figures 69 and 73). An additional 17.4 m of Berriasian mudstone belonging to the Kimmeridge Clay Formation is also present in this well. According to Dean et al. (1999), the sequence in adjacent wells 205/23-1 and 205/30-1 is punctuated by an unconformity of Aptian age. A thickness of at least 201.5 m was drilled in well 205/23-1. In the north-east part of the basin, the thickness of the Cromer Knoll Group is generally much reduced, ranging from 17.4 m in well 208/23-1 to a maximum of 209.5 m in well 207/01-2. Moreover, there is some dispute as to the age range of the preserved sequence in this region. Dean et al. (1999) and Grant et al. (1999) indicate a predominantly Aptian to Albian age for the preserved Cromer Knoll Group in well 207/01-2, supported by Goodchild et al. (1999) who cite a late Aptian age, though they also state that accurate dating is problematic. In contrast, Ritchie et al. (1996) imply a late Berriasian to Albian age on the basis of biostratigraphical data. The occurrence of Berriasian to Barremian strata in the lower part of the section is suggested by a foraminiferal microfauna that includes *Textularia foeda*, *Glomospira* spp., *Uvigerinammina* sp., and *Lenticulina* sp., and debris of the bivalve *Inoceramus*, whereas the occurrence of the dinoflagellate cyst

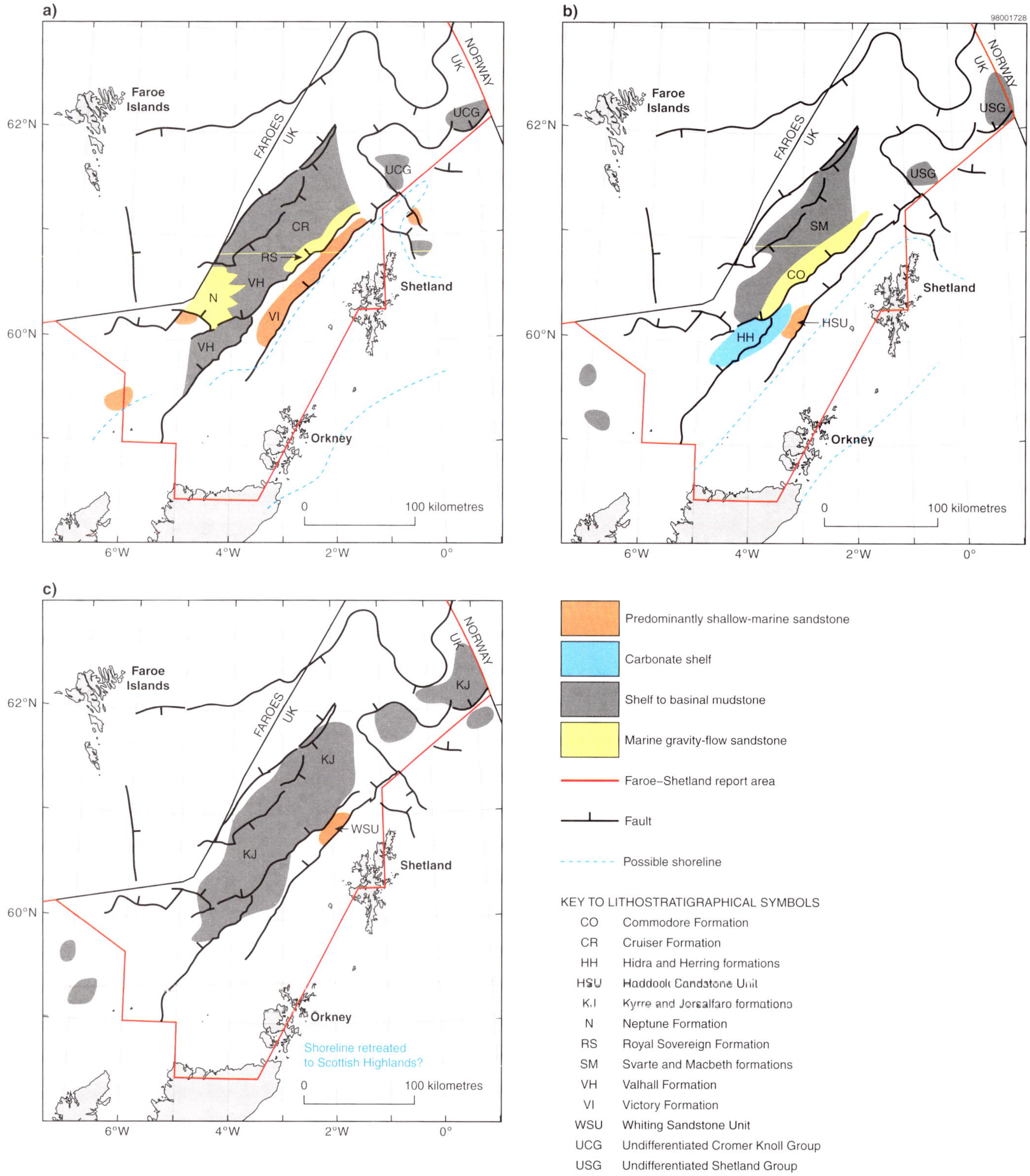

Figure 72 Generalised Cretaceous palaeogeography of the Faroe–Shetland region for a) late Berriasian to early Aptian; b) Cenomanian to Turonian; c) Coniacian to Maastrichtian. Location of palaeo-shoreline based on Harker, (2002). Structural elements derived from Figure 7.

Tubotuberella apatela is indicative of an age no younger than mid Valanginian.

The whole Lower Cretaceous succession in the West Shetland Basin is assigned to the Victory Formation of Ritchie et al. (1996), which is dominated by sandstone that locally grades into, and is interbedded with, claystone and limestone, particularly in the south-west part of the basin (Figure 73). The Victory Formation is locally divided into five informal units based on wireline well log response (Figures 70 and 74). Unit I contains a basal conglomerate of variable age, usually present where the Victory Formation overlies Precambrian crystalline basement. Units II and V consist of relatively low gamma-ray sandstone, whereas higher gamma-ray

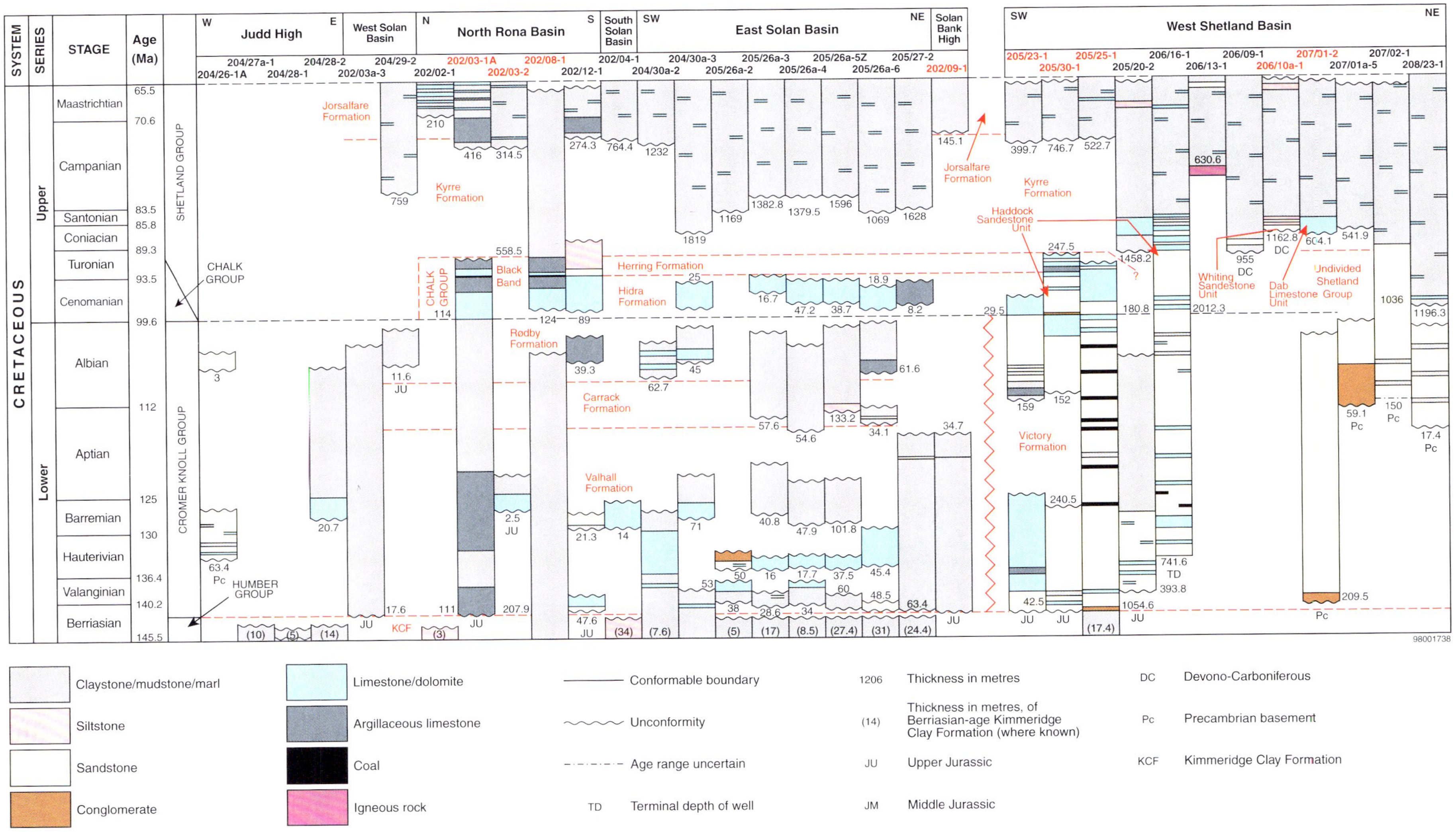

Figure 73 Stratigraphical range chart showing generalised lithology and thickness of Cretaceous rocks in the North Rona, East Solan, West Solan, South Solan and West Shetland basins, and on the Judd and Solan Bank highs (partly modified from Ritchie et al., 1996 and Dean et al., 1999). Timescale is based on 'A Geologic Time Scale 2004' by F M Gradstein, J G Ogg, A G Smith, et al. (2004) and the International Stratigraphical Chart, 2006 (ICS) with additions. For locations of wells and lithostratigraphical nomenclature, see Figures 69 and 70, respectively. Guide to lithostratigraphical framework and key wells shown in red.

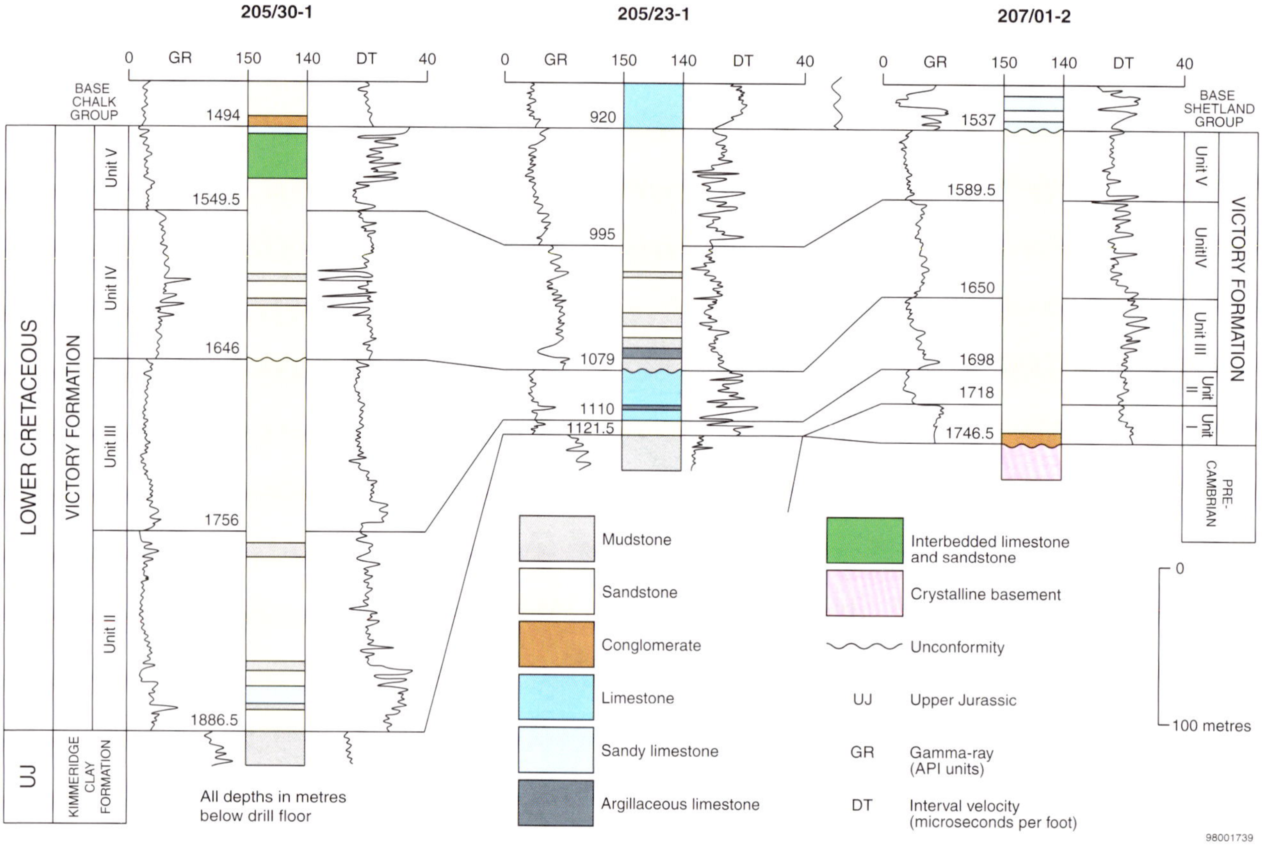

Figure 74 Well log correlation of the Lower Cretaceous Victory Formation (Cromer Knoll Group) in wells 205/30-1, 205/23-1 and 207/01-2 within the West Shetland Basin (modified from Ritchie et al., 1996). For location of wells, see Figure 69.

values characterise the sandstones of Units III and IV. Unit IV has the highest gamma-ray values of all the units, as the sandstone is generally more argillaceous and carbonaceous. Locally, a spike at the base of Unit IV has been tentatively interpreted to correlate either to the intra-Barremian Munk Marl Bed or the intra-Aptian, organic-rich, Fischschiefer (Fish Shales) mudstone (Johnson and Lott, 1993; Ritchie et al., 1996).

A crudely stratified conglomerate facies of Unit 1 occurs at the base of the section in wells 205/25-1, 207/01-2 and 01a-5 (Figures 73 and 74) and generally consists of subrounded to angular quartz and gneiss fragments of pebble to cobble grade. The clasts are either framework supported or embedded within a poorly sorted, well-cemented sandstone matrix (Goodchild et al., 1999). Towards the top of the conglomerate facies the beds become thinner and the clast size decreases, as the proportion of interbedded sandstone associated with Units II to V increases. In wells 205/23-1, 205/30-1, 207/01-2 and 01a-5, the sandstone is generally white to grey, buff and brown, fine to medium-grained, commonly calcareous, micaceous, glauconitic and carbonaceous (Ritchie et al., 1996). The sandstone is commonly bioturbated and contains foraminifera and bioclastic debris. Locally, granule to pebble-sized clasts occur in medium to coarse-grained sandstone beds, and depict crude high-angle cross-stratification (Goodchild et al., 1999).

Interbedded argillaceous rocks within the Victory Formation are variably coloured, ranging from dark grey carbonaceous mudstone or shale to grey-green to red-brown mudstone and claystone. They are variably calcareous and locally grade to siltstone. Interbedded limestone is generally white and includes both micrite and sandy, shelly limestone with rare grainstone. Beds of micrite up to 25 m thick, described as wackestone by Ritchie et al. (1996), occur in well 205/23-1. Limestone is common in the Valanginian to Barremian interval, and also occurs in the Albian to Cenomanian (Upper Cretaceous) section (Figure 73). Several thin, black, argillaceous coal beds are recorded in well 205/25-1 and also disseminated in the lower part of well 206/16-1.

The Victory Formation was deposited in a paralic to shallow marine environment as the West Shetland Basin subsided between the West Shetland and Rona highs, largely due to growth along the Shetland Spine Fault during Early Cretaceous rifting (Duindam and van Hoorn, 1987; Hitchen and Ritchie, 1987; Meadows et al., 1987; Ritchie et al., 1996; Goodchild et al., 1999; Figure 30). The conglomerate facies represents fan-delta deposits that built out from the emergent West Shetland and Rona highs. The overlying bioturbated sandstone beds are interpreted as shoreface to inner shelf deposits that accumulated as the fan deltas and the Rona High became gradually submerged in response to a rise in eustatic sea level (Goodchild et al., 1999).

Episodic, tectonically induced emergence and nonmarine deposition is recorded by the presence of thin coal beds in well 205/25-1 adjacent to the Shetland Spine Fault.

North Rona Basin, the Solan basins, Papa Basin and adjacent highs

Lower Cretaceous rocks occur in the north-east part of the North Rona Basin and throughout the West Solan, South Solan and East Solan basins (Figure 69). They are also present on the extreme north-east end of the Solan Bank High and across the north-east Judd High. Mostly, though not exclusively, the rocks of the Cromer Knoll Group are separated from the underlying Kimmeridge Clay Formation by an unconformity (Figure 73). The upper boundary of the Lower Cretaceous sequence is also largely represented by an unconformity. The preserved sequence spans the Early Cretaceous interval, though there are a number of apparent unconformities, especially in the East Solan Basin, which punctuate the succession.

Maximum drilled sequences range from approximately 200 m in the North Rona Basin to 330 m in the East Solan Basin (Figure 73), where the Cromer Knoll Group can be divided into three formations, the Valhall, Carrack and Rødby formations (Ritchie et al., 1996). The Cromer Knoll Group is a predominantly argillaceous sequence of variably calcareous mudstone grading to argillaceous limestone, with sporadic limestone, siltstone, sandstone and rare conglomerate. In areas where the Valhall Formation is recognised, the Lower Cretaceous succession is relatively thin and represents deposition in an aerobic, open marine, shelf to basinal setting that was essentially starved of coarser-grained clastic input (Stoker et al., 1993; Ritchie et al., 1996). This resulted in the local development of limestone and condensed sequences (Booth et al., 1993), the presence of early diagenetic calcareous and siliceous nodules, and the replacement of bioclasts by siderite and phosphate (Meadows et al., 1987). Sporadic beds of sandstone and conglomerate in the North Rona and East Solan basins may reflect redeposition from the West Shetland Basin, or from localised fault activity (Booth et al., 1993). The overlying Carrack Formation comprises organic-rich mudstone indicative of basinal restriction, with bottom-water oxygen depletion (Ritchie et al., 1996). This change in basin oceanography in the North Sea has been linked to possible rejuvenation of intrabasinal relief driven by the culmination of the Austrian intra-plate tectonic activity (Ziegler, 1982; Crittenden et al., 1997; Oakman and Partington, 1998). The ensuing forced regression is envisaged to have resulted in the facies change from calcareous mudstone of the Valhall Formation to dark mudstone of the Carrack Formation. In the report area, this tectonic activity may have been compounded by the later Early Cretaceous phase of main Atlantic rifting (Figures 8 and 13). The return to a more open marine, well-oxygenated, shelfal environment generally accompanied deposition of the Rødby Formation.

North Rona Basin

Lower Cretaceous strata have been proved in wells 202/02-1, 202/03-1A, -2, 202/08-1 and 202/12-1 (Figure 73), where they are preserved as part of a series of thin Upper Jurassic to Lower Cretaceous wedges within the half-graben complex that is the North Rona Basin (Figure 28). All wells include organic-rich mudstone of the Kimmeridge Clay Formation, though only in 202/08-1 do these strata extend into the Berriasian where they are conformably overlain by rocks of the Valhall Formation. In well 202/02-1, a 3 m thick Berriasian siltstone is unconformably overlain by Upper Cretaceous strata. An unconformity separates Upper Jurassic and Lower Cretaceous rocks in the remaining wells (Figures 60a and 73).

In the central part of the North Rona Basin, well 202/03-1A penetrated the most complete section of Lower Cretaceous rocks, incorporating the Valhall, Carrack and Rødby formations (Figure 73). This well proved 111 m of Berriasian to Albian pale to dark grey, red-brown and green, silty, calcareous mudstone and argillaceous limestone, the latter restricted to the Valhall Formation. The Valhall Formation has been divided into a number of informal units, V1-V7, based on lithological and wireline well log characteristics as defined by Johnson and Lott (1993) (Figures 70 and 75). Although the wireline well log displays a variable serrated response, there is an overall upward decrease in gamma-ray value observed in well 202/03-1A. However, a high gamma-ray and low velocity spike associated with Unit V5 has been tentatively correlated with the organic-rich Fischschiefer mudstone. The mudstone of the overlying Carrack Formation is generally darker and more carbonaceous, and its boundary with the Valhall Formation is marked by a downward decrease in gamma-ray values and an increase in velocity (Figure 75). The grey to red-brown mudstone of the Rødby Formation generally display lower gamma-ray and higher interval velocity responses than the Carrack Formation. This sequence is conformable with the overlying Upper Cretaceous Hidra Formation. Valhall Formation Units V1 to V7 are also reported in well 202/08-1 to the south (Ritchie et al., 1996), which proved 207.9 m of grey to locally red-brown, calcareous, glauconitic claystone of Berriasian to Albian age, although the upper part of the section (Rødby Formation) in this well is eroded and unconformably overlain by carbonates of the Upper Cretaceous Hidra Formation.

In the north of the North Rona Basin, well 202/03-2 revealed a 2.5 m thick section of Barremian to Aptian red-brown mudstone and white to grey argillaceous limestone (Figure 73). On the southern margin of the basin, well 202/12-1 drilled three separate unconformity-bounded units. At the base, there is a 47.6 m thick section of upper Berriasian to lower Valanginian grey bioclastic limestone, grading downwards to sandy limestone and green to grey, very fine to fine-grained, well-sorted sandstone. This unit is unconformably overlain by 21.3 m of green, fine to medium-grained, well-sorted, glauconitic sandstone of early Barremian age, which becomes marly towards the base. Both these units correlate with the Valhall Formation. The uppermost unit consists of 39.3 m of middle Albian argillaceous limestone that correlates with the Rødby Formation, and is red-brown, soft to firm and locally grades to claystone.

East Solan Basin

Lower Cretaceous strata have been proved in wells 204/30a-2, -3, 205/26a-2, -3, -4, -5Z, -6 and 205/27-2 in the East Solan Basin (Figures 39 and 73). The continuity of the record is punctuated by several unconformities of intra-Berriasian, Valanginian to Hauterivian, Hauterivian to Barremian, intra-Aptian and Aptian

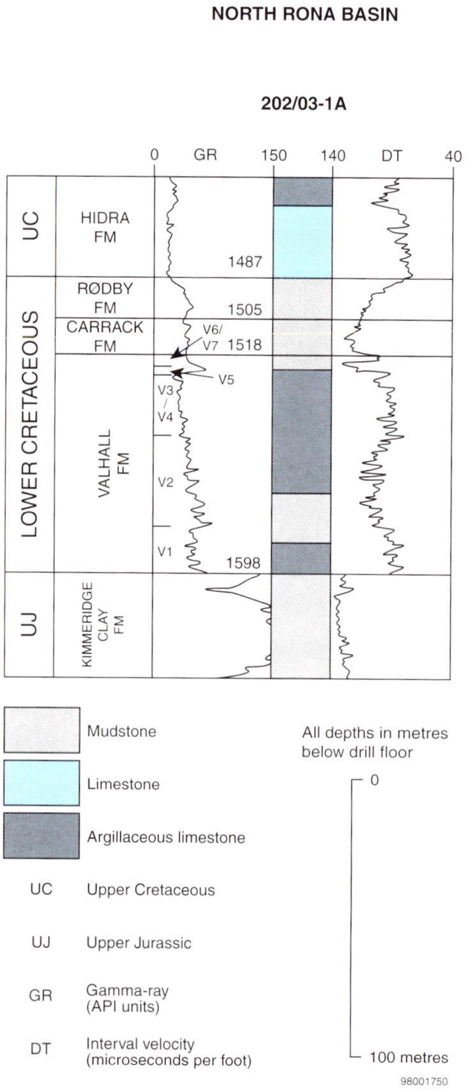

Figure 75 Well log summary of the Lower Cretaceous Valhall, Carrack and Rødby formations (Cromer Knoll Group) in well 202/03-1A within the North Rona Basin (modified from Ritchie et al., 1996). For location of well, see Figure 69.

to Albian age. The intra-Berriasian break largely separates the Kimmeridge Clay Formation from the overlying Valhall Formation, though a conformable boundary is recorded in wells 204/30a-2 and -3. The boundary between the Lower and Upper Cretaceous is marked by an unconformity in all wells.

Dark brown to yellow-brown to dark grey, noncalcareous, carbonaceous claystone belonging the Kimmeridge Clay Formation occurs in all of the wells (Figures 73). The Berriasian part of this unit ranges from 5 to 31 m thick. The overlying succession is dominated by grey to red-brown and locally black, calcareous claystone of the Valhall, Carrack and Rødby formations. The claystone locally grades to argillaceous limestone, and is commonly interbedded with white to red-brown and grey-green limestone in the Valanginian to Barremian (Valhall Formation) interval in most wells (Figures 73). In well 205/26a-2, the Hauterivian section also includes grey-brown, very fine to medium-grained, well-sorted sandstone and pale grey conglomerate. Grey to dark green, silty, calcareous sandstone is recorded in the upper Berriasian to lowermost Valanginian sections in well 205/26a-6, whereas sandstone and red to green siltstone of the Carrack Formation was proved in the Aptian section in well 205/26a-5Z.

Papa Basin

At the north-west end of the Papa Basin, BGS borehole 82/02 (Figure 69) recovered a 9 m thick section of greenish-grey sandstone with heavy mineral banding and mudstone rip-up clasts of possible Early Cretaceous age.

West and South Solan basins

Lower Cretaceous rocks have been proved in wells 202/03a-3 and 204/29-2 in the West Solan Basin (Figures 73). Well 202/03a-3 proved 17.6 m of Cromer Knoll Group, with the upper 7.2 m consisting of grey to greyish-green, moderate to highly calcareous mudstone assigned to the Rødby Formation, overlying 10.4 m of reddish-brown very slightly silty, calcareous mudstone, mainly assigned to the Valhall Formation. The latter appears to rest unconformably on the Upper Jurassic Kimmeridge Clay Formation (Figures 60a and 73). In the north-east part of the West Solan Basin, well 204/29-2 proved 11.6 m of orange-brown to red-brown, locally very silty, noncalcareous claystone, with sporadic intercalations of siltstone, assigned to the Rødby Formation. The claystone rests unconformably on Upper Jurassic sandstone. In both wells, the top of the Cromer Knoll Group is marked by an unconformity.

In the South Solan Basin, Lower Cretaceous strata have been tested in well 202/04-1, which proved 14 m of Barremian limestone unconformable on a 34 m thick siltstone of the Lower Cretaceous (Berriasian) Kimmeridge Clay Formation (Figures 73).

Solan Bank and Judd highs

Lower Cretaceous rocks are present on the Solan Bank and Judd highs (Figures 73). The Solan Bank High forms the faulted eastern margin of the North Rona and South Solan Basin (Figure 69). Lower Cretaceous strata have been proved in well 202/09-1 on the northern end of the high, which penetrated 34.7 m of red-brown, calcareous, silty mudstone and pale grey argillaceous limestone of Valanginian to Aptian age. These rocks rest unconformably on dark grey to grey-black mudstone of the Upper Jurassic Kimmeridge Clay Formation and are unconformably overlain by Campanian to Maastrichtian strata (Figures 60a and 73).

The Judd High forms the faulted western margin of the West Solan Basin (Figure 69). Lower Cretaceous strata assigned to the Cromer Knoll Group have been tested in wells 204/26-1A and 204/28-2 (Figure 73). Well 204/26-1A proved 3 m of Albian sandstone unconformable on 63.4 m of Hauterivian to Barremian sandstone with sporadic interbedded claystone and limestone, which increase toward the base of the section. These rocks rest unconformably on Precambrian crystalline basement. Well 204/28-2 proved a 20.7 m thick unconformity bounded section of Barremian to Albian. The basal 5.7 m consists of limestone assigned to the Valhall Formation, whereas the overlying 15 m of claystone belongs mainly to the Carrack and Rødby formations. The Cromer Knoll Group at this site is underlain by the Kimmeridge Clay Formation, which includes a 14 m thick section of Berriasian claystone and siltstone. Wells 204/27a-1 and 204/28-1 proved 10

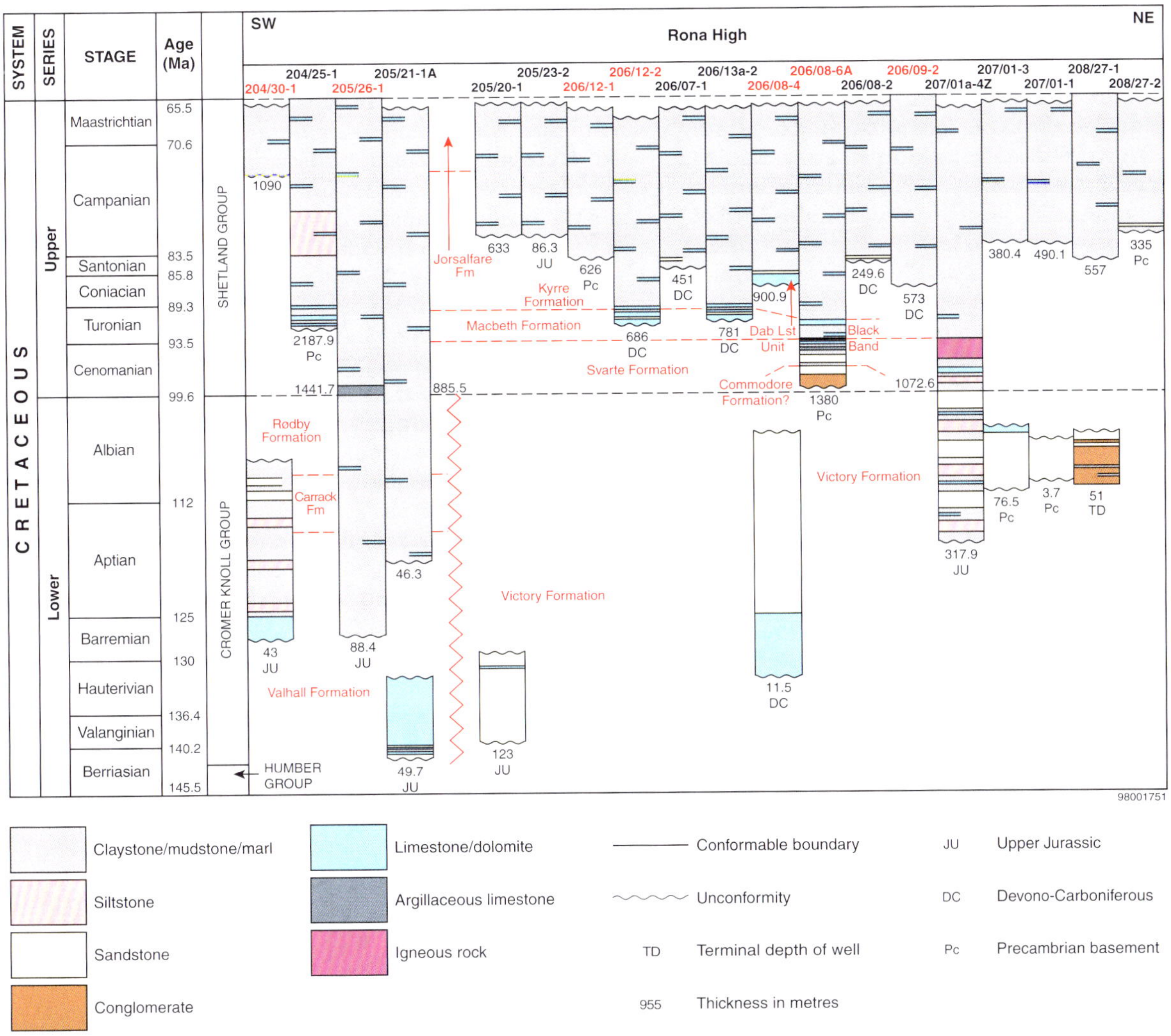

Figure 76 Stratigraphical range chart showing generalised lithology and thickness of Cretaceous rocks on the Rona High (partly modified from Ritchie et al., 1996; Dean et al., 1999 and Grant et al., 1999). Timescale is based on 'A Geologic Time Scale 2004' by F M Gradstein, J G Ogg, A G Smith, et al. (2004) and the International Stratigraphical Chart, 2006 (ICS) with additions. For locations of wells and lithostratigraphical nomenclature, see Figures 69 and 70, respectively. Guide to lithostratigraphical framework and key wells shown in red.

and 5 m respectively of pale grey to olive-black, slightly silty and moderately calcareous Berriasian mudstone assigned to the Kimmeridge Clay Formation (Ritchie et al., 1996). In well 204/27a-1, the Berriasian appears to be conformable on Upper Jurassic strata, whereas in well 204/28-1 an unconformity separates upper Berriasian and Tithonian (Upper Jurassic) strata (Figures 60a and 73). Younger Cretaceous strata are absent at these sites (e.g. Figure 27).

Rona High

On the Rona High, rocks belonging the Cromer Knoll Group have been proved in nine wells mainly at the south-west and north-east ends (Figure 76) Over much of the high, however, Lower Cretaceous strata are absent (Figure 69). The bulk of the drilled sequence is Barremian to Albian in age, though isolated occurrences of Berriasian to Hauterivian rocks have also been proved. Most of the drilled sections are less than 100 m thick; however, over 300 m of Aptian to Albian strata was proved in well 207/01a-4Z. In the majority of wells, the Lower and Upper Cretaceous sequences are separated by an unconformity. The base of the Cretaceous is also predominantly erosional and variably underlain by Upper Jurassic, Devono-Carboniferous and Precambrian crystalline basement rocks.

The lithology of the Cromer Knoll Group varies along the length of the Rona High. For example, fine-grained argillaceous rocks characteristic of the Valhall, Carrack and Rødby formations occur at the south-west end of the high are replaced by sandstone and conglomerate of the Victory Formation on the central and north-east parts (Ritchie et al., 1996; Figure 76). Grey, red-brown and green, slightly silty and calcareous claystone dominates the Cromer Knoll Group at the south-west end of the Rona High, though interbedded siltstone, sandstone and limestone are present mostly within the Valhall Formation. Well 204/30-1 proved a 43 m thick sequence of Aptian to Albian claystone locally grading to greenish siltstone and white sandstone, overlying a basal Barremian limestone. A 49.7 m thick, unconformity-bounded unit of grey, Berriasian to Hauterivian limestone with thin sporadic interbeds of claystone and sandstone at its base is reported in well

205/21-1A. These limestones form part of a basal unit recognised within the Valhall Formation (Ritchie et al., 1996). Unconformably overlying the basal limestone in well 205/21-1A is 46.3 m of Aptian to Albian claystone. A comparable claystone sequence, 88.4 m thick and spanning the Barremian to Albian interval, was proved in well 205/26-1. The sections in both these wells continue without any obvious discontinuity into the Upper Cretaceous.

On the central part of the Rona High, well 205/20-1 drilled 123 m of Valanginian to lower Barremian sandstone with sporadic thin limestone beds in the upper part of the section (Figure 76). The sandstone beds varied from very fine to coarse-grained, and well to poorly sorted, are calcareous and bioturbated, and contain localised inclusions (rip-up clasts?) of black lignite. The sandstone appears to unconformably overlie siltstone of Upper Jurassic Kimmeridge Clay Formation (Figure 60b). Farther to the north-east, well 206/08-4 penetrated 7.4 m of buff to dark brown, very fine-grained, well-sorted, calcite cemented sandstone of Aptian to Albian age overlying a basal 4.1 m thick nodular limestone of Hauterivian to Barremian age. The sandstone and limestone units in both of these wells are assigned to the Victory Formation (Ritchie et al., 1996).

At the north-east end of the Rona High, well 207/01a-4Z proved a 317.9 m thick Aptian to Albian sequence of interbedded sandstone, siltstone, and mudstone with thin bands and stringers of limestone (Figure 76). Farther to the north-east, only Albian sediments are preserved (Dean et al., 1999) in wells 207/01-1, -3 and 208/27-1, although older strata are recorded in the intervening well 207/01-2 in the adjacent West Shetland Basin (Ritchie et al. 1996; Figure 73). Wells 207/01-1 and -3 proved 3.7 m and 69.8 m respectively of grey-brown, medium to fine-grained, moderately well-sorted, glauconitic sandstone. In addition, a 6.7 m thick sandy limestone unit occurs at the top of the section in the latter well. Grant et al. (1999) include this limestone unit within the Victory Formation, although Ritchie et al. (1996) assigned an equivalent limestone in the nearby well 207/01-2 to the Dab Limestone Unit of the Kyrre Formation (Figure 70). In contrast, a conglomeratic unit, at least 51 m thick, was drilled in well 208/27-1, and is probably equivalent to Victory Formation Unit 1 of Ritchie et al. (1996). An age of 'not older than Albian' is indicated on the original well composite log. The conglomerate is composed predominantly of gneissose clasts, most likely derived from nearby basement rocks. Two thin mudstone beds are present in the lower part of the cored section, whereas the top of the unit is interbedded with yellow-grey, very fine to fine-grained, well-sorted, slightly calcareous and friable sandstone.

Faroe–Shetland Basin

Lower Cretaceous rocks of the Cromer Knoll Group have been proved throughout the UK national sector of the Faroe–Shetland Basin and are preserved in the Judd, Foula, Flett and Erlend sub-basins and on intrabasinal highs, such as the Westray, Flett and Corona highs (Figure 69). The Cromer Knoll Group within the basin is separated from underlying strata by an unconformity (Figures 77, 79, 81 and 82). In contrast, the Lower–Upper Cretaceous boundary is a predominantly conformable transition, although local unconformities are noted in sections adjacent to the faulted eastern margin of the Faroe–Shetland Basin, in particular on the Rona and East Shetland highs, as well as on the intrabasinal Westray High. The preserved sequence almost spans the Early Cretaceous interval, although there are a number of unconformities that punctuate the succession. In particular, several wells in the Foula Sub-basin terminated within the Cromer Knoll Group.

The thickness of the Lower Cretaceous succession in the Faroe–Shetland Basin is highly variable (Figures 77, 79, 81 and 82). The maximum drilled thickness of Lower Cretaceous rock occurs in wells from the Foula Sub-basin, where nearly 1.5 km has been proved (Figure 79). In the Judd, Flett and Erlend sub-basins, the drilled thickness is mostly less than 200 m (Figures 77, 81, and 82). The argillaceous Valhall, Carrack and Rødby formations extend northwards from the North Rona Basin and the Solan basins into the south-west part of the Faroe–Shetland Basin, within the Judd Sub-basin and south-west Flett Sub-basin, but are replaced in the central part of the basin, on the Flett High and in the Foula sub-basin, by the similarly argillaceous Cruiser Formation (Ritchie et al., 1996; Harker, 2002; Figures 70 and 72a). Coarser clastic units of sandstone and conglomerate fringe the Faroe–Shetland Basin as a result of sporadic fault activity. These include the Royal Sovereign and Commodore formations in the Foula Sub-basin, the Neptune Formation in the Judd Sub-basin and the Commodore Formation in the Flett Sub-basin (Figures 72a and b, 77, 79 and 81). Although deposition of the Commodore Formation continued into the Cenomanian, for descriptive purposes it is included within the Cromer Knoll Group. In the north-east part of the Faroe–Shetland Basin, the Cromer Knoll Group remains undivided within the Erlend Sub-basin (Figure 82).

In the following description, the Lower Cretaceous succession is described from the various sub-basins and intrabasinal highs within the UK national sector of the Faroe–Shetland Basin, in a general south-to-north transect across the basin. At the present time there is very little information on the occurrence and distribution of Lower Cretaceous rocks within the Faroese national sector of the Faroe–Shetland Basin, although a Lower Cretaceous section has been interpreted to occur within the Guðrun Sub-basin, west of the Corona High (Figure 37), and possibly as part of inferred Mesozoic successions within the Annika, Brynhild, Grimhild and Steinvør sub-basins and associated highs (e.g. Keser Neish, 2003; see Chapter 2).

Judd Sub-basin

The Cromer Knoll Group has only been proved in well 204/23-1 within the Judd Sub-basin where the drilled section is approximately 140 m thick (Figures 69, 77 and 78). Here, the base of the Lower Cretaceous succession rests unconformably on Upper Jurassic Kimmeridge Clay Formation (Figure 78) whereas the top of the Cromer Knoll Group is conformable with the overlying Shetland Group. In well 204/29-1, the Kimmeridge Clay Formation may extend into the lowest part of the Cretaceous where claystone of 'Ryazanian' age is noted on the composite well log, but no younger rocks of Early Cretaceous age are preserved at this site.

Between the Judd and Westray highs, the Lower Cretaceous is preserved as a series of asymmetric wedges associated with half-graben development in the Judd Sub-basin (Figure 27). Within the basin, well 204/23-1

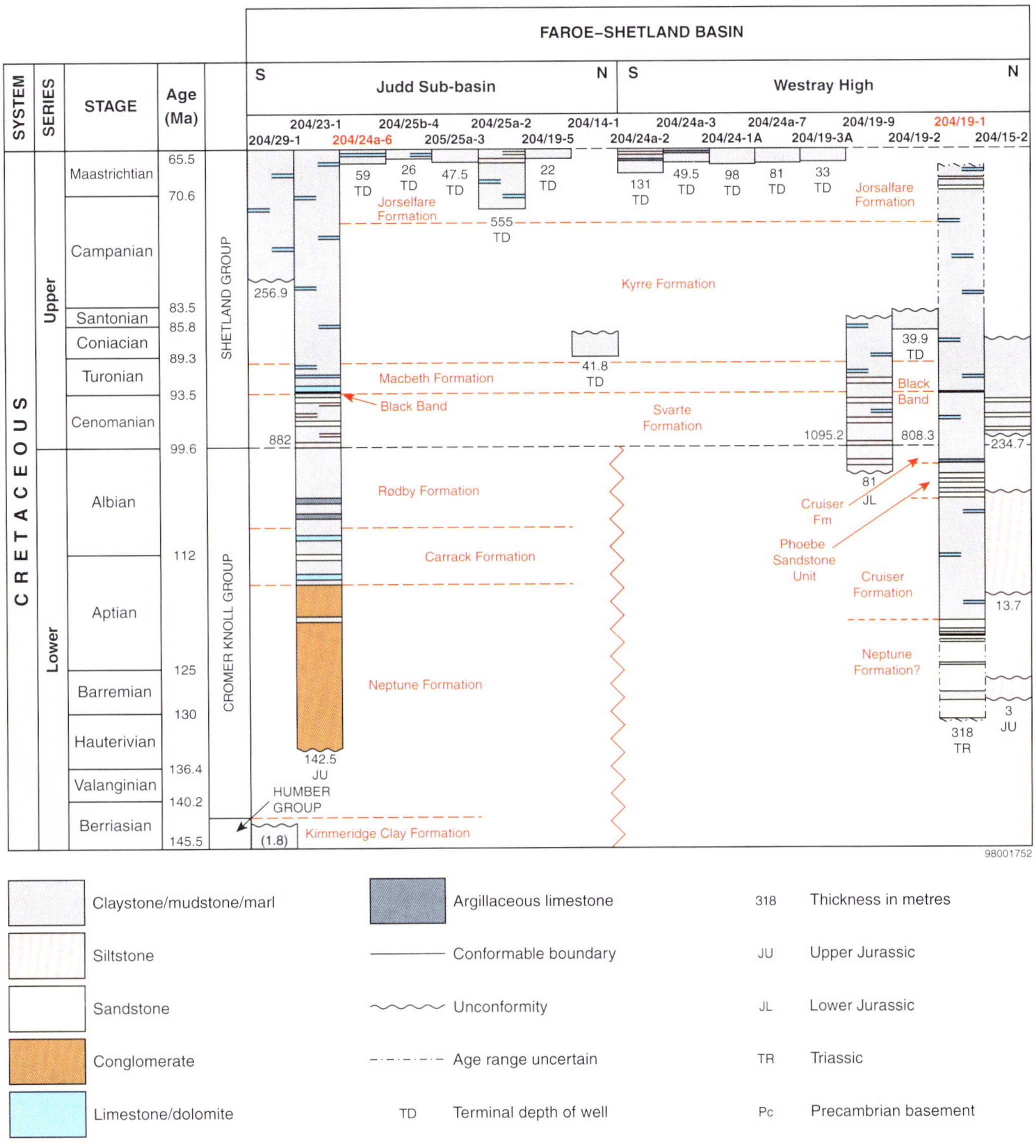

Figure 77 Stratigraphical range chart showing generalised lithology and thickness of Cretaceous rocks in the Judd Sub-basin and on the Westray High (partly modified from Ritchie et al., 1996 and Dean et al., 1999). Timescale is based on 'A Geologic Time Scale 2004' by F M Gradstein, J G Ogg, A G Smith, et al. (2004) and the International Stratigraphical Chart, 2006 (ICS) with additions. For locations of wells and lithostratigraphical nomenclature, see Figures 69 and 70, respectively. Guide to lithostratigraphical framework and key wells shown in red.

proved a dark green to grey, clast-supported conglomeratic unit, 42.5 m thick, composed largely of metamorphic clasts, with sporadic mudstone beds, 2 to 3 m thick. The mudstone is dark grey to brown-black, silty, slightly calcareous and carbonaceous. The conglomerate section ranges from Hauterivian to Aptian in age, and is assigned to the Neptune Formation. This is followed by a 10.5 m thick unit of mudstone with interbedded dark grey to green, fine to medium-grained, moderately sorted sandstone and sandy limestone of the Carrack Formation and 89.5 m of mudstone with sporadic siltstone and argillaceous limestone bands of the Rødby Formation. According to Ritchie et al. (1996), the Rødby Formation in the well can be divided into three sub-units, R1 to R3, primarily on the basis of wireline well log signature (Figure 78). In common with the North Sea (Johnson and Lott, 1993) the middle unit (R2) has a higher gamma-ray and lower velocity response compared to the lower (R1) and upper (R3) units, and may have a darker grey colour.

Westray High

The Cromer Knoll Group was tested in wells 204/19-1 and 204/19-9 on the southern part of the Westray High, and in well 204/15-2 on the northern part (Figures 69 and 77). The thickness of the succession is variable, ranging from approximately 15 to 320 m. The base of the Cromer Knoll Group is an unconformity that is generally underlain by Lower to Upper Jurassic and Triassic rocks (Ritchie et al., 1996; Figures 55a and 60b). The nature of the top of the group is variable; seemingly conformable with the Shetland Group on the southern part of the Westray High, but an unconformity between the Lower and Upper Cretaceous successions is recorded in well 204/15-2 on the northern high.

On the southern Westray High, the upper Albian section in wells 204/19-1 and 204/19-9 consists of 87.5 and 81 m, respectively, of interbedded mudstone and siltstone with traces of limestone, which are assigned to the Cruiser Formation (Figure 77). In well 204/19-1, the Cruiser Formation extends down into the Aptian, and has a total thickness of about 160 m, though this is split in the middle Albian interval by a 49 m thick unit of interbedded sandstone and mudstone termed the Phoebe Sandstone unit (Figure 70), which on wireline well logs is marked by a general decrease in gamma-ray and an increase in the velocity values (Figure 78). This unit is regarded as a lateral equivalent of the Commodore Formation that is developed farther to the north-east in the area of the Foula Sub-basin and Flett High (Figures 70 and 79). According to Ritchie et al. (1996), the Cruiser Formation in well 204/19-1 is underlain

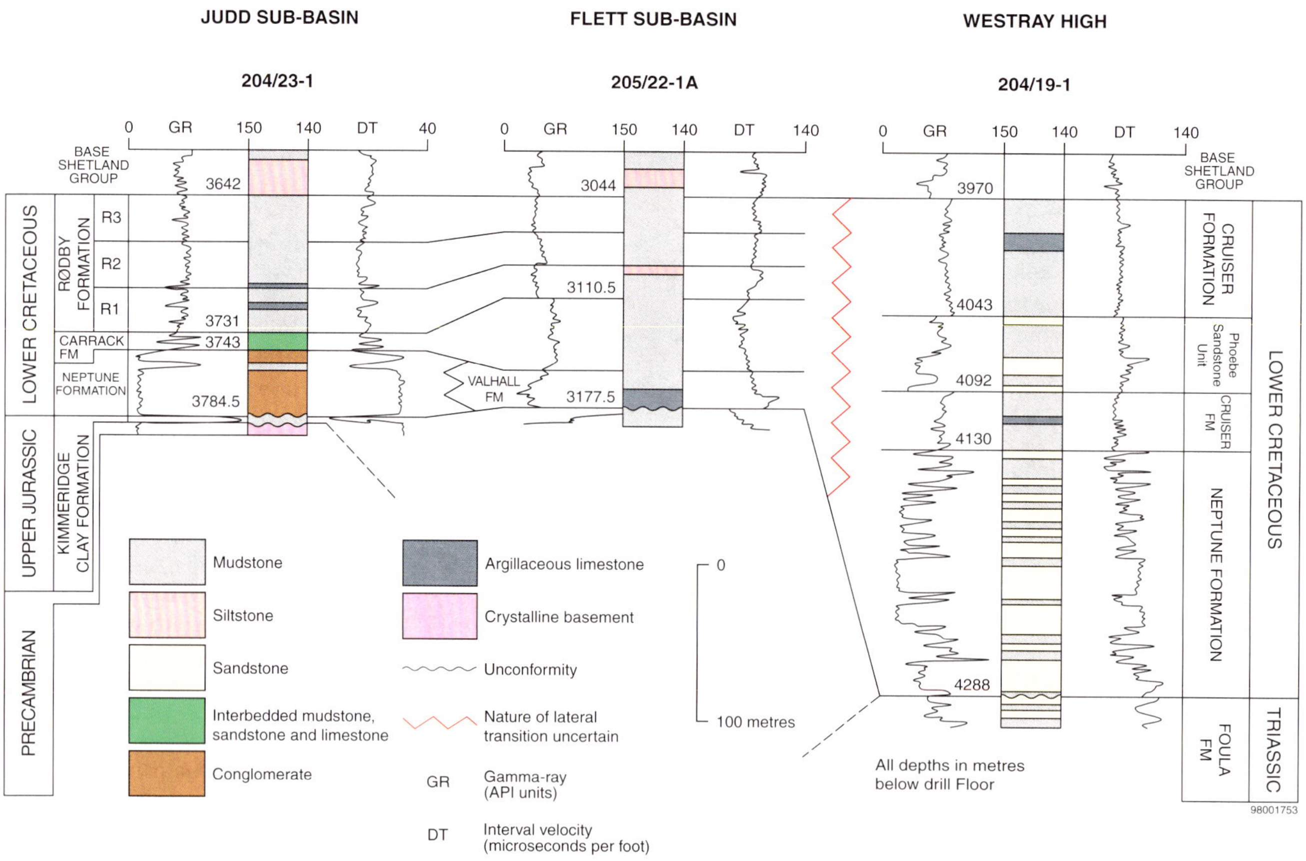

Figure 78 Subdivision and well log correlation of the Lower Cretaceous Cromer Knoll Group in wells 204/23-1, 205/22-1A and 204/19-1 within the Judd Sub-basin, Flett Sub-basin and on the Westray High (modified from Ritchie et al., 1996). For location of wells, see Figure 69.

by a 158 m thick sandstone-dominated unit that they correlate to the Hauterivian to Aptian Neptune Formation, which itself is interpreted to unconformably overlie Triassic strata (Figures 27 and 77). However, this contrasts with the composite well log that assigns the bulk of this interval to the Upper Jurassic, with only the upper 20.1 m of interbedded sandstone and mudstone regarded as undifferentiated Lower Cretaceous strata.

On the northern part of the Westray High, well 204/15-2 proved a lithostratigraphically unassigned 16.7 m thick sequence of brown-grey to brown-black carbonaceous, micaceous siltstone (Figure 77). This section is cut by an unconformity that separates a 3 m thick Barremian unit from a 13.7 m thick Aptian to Albian unit.

Foula Sub-basin

Rocks of the Cromer Knoll Group have been proved in wells 206/11-1, 206/03-1, 206/04-1, 206/05-1, 2, 214/29-1 and 208/26-1 (Figures 69 and 79). The drilled sequence ranges from about 30 m to nearly 1.5 km, with the thickest sequences generally including the Commodore Formation that straddles the Lower–Upper Cretaceous boundary. The base of the Lower Cretaceous succession has only been penetrated adjacent to the Rona High where it is marked by an unconformity that is variably underlain by Jurassic and Precambrian crystalline basement rocks (Figures 30 and 31). The top of the Cromer Knoll Group is conformable with the Shetland Group in most wells excepting 206/05-2 and 208/26-1 close to the Rona Fault, where an unconformity separates Albian and Cenomanian strata.

The lithology of the Cromer Knoll Group in the Foula Sub-basin is highly variable with the predominantly coarse-grained clastic lithofacies of the Royal Sovereign and Commodore formations separated by the argillaceous rocks of the Cruiser Formation (Figure 79). The Royal Sovereign and Commodore formations are best developed adjacent to the Rona Fault and a causal relationship between these clastic units and movement along the fault is envisaged (Figure 72a and b; Ritchie et al., 1996; Dean et al., 1999; Grant et al., 1999).

The Aptian to Albian Royal Sovereign Formation forms the basal Cretaceous unit penetrated in the Foula Sub-basin, and consists predominantly of conglomerate and sandstone (Figure 79). In the extreme north-east part of the Sub-basin, well 208/26-1 proved a 60.5 m thick basal conglomerate, overlying Precambrian crystalline basement, dominated by clasts of quartz, schist and gneiss that is itself overlain by 133.5 m of pebbly sandstone, which has the same clast types set in a matrix of very fine to fine-grained sandstone. A basal conglomerate facies has also been proved in wells 206/04-1 and 206/11-1 (Figure 80) located in the central and south-west parts of the sub-basin, respectively. Well 206/04-1 cored an 85.3 m thick pebble to boulder grade conglomerate that contained igneous and metamorphic rock clasts, which are very angular and commonly shattered, with sporadic interbedded sandstone. In well 206/11-1, the conglomerate is 122 m thick and, in addition to metamorphic clasts, includes clasts of shale and siltstone. These wells are located adjacent to

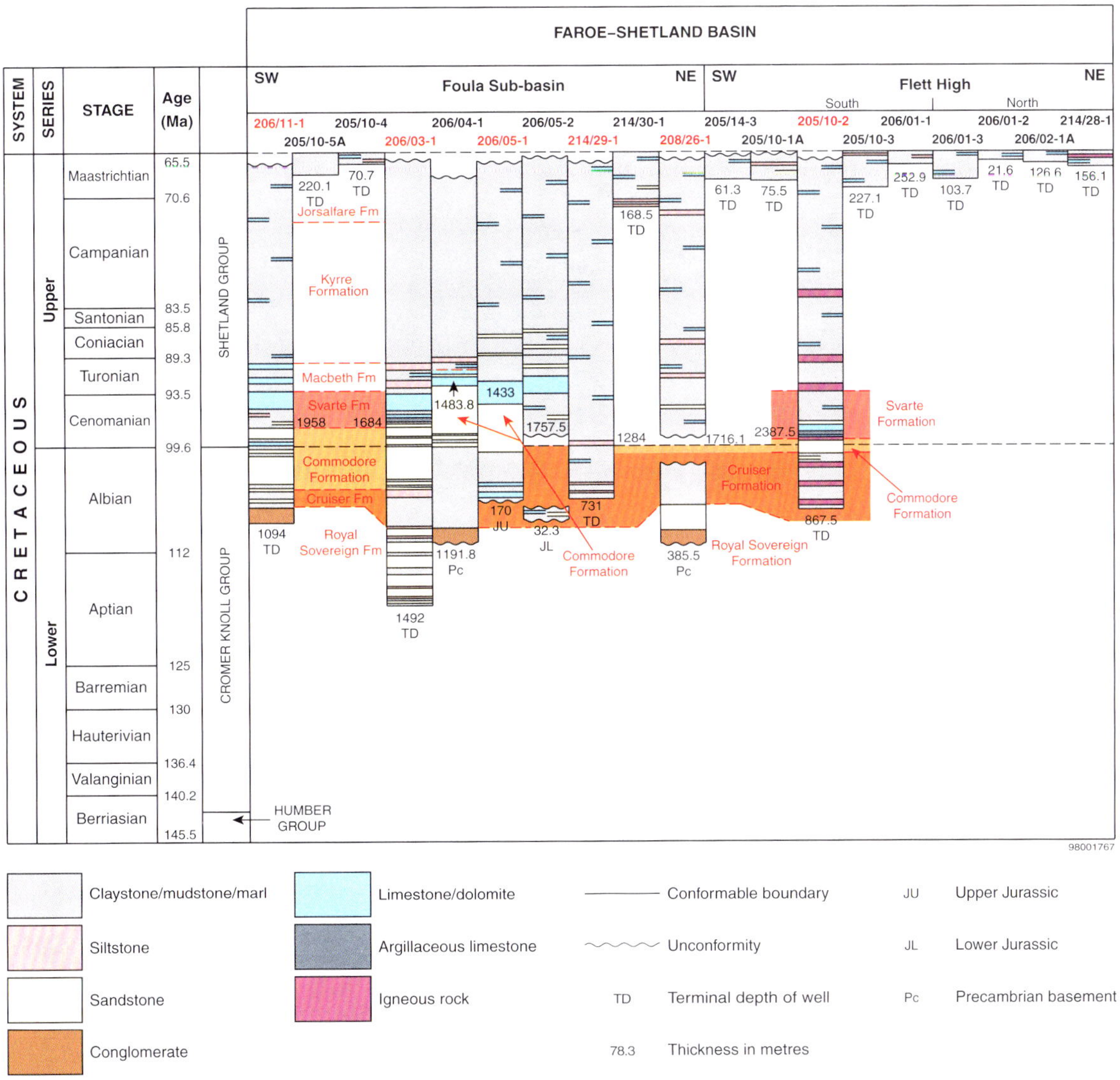

Figure 79 Stratigraphical range chart showing generalised lithology and thickness of Cretaceous rocks in the Foula Sub-basin and on the Flett High (partly modified from Ritchie et al., 1996 and Grant et al., 1999). Timescale is based on 'A Geologic Time Scale 2004' by F M Gradstein, J G Ogg, A G Smith, et al. (2004) and the International Stratigraphical Chart, 2006 (ICS) with additions. For locations of wells and lithostratigraphical nomenclature, see Figures 69 and 70, respectively. Guide to lithostratigraphical framework and key wells shown in red.

the Rona Fault (Figure 69). By way of contrast, well 206/03-1 is situated farther into the central part of the Foula Sub-basin where a 401.5 m-thick sequence of interbedded sandstone, claystone/mudstone and sporadic siltstone is preserved. This section has also been assigned to the Royal Sovereign Formation (Ritchie et al., 1996). The sandstone is fine to medium-grained, locally coarse-grained, quartz-rich, poor to moderately well-sorted, and forms beds up to several tens of metres thick that have sharp bases. The dark grey claystone/mudstone is laminated and generally noncalcareous.

The overlying Cruiser Formation occurs in all wells that penetrate Lower Cretaceous strata within the sub-basin and is composed mainly of claystone that ranges in thickness from 26 m in 206/11-1 to 772 m in 206/03-1 (Figures 80). In the latter well the Cruiser Formation consists predominantly of grey, laminated and bioturbated claystone interbedded with sporadic thin siltstone, which increases near the top of the sequence. The claystone is carbonaceous (with plant fragments) and displays soft-sediment deformation structures, including load casts and slumped bedding (Ritchie et al., 1996). Interbedded thin sandstone and siltstone occur in well 206/11-1, whereas thin bands of limestone occur locally in well 206/05-1. In most of the wells, the Cruiser Formation occurs in continuity with the Royal Sovereign and Commodore formations. However, in well 206/05-2, the Cruiser Formation is preserved as an isolated, unconformity bounded unit, 32.3 m thick, which rests on Lower Jurassic strata and is overlain by the Upper Cretaceous Svarte Formation (Figure 79). The Svarte Formation also unconformably overlies the Cruiser Formation in well 208/26-1. Both wells 208/26-1 and 206/5-2 are located adjacent to the hanging-wall block of the Rona Fault.

The sandstone-dominated Commodore Formation occurs in wells 206/11-1, 206/03-1, 206/04-1 and 206/05-1 in the south-west and central part of the Foula Sub-basin, where it ranges in thickness from 946 m in 206/11-1 to 15 m in 206/05-1 (Figure 79). The

Figure 80 Well log correlation of the Lower Cretaceous Cruiser and Commodore formations (Cromer Knoll Group) in wells 206/11-1 and 206/03-1 within the Foula Sub-basin (modified after Ritchie et al., 1996). For location of wells, see Figure 69.

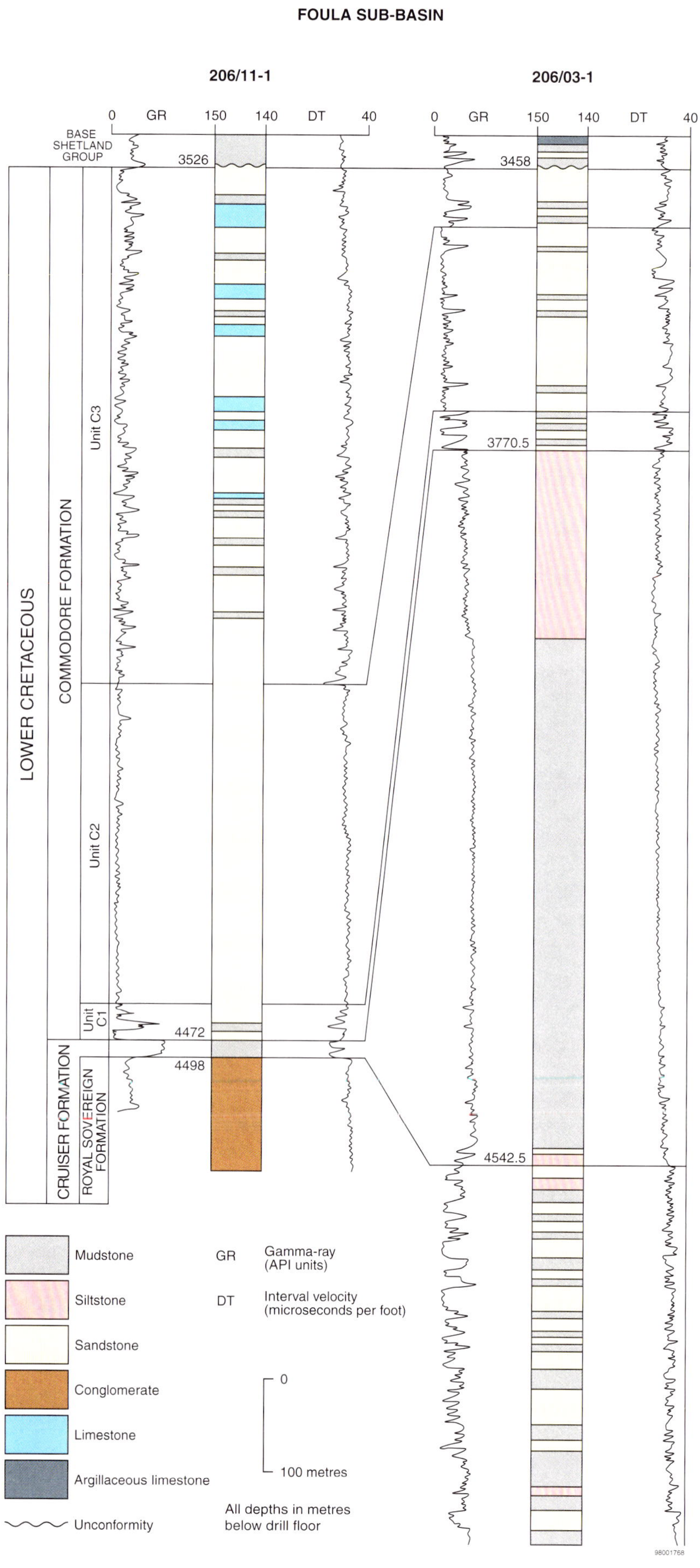

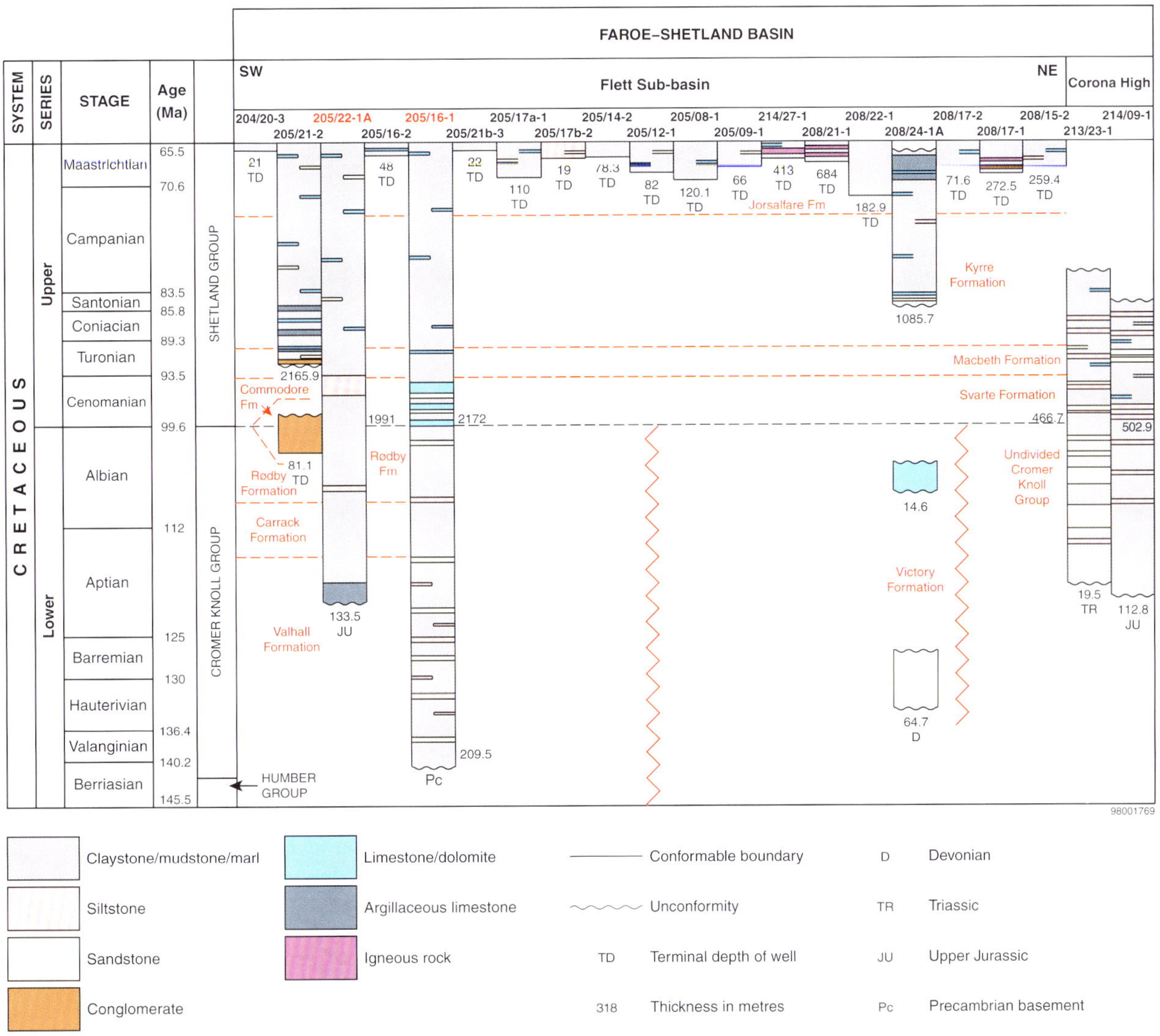

Figure 81 Stratigraphical range chart showing generalised lithology and thickness of Cretaceous rocks in the Flett Sub-basin and on the Corona High (partly from Ritchie et al., 1996). Timescale is based on 'A Geologic Time Scale 2004' by F M Gradstein, J G Ogg, A G Smith, et al. (2004) and the International Stratigraphical Chart, 2006 (ICS) with additions. For locations of wells and lithostratigraphical nomenclature, see Figures 69 and 70, respectively. Guide to lithostratigraphical framework is shown in red.

age range of the Commodore Formation is generally taken as Albian to Cenomanian (Ritchie et al., 1996), though in well 206/04-1 the well log data extend the upper age of the unit into the Turonian. In well 206/03-1, the Commodore Formation is 312.5 m thick and consists mainly of white to pale grey, fine to medium and locally coarse-grained, variably sorted calcareous sandstone with sporadic thin grey mudstone beds. The sandstone locally contains intraformational pebbles, glauconite and comminuted shell fragments (Meadows et al., 1987). Within this section, the Commodore Formation has been divided into three units, C1 to C3, on the basis of wireline well log response (Ritchie et al., 1996; Figure 80). The lower unit, C1, is 46 m thick and is characterised by sandstone beds <5 m thick, interbedded with many thin mudstone beds and this generates a highly variable gamma-ray well log signature. The middle unit, C2, forms the bulk of the formation in well 206/03-1 and consists of uniform low-gamma sandstone with rare thin mudstone beds that form sharp gamma-ray log peaks. The upper unit, C3, is a high and variable gamma-ray sandstone with occasional thick mudstone beds. A comparable subdivision of the Commodore Formation is also reported in well 206/11-1 (Figure 80), where unit C3 is the thickest of the sandstone-dominated units (564 m within a 946 m thick section) that also includes metre-thick interbeds of beige and brownish, sandy and chalky limestone. In well 206/04-1, the Commodore Formation is 544 m thick and includes thin intraformational conglomerate beds of Turonian age, with clasts of limestone and mudstone. A 15 m thick fine-grained sandstone of Albian to Cenomanian age proved in well 206/05-1 is assigned by Ritchie et al. (1996) to the Commodore Formation, though Grant et al. (1999) do not recognise the occurrence of this sandstone in their appraisal of the Turonian play in the Foula Sub-basin. Although the presence of the Commodore Formation is proved locally on the Flett High (Figure 79), the westerly extent of this formation remains largely unknown. It may correlate with the Phoebe Sandstone Unit on the Westray High (Ritchie et al., 1996; Figure 70).

Flett High

Lower Cretaceous strata have been proved only in well 205/10-2 on the southern part of the Flett High (Figures 69 and 79). The well penetrated 867.5 m of the Cromer Knoll Group before terminating within the Lower Cretaceous succession. The section also includes numerous Palaeogene sills. The bulk of the sequence in this well is assigned to the Cruiser Formation, which is 834 m thick, and consists of dark grey to grey-black, silty, noncalcareous claystone with thin interbedded, pale grey, very fine-grained sandstone. The overlying Commodore Formation is 33.5 m thick and consists of fine to medium-grained, calcareous sandstone.

Flett Sub-basin

In the north-east part of the Flett Sub-basin, Lower Cretaceous strata have only been drilled in well 208/24-1A (Figures 69 and 81). In this well, the Cromer Knoll Group rests unconformably on Devonian strata, and is unconformably overlain by the Upper Cretaceous Shetland Group. The preserved section is 79.3 m thick and consists of two unconformity-bounded units that have much in common with the depositional environment of the West Shetland Basin, with the rocks correlatable with the Victory Formation (Ritchie et al., 1996). A lower unit, 64.7 m thick, is composed of grey, fine to very coarse-grained, moderately sorted, quartzose, glauconitic, calcite-cemented sandstone of Hauterivian to Barremian age, which may be correlated to Unit III of the Victory Formation. The upper unit is of Albian age and consists of a 14.6 m thick sequence of hard, white, chalky limestone that becomes increasingly sandy and glauconitic with depth. This unit is tentatively correlated with Unit V of the Victory Formation.

Within the south-west part of the Flett Sub-basin, The Cromer Knoll Group has been proved in wells 205/22-1A, 205/16-1 and 205/21-2 with thicknesses of approximately 80 to 210 m proved (Figure 81). A 54.5 m thick sequence of dark grey to brown mudstone with common interbeds of white to buff, very fine to fine-grained sandstone and dark grey to brown siltstone of late Berriasian to Aptian age characterise the Valhall Formation in well 205/16-1. In the south-west corner of the sub-basin, the Valhall Formation in well 205/22-1A is represented by a basal argillaceous limestone unit, about 12 m thick, of Aptian age overlain by a 10.5 m thick section of mudstone. In wells 205/16-1 and 205/22-1A the Carrack Formation is composed mainly of mudstone that forms a unit 40 to 45 m thick. The overlying Rødby Formation is thickest in well 205/16-1, where a 115 m thick sequence of dark grey to dark brown mudstone with sporadic beds of white to grey, very fine to fine-grained sandstone and dark grey siltstone was penetrated. Wells 205/22-1A drilled 66 m of mudstone with sporadic siltstone and argillaceous limestone bands. According to Ritchie et al. (1996), the Rødby Formation in these two wells can be divided into three sub-units, R1 to R3, primarily on the basis of wireline well log signature (e.g. Figure 78). The top of the Cromer Group in wells 205/16-1 and 205/22-1A is conformably overlain by Shetland Group Svarte Formation, whereas its base is defined by either Upper Jurassic Kimmeridge Clay Formation or Precambrian crystalline basement (Figures 60b and 81). Adjacent to the Rona High, well 205/21-2 terminated in an 81.1 m thick sequence of upper Albian to lower Cenomanian conglomerate. The highly fracture conglomerate is clast supported and poorly-sorted, with subangular to rounded clasts of mainly diorite, to cobble grade, set in a matrix of black to greenish grey clay, silt-size quartz, feldspar and mafic minerals. This unit has been correlated to the Commodore Formation (Ritchie et al., 1996) by comparison with a similar sequence in well 206/08-6A on the Rona High.

Corona High

On the Corona High, Lower Cretaceous rocks have been proved in wells 213/23-1 and 214/09-1, on the central and north-eastern parts of the high, respectively (Figures 69 and 81). In well 213/23-1 a 19.5 m thick section of Aptian to Albian strata was drilled unconformably overlying Triassic rocks (Figure 55a). Proven Albian strata were also cored in well 214/09-1 though the lower age limit of this 112.8 m thick sequence is unknown. The section in this well rests unconformable on the Upper Jurassic Kimmeridge Clay Formation (Figure 60b). On seismic data, the Cretaceous section is commonly not easily distinguished from older strata on the Corona High (e.g. Figures 30, 31 and 33). In both wells, Lower Cretaceous strata appear to pass conformably upwards into the Upper Cretaceous succession. Claystone predominates in both wells, and is generally pale to dark grey, occasionally orange-brown, slightly to noncalcareous, firm, blocky and carbonaceous. However, interbedded sandstone is common in 213/23-1 whereas siltstone is more prevalent in 214/09-1. The sandstone is generally brown, very fine-grained, well-sorted and cemented with silica. The siltstone is commonly grey and locally sandy. The lithostratigraphical assignment of these sections remains undefined.

Erlend Sub-basin

The Erlend Sub-basin occurs in the north-east corner of the Faroe–Shetland Basin where the nature and distribution of Cretaceous strata is poorly resolved (Figure 69). On seismic profiles, Palaeogene lavas and igneous sills obscure much of the pre-Cenozoic succession (e.g. Figures 24 and 26). Lower Cretaceous rocks have only been tested in well 219/28-1, adjacent to the Møre Basin and East Shetland High, which recovered 15.2 m of dark brownish grey, variably calcareous, slightly silty claystone with stringers of pale greenish grey, red-brown and pale grey limestone of Hauterivian to Barremian age (Figure 82). These rocks rest unconformably on a Palaeozoic or Precambrian crystalline basement and are unconformably overlain by Upper Cretaceous strata. This claystone is assigned an undifferentiated status within the Cromer Knoll Group (Ritchie et al., 1996; Harker, 2002).

Møre, Unst and Magnus basins, and the Erlend High

Within the report area, rocks belonging to the Cromer Knoll Group have been proved in three wells and two BGS boreholes on the flanks of the Møre and Unst basins, and on the Erlend High (Figures 69 and 82). These are largely of Hauterivian to Barremian in age, and lithostratigraphically undivided (Ritchie et al., 1996). In contrast, Lower Cretaceous rocks are interpreted to be absent on the East Shetland High and along

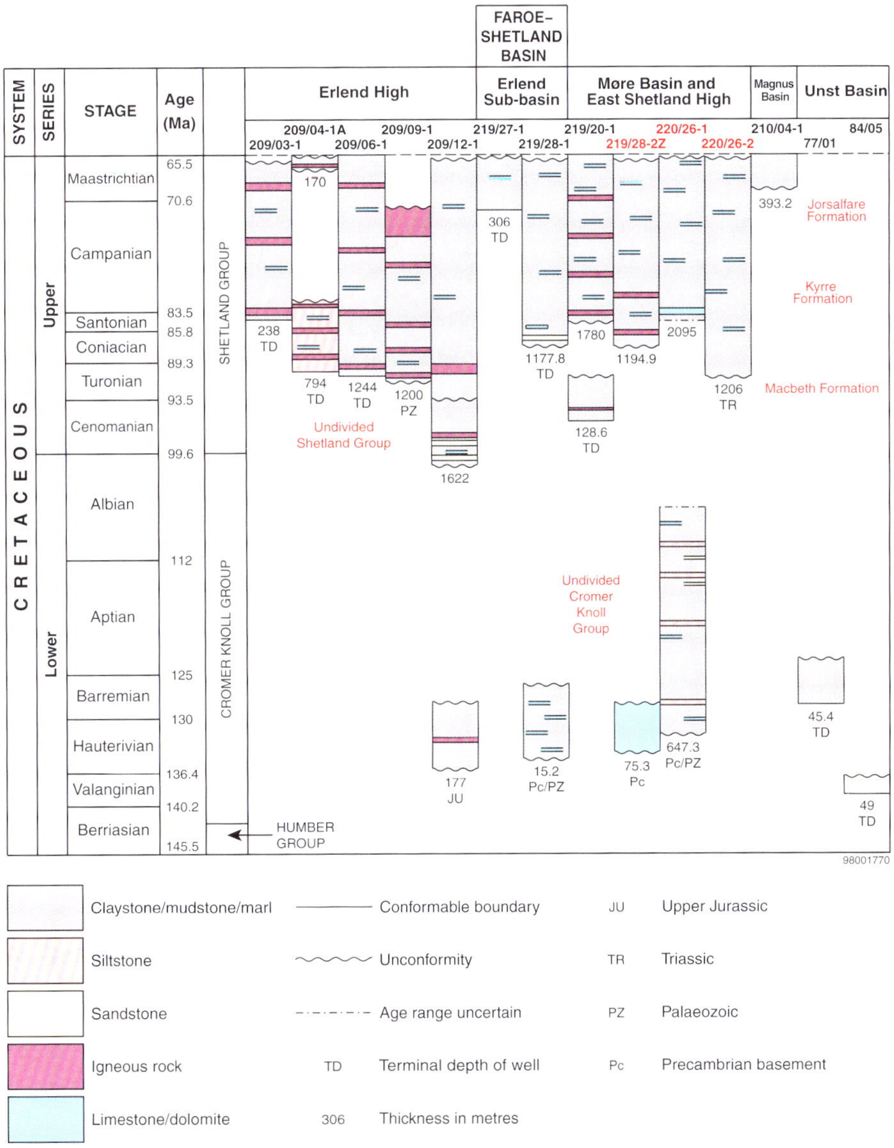

Figure 82 Stratigraphical range chart showing generalised lithology and thickness of Cretaceous rocks in the Erlend Sub-basin, Møre, Magnus and Unst basins and the Erlend and East Shetland highs (partly from Ritchie et al., 1996). Timescale is based on 'A Geologic Time Scale 2004' by F M Gradstein, J G Ogg, A G Smith, et al. (2004) and the International Stratigraphical Chart, 2006 (ICS) with additions. For locations of wells and lithostratigraphical nomenclature, see Figures 69 and 70, respectively. Guide to lithostratigraphical framework and key wells shown in red.

the northern margin of the Magnus Basin (Figure 69), though up to 1.5 km of Lower Cretaceous strata have been reported close to the southern bounding fault of the basin (Johnson et al., 1993). The Erlend, East Shetland and Manet highs probably had considerable inherited (from the Jurassic) topographical relief, against which the Lower Cretaceous sediments onlapped (Nelson and Lamy, 1987; Figures 23 and 24).

Møre Basin

The Møre Basin may have accumulated between 2 and 4 km of Lower Cretaceous sediment (Nelson and Lamy, 1987; Brekke, 2000; Færseth and Lien, 2002). Seismic evidence presented here suggests that up to nearly 2.5 s TWTT of Lower Cretaceous is present on the southern flank of the basin (Figure 23). Well 219/28-2Z proved 75.3 m of white to pale brown, micritic limestone of Hauterivian to Barremian age incorporating coarse-grained sand to boulder grade clasts of pale grey to green psammitic gneiss (Figure 82). This contrasts with well 220/26-1 which recovered a 647.3 m thick mudstone-dominated section, grading to siltstone, and with sporadic traces of sandstone and limestone. The basal 3 m of this section are assigned a Barremian to Hauterivian age, though the bulk of the sequence is poorly dated as mid Albian to early Santonian (Early to Late Cretaceous). These rocks are assigned an undifferentiated status within the Cromer Knoll Group (Ritchie et al., 1996). In both wells the rocks unconformably overlie Palaeozoic or Precambrian crystalline basement.

Unst Basin

The Cromer Knoll Group extends into the Unst Basin where BGS boreholes 77/01 and 84/05 proved an undifferentiated section of Valanginian to Aptian age (Johnson et al., 1993; Figures 69 and 82). Borehole 77/01 recovered 45.4 m of pale grey to black claystone with sporadic rip-up clasts of pyritic claystone. A rich calcareous nannoflora includes *Chiastozygus litteracius*, *Micrantholithus obtusus*, *Calcialathina oblongata*, *Lithraphidites bolli*, *Nannoconus steinmanni* and *Nannoconus wassalli*, indicative of an Early Barremian to Early Aptian age. Quaternary strata unconformably overlie this section. Farther north, borehole 84/05 proved a 49 m thick sandstone unit that is slightly muddy, greenish grey, moderately well-sorted, very fine to fine-grained, is bioturbated, and includes shell fragments, rare plant fragments, and has sporadic interbeds of dark green claystone and brighter green glauconite-rich intervals. Sporadic ash bands may also be present. The sandstone preserved a dinoflagellate cyst assemblage that includes *Lagenorhytis delicatula, Cymososphaeridium validum, Scriniodinium pharo, Pseudoceratium brevicornutum, Ctenidodinium elegantulum* and *Nelchinopsis kostromiensis*, which are all part of the *Spiniferites ramosus* zone, indicative of a Valanginian age. Pliocene and Quaternary strata overlie the sandstone.

Erlend High

Lower Cretaceous rocks have been proved only on the southern part of the Erlend High in well 209/12-1, which recovered 177 m of dark brown claystone with sporadic thin sandstone of 'Neocomian' (Berriasian to Hauterivian) to Barremian age (Figures 69 and 82). The claystone is cut by thin acidic intrusions. These rocks rest unconformably on Upper Jurassic sandstone (Figure 60b), and are assigned an undifferentiated status within the Cromer Knoll Group.

West Lewis and north-east Rockall basins

Lower Cretaceous rocks have been tested in a commercial well and two BGS boreholes in the West Lewis Basin (Figure 69), where a thickness of at least 175 m has been proved adjacent to the footwall of the basin-bounding block. On the eastern, up-dip flank of the basin, Lower Cretaceous strata occur at subcrop below the Quaternary. In the north-east Rockall Basin, the occurrence of Lower Cretaceous strata remains poorly resolved. The Cromer Knoll Group in both these basins remains undifferentiated.

West Lewis Basin

Lower Cretaceous strata have been recovered in well 164/25-1Z and BGS boreholes 88/01 and 90/05 (Figure 69). Well 164/25-1Z penetrated 176.5 m of reddish brown siltstone and sandstone, which is argillaceous, very fine to fine-grained, poorly-sorted, and quartzose. An abundant and diverse late Albian microflora includes the dinoflagellate cyst *Litosphaeridium siphoniphorum* and the calcareous nannoplankton *Watznaueria barnesae, Biscutum constans, Tegumentum stradneri, Rhagodiscus splendens, Zygodiscus sisyphus* and *Tranolithus orionatus*. On the composite well log this interval had previously been assigned to the underlying Permo-Triassic section, but the fossil assemblage confirms an upper Albian clastic package. The section is overlain by Upper Cretaceous rocks, and unconformably overlies Permo-Triassic strata (Hitchen and Stoker, 1993; Stoker et al., 1993).

BGS borehole 88/01 proved 4 m of coarse-grained, poorly-sorted, calcareous and glauconitic sandstone with abundant lithic and shell fragments. Marine and terrestrially derived palynomorphs indicate a Barremian age. The sandstone rests unconformable on Middle Jurassic mudstone. Borehole 90/05 tested 21 m of dark grey to black, massive to poorly bedded, organic-rich Berriasian mudstone, with lignite bands and fragments. This sequence has been assigned to the Kimmeridge Clay Formation. Quaternary deposits unconformably overlie the sequences in both boreholes.

North-East Rockall Basin

Well 164/07-1 is reported to have terminated in Albian to Cenomanian claystone (Figure 69; Archer et al., 2005). These claystones are intruded by over seventy dolerite sills, which make up around half of the total Lower Cretaceous section.

DISTRIBUTION OF UPPER CRETACEOUS

The distribution of Upper Cretaceous rocks within the Faroe–Shetland region is described firstly from the marginal basins of the West Shetland, North Rona and Solan basins, and the adjacent Solan Bank High (Figure 69). The Rona High separates the West Shetland and Faroe–Shetland basins, and is described separately. This is followed by the Faroe–Shetland Basin, which is divided into a number of sub-basins and intrabasinal highs. Peripheral basins and highs, such as the Møre and Magnus basins, and the Erlend High in the north-east and the West Lewis and north-east Rockall basins in the south-west are described at the end of the section.

West Shetland Basin

Upper Cretaceous strata occur throughout the West Shetland Basin (Figures 69 and 73). The West Shetland High marks the eastern boundary of the succession whereas to the west, it extends across the Rona High and into the Faroe–Shetland Basin (Figure 31). The Chalk and Shetland groups are both preserved within the West Shetland Basin, although the Chalk Group is restricted to the south-west part of the basin, as well as the Cenomanian to Turonian interval (Figures 70 and 72b). In most of the wells, the Chalk and Shetland groups are separated by an unconformity, though regionally they are partly laterally equivalent (Figures 70 and 73). In well 206/16-1, the boundary between the groups appears more transitional, however there is a marked change from the interbedded sandstone and limestone of the Chalk Group into the claystone to mudstone dominated Shetland Group. The top of the Shetland Group is mostly (though not exclusively) marked by an unconformity with the overlying Cenozoic section.

The Chalk Group was tested in wells 205/23-1, 205/25-1, 205/30-1 and 206/16-1, which proved a predominantly carbonate-rich sequence, up to 247.5 m thick in well 205/30-1, belonging the Hidra and Herring formations (Figures 73 and 83). The limestone varies from fine-grained to cryptocrystalline, is white to grey and occasionally buff, pink, red-brown, pale green, and grades locally to calcareous sandstone (Ritchie et al., 1996). In wells 205/25-1, 205/30-1 and 206/16 1, adjacent to the Shetland Spine Fault, the limestone of the Hidra Formation passes laterally to a calcareous sandstone-rich facies referred to as the Haddock Sandstone Unit (Figure 72b). This low-gamma sandstone (e.g. Figure 83) is variably coloured white and buff to grey-green and orange-brown, fine to very coarse-grained, well to poorly sorted and compact to friable. It is commonly cemented by calcite and pyrite, contains glauconite, mica and carbonaceous fragments and interbedded with sandy, sometimes cherty limestone towards the top of the section. The Haddock Sandstone Unit reaches a maximum drilled thickness of 139 m in well 205/30-1, with a thin conglomerate composed of clasts of reworked limestone marking the base.

In well 205/30-1, the base of the overlying Herring Formation is marked by a thin, dark grey to black, locally dark green or varicoloured, noncalcareous, pyritic mudstone that displays very high gamma-ray and very low velocity well log responses (Figures 73 and 83). This pyritic mudstone represents the Black Band as formally recognised at the base of the Herring Formation (Johnson and Lott, 1993).

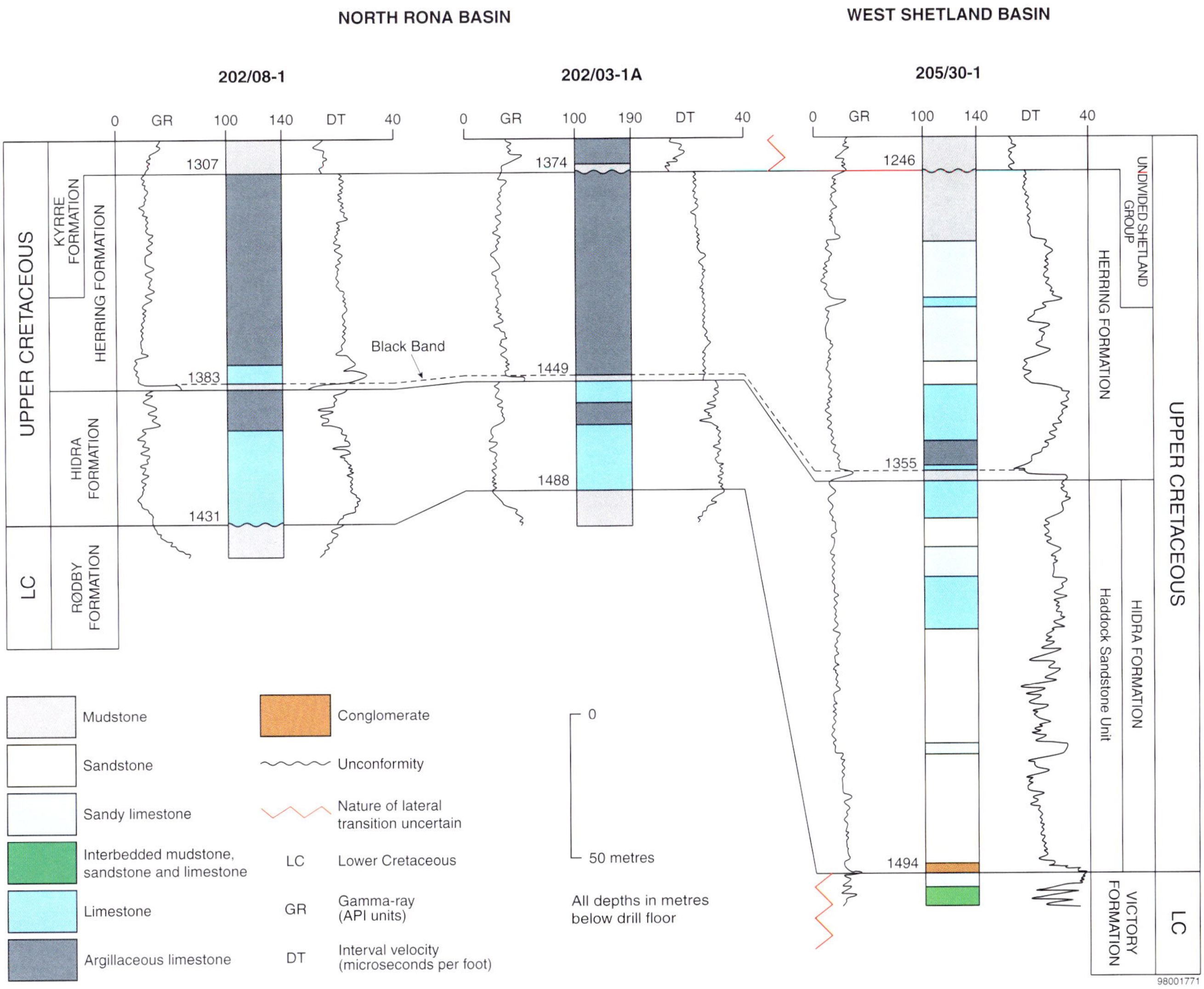

Figure 83 Well log correlation of the Upper Cretaceous Hidra and Herring formations (Chalk Group) in wells 202/08-1, 202/03-1A and 205/30-1 within the North Rona and West Shetland basins (modified from Ritchie et al., 1996). For location of wells, see Figure 69.

The Shetland Group was tested in all wells in the West Shetland Basin, which proved an approximately 400 m to 1.7 km thick monotonous sequence of grey, calcareous claystone with sporadic stringers and thin beds of limestone, dolomite and rare sandstone and siltstone. The Kyrre and Jorsalfare formations comprise the Shetland Group (Figure 70), though in much of the basin the succession remains undivided (Figure 73). In well 206/16-1, the Kyrre Formation consists of 657 m of grey, moderately calcareous claystone that is interbedded with pale grey to green-grey, very fine to fine-grained, locally medium to coarse-grained, calcareous sandstone and white to grey limestone in the lower part of the section. This is overlain by a 1013.8 m thick sequence of grey, calcareous claystone of the Jorsalfare Formation, with disseminated limestone and dolomite stringers, although thicker limestone beds are present in the lower Maastrichtian section.

In well 207/01-2, the Kyrre Formation is represented by an 11 m thick limestone-rich facies referred to as the Dab Limestone Unit (Ritchie et al., 1996; Figure 73), which consists of white, pale grey, brown, cream and buff, fine to medium-grained, glauconitic, pyritic, silty and sandy limestone with thin interbeds of grey, calcareous, carbonaceous mudstone. It should be noted that Grant et al. (1999) included the limestone in this well as part of the Victory Formation. In well 206/10a-1, the lower part of the Kyrre Formation is marked by a 154 m thick sequence of siltstone and sandstone informally termed the Whiting Sandstone Unit (Ritchie et al., 1996), which on wireline well logs is defined by low gamma-ray and high velocity log responses (Figures 70, 72c and 84). The siltstones are grey to dark grey, green or red to brown, partly calcareous and carbonaceous; the sandstones are grey or grey to brown or occasionally orange to dull yellow, silty to very fine-grained, moderately sorted and argillaceous.

In the north-east part of the West Shetland Basin, the Shetland Group in undivided. Wells 207/02-1 and 208/23-1 recorded a 1 to 1.2 km thick undivided mixed clastic assemblage of sandstone and claystone with sporadic limestone beds and stringers (Figure 73).

North Rona Basin, the Solan basins and the Solan Bank High

Upper Cretaceous rocks have been proved in 16 wells within the North Rona and Solan basins, and the

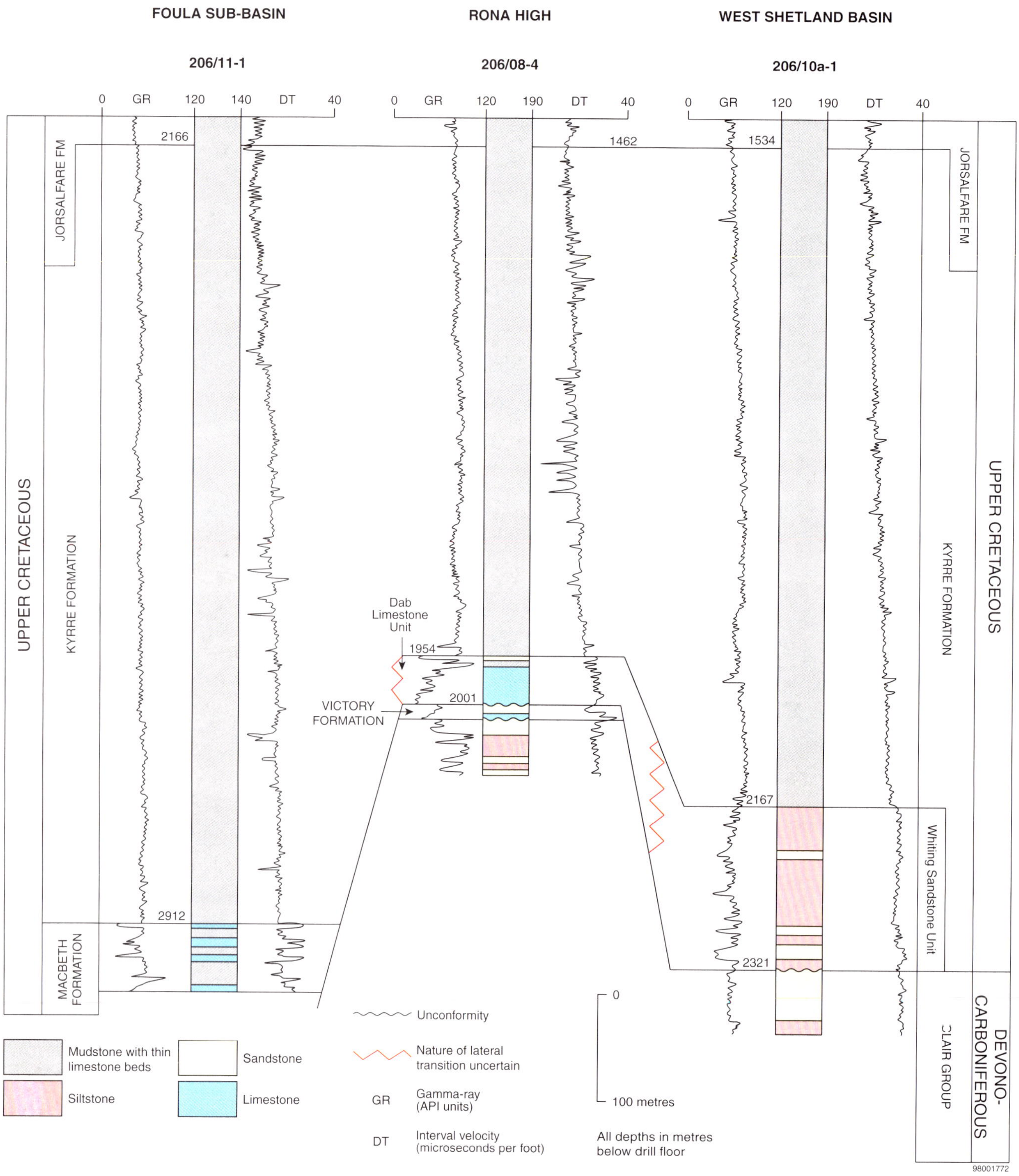

Figure 84 Well log correlation of the Upper Cretaceous Kyrre Formation (Shetland Group) from wells 206/11-1, 206/08-4 and 205/10a-1 within the eastern margin of the Faroe–Shetland Basin, Rona High and the West Shetland Basin (modified from Ritchie et al., 1996). For location of wells, see Figure 69.

Solan Bank High immediately south of the Faroe–Shetland Basin (Figures 69 and 73). Limestone and argillaceous limestone of the Chalk Group are present in the North Rona and East Solan basins. In the North Rona Basin, a maximum drilled thickness of 124 m incorporates both the Hidra and Herring formations, whereas a thinner drilled sequence up to 47 m thick in the East Solan Basin is all assigned to the Hidra Formation. The Chalk Group is largely separated from the overlying Shetland Group by an unconformity. The Shetland Group reaches a maximum drilled thickness of approximately 1.8 km in the East Solan Basin and 560 m in the North Rona Basin. A 759 m thick section of Shetland Group strata was drilled in the West Solan Basin, unconformably overlying Lower Cretaceous rocks. Rocks of the Shetland Group extend onto the Solan Bank High but are absent from the Judd High (Figure 27).

NORTH RONA BASIN

The Upper Cretaceous succession forms a partially eroded wedge that infills and buries the rifted Upper Jurassic to Lower Cretaceous basinal topography (Figure 28). Upper Cretaceous rocks have been drilled in wells 202/02-1, 202/03-1A, -2, 202/08-1 and 202/12-1, which tested both the Chalk and Shetland groups (Figures 69 and 73). The most complete section was recovered in well 202/08-1 which proved 682.5 m of Cenomanian to Maastrichtian rocks. The Cenomanian to Turonian Hidra and Herring formations (Chalk Group) consist of 124 m of predominantly white to pale grey limestone interbedded with argillaceous limestone and varicoloured, carbonaceous, laminated mudstone (Figure 83). This is overlain by a thicker succession of grey claystone of the Kyrre and Jorsalfare formations. A comparable Chalk Group sequence is present in well 202/03-1A (Figure 83). In both well 202/08-1 and 202/03-1A, the base of the Herring Formation is marked by a dark grey to black, occasionally dark green noncalcareous, pyritic mudstone that is correlated with the Black Band (Ritchie et al., 1996). On wireline well logs, the Black Band is recognised by a single peak with high gamma-ray and very low velocity log responses (Figure 83).

The top of the Chalk Group in well 202/03-1A is an unconformity, which is overlain by 416 m of pale to dark grey and red-brown calcareous mudstone interbedded with thin sporadic white and buff-brown limestone, argillaceous limestone and dolomite assigned to the Kyrre and Jorsalfare formations. The latter has been informally divided into units J1 and J2 (Figure 85). Unit J1 forms a relatively thin succession of sometimes-reddened calcareous mudstone and argillaceous limestone. Unit J2 forms the bulk of the formation and is made up of grey mudstone and sporadic argillaceous limestone beds. Unit J2 commonly becomes more calcareous towards its top. The boundary between the two units is marked on wireline well logs by a downward decrease in gamma-ray log values and an increase in velocity.

In well 202/12-1, the Cenomanian Hidra Formation is represented by a 12.8 m thick, white, chalky limestone, which is overlain by 76.2 m of Turonian to ?Coniacian pink to red-brown siltstone that grades to fine to medium-grained sandstone at the base (Figure 73). This equivalent of the Herring Formation is coarser-grained than that represented in wells 202/03-1A and 202/08-1. An unconformity separates the Turonian to ?Coniacian strata from the overlying upper Campanian to Maastrichtian sequence, which is 274.3 m thick and consists predominantly of grey, soft to firm, calcareous claystone though the Campanian section is dominated by a 51.8 m thick argillaceous limestone.

Wells 202/02-1 and 202/03-2 proved 210 and 314.5 m respectively of Campanian to Maastrichtian rocks, which consist predominantly of grey to dark grey and greenish grey, carbonaceous, calcareous and locally glauconitic mudstone with sporadic interbeds or stringers of white and grey limestone and orange brown dolomite. These rocks are assigned largely to the Jorsalfare Formation.

The Shetland Group in wells 202/02-1, 202/03-1A and -2 appears to be conformable with the overlying Paleocene section (Figure 73). In the southern part of the North Rona Basin, an unconformity separates the Maastrichtian from younger strata in wells 202/08-1 and 202/12-1 adjacent to the Solan Bank High.

EAST SOLAN BASIN

In excess of 1 s TWTT of Upper Cretaceous strata are preserved in the East Solan Basin, where they appear to form a gentle syncline bounded by the Otter Bank Fault and the Rona High (Figure 39). Upper Cretaceous rocks have been drilled in wells 204/30a-2 and 3, 205/26a-2, -3, -4, -5Z, -6, and 205/27-2, where the Chalk and Shetland groups are separated by an unconformity (Figures 69 and 73). Only the Cenomanian Hidra Formation of the Chalk Group is preserved, whereas both the Kyrre and Jorsalfare formations of the Shetland Group are represented. The age of the Shetland Group rocks recovered ranges from the Campanian to Maastrichtian, though well 204/30a-3 ranges from the Coniacian, and wells 205/26a-2, -6 and 205/27-2 extend from the Santonian.

The Chalk Group is present in wells 204/30a-3, 205/26a-3, -4, -5Z and -6, which proved a white to pale olive-grey and pale brown limestone, locally glauconitic and with scattered sandy grains or interbeds of red-brown claystone (Ritchie et al., 1996; Figure 73). Argillaceous limestone is recorded in well 205/27-2. The drilled section in the East Solan Basin does not exceed 50 m in thickness. The overlying Shetland Group was tested in all wells and ranged from approximately 1.1 to 1.8 km in thickness. The wells proved predominantly pale to dark grey, occasionally green and black, calcareous claystone, which is soft, sticky and blocky, locally silty, with traces of pyrite and carbonaceous material, and sporadic interbeds or stringers of brown dolomite and white argillaceous limestone. Danian strata conformably overlie the Shetland Group in all wells.

WEST AND SOUTH SOLAN BASINS AND THE SOLAN BANK HIGH

In the West Solan Basin, Upper Cretaceous rocks were drilled in well 204/29-2 (Figures 69 and 73), which proved 759 m of pale to dark grey and greenish grey, moderate to very calcareous mudstone with traces of sand and sporadic interbeds/stringers of white to pale brown and yellow-brown limestone and dolomite. The section spans the early Campanian to Maastrichtian interval. In the South Solan Basin, well 202/04-1 penetrated 764.4 m of predominantly claystone with sporadic limestone and dolomite stringers of Campanian to Maastrichtian age. In both wells, Danian strata conformably overlie the Upper Cretaceous rocks. On the Solan Bank High, well 202/09-1 drilled 145.1 m of Maastrichtian to Campanian mudstone, which is dark grey, calcareous, and with traces of glauconite. The nature of the Cretaceous–Paleocene boundary is uncertain in this well, though adjacent wells in the North Rona Basin suggest that it might be an unconformity.

Rona High

The Rona High is completely buried by Upper Cretaceous rocks (Figures 30, 31 and 37), which have been proved in a large number of wells (Figures 69 and 76). The rocks mostly range from Turonian to Maastrichtian in age, though four wells, 205/26-1, 205/21-1A, 206/08-6A and 207/01a-4Z, extend into the Cenomanian. The rocks all belong to the Shetland Group, although the Black Band, more commonly associated with the Chalk Group,

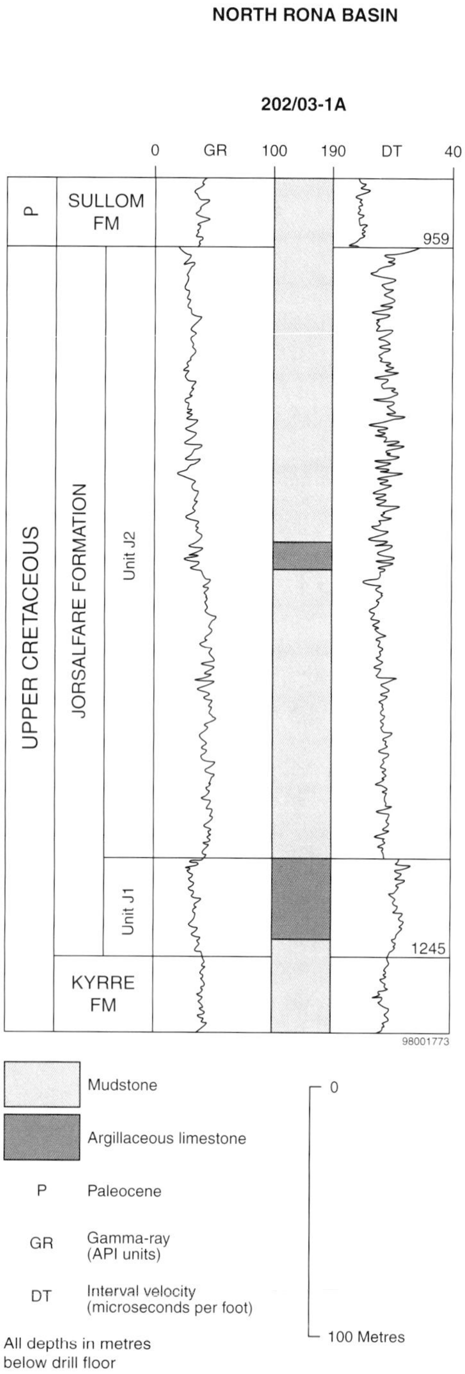

Figure 85 Well log summary of the Upper Cretaceous Jorsalfare Formation (Shetland Group) from well 202/03-1A within the North Rona Basin (modified from Ritchie et al., 1996). For location of well, see Figure 69.

is recorded in well 206/8-6A (Figure 86). The drilled sequences are thickest at the south-west end of the Rona High, and range from 880 m to nearly 2.2 km whereas on the central and north-east part of the high, the sequence varies from approximately 250 m to 1.4 km thick.

Most of the Shetland Group is bounded by unconformities at its base and top, though continuity from the Lower Cretaceous and into the Danian occurs in several wells (Figure 76). In all wells, the Shetland Group consists of a monotonous sequence of pale to dark grey, brown, green and occasionally red, blocky to subfissile, calcareous to noncalcareous, slightly silty claystone, with traces of glauconite and pyrite, and common stringers of dolomite and limestone (Ritchie et al., 1996). The Kyrre and Jorsalfare formations comprise much of this argillaceous succession, though in many wells the Shetland Group remains undifferentiated.

Most lithological variation occurs in the lower part of the succession, within the rocks of the Svarte and Macbeth formations (Figure 76). In well 205/26-1, the Svarte Formation is marked by a 10 m thick sequence of calcareous mudstone and argillaceous limestone that is distinguished from the overlying undivided Shetland Group by a marked downward decrease in the gamma-ray log values (Ritchie et al., 1996). On the central part of the Rona High, the Svarte Formation is 117 m thick in well 206/08-6A, and consists of mudstone with interbedded dolomite and dolomitic sandstone, and a 20 m thick unit of buff, very fine to fine-grained, calcareous sandstone near the base. Underlying the Svarte Formation in this well is an undated conglomeratic and pebbly sandstone unit, 40 m thick. The conglomerate includes clasts of basement, and is interbedded with green-brown, fine to medium-grained sandstone and pebbly sandstone, and sporadic thin mudstone that has been assigned by Ritchie et al. (1996) to the early Albian to Cenomanian Commodore Formation, which at this location may be laterally equivalent to the lower part of the Svarte Formation. On the north-east part of the Rona High, well 207/01a-4Z proved at least 62.5 m of interbedded siltstone, mudstone and limestone belonging the Svarte Formation, which is cut by a Palaeogene dolerite intrusion.

Rocks of the Macbeth Formation are best preserved on the south-west and central parts of the Rona High, with a key section illustrated by well 206/08-6A that proved 205 m of mudstone with interbedded limestone, dolomite and sporadic sandstone (Figures 76 and 86). The mudstone is pale to dark grey, variably calcareous, glauconitic, pyritic and microcarbonaceous. The base of the unit in this well is marked by Black Band noncalcareous mudstone, which displays the typical high gamma-ray wireline well log response. The limestone and dolomite is white or brown to pale grey, argillaceous, sandy and glauconitic. The interbedded nature of the succession in well 206/08-6A produces highly serrated gamma-ray and velocity log responses (Figure 86). A comparable sequence of interbedded mudstone, limestone and sandstone is also present in wells 204/25-1, 206/12-2 and 206/13a-2. This association of limestone and the Black Band strongly implies that the Macbeth Formation is the lateral equivalent of the Herring Formation. In other wells, such as 205/26-1 and 205/21-1A that incorporate Turonian mudstone, the Shetland Group remains as undivided.

There is some local variation within the overlying Kyrre Formation on the central part of the Rona High, where the predominantly mudstone-rich succession grades into a basal limestone facies. In well 206/08-4, the Kyrre Formation consists of 492 m of mudstone that overlies a 47 m thick unit of limestone with thin interbeds of mudstone (Figure 84). This limestone unit has been informally termed the Dab Limestone Unit, as described from the West Shetland Basin (Ritchie et al., 1996). Similar limestone units have been proved in well 207/01-3 on the north-east Rona High, and in wells 207/01-2 and 205/20-2 in the West Shetland Basin, though, as previously noted, there is locally some dispute as to their stratigraphical assignment. In con-

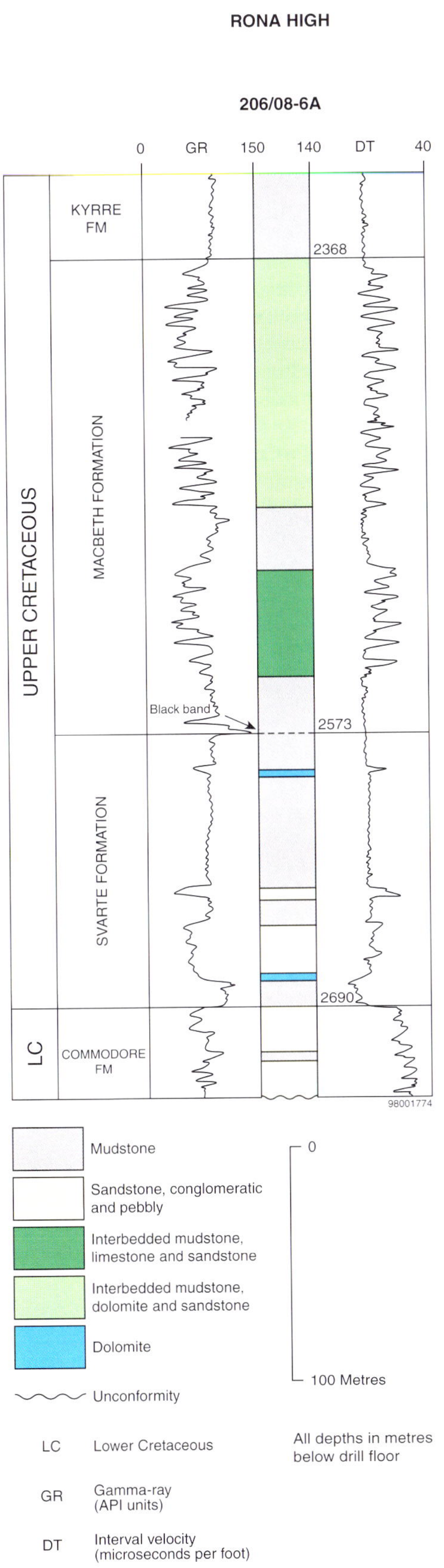

Figure 86 Well log summary of the Upper Cretaceous Svarte and Macbeth formations (Shetland Group) in well 206/08-6A within the Rona High (modified from Ritchie et al., 1996). For location of well, see Figure 69.

trast to Ritchie et al. (1996), Grant et al. (1999) assign the limestone in wells 207/01-2 and 207/01-3 to the top of the Victory Formation.

Faroe–Shetland Basin

Upper Cretaceous rocks of the Shetland Group occur extensively within the UK national sector of the Faroe–Shetland Basin (Figures 69, 77, 79, 81 and 82). As noted above (see Lower Cretaceous), the Lower–Upper Cretaceous contact is a mainly conformable transition, albeit unconformities are sporadically developed on the faulted eastern margin of the Faroe–Shetland Basin where the Cenomanian to Santonian section is locally absent. In many wells, the preserved sequence spans the Late Cretaceous interval, though the majority in the Faroe–Shetland Basin terminated within the Maastrichtian sequence.

The maximum-drilled thickness of Upper Cretaceous rocks occurs in the Flett Sub-basin and High in particular, where wells proved up to approximately 2.4 km of section, although this is locally inflated due to the presence of igneous sills (Figure 79). Although the Chalk Group is not reported to extend into the Faroe–Shetland Basin (Figure 72b), a Cenomanian to Turonian carbonate facies is well developed in the Judd, Flett and Foula sub-basins associated with the Svarte and Macbeth formations (Figures 77, 79 and 81). However, the bulk of the Shetland Group consists of argillaceous rocks that dominate the Coniacian to Maastrichtian section, and comprise the Kyrre and Jorsalfare formations, though much of this section remains locally undivided. Whilst at the present time there is very little information on the occurrence and distribution of Upper Cretaceous rocks within the Faroese national sector of the Faroe–Shetland Basin, an Upper Cretaceous sequence has been interpreted to occur within the Guðrun, Annika, Brynhild, Grimhild, Corona and Steinvør sub-basins and associated highs (Figures 32, 37, 38 and 69).

Judd Sub-basin

Rocks of the Shetland Group have been proved in eight wells in the Judd Sub-basin, although the entire preserved sequence was only tested in well 204/23-1, with the remainder terminating within the Campanian to Maastrichtian (Figures 69 and 77). Within well 204/23-1, the thickness of the Shetland Group is approximately 900 m. In well 204/14-1, adjacent to the northern Westray High, the Santonian to Maastrichtian section is missing.

The Shetland Group is characterised by pale to dark grey, locally greenish grey and pale brown, soft to firm, blocky, locally silty, calcareous, glauconitic and carbonaceous mudstone and claystone with common stringers and bands of white to grey and yellow-brown dolomite and limestone (Ritchie et al., 1996). Within well 204/23-1, greater lithological variation occurs in the Cenomanian to Turonian section that correlates with the Svarte and Macbeth formations (Figure 77). These formations comprise a 216 m thick section of interbedded mudstone and dolomitic limestone with the noncalcareous mudstone of the Black Band at the base. This unit is assigned to the Macbeth Formation and is underlain by a 26 m thick sequence of interbedded mudstone, siltstone and sandstone correlated to the Svarte Formation.

Westray High

The Westray High is completely buried by Upper Cretaceous strata (Figure 27). On the southern part of the south Westray High, a Maastrichtian section up to at least 131 m thick has been proved in wells 204/24a-2 and 3, 204/24-1A, 204/24a-7 and 204/19-3A (Figures 69 and 77). This is characterised by grey and greenish black to pale brown, firm, blocky mudstone with sporadic dark grey siltstone and yellow-grey limestone (Ritchie et al., 1996). The nature of the underlying section has not been tested in these wells, which were terminated within the Maastrichtian. In all these wells Danian strata conformably overlie the Maastrichtian. In contrast, in wells 204/19-2 and -9 on the northern part of the south Westray High, and in well 204/15-2 on the north Westray High, the Upper Cretaceous section is no younger than Santonian in age. Although a Campanian to Maastrichtian section has been reported by Ritchie et al. (1996) on the basis of the wireline well log signature from well 204/19-1, and is cited by these authors as a key reference section for the Jorsalfare Formation. Their interpretation contradicts the composite well log information for this well that indicates Santonian strata are unconformably overlain by Upper Paleocene rocks. The biostratigraphical data for the adjacent wells 204/19-2 and -9 supports the absence of the Jorsalfare Formation.

In well 204/19-9, the Cenomanian to Santonian sequence is 1095.2 m thick and consists predominantly of mudstone and thin limestone bands, though sporadic siltstone beds occur in the Cenomanian to Turonian interval, equivalent to the Svarte and Macbeth formations (Figure 77). Separated by the Black Band, these two formations are represented in well 204/19-1, where they form 58.5 m (Svarte Formation) and 192 m (Macbeth Formation) thick units of claystone with thin limestone bands and stringers. The Macbeth Formation is overlain by at least 140 m of Kyrre Formation claystone, which may also include the overlying 417.6 m of claystone with sporadic sandstone, siltstone and limestone assigned by Ritchie et al. (1996) to the Jorsalfare Formation, though, as noted above, this issue remains to be resolved. On the northern Westray High, well 204/15-2 proved a Cenomanian to Coniacian section dominated by dark grey, pale grey-green and brown-grey, moderate to very calcareous, silty, micaceous claystone. Beds of pale grey and pink-grey, very fine-grained, well-sorted, slightly carbonaceous and glauconitic sandstone, locally grading to siltstone, occur within the Cenomanian section. An unconformity separates Lower and Upper Cretaceous rocks in this well.

Foula Sub-basin

In the Foula Sub-basin, rocks of the Shetland Group have been proved in 10 wells although the entire preserved sequence was only tested in wells 206/11-1, 206/03-1, 206/04-1, 206/05-1, -2, 214/29-1 and 208/26-1 on the eastern side of the Sub-basin, adjacent to the Rona High (Figures 69 and 79). By way of contrast, wells 205/10-4, -5A and 214/30-1 located on the western side of the Foula Sub-basin, adjacent to the Flett High, were terminated in late Campanian to Maastrichtian claystone. These wells pass conformably into the Danian, whereas the wells in the eastern part of the sub-basin are separated from Paleocene to Lower Eocene strata by an unconformity. The thickness of the Upper Cretaceous succession in the Foula Sub-basin is variable, and on seismic profiles appears to be greatest in the south central part of the sub-basin (Figures 30 and 31). In the wells that totally penetrated the Upper Cretaceous succession, the preserved thickness ranges between approximately 1.3 and 2 km.

The bulk of the Shetland Group is undivided in the Foula Sub-basin, particularly the upper Turonian to Maastrichtian section (equivalent to the Kyrre and Jorsalfare formations). The thickness of this section ranges from approximately 1.1 to 1.7 km, and is characterised by pale grey, olive-grey and dark greenish grey, firm to hard, variably calcareous, silty, blocky claystone, locally grading to siltstone, with common stringers of dolomite and limestone (Ritchie et al., 1996; Figure 79). Rare, thin, very fine to fine-grained, locally coarse-grained, moderately to poorly sorted sandstone is present in the upper Turonian to Coniacian section in wells 206/05-1 and -2. Interbedded siltstone is common in the Campanian and Maastrichtian section in wells 214/30-1 and 208/26-1. A possible distinction between the Kyrre Formation and the overlying Jorsalfare Formation in the Foula Sub-basin is best defined in well 206/11-1, where a downward transition from varicoloured, calcareous mudstone characteristic of the Jorsalfare Formation to dark grey, noncalcareous mudstone common to the Kyrre Formation is marked on wireline well logs by a downward increase in gamma-ray and a decrease in velocity (Ritchie et al., 1996; Figure 84).

In wells 206/11-1, 206/03-1, 206/04-1, 206/05-1 and -2, the Cenomanian to Turonian section displays greater variation in lithology with a significant component of limestone in both the Macbeth and Svarte formations (Figure 79). In the south-west part of the Foula Sub-basin, well 206/11-1 proved a 510 m thick sequence of interbedded, beige to brown, slightly dolomitic limestone and black to grey-brown silty claystone assigned to the Macbeth Formation (Ritchie et al., 1996). A similarly interbedded limestone and claystone sequence, up to 104 m thick, which underlies the Macbeth Formation, is correlated to the Svarte Formation, which also contains sporadic lenses of siltstone and sandstone. In the central part of the Foula Sub-basin, the combined thickness of the Macbeth and Svarte formations ranges from approximately 270 to 450 m. In this region, the Macbeth Formation consists mainly of claystone and siltstone 365 to 380 m thick in wells 206/03-1 and 206/04-1. In well 206/05-2, a 191 m thick sequence of claystone with interbedded limestone and thin sandstone is present. In contrast, limestone dominates the Svarte Formation in the central Foula Sub-basin. Well 206/03-1 contains a 78 m thick section of white to pale grey, hard, variably sparry and argillaceous limestone interbedded with claystone and calcareous sandstone. Comparable sequences are also present in wells 206/04-1 and 206/05-1, which proved 27.4 m and 175 m, respectively, of Svarte Formation. In well 206/05-2, a 78.4 m thick claystone-dominated sequence was proved. In the north-east part of the Foula Sub-basin, the Shetland Group is undivided, and in wells 214/29-1 and 208/26-1 the section is claystone dominated with sporadic beds of siltstone and sandstone.

Flett High

Upper Cretaceous rocks overlie the Flett High (Figures 30 and 31), although the only well to penetrate through the Shetland Group is 205/10-2 on the southern part of the high (Figures 69 and 79). This well drilled an approximately 2 km thick sequence of undivided Turonian to Maastrichtian claystone with traces of limestone, overlying a 332.5 m thick section of Svarte Formation claystone, limestone and sandstone. The claystone throughout the succession is varicoloured (brown-black, olive-grey, brownish grey), variably calcareous, blocky and silty with traces of yellow-brown, argillaceous and dolomitic limestone. In the Svarte Formation, the interbedded sandstone is white, very fine-grained and calcareous, and the limestone is pale grey and locally argillaceous. The section in this well is cut by numerous Palaeogene intrusions. All of the remaining wells drilled on the Flett High, i.e. 205/14-3, 205/10-1A, -3, 206/01-1, -2, -3, 206/02-1A and 214/28-1, terminated within the Maastrichtian, which is dominated by claystone with subordinate siltstone and traces of limestone and sandstone. The maximum drilled thickness is 252.9 m in well 206/01-1. In wells 205/14-3, 205/10-1A and -2, on the southern part of the Flett High, Paleocene strata unconformably overlie the Upper Cretaceous rocks. This contrasts with well 205/10-3 together with all the wells from the northern part of the high, i.e. 206/01-2 and 3, 206/02-1A and 214/28-1, where Upper Cretaceous strata pass conformably into the Danian succession.

Flett Sub-basin

Upper Cretaceous rocks have been drilled in 19 wells throughout the Flett Sub-basin (Figures 69 and 81). Within the north-east part of the sub-basin, well 208/24-1A recovered an unconformity-bounded Santonian to Maastrichtian section approximately 1.1 km thick. Towards the south-west part of the sub-basin, only wells 205/16-1 and 205/22-1A have penetrated the full succession with almost 2.2 km proved in the former.

Within the north-east part of the Flett Sub-basin, the Kyrre and Jorsalfare formations have been tentatively defined in well 208/24-1A, with the bulk of the Maastrichtian section comprising pale grey to greenish grey argillaceous limestone with rare carbonaceous streaks, though claystone is prevalent in the underlying Campanian to Santonian section, whereas interbedded thin limestone and sandstone increase in abundance towards the base of the section.

Within the south-west part of the basin, the Kyrre and Jorsalfare formations are interpreted to be present and characterised by pale to dark grey, locally greenish grey and pale brown, soft to firm, blocky, locally silty, calcareous, glauconitic and carbonaceous mudstone and claystone with common stringers and bands of white to grey and yellow-brown dolomite and limestone and also sandstone (Ritchie et al., 1996). Argillaceous limestone beds up to 10 m thick have been proved in the Coniacian to Santonian section in well 205/21-2 (Figure 81), and traces of fine to medium-grained and moderately sorted sandstone and siltstone occur in wells 205/21-2 and 205/22-1A. In this part of the basin, greater lithological variation occurs in the Cenomanian to Turonian section that correlates with the Svarte and Macbeth formations. Adjacent to the Rona High, well 205/21-2 proved a 225 m thick sequence of the Macbeth Formation, which includes a 5 m thick conglomeratic unit, composed of angular clasts of metaquartzite, within the interbedded mudstone and limestone sequence. The Svarte Formation in this part of the Flett Sub-basin is best represented in wells 205/22-1A and 205/16-1, which proved 16 m of mudstone with interbedded siltstone and 35 m of sandstone and limestone, respectively.

With the exception of well 208/22-1, all remaining wells within the Flett Sub-basin terminated within the Maastrichtian succession, which varies between 19 and 684 m thick, and is conformably overlain by Danian strata (Figure 81). Although the thickness of the Upper Cretaceous sequence in wells 214/27-1, 208/21-1 and 208/17-1 exceeded 250 m, this is in part due to section thickness being inflated by the presence of dolerite sills. The Shetland Group is dominated by pale to dark grey, and locally orange-brown, firm to hard, blocky to sub-fissile, calcareous claystone, though siltstone is recorded in well 205/17b-2. Traces of grey, very fine to coarse-grained sandstone and pale grey to buff, limestone are common.

Corona High

On seismic profiles, the distinction between Upper Cretaceous and older strata on the Corona High is ambiguous (e.g. Figures 30 and 31), although its presence overlying the high has been proved in wells 213/23-1 (Figure 52b) and 214/09-1 (Figure 33) where thicknesses of 466.7 m and 502.9 m, respectively, have been drilled (Figures 69 and 81). However, the preserved section is incomplete; well 213/23-1 cored a Cenomanian to Campanian sequence, whereas Cenomanian to Santonian strata were recovered in well 214/09-1. The top of the sequence is an unconformity overlain by Upper Paleocene strata.

The successions in both wells are similar, consisting predominantly of pale to dark grey, olive-grey, olive-black and brownish grey claystone that is generally firm, blocky, moderately to noncalcareous, carbonaceous, with traces of glauconite and pyrite. The claystone locally grades to pale grey, calcareous, locally glauconitic siltstone. In well 214/09-1, the middle Turonian section is marked by a brownish grey, fine to coarse-grained, moderately sorted, quartzose sandstone that is friable, slightly glauconitic and calcite cemented, and contains thin beds of claystone and siltstone. Thin sandstone beds also occur in the Cenomanian section in well 213/23-1. Traces of olive-grey to grey-brown limestone and dolomite occur throughout the succession. In well 214/09-1, the well log correlates the upper 258.7 m of the mid Turonian to Santonian section with the Kyrre Formation. Although the underlying part of the section is stratigraphically equivalent to the Macbeth and Svarte formations, there is a general absence of limestone that otherwise characterises these units elsewhere in the Faroe–Shetland region.

Erlend Sub-basin

In the Erlend Sub-basin, the occurrence of Upper Cretaceous strata is poorly understood as the sub-Cenozoic section is largely obscured by Palaeogene volcanic rocks (e.g. Figures 24 and 26). However, rocks of the Shetland Group have been proved in wells 219/27-1 and 219/28-1, adjacent to the Møre Basin, with a maximum thickness of approximately 1.2 km proved in 219/28-1 (Figures 69 and 82). This well recovered a Coniacian to Maastrichtian sequence dominated by dark grey to dark brownish grey, moderately to noncalcareous claystone with sporadic stringers of white, cream and buff-orange limestone, and dark grey, fine to very fine-grained, variably sorted, glauconitic sandstone. The sandstone increases in abundance towards the base of the Upper Cretaceous section with beds up to 5 m thick. The rocks have been assigned to the Kyrre and Jorsalfare formations (Ritchie et al., 1996).

Møre and Magnus basins, and the Erlend and East Shetland highs

In the north-east part of the report area, Upper Cretaceous rocks have been proved in 10 wells that record a predominantly Turonian to Maastrichtian age history of sedimentation (Figures 69 and 82). The preserved succession is largely assigned to the Kyrre and Jorsalfare formations of the Shetland Group, though rocks of the Middle Turonian Macbeth Formation are present. In the Møre Basin, the Upper Cretaceous section may be up to 3 to 4 km thick (Nelson and Lamy, 1987). Seismic evidence presented here suggests that up to 1.75 s TWTT is present (Figure 23). In the Magnus Basin, the thickness of the Shetland Group exceeds 2 km (Johnson et al., 1993). On the Erlend High, the drilled thickness ranges from approximately 240 m to 1.6 km. Whilst well 209/12-1 penetrated the entire Upper Cretaceous sequence on the Erlend High, the seismic expression of the Shetland Group is poorly imaged on seismic data due to later, Palaeogene volcanism (Figure 26). The adjacent East Shetland High, which is almost devoid of Lower Cretaceous rocks, became fully buried along its northern prolongation, which separates the Magnus and Møre basins, during the Late Cretaceous (Figures 23 and 69).

Møre Basin and East Shetland High

In the Møre Basin, wells 219/28-2Z and 220/26-1 are sited on the flank of the basin, whereas well 219/20-1 occurs in a more basinal location (Figure 69). The Upper Cretaceous section in the marginal wells appears to be unconformity bounded at the top and base of the section, whereas the more central well 219/20-1 is conformably overlain by Lower Paleocene Danian strata (Figure 82). A maximum drilled thickness of approximately 2.1 km was proved in well 220/26-1, and assigned a Santonian to Maastrichtian age, though the accurate location of the Lower–Upper Cretaceous boundary in this well remains ambiguous. In the adjacent well on the north-west flank of the East Shetland High, 220/26-2, a 1.2 km thick sequence of Turonian to Maastrichtian rocks was recovered, unconformable on Triassic strata (Figure 55a). This is consistent with seismic profile data that show the base of the Upper Cretaceous section to be irregular and erosional at this location (Figure 23). The maximum proven stratigraphical range was tested in well 219/20-1 where approximately 1.8 km of Santonian to Maastrichtian rocks unconformably overlie 128.6 m of Cenomanian to Turonian strata.

In general terms, the sequence in the Møre Basin and on the north-west margin of the East Shetland High is dominated by claystone and mudstone, which is mainly grey to brown and occasionally black in the Cenomanian to Turonian section (Ritchie et al., 1996). The claystone and mudstone is slightly to noncalcareous, and commonly has sporadic interbeds of siltstone and stringers of cream to grey-brown limestone, although 5 to 10 m thick beds may occur near the base of the section in well 220/26-1 (Figure 82). The rocks have been assigned to the Shetland Group by Ritchie et al. (1996), predominantly the Kyrre and Jorsalfare formations, though representatives of the Macbeth and Svarte formations may be present in wells 219/20-1 and 220/26-2. Numerous igneous intrusions, including Campanian sills in well 219/28-2Z (see Chapter 9) cut the succession.

Magnus Basin

Outwith the report area, Upper Cretaceous rocks have been drilled in well 210/04-1 in the north-west part of the Magnus Basin. This well drilled 393.2 m of Maastrichtian claystone resting unconformably on Permo-Triassic strata, and conformably overlain by Lower Paleocene strata. The claystone is tentatively assigned to the Jorsalfare Formation, which, in more central parts of the basin, reaches over 580 m thick (Johnson et al., 1993).

Erlend High

On the Erlend High, Upper Cretaceous rocks have been recovered in wells 209/03-1, 209/04-1A, 209/06-1, 209/09-1 and 209/12-1, though only the latter well penetrated the entire preserved succession, the other wells being terminated in the Turonian to Santonian section (Figures 69 and 82). The drilled thickness of the Upper Cretaceous ranges from 238 m in well 209/03-1 to approximately 1.6 km in well 209/12-1, though the latter includes an undifferentiated, 142 m thick basal section of Albian (Lower Cretaceous) to Cenomanian strata.

The Upper Cretaceous section on the Erlend High is dominated by pale to dark grey, dark grey-brown and brown-black, occasionally silty and glauconitic claystone, although well 209/04-1A proved 794 m of grey siltstone (Figure 82). Limestone and dolomite stringers are common throughout the sequence, with sporadic interbeds of sandstone. The latter may form beds up to several metres thick within the upper Albian to Cenomanian section in well 209/12-1, near the Lower–Upper Cretaceous unconformity. A break in the section is also noted in well 209/04-1A with most of the Campanian missing, and a break is also indicated on the well log for well 209/12-1 separating the Cenomanian and Turonian. Whilst the rocks are assigned to the Shetland Group, it remains largely undivided at present. Everywhere on the Erlend High, the sequence is cut by numerous basic to acidic intrusions, probably of Palaeogene age (e.g. Ritchie and Hitchen, 1996).

West Lewis and North-east Rockall basins

Upper Cretaceous rocks have been proved in two wells in the West Lewis and North-east Rockall basins (Figure 69). In the West Lewis Basin, well 164/25-1 was drilled close to the eastern margin of the West Lewis High whereas well 164/07-1 is located in the central part of the North-east Rockall Basin. The rocks are tentatively assigned to the Shetland Group, which is intruded by Palaeogene sills and remains undivided in both basins.

West Lewis Basin

Well 164/25-1 penetrated 225 m of medium to dark grey, moderately calcareous mudstone that was deposited in an open-shelf marine environment. Although the section spans the Cenomanian to Maastrichtian interval, the bulk of the sequence is of Campanian to Maastrichtian age, with the Cenomanian to Santonian section being highly condensed (approximately 14 m thick). The Turonian section is considered to be missing with an unconformity separating reddish lithologies below from grey mudstone above.

North-East Rockall Basin

Well 164/07-1 proved 1.8 km of latest Albian (Lower Cretaceous) to Coniacian claystone unconformably overlain by Upper Paleocene strata (Archer et al., 2005). The claystone is dark grey to grey-black, slightly silty and noncalcareous, with hornfelsed contacts associated with dolerite sills.

8 Cenozoic (sedimentary)

Martyn Stoker[1] and Thomas Varming[2]
[1] *British Geological Survey*
[2] *Jarðfeingi (Faroese Earth and Energy Directorate)*

The Cenozoic Era began at 65.5 Ma (Figure 6) and marked the onset of an interval of major global change, including the rearrangement of tectonic plates in the Arctic–North Atlantic region, the realignment of oceanic circulation and the transition from a warm 'greenhouse' to a cold 'icehouse' climate. This legacy of change is reflected in the geological record of the Faroe–Shetland region where, following the Early Eocene initiation of sea-floor spreading between north-west Europe and Greenland, a succession of post-rift tectonic episodes drove regional changes in the patterns of sedimentation and deep-ocean circulation along the ocean margin (Praeg et al., 2005; Stoker et al., 2005b and c). Although the post-rift development of the continental margins bordering the north-east Atlantic is generally classed as passive (i.e. tectonically quiescent), the evidence for a pattern of stepwise subsidence and recurrent uplift preserved in the Cenozoic succession of the Faroe–Shetland region indicates a margin evolution that has been anything but passive.

The Cenozoic is divided into Palaeogene and Neogene periods (Berggren et al., 1995a, b; Gradstein et al., 2004), a subdivision that is adopted in this report. In this scheme, the Neogene period is taken to incorporate the Miocene to Holocene interval. The status of the Quaternary as a formal stratigraphical unit, which historically incorporates the Pleistocene and Holocene series, is currently the subject of a major debate (Gibbard et al., 2005) that is beyond the scope of this report.

The Faroe–Shetland region, including the Faroe Bank Channel–Wyville Thomson Ridge area, is the main focus of this chapter (Figure 87). The Møre Basin is a major structural domain offshore mid Norway that influenced Palaeogene sedimentation in the north-east part of the Faroe–Shetland region, and general information together with well data from within the report area are incorporated where appropriate. Information from the area to the north and west of the Faroe Islands is scarce but is incorporated where available, including Paleocene data from the Kangerlussuaq Basin of East Greenland.

The bathymetry of the Faroe–Shetland region (Figure 87), including the Iceland–Faroe Ridge, is ultimately a legacy of crustal thickness variations due to syn- to post-rift tectonic and magmatic activity. Although deep-water basins (>500 m water depth) existed during the syn- to early post-rift (mid Cretaceous to Early Palaeogene) development of the region (e.g. Lamars and Carmichael, 1999; Smallwood and Gill, 2002), these largely evolved in a compartmentalised (segmented) manner, controlled to some extent by north-west-trending lineaments or transfer zones (Figure 7). The Brendan Lineament separates the Faroe–Shetland Basin from the Møre Basin to the north-east though its boundary with the Munkur Basin to the south-west is less well understood. The present day physiographical configuration of the continental margin, whereby the bathymetrically shallow Faroe Shelf and adjoining Munkagrunnur and Fugloy ridges are separated from the West Shetland Shelf, Wyville Thomson Ridge and Faroe Bank High by the deep-water conduit of the Faroe–Shetland and Faroe Bank channels (Figure 87), was largely established during Late Palaeogene subsidence and latest Palaeogene to Early Neogene compression (Davies et al., 2004; STRATAGEM Partners, 2002, 2003; Stoker et al., 2005b and c; Figure 14). In particular, Early to Mid Miocene intensification of contractional deformation resulted in the enhanced growth of inversion structures, such as the Wyville Thomson, Munkagrunnur and Fugloy ridges (Boldreel and Andersen, 1993; Johnson et al., 2005b), and may have been complemented by differential subsidence in adjacent basinal areas, thus forming and shaping the modern Faroe–Shetland and Faroe Bank channels (Stoker et al., 2005b and c). In the subsequent late post-rift (Neogene) period, the region evolved as a more integrated margin.

The structural terminology used in this chapter reflects this change in structural development. Structural elements such as the Faroe–Shetland Basin, West Shetland Basin, Faroe Bank Channel Basin and West Shetland Platform are more applicable to the syn- to early post-rift (Early Palaeogene) phase of margin development rather than the late post-rift (Late Palaeogene to Neogene) expression of the margin. For the latter, the more unified structural/physiographical terms such as Faroe–Shetland and Faroe Bank channels and Faroe and West Shetland shelves apply to an area that is still being shaped today by bottom-current and sporadic mass-flow processes.

Palaeogene and Neogene strata occur throughout most of the report area (Figure 87). Their absence from parts of the West Shetland Shelf is largely the result of Mid to Late Pleistocene glacial erosion. The outcrop of Palaeogene strata on the Faroe Shelf, the Fugloy and Munkagrunnur ridges, the Wyville Thomson and Ymir ridges, and the Faroe Bank High contrasts with the West Shetland region where its landward limit is marked by erosional truncation on the outer part of the modern-day West Shetland Shelf. The latter broadly coincides with the hinge zone marking the transition from 'normal' thickness continental crust beneath the shelf to the thinner, stretched, crust beneath the Faroe–Shetland Basin (Stoker et al., 1993 and references therein; Figure 17). The distribution of the Palaeogene inliers in the Fugloy Ridge–Wyville Thomson Ridge region may partly relate to the axes of anticlinal folds that were formed/enhanced during latest Palaeogene to early Neogene compression (Boldreel and Andersen, 1993; Johnson et al., 2005b; Stoker et al., 2005b; Figure 14).

The base of the Cenozoic succession is well defined on the basis of seismic reflection profiles calibrated by biostratigraphical and lithostratigraphical data from wells on the West Shetland margin and in the eastern part of the Faroe–Shetland Basin, beyond the south-eastern limit of the lower Palaeogene volcanic strata. In this area, the succession reaches a maximum thickness of about 4 km in the Flett Sub-basin, composed predominantly of Paleocene and Eocene sediments (Figure 88a). The Upper Cretaceous/Palaeogene boundary

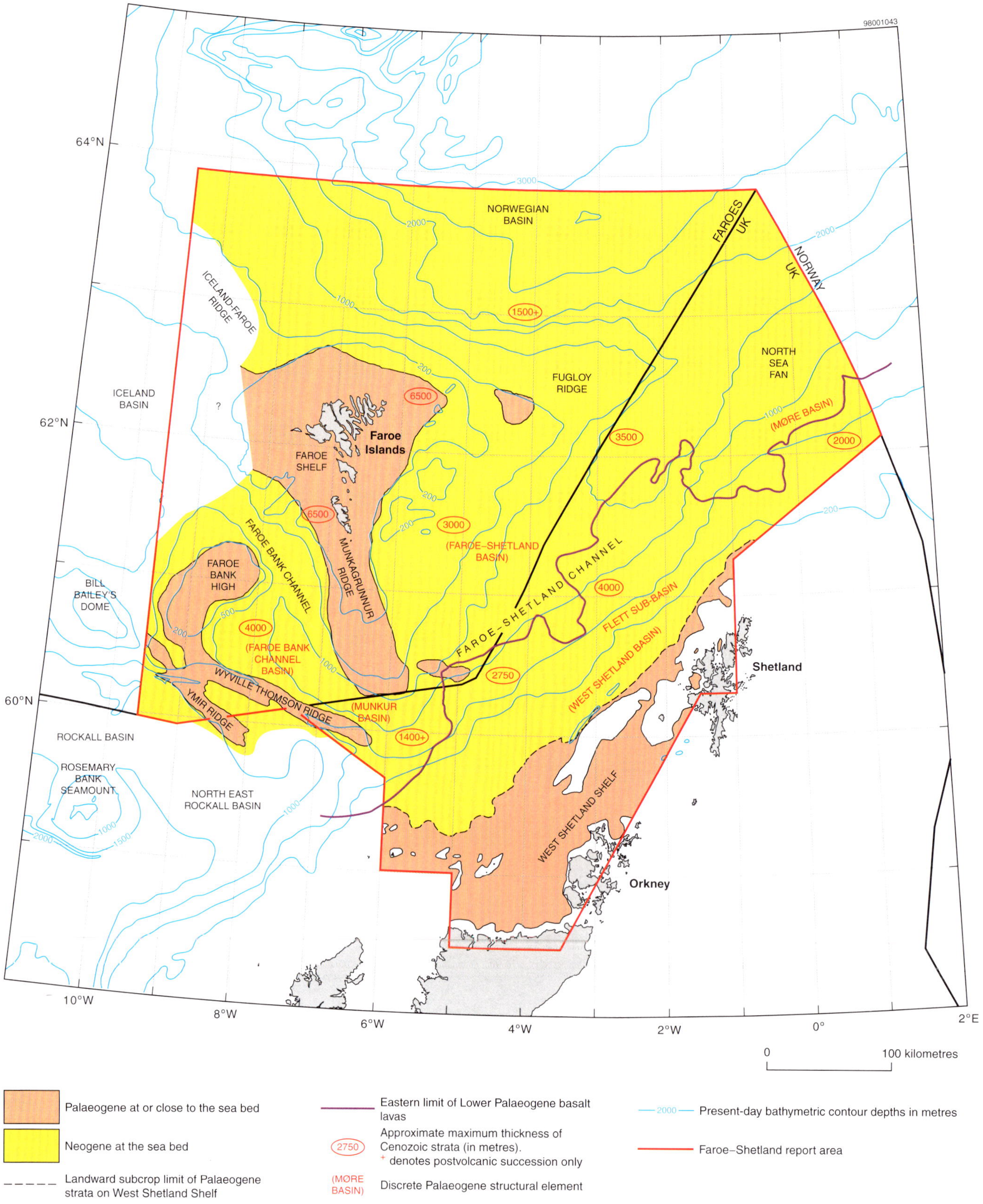

Figure 87 Distribution of Palaeogene and Neogene strata within the Faroe–Shetland report area.

Figure 88 *(opposite)* Geoseismic cross-sections showing the general structural disposition of Cenozoic strata within the Faroe–Shetland Basin and adjacent areas. Section a) modified from Lamars and Carmichael, (1999); Section b) constructed from seismic profiles (courtesy of Fugro Multi Client Services). Sub-basalt information in both sections derived from Keser Neish, (2003), Archer et al., (2005) and Keser Neish and Ziska, (2005).

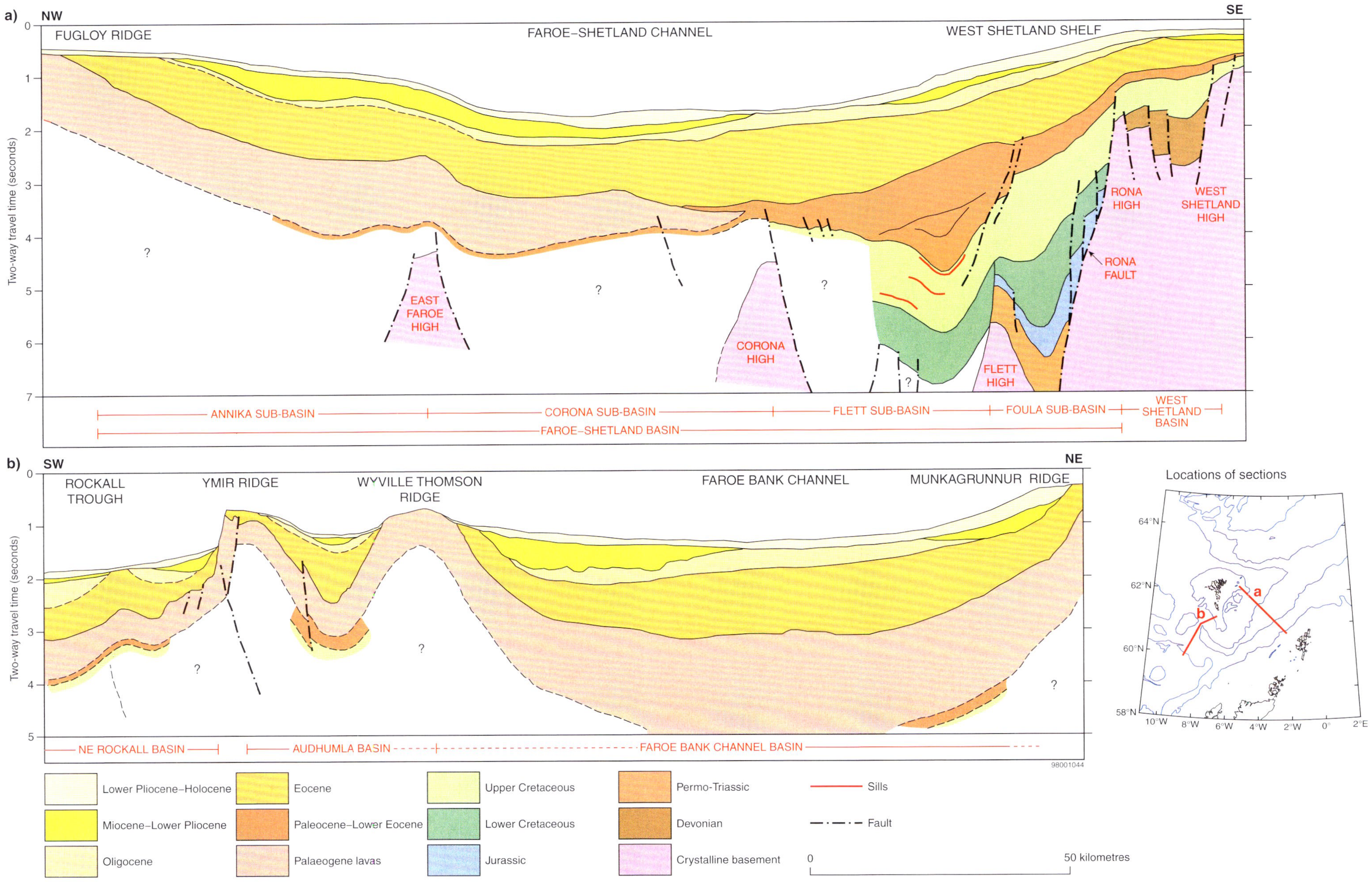

a)
NW
SE
FUGLOY RIDGE
FAROE–SHETLAND CHANNEL
WEST SHETLAND SHELF
Two-way travel time (seconds)
0
1
2
3
4
5
6
7
EAST FAROE HIGH
CORONA HIGH
FLETT HIGH
RONA HIGH
RONA FAULT
WEST SHETLAND HIGH
ANNIKA SUB-BASIN
CORONA SUB-BASIN
FLETT SUB-BASIN
FOULA SUB-BASIN
WEST SHETLAND BASIN
FAROE–SHETLAND BASIN
b)
SW
NE
ROCKALL TROUGH
YMIR RIDGE
WYVILLE THOMSON RIDGE
FAROE BANK CHANNEL
MUNKAGRUNNUR RIDGE
NE ROCKALL BASIN
AUDHUMLA BASIN
FAROE BANK CHANNEL BASIN
98001044
Locations of sections
64°N
62°N
60°N
58°N
10°W
8°W
6°W
4°W
2°W
0°
2°E
a
b
Lower Pliocene–Holocene
Miocene–Lower Pliocene
Oligocene
Eocene
Paleocene–Lower Eocene
Palaeogene lavas
Upper Cretaceous
Lower Cretaceous
Jurassic
Permo-Triassic
Devonian
Crystalline basement
Sills
Fault
0
50 kilometres

is locally marked by intrusive sills (Lamars and Carmichael, 1999) (Figure 31). Farther west, in the basalt-dominated Faroe region, recognition of the base of the Cenozoic is more problematic due to the high impedance contrast of the lower Palaeogene volcanic rocks, which commonly obscures or reduces the quality of the image of the sub-basalt geology on seismic profiles (e.g. Figures 30, 31, 33 and 88).

Deep drilling on the Faroe Islands indicates a thickness in excess of 6 km for the volcanic succession (see Chapter 9), whilst a total Cenozoic thickness of 3 to 4 km has been inferred from seismic interpretation in the adjacent Faroe Bank Channel Basin and western Faroe–Shetland Basin (Keser Neish, 2003; Keser Neish and Ziska, 2005; Figure 87). North of the Faroe Islands, up to 1.5 km of Eocene and younger strata have been imaged on seismic profiles overlying the top of the volcanic succession, above the continent–ocean boundary and the oceanic basement of the Norwegian Basin (Nielsen and van Weering, 1998; Roberts et al., 2005). The Møre Basin was the site for major Cenozoic deposition, with 2.5 to 3 km of Cenozoic sediment preserved at its centre, and about 2 km on its south-west margin (Hamar and Hjelle, 1984; Blystad et al., 1995; Figure 87).

The nature of the basal contact is variable. It is characteristically conformable in the deepest parts of the Faroe–Shetland Basin, where mudstone deposition associated with Early Paleocene submarine fan development was influenced by an end Cretaceous fault-induced topography (Smallwood and Kirk, 2005). In contrast, an unconformable relationship is preserved on basin-floor highs (e.g. Figures 30 and 31), which were progressively onlapped and covered by the fans. In the Faroe Bank Channel Basin, Keser Neish and Ziska (2005) interpret the base of the Cenozoic as a surface of nondeposition. An erosional unconformity separates Cretaceous and Cenozoic strata along the flanks of the Møre Basin (Ritchie et al., 1996; Brekke, 2000). The variable nature of this surface is most likely attributed to the creation of significant end-Cretaceous submarine topography, as a result of Late Cretaceous rifting in the proto-north-east Atlantic region (e.g. Dean et al., 1999; Roberts et al., 1999; Spencer et al., 1999; Figures 8 and 89). Extension and inversion linked to strike-slip tectonics were driven in response to the northward propagation of the Atlantic rift-spreading system into the Labrador Sea (Coward et al., 2003), which would have placed the entire region between the Rockall Plateau and the Vøring margin in an area of oblique strike-slip motion (Roberts et al., 1999; Figure 90b).

The Labrador Sea opened in the Paleocene (Chalmers, 1997), and an extensional regime persisted in the proto-north-east Atlantic region. In the Faroe–Shetland region, the Early Paleocene has been interpreted as a time of localised rifting largely followed by significant regional uplift (Dean, et al., 1999; Doré, et al., 1999; Roberts et al., 1999). The highest Paleocene sedimentation rates accompanied rifting and extension; the input of a series of major turbidite fans into the Faroe–Shetland Basin peaked in the Mid to Late Paleocene T35 to T36 interval (the BP sequence-stratigraphical 'T' terminology is from Ebdon et al. (1995) and Lamers and Carmichael (1999), after which rifting was terminated (Dean et al., 1999; Smallwood and Gill, 2002). Whilst much of this sediment may have been sourced from the east and south (Scotland) (Ebdon et al., 1995; Lamers and Carmichael, 1999; Andersen et al., 2002), a westerly source (from Greenland) has also recently been proposed (Sørensen, 2003; Jolley et al., 2005; Larsen and Whitham, 2005; Smallwood, 2005a). The subsequent build-out of major prograding shelf-margin systems during the Late Paleocene T38 to T40 intervals, sourced from the North Scottish region, together with some base sea level fall, resulted in the establishment of nonmarine conditions over a large part of the Faroe–Shetland region (Lamers and Carmichael, 1999; Figure 90a). A comparable history of turbidite sedimentation and palaeogeographical development is also recorded from the Møre Basin (Brekke et al., 1999; Gjelberg et al., 2005).

This major Late Paleocene clastic input into the Faroe–Shetland Basin, and widespread emergence of adjacent areas, is commonly linked to uplift induced by a thermal anomaly (hot spot) beneath a proto-Iceland, though there are conflicting views on the nature of the thermal anomaly. Some workers (e.g. White and Lovell, 1997; Jones et al., 2002; Maclennan and Lovell, 2002; Smallwood and Gill, 2002) interpret the Iceland hot spot as a mantle plume beneath the lithosphere, which provided support until continental break-up. The extensive and coeval volcanism has also been linked to the interaction between rifting and an Iceland mantle plume (Smallwood and White, 2002 and references therein; see Chapter 9). An alternative view is that the hot spot anomaly is an upper mantle response to plate break-up, of which the volcanism is a by-product of this extension (Doré et al., 1999; Foulger and Anderson, 2005; Lundin and Doré, 2005). Whatever the cause, this tectonic episode was the precursor to continental break-up between Greenland and north-west Europe, which occurred in the Early Eocene, at approximately 55 to 54 Ma (Chron 24r) (e.g. Ritchie et al., 1999a), forming the Iceland and Norwegian basins (Figure 91). The onset of sea-floor spreading was marked by widespread phreatomagmatic eruptions that produced large volumes of basaltic tephra, including those of sequence T50 (the Balder Formation) in the Faroe–Shetland and North Sea regions (Jolley and Bell, 2002a). As the Iceland and Norwegian basins widened, the Iceland Volcanic Province was formed, including the Greenland–Scotland Ridge and Iceland itself (Figures 91, 92 and 93; Foulger and Anderson, 2005).

A pronounced warming trend also peaked in the Early Eocene with a climatic optimum between about 50 and 52 Ma, this trend having been developing since the Mid Paleocene (59 Ma) (Zachos et al., 2001). Superimposed on this general warming trend is a unique, 100 000 year-long episode starting at the Paleocene–Eocene boundary, when high latitudes and global deep waters were subjected to a 6–8°C warming (Schmitz and Pujalte, 2003). Floral migration was associated with this Paleocene–Eocene Thermal Maximum (PETM), and resulted in closely similar coeval floras in the East Greenland and Faroe–Shetland regions, which have been used for stratigraphical correlation purposes (Jolley et al., 2005).

The post-rift development of the Faroe–Shetland region resulted in a general deepening of the basinal area albeit punctuated by a number of tectonic events, which are manifest as semiregional to regional unconformities (Figure 89). Following the onset of sea-floor spreading in the Early Eocene, deep-marine conditions were rapidly re-established in the sub-basins (Lamers and Carmichael, 1999; Smallwood and Gill, 2002; Smallwood

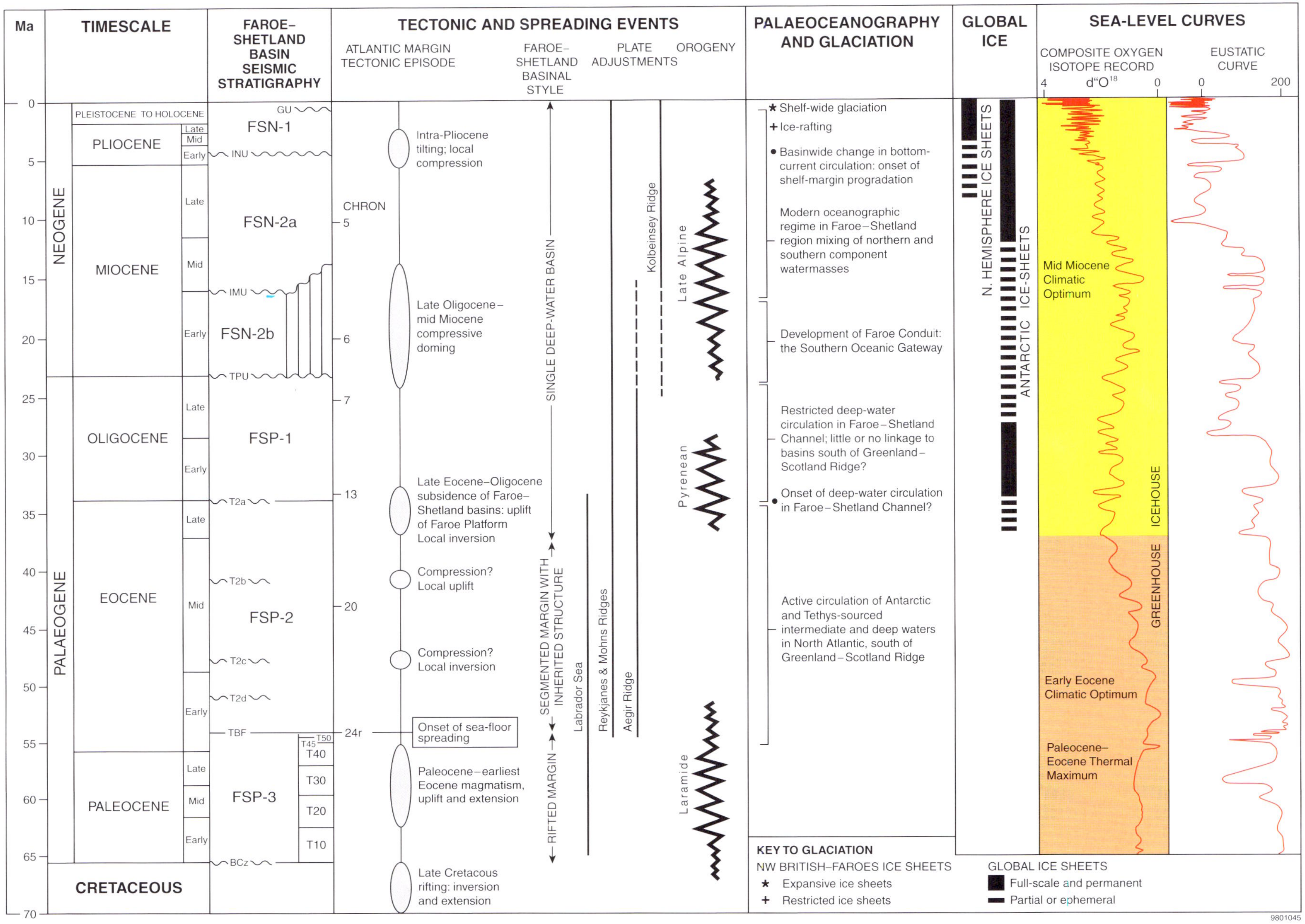

Figure 89 Cenozoic event stratigraphy for the Faroe–Shetland region. Timescale is based on 'A Geologic Time Scale 2004' by F M Gradstein, J G Ogg, A G Smith, et al. (2004) and the International Stratigraphical Chart, 2006 (ICS) with additions; the sea-level curves and indication of global ice extent are modified from Abreu and Anderson, (1998); tectonic and other events derived from a variety of sources (see text for details). Abbreviations: see Figures 14 and 94 and text for details of seismic-stratigraphical notation except, GU=Glacial Unconformity and BCz=Base Cenozoic.

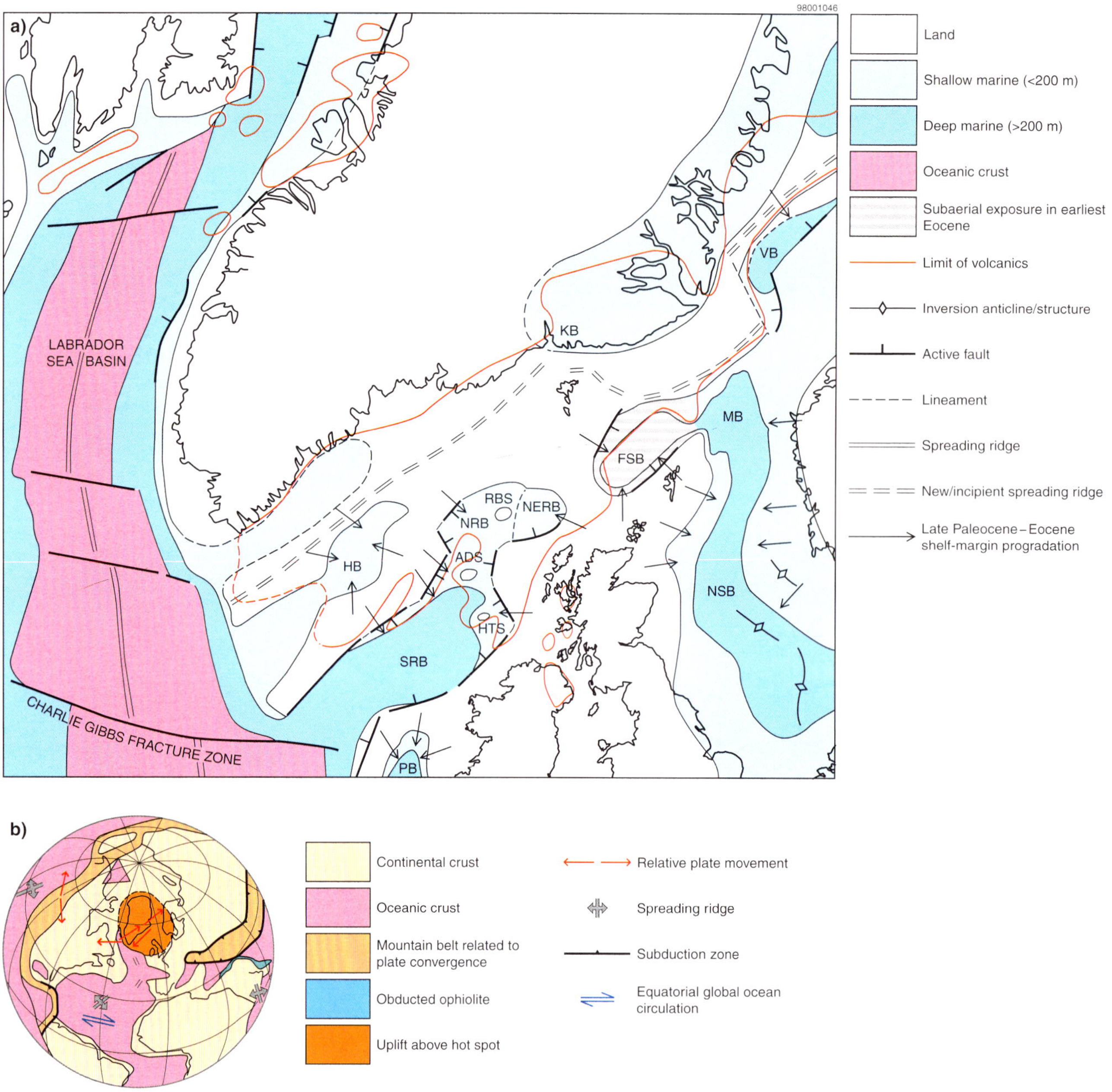

Figure 90 a) Palinspastic map for the Late Paleocene to Early Eocene interval. Abbreviations: see Figures 7, 8, 10 and 11 except, ADS=Anton Dohrn Seamount; HB=Hatton Basin; HTS=Hebrides Terrace Seamount KB=Kangerlussuaq Basin; NSB=North Sea Basin; RBS=Rosemary Bank Seamount; SRB=South Rockall Basin; b) Global view of Paleocene tectonics and ocean circulation at about 60 to 65 Ma. Maps based on information derived from Saunders et al., (1997), Naylor et al., (1999), Andersen et al., (2000), Brekke, (2000), Redfern, (2000), BGS and PAD, 2002, Faleide et al., (2002), Mosar et al., (2002), Coward et al., (2003), McInroy et al., (2006) and BGS (2007) (slightly modified reproduction from Coward et al. (2003) by permission of The Geological Society, London).

and Kirk, 2005; Figure 91). Robinson et al. (2004) describe a system of pulsed sedimentation throughout the Eocene in the eastern part of the Faroe–Shetland Basin, with intervals of shelf-margin progradation, sourced from the south and east (Andersen et al., 2002), interrupted by at least two phases of Mid Eocene channelised incision. The formation of these channel systems has been linked by Robinson et al. (2004) to compressional movements that are known to have caused basin-inversion at this time (Ritchie et al., 2003; Sørensen, 2003; Smallwood, 2004; Johnson et al., 2005b).

Although the thickest accumulations of Eocene strata overlie existing Paleocene sub-basins in the Faroe–Shetland Basin (e.g. Figure 30) and the Faroe Bank Channel Basin, the lack of extensional faulting in the Eocene section implies that the basinal structural framework inherited from the Paleocene controlled the location of the depocentres (Smallwood and Gill, 2002; Davies et al., 2004). To the north of the Faroe Islands, the spreading ridge continued to subside (Andersen et al., 2000). In the Møre Basin, post-Lower Eocene sediments are distributed across the basin though they are thickest in the west, adjacent to the

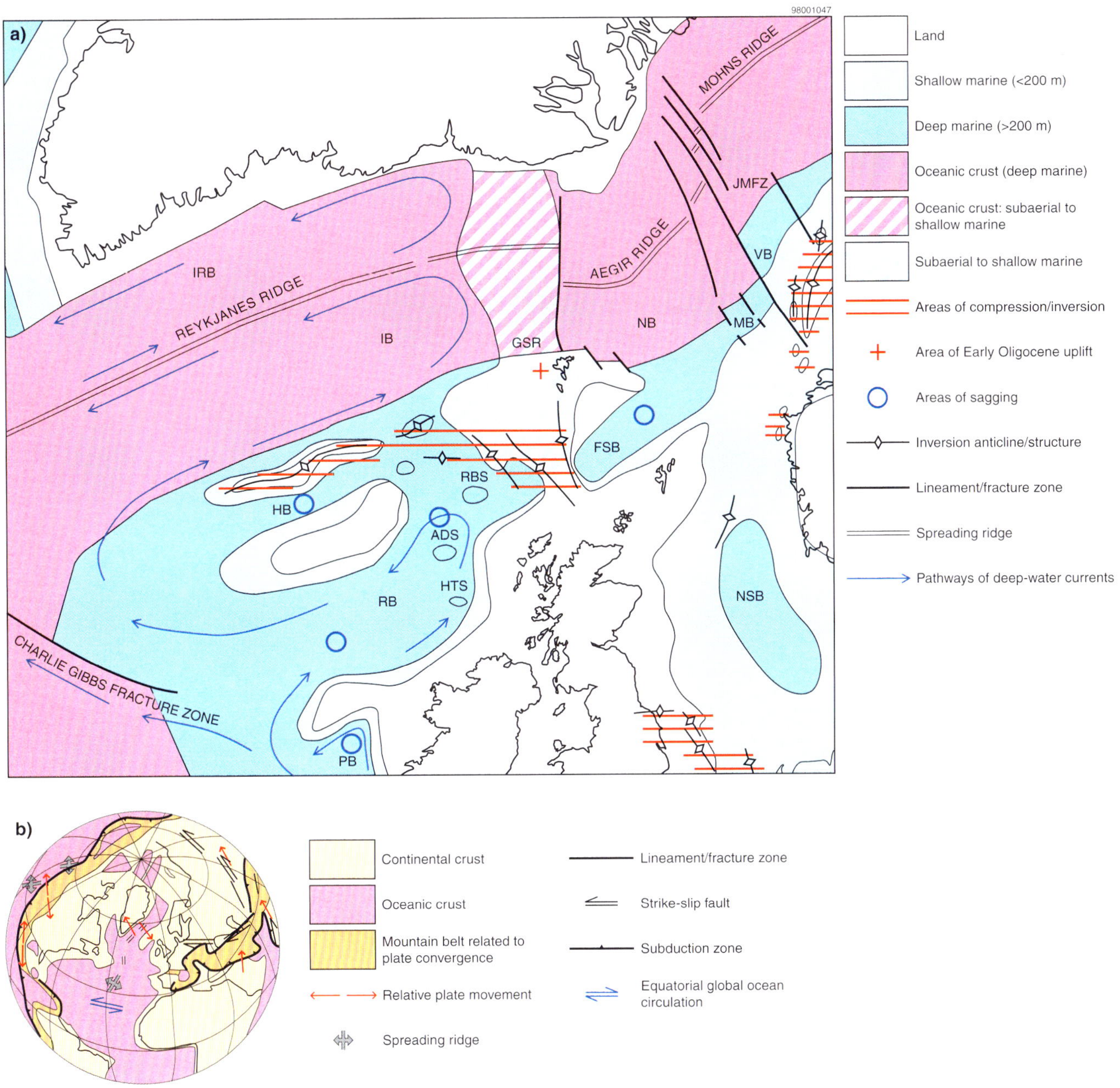

Figure 91 a) Palinspastic map for the Late Eocene to Oligocene interval. Abbreviations: see Figures 7, 8, 10, 11, 15 and 90 except, GSR=Greenland–Scotland Ridge and IRB=Irminger Basin; b) Global view of Oligocene tectonics and ocean circulation at about 30 Ma. Maps based on information derived from Thiede and Eldholm, (1983), Jansen and Raymo, (1996), Andersen et al., (2000), Redfern, (2000), Faleide et al., (2002), Coward et al., (2003), Johnson et al., (2005b) and Stoker et al., (2005a, b and c) (slightly modified reproduction from Coward et al. (2003) by permission of The Geological Society, London).

Møre Marginal High (Brekke et al., 1999; Martinsen et al., 1999). Basinal sedimentation changed at the onset of the Oligocene from a mud-dominated clastic system in the Eocene to deposition of biosiliceous ooze in the Early Oligocene (Davies and Cartwright, 2002). This implies a major reduction in the amount of shelf-derived material transported into the basins (Kershaw, 2000), and has been linked to the initiation of a deep-water current system in the Faroe–Shetland region (Davies et al., 2001). This most likely represents a major change in basin geometry from the Paleocene–Eocene segmented margin of the Faroe–Shetland Basin to the precursor of the single deep-water basin that is the present day Faroe–Shetland Channel.

The Møre Basin was also losing its discrete expression, becoming integrated into the continental margin developing offshore Norway, and flanking the larger Norwegian Basin (Blystad et al., 1995; Martinsen et al., 1999; Figures 91 and 92). A comparable change in basin hydrodynamics has been reported from the Rockall Basin to the south, where rapid, kilometre-scale differential subsidence (sagging) is envisaged to have ended Mid to Late Eocene shelf-margin progradation and created the present day deep-water basin (Stoker et al., 2005b; Praeg et al., 2005). Up to 1 km of basinal subsidence is reported to have occurred in the Faroe–Shetland region in the Oligocene to Holocene interval, though accurate timing of this has not been established

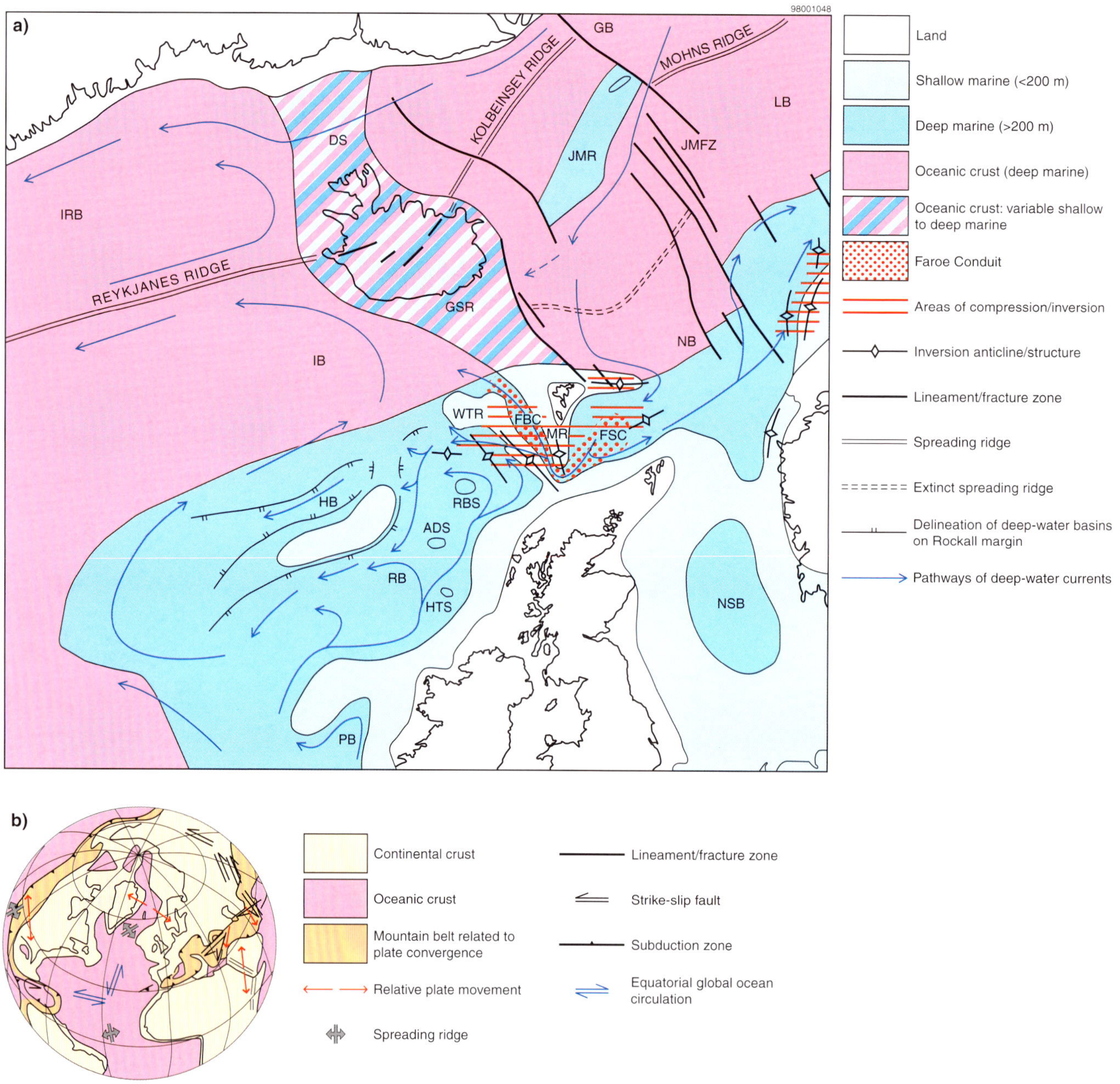

Figure 92 a) Palinspastic map for the Mid to Late Miocene interval. Abbreviations: see Figures 7, 15, 90 and 91 except, DS=Denmark Strait; FBC=Faroe Bank Channel; FSC=Faroe–Shetland Channel; GB=Greenland Basin, JMR=Jan Mayen Ridge and LB=Lofoten Basin; b) Global view of Late Miocene tectonics and ocean circulation at about 10 Ma. Maps based on information derived from Jansen and Raymo, (1996), Andersen et al., (2000), Redfern, (2000), Galloway, (2001), Faleide et al., (2002), Huuse, (2002), Coward et al., (2003), STRATAGEM Partners, (2003) and Stoker et al., (2005a, b and c) (slightly modified reproduction from Coward et al. (2003) by permission of The Geological Society, London).

(Davies et al., 2001; Sørensen, 2003), and it may even have been initiated in the Late Eocene (Andersen et al., 2000). The mechanism driving the change in sedimentation has similarly not been established, and a link to Late Eocene–Early Oligocene uplift of the Faroe Platform, and compression in the area of the Munkagrunnur and Wyville Thomson ridges, cannot be discounted (Waagstein and Heilmann-Clausen, 1995; Andersen et al., 2000; Johnson et al., 2005b) (Figure 91). Additionally, increased basinal subsidence from the Early Oligocene has been suggested from analyses of sediment accommodation on the northern North Sea margin (Jordt et al., 2000).

In common with the North Atlantic region in general, two main tectonic episodes have had an important influence on the Neogene development of the region. Firstly, in the Late Oligocene to Mid Miocene, compressive tectonism resulted in the formation/enhancement of inversion structures, including the Wyville Thomson, Ymir, Munkagrunnur and Fugloy ridges, and complementary synclines, such as the present day Faroe Bank Channel (Boldreel and Andersen, 1995; Lundin and Doré, 2002; Johnson et al., 2005b; Stoker et al., 2005b and c; Figures 8, 14, 89 and 92). Secondly, from the Early Pliocene, onshore uplift and tilting of the continental margin was accompanied by accelerated offshore subsidence and

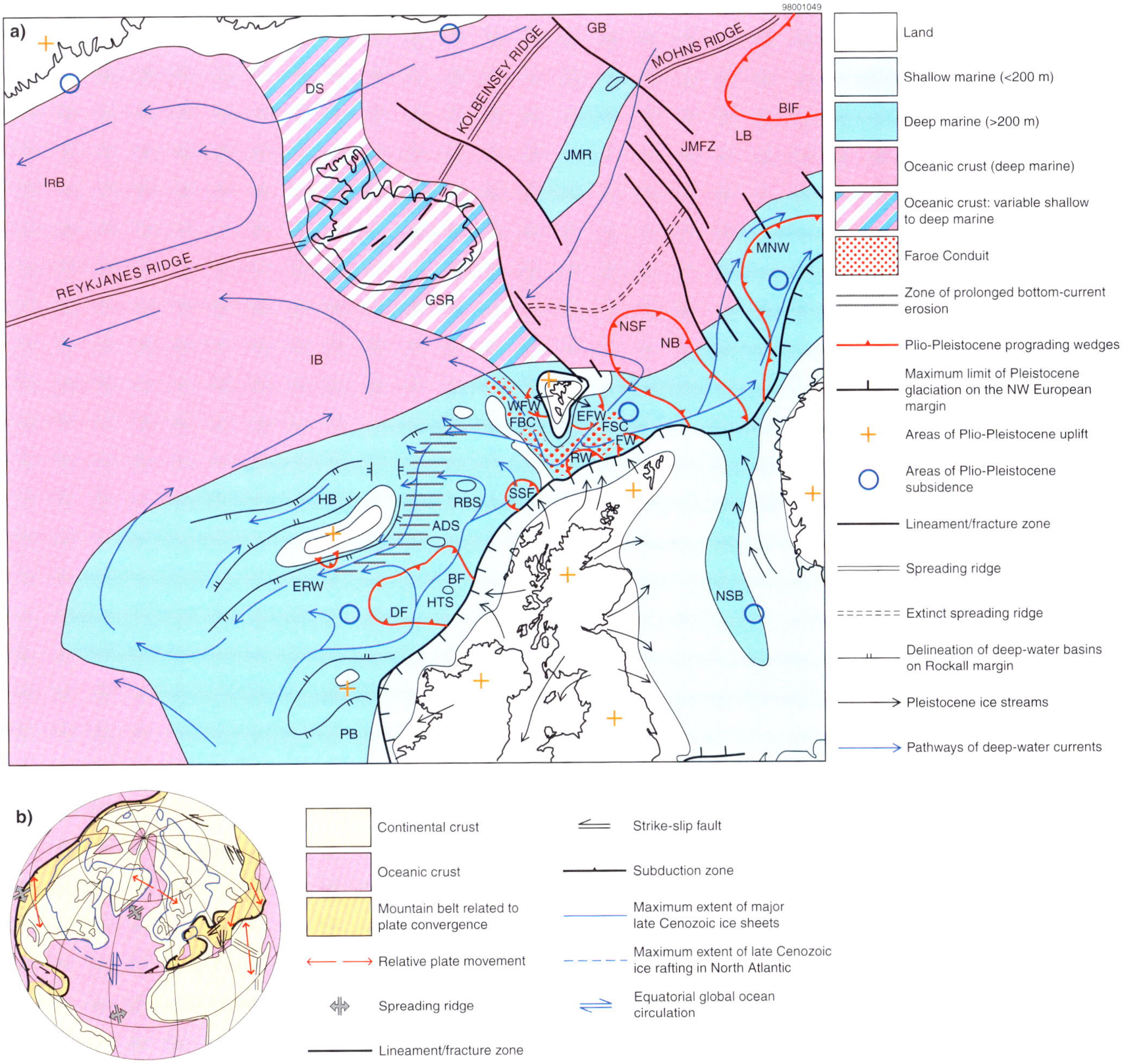

Figure 93 a) Palinspastic map for the Plio-Pleistocene interval. Abbreviations: see Figures 7, 15, 90, 91 and 92 except, BIF=Bear Island Fan; BF=Barra Fan; DF=Donegal Fan; EFW=East Faroe Wedge; ERW=East Rockall Wedge; FW=Foula Wedge; MNW=mid Norwegian Wedge; NSF=North Sea Fan; RW=Rona Wedge; SSF=Sula Sgeir Fan and WFW=West Faroe Wedge; b) Global view of Pliocene tectonics and ocean circulation at about 4 Ma, but incorporating the maximum extent of northern hemisphere glaciation since c. 2.74 Ma. Maps based on information derived from CLIMAP Members, (1981), Eiriksson, (1981), Thiede and Eldholm, (1983), Jansen and Raymo, (1996), Japsen and Chalmers, (2000), Redfern, (2000), Faleide et al., (2002), Huuse, (2002), Coward et al., (2003), STRATAGEM Partners, (2003), and Stoker et al., (2005a, b and c).

the progradation of sedimentary wedges (Andersen et al., 2000, 2002; Stoker, 2002; Davies et al., 2004; Stoker et al. 2005a and b; Figures 89 and 93). Localised compressional deformation may have continued to affect the north-east Faroe–Shetland Channel (Johnson et al., 2005b) and in other areas too (Ritchie et al., 2008).

The variations in subsidence and sediment supply associated with these tectonic movements had a pronounced effect on the patterns of shelf-margin and deep-water sedimentation and palaeoceanographical circulation. Although deep-water basins have existed in the Faroe–Shetland region since the Early Palaeogene, its present morphological expression as the Faroe–Shetland and Faroe Bank channels is probably a result of Late Eocene to Oligocene subsidence combined with Early to Mid Miocene compression (Davies et al., 2002; Stoker et al., 2005b and c). The linkage between the Faroe–Shetland and Faroe Bank channels instigated a passageway (herein referred to as the Faroe Conduit) for the persistent exchange of intermediate and deep-water masses between the Nordic Seas and the Atlantic Ocean, across the Greenland–Scotland Ridge

(Johnson et al., 2005b; Stoker et al., 2005c; Figure 92). The other significant deep-water passageway across the Greenland–Scotland Ridge, the Denmark Strait (Figure 92), may also have developed at about this time, as the Greenland–Scotland Ridge became fully submerged (Thiede and Eldholm, 1983).

The Early Pliocene event included uplift of the Faroe and West Shetland shelves, the latter being tilted seaward by up to 0.5° (Andersen et al., 2000, 2002; Stoker, 2002). This resulted in basinal progradation of the shelf margins bordering the Faroe–Shetland region, as well as deep-marine erosion during a reorganisation of bottom current patterns (Stoker et al., 2005b). Progradation and development of the West Shetland and Faroe margins was enhanced by Late Pliocene to Pleistocene denudation and glaciation (Figure 93). Glaciation of this region represents the culmination of a progressive cooling of northern hemisphere climate since the Mid Miocene, although global cooling and the change from greenhouse to icehouse conditions began earlier, at the beginning of the Late Eocene (Zachos et al., 2001; Figure 89).

STRATIGRAPHICAL FRAMEWORK OF THE FAROE–SHETLAND REGION

Several stratigraphical schemes have been applied to the Cenozoic succession in the Faroe–Shetland region including sequence-stratigraphical, lithostratigraphical and seismic-stratigraphical schemes. The use of different stratigraphical procedures largely reflects the varied objectives of commercial, government and academic institutions. The resultant scheme presented here (Figure 94) represents a hybrid framework that is based on the most commonly used subdivisions and nomenclature, and which are summarised below. Detailed stratigraphical information will be presented in the relevant sections.

Sequence stratigraphy

Arguably, the discovery of the Foinaven oil field in 1992 (Cooper et al., 1999) acted as the catalyst for the development of the current Cenozoic stratigraphical framework in the Faroe–Shetland region. This discovery revitalised the Paleocene fan play in this region, and led to Paleocene-age reservoirs and combined structural–stratigraphical traps becoming a focus for exploration companies (Brooks et al., 2001). The subtle nature of this play necessitated a means whereby potentially similar stratigraphical targets could be identified. Utilising a sequence-stratigraphical methodology, Mitchell et al. (1993) and Ebdon et al. (1995) both subdivided the Paleocene succession into distinct shelf, slope and basinal systems, which provided a context for the sandstone reservoir and the intraformational mudstone seal, both of which are critical to this play. However, the complex cyclical nature of the succession is borne out by the fact that Mitchell et al. (1993) recognised nine sequences within the Paleocene on the basis of type-1 unconformity recognition, whereas Ebdon et al. (1995) identified only eight sequences within the same interval using maximum flooding surfaces as their criteria for subdivision. The use of regional flooding surfaces to identify mudstone seals has proved very effective in the Faroe–Shetland region (Brooks et al., 2001).

The scheme presented by Ebdon et al. (1995) is commonly referred to as the T-sequence scheme (the letter 'T' abbreviated from 'Tertiary'), and has subsequently developed into common usage by exploration companies. This T-sequence scheme was formulated by the integration of well, seismic and log data constrained by the regional biostratigraphical zonation. In the Faroe–Shetland region, the key units are the T10 to T50 sequences (Figure 94), which represent a series of regionally mappable Paleocene to lowermost Eocene sand-rich units separated by the more shaly intervals of the maximum flooding surfaces. The identification of local bioevents has improved the resolution of the scheme, particularly the subdivision of the T20 and T30 intervals into 10 biostratigraphically constrained packages (Lamers and Carmichael, 1999), though not all of these higher-resolution packages are regionally distinguishable (Ebdon et al., 1995). It should be noted that in Figure 94, the positioning of the T-sequence boundaries is largely based on Lamers and Carmichael (1999), though the top of T40 is modified according to Jolley and Bell (2002a) and Jolley et al. (2005).

Lithostratigraphy

The regionally extensive T10–T50 sequences are broadly equivalent to Paleocene to lowermost Eocene lithostratigraphical units, originally correlated in the North Sea to formations within the Shetland, Montrose and Moray groups (Jones and Milton, 1994). The North Sea lithostratigraphy was extended into the Faroe–Shetland region by Knox et al. (1997) with only slight modification (replacement of the Montrose Group with the Faroe Group). A number of formations comprise these groups (Figure 94), the most regionally distinctive being the highly tuffaceous Balder Formation that forms the top of the Moray Group (Knox et al., 1997). The top of this formation has been correlated with the top of the T50 sequence (Jones and Milton, 1994; Ebdon et al., 1995). Individual formations have been further subdivided, both formally and informally, and details of these are presented elsewhere in this chapter.

The lithostratigraphy of the Paleocene to lowermost Eocene basaltic lava pile that forms the Faroe Islands has recently been formalised (Passey and Jolley, 2009). The Faroe Islands Basalt Group consists of a number of formations and members that replace the long-standing Lower, Middle and Upper Basalt subdivision (see Figure 122).

Unlike the sequence-stratigraphical scheme, the lithostratigraphical framework of Knox et al.,(1997) was applied to the entire Cenozoic succession, with the Stronsay, Westray and Nordland groups representing the post-rift Eocene to Holocene succession. The Stronsay and Westray groups represent the Eocene (post-Balder Formation) and Oligocene successions, respectively, and remain lithostratigraphically undivided in the Faroe–Shetland region (Figure 94). A detailed study of the Late Palaeogene to Neogene succession by Stoker (1999) subsequently refined the age of the Westray/Nordland group boundary in this region, proposing a Late Oligocene to Early Miocene age in contrast to the Mid Miocene assignation of Knox et al. (1997). Thus, the Nordland Group is taken to represent the entire Neogene succession and has been informally divided into lower, middle and upper units that correspond to distinct changes in patterns of sedimentation (Stoker, 1999).

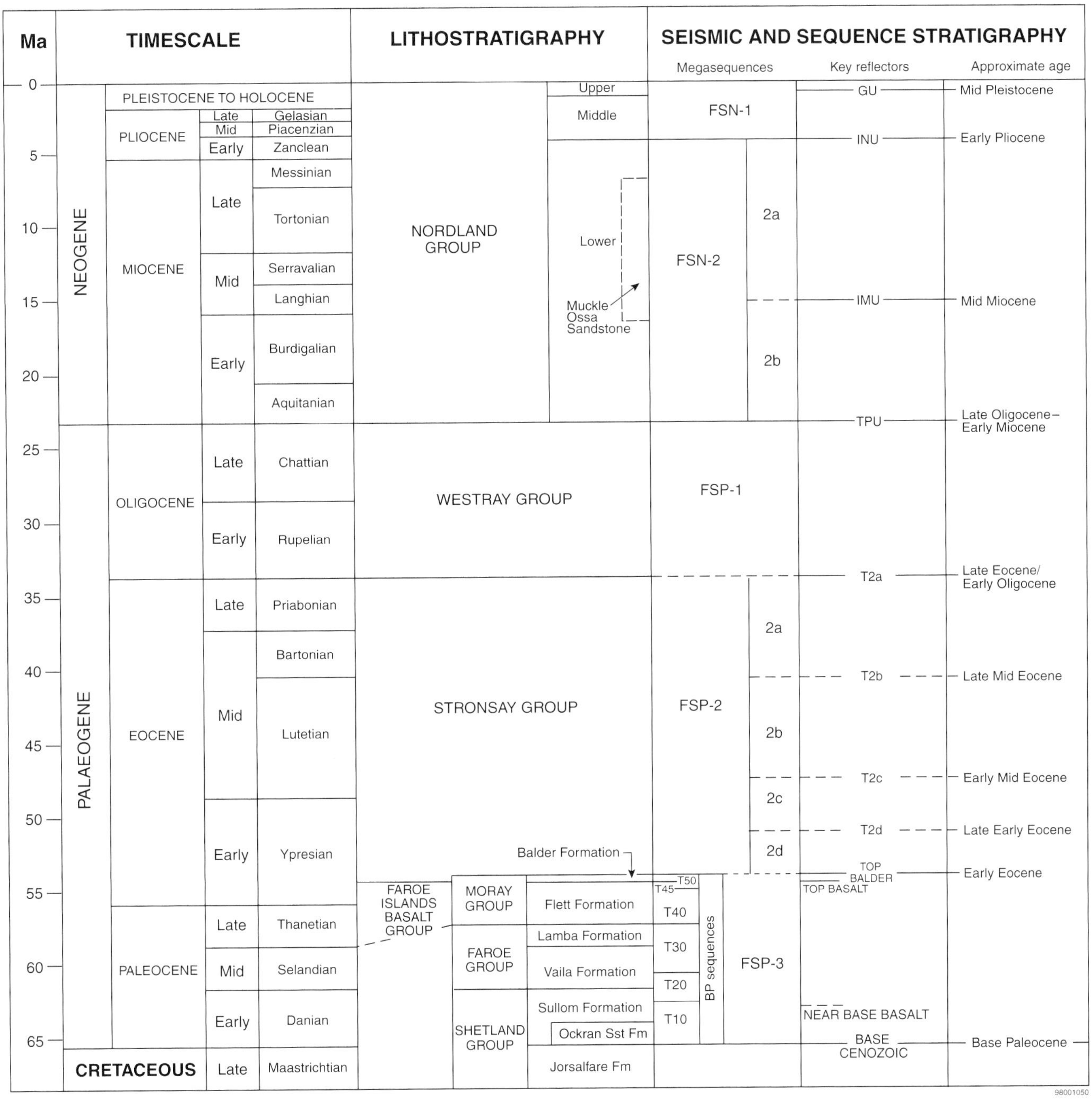

Lithostratigraphy			BP sequence
Balder Formation		B2	T50
		B1	
Flett Formation	Hildasay Sandstone Mbr	F3	T45
		F2	
	Colsay Sandstone Member	F1B	T40
		F1A	
Lamba Formation		L2	T38
	Westerhouse Sandstone Mbr	L1	T36
	Kettla Member		
Vaila Formation		V4	T35
		V3	T31-34
		V2	T25-28
		V1	T22
Sullom Formation		S2	
	Ockran Sst Fm	S1	T10

Figure 94 Summary of lithostratigraphical, seismic- and sequence-stratigraphical schemes applied to the Cenozoic succession in the Faroe–Shetland region (see text for details). Lithostratigraphical data from Ritchie et al., (1996), Knox et al., (1997), Naylor et al., (1999), Stoker, (1999), Jolley and Bell, (2002a), Waagstein et al., (2002), Sørensen, (2003) and Chapter 9. Seismic-stratigraphical data from Andersen et al., (2000), STRATAGEM Partners, (2002), Robinson, (2004), Robinson et al., (2004) and Stoker et al., (2005a). The BP sequence-stratigraphical terminology is from Ebdon et al., (1995) and Lamers and Carmichael, (1999). Timescale is based on 'A Geologic Time Scale 2004' by F M Gradstein, J G Ogg, A G Smith, et al. (2004) and the International Stratigraphical Chart, 2006 (ICS) with additions. Abbreviations: see Figures 14 and 89. It should be noted that the age range of the Faroe Islands Basalt Group remains ambiguous. Palynological data imply that a large part of the group section correlates with the Flett Formation, whereas radiometric data suggest an age range extending back into the Early Paleocene (see Chapter 9). Inset shows higher resolution subdivision of, and correlation between, the early Palaeogene lithostratigraphy and BP sequence stratigraphy based on Jolley et al., (2005).

Seismic stratigraphy

The contrast in sequence-stratigraphical and lithostratigraphical detail between the Paleocene to lowermost Eocene and the rest of the Cenozoic succession reflects, to a large degree, the fact that the Paleocene has been the major petroleum exploration play fairway in this region. However, an increasing need to better understand the post-rift development of the area is warranted by an increasing realisation that the post-rift history of the Faroe–Shetland Basin has not followed a simple McKenzie (1978) type model of thermal subsidence following extension (Ceramicola et al., 2005), and compression has had a major influence on its structural and stratigraphical development (Andersen et al., 2000; Ritchie et al., 2003; Davies et al., 2004; Stoker et al., 2005b). This is relevant both to a general understanding of this passive margin, as well as an improved assessment of hydrocarbon prospectivity in the Faroe–Shetland Basin, including the higher risk Middle Eocene basin-floor fan plays (Brooks et al., 2001; Davies et al., 2004). Thus, a seismic-stratigraphical approach has been utilised in several studies of the post-rift succession in order to define a number of regionally significant marker horizons and constituent megasequences. A key objective of this methodology has been to ensure that the megasequences represent physically mappable units throughout the Faroe–Shetland region.

Palaeogene

The mainly post-rift Palaeogene succession is poorly resolved beyond Series level. Andersen et al. (2000) and Sørensen (2003) both recognised an intra-Eocene reflector to the north and east of the Faroe Islands, to which they assigned an arbitrary 'Mid Eocene' age (Table 5), though stratigraphical correlation in the Faroe region is a major problem due to lack of well and borehole data. More recently, Robinson (2004), Robinson et al. (2004) and Smallwood (2004) have proposed a more detailed subdivision of the Eocene succession that identifies a number of intra-Eocene reflectors between the 'Top Balder' and 'Top Eocene' reflectors. The four-fold scheme of Robinson (2004) is the most regional in extent and matches well with the stratigraphical record proved by BGS borehole 99/03 (Table 5) within the Judd Sub-basin (Figure 2). Consequently, this scheme has been utilised as the basis of the informal subdivision of the Eocene succession presented in this report, though it is stressed that this subdivision remains provisional.

The stratigraphical notation used in this report is adopted from STRATAGEM Partners (2002, 2003) where 'FSP' is an abbreviation for 'Faroe–Shetland Palaeogene'. Three Palaeogene megasequences are proposed consisting, in ascending stratigraphical order, of: 1) FSP-3 (Paleocene to Early Eocene); 2) FSP-2 (Early to Late Eocene); and, 3) FSP-1 (Oligocene) (Figure 94). Although this is the first time that the FSP succession has been subdivided, previous work on the Faroese margin has resulted in a number of seismic units being recognised within the Cenozoic section (Nielsen and van Weering, 1998; Andersen et al., 2000, 2002). The correlation between these units and the nomenclature proposed below, in this report, is indicated in Table 6.

Megasequence FSP-3 incorporates the detailed sequence and lithostratigraphical subdivisions defined in the eastern part of the study area, outside of the basalt limit (Figure 94). A regional intra-Paleocene unconformity of 'Mid Paleocene' age is commonly reported (Ritchie et al., 1996; Dean et al., 1999), and appears to mark an upward change from the generally calcareous mudstone of the upper part of the Shetland Group (Sullom Formation) to interbedded mudstone, siltstone and sandstone of the Faroe and Moray groups. This broadly correlates with the Danian/Selandian boundary (Knox et al., 1997). Where basalt is present, it largely dominates FSP-3, with much of the sub-basalt detail commonly obscured. However, in the Faroe region, a two-fold subdivision of the sub-basalt Paleocene sedimentary section has been proposed (Keser Neish and Ziska, 2005), though this has not been calibrated by well data. Due to its prominence as a regional marker (cf. Ebdon et al., 1995), the Top Balder Formation reflector is herein provisionally taken as the top of FSP-3. However, as the polarity of the reflection varies laterally with lithology, this horizon is not always recognisable, particularly in basinal areas (Smallwood and Gill, 2002). Instead, the high-amplitude top basalt surface is locally the strongest Late Paleocene to Early Eocene event (cf. Fig. 9 of Andersen et al., 2000; Davies et al., 2004) thereby providing an approximate top to FSP-3. Indeed Smythe (1983), Ritchie et al. (1999a) and Smallwood and Gill (2002) report that the Top Balder Formation onlaps the basalt in the Faroe region.

Megasequence FSP-2 corresponds to the Stronsay Group of Knox et al. (1997), and has been informally divided into four units, 2a–2d, which are bounded by semiregional reflectors (Robinson, 2004; Robinson et al., 2004) the tops of which are herein referred to as T2a–T2d (Table 5 and Figure 94). Correlation with BGS borehole 99/03 suggests that FSP-2d is Ypresian (NP12) in age, FSP-2b is Lutetian (NP14-15), and FSP-2a is Priabonian (Figure 95). FSP-2c is not present at the core site, cut out by the early Mid Eocene T2c unconformity. However, a substantial thickness of FSP-2c occurs to the north of 99/03, which constrains the age of this unit to the late Ypresian to earliest Lutetian (NP12–14) interval. This is consistent with information from commercial well data that have tested the Eocene succession, such as 204/22-1 (Figure 95). This subdivision is broadly supported by other seismic-stratigraphical studies (e.g. Faroes GEM Network, 2001a, b, c and d; Andersen et al., 2002; Smallwood and Gill, 2002; Austin, 2004; Smallwood, 2004), though ambiguity remains over the age assignment of FSP-2c (Table 5).

Megasequence FSP-1 corresponds to the Westray Group of Knox et al. (1997) (Figure 94). In areas of well control, the base of FSP-1 (T2a) is reportedly marked by a high-amplitude reflector (Knox et al., 1997), and is locally observed to represent an angular unconformity (Stoker, 1999; Davies and Cartwright, 2002). However, in regional terms, the Eocene–Oligocene boundary remains poorly defined (Figure 95). The top of FSP-1 is marked by the angular Top Palaeogene Unconformity (STRATAGEM Partners, 2002; Stoker et al., 2005a and b). This reflection may be difficult to interpret where the upper Palaeogene to lower Neogene succession is locally cross-cut by a distinctive high-amplitude seismic reflector generated by the diagenetic transformation of Opal A to Opal C/T (Cristobalite/Tridymite) (Davies et al., 2001; Davies and Cartwright, 2002; Figure 26, see BSR). In the shallower, southern

Table 5 Correlation of regionally important Eocene seismic reflectors.

Andersen et al. (2000)	Sørensen (2003)	Smallwood (2004)	Robinson et al. (2004)	Robinson (2004)	BGS borehole 99/03	Common age range	This report		
CP100	Top Eocene	Top Eocene		Top Eoc 4		Late Priabonian/ early Rupelian	T2a		
			Leoc1		Priabonian			2a	FSP-2
		Mid Eocene	Meoc2	Top Eoc 3		Lutetian/Bartonian	T2b		
					Lutetian NP14-15			2b	
CP060		Lutetian 3	Meoc1	Top Eoc 2		Early Lutetian	T2c		
	Mid Eocene	Lutetian 2						2c	
		Lutetian 1							
		Lower Eocene		Top Eoc 1		Late Ypresian	T2d		
					Ypresian NP12			2d	
CP030	Top Balder	Top Balder	Eeoc1	Top T50		Early Ypresian	Top Balder		

98001175

part of the report area, the Palaeogene/Neogene boundary is commonly a composite unconformity combining the Top Palaeogene, intra-Miocene and intra-Neogene unconformities (Johnson et al., 2005b; Figure 95).

NEOGENE

A Neogene seismic stratigraphy has recently been published for the bulk of the Faroe–Shetland region (Stoker et al., 2005a and b) based on the results of the STRATAGEM project (STRATAGEM Partners, 2002, 2003). This framework has been built upon the foundation of a number of earlier independent studies, including the BGS 1:250 000 scale mapping programme in the West Shetland region (cf. Stoker et al., 1993; Stoker, 1999), and the work of the Geological Survey of Denmark and Greenland and co-workers in the Faroe region (e.g. Boldreel and Andersen, 1995; Nielsen and van Weering, 1998; Andersen et al., 2000, 2002; Sørensen, 2003).

The Neogene succession has been divided into two megasequences consisting, in ascending stratigraphical order, of: 1) FSN-2 (Miocene to Early Pliocene); and, 2) FSN-1 (Early Pliocene to Holocene) (Figure 94). The 'FSN' notation was introduced by STRATAGEM Partners (2002) as an abbreviation for 'Faroe–Shetland Neogene'. Correlation with the seismic units recognised by Nielsen and van Weering (1998) and Andersen et al. (2000, 2002) on the Faroese margin is indicated in Table 6.

The base of the FSN-2 megasequence is marked by the TPU, and is separated from the FSN-1 megasequence by the INU (Figure 94). The FSN-2 megasequence corresponds to the Lower Nordland unit of Stoker (1999). In deep water, the IMU locally divides FSN-2 into units' 2a and 2b. In the southern part of the region, especially on the flanks of the Wyville Thomson Ridge, the TPU and IMU are locally composite (Johnson et al., 2005b; Stoker et al., 2005a and b). The FSN-1 megasequence represents the sediment wedges

Table 6 Correlation of the Cenozoic seismic-stratigraphical units and key reflectors utilised in this report with those previously recognised on the north (Nielsen and van Weering, 1998) and the south and east Faroe margin (Andersen et al., 2000 and 2002). The North Faroe margin borders the Norwegian Basin, the South Faroe margin borders the Faroe Bank Channel and the East Faroe margin borders the Faroe–Shetland Basin.

FAROE–SHETLAND REGION						Approximate age
North Faroe Margin	South and east Faroe margin	Key reflectors		This report		
Unit 9	FPC-D.3					
		CN-050	GU	FSN-1		Mid Pleistocene
Sequence 3	FPC-D.2					
		CN-040	INU			Early Pliocene
Sequence 2	FPC-D.1			2a	FSN-2	
		CN-030	IMU			Mid Miocene
Sequence 1	FPC-C			2b		
		CN-010	TPU			Late Oligocene/Early Miocene
Unit A	FPC-B.4			FSP-1		
		CP-100	T2a			Late Eocene/Early Oligocene
Sequence 0	FPC-B.3			2a	FSP-2	
			T2b			Late Mid Eocene
		CP-060		2b		
	FPC-B.2		T2c			Early Mid Eocene
		CP-030		2c		
			T2d			Late Early Eocene
	FPC-B1			2d		
		CP-010	Top Balder Fm			Early Eocene
Basalt basement	FPC-A			FSP-3		

98001051

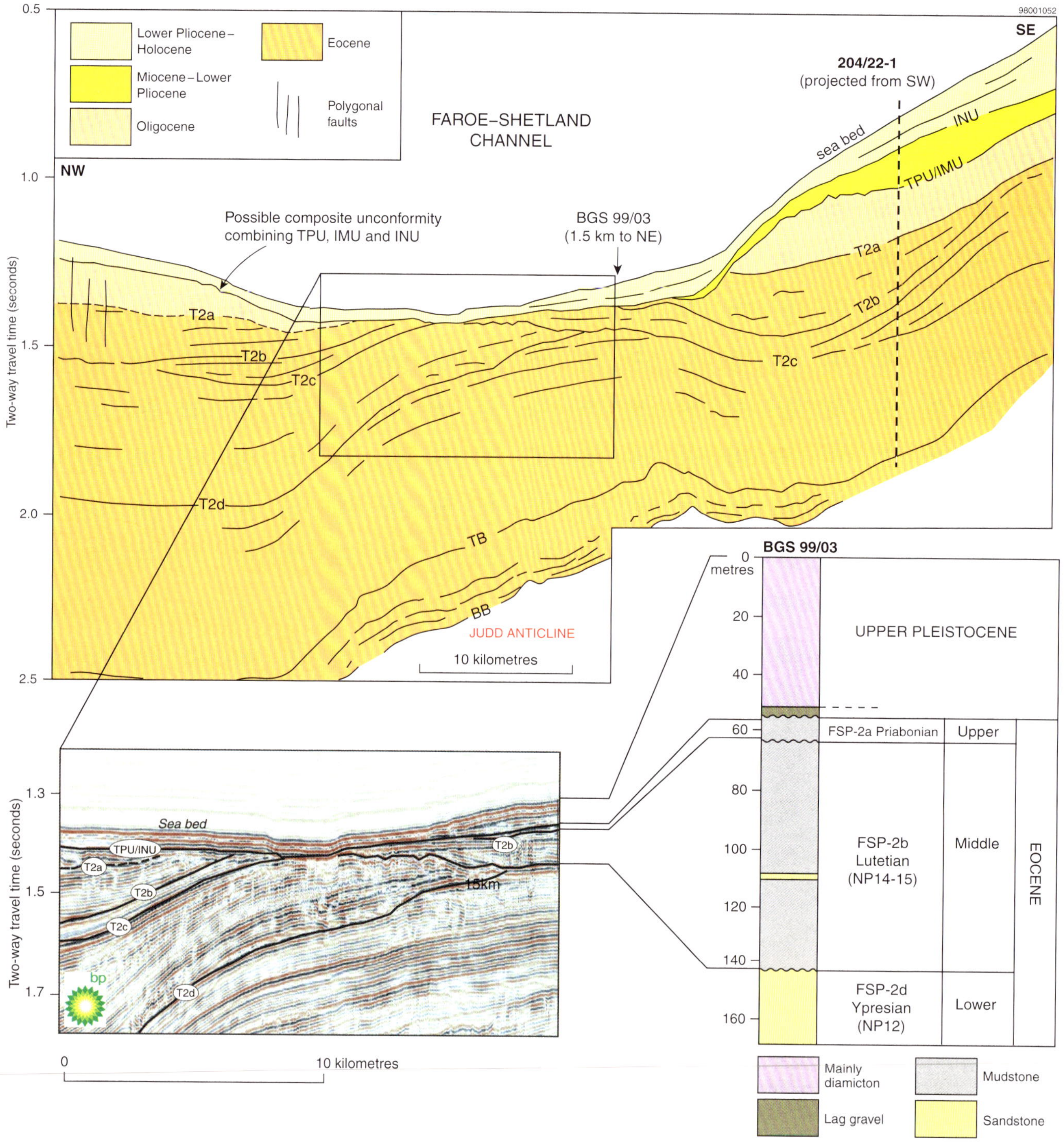

Figure 95 Preliminary interpreted geoseismic cross-section of a high-resolution seismic profile across the Faroe–Shetland Channel and the Judd Anticline, showing the Eocene succession calibrated to BGS borehole 99/03. Inset shows stratigraphical detail adjacent to borehole site. Abbreviations: see Figure 14 and text for details of seismic-stratigraphical notation except, BB=base Balder Formation. Note the possible composite unconformities marking the Palaeogene–Neogene boundary. For location of geoseismic cross-section, see Figure 96. The stratigraphical subdivision is supported regionally by nearby well 204/22-1 (see Figure 101). Whilst the projected position of well 204/22-1 shows its general seismic-stratigraphical context, it should be noted that the T2c–T2d interval (i.e. subunit 2c of megasequence FSP-2) might be present at the well site. Seismic data courtesy of BP.

that prograde seaward off the West Shetland and Faroe Shelf margins. An intra-Pleistocene (early Mid Pleistocene) Glacial Unconformity, best expressed in the shelf areas, separates ice-contact glacial deposits (above) from deltaic sediments (below). This subdivision corresponds to the Middle and Upper Nordland units of Stoker (1999).

PALAEOGENE

Sedimentary and volcanic rocks of Palaeogene age occur throughout most of the report area, and are absent only on the West Shetland and Orkney–Shetland highs (Figure 96). Subdivision of the Palaeogene succession to Series level, as shown on Figure 96, remains tentative

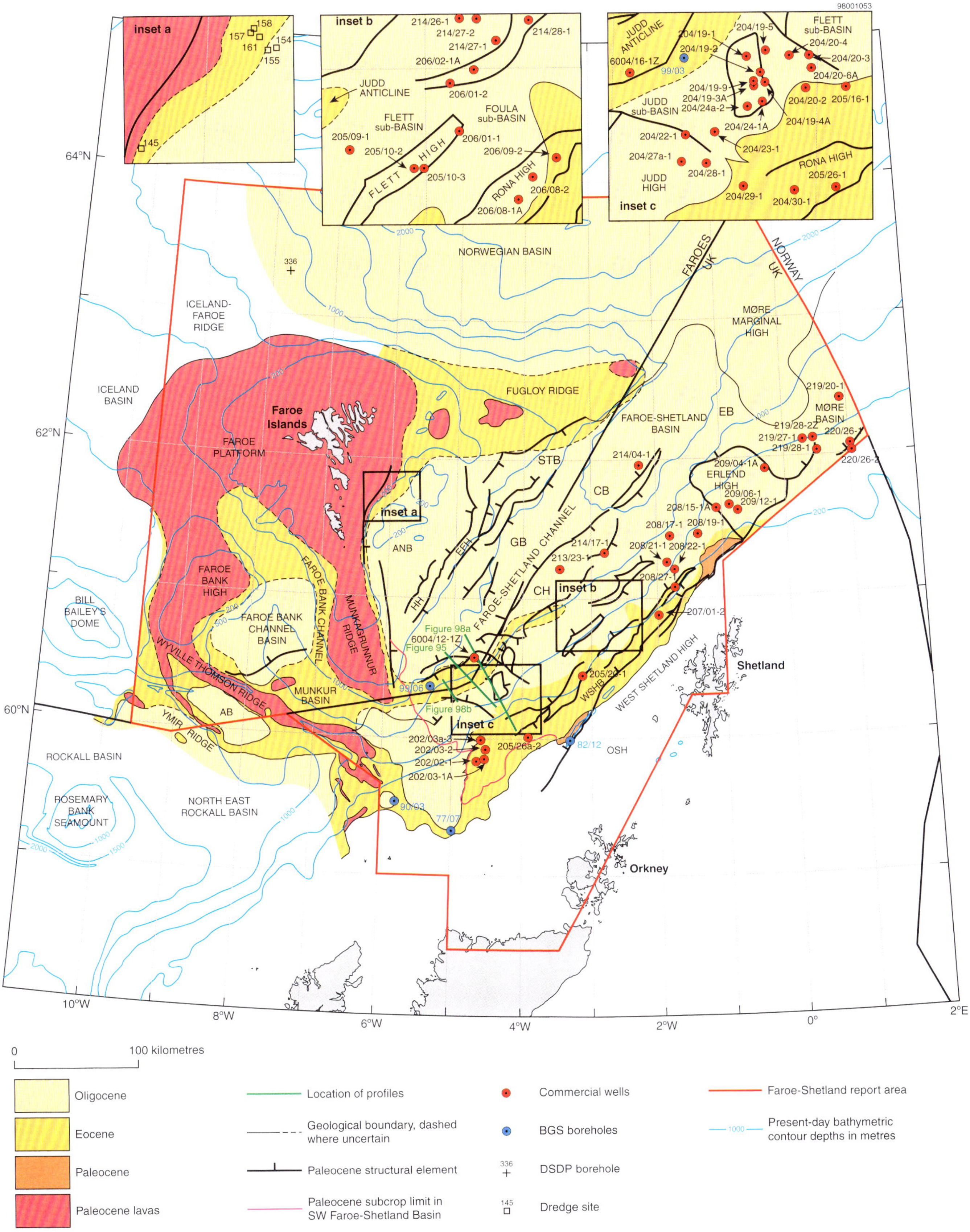

Figure 96 Distribution of Palaeogene sedimentary and volcanic rocks. Structural elements derived from Figure 7. Abbreviations: see Figure 7.

over much of the area, as is evident from the assignment of these strata to three regional megasequences (FSP-1 to 3) (Figure 94). Series definition is best established in the West Shetland region where most shallow boreholes and commercial wells are sited (Figure 96), though it should be noted that in many of the wells the Eocene–Oligocene boundary is poorly defined.

For the purpose of this report, the description of the Paleocene in the Faroe–Shetland region incorporates the interval bounded by the lithostratigraphical units of the Sullom/Ockran Sandstone to Flett formations (Shetland to Moray groups) (the T10 to T45 sequences), whilst the Eocene section includes the Balder Formation of the Moray Group (sequence T50) and the Stronsay Group (Figure 94). From a seismic-stratigraphical perspective, the Paleocene includes the bulk of megasequence FSP-3, with the uppermost part of FSP-3 and the whole of FSP-2 comprising the Eocene. The Oligocene is represented by the Westray Group (megasequence FSP-1). In the Møre Basin, the Tang Formation of the Rogaland Group comprises the Paleocene, the Tare Formation (also Rogaland Group) is equivalent to the Early Eocene Balder Formation, with the overlying Eocene to Oligocene succession represented by the Brygge Formation (Hordaland Group) (Dalland et al., 1988; Martinsen et al., 2005). The succession in the Norwegian Basin, north-east of the Iceland–Faroe Ridge, remains undivided.

During the Paleocene, renewed tectonic movement on some existing faults enhanced an inherited end-Cretaceous fault-induced topography. In the Faroe–Shetland Basin, this topography strongly influenced Early Paleocene submarine fan development, which was focused in the basin-floor deeps (Smallwood and Kirk, 2005). Progradation of the deep-water fan system continued during the Mid Paleocene, with the episodic input of sand punctuating the background accumulation of hemipelagic mudstone (Lamers and Carmichael, 1999; Sullivan et al., 1999). The Late Paleocene prograding shelf-margin systems accumulated outer shelf to basinal sandstones and mudstones, locally reworked by mass-flow processes (Knox et al., 1997). The major extrusive activity, which began in the Late Paleocene (see Chapter 9), culminated in the earliest Eocene, concomitant with regional uplift and widespread emergence (Figure 90) manifest as a regional unconformity at the base of the Balder Formation that records subaerial fluviatile erosion (Lamers and Carmichael, 1999; Smallwood and Gill, 2002). The topography of this surface was subsequently infilled following marine transgression, and deeper water conditions returned to the Faroe–Shetland Basin in the Early Eocene.

In the north of the area, the Møre Marginal High was emergent and separated the Paleocene basins of East Greenland and mid Norway (Brekke et al., 1999). As with the Faroe region, large volumes of extrusive volcanic rocks flowed across the exposed platform. The margin of the Møre Basin was also locally exposed (Gjelberg et al., 2005), though a narrow marine connection with the Faroe–Shetland Basin may have existed throughout Paleocene to Early Eocene times (Brekke et al., 1999; Figure 90). Following continental break-up to the west of the Møre Marginal High, this platform gradually subsided and was subsequently onlapped and overstepped by north-westerly prograding Eocene and younger strata (Hamar and Hjelle, 1984; Blystad et al., 1995).

In the Early Eocene, deltaic and shallow-marine sandstones sourced from the south and east built out into the Faroe–Shetland Basin, and probably the Faroe Bank Channel Basin, with silty mudstones in the more basinal areas (Knox et al., 1997; Knox, 2002). Limestones were deposited to the east of the Faroes, whilst shallow-marine clastic rocks deposited in the area of the continent–ocean boundary were sourced from the Faroe region, albeit to the north of an inferred watershed (Andersen et al., 2000). A series of overlapping basin-floor fans fringed by prograding shelf-margin deltaic systems developed in the central part of the Faroe–Shetland Basin during Mid Eocene times, sourced from the West Shetland area (Brooks et al., 2001; Davies et al., 2004; Robinson et al., 2004). Subaerial incision of the deltatop sediment and subsequent draping of this surface by prograding Late Eocene clinoforms is indicative of fluctuating relative sea level, though it remains uncertain if this reflects a tectonic or eustatic control (Robinson et al., 2004). Palaeogeographical reconstruction indicates that the contemporary Mid to Late Eocene shelf edge, west of Shetland, was located seaward of the Rona Fault that bounds the eastern margin of the Faroe–Shetland Basin (Robinson et al., 2004, fig. 8; Figure 88a). The local absence of Eocene sediments over the Møre Marginal High indicates that this platform remained an important source area for the Møre Basin (Martinsen et al., 1999). Myhre et al. (1992) have suggested that this structural high influenced sedimentation until it was buried in the Late Oligocene.

Late Eocene to Early Oligocene uplift of the Faroe Platform forced a westward shift in the watershed on the platform, and resulted in Oligocene clastic material being deposited both to the north and east of the Faroe region (Waagstein and Heilmann-Clausen, 1995; Andersen et al., 2000; Johnson et al., 2005b). Late Oligocene paralic to shallow-marine sedimentation in the West Shetland area has been interpreted as a transgressive systems tract (Evans et al., 1997; Knox et al., 1997; Stoker, 1999). This may be indicative of a general subsidence in the area, replacing the rifted topography of the Faroe–Shetland Basin with the broader deep-water trough that is the Faroe–Shetland Channel (Figures 87 and 96). The greater Møre Basin area was similarly developing into a continental margin with a shelf–slope–basin floor physiography (Martinsen et al., 1999). In deep water, biosiliceous oozes deposited in association with contourite sediment drifts reflect a major change in the circulatory regime in the Faroe–Shetland region (Davies et al., 2001).

This change has been interpreted by Davies et al. (2001) to mark the initiation of North Atlantic Deep Water formation flowing southerly from the basins to the north of the Greenland–Scotland Ridge (e.g. Norwegian Basin) via the Faroe Conduit into those to the south of the ridge (e.g. Iceland, Hatton, Rockall basins). However, isotopic, taxonomic and sedimentological data from DSDP and ODP sites on either side of the ridge suggest that these basins remained essentially isolated in terms of deep-water exchange during the Oligocene (Blanc et al., 1980; Thiede and Myhre, 1996; Kaminski and Austin, 1999). This is consistent with the suggestion that the Faroe Bank Channel (part of the Faroe Conduit) was shaped by early Neogene tectonic processes, and thus a proper deep-water connection is possibly a Neogene phenomenon (Stoker et al., 2005b and c; Figures 92 and 93).

Paleocene to earliest Eocene

Paleocene to lowermost Eocene sedimentary rocks are widely distributed within the Faroe–Shetland Basin, with a proven south-west limit extending onto the Judd High (Smallwood, 2005a; BGS 1986a, 1991c). Syndepositional movement on faults bounding marginal and intrabasinal highs, such as the Judd, Westray, Flett and Corona highs and Sjúrður Ridge, resulted in considerable accumulation of up to 2 to 3 km of marine shale and turbiditic sandstone in the adjacent Judd and Flett sub-basins, shed from the highs as well as the hinterland (i.e. Greenland and Scotland) (Hitchen and Ritchie, 1987; Duindam and van Hoorn, 1987; Dean et al., 1999; Lamers and Carmichael, 1999; Jones et al., 2002; Keser Neish, 2003; Smallwood et al., 2004; Jolley et al., 2005; Larsen and Whitham, 2005; Smallwood and Kirk, 2005; Smallwood, 2005a; Figures 88, 97 and 98). On the eastern margin of the Faroe–Shetland Basin, rejuvenation of the Rona and West Shetland highs similarly sourced the Foula Sub-basin and the West Shetland Basin, though the sequence is generally thinner (<1 km thick). Over the intrabasinal highs, the Paleocene sequence is generally thinner than in the basins and locally absent (Figures 88 and 98a). Within the basalt-covered region of the Faroese area, no firm evidence of Paleocene sedimentation is available. West of the Corona High, Smallwood (2005a) has interpreted an average Paleocene sediment thickness of 1.4 km in the Corona Sub-basin. Farther to the south-west, in excess of 1 km of Paleocene sediment is inferred for the Faroe Bank Channel Basin (Keser Neish and Ziska, 2005), though their distribution remains unknown.

The Møre Basin accumulated a thick sandy turbidite sequence along the eastern margin of the basin, though sediments also prograded from the west, derived from the Møre Marginal High (Martinsen et al., 1999, 2005; Brekke, 2000). This clastic input may reflect contemporary uplift of adjacent structural highs (Gjelberg et al., 2001). Sediment accommodation was probably controlled by differential subsidence within the basin above deep-seated Jurassic faults (Martinsen et al., 2005). Pre-drift reconstructions of the area north of the Faroe–Shetland region suggest that the Kangerlussuaq Basin of southern East Greenland lay in close proximity to the Faroe–Shetland Basin (Larsen et al., 1999; Larsen and Whitham, 2005; Figure 90). The possibility that the Kangerlussuaq region acted both as a source and a conduit for sediment transfer into the Faroe–Shetland Basin (Jolley et al., 2005; Larsen and Whitham, 2005; Smallwood, 2005a) may have implications for the occurrence of Paleocene sandstone in offshore areas around the Faroe Islands, including the area to the north of the islands.

In the report area, lithological and stratigraphical information is based on commercial well data and biostratigraphical reports, and BGS boreholes located within, and on the eastern and southern flanks of, the Faroe–Shetland Basin, and from the south-west Møre Basin (Figure 96). For the purpose of this section, the Paleocene to earliest Eocene succession in the Faroe–Shetland region is divided, in ascending stratigraphical order, into the Sullom and Ockran Sandstone formations of the Shetland Group, the Vaila and Lamba formations of the Faroe Group, and the Flett Formation of the Moray Group. This succession corresponds to the T10 to T45 BP sequence nomenclature, and the seismic-stratigraphical FSP-3 megasequence (Figures 94 and 97). Information from the Møre and Kangerlussuaq basins is described separately.

Faroe–Shetland region: Shetland Group, Sullom Formation (FSP-3 Megasequence, T10–T22)

The Sullom Formation (Figures 94 and 97) forms the bulk of the Paleocene section of the Shetland Group within the Faroe–Shetland Basin, where it is dominated by green- grey to dark grey calcareous mudstone and subordinate sandstone. It has a variable thickness that reaches up to approximately 1 km in well 208/17-1 (Figure 96). The occurrence of sandstone increases along the margin of the basin, and in the extreme south-east of the region the Sullom Formation passes laterally into the sandstone-dominated Ockran Sandstone Formation (Knox et al., 1997). This probably represents a shelf terrace that was structurally controlled by the Judd Lineament along the southern edge of the basin (Figure 7).

The base of the Sullom Formation occurs immediately above the highly reflective upper part of the Late Cretaceous Jorsalfare Formation (Shetland Group); a boundary also marked by a sharp downward increase in gamma-ray log values, and a decrease in velocity (Ritchie et al., 1996). This boundary is interpreted by Ebdon et al. (1995) as a maximum flooding surface, and is associated with earliest Paleocene (Danian) marine microfossils, such as *Globigerina simplicissima, Globorotalia planocompressa, Globorotalia eobulloides, Globigerina trivialis, Senoniasphaera inornata, Spongodinium delitiense, Alisocysta reticulata, Palaeoperidinium pyrophorum, Palaeocystodinium* cf. *australinum* and *Carpatella cornuta* (Ebdon et al., 1995; Knox et al., 1997; Mudge and Bujak, 2001). According to Knox et al. (1997), the top of the Sullom Formation is characterised by an increase in sandstone interbeds, though this is locally difficult to define in the south of the area where sandstone occurs both in the Sullom Formation and the basal Vaila Formation (Figure 97). On structural highs, such as the Rona, Judd and Westray highs, the Sullom Formation is locally thin to absent, with erosion occurring prior to the deposition of younger Paleocene units.

Shelf, slope and basinal sediments comprise the Sullom Formation. A thin sequence of shelf deposits has been proved around the margin of the Faroe–Shetland Basin (Ebdon et al., 1995; Knox et al., 1997). On its south-eastern margin, well 204/28-1 proved an 18.9 m thick sequence of siltstones, whereas well 205/26-1 to the east on the Rona High indicated a condensed carbonate shelf deposit (Figure 96). In the north-east of the Faroe–Shetland Basin, wells 208/15-1A and 209/06-1 proved the occurrence of shelf sandstones.

The transition to deeper water is marked by an interbedded slope-to-basin floor sequence of sandstones and mudstones, with evidence of mass-flow reworking. This reflects uplift and denudation of the structural highs causing sediments to be shed into the basin (Smallwood, 2005a). On the flank of the Westray High, well 204/19-3A cored a 27 m thick, quartzose, pale grey, medium to coarse-grained sandstone that included a thin (<1 m thick) pebbly section towards the base of the 101 m thick Sullom Formation (Figures 96 and 97). In the adjacent well 204/19-4A, a 307 m thick section of Sullom Formation consists mainly of quartzose, slightly glauconitic, sandstone with several conglomeratic

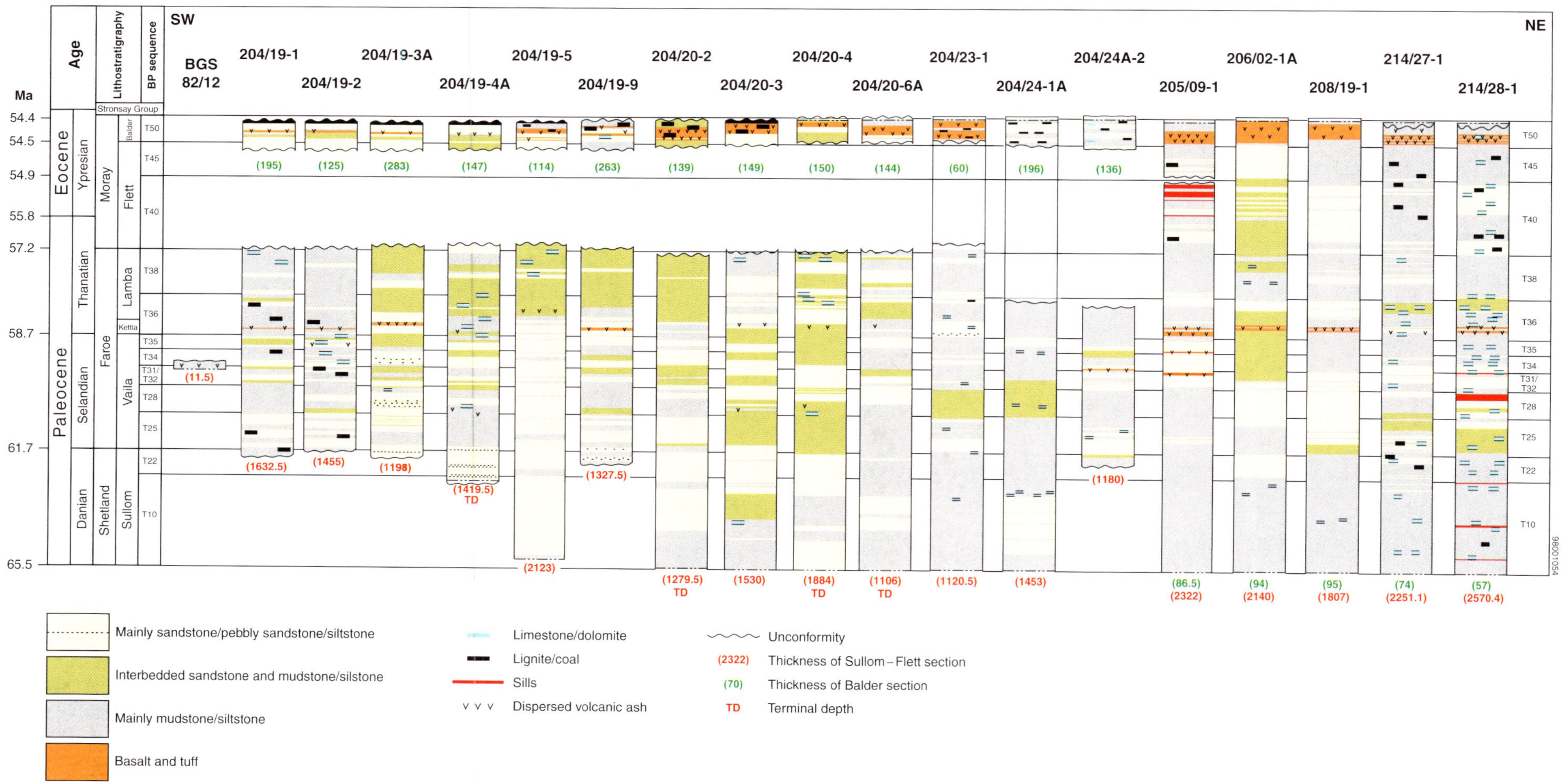

Figure 97 Stratigraphical range chart showing thickness (m) and generalised lithology for the Paleocene to Lower Eocene succession from selected wells and a BGS borehole within the Faroe–Shetland Basin. This information is mainly derived from released well logs with added information from Morton et al., (1988b), Ebdon et al., (1995), Knox et al., (1997), Cooper et al., (1999), Jowitt et al., (1999), Lamers and Carmichael, (1999), Leach et al., (1999), Naylor et al., (1999), Ritchie et al., (1999a), Mudge and Bujak, (2001), Jolley and Bell, (2002a), Carruth, (2003), Jolley and Morton, (2004). Timescale is based on 'A Geologic Time Scale 2004' by F M Gradstein, J G Ogg, A G Smith, et al. (2004) and the International Stratigraphical Chart, 2006 (ICS) with additions. For locations of wells, see Figure 96.

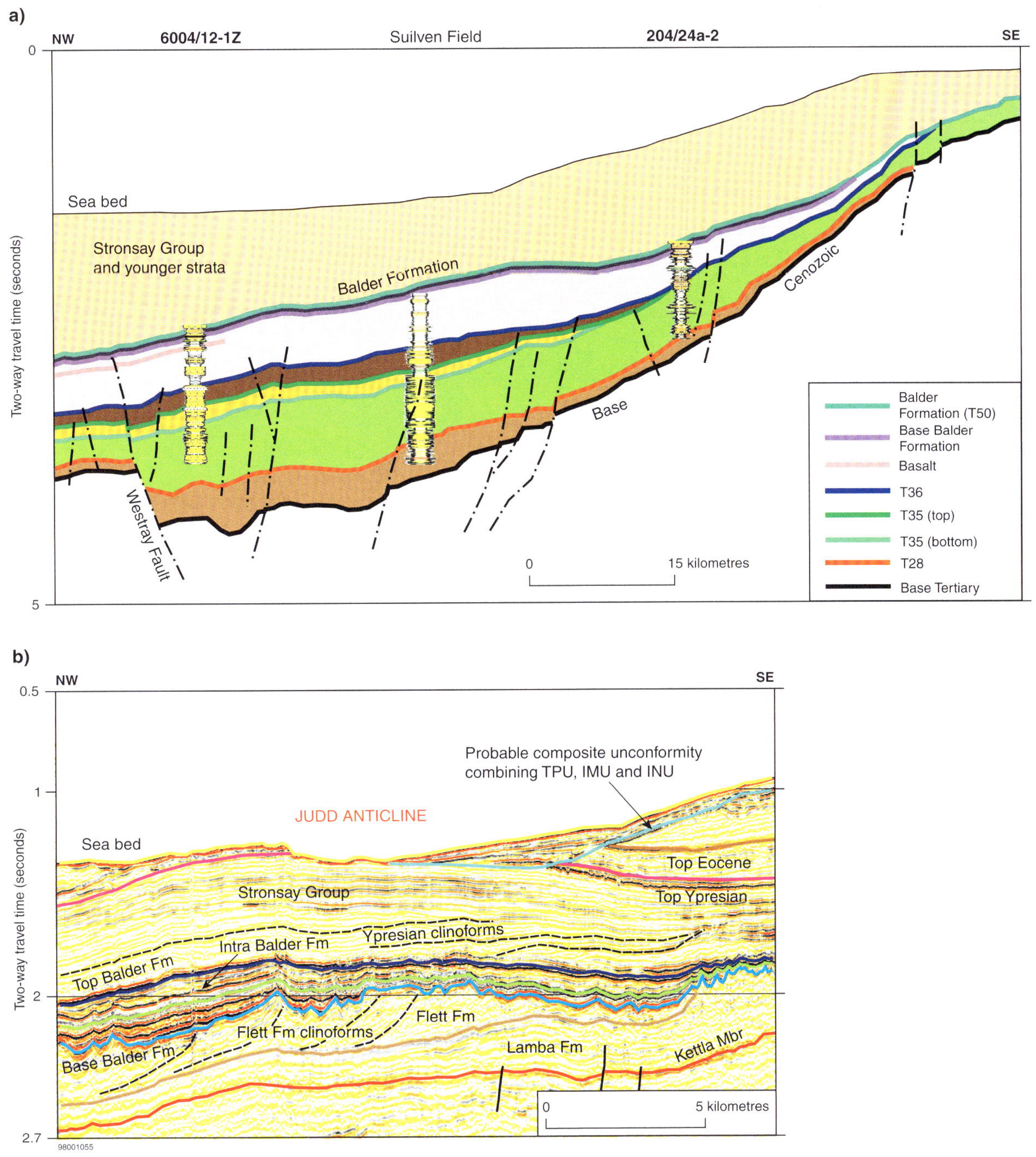

Figure 98 a) Geoseismic cross-section across the Judd Sub-basin. Representative well sticks from Hollingsworth, (2002) display the increase in sand proportion basinward in the Judd Sub-basin, towards the Westray Fault (modified from Smallwood, 2005a); b) Seismic profile across the Judd Anticline showing the character of the Balder Formation and the prograding clinoforms within the Flett Formation and Stronsay Group (modified from Smallwood and Gill, 2002). Abbreviations: see Figure 14. For location of profiles, see Figure 96. Seismic data courtesy of CGG Veritas, and both images reproduced by permission of The Geological Society, London.

beds, up to 6 m thick, incorporating flakes of two different mudstones. One of the populations of mudstone flakes in the conglomerate is dark grey, calcareous and slightly glauconitic, whereas the second population of flakes consists of an olive to black, calcareous and carbonaceous mudstone. The latter set of mudstone flakes may be derived from Kimmeridge Clay Formation rocks. A slope-to-basin sandstone was also proved in well 204/19-5, which penetrated 269.9 m of the Sullom Formation consisting of very fine to fine-grained, occasionally coarse-grained, sandstone with rare lithic grains, which becomes interbedded with siltstone and mudstone near the top of the section.

In the area of the Judd and southern Flett sub-basins, including the Westray High, deep-marine mudstone was deposited, e.g. wells 204/23-1, 204/19-1 and 205/16-1, though an association of mixed sandstone and mudstone occurs in wells 204/24-1A, 204/20-2 and -3 (Figures 96 and 97). The sandstone is very fine to fine-grained, quartzose, white to pale grey, and variably glauconitic; the mudstone is pale to dark grey, and rarely green, and variably calcareous and carbonaceous. This accumulation of basinal sandstone marks the onset of deep-water fan progradation into the Faroe–Shetland Basin (Smallwood, 2005a). Heavy mineral analysis of wells in the Judd Sub-basin indicates derivation of sediment from the surrounding structural highs, especially to the south and east of the basin (Morton et al., 2002a). However, pollen analysis indicates an additional westerly source of input; the angiosperm pollen *Momipites dilatus* has only previously been recorded from North America (Naylor et al., 1999; Jolley et al., 2005). Farther north, in the Flett Sub-basin and High, wells 214/27-1, 214/28-1, 205/09-1 and 206/02-1A proved up to 429 m of black to dark grey mudstone, though the sections locally coarsen upward to include thin sandstone interbeds (e.g. 214/27-1) (Figures 96 and 97). Planktonic foraminifera indicate deposition in a fully marine environment with access to open oceanic circulation (Jowitt et al., 1999). In well 214/28-1, the mudstones are intruded by dolerite sills.

Faroe–Shetland region: Shetland Group, Ockran Sandstone Formation (Fsp-3 Megasequence, T10).

The Ockran Sandstone Formation (Figure 94) represents a unit of bioclastic sandstones restricted to the East Solan Basin and south-west Rona High, close to the south-western margin of the Faroe–Shetland Basin (Booth et al., 1993; Figure 96). They were deposited in an inner shelf setting (Knox et al., 1997), and are equivalent to the 'Bryozoan Sands' first reported by Hitchen and Ritchie (1987). The rocks consist of medium to coarse-grained, variably argillaceous, bioclastic sandstones that grade into sandy limestones towards the base of the formation. The bioclastic debris consists mainly of bryozoans, but also includes bivalves and gastropods. Wireline well logs indicate a consistently sandy and well-bedded formation, with high-velocity spikes indicative of calcite-cemented beds. The latter are particularly abundant in the lower part of the section, such as that encountered in the well 205/26a-2 (Figure 96), which reveals a 260 m thick section of calcareous sandstone that grades into a sandy limestone. On the Rona High, well 204/30-1 proved an 82.5 m thick section of predominantly sandstone with a thin basal mudstone. Correlation from wireline well logs and biostratigraphical correlation indicates that the Ockran Sandstone Formation equates to the lower part of the Sullom Formation of more basinal sections (Knox et al., 1997).

Faroe–Shetland region: Faroe Group, Vaila Formation (Fsp-3 Megasequence, T25–T35)

The Vaila Formation (Figures 94 and 97) occurs widely throughout the Faroe–Shetland Basin, ranging in thickness from approximately 600 m to 1 km in the southern and central part of the basin (Knox et al., 1997), whereas an approximately 1.5 km thick succession of the T28–T35 section has been encountered on the Faroese side of the Judd Sub-basin (Smallwood and Kirk, 2005). In the northern part of the Faroe–Shetland Basin, the formation thins to about 200 m thick, and it is commonly thin to absent on the margin of the basin and over intrabasinal highs, such as the Corona High. The system of deep-water fan sand progradation initiated within the Sullom Formation continued during the deposition of the Vaila Formation, with thick basinal sandstones punctuating the background hemipelagic mudstone accumulation (Ebdon et al., 1995; Lamers and Carmichael, 1999; Sullivan et al., 1999; Figure 99), which culminated in peak Paleocene sediment accumulation rates in the upper T35 to T36 interval (Jolley et al., 2002). Heavy mineral and pollen analyses indicate source regions located in the Scottish and Greenland hinterlands (Morton et al., 2002a; Jolley et al., 2005). This depositional system has become a major hydrocarbon play in the Faroe–Shetland region, and includes the Suilven, Loyal, Schiehallion and Foinaven discoveries (see Chapter 12).

The lithology of the Vaila Formation shows considerable variability, ranging from mudstone to interbedded mudstone and sandstone and thick-bedded sandstone. These sediments accumulated in depositional environments ranging from outer shelf shallow-marine to deep-water basin settings (Figure 99). Basinal sandstones are interpreted to represent lowstand fan deposits with sedimentation having occurred in the form of sandy slumps and mass-flows, grading into turbidites (Mitchell et al., 1993; Lamers and Carmichael, 1999; Naylor et al., 1999). Sporadic tuff intervals have been cored in some of the mudstone sections, and most probably represent distal accumulation products of eruptions in the Faroe–Greenland area, though more localised volcanic sources cannot be discounted (Morton et al., 1988b; Knox et al., 1997; Jolley and Bell, 2002a).

In the mudstone-dominated basinal sections, Knox et al. (1997) divided the Vaila Formation into four informal subdivisions, Vaila units 1 to 4 (V1–V4) (Figure 94), largely on the basis of the gamma-ray wireline well log signature. Units V1 and V2 are equivalent to T22 and T25–28, respectively; unit V3 spans the T31–T35 interval, with V4 equivalent to the upper part of T35 (Jolley et al., 2005). Biostratigraphically, the formation as a whole is characterised by the dinoflagellate cysts *Palaeocystodinium pyrophorum*/*Palaeocystodinium bulliforme* and the *Palaeocystodinium pyrophorum* acme, and the radiolaria *Cenodiscus* spp. and *Cenosphaera lenticularis* acme, indicative of a mainly Selandian age (Knox et al., 1997). The dinoflagellate cyst *Isabelidinium viborgense* biomarker became extinct during the T31 interval in the Faroe–Shetland Basin (Jolley and Bell, 2002a).

Seismic data from the Judd and Flett sub-basins reveal channelised and chaotic seismic facies associated with the Vaila Formation; these are attributed to a submarine slope depositional environment subject to mass failure (Cooper et al., 1999; Leach et al., 1999). The vertical and lateral lithofacies variation observed in well logs from these basins (e.g. wells 204/20-2, -3 and -6A; Figure 97) is consistent with a channelised slope complex. Detailed study of the T31 and T34 sandstone intervals, which form part of the Schiehallion and Loyal reservoirs, indicates a series of north-westerly trending, laterally amalgamated, channel complexes, locally meandering, with thicknesses up to 70 m and widths ranging from 100m to 1 km (Leach et al., 1999).

Sedimentological analyses of cored sections penetrating this channelised seismic facies indicate that it is dominated by beds of massive structureless sandstone, albeit locally laminated at the tops of the beds. Subordinate facies include mud-clast conglomerates, which are particularly common towards the base of the reservoir units, and thin-bedded sandstone that is more common towards the top of these units. This is illustrated in well 204/19-1 (Figure 96), where a 7 m cored section from the most sand-rich facies of the well displays medium to coarse-grained

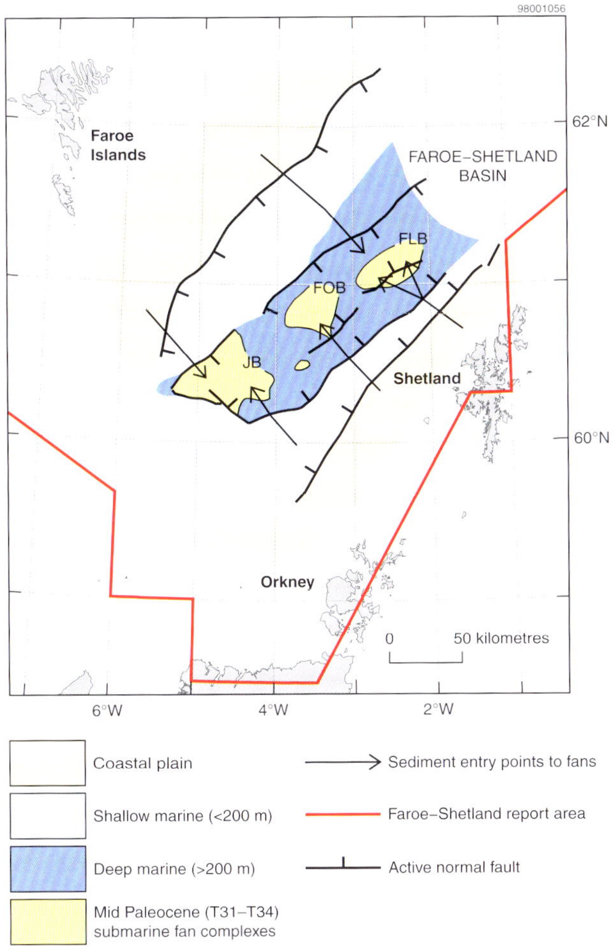

Figure 99 Gross palaeogeographical reconstruction during Mid Paleocene fan development in the Judd (JB), Foula (FOB) and Flett (FLB) sub-basins, based on information derived from the Vaila Formation (T31–T34) (modified from Lamers and Carmichael, 1999 and Jolley et al., 2005).

sandstone with bed thicknesses ranging from 0.3 m to 3 m (Knox et al., 1997). Upward fining is locally observed with coarse-grained sand to pebble grade material occurring at the base of some sandstone beds. Mudstone flakes and laminations, and convolute lamination indicative of dewatering are also observed. The channel complexes are envisaged as scours into the submarine slope, which were subsequently back-filled with stacked, mass-flow sandstone (Leach et al., 1999). Mudstone units, 10 to 20 m thick, separate these sandstone intervals.

A comparable sandstone facies has been proved in the north-eastern part of the Flett Sub-basin. Here, a 10 m cored section in well 208/19-1 proved medium to very coarse-grained, upward-coarsening to upward-fining sandstone, 0.25 to 3 m thick, with variable proportions of pebbles and mudstone clasts. Thinner units include medium-grained sandstone with horizontal bedding picked out by carbonaceous laminae and coarse matrix-supported conglomeratic sandstone with pebbles and mudstone or sandstone clasts. Thin beds of micaceous, carbonaceous mudstone and siltstone are also present (Knox et al., 1997). The gamma-ray well log response of the sand-rich facies shows a subdued serrated signature giving the unit an overall massive appearance.

A strongly serrated gamma-ray well log signature in wells 214/27-1, 214/27-2 and 214/28-1 is indicative of a more interbedded sandstone/mudstone succession in the east central Flett Sub-basin, and on the Flett High (Figure 96). Sandstones are the dominant lithology in these sections, with individual units ranging from 0.25 to 2.5 m in thickness. The thinner beds (0.25 to 1.0 m) are dominated by clean, well-sorted, fine to medium-grained sand. Many beds fine upwards, with generally structureless sandstone passing into muddy sandstone, siltstone or silty mudstone with parallel lamination or ripple cross-lamination, or micaceous sandstone. The thinnest beds consist entirely of laminated or cross-laminated sandstone, which is variably carbonaceous and commonly affected by burrowing. Mudstone fragments only occur rarely in these thin sandstone beds. The associated mudstone (0.1 to 0.3 m) is typically silty, micaceous, and carbonaceous, commonly laminated or cross-laminated, bioturbated and with local slumping. Less commonly the mudstone is finer grained, grading to micaceous claystone (Knox et al., 1997).

A cored section in well 206/01-2 (Figure 96) shows sandstone units with subordinate sections in which mudstone predominates over sandstone. The mudstone is dark grey, micaceous and commonly bioturbated, and is interbedded with sporadic, thin, laminated sandstone layers displaying convolution and disruption due to slumping. By way of contrast, thicker-bedded (4 to 15 m), sharply defined sandstone and mudstone units with a blocky gamma-ray wireline well log response were penetrated in wells 206/01-2 and 205/09-1. Lags of coarse-grained sand or fragments of carbonaceous material locally occur at the base of the sandstone beds, whereas the mudstone is typically micaceous and bioturbated, silty, and with disseminated sand grains and mudstone clasts. Slumped intervals of laminated to thin-bedded siltstone and sandstone are common.

Thin layers of tuff, clearly distinguishable by their green-grey to green colour are sporadically present throughout the cored mudstone section in well 206/01-2. A thicker unit of tuffaceous sandstone is present in well 205/09-1 (Figure 97), where the cored section reveals tuffaceous, locally pebbly, sandstone grading into sandy tuffite. Additional thinner tuffaceous sandstone units are identified lower in the section based on their gamma-ray wireline well log character and sidewall cores. Further evidence of volcanic activity occurring within the deposition of the Vaila Formation is found in wells 204/20-3, -4 and 204/24a-2, on and adjacent to the Westray High (Figures 96 and 97), where tuffaceous rock has been found in Vaila Formation units V2 and V3. Wells 204/20-3 and 204/20-4 both encountered thin sequences of tuffaceous mudstone. In well 204/24a-2, the base of the section consists of about 13 m of pale grey, sub-fissile, homogenous mudstone, which is overlain by a 28 m thick carbonaceous siltstone that shows an upwardly increasing abundance of tuffaceous inclusions, and is itself capped by a 4 m thick, pale grey, silty tuff.

In the Papa Basin, BGS borehole 82/12 cored an 11.5 m thick section of very dark grey, faintly laminated, tuffaceous claystone (Figures 96 and 97). Although the sediment contains a rich Danian palynomorph assemblage (Morton et al., 1988b), the common occurrence of the dinoflagellate cyst *Palaeocystodinium bulliforme* implies a Selandian age (Jolley and Bell, 2002a), and places it just below the top of the Vaila Formation (Knox et al., 1997). Consequently, the tuffs in borehole

82/12 are interpreted as being equivalent to the V2/V3 ashes farther west (Jolley and Bell, 2002a).

Faroe–Shetland region: Faroe Group, Lamba Formation (Fsp-3 Megasequence, T36–T38)

The Lamba Formation (Figures 94 and 97) consists of a progradational package of mudstone and sandstone that built out from south-east to north-west across the Faroe–Shetland Basin (Knox et al., 1997; Smallwood and Gill, 2002). Seismic data from the Judd Sub-basin images delta-front clinoforms up to 500 m high, indicating a minimum water depth for the basin. The Lamba Formation exceeds 700 m in thickness and, in basinal sections, can be divided into Lower Lamba (L1) and Upper Lamba (L2) units (Figure 94). Whereas the L1 and L2 units are informal subdivisions based on gamma-ray wireline well log signatures that reveal upward-fining profiles, the Kettla and Westerhouse Sandstone members recognised within L1 are formally defined (Knox et al., 1997). The L1 and L2 units correlate with T36 and T38 sequences of Ebdon et al. (1995), and are assigned a Thanetian age (Jolley et al., 2005). The Kettla Member is a distinctive volcaniclastic interval that locally forms a prominent seismic reflector at the base of the Lamba Formation (Smallwood and Gill, 2002; Figures 97 and 98b). The first downhole occurrence of the foraminiferal biomarker *Spiroplectammina spectabilis* marks the top of the formation (Knox et al., 1997). The acme occurrence of the dinoflagellate cysts *Areoligera senonensis* and *Areoligera gippingensis* (Mudge and Bujak, 2001) close to the L1/L2 boundary is an additional distinctive biomarker of the Lamba Formation.

The Kettla Member (Figure 94) is a dark grey to green tuffaceous siltstone grading to silty tuffite, and consists of degraded volcanic ash and volcanogenic lithoclasts that occur as two units separated by a thin unit of mudstone (Jolley et al., 2005; see Figure 129 and Chapter 9). The tuffaceous interval displays a characteristic double peak on the gamma-ray well log response (Sørensen, 2003). This unit has been identified within the Judd, Flett and Foula sub-basins, and on the Westray and Flett highs, where it is encountered in numerous wells, e.g. 204/19-1, -2, 204/20-3, 6004/16-1Z (Smallwood and Kirk, 2005), 205/09-1, 206/02-1A, 208/19-1 and 214/27-1 (Figure 96). The derivation of the volcaniclastic rock remains uncertain; heavy mineral and pollen assemblages indicate potential multiple source areas to the east and west of the Faroe–Shetland Basin (Jolley et al., 2005).

There is an absence of Greenland-derived flora in the post-Kettla Member sediments of the Lamba Formation (Jolley et al., 2005). The ensuing north-westerly progradation of the shelf margin is envisaged to have been driven by uplift and denudation of the Scottish hinterland (Jones et al., 2002; Smallwood, 2005a). Deposition occurred in environments ranging from outer shelf to basin floor, with the shelf slope break having advanced to the north-west, and basinal sandstones having accumulated beyond the Westray High (Naylor et al., 1999).

The Westerhouse Sandstone Member (Figure 94) of the L1 unit of the Lamba Formation occurs in a restricted area in the Flett Sub-basin, and consists of a succession of discrete sandstone and mudstone units (Knox et al., 1997). Bed thicknesses are mostly between 1 and 5 m thick, but locally reach 10 to 15 m in thickness, producing a distinctly serrated wireline well log response.

A cored section from the middle of the Westerhouse Sandstone Member in well 214/27-2 (Figure 96) indicates a sequence of medium-grained, well-sorted, pale grey to grey-green sandstone separated by units of thin, grey claystone, silty micaceous mudstone and siltstone. The sandstone is predominantly massive, with crude upward fining of the sand fraction and concentrations of mudstone rip-up clasts in the upper part of the unit. In this well, the Westerhouse Sandstone Member is 403 m thick, and extends downwards to the top of the Kettla Member. In other wells (e.g. 214/28-1; Figure 97) the two members are separated by several tens of metres of mudstone. The sandstone is interpreted as a debris-flow deposit, associated by Mitchell et al. (1993) with a basinal lowstand systems tract. The interbedded mudstone is typically pale to medium grey, silty and generally displays a monotonous lithology (Knox et al., 1997).

More thickly bedded sandstone units, presumably equivalent to the Westerhouse Sandstone Member, occur locally in wells in more marginal sections e.g. 204/19-1 and 205/09-1. A cored section in well 205/09-1 reveals poorly bedded sandstone with discontinuous laminae associated with dish structures, local concentrations of mudstone flakes and carbonaceous fragments (Knox et al., 1997). Medium to coarse-grained, locally pebbly sandstone also occurs higher in the Lamba Formation in these settings e.g. well 204/23-1 in the Judd Sub-basin and well 208/19-1 in the north-eastern part of Faroe–Shetland Basin, with thicker sections displaying an overall upward-fining profile. Marine palynofloras sharply decline and disappear towards the top of the Lamba Formation (Naylor et al., 1999).

Faroe–Shetland region: Moray Group, Flett Formation (Fsp-3 Megasequence, T40–45)

The north-westerly prograding system that deposited the Lamba Formation was replaced by the northerly prograding Flett Formation (Figures 94 and 97) that is dominated by sandstone and mudstone. The sediments are predominantly of progradational shallow-marine facies, ranging from paralic 'delta-top' facies in proximal sections to 'prodelta' facies in more distal settings (Ebdon et al., 1995; Knox et al., 1997; Smallwood and Gill, 2002). Prodelta clinoforms (Figure 98b) indicate water depths of at least 300 m. The Flett Formation was largely deposited beyond the Lamba Formation shelf edge, resulting in shelf bypass, erosion and nondeposition along the southern margin of the Faroe–Shetland Basin (Ebdon et al., 1995). This shift in depositional focus may have been, at least in part, linked to uplift associated with the initiation of widespread extrusive volcanic activity during this interval (Jolley and Bell, 2002a; Smallwood and Gill, 2002). This resulted in a significant unconformity in this area, where the Flett Formation is commonly absent, and the younger Balder Formation rests unconformably on the Lamba Formation (Figure 97).

The Flett Formation exceeds 500 m in thickness in the Flett Sub-basin (Ellis et al., 2002), and can be divided into three units, the Lower, Middle and Upper Flett units (F1–F3) (Figure 94). These are informal subdivisions based on wireline well log characteristics and biostratigraphy, but the Colsay Sandstone and Hildasay Sandstone members within the F1 and F2 units, respectively, are formally defined (Knox et al., 1997). The F1 unit correlates with sequence T40, whereas F2 and F3 correlate with T45 (Jolley et al., 2005). Unit

F1 is further divided into 1a and 1b, also referred to as the Lower and Upper Colsay Sandstone units (Knox et al., 1997), the boundary of which is correlated to the Paleocene/Eocene boundary (Jolley and Bell, 2002a; Jolley et al., 2005). This subdivision of the Flett Formation largely represents the recognition of gross upward-coarsening cycles within units F1a, F1b and F2, though unit F3 is characterised by high-gamma mudstone. Although marine dinoflagellate cysts, especially the genus *Apectodinium*, are abundant in unit F1, assemblages in the overlying F2 and F3 units are dominated by terrestrial forms (Knox et al., 1997). In mudstone-dominated sections, the boundary with the overlying Balder Formation is marked by a major increase in tuffs.

In the Colsay Sandstone and Hildasay Sandstone members, paralic delta-top facies recovered in wells 206/02-1A and 208/22-1 (Figure 96) include varicoloured, shelly (gastropods and bivalves), bioturbated and carbonaceous, laminated and rippled claystone, mudstone and siltstone, with grey to pale brown, fine to coarse-grained and pebbly sandstone, and brown to black lignite (Knox et al., 1997). The sandstone is variably cemented by kaolinite with local tight cementation by calcite. The lignite locally overlies in situ rootlets. In proximal sections, e.g. wells 205/09-1 and 206/01-1 (Figure 96), the Lower Colsay Sandstone unit (F1) commonly consists of a thin basal sandstone overlain by a relatively thick upward-coarsening silt/sandstone unit, frequently with lignite in the upper part. The distal progradational settings farther north are dominated by grey laminated or bioturbated mudstone and siltstone interbedded with fine to coarse-grained sandstone. However, the depositional slope was subjected to mass wasting and slumping. In well 214/28-1 (Figure 96), a 367.6 m thick deep-water sandstone deposit displaying a distinctive serrated wireline gamma-ray well log signature implies a stacked succession of relatively thin-bedded sandstone interpreted as a series of debris-flow deposits (Shanmugan et al., 1995). Units of thin deep-water sandstone of probable turbidite origin are also preserved in wells 206/02-1A, 208/21-1 and 208/17-1 (Knox et al., 1997; Figure 96).

In the Flett Sub-basin, well 205/09-1 penetrated about 70 m of basaltic rocks within the lower part of the Flett Formation. The palynoflora sampled directly beneath and between the lava flows shows a population of aquatic and aquatic-marginal pollen and spore taxa that include *Sparganiaceaepollenites* species, *Azolla* massulae, *Carayapollenites veripites, Platycaryapollenites platycaryoides, Intratriporopollenites microreticulatus* and *Monocolpopollenites tranquilus* (Ellis et al., 2002). This assemblage is typical of an aquatic and swamp community, and is comparable to assemblages recovered from the Prestfjall Formation on Suðuroy. The palynomorph assemblages in well 205/09-1 have been assigned to the upper Flett unit 1b to lowermost Flett unit 2 (Ellis et al., 2002).

Møre Basin

In the Møre Basin (Figure 96), Paleocene to earliest Eocene strata equivalent to the Sullom/Ockran Sandstone to Flett formations are correlated with the Tang Formation of the Rogaland Group (Dalland et al., 1988). Throughout the basin, these strata are mainly composed of deep-water argillaceous sediments, though turbidite sandstone is present along the margin of the basin (Martinsen et al., 2005), and tuffaceous rock is common in the upper part of the formation. In excess of 1.2 km of Paleocene sediment has been recorded from the eastern part of the Møre Basin (Gjelberg et al., 2005).

Close to the south-west margin of the Møre Basin, well 220/26-2 penetrated 7 m of Danian limestone lying unconformably on Upper Cretaceous rocks (Figure 96). The Danian limestone is unconformably overlain by 203 m of grey-brown mudstone of Thanetian age, sandy at the base, tuffaceous near the top, and with sporadic, thin, interbedded limestone. A comparable sequence was cored in the adjacent well, 220/26-1, where an 18 m thick section of red Danian mudstone is overlain by 152 m of pale grey to grey-brown Thanetian mudstone and siltstone, becoming tuffaceous at the top of the section. Farther west, well 219/28-1 proved a basal 24 m thick section of interbedded claystone and limestone of Danian/Selandian age overlain by 152 m of Thanetian claystone with subordinate limestone, sandstone, and tuff. An equivalent 245 m thick Thanetian succession was cored in well 219/27-1 resting unconformably on Upper Cretaceous claystone and limestone. To the north, well 219/20-1 drilled 349 m of undifferentiated Paleocene sediment consisting of dark grey, green and blue-grey mudstone, that is slightly silty and tuffaceous. The Møre Marginal High was an area of positive structural relief (Figure 96), and Paleocene deposits are uniformly thin, typically 200 to 350 m, with lithologies consisting mainly of lavas and shallow-water sedimentary rocks (Nelson and Lamy, 1987).

Kangerlussuaq Basin

The Kangerlussuaq Basin is located on the southern East Greenland continental margin (Figure 90), and includes a 200 m thick sequence of Paleocene to lowest Eocene mudstone, sandstone and conglomerate lying unconformably on Cretaceous mudstones (Larsen and Whitham, 2005). The sequence is capped by a thick succession of Eocene basalt and basaltic sediment. The Paleocene to lowermost Eocene unit shows an upward-shallowing trend from deep-marine mudstone and channelised turbidite sandstone to shallow-marine and deltaic sandstone and silty mudstones to fluvial sandstone and conglomerate. On the basis of macrofauna and palynological studies, Larsen et al. (1999) assigned the entire unit an Early to Mid Paleocene age. However, Jolley and Whitham (2004) suggest that the pollen floras are indicative of deposition in late T40 time, immediately following the PETM.

Eocene

Eocene sedimentary rocks occur widely in the basinal areas of the Faroe–Shetland region, where they locally exceed 1 km in thickness, but are largely thin to absent on the shallow platforms and ridges due to subsequent erosion (Figures 88 and 96). For the most part, the Eocene rocks are concealed beneath an upper Palaeogene to Neogene cover, although they are locally exposed on the flanks of the basins and, more unusually, within deep-water channels (Waagstein and Heilmann-Clausen, 1995; Stoker et al., 2003; Smallwood, 2004).

Following Late Paleocene to earliest Eocene uplift and volcanism, much of the Eocene sedimentary record in the Faroe–Shetland Basin indicates phases of coastal/shallow marine to deep marine deposition interspersed with periods of local and/or regional uplift (Smallwood and Gill, 2002; Sørensen, 2003; Robinson et al., 2004; Figure 89). This has resulted in a predominantly northerly prograding clastic succession, commonly tuffaceous in Lower Eocene strata, and characterised by the development of shelf margin deltaic systems interdigitating with deep-water submarine fans, especially in the Middle Eocene succession. Shallow-marine clastics and the northward and southward progradation of deltas from the Wyville Thomson–Ymir ridge system into the Faroe Bank Channel Basin and north-east Rockall Basin, respectively, implies the early stages in the development of this ridge system high (Stoker et al., 1988; Tate et al., 1999).

In the Møre Basin, subsidence during the Eocene allowed thick sequences to accumulate mainly along the western margin of the basin (Figure 23), sourced from the adjacent Møre Marginal High (Brekke et al., 1999; Martinsen et al., 1999). North of the Faroe Platform, the subsiding continental margin is envisaged to have been continuously transgressed throughout the Eocene (Nielsen and van Weering, 1998), though the adjacent Iceland–Faroe Ridge probably remained above sea level (Thiede and Eldholm, 1983).

In the report area, lithological and stratigraphical information is derived from wells and BGS boreholes located in the eastern half, and on the southern flank, of the Faroe–Shetland Basin, and in the West Shetland Basin (Figure 96). Commercial wells in the south-west Møre Basin and DSDP site 336 provide information from the northern flank of the Iceland–Faroe Ridge. For the purpose of this section, the data that comprise the Eocene succession in the Faroe–Shetland region are divided into the Balder Formation of the Moray Group (uppermost megasequence FSP-3) and the Stronsay Group (FSP-2); the Møre Basin and DSDP site 336 are described separately.

Faroe–Shetland region: Moray Group, Balder Formation (uppermost Fsp-3 megasequence)

The Balder Formation (Figures 94, 97 and 100) has been extensively mapped throughout the Faroe–Shetland Basin, though it is best developed and thickest (more than 200 m in places) in the southern part of the basin between 60° and 61°N (Smallwood and Gill, 2002; Figure 100). Whilst the base of the formation may appear conformable in the central part of the Faroe–Shetland Basin (Smallwood and Gill, 2002), it becomes increasingly unconformable to the south and east where an angular unconformity is incised into the underlying Flett and Lamba formations (Ebdon et al., 1995; Figure 98b). This surface has been interpreted as a northward-draining subaerial valley system (Lamers and Carmichael, 1999; Smallwood and Gill, 2002). The top of the Balder Formation equates to the top of sequence T50 (Figure 94), which is a prominent regional marker (Ebdon et al., 1995). This surface is interpreted to represent a maximum flooding event, with its high-amplitude reflective character attributed largely to the presence of coals, tuffs or porous sandstones (Smallwood and Gill, 2002).

Knox et al. (1997) characterised the Balder Formation as consisting predominantly of grey, variably silty and carbonaceous mudstone with abundant layers of grey-green tuff, with interbedded sandstone units and thin lignitic coal in and adjacent to the southern part of the basin (Figure 100). Tuffaceous rock is most prevalent in the lower part of the formation, which has been informally subdivided into lower B1 (tuff-rich) and upper B2 (tuff-poor) units (Figure 94). The latter is generally indicated on wireline well logs by high gamma-ray values. The occurrence of anomalous sections, such as that penetrated in well 204/28-1, which proved two units of coarse-grained sandy tuffite separated by a unit of relatively high-gamma mudstone, has been attributed to reworking (Figure 100), an interpretation supported by the absence of tuffaceous material from the upper part of the section in basinal wells (Smallwood and Gill, 2002).

Biostratigraphically, the formation as a whole is dominated by an abundance of terrestrially derived kerogen and the predominance of miospores including acme occurrences of *Inapaturopollenites* spp. and *Caryapollenites simplex* at the top and within the formation, respectively (Ebdon et al., 1995; Knox et al., 1997). Marine fossils are sparse but include a downward incoming of the diatom *Coscinodiscus* spp. 1 and 2 at the top of the section, whilst the dinoflagellate cyst *Deflandrea oebisfeldensis* displays an acme occurrence within the formation, and *Ceratiopsis wardenensis* records a downward influx. These biomarkers are indicative of a Ypresian (Early Eocene) age.

Seismic reflections within the lower part of the Balder Formation onlap the incised basal surface (Figure 98), recording the transgressive infill of the pre-existing subaerial valley network (Ebdon et al., 1995; Smallwood and Gill, 2002). The build-out of northward-prograding deltaic systems across this surface is indicated by the interbedded clastic succession in the southern part of the report area (Figure 100).

BGS borehole 90/03 (Figure 100) proved a 91.5 m thick section of mudstone and sandstone with lignite present both in thin bands and as derived clasts (Hitchen et al., 1995a). The sandstone units are dark grey to black, poorly sorted, slightly gravelly and muddy, and are predominantly volcaniclastic in composition. They display upward-coarsening units and reworking is evident throughout the section incorporating mud rip-up clasts, contemporaneously derived organic, plant and volcanic debris, and a Late Cretaceous palynoflora. The sediments preserve a rich assemblage of terrestrially derived palynomorphs including *Retitricolpites retiformis, Inapaturopollenites hiatus, Intratriporopollenites microreticulatus/pseudoinstructus, Alnipollenites verus, Caryapollenites circulus, Triatriopollenites subtriangulus, Laevigatosporites haardti* and *Tricolpites hians*. Together with the sedimentology, the composition of the palynoflora is consistent with deposition in a deltaic setting.

Well 204/27-1 penetrated one of the valleys in the base-Balder surface over which the deltas migrated and proved an 112 m thick channel-fill deposit of pebbly sandstone overlain by tuffite, lignite and tuffaceous mudstone (Robinson, 2004; Figure 100). The sandstone is fine to medium-grained, poorly sorted, with common lithic clasts. Well 204/22-1, also located in a channel albeit in a more basinal location, consists of a tuffaceous-rich unit with sporadic thin sandstone and mudstone beds (Figure 101). In both of these wells, the degree of reworking of the tuffaceous material remains uncertain. In the interfluve areas between the channels, the Balder Formation is commonly thinner and more argillaceous. Well 204/23-1 revealed a 60 m thick sequence of mudstone overlying tuffaceous mudstone with thin interbedded lignite (Figure 100).

On the eastern margin of the Faroe Platform, yellowish brown, yellowish grey or greenish grey tuff has been

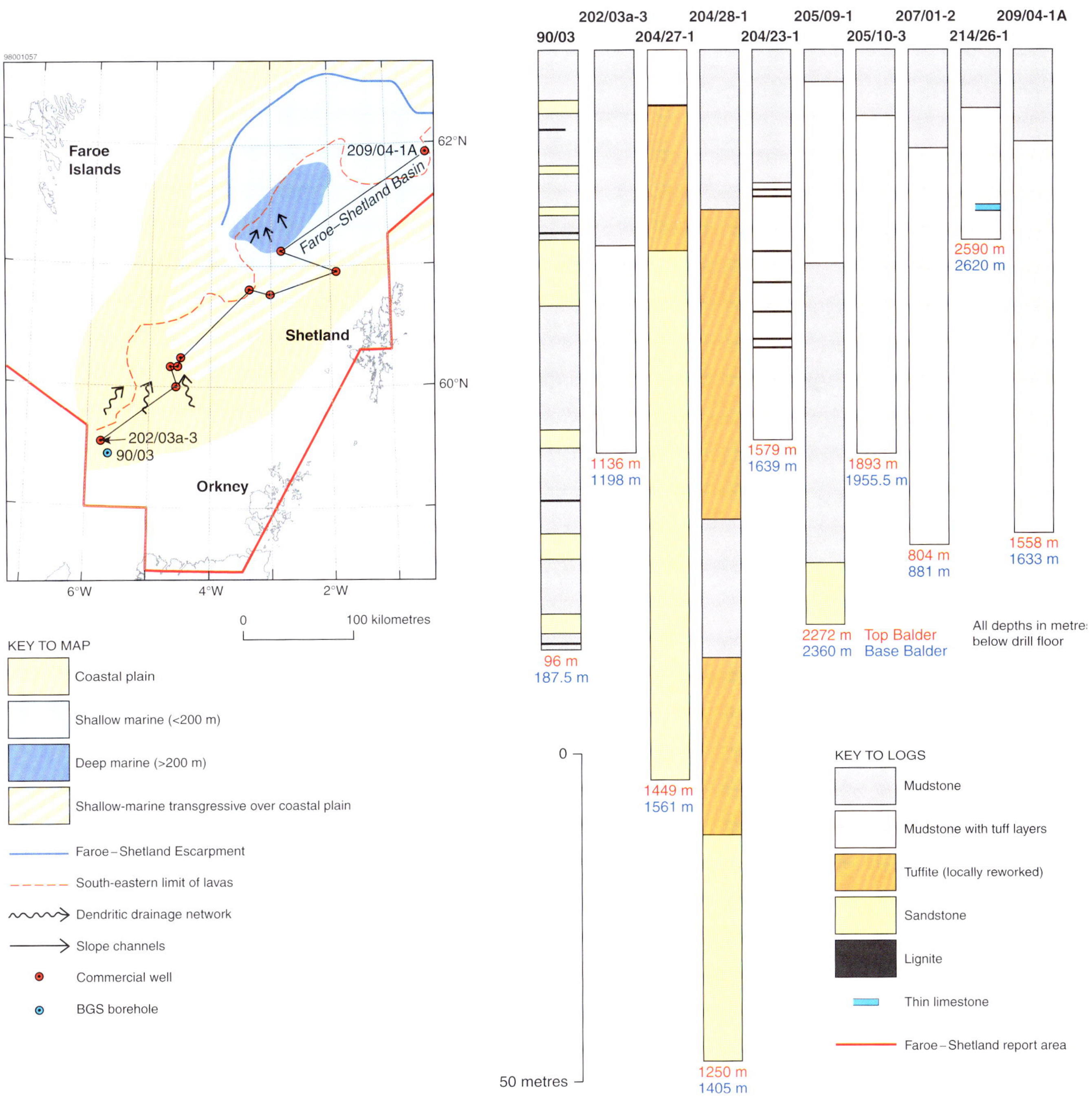

Figure 100 Selected BGS borehole and commercial wells showing lithological variation within the Balder Formation. All depths are to top (red) and base (blue) of Balder Formation. Note that depths in BGS borehole 90/03 are below sea bed. Inset map shows gross palaeogeographical reconstruction during deposition of the Balder Formation (earliest Eocene; sequence T50 time): widespread emergence at the beginning of Balder Formation deposition was followed by flooding of the coastal plain at the end of this interval. Well logs and reconstruction are based on information derived from Ebdon et al., (1995), Hitchen et al., (1995a), Lamers and Carmichael, (1999), Smallwood and Gill, (2002) and Robinson, (2004).

been collected in dredge hauls 145, 157, 158 and 161 (Figure 96) recovered from in situ, easterly dipping, outcrops (Waagstein and Heilmann-Clausen, 1995). The tuff from sites 145, 157 and 158 is largely nonmarine (fluvial or paralic) with spores, pollen, plant fragments and lignite. In contrast, sample 161 recovered both nonmarine and bioturbated marine tuff, from which the dinoflagellate cyst assemblage, including *Hystrichosphaeridium tubiferum*, *Paralecaniella indentata*, *Eatonicysta ursulae* subsp. *ursulae*, *Deflandrea phosphoritica* and common representatives of the *Apectodinium homomorphum* group suggest a latest Paleocene to Early Eocene (mid Ypresian) age.

The edge of the delta front is envisaged to have been located at about 61°N where the Balder Formation becomes increasingly dominated by mudstone (Knox et al., 1997; Lamers and Carmichael, 1999; Figure 100). Basal sandstone overlain by a predominance of mudstone

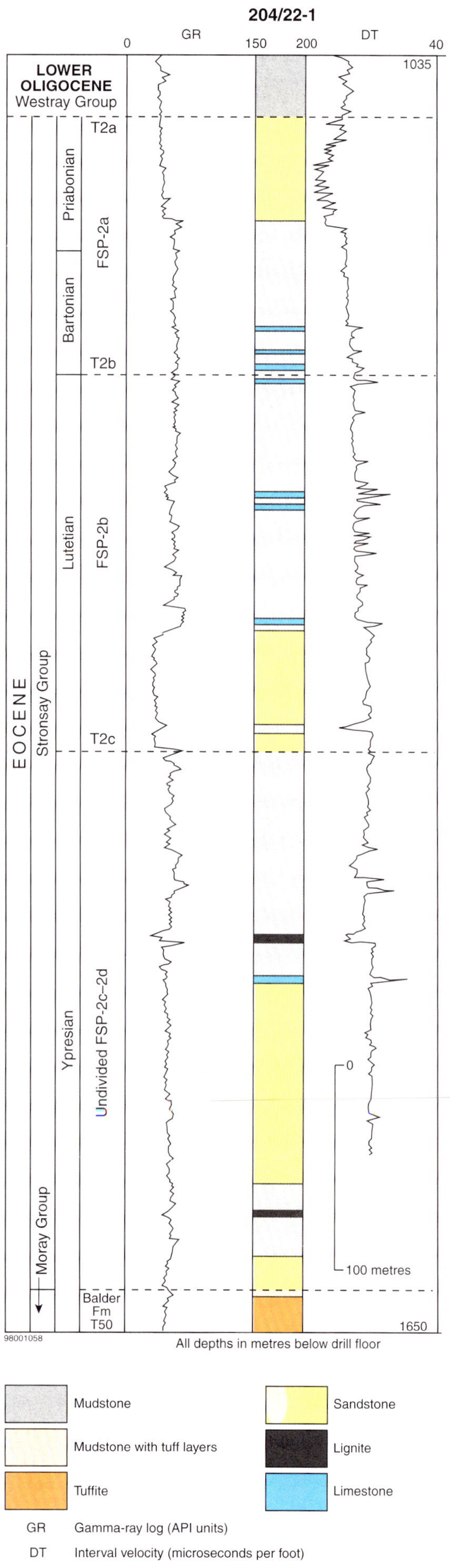

Figure 101 Simplified well log summary of the Eocene succession within well 204/22-1 from the southern margin of the Faroe–Shetland Basin. Seismic-stratigraphical context of well shown on Figure 95. For location, see Figure 96.

and tuffaceous mudstone comprised the 88 m thick section of Balder Formation in well 205/09-1 at the northern edge of the coastal plain. The biostratigraphical signature of these sediments indicates the influence of both freshwater and restricted marine conditions (Robinson, 2004). A mixed environmental assemblage is also noted from wells 208/19-1 and 214/27-1 (Figure 96), which penetrated about 75 and 74 m, respectively, of mudstone and tuffaceous mudstone containing the marine dinoflagellate cysts *Deflandrea oebisfeldensis* and *Hystrichosphaeridium tubiferum* and planktonic microfossil *Coscinodiscus* together with abundant *Taxodiaceaepollenites hiatus*, a pollen derived from intertidal *Taxodium* swamps (Knox et al., 1997; Mudge and Bujak, 2001). In contrast, well 214/26-1 (Figures 100), which proved a 30 m thick sequence of tuffaceous mudstone, mudstone and thin limestone, is reported to record a consistent upper bathyal (outer shelf to upper slope) environment of deposition (Robinson, 2004). Similar conditions are also recorded in wells 205/10-3 and 207/01-2 (Figure 100) from the upper part of the Balder Formation, consistent with eventual marine flooding of the delta system (Ebdon et al., 1995). Reworking processes are envisaged within the marine environment, and it has been suggested that relatively coarse-grained tuffaceous rock reported from well 209/04-1A (Figure 100) may be of local derivation (Knox et al., 1997).

Faroe–Shetland region: Stronsay Group (Fsp-2 Megasequence)

The sediment of the Stronsay Group (Figure 94) marks a return to deeper water conditions. Although coal continued to develop within the basal part of the group on the margins of the Faroe–Shetland Basin during marine transgression, palynomorphs and 250 m high prograding clinoforms within the Ypresian section imply an overall deepening of the basinal area (Knox et al., 1997; Smallwood and Gill, 2002). Knox et al. (1997) described a gross south-to-north facies change within the Stronsay Group of the Faroe–Shetland region, from sandstone-dominated along the prograding southern margin of the basin to a mudstone-dominated deeper basinal area north of 61°N. This simplified representation of the facies belts is complicated in the Middle Eocene section by the incursion of thick, deep-water, basin-floor sandstones as far north as 62°N (Brooks et al., 2001; Davies et al., 2004). Indeed, deposition throughout the Eocene was punctuated by a number of local and regional tectonic events linked to compression and uplift (Waagstein and Heilmann-Clausen, 1995; Andersen et al., 2000; Sørensen, 2003; Robinson, 2004; Robinson et al., 2004; Smallwood, 2004), the effects of which are manifest in the subdivision of the Eocene succession into four unconformity-bounded seismic-stratigraphical units (Figures 89 and 94). Thus, for the purposes of this account, the Eocene is described on the basis of the seismic-stratigraphical division of the Stronsay Group. The four units (2a–2d) of the FSP-2 megasequence (Table 5 and Figure 94) have been provisionally applied to the succession in the Faroe–Shetland Basin.

This stratigraphical scheme is best developed in the southern part of the Faroe–Shetland Basin, where BGS borehole 99/03 and well 204/22-1 provide useful reference sections for the calibration of the seismic data (Figures 95 and 101). Whilst a significant thickness of Eocene strata is preserved in the Faroe Bank Channel Basin (Figure 88b), this remains largely undivided, although a basal sequence with comparable seismic character to the Balder Formation together with the occurrence of a possible 'Middle Eocene' unconformity have been reported (Sørensen, 2003; Keser Neish and Ziska, 2005). The basinal sediments are locally intensely deformed by complex polygonal fault systems (Davies et al., 1999; Davies and Cartwright, 2002 see Chapter 11; Figure 102). North of the Faroe Platform, the Eocene is interpreted as a transgressive systems tract though no subdivision of the sequence has been undertaken (Nielsen and van Weering, 1998).

Eocene units FSP-2c and FSP-2d are predominantly of Ypresian age, with FSP-2c possibly extending into the earliest Lutetian (Figure 94). They are separated from younger strata by an angular unconformity that is overlain by sediments of early Lutetian (NP14 to 15) age belonging to unit FSP-2b (Figure 95). In the southern part of the Faroe–Shetland Basin, well 204/22-1 penetrated an undivided section of units FSP-2c and 2d, and proved a 260 m thick Ypresian succession of interbedded sandstone, mudstone, thin lignite and occasional limestone (Figure 101). The sandstone is typically fine to medium-grained, quartzose, with scattered mudstone intraclasts, and commonly glauconitic; the mudstone is carbonaceous, varies from pale olive-grey to blue-grey, and contains tuffaceous fragments near the base of the section. Comparable sequences are recorded in wells such as 202/03-1A, -2, 204/23-1, 204/28-1 and 204/29-1 (Figure 96) where distinctive units of high-gamma ray, green to blue-green mudstone are interbedded with sharply defined units of low-gamma ray sandstone with local lignite (Knox et al., 1997; Robinson, 2004).

BGS borehole 99/03, located about 20 km north of well 204/22-1 (Figure 96), cored 23.8 m of fine to medium-grained, greenish grey, calcite cemented, quartzose sandstone from unit FSP-2d (Figure 95). The palynomorph assemblage in this borehole proved a mixed marine and terrestrial flora typical of a deltaic/shelf margin environment, and include the dinoflagellate cysts *Charlesdowania coleothrypta, Dracodinium vareilongitudum, Heteraulacacysta leptaela* and *Pthanoperidium echinatum* which are indicative of a Ypresian (NP12) age. This dating is supported by the foraminifera *Cancris subconicus, Vaginulian decorats* and *Osangularia expansa*, and the ostracods *Leguminocythereis biocostata* and *Trachyleberis spiniferrima*. Farther west, BGS borehole 99/06 proved 14 m of dark greenish grey, Lower Eocene mudstone. The green colour of the sediment may be due to chlorite derived from the tuffaceous content. The dinoflagellate cysts *Dracodinium condylos, Heteraulacacysta leptaela, Achilleodinium biformoides, Glaphyrocysta ordinata, Wetzeliella meckalfeldensis* and *Wetzeliella samlandica* imply a mid Ypresian (intra-NP12 biozone) age. In both boreholes, the biostratigraphical age is further supported by the absence of *Apectodinium*, which confirms that these sediments are younger than the NP10 biozone.

Farther north along the basin margin, well 205/20-1 (Figure 96) drilled very fine to coarse-grained sandstone throughout the FSP-2c to 2d section, with traces of limestone and lignite beds near the base of the sandstone (within unit FSP-2d) (Robinson, 2004). Early Eocene forms dominate the biostratigraphical assemblage of pollen and dinoflagellate cysts, though the occurrence of *Wetzeliella ovalis* and *Wetzeliella homomorpha* may indicate that part of the sand-rich succession extends into the Middle Eocene. A deltaic setting with restricted marine influence is envisaged.

In the central part of the Faroe–Shetland Basin and on the Rona High, wells 205/09-1, 205/10-2 and 206/09-2 (Figure 96) proved marine siltstone, mudstone and sporadic limestone from strata equivalent to unit FSP-2d (Knox et al., 1997; Robinson, 2004; Robinson et al., 2004). The occurrence of biomarkers, such as *Eatonicysta ursulae, Cibicidoides gr. eocaenus* and *Spiroplectammina navoarroanna* indicate an Early Eocene age, and an outer shelf environment of deposition. According to Robinson (2004), this Lower Eocene section is pervasively cut by polygonal faults. In well 205/09-1, a sand-dominated section with interbeds of siltstone and mudstone that comprise unit FSP-2c overlies this argillaceous sequence. The sandstone exceeds 400 m in thickness, is generally fine to medium-grained, moderate to well-sorted, and is commonly pale, occasionally pinkish or yellowish. A diverse microfossil assemblage that includes *Pseudohastigerina micra, Truncorotaloides cf. rohri, Neoeponides karsteni* and *Cyclammina amplectens* together with occurrences of *Cenosphaera* spp. and *Rhabdammina robusta* imply an early Lutetian age (Robinson, 2004). Deposition occurred in a marine outer shelf to upper bathyal setting.

In the northern, deeper water part of the basin, units FSP-2c and 2d are generally thinner (a combined thickness of mostly less than 200 m) and are dominated by outer shelf to bathyal mudstone in wells such as 208/15-1A, 214/28-1, 214/26-1 and 214/04-1 (Knox et al., 1997; Figures 96 and 103). According to Knox et al. (1997), the mudstone is typically pale to grey, grading to grey-green, but with red-brown and variegated tuffaceous mudstone near the base of the Eocene section. Sandier units between 20 and 100 m thick are reported in wells such as 214/17-1 and 213/23-1 (Robinson, 2004; Figure 103).

Tuffaceous limestone and phosphatic sediment have been inferred to crop out on the western flank of the Faroe–Shetland Basin, based on the composition of glacial erratics recovered in dredge samples from the modern-day Faroe Shelf (Waagstein and Heilmann-Clausen, 1995). The tuffaceous sediment is matrix supported with recrystallised calcium carbonate, displays evidence of bioturbation and carrying pellets of collophane, a yellowish isotropic phosphate. Marine dinoflagellate cysts reveal a range of dates from Ypresian to Lutetian. Whilst these sediments broadly correlate with units FSP-2c and 2d, some of the younger samples (late Lutetian) described by Waagstein and Heilmann-Clausen (1995) in this suite of sediments are time equivalents of the overlying Eocene unit FSP-2b.

The top of unit FSP-2c was locally eroded prior to the deposition of the younger Eocene strata. In the southern part of the Faroe–Shetland Basin, reflector T2c and the underlying strata may have been folded and eroded prior to the deposition of the overlying basin-floor fans and shelf margin clinoforms of Eocene unit FSP-2b (see below), which prograded and downlapped onto T2c (Figure 95). This tectonic movement

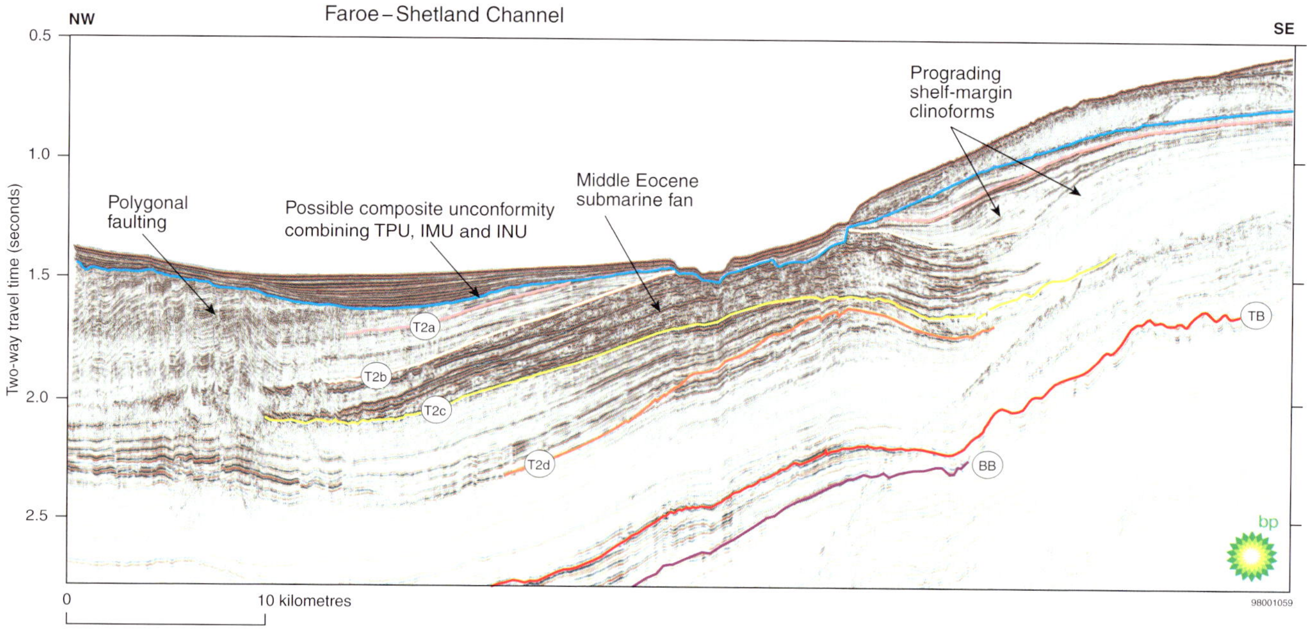

Figure 102 Seismic profile showing the expression of shelf-margin and submarine fan deposits of the Stronsay Group, bounded by the Top Balder Formation (TB) and Top Stronsay Group (T2a) unconformities. Polygonal faulting in the axial region of the channel disrupts the bedding of the Eocene and Oligocene (Westray Group) strata, the latter occurring between the T2a Unconformity and the composite Top Palaeogene (TPU), intra-Miocene (IMU) and Intra-Neogene (Early Pliocene) (INU) unconformities. Abbreviations: BB=Base Balder Formation. For location of profile, see Figure 103. Seismic data courtesy of BP.

may be linked to incipient inversion of the 'Judd Anticline' (Sørensen, 2003; Robinson, 2004). The erosion is attributed here to submarine processes, perhaps in advance of the overlying submarine fan development. Submarine erosion is also recorded from the central part of the basin where the sandstone package in well 205/09-1 has been incised by a broad channel, part of a broader north to north-east shelf-parallel channel network that has been imaged between blocks 205/13 and 205/05 (Robinson et al., 2004). The channel system is up to 2.5 km wide, and has incised between 80 and 200 m into the underlying sandstone package. At its northern end, the channel turns abruptly downslope to the west. This feature is interpreted as a submarine channel that has a deeper incised canyon at its northern end. The trigger mechanism for channel formation has been postulated to be due either to a lowering in relative sea level over the shelf or that it was controlled by uplift or faulting of the Flett High (Robinson et al., 2004; Figure 96).

Eocene unit FSP-2b (Figure 94) records an expansion in the area of marine influence in the Faroe–Shetland region coupled with a major change in the style of deep-water sedimentation, expressed by several areas of focused fan-sand deposition within the eastern half of the Faroe–Shetland Basin. In the southern part of the basin, well 204/22-1 penetrated a 63 m thick basal sandstone, medium to coarse-grained, well-sorted and quartzose with thin mudstone interbeds near the base, overlain by a 120 m thick section of olive grey to pale brown mudstone grading to siltstone, with sporadic thin brown limestone (Figure 101). The mudstone preserves a diverse and abundant marine microfauna, including the benthic foraminifera *Planulina costata*, the planktonic species *Truncorotaloides pseudodubia* and *Globigerinatheka index*, and the ostracod *Leguminocythereis oertlii*, all of which indicate a Lutetian (Mid Eocene) age.

The change to a predominantly marine environment in this area is consistent with the general seismic architecture of the margin of the basin that shows a prograding sequence of shelf margin clinoforms sourced from the Orkney–Shetland High interdigitating with a basin-floor fan (Figures 95 and 102). The fan may be up to 100 km long and 50 km wide, and may attain a thickness of 200 m. In common with Middle Eocene fan development farther north (Figure 103), the basal FSP-2b sandstone in well 204/22-1 may form part of the fan complex. The sporadic occurrence throughout this section of noncalcareous agglutinating foraminiferids of the *Rhabdammina* biofacies (cf. Gradstein and Berggren, 1981), including *Bathysiphon cocenicus* and *Cyclammina* gr. *placenta* implies that water circulation in the basin remained restricted despite marine expansion.

A Lutetian age for FSP-2b is also supported by BGS borehole 99/03, which cored a mud-dominated basinal section (Figure 95). The borehole proved a 79.7 m thick sequence of dark greenish grey marine mudstone with sporadic thin, intermittently graded, interbeds of quartzose sandstone. The unconformable base of the section is marked by a rubbly lag that is characterised by abundant agglutinating foraminifera, sharks teeth and glauconite. Marine palynomorphs are characteristic of the early to mid Lutetian, with key biomarkers such as the dinoflagellate cysts *Areoligera senonensis*, *Pthanoperidinium comatum* and *Selenopemphix coronata* indicative of the NP14 to 15 biozones. This is supported by the occurrence of the diatom *Coscinodiscus* sp. 1, the radiolarian *Cenosphaera*, and the foraminifera *Uvigerina germanica* and *Pullenia osloensis*.

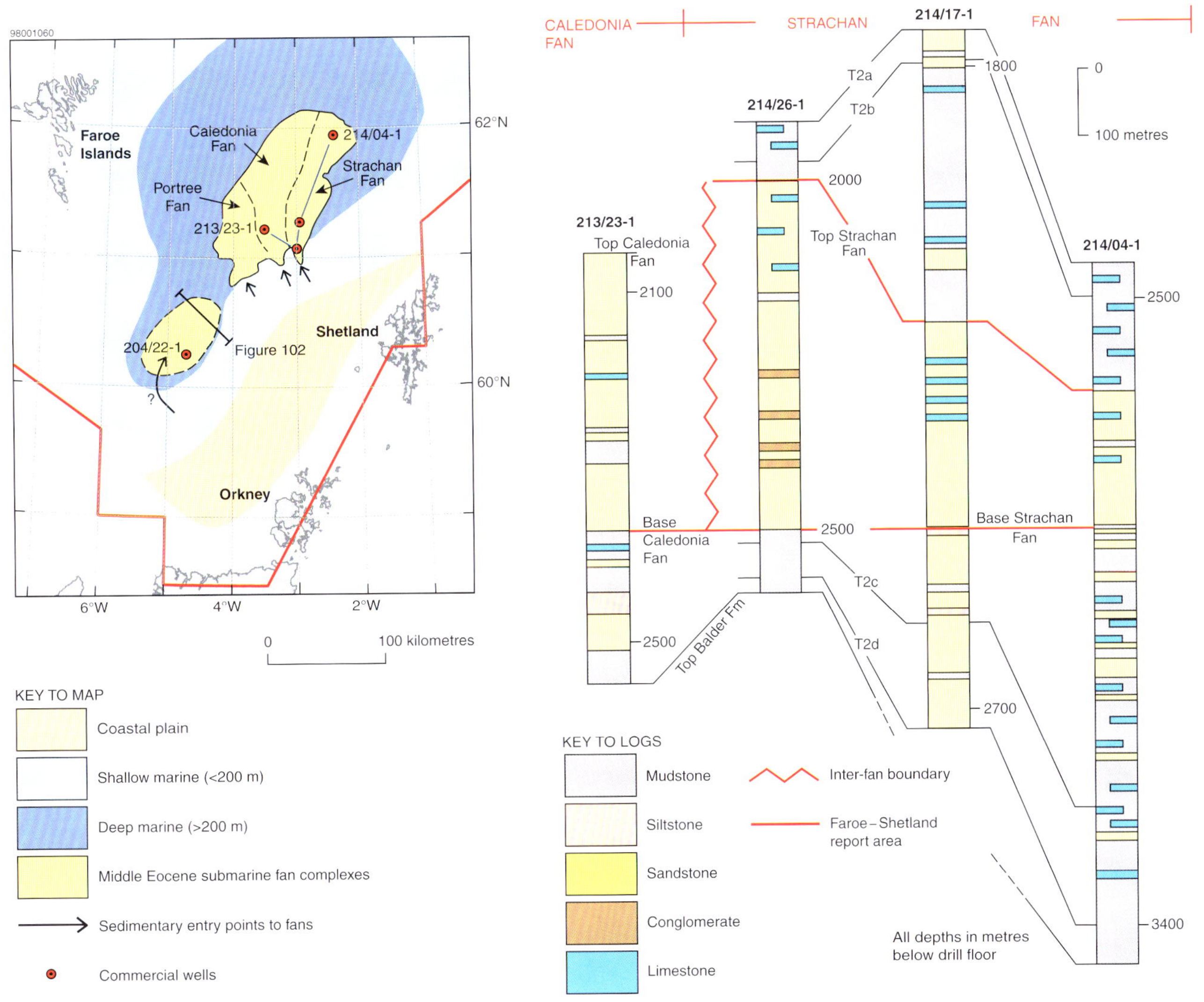

Figure 103 Selected well logs showing the nature of the Stronsay Group sediments in the areas of the Middle Eocene submarine fan complexes within the Faroe–Shetland Basin (modified from Robinson, 2004). Inset map shows gross palaeogeographical reconstruction during Middle Eocene fan development, based on information derived from Brooks et al., (2001) and Robinson, (2004). Abbreviations: see Figure 94. For well log of 204/22-1, see Figure 101.

Middle Eocene shelf margin progradation sourced from the West Shetland High is reported in the area of wells 205/10-2 and 205/09-1 (Robinson, 2004; Robinson et al., 2004; Figure 96). The more landward of these wells, 205/10-2, records a high proportion of sand that contrasts with siltstone proved in the more basinal well 205/09-1. The former penetrated the topsets and foresets of the prograding shelf-margin, whilst the siltstone of the latter probably represents their distal basinal equivalent (e.g. bottomset) (Robinson, 2004). Biostratigraphical information from the siltstone is indicative of an outer shelf to upper bathyal setting, with water depths up to a maximum of 400 m at this location.

The most significant development of Middle Eocene fans occurs between 61° and 62°N, where a series of overlapping, linear basin-floor fans can be traced across the northern part of the Faroe–Shetland Basin (Figure 31 and 103). Three main submarine canyon entry points that fed the fan complex have been identified on the eastern edge of the palaeoshelf, and the fan deposits have flowed northwards across the basin floor (Brooks et al., 2001). The individual components of the fan complex have been informally termed the Strachan, Caledonia and Portree fans by Davies et al. (2004) (Figure 103). The fan complex is more than 100 km long, exceeds 50 km in width, and attains a maximum thickness of 550 m (Brooks et al., 2001). The internal geometry of the fans is complex, with zones of high-amplitude layered seismic reflectors alternating with zones of more chaotic reflectors. Internal unconformities are common. The more proximal, confined, southern portions of both the Strachan and Caledonia fans have erosional bases, which contrasts with their unconfined northern ends, located beyond the toe-of-slope, where the bases are not erosional (Davies et al., 2004). On the basis of clinoform geometries along the eastern margin of the basin, water depths are envisaged to have been between 100 m and 1 km (Robinson, 2004).

The Strachan and Caledonia fans have been tested by wells 213/23-1, 214/04-1, 214/17-1 and 214/26-1, which revealed their sand-rich nature (Figure 103).

Individual sandstone beds range from a few metres to tens of metres in thickness locally separated by siltstone or mudstone units up to 10 m thick (Robinson, 2004). Generally, the sand-rich section is encased in mudstone, though in well 214/17-1 sandstone and interbedded siltstone and mudstone form a major component of unit FSP-2b below the base of the Strachan Fan. The fan sandstones are commonly medium to coarse-grained, clean, translucent, pale orange or pale yellow, though sometimes reddish, pinkish and greenish in colour. They are quartz dominated with feldspathic and lithic components, and are commonly glauconitic. Although they display moderate to poor sorting with sporadic mudstone intraclasts, there is generally good to moderate porosity throughout the sandstone. In well 214/26-1, sporadic beds of conglomerate, up to 5 m thick, occur towards the base of the Strachan Fan. This well occurs at the narrow proximal end of the fan (Figure 103), and was drilled through an incised feature that may form part of the feeder canyon (Robinson, 2004). Thin calcareous bands grading to sandy limestone occur in all wells. Positive relief on the top of all the fans is interpreted as depositional in origin, and represents mudstone draping of the lobes (Davies et al., 2004).

Incision into the top of the FSP-2b shelf margin sediments by a system of braided and meandering channels trending orthogonal to the shelf edge, prior to the deposition of Eocene unit FSP-2a (Figure 94) implies a relative fall in sea level in the late Mid Eocene, though it is uncertain whether this relates to eustacy, changes in sediment supply or uplift of the Flett High (Robinson et al., 2004). However, this relative sea level fall was short lived. The Middle Eocene submarine fan is downlapped by shelf margin clinoforms that indicate a renewed build-out of the Orkney–Shetland and West Shetland highs (Figure 102). This prograding sediment wedge together with basinal strata that onlap the fan comprise Eocene unit FSP-2a in the southern part of the Faroe–Shetland Basin. Thin basinal sediments characterise FSP-2a in the northern, deep-water part of the basin, but Upper Eocene strata are missing from the Faroe Platform where Oligocene sediments unconformably overlie sediments equivalent to Eocene unit FSP-2b (Waagstein and Heilmann-Clausen, 1995). The occurrence of reworked Mid to Late Eocene dinoflagellate cysts within the Oligocene sandstones suggests that they were formerly present in this area, but may have been removed as a result of late Mid to Late Eocene uplift of the Faroe Platform (Waagstein and Heilmann-Clausen, 1995; Andersen et al., 2000).

In the southern part of the Faroe–Shetland Basin, the FSP-2a shelf margin succession was tested by well 204/22-1, which drilled a 77 m thick section of pale brown, glauconitic, mudstone with sporadic thin sandy limestone beds, overlain by 48 m of medium to coarse-grained sandstone (Figure 101). The occurrence of *Nummulites* gr. *prestwichianus* in the sandstone, and the benthic foraminifera *Arenobulimina* sp., *Lamarckina halkyardi* and *Cibicides truncanus*, together with planktonic forms, such as *Globigerinatheka index* and *Turborotalia cerroazuluensis pomeroli*, common in the mudstone, suggests a Mid to Late Eocene age for FSP-2a (Figure 94). A Priabonian (Late Eocene) age is assigned to a 7 m thick section of dark yellow-brown to olive- grey/dark olive-grey, basinal mudstone cored in BGS borehole 99/03 (Figure 95), which included the foraminifera *Cassidulina carapitana*, *Gyroidina girardana* and *Reticulophragmium amplectens*. The upper part of the FSP-2a unit has been eroded at this site. Although the *Rhabdammina* biofacies is present both in well 204/22-1 and borehole 99/03, the presence of *Nummulites* in the former suggests an increasing oceanic influence along the Atlantic margin.

Farther north, the FSP-2a shelf–margin sequence has built out across the incised top of unit FSP-2b, though the prograding clinoforms appear to have a lower angle of slope (<1°) relative to the Middle Eocene clinoforms (~1.5°) (Robinson et al., 2004). This may be indicative of progradation into shallower water depths and/or a lower rate of sediment supply. The rocks in well 205/09-1 are predominantly argillaceous and were deposited in an oxygenated outer-shelf setting with open-marine influence. An abundant and diverse benthonic foraminiferal assemblage, including *Cibicides westi*, *Bolivina cookie* and *Uvigerina eocaena* is indicative of a Late Eocene age.

In deeper water, the nature of the boundary between FSP-2a and 2b is unclear; the basinal seismic reflections have lower amplitude and are more discontinuous in character (Robinson, 2004). Wells 214/04-1 and 214/26-1 proved a thin (<50 m) mudstone-dominated sequence with thin limestone bands similar to the underlying FSP-2b unit, whereas sandstone and thinner mudstone was drilled in well 214/17-1 (Figure 103). Although the top of FSP-2a has been locally removed by submarine erosion during the Oligocene, the occurrence of key dinoflagellate cyst biomarkers in well 214/04-1, such as *Areosphaeridium michoudii* and *Heteraulacacysta porosa* indicates a late Mid to Late Eocene age for the upper part of unit FSP-2a in this well (Davies et al., 2002).

Møre Basin

In the Møre Basin, Lower Eocene strata equivalent to the Balder Formation are correlated with the Tare Formation, whereas the overlying strata, equivalent to the Stronsay Group, form part of the Brygge Formation of the Hordaland Group (Dalland et al., 1988; Martinsen et al., 2005). Throughout the basin, these strata are mainly composed of deep-water, slope to basin-floor, fine-grained, marine rocks, though turbidite sandstone is locally present (Martinsen et al., 1999). The bulk of the sediment was sourced from the west or south.

Towards the south-west margin of the Møre Basin, wells 219/28-1 and 219/28-2Z (Figure 96) penetrated 867 and 993 m respectively of Lower to Middle Eocene strata above which the Oligocene rests unconformably. The section consists predominantly of claystone with thin limestone and sporadic sandstone, and becomes increasingly tuffaceous towards the base. Farther east, well 220/26-1 proved 108 m of tuffaceous mudstone, most probably equivalent to the Tare/Balder formations, overlain by at least 40 m of Middle to Upper Eocene, grey-green, mudstone and thin interbedded limestone. The top of the Eocene section in this well is undefined on the composite well log. Similarly, adjacent well 220/26-2 on the north-west flank of the East Shetland High penetrated a 50 m thick section of Lower Eocene, grey, blue-green and brown mudstone overlain by undifferentiated Middle Eocene to Lower Miocene strata. To the north, well 219/20-1 is located slightly farther into the basin and proved a 925 m thick sequence of Lower to Upper Eocene sediments. The basal 45 m consists of grey silty claystone assigned on the composite well log to be equivalent to the Tare/Balder formations. The overlying sediments consist of yellow-brown and red-brown claystone becoming grey, grey-brown and grey-green higher in the section. The claystone occasionally grades into siltstone, contain specks and streaks of carbonaceous material, and is locally interbedded with thin beds of buff to pale brown limestone.

ICELAND–FAROE RIDGE: DSDP SITE 336

DSDP site 336 is located at about 800 m water depth on the northern flank of the Iceland–Faroe Ridge (Figure 96). This site penetrated to 515 m below sea bed, and proved Middle Eocene basalt, radiometrically dated by K–Ar as 43 to 40 Ma (Talwani et al., 1976), forming the basement to the ridge (Figure 104). The basalt grades into, and is overlain by, an 8 m thick unit of volcanic conglomerate (basaltic rubble), which in turn is overlain by a 13 m thick unit of red claystone. This entire section is interpreted as a ferruginous lateritic palaeosol formed by the in situ weathering of the basaltic basement (Nilsen, 1978a; Nilsen and Kerr, 1978). The palaeosol is overlain by 295 m of Middle Eocene to Upper Oligocene grey to olive marine mudstone, the bulk of which is probably Mid to Late Eocene in age. Micropalaeontological data indicate submergence of the Iceland–Faroe Ridge at site 336 from a subaerial setting during palaeosol formation to a neritic to upper bathyal (shelf to upper slope) environment during the Late Eocene (Talwani et al., 1976; Schrader et al., 1976; Berggren and Schnitker, 1983). However, the crest of the Iceland–Faroe Ridge (sited about 400 m higher than the sea bed at site 336, which itself is 463 m above the palaeosol) possibly remained above sea level (Talwani et al., 1976; Nilsen, 1978b).

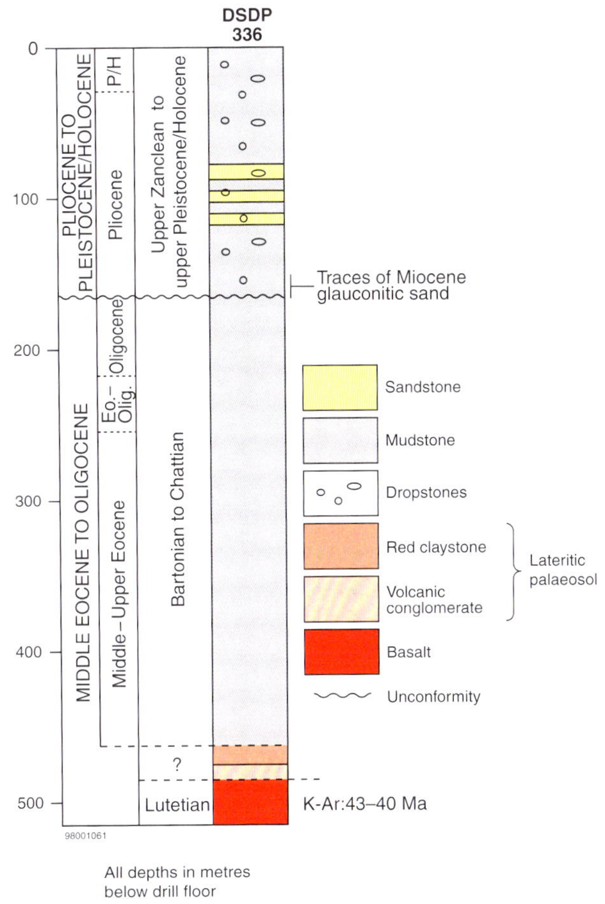

Figure 104 Simplified log of borehole DSDP 336 from the Iceland–Faroe Ridge (modified from Talwani et al., 1976). For location of borehole, see Figure 96.

Oligocene

Oligocene rocks are widely distributed throughout the Faroe–Shetland region, but are locally absent at its south-west end and in the Faroe–Wyville Thomson area (Figure 96). This distribution pattern reflects the Late Eocene to Early Oligocene inception of a single deep-water basin in the Faroe–Shetland region, which was further shaped by early Neogene compression. The latter formed or reactivated inversion structures, such as the Judd Anticline (Figures 14 and 95), from which Oligocene strata were eroded prior to the Mid Miocene. A patchy record of paralic to shallow-marine sedimentation is preserved on the shallow shelves and ridges flanking the present day Faroe–Shetland and Faroe Bank channels, a legacy of late Neogene erosion due to uplift and glaciation.

A more complete deep-water succession is preserved in the northern part of the report area that deepens into the Norwegian Basin. The Møre Basin still retained some expression as Oligocene strata are locally thin or absent over structural highs, such as the Møre Marginal High (Hjelstuen et al., 1999; Brekke, 2000; Lundin and Doré, 2002). For the most part, however, the greater Møre Basin area reflects shelf-margin progradation concomitant with the developing physiography of the northern North Sea to mid Norwegian continental margin (Martinsen et al., 1999).

The thickness of the preserved Oligocene rocks remains uncertain due to ambiguity in its recognition within, and correlation between, individual commercial wells, and its correlation to the seismic stratigraphy (cf. Stoker, 1999). In general, the Oligocene appears to thicken north-eastwards within the Faroe–Shetland region (Damuth and Olson, 1993), and may be about 250 to 300 m thick in deep-water axial areas of the Faroe–Shetland Channel. A broadly comparable thickness is inferred for the Faroe Bank Channel (assuming an interval velocity of 2.0 km/s in Figure 88b). In the Møre Basin, Oligocene strata are thickest (300 to 400 m) on the eastern margin of the basin, thinning westwards onto the Møre Marginal High (Martinsen et al., 1999). In the report area, the bulk of the available information on the Oligocene is derived from BGS boreholes and commercial wells located on the shelf and slope west and north of Shetland, and in the Faroe–Shetland Channel. Dredge samples provide information from the east Faroe Shelf, whereas DSDP site 336 presents data from the northern flank of the Iceland–Faroe Ridge. For descriptive purposes, the Westray Group (FSP-1 megasequence) is here taken to comprise the Oligocene section in the Faroe–Shetland region, including the Faroe–Wyville Thomson area; information from the Møre Basin and site 336 on the Iceland–Faroe Ridge is presented separately.

FAROE–SHETLAND REGION: WESTRAY GROUP (FSP-1 MEGASEQUENCE)

The Westray Group (Figure 94) has been extensively mapped in the Faroe–Shetland region (Knox et al., 1997; Stoker, 1999; Sørensen, 2003), though its distribution in the Faroe–Wyville Thomson area remains tentative. West of Shetland, the Westray Group is restricted to the outer part of the modern shelf, where it unconformably overlies rocks ranging from Precambrian to Eocene in age.

Its landward limit is marked by a well-defined erosional pinchout (Figure 88a). In the south of the report area, the Westray Group is preserved as an eroded, sheet-like deposit, generally <50 m thick, and locally occupying basinal hollows in the underlying surface, which is commonly onlapped by flat-lying to undulatory reflections.

Boreholes and commercial wells have proved a paralic to shallow-marine sedimentary sequence preserved on the outer West Shetland High. On the Solan Bank High, BGS borehole 77/07 (Figures 96 and 105) recovered a 13.3 m thick sequence of highly carbonaceous, fine-grained sandstone and silty mudstone with lignite beds, capped by a thin unit of hard, dark grey, homogenous mudstone and silty mudstone with thin silty and sandy partings (Evans et al., 1997). Over 30% of the sediment in parts of the core consists of very dark brown to black comminuted carbonaceous matter. The carbonaceous sequence was deposited in nonmarine, floodplain and forest swamp environments, which gave way to brackish marine conditions with the deposition of the overlying mudstone. The nonmarine portion contains a high proportion of terrestrially derived spore and pollen taxa indicative of a Late Oligocene age (Evans et al., 1997). Sporadic dinoflagellate cysts from the overlying brackish-marine strata included the *Areoligera semicirculata* biomarker, indicative of an age no younger than late Chattian.

Shallow-marine sandstone units predominate in wells located along the edge of the West Shetland and Orkney–Shetland highs e.g. 202/02-1, 202/03a-3, 206/08-1A and -2 (Figure 96), and are characteristically white to pale grey and grey-brown, silty to coarse-grained, poorly sorted, bioclastic, variably calcareous and glauconitic, with common fragments of soft, brown to black, lignite (Knox et al., 1997; Stoker, 1999). Subordinate interbedded lithologies include pale to dark grey, green and red-brown, firm to hard, silty, slightly calcareous and glauconitic mudstone, and white, hard, sandy, glauconitic limestone. The marine sequence contains a fairly rich and diverse microfossil content, including calcareous foraminiferal assemblages typical of the mid Oligocene *Rotaliatina bulimoides* bioevent.

Wells located on the present day slope reveal a transition into finer-grained lithologies. On the uppermost slope, mixed sandstone/siltstone units with thin limestone beds were proved in wells 204/28-1, 208/27-1 and 209/12-1 (Figures 96 and 105). Knox et al. (1997) have documented evidence of upward-coarsening cyclicity within the shallow-marine sediments, with siltstone to fine-grained sandstone couplets described from wells such as 202/03a-3 and 208/27-1. Farther down the present day slope, wells 208/15-1A, 209/06-1, 214/28-1 and 219/20-1 recovered predominantly mudstone, commonly grey to pale brown, soft to firm, bioclastic and variably calcareous and glauconitic. The seismic character of the marine sediments includes both acoustically layered and chaotic facies, with some evidence of progradation into the Faroe–Shetland Channel (Stoker, 1999).

East of the Faroe Islands, dredge hauls 154 and 155 of Waagstein and Heilmann-Clausman (1995) (Figure 96) sampled essentially in situ exposures, and proved feldspathic and lithic marine sandstones of Oligocene age associated with a 250 m thick section of easterly dipping rocks. The sandstone is fine-grained, well-sorted, poorly cemented and bioturbated. Abundant *Palaeophycus tubularis* burrows observed in sandstone from dredge 155 are indicative of a high-energy, oxygen-rich environment of deposition. The mineral composition of the sandstone indicates derivation from basalt to the west. Palynological analysis of the sandstone revealed a predominance of Early Oligocene (Rupelian) species, including the dinoflagellate cysts *Thalassiphora fenestrata* and *Reticulatosphaera 'pseudoursulae'*. The sediment also contained reworked dinoflagellate cysts of Early and Mid Eocene age, which have been interpreted to indicate the presence of an unconformity separating the Oligocene and Lower to Middle Eocene strata in this area (Waagstein and Heilmann-Clausman, 1995).

Little is known of the Westray Group sediments in the deeper waters of the Faroe–Shetland region. On seismic profiles they are generally acoustically layered albeit commonly disrupted in reflector continuity by intense polygonal faulting (Davies et al., 1999; Davies and Cartwright, 2002; Figure 102). The basinal Oligocene sediments are interpreted to unconformably overlie Eocene strata (Davies and Cartwright, 2002), with the top of the unit marked by a widespread angular unconformity, the Top Palaeogene Unconformity, which in places is highly irregular in form (Stoker, 1999; Andersen et al., 2000; Stoker et al., 2005b). The deep-water stratigraphy has been tested by well 214/04-1 (Figure 96), which proved about 250 m of silty mudstone rich in biogenic silica (diatoms) and carbonate (foraminifera), described as a biosiliceous ooze by Davies et al. (2001). An Oligocene age is indicated by the occurrence of a calcareous benthonic foraminiferal and dinoflagellate cysts assemblage that includes *Labrospira scitulus, Areoligera semicirculata, Gyroidina soldani, Elphidium latidorsatum* and *Turrilina alsatica* (Davies and Cartwright, 2002). The diatom assemblage contains between 50 and 70% of nearshore benthic genera, such as *Sceptroneis* and *Paralia*, the incorporation of which into deep-water sediments is attributed to reworking by bottom currents (Davies et al., 2001).

A consequence of the high biogenic silica fraction within the basinal sediments is the later (Mid to Late Miocene) diagenetic transformation of the biogenic silica (Opal A) to Opal C/T, which resulted in the formation of a porcellanite (Davies and Cartwright, 2002) known as a BSR that appears to crosscut stratal reflections (Figure 26). At well site 214/04-1, the porcellanite lies within sediments of the Westray Group.

Farther south, a 305 m thick sequence of undifferentiated Oligocene claystone with occasional thin sandstone interbeds is recorded from well 213/23-1 (Figure 96). Oligocene sediments have also been interpreted to occur in the synclines between the Ymir, Wyville Thomson and Munkagrunnur ridges (BGS, 2007; Figure 96). Between 350 and 400 m of Oligocene sediment has been inferred from the latter syncline, which forms the modern Faroe Bank Channel (Sørensen, 2003; Stoker et al., 2005c; Figure 88b).

Møre Basin

In the Møre Basin, Oligocene strata form part of the Brygge Formation of the Hordaland Group (Dalland et al., 1988). The strata are mainly composed of outer shelf to basin-floor, fine-grained, marine rocks, including turbidite sandstone (Martinsen et al., 1999). The bulk of the sediment was sourced from the east.

In the report area, well 219/20-1 (Figure 96) penetrated 235 m of Lower to Upper Oligocene rocks overlying Eocene strata (Stoker, 1999). The Oligocene rocks consist predominantly of pale grey, slightly silty, claystone

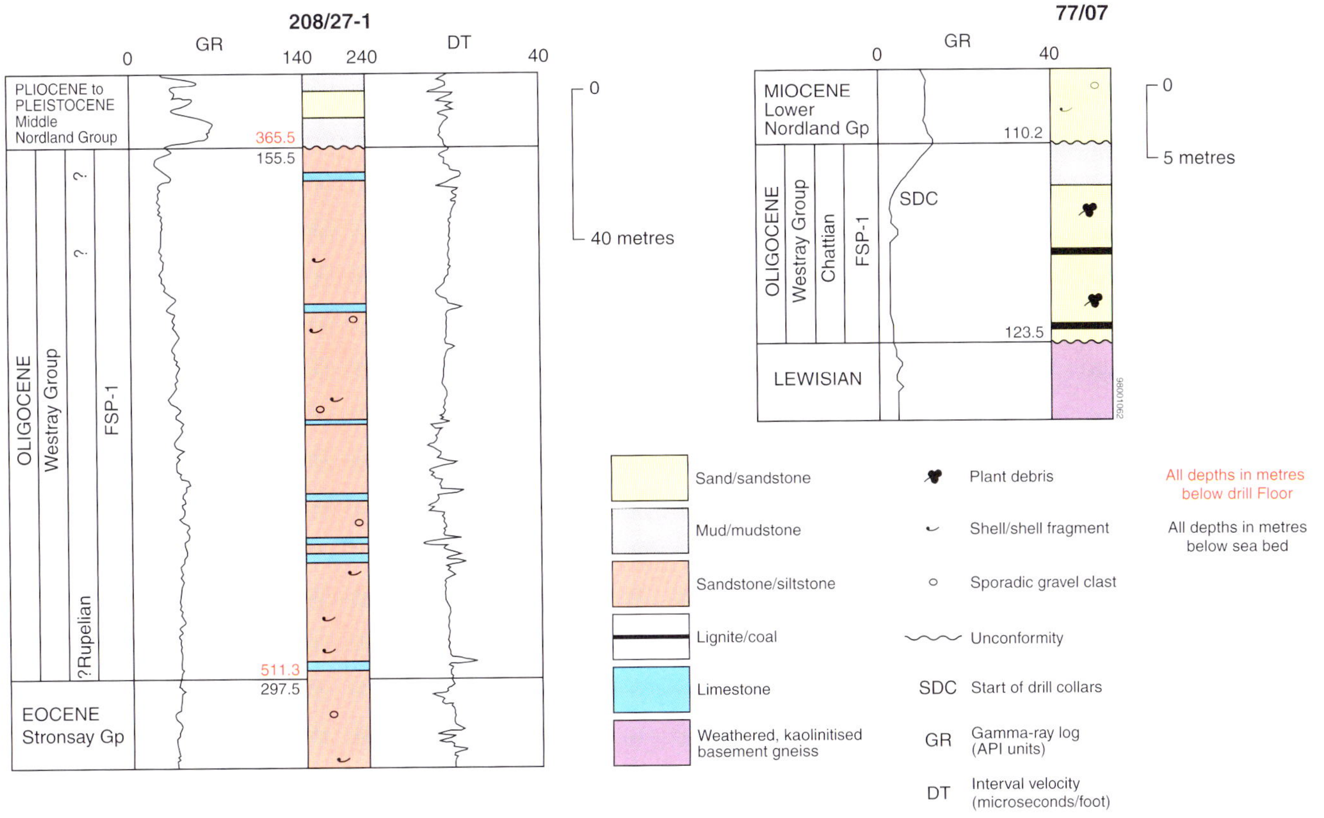

Figure 105 Simplified well logs of Oligocene strata in well 208/27-1 and BGS borehole 77/07 from the West Shetland Shelf and upper slope (modified from Stoker, 1999). For location of the well and borehole, see Figure 96.

with sporadic thin interbeds of white to grey-brown limestone. Rare coal bands set within a sandy interval occur near the top of the Oligocene section, though the latter is marked by an angular unconformity overlain by Plio-Pleistocene deposits. On the flank of the basin, well 219/28-1 proved 110 m of Lower to Upper Oligocene claystone with interbedded limestone and sandstone. The base and top of the Oligocene section in this well are both marked by unconformities. Rocks in adjacent wells (e.g. 220/26-1, 220/26-2) are marked as undifferentiated Eocene to Neogene on composite well logs.

Iceland-Faroe Ridge: DSDP Site 336

DSDP site 336 proved at least 48 m of Oligocene rocks underlain by a 38 m thick section of undifferentiated Eocene–Oligocene age (Talwani et al., 1976; Figure 104). The Oligocene sediments consist of grey to olive-grey, bioturbated, stiff, marine mudstone with increased amounts of biosiliceous material and glauconite relative to the underlying Eocene section. A disparate, Oligocene, microfossil assemblage at site 336 contrasts markedly with rich and diverse nannofossil and foraminiferal assemblages at DSDP site 552 (63° 38.97' N, 12° 28.26' W) on the southern flank of the Iceland–Faroe Ridge. This suggests that the deep-water basins north and south of the Iceland–Faroe Ridge remained unconnected during the Oligocene, and the ridge itself may still have been largely emergent (Talwani et al., 1976; Figure 91). This is consistent with the sequence-stratigraphical interpretation of the Oligocene section preserved on the north Faroe Slope, to the east of DSDP 336, where a shelf-margin systems tract prograding into the Norwegian basin is overlain by a transgressive systems tract (Nielsen and van Weering, 1998).

NEOGENE

Neogene strata are widespread throughout much of the Faroe–Shetland region, being absent mainly on the Faroe Shelf and adjacent highs and ridges (Figure 87). The total thickness of the Neogene succession locally exceeds 1.4 s TWTT in the northern part of the report area, on the North Sea Fan (Stoker et al., 2005a). This is equivalent to a maximum estimated thickness of 1.4 km based on an interval velocity of 2.0 km/sec, though a range of 1.5 to 2.0 km/sec (1.05 to 1.4 km) is more generally applicable. In the southern part of the area, the Neogene succession is generally less than 500 ms TWTT in thickness.

The bulk of the Neogene succession is assigned to the two regional megasequences of FSN-2 (Miocene to Lower Pliocene) and FSN-1 (Lower Pliocene to Holocene) (Figure 94), whose distribution and thickness are shown in Figures 106 and 107, respectively. The available lithological and age data for these megasequences are summarised in Figure 108 and Table 7. The age data are based mainly on biostratigraphical information, though strontium isotope data are also used for some of the Miocene sediments. On the West Shetland margin, the Pleistocene to Holocene (Quaternary) component of the FSN-1 megasequence has previously been subdivided into a number of informal, higher-resolution, seismic-stratigraphical sequences during the course of the BGS 1:250 000 offshore mapping programme. The details of this scheme have been presented elsewhere (cf. Stoker et al., 1993); however, its correlation to the megasequence framework presented in this report is indicated in Figure 109, together with a tentative correlation to recently developed schemes for the East Faroe margin (Andersen et al., 2000, 2002; Faroes GEM Network, 2001a, b, c and d; Austin, 2004).

The Neogene tectonic development of the Faroe–Shetland region has involved both compressive and epeirogenic movements, in the early (Miocene) and late (Plio-Pleistocene) Neogene respectively, which have influenced patterns of palaeoceanographical circulation and sedimentation (Stoker et al., 2005a, b, c; Figures 89, 92 and 93). The most obvious expression of early Neogene compression is the development of spectacular elongate anticlinal domes, including the Fugloy, Munkagrunnur, Wyville Thomson and Ymir ridges (Figures 26, 34, 88b, 106 and 110). The latter two ridges are sometimes referred to collectively as the Wyville Thomson Ridge Complex (Boldreel and Andersen, 1993), which may also include the anticlinal domes of the Faroe Bank High and Bill Bailey's Dome (Kimbell et al., 2005; Figure 106). These anticlinal structures are up to 4 km in amplitude and tens of kilometres across, with axial traces that extend for up to a few hundred kilometres (Boldreel and Andersen, 1998; Johnson et al., 2005b). In addition, the formation of the Judd Anticline and the reactivation of early Palaeogene structural highs, such as the Corona and East Faroe highs, created an undulating palaeobathymetry. Although these anticlinal structures may have been developing since the Paleocene/Eocene (Figure 89), an intensification of contractional deformation is interpreted to have occurred during the Early to Mid Miocene (Johnson et al., 2005b; Stoker et al., 2005c). Regional flexure across the continental margin may have preceded this phase of deformation, reflecting the build-up of stress (Stoker et al., 2005c). Although the amplitude of flexural deflection may have been relatively small (tens to a few hundreds of metres), it may have been sufficient to change the bathymetric threshold in the Faroe–Shetland region (Cloetingh et al., 1990). A modification to the pattern of deep-water circulation is expressed by erosion associated with the formation of the Top Palaeogene Unconformity (Figure 111).

This discrete phase of early Neogene compressional tectonism was coeval with a local reorganisation of the adjacent north-east Atlantic plate system, which accommodated the transfer of the spreading ridge across the Jan Mayen area between about 23 and 15 Ma (Brekke, 2000; Mosar et al., 2002). Regional flexure is envisaged to be a response to the prolonged build-up of compressive stresses during this interval, which was followed by the rapid release of stress in specific areas in the form of the anticlinal structures. The formation of deformation structures initiated a change in deep-water circulation patterns in the early Mid Miocene, expressed by the development of the intra-Miocene Unconformity. Differential vertical movements in the southern part of the report area, notably the development of the Wyville Thomson and Munkagrunnur ridges and the complementary syncline of the Faroe Bank Channel, are envisaged to have created the Faroe Conduit (Stoker et al., 2005c).

The inferred period of compression is bracketed by the formation of the Top Palaeogene and intra-Miocene unconformities, which are separate unconformities in the deeper, northern part of the report area, but may merge in the shallower, southern Faroe–Shetland Channel and the Faroe Bank Channel to form a composite unconformity (Figures 110 and 112b). The formation of this deep-water passageway, concomitant with a general deepening of the Greenland–Scotland Ridge (Thiede and Eldholm, 1983), facilitated the onset or acceleration of the exchange of intermediate and deep waters across the ridge after about 15 Ma. The onlapping, Middle to Upper Miocene, basinal strata represent an expansion of contourite drift accumulation in the Faroe–Shetland region, as well as the north-east Atlantic region in general (Laberg et al., 2005). The preservation of upslope-retreating lowstand wedges (Figure 113) and Middle to Upper Miocene transgressive sandstones on the West Shetland margin is further indication of a general subsidence of the region in Mid Miocene to earliest Pliocene time (Stoker, 1999; Stoker et al., 2005a).

Compression may have continued to affect the north-east Faroe–Shetland Channel into Pliocene–Holocene time, including the formation of the Pilot Whale Anticline and associated mud volcanoes/diapirs (Ritchie et al., 2003; Johnson et al., 2005b; Figures 24). However, the main instigation of late Neogene change was the onset of epeirogenic movements from about 4 Ma (Early Pliocene). This modified the patterns of contourite sedimentation, and facilitated the onset of rapid progradation of the continental margins bordering Shetland and the Faroe Islands by up to 50 km (Andersen et al., 2000; Stoker, 2002; Dahlgren et al., 2005; Nielsen et al., 2005; Stoker et al., 2005a and b).

The sedimentary response to change is well illustrated by comparing the distribution of the FSN-2 and FSN-1 megasequences in Figures 106 and 107, which indicate a late Neogene shift in the focus of sedimentation from the basinal areas (FSN-2) to the shelf and slope (FSN-1). The late Neogene growth of the slope apron is especially well developed in five depocentres — the Rona, Foula, East Faroe and West Faroe wedges and the North Sea Fan — where Pliocene to Holocene sediment thicknesses commonly range from 200 to 600 ms TWTT, but locally exceed 1.2 s TWTT on the North Sea Fan (Figures 107 and 114). The build-out of the shelf and slope is inferred to record a marked increase in sediment supply to the margin in response to its uplift and tilting (Andersen et al., 2000; Praeg et al., 2005; Stoker et al., 2005b). The West Shetland margin was titled seaward by about 0.5° (Stoker, 2002; Figure 112a).

Tectonic and isostatic mechanisms are both invoked as driving the uplift, the latter in response to late Neogene glaciation. However, although glacially derived sediments form a significant component of the prograding sediment wedges, the initiation of the wedges on the West Shetland margin has been proved to predate the onset of glaciation (at about 2.74 Ma) by up to 1 Ma (Stoker, 2002). This may indicate that tectonic uplift was the initiating mechanism of change, whereas isostatic uplift may have been a sustaining factor. The prograding sediment wedges form part of a larger domain of shelf-margin progradation extending the length of the north-west European Atlantic margin, indicative of the regional extent of this tectonic pulse (Figures 89 and 93).

Uplift of onshore and shallow shelf areas bounding north-west Britain and the Faroe Islands may have been

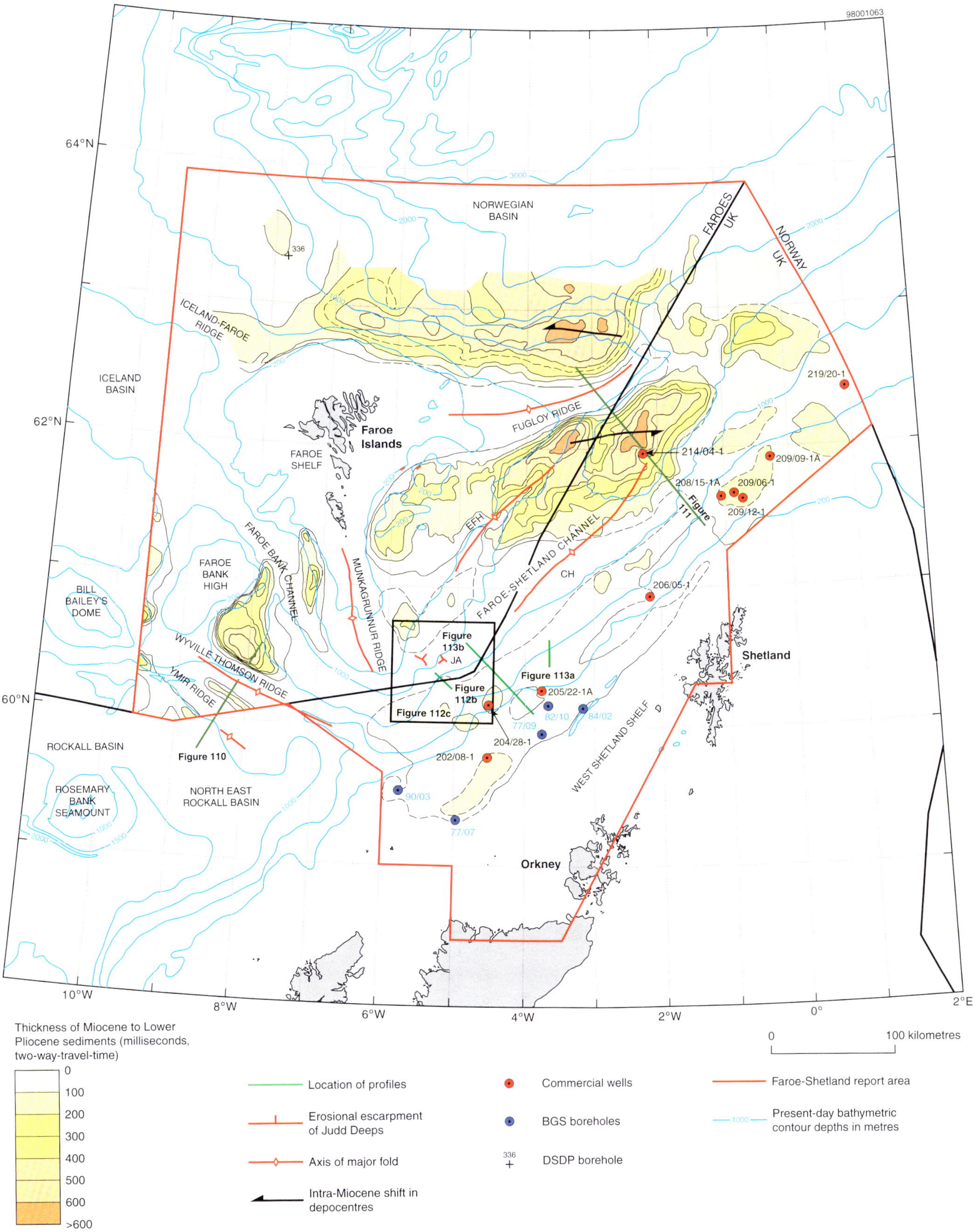

Figure 106 Distribution and thickness of Miocene to Lower Pliocene strata (FSN-2). Abbreviations: see Figure 7 except, JA=Judd Anticline.

accompanied by accelerated subsidence of several hundred metres in the surrounding basins (e.g. Cloetingh et al., 1990; Praeg et al., 2005). This change in palaeobathymetry may have been a contributory factor in the widespread development of the intra-Neogene Unconformity in deep water (Figures 110, 112b and 114c). On a global scale, Early Pliocene submarine erosion may have been driven by an increase in the volume and

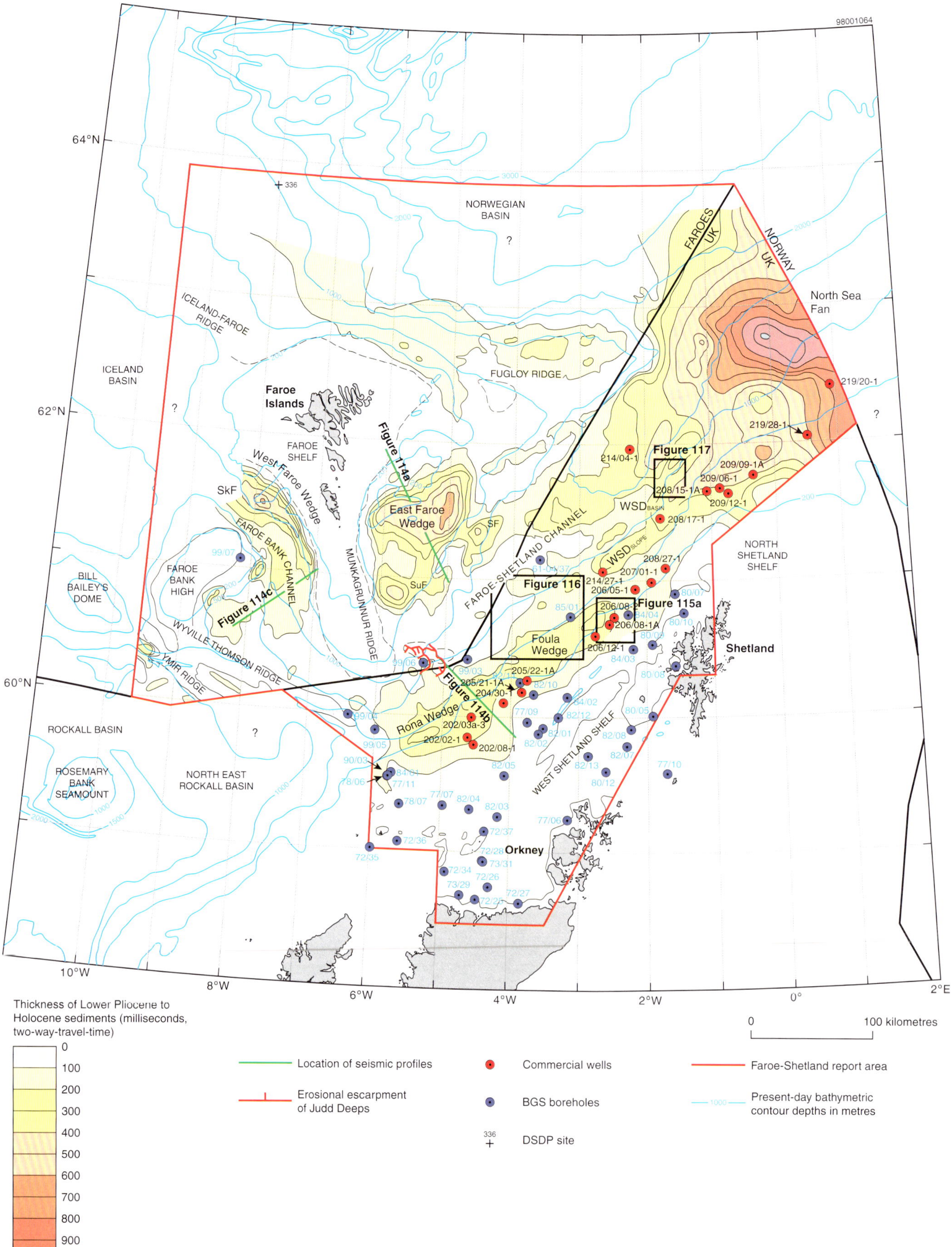

Figure 107 Distribution and thickness of Lower Pliocene to Holocene strata (FSN-1). Abbreviations: SF= Sandoy Fan; SuF=Suðuroy Fan; SkF=Skeivi Fan; WSDSLOPE and WSDBASIN = West Shetland Drift (modified from Knutz and Cartwright, 2003 and 2004).

strength of Norwegian Sea Deep Water (NSDW) being transported into the Atlantic Ocean via the Greenland–Scotland Ridge (including the Faroe Conduit), following the closure of the Central American Seaway (Haug and Tiedemann, 1998; Lear et al., 2003). In the Faroe Conduit, this pulse of vigorous deep-marine erosion is likely to have been enhanced by the change in its palaeobathymetry. Subsequent patterns of deep-water sedimentation, including the modern-day setting, continued to be dominated by bottom-current activity (Masson et al., 2004).

The interaction between downslope and alongslope processes during late Neogene time has been very influential in the present day shaping of the continental margin. The erosional and depositional effects of this dynamic sedimentary environment are strongly illustrated on sea-bed imagery (Bulat and Long, 2001; Masson, 2001; Long et al., 2004; Bulat and Long, 2005; Masson et al., 2004; see Figure 138), and are discussed in more detail in chapters 10 and 11.

In the following sections, the Neogene is described broadly in terms of its Miocene and Pliocene to Holocene successions, though the boundary between the FSN-2 and FSN-1 megasequences is probably of Early Pliocene age (Figure 108). This subdivision most aptly corresponds to the two-stage tectonic evolution of the region summarised above.

Miocene

Miocene strata are best preserved in the northern part of the Faroe–Shetland region where they have accumulated in two discrete depocentres either side of the Fugloy Ridge (Figure 106). These rocks comprise the FSN-2 megasequence (broadly equivalent to the Lower Nordland unit of the Nordland Group) (Figure 94), and locally exceed 600 ms TWTT in thickness. Well 214/04-1 indicates that the top of the FSN-2 megasequence extends into Lower Pliocene strata in the Faroe–Shetland Basin (Davies and Cartwright, 2002). Farther south, the distribution and thickness of the FSN-2 megasequence is more variable. On the West Shetland margin, a widespread succession of Miocene sandstones rarely exceeds 100 ms TWTT (Stoker, 1999), whereas localised accumulations in excess of 500 ms TWTT occur in the Faroe Bank Channel. The latter contrasts with elsewhere within the Faroe–Shetland and Faroe Bank channels where Miocene strata are absent. They are also absent from the crest of the Fugloy and Munkagrunnur ridges, and from adjacent ridges in the Wyville Thomson area.

The deep-marine sedimentary system that was established in the Oligocene continued into the Early Miocene. Although there was no obvious change in the style of deep-water sedimentation, the pattern of deposition and erosion was strongly influenced by the early Neogene deformation. Immediately south of the Fugloy Ridge, the FSN-2 megasequence has been divided into two units, the FSN-2a and FSN-2b units, which are separated by intra-Miocene and Mid Miocene unconformities (Stoker et al., 2005a; Figures 14 and 94). The former unconformity is the more important and is interpreted to have formed by deformation linked to compression and inversion in this region (Ritchie et al., 2003; STRATAGEM Partners, 2003; Sørensen, 2003; Stoker et al., 2005a). Biostratigraphical data from well 214/04-1 indicate an approximate Mid Miocene age for the formation of the unconformity (Davies and Cartwright, 2002; Figure 108), which has resulted in Middle Miocene to Lower Pliocene sediments onlapping folded Lower Miocene and older strata (Ritchie et al., 2003; Figure 111). A Mid Miocene age is consistent with its regional correlation to comparable unconformities off Norway and in the Rockall–Porcupine region, which developed in the interval between 16 and 11 Ma (cf. Stoker et al., 2005c and references therein). An intra-Miocene Unconformity is also recognised on the North Faroe margin, though the stratigraphical terminology of the Miocene units is different (Nielsen and van Weering, 1998; Andersen et al., 2000; cf. Table 6).

The two-fold subdivision of the FSN-2 megasequence is less clear in the southern part of the Faroe–Shetland region. In the Faroe–Shetland Channel, compressional deformation combined with vigorous Early Miocene bottom-current erosion largely curtailed the accumulation of Lower Miocene sediments at its shallower, south-west end (STRATAGEM Partners, 2003; Stoker et al., 2005a). On the West Shetland margin, the early Neogene record is largely expressed by a composite Top Palaeogene to intra-Miocene unconformity overlain by Middle to Upper Miocene (equivalent to FSN-2a) and younger strata (Figure 114b). The intensity of the deep-water erosion is best expressed where the composite unconformity is highly irregular and shaped into a series of erosional deeps, termed the Judd Deeps (also known as the Munkagrunnur Falls) that have a relief up to 200 m (Stoker et al., 2003; Figures 112b and c). Their formation has been attributed to bottom-current activity that was strongly focused by contemporary deformation of the sea bed due to inversion of the Judd Anticline (Stoker et al., 2002; Smallwood 2004; Figure 106). Subdivision of the FSN-2 megasequence in the Faroe Bank Channel remains uncertain (STRATAGEM Partners, 2003; Sørensen, 2003; Johnson et al., 2005b; Stoker et al., 2005c; Figure 110).

BGS boreholes from the adjacent West Shetland Shelf proved a predominantly late Early to Late Miocene age for a sandstone-dominated section within the Lower Nordland unit, termed the Muckle Ossa Sandstone (Stoker, 1999; Figures 94 and 108). Although the stratigraphical range of several of these boreholes appears to extend back to the NN2 biozone (Aquitanian), the overlapping range of key biostratigraphical indicators suggests that those species older than biozone NN4 (Burdigalian/Langhian) may be reworked (Stoker, 1999). Lower Miocene rocks are, however, preserved on the slope to the north of Shetland, proved in wells such as 208/15-1A (Figure 108).

For ease of description, the Miocene to Lower Pliocene succession is presented separately from four regions: i) the East Faroe margin, between the Munkagrunnur and Fugloy ridges; ii) the North Faroe margin, which slopes into the Norwegian Basin; iii) the West Shetland margin, predominantly the outer shelf and slope; and, iv) the Faroe Bank Channel. These regions best reflect the variable distribution of these strata (Figure 106).

East Faroe margin

On the East Faroe margin, Miocene to Lower Pliocene strata are preserved in an elongate, north-north-east-trending basin that extends from beneath the present day Faroe Shelf into the northern Faroe–Shetland Channel, and possibly continues below the slope north of Shetland (Figure 106). The development of this

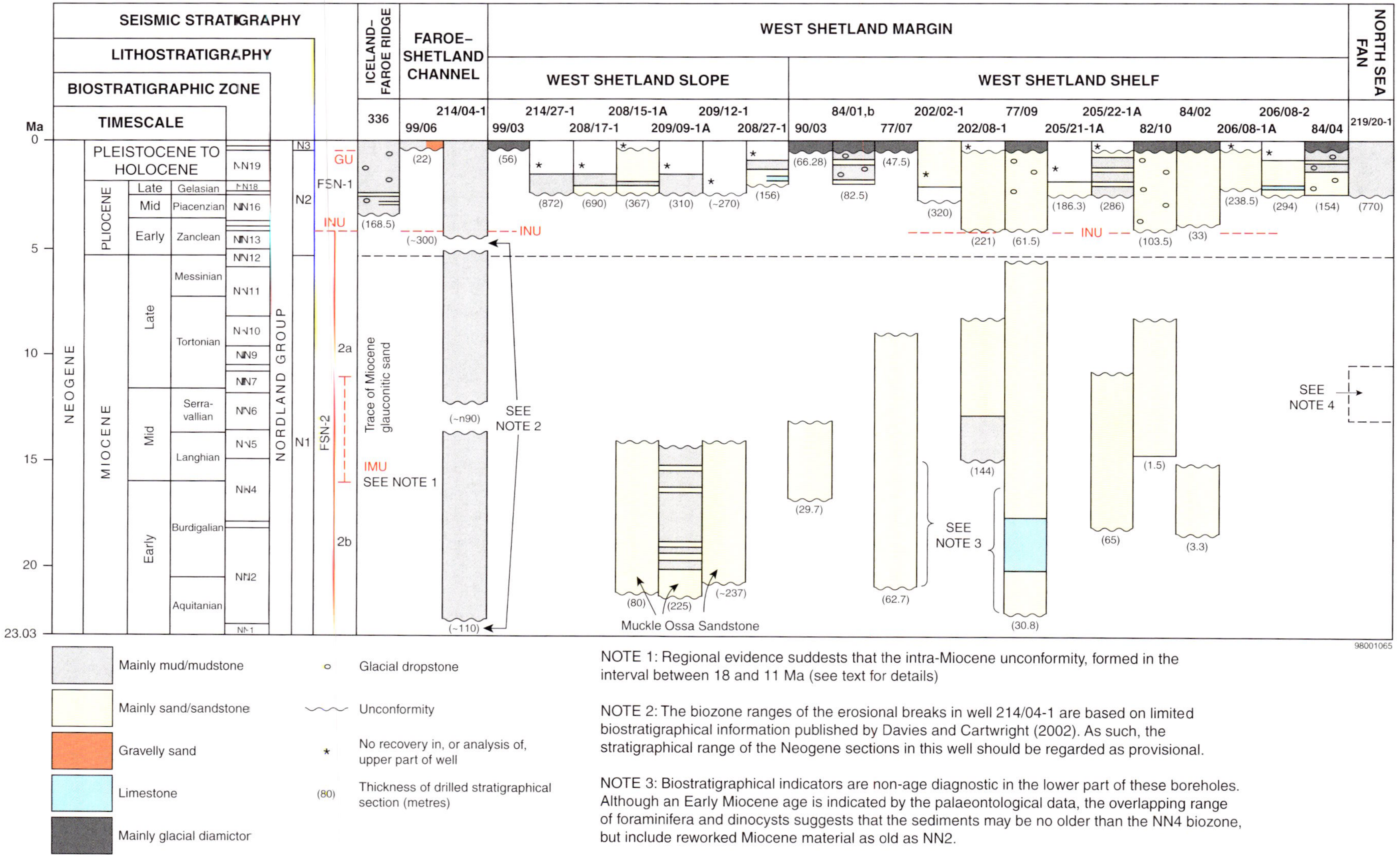

NOTE 1: Regional evidence suddests that the intra-Miocene unconformity, formed in the interval between 18 and 11 Ma (see text for details)

NOTE 2: The biozone ranges of the erosional breaks in well 214/04-1 are based on limited biostratigraphical information published by Davies and Cartwright (2002). As such, the stratigraphical range of the Neogene sections in this well should be regarded as provisional.

NOTE 3: Biostratigraphical indicators are non-age diagnostic in the lower part of these boreholes. Although an Early Miocene age is indicated by the palaeontological data, the overlapping range of foraminifera and dinocysts suggests that the sediments may be no older than the NN4 biozone, but include reworked Miocene material as old as NN2.

NOTE 4: Re-analysis of biostratigraphical data from well 219/20-1 has suggested that in situ Middle Miocene sediments, NN8 at youngest, may be present. Seismic records in the area show a Miocene section pinching out to the SW, just beyond the well location. Consequently, a potential Mid Miocene section cannot be discounted in this well.

Figure 108 Stratigraphical range chart showing thickness and generalised lithology for the Neogene succession from selected wells, BGS boreholes and a DSDP site within the Faroe–Shetland region. Stratigraphical range bars based on biostratigraphical and strontium isotope data derived from Stoker, (1999), STRATAGEM Partners, (2002) and Stoker et al., (2005a). Abbreviations: see Figures 14 and 89. Timescale, including biostratigraphical zonation (based on calcareous nannoplankton), is based on 'A Geologic Time Scale 2004' by F M Gradstein, J G Ogg, A G Smith, et al. (2004) and the International Stratigraphical Chart, 2006 (ICS) with additions. For location of sites, see Figure 107.

Table 7 Summary of predominantly Middle to Upper Pleistocene (<0.5 Ma) glacial sediments recovered in BGS boreholes not included in Figure 108. For location, see Figure 107.

Borehole	Thickness (metres)	Lithology
72/25	12.5	Sand on diamicton
72/26	8.0	Sand on diamicton
72/27	18.0	Sand on diamicton
72/28	2.0	Sand on diamicton
72/34	8.5	Sand on mud on diamicton
72/35	16.0	Pebbly sand on diamicton
72/36	40.0	Pebbly sand on diamicton with thin interbedded mud
72/37	23.0	Pebbly sand and muddy gravel
73/29	28.5	Pebbly sand on diamicton
73/31	28.0	Pebbly sand
77/06	10.4	Sand on diamicton
77/11	8.0	Gravel
78/06	7.0	Muddy pebbly sand on basal gravel
78/07	84.0	Diamicton
80/05	4.3	Sand
80/07	10.0	Diamicton
80/08	34.6	Sand on mud on diamicton
80/09	33.6	Gravelly sand on diamicton with interbedded mud
80/10	10.1	Sand on gravel
80/12	10.8	Sand and gravel
82/01	74.5	Thin mud on diamicton
82/02	27.0	Mud on basal diamicton
82/03	65.86	Pebbly sand on mud on diamicton
82/04	69.25	Diamicton
82/05	47.5	Interbedded sand and diamicton
82/07	2.85	Sandy gravel on thin basal diamicton
82/08	8.2	Diamicton
82/11	46.15	Mud on diamicton, locally sandy
82/12	16.5	Diamicton
82/13	2.0	Gravel
84/03	25.4	Mud on diamicton
85/01	13.9	Pebbly mud
99/04	15.17	Sandy mud on sand and gravel
99/05	33.7	Pebbly sand and sandy gravel; muddy at base
99/07	0.25	Gravelly sand

98001176

basin was structurally controlled by the early Neogene reactivation of the Fugloy Ridge and the Corona High, which form its northern and southern boundaries, respectively. The variation in sediment thickness within the basin is a consequence of intrabasinal folding above structures such as the East Faroe High (Ritchie et al., 2003; Keser Neish, 2005), though the upper surface of the FSN-2 megasequence is everywhere erosional (the intra-Neogene Unconformity) (Figures 111 and 114a).

The sediment fill in the deeper part of the basin is characterised by a seismically layered configuration of parallel, undulatory reflections that commonly display low-angle onlap onto the margin of the basin (Stoker et al., 2005a; Figure 111). This seismic facies is consistent with axial sedimentation in the basin driven mainly by bottom currents (Stoker et al., 1998; Howe et al., 2002). The undulations represent mounded, hummocky structures formed by deformation of the sediments due to density inversion during early burial, which was initiated by differential loading above polygonal faults (Davies et al., 1999). The diagenetic reflector (see the Oligocene section) locally crosscuts the lower part of the section (Figure 111).

On the Faroe Shelf, the strata display a more acoustically structureless character, and their distribution is locally terminated by erosional pinchout on the margin of the basin, and adjacent to intrabasinal highs (Figure 114a). Some of this erosion probably occurred during the Mid Miocene, concomitant with the formation of the intra-Miocene Unconformity that resulted from compressional deformation (Andersen et al., 2000). In the Faroe–Shetland Channel, the intra-Miocene Unconformity forms a distinct angular surface where the FSN-2b deposits have been folded and deformed, and subsequently overlain and onlapped by sediments of

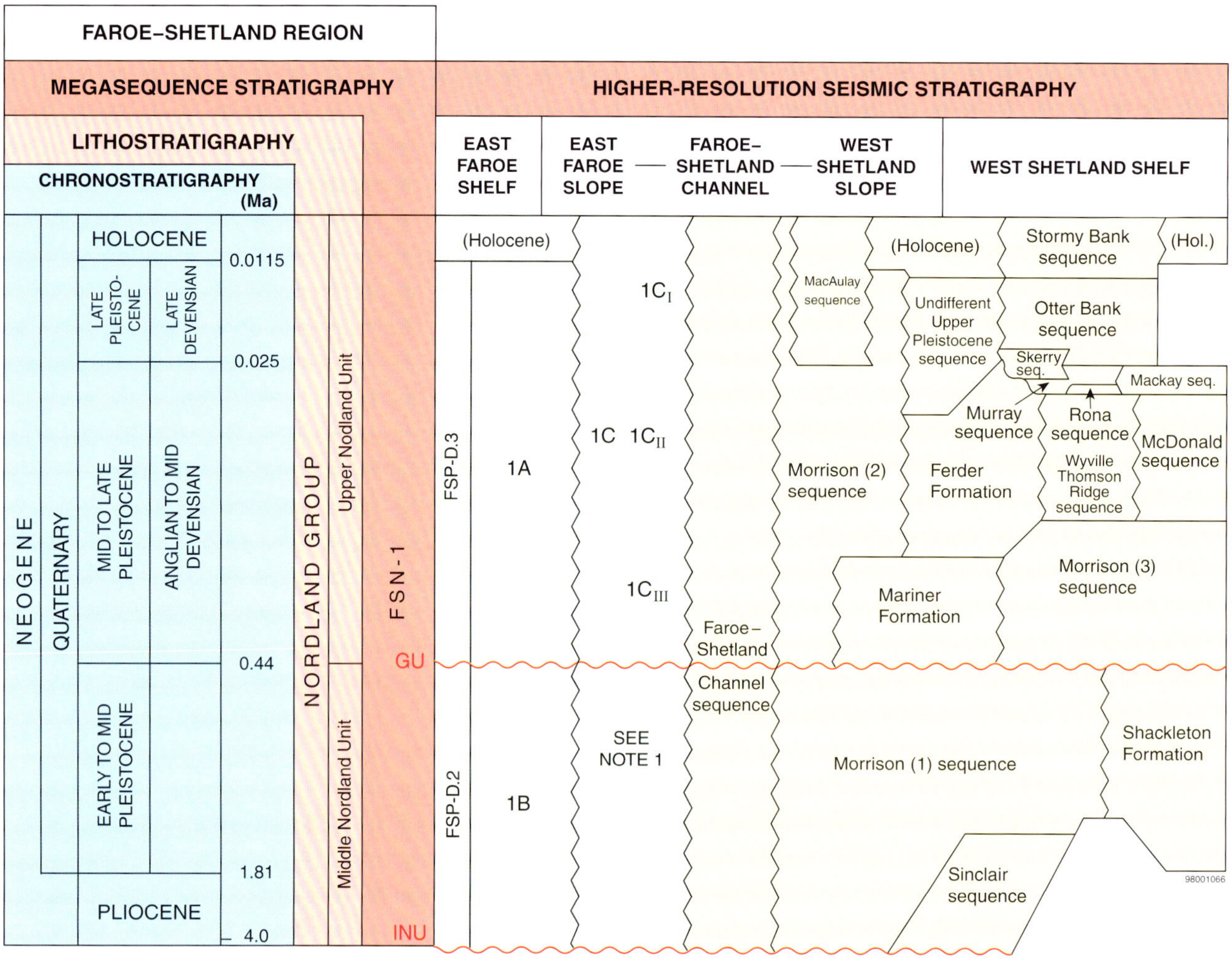

NOTE 1: Stratigraphic range of subunits $1C_{II}$ and $1C_{III}$ remains uncertain. Well 214/04-1 indicates on Early Pliocene–Holocene age range for the Faroe–Shetland Channel Sequence, which implies a comparable age range for the stratigraphically-equivalent Unit 1C

Figure 109 Late Neogene stratigraphical framework for the Faroe–Shetland region, showing tentative correlation between the West Shetland (Stoker et al., 1993) and East Faroe (modified from Faroes GEM Network, 2001a, b, c and d and Austin, 2004) high-resolution stratigraphical schemes. The FSP-D.2 and D.3 notation is from Andersen et al., (2000 and 2002). See text for details. Abbreviations: see Figures 14 and 89.

the FSN-2a unit (Figure 111). A north-eastward shift in the focus of sedimentation within the basin followed the phase of deformation (STRATAGEM Partners, 2003; Stoker et al., 2005a; Figure 106).

Lithological information from this basin is restricted to well 214/04-1, in the north-east Faroe–Shetland Channel. This well proved about 110 m of Lower to lower Middle Miocene diatomaceous mud, which included foraminifera such as *Globorotalia praescitula* and *Martinotiella bradyana*, unconformably overlain by about 90 m of Middle Miocene to Lower Pliocene sediment of comparable facies, with common *Neogloborigerina continuosa* and the Lower Pliocene *Orbulina universa* immediately below the intra-Neogene Unconformity (Davies and Cartwright, 2002; Figures 108 and 111). Farther east, well data on the north Shetland Slope may represent a marginal facies to the main basin. Well 208/15-1A (Figure 106) penetrated an acoustically structureless section, and proved an 80 m thick sandstone-dominated sequence of Early to early Mid Miocene age, which is yellow-brown, medium to coarse-grained, poorly sorted, bioclastic and glauconitic (Stoker, 1999; Figure 108). Similar sandstone was recorded in nearby wells 209/06-1 and 209/09-1A, although the upper part of the section in 209/06-1 was dominated by pale brown to grey mudstone with thin interbeds of medium to coarse-grained shelly sandstone. In contrast, well 209/09-1A revealed an interbedded sequence of grey to dark grey, soft to firm, calcareous mudstone and olive grey to brown-black, well-sorted glauconitic sandstone with occasional shell fragments, lignitic and shale clasts. To the north-east, it remains uncertain whether or not well 219/20-1 recovered Miocene strata (cf. Figure 108).

North Faroe margin

On the North Faroe margin, Miocene sediments are thickest in the midslope region, in excess of 600 ms TWTT, thinning upslope and interpreted to form part of a wedge-shaped shelf-margin succession that onlaps the Fugloy Ridge (Nielsen and van Weering, 1998; Figure 111). The Miocene succession is poorly resolved due to late Neogene mass failure on the lower slope (van Weering et al., 1998; Kuijpers et al., 2001). An

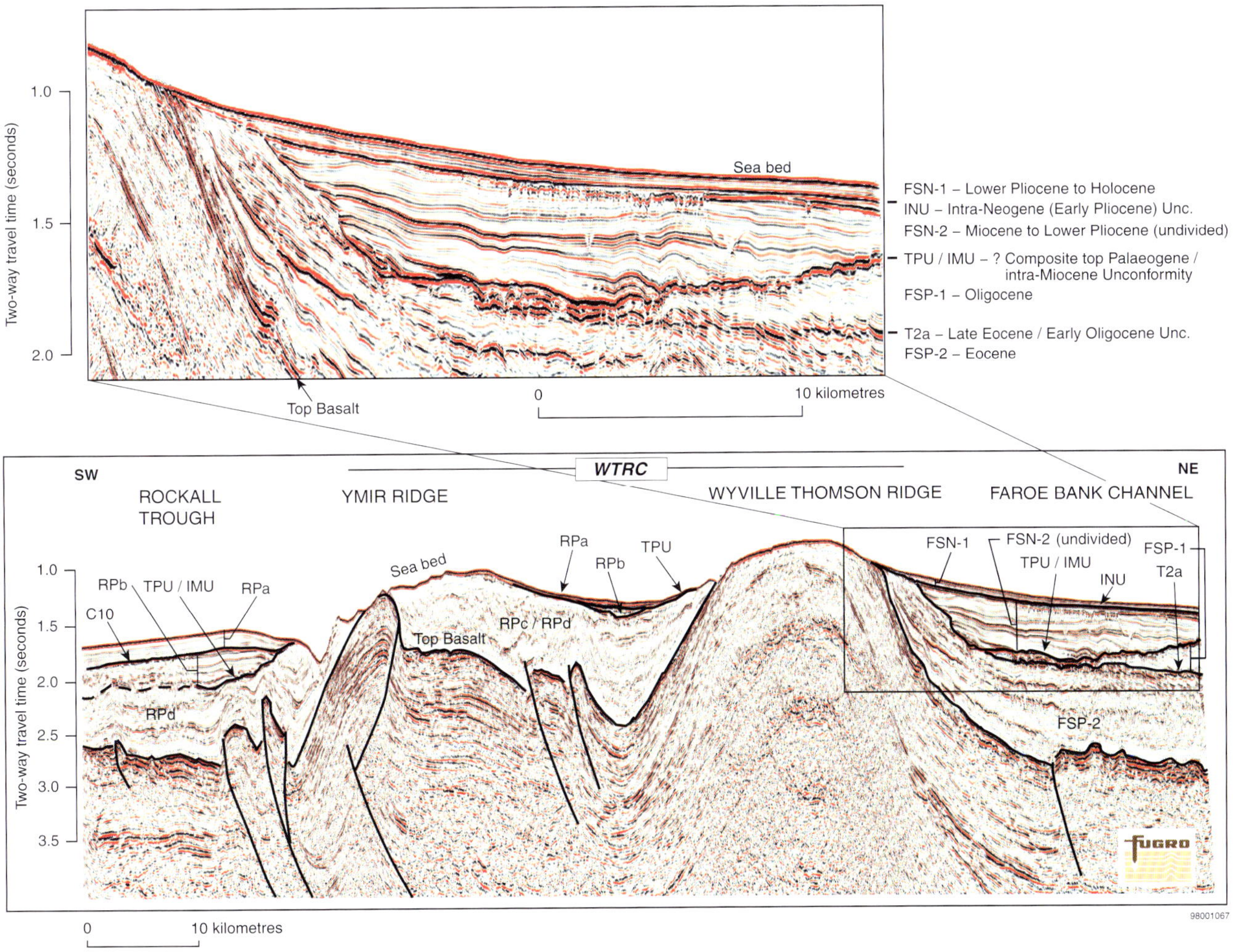

Figure 110 Seismic profile across the Wyville Thomson Ridge Complex (WTRC) showing deformed Palaeogene sediments (FSP-1, FSP-2, RPd and RPd/RPc) onlapped by Neogene (FSN-1, FSN-2, RPa and RPb) strata (modified from Stoker et al., 2005c). The RP notation relates to the Rockall region (cf. Stoker et al., 2005a): RPa, RPb and RPd correlate with FSN-1, FSN-2 and FSP-1 to 2, respectively. Reflector T2a marks the Eocene/Oligocene boundary. The C10 Rockall reflector correlates with the INU. Inset shows part of seismic profile in greater stratigraphical detail. For location of profile, see Figure 106. Seismic data courtesy of Fugro Multi Client Services.

angular unconformity separates Lower Miocene strata, which form part of the North Faroes sequence 1 (as defined by Nielsen and van Weering, 1998), from Middle to Upper Miocene strata that correlate with sequence 2 (Table 6), and extend higher onto the north Faroe Slope than the Lower Miocene section. Andersen et al. (2000) suggest that this unconformity may be indicative of a discrete tectonic phase in late Early Miocene times, which was accompanied by a shift in the focus of sedimentation (STRATAGEM Partners, 2003; Stoker et al., 2005a) depicted by the black arrows in Figure 106.

The Lower Miocene section displays subparallel, continuous reflections that onlap eroded Oligocene strata on the upper slope, whereas the basinal part of the section has been disturbed or removed by later mass movement (Nielsen and van Weering, 1998). According to Nielsen and van Weering (1998), this section represents a highstand systems tract. The overlying Middle to Upper Miocene package is similarly characterised by subparallel, continuous internal reflections, which preserve an overall sigmoidal depositional geometry on the middle and upper slope. An upwards transition from shelf-margin through transgressive to highstand systems tract is inferred by Nielsen and van Weering (1998), though much of the character of the lower slope-basin floor section has been disrupted by mass wasting. Numerous rotated fault blocks are observed in the upper part of the sequence.

Lithological information from this region is restricted to DSDP site 336 on the Iceland–Faroe Ridge, where traces of glauconitic sand, in the interval between 150 and 168 m below sea bed (Figure 104), contain fragments of the Miocene foraminifera *Martinotiella communis* (Talwani et al., 1976). However, this material is mixed with Upper Oligocene and Plio-Pleistocene sediment, which strongly suggests reworking at the base of the Plio-Pleistocene section. Nevertheless, the occurrence of Miocene glauconitic sand, albeit reworked, may indicate that the interval between the Late Oligocene and the Pliocene was characterised by nondeposition (except for glauconite) at this site (Talwani et al., 1976).

West Shetland margin

On the West Shetland margin, Miocene strata are mainly restricted to the shelf and slope (Figure 106).

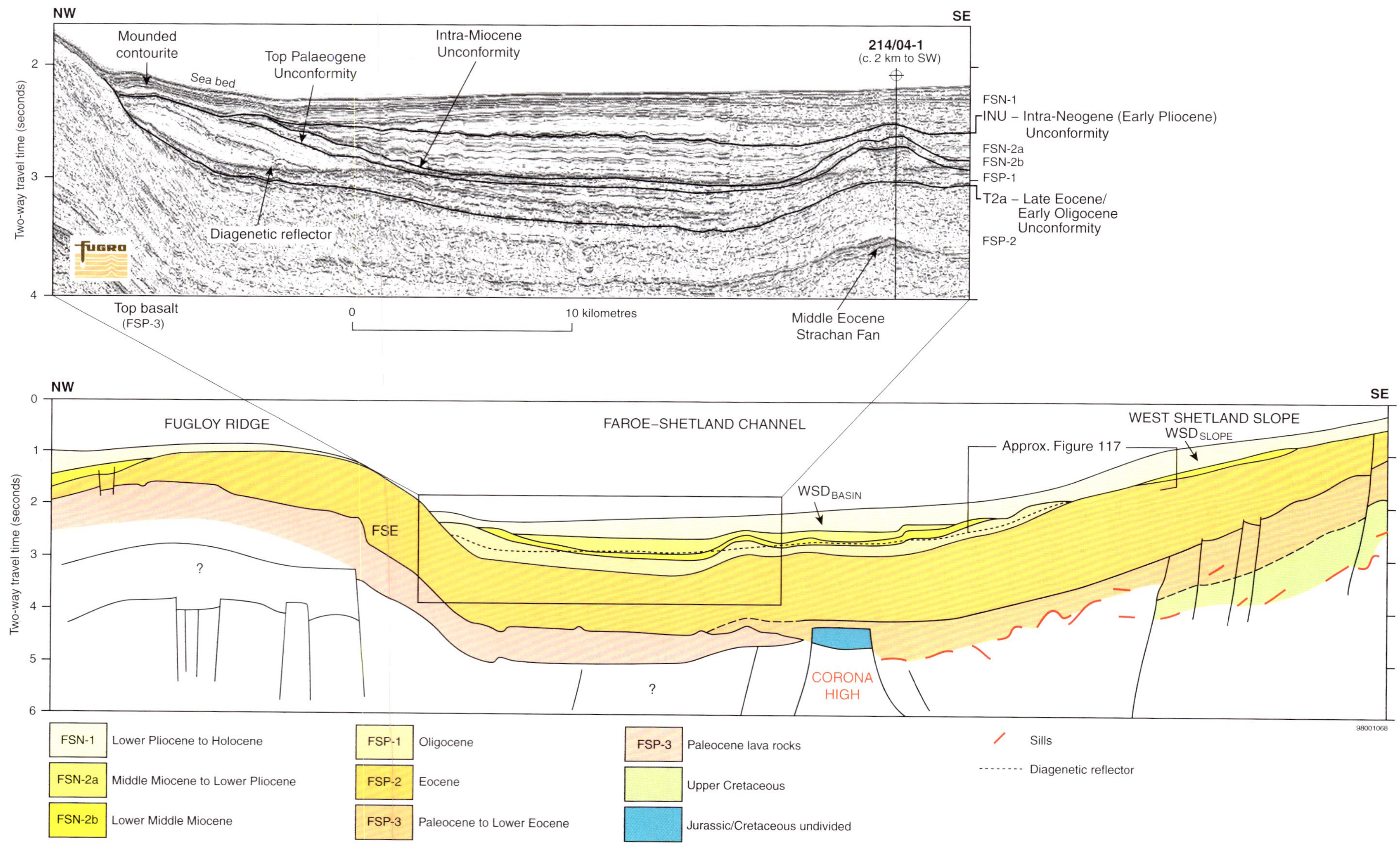

Figure 111 Geoseismic cross-section showing general disposition of Cenozoic strata across the north Faroe–Shetland Channel, including the Fugloy Ridge. Inset shows seismic profile that illustrates stratigraphical detail of upper Palaeogene to Neogene section with emphasis on the subdivision of the Miocene sequence by the intra-Miocene Unconformity. Abbreviations: see Figures 24, 94 and 107. For location of geoseismic cross-section, see Figure 106. Seismic data and geoseismic interpretation of data courtesy of Fugro Multi Client Services.

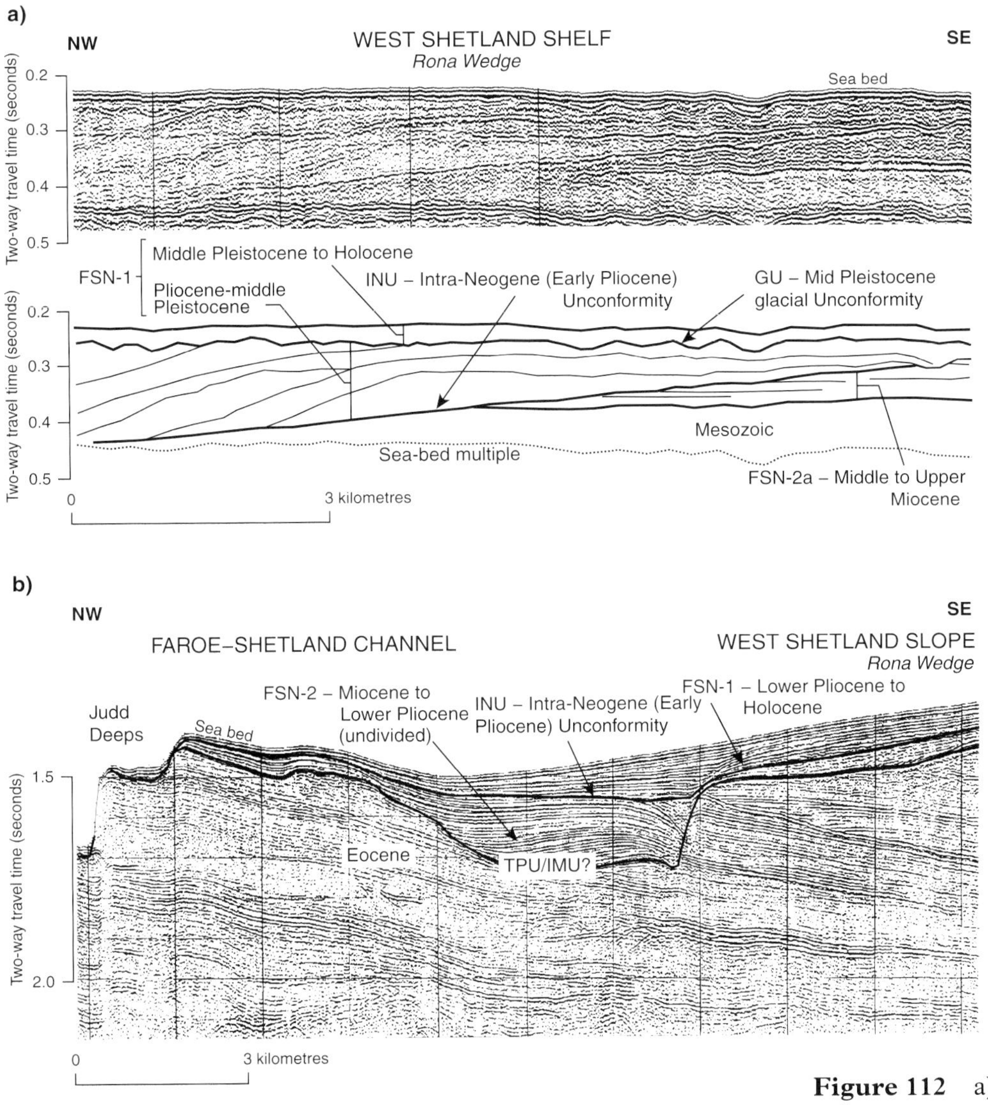

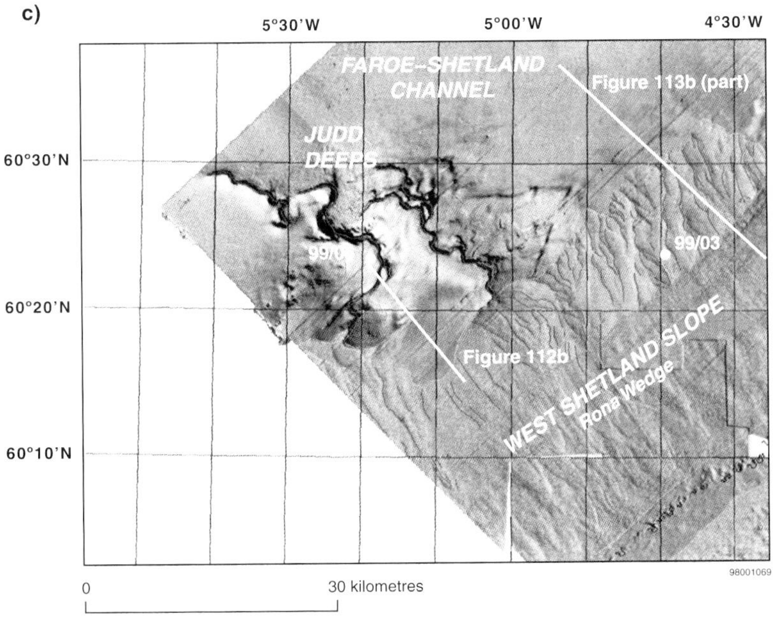

Figure 112 a) BGS airgun profile on the West Shetland Shelf showing Miocene strata (FSN-2a-Muckle Ossa Sandstone) truncated by the intra-Neogene Unconformity (INU), which is overlain by prograding Plio-Pleistocene deposits (FSN-1) of the Rona Wedge. The shelf-wide Glacial Unconformity (GU) marks the onset of extensive glaciation (modified from Stoker, 2002). For context of profile, see Figure 113b; b) BGS airgun profile showing the irregular nature of the Top Palaeogene Unconformity (TPU), which may be locally composite with the intra-Miocene Unconformity (IMU). Part of this surface is buried beneath an FSN-2 infilling sediment drift and the lower part of the Rona Wedge slope apron (FSN-1), part remains open – the Judd Deeps (modified from Stoker et al., (2005b). For location of profile, see figures 106 and 112c; c) Sea-bed shaded relief image (illuminated from the north-east) (Long et al., 2004) showing the Judd Deeps depicted by the scarp slopes, which are in shadow (black), the debris flows of the Plio-Pleistocene Rona Wedge, the smoother sea bed of the Faroe–Shetland Channel, and the locations of BGS boreholes 99/03 and 99/06. For general location of image, see Figure 106 (modified after Stoker et al., 2003). Sea-bed image derived from Figure 138.

a)

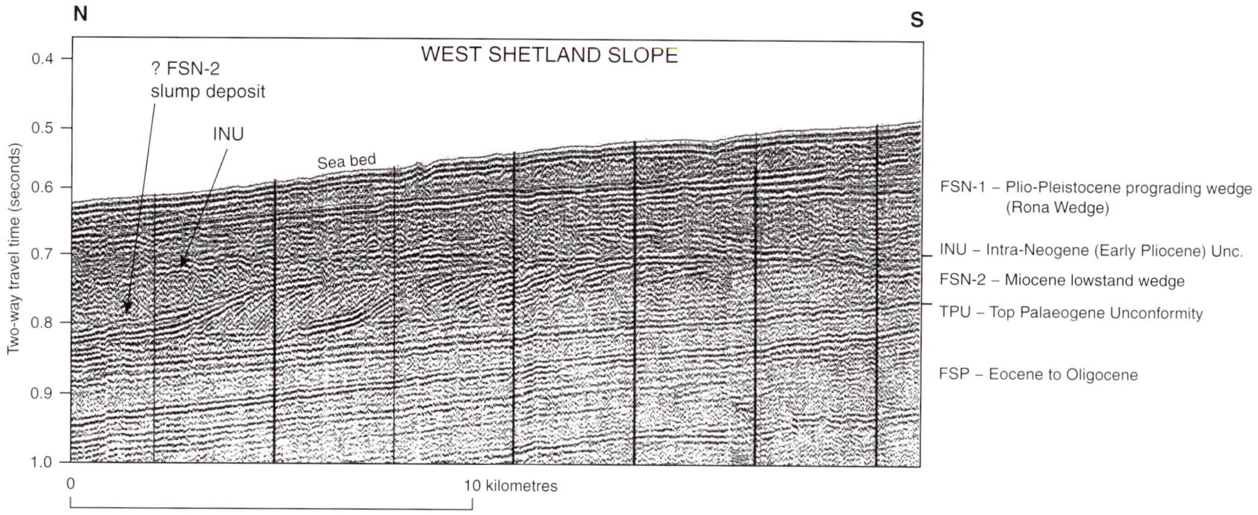

b)

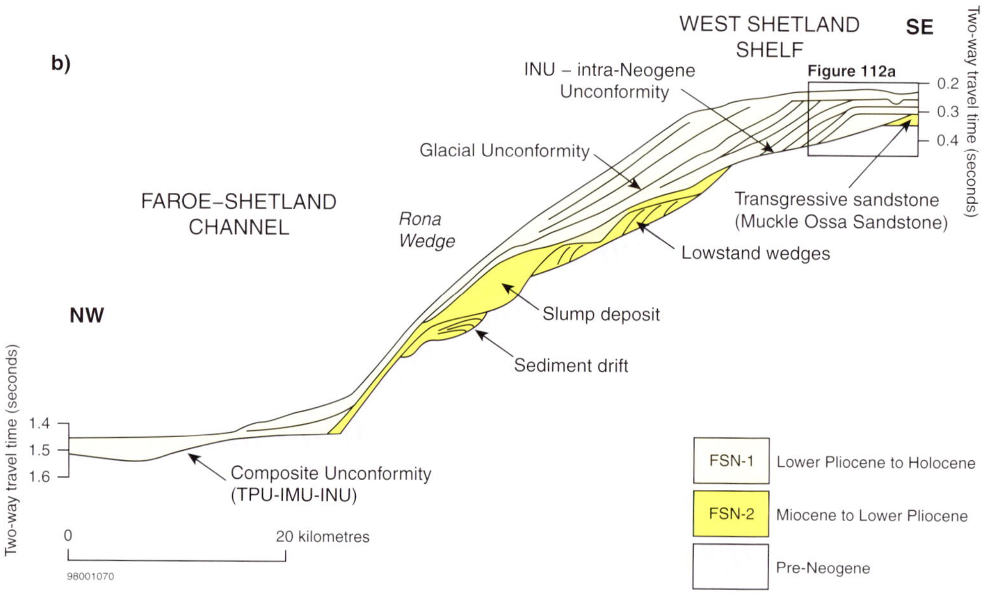

Figure 113 a) BGS sparker profile from the middle to upper West Shetland Slope showing Miocene lowstand wedge and ?latest Miocene to Early Pliocene slump deposit; b) Geoseismic section across the West Shetland Shelf showing preserved Miocene depositional systems on the shelf and slope, below the Plio-Pleistocene prograding wedge. See text for details. Abbreviations: see Figure 14. For location of profile and geoseismic cross-section, see Figure 106; part of b also located in Figure 112c.

On the basis of biostratigraphical data, these strata are tentatively correlated with unit FSN-2a though the effect of reworking in a predominantly sandy succession reduces the biostratigraphical precision (Figure 108). On the West Shetland Slope, the Miocene section is interpreted to overlie a composite Top Palaeogene to intra-Miocene Unconformity, whereas the top of the section is everywhere eroded by the intra-Neogene Unconformity (Figures 112b and 114b). Miocene strata are mostly absent from the south-west Faroe–Shetland Channel (Figure 106).

On the West Shetland Shelf, the Miocene deposits are preserved as an eroded, sheet-like unit up to 150 m thick (Stoker, 1999) characterised by a seismically layered configuration of subparallel, flat-lying to undulatory reflections that display landward onlap. Indications of internal channels are locally observed. The shelf succession has been tested in a number of BGS boreholes and commercial wells, which proved a sandstone-dominated facies, termed the Muckle Ossa Sandstone (Stoker, 1999; Figures 94 and 108). BGS boreholes 77/07, 77/09 and 90/03 penetrated between 29.7 and 62.7 m of sandstone that is characteristically green to dark green, occasionally black, fine to coarse-grained, poorly sorted, generally friable but partly calcite cemented, muddy, glauconitic and bioclastic, with sporadic gravel clasts. The bioclasts include molluscs, bivalves and corals. Similar greensand was also recovered in BGS boreholes 82/10 and 84/02, and in wells 202/08-1 and 206/05-1, which proved thicknesses ranging from 3.3 m (84/02) to 144 m (202/08-1). Subordinate interbedded lithologies include dark green mud-

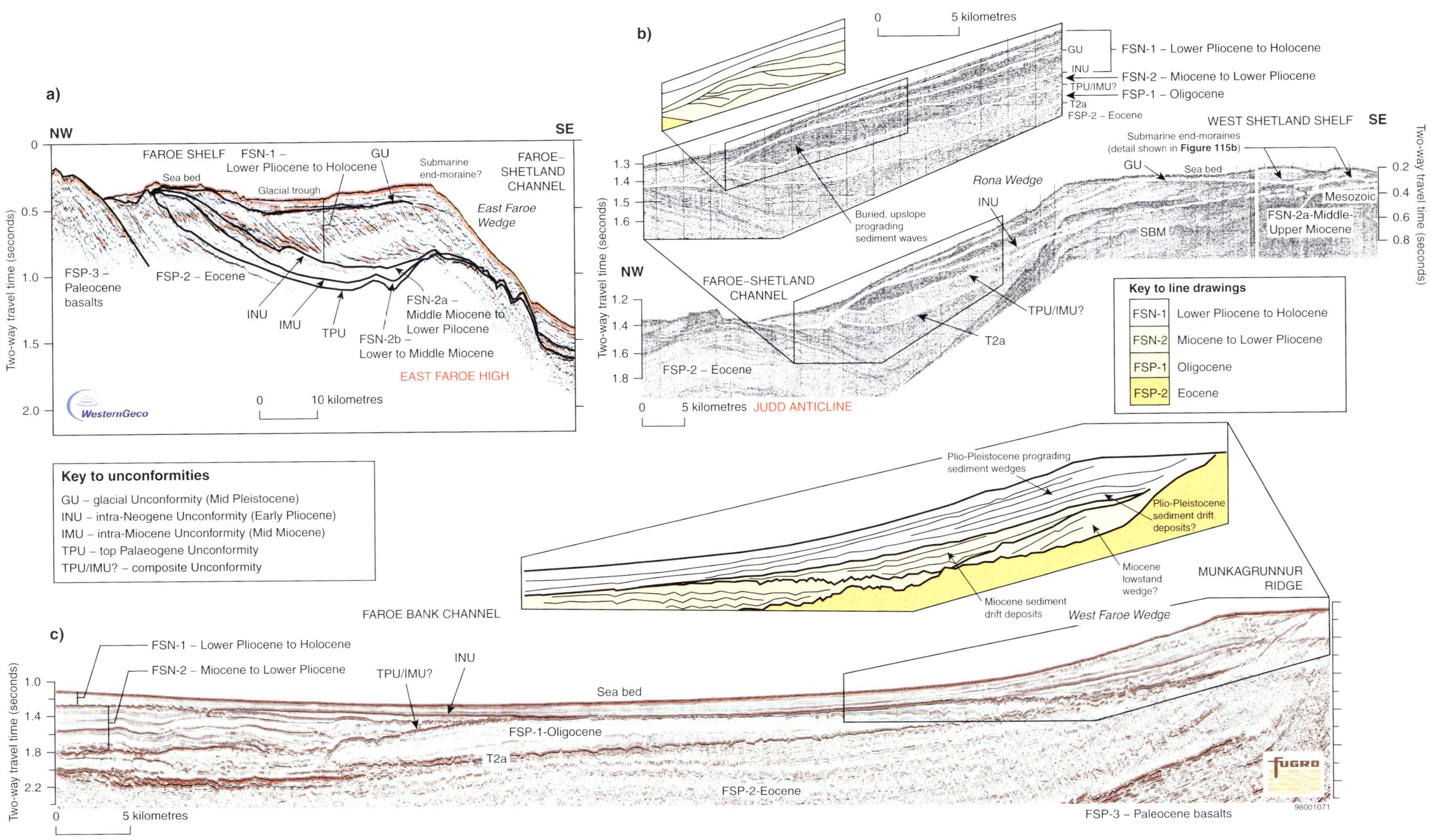

Figure 114 Seismic profiles showing examples of Lower Pliocene to Holocene prograding sediment wedges; a) East Faroe Wedge; b) Rona Wedge and; c) West Faroe Wedge. Note also the expression of key Neogene unconformities, especially the erosive character of the Early Pliocene intra-Neogene Unconformity (INU) in the Faroe Bank Channel (profile c) and the Mid Pleistocene Glacial Unconformity (GU) (profiles a and b). Detail of moraines (profile b) expanded in Figure 115b (see text). Profile (a) modified from STRATAGEM Partners, (2002) and profile (b) modified from Stoker et al., (2005b). For locations of profiles, see Figure 107. Seismic data courtesy of WesternGeco (profile a) and Fugro Multi Client Services (profile c).

stone, sandy siltstone, sandy limestone, lignite and thin mudstone. In well 202/08-1, a basal mudstone facies underlies the sandstone. This variation is commonly reflected in the gamma-ray well log signature of this unit (Stoker, 1999), although the fluctuating trace may also reflect more subtle changes in the content of the glauconite and clay mineral assemblages.

In borehole 77/07, an upward increase in gamma-ray well log values reported by Evans et al. (1997) marks a change from a lower, glauconite-rich, sandstone unit, to an upper unit consisting of interbedded fine-grained sandstone and sandy siltstone, with a greatly reduced glaucony content. There is also a change in clay mineral assemblages with the lower sandstone unit dominated by kaolinite and mica, and the upper interbedded unit distinguished by high smectite and mica, with lesser kaolinite and chlorite. Uncemented samples have a high proportion of mud matrix but remain relatively porous, with 15 to 18% porosity. Petrological and geochemical analysis of the glaucony in borehole 77/07 revealed that it is comparatively well evolved, with a K_2O content of about 7 to 8% (Evans et al., 1997), compared with maximum-recorded K_2O contents of 9% (Odin and Morton, 1988). A variety of initial components have been replaced by glauconite, including detrital minerals and carbonate bioclasts. Some grains show a poorly defined oolitic texture.

The Muckle Ossa Sandstone is characterised by a rich and diverse microfossil assemblage (Stoker, 1999). The calcareous microfauna is indicative of an Early to Mid Miocene stratigraphical range, whereas the microflora implies a late Early to Late Miocene age range. In the lower part of the unit, the foraminiferal and calcareous algal assemblages are characterised by the earliest Miocene *Bolboforma spinosa*/*Bolboforma rotunda* bioevent, with the earliest Mid Miocene *Heteropleura dutemplei peelensis* bioevent identified in the upper part of the section. The dinoflagellate cyst assemblage consists of several distinct biomarkers, including *Labyrinthodinium truncatum, Unipontidinium aquaeductum, Palaeocystodinium golzowense, Systematophora placantha* and *Apteodinium spiridoides*. These dinoflagellate cyst species are indicative of a warm, marine environment. The common overlap range of the foraminiferal and dinoflagellate cyst markers in BGS boreholes 77/07, 77/09, 84/02 and 90/3 corresponds to the NN4/5 biozones (Stoker, 1999), which suggests that species with a range older than NN4 may be reworked. Strontium isotope ($^{87}Sr/^{86}Sr$) analyses were undertaken on bivalve fragments and foraminifera from the Muckle Ossa Sandstone in BGS boreholes 77/07, 77/09, 84/02 and 90/03 (Stoker, 1999). The measured $^{87}Sr/^{86}Sr$ ratios gave the following ages: 77/07, 18.35–20.73 Ma; 77/09, 5.5–8.0 Ma; 84/02, 16.95–18.45 Ma; 90/03, 16.34–16.78 Ma. All of the ratios confirm a Miocene age, but the wide variation in age estimate further implies that reworking processes may have been operative.

On the West Shetland Slope, the Miocene section displays a more variable character. Small, lowstand wedges and mounded slump deposits, up to 180 m thick, occur on the upper to midslope (Stoker, 1999). The wedges are characterised by a seismically layered, parallel-to-sigmoidal, prograding reflection configuration (Figure 113), with indications of toplap and downlap. At least two phases of wedge development are preserved on the slope, with the younger wedge stacked on top of the older wedge. The slump deposits are seismically structureless and commonly mounded, and generally overlie the wedges. Well 204/28-1 recovered traces of glauconitic sand from the slump package (Figure 106).

On the lower slope, eroded remnants of upslope-migrating sediment drifts, up to 120 m thick, are preserved. These deposits are seismically layered, commonly exhibit mounded waveforms, and show downlapping and onlapping reflection terminations (Figures 113b and 114b). The drift deposits appear to overlap with the wedges (Stoker, 1999). At the base of the slope, the highly irregular, sculpted morphology of the composite Top Palaeogene to intra-Miocene Unconformity is locally infilled by sediment drift deposits, which may have an age range that extends into the Early Pliocene (Stoker et al., 2005b; Figure 112b).

The Muckle Ossa Sandstone is interpreted to represent a marine transgression that flooded the paralic environment characteristic of Late Oligocene time (Evans et al., 1997). This rise in relative sea level is consistent with the stacking geometry of the upper slope, prograding, lowstand wedges which backstepped upslope in response to relative sea level change. The deeper water sediments were greatly influenced by bottom currents that created the mounded, upslope-accreting sediment drifts on the slope. Overlap between the wedges and the drifts indicate that they developed simultaneously as companion systems. Deposition on the shelf was terminated by a major phase of erosion, associated with the development of the intra-Neogene Unconformity, which truncated the underlying Middle to Upper Miocene strata prior to seaward tilting of the margin and the subsequent progradation of the FSN-1 megasequence (Stoker 2002; Stoker et al. 2005b; Figures 112a and 114b). The mounded, mass-flow slump deposit preserved on the slope, above the Miocene wedges and sediment drift (Figure 113), may have accumulated as a latest Miocene to Early Pliocene slump in response to erosion of the Muckle Ossa Sandstone (Stoker, 1999).

Faroe Bank Channel

Two discrete accumulations of Miocene sediments are interpreted within the Faroe Bank Channel (Figure 106). The thickest accumulation, in excess of 500 ms TWTT, occurs on the southern flank of the channel adjacent to the Wyville Thomson Ridge and Faroe Bank High, whereas a thinner sequence, up to 300 ms TWTT, is located on the western flank of the Munkagrunnur Ridge. Both accumulations overlie a highly irregular unconformity cut into Oligocene and older strata (Figures 88b, 110 and 114c). The age range of these proposed Miocene sections remains uncertain. STRATAGEM Partners (2003) inferred that sediments belonging to the FSN-2a and 2b units were present in both of the accumulations, whereas Sørensen (2003) suggested that Lower Miocene deposits were only present in the Wyville Thomson Ridge–Faroe Bank High region. In contrast, Johnson et al. (2005b) and Stoker et al. (2005b, c) have suggested that the basal unconformity may represent a composite Top Palaeogene to intra-Miocene surface. This would imply that the overlying sediments are all predominantly of Mid Miocene and younger age. In light of this ambiguity, the Faroe Bank Channel sediments remain largely undivided. The eroded top of both accumulations represents truncation at the Early Pliocene intra-Neogene Unconformity, and suggests that these strata were formerly more extensively developed in the Faroe Bank Channel.

The Miocene deposits display a predominantly seismically layered reflection configuration that onlaps the basal unconformity (Figures 110 and 114c). The reflections are predominantly subparallel and continuous, though localised convergence occurs adjacent to the Wyville Thomson Ridge, which may be indicative of intraformational unconformities. A localised wedge-shaped deposit is observed in the lower part of the section adjacent to the Munkagrunnur Ridge, and has been interpreted as a lowstand fan deposit (STRATAGEM Partners, 2003; Stoker et al., 2005a; Figure 114c). This fan is onlapped and buried by younger Miocene strata. The basinal reflections are commonly undulatory, which may be a consequence of the irregular nature of the underlying surface, or represent some degree of deformation during burial akin to that observed on the East Faroe margin (see above). No lithological data exist for these deposits.

The association of a highly sculpted basal unconformity and the marked change in onlap pattern adjacent to the Wyville Thomson Ridge have been linked to a major growth phase of the ridge, concomitant with a change (most probably a deepening) in the bathymetric threshold of the complementary syncline that is the Faroe Bank Channel (Johnson et al., 2005b; Stoker et al., 2005b and c). Although Palaeogene strata onlap the Wyville Thomson Ridge, indicating growth of the ridge throughout that interval, the angle of onlap is most extreme where the Miocene strata abut the ridge. This has been interpreted to indicate a Miocene growth of the anticlinal ridge of the order of 1 km (Stoker et al., 2005c). In contrast, the strongly erosional nature of the basal unconformity is not replicated in the underlying Palaeogene section. This is taken to indicate that its formation most probably was a consequence of a vigorous pulse or series of pulses of bottom-current erosion during the early Neogene, when the barrier that was the Greenland–Scotland Ridge was breached by deep-water flow through the Faroe Conduit (Figure 92).

It has been suggested that during periods of strong bottom flow through the Faroe Conduit, the bottom water was piled up in the southern part of the Faroe–Shetland and Faroe Bank channels, thus allowing some of the deep water to overflow the Wyville Thomson and Ymir ridges into the north Rockall Basin (Andersen and Boldreel, 1995). An outlier of inferred Miocene sediment in the syncline between the Wyville Thomson and Ymir ridges (Figures 106 and 110), correlated to the RPb megasequence of the Rockall Basin (Stoker et al., 2005a), may represent a remnant of this overflow. The deep-water overflow water continued to move northwards into the Iceland Basin depositing Miocene strata within the passageway between the Faroe Bank High and Bill Bailey's High (Boldreel et al., 1998).

Pliocene to Holocene

The sediments deposited during the late Early Pliocene to Holocene interval comprise the FSN-1 megasequence, which is widely distributed throughout the report area (Figures 94 and 107). The thickest accumulation is located in the north-east part of the report area, where over 1.2 s TWTT of sediment is preserved on the North Sea Fan. Similar albeit smaller and thinner accumulations occur along the outer part of the West Shetland and Faroe margins, where the slope apron locally exceeds 300 and 600 ms TWTT in thickness, respectively, reflecting the development of the Rona, Foula, East Faroe and West Faroe prograding sediment wedges (Figure 107). By way of contrast, the Lower Pliocene to Holocene succession is commonly <100 ms TWTT thick on the inner parts of the margins, being thin to absent on rock platforms on the shallowest parts of shelves and banks where only a veneer of sediment is present. In the Faroe–Shetland and Faroe Bank channels, the deep-water succession is generally less than 200 ms TWTT in thickness, though in the north-east Faroe–Shetland Channel the sequence locally exceeds 400 ms TWTT in thickness in association with sediment drift development. On the North Faroe margin, an outer shelf wedge up to 200 ms TWTT thick thins to less than 100 ms TWTT in the Norwegian Basin.

The change in sedimentation style in the Early Pliocene represents a regional response to differential uplift and subsidence within the Faroe–Shetland region, instigated by epeirogenic movements (at about 4 Ma), which changed the sediment supply to the West Shetland and Faroese margins and modified the deep-water bottom-current system through submarine erosion (Stoker et al., 2005a and b). On seismic profiles, the intra-Neogene Unconformity marks the stratigraphical expression of this tectonic phase (Figures 110, 111, 112 and 114). Significantly, the onset of progradation of the West Shetland margin predates the widespread glaciation of Britain and its offshore region, which is dated from the early Mid Pleistocene at about 0.44 Ma (Stoker et al., 1994; Stoker, 2002), though glacial sediments do contribute markedly to the upper part of the FSN-1 megasequence above the Glacial Unconformity (Figures 94 and 108).

A previous subdivision of the West Shetland margin succession resulted in the plethora of seismic-stratigraphical units that is indicated in Figure 109. The abundance of stratigraphical units comprising the succession above the Glacial Unconformity presents a complexity that is more apparent than real, and is largely a reflection of the diverse sediment–landform association inherent within the glacial depositional system. This stratigraphical scheme has not been updated since its publication (Stoker et al., 1993), and a revision of the scheme is beyond the scope of this report. Although the scheme was broadly applied to the Faroe–Shetland Channel succession, no high-resolution seismic-stratigraphical subdivision of the basinal succession, which was treated as a single sequence, was possible beyond the limit of the slope apron. On the East Faroe margin, the section above the Glacial Unconformity was more recently divided into a number of sub-units, with particular emphasis on the basinal succession (Faroes GEM Network, 2001a, b, c and d; Austin, 2004). A tentative correlation with the West Shetland margin is shown in Figure 109. On both the West Shetland and East Faroe margins, reference to specific units will be made where appropriate for adding detail to the interpretation of megasequence FSN-1; however, for a fuller definition of the various units, the reader is referred to Stoker et al. (1993), Faroes GEM Network (2001a, b, c and d) and Austin (2004).

For ease of description, the Pliocene to Holocene succession is presented separately from the West Shetland margin, the Faroe–Shetland Channel, the Faroe margin, the Faroe Bank Channel and adjacent banks, and a brief summary from the North Sea Fan. This is followed by a general discussion on glacial history and ice-sheet reconstruction in the Faroe–Shetland region. For this, reference is made to both British and North West European chronostratigraphic stage names (see below); the latter nomenclature is adopted by Faroese Quaternary workers. As the emphasis throughout this section has been to summarise the Cenozoic succession to series level, ultra-high-resolution studies, such as those focused on latest Pleistocene to Holocene millennial-scale events associated with the deglaciation of the last British and Faroese ice sheets, are beyond the scope of this report. For more information on the latter, with respect to the report area, the reader is directed towards the work of Rasmussen et al. (1996), Abrantes et al. (1998), Kuijpers et al. (1998), Lassen et al. (1999, 2001), Rasmussen et al. (2002), STRATAGEM Partners (2002, and references therein) and Andresen et al. (2006).

West Shetland margin

The prograding, Rona and Foula sediment wedges, which dominate the West Shetland margin south of 61°N, overlap to form an elongate, outer shelf to upper slope, depocentre (Figure 107). To the north of 61°N, the prograding shelf-margin deposits are thinner, with the major accumulation developed on the lower slope as part of a sediment drift succession preserved in the northern Faroe–Shetland Channel. North of Shetland, the shelf-margin succession merges with the North Sea Fan.

The development of these prograding wedges has resulted in a seaward build-out of the West Shetland margin by up to 40 km since the Early Pliocene, with the main advance of the shelfbreak having occurred prior to the onset of Mid Pleistocene glaciation (Stoker et al., 1993). The construction of the prograding wedges can be divided into two main phases separated by the formation of the Glacial Unconformity, which marks the onset of extensive shelf-wide glaciation at about 0.44 Ma (Stoker, 1995). In the late Early Pliocene to Mid Pleistocene interval, progradation of the shelf-margin was initially restricted to the contemporary outer shelf, though with time extended by downlap into deeper water (Stoker, 1999). Since the Mid Pleistocene, progradation has extended from the shelf to the floor of the Faroe–Shetland Channel (Figure 114b), interspersed with sporadic episodes of mass failure on the West Shetland Slope, such as the Afen, Palaeo-Afen and Miller slides (Wilson et al., 2004; Evans et al., 2005; see Chapter 11).

On seismic profiles, the Rona and Foula wedges display a variable sigmoid-oblique to oblique-parallel reflection configuration (Figures 112, 113 and 114). These patterns are expressed by clinoforms that form continuous, medium to high amplitude reflectors separating thicker, acoustically structureless to chaotic packages of mass-flow deposits (Stoker, 1995; Davison and Stoker, 2002). On the outer shelf, topsets are locally preserved below the Glacial Unconformity, and the occurrence of lowstand–highstand couplets implies some degree of relative sea level change during the early phase of wedge construction (Stoker, 2002). Above the Glacial Unconformity, oblique-parallel clinoforms are predominant on the West Shetland Slope, whereas the shelf succession displays an aggradational character that internally consist of stacked sheet-like sequences interpreted as composite glacial units containing ice-contact, ice-proximal and glacimarine deposits (Stoker et al., 1993; Davison, 2005; Figure 115). It is the glacial geomorphic shaping of these sequences into sub to proglacial bedforms and deposits, such as moraines and ice-contact fans (Stoker and Holmes, 1991; Long et al., 2004; Davison, 2005; Stoker et al., 2006), which has contributed to the complexity of the seismic stratigraphy above the Glacial Unconformity (Figure 109).

The nature of the West Shetland margin succession has been tested in BGS boreholes, and a few commercial wells, though the bulk of the information relates to the section above the Glacial Unconformity (Figure 108 and Table 7). In general terms, the Glacial Unconformity separates underlying sand-dominated strata from overlying mud-dominated sediments. On the West Shetland margin, the sediments below the Glacial Unconformity were previously assigned to the Sinclair and Morrison (unit 1) sequences (Stoker et al., 1993; Figure 109). These sequences comprise the main part of the Rona and Foula wedges below the Glacial Unconformity, with the Sinclair sequence representing the oldest, innermost part of the wedges (Figure 115b).

The Sinclair sequence locally exceeds 100 m in thickness, and has been sampled in BGS boreholes 77/09, 82/10, 82/11, 84/01 and 1b, 84/02 and 04 (Figures 107 and 108), which recovered pebbly, shelly, glauconitic sand with subordinate, thin, interbedded mud of deltaic and shallow-marine origin (Cockcroft, 1987; Stoker, 1999; STRATAGEM Partners, 2002; BGS, 1986b; 1991b). The sand is compact and firm, varies in colour from grey to brown to olive, is generally poorly sorted, very fine to very coarse-grained, and slightly muddy with common scattered pebbles and shell debris. The interbedded mud varies from very dark grey to yellow-brown, is commonly firm to hard, and also contains scattered pebbles and shell fragments. Well 202/02-1 proved at least 65 m of sandy deltaic deposits, whereas about 80 m of interbedded sand and mud were tested in well 205/22-1A (Figure 108). Shells, pebbles and sand were recovered in wells 202/03a-3, 205/21-1A, 206/08-1A, -2, 206/12-1, 206/05-1 and 207/01-1 (Figure 107). There are also subordinate interbeds of laminated to thin-bedded sand and mud, rare gravel beds (borehole 84/04), sporadic lignite (wells 202/08-1 and 204/30-1) and thin limestones (well 206/08-2). Micropalaeontological data indicate that cold climatic conditions prevailed during deposition (Cockcroft, 1987).

The Sinclair sequence has been dated on the basis of biostratigraphical data as Pliocene to Early Pleistocene in age (BGS 1986b, 1991b). In borehole 77/09, the presence of the Early Pliocene dinoflagellate cysts *Amiculosphaera umbracula* and *Spiniferites splendicus*, and the planktonic foraminifer *Globorotalia crassaformis*, indicate that the proximal part of the Rona Wedge correlates with the NN13–14 biozone, which implies an age of about 4 Ma for the onset of wedge development (Stoker, 2002).

The Morrison (unit 1) sequence is up to 200 m thick on the outer shelf and upper slope, but commonly thins downslope and is locally absent in the Faroe–Shetland Channel. On the northern part of the margin, it passes laterally into the Faroe–Shetland Channel sequence on the mid to lower slope. This part of the shelf-margin succession has been tested in a number of wells and boreholes (Stoker et al., 1993; Stoker, 1999; BGS 1990b, 1991b; Figures 107 and 108). In the area of the Rona Wedge, well 204/30-1 proved a 20 m thick basal unit of pale grey to white mud overlain by at least 140 m of diamicton (a poorly sorted admixture of mud, sand and gravel), which may represent prodeltaic

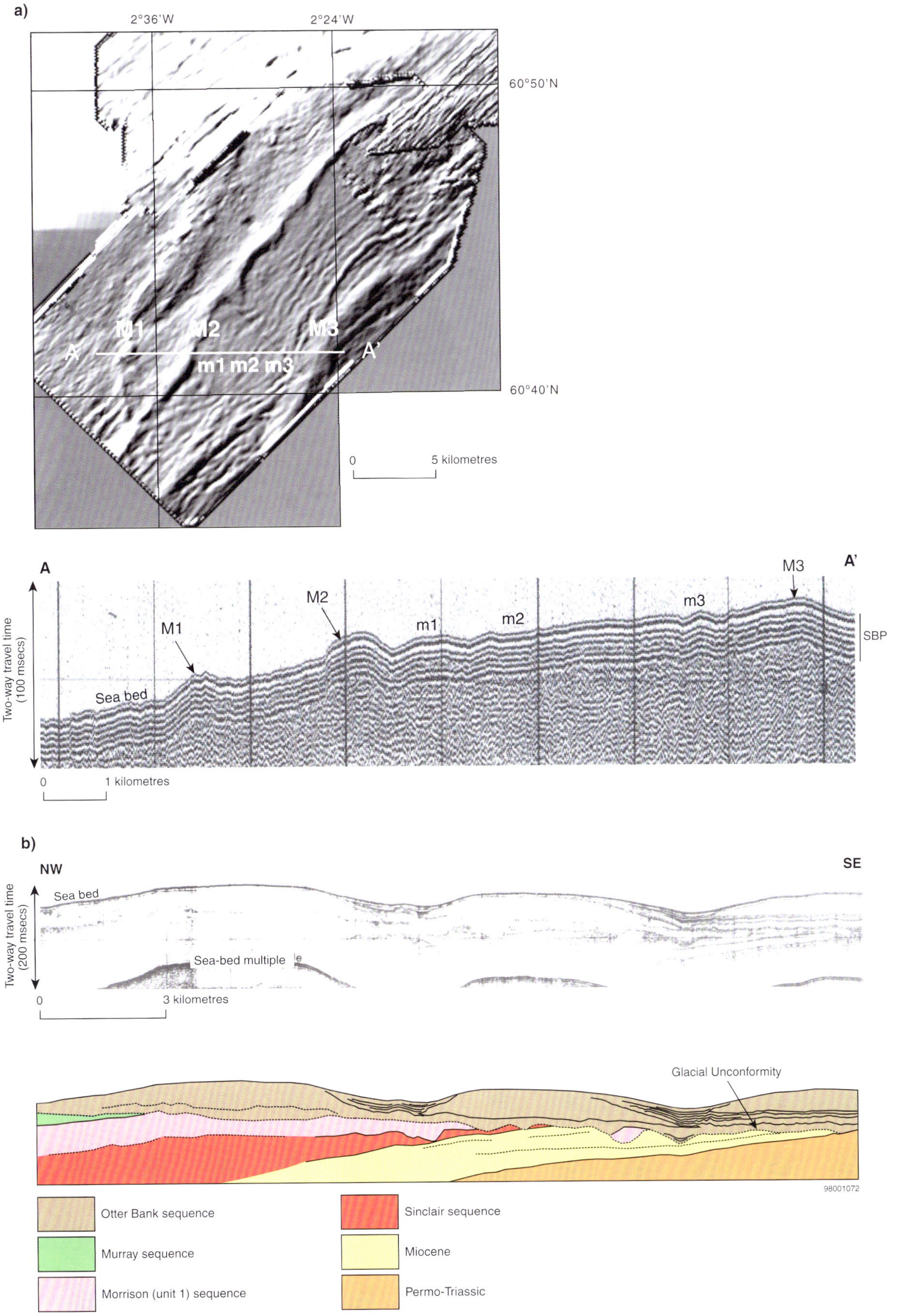

Figure 115 Moraines preserved in the Otter Bank sequence on the West Shetland Shelf; a) Sea-bed image (illuminated from the north-west) and BGS sparker profile showing moraine ridges at various scales (larger, MI–M3 and smaller, m1–m3). For location, see Figure 107. Sea-bed image derived from Figure 138; b) BGS sparker profile showing large moraines, which become younger to the south-east, and are associated with acoustically layered, ponded, stillstand deposits (modified from Stoker et al., 2006). For context of profile, see Figure 114b. Abbreviations: SBP= sea-bed pulse.

sedimentation followed by extensive mass-flow deposition as the shelf-margin prograded (Stoker et al., 1993). The diamicton consists of pebbles and cobbles of quartzite, igneous and metamorphic lithologies set in a pale brown to grey, shelly, clay matrix. Well 205/22-1A penetrated about 120 m of mud with sporadic, thin, pebbly and shelly sands.

From the Foula Wedge, borehole 84/04 (Figure 107) proved 57 m of predominantly pebbly, shelly, muddy, diamicton with sporadic sands. Dinoflagellate cysts preserved in these sediments, such as *Bitectatodinium tepikiense* and *Operculodinium centrocarpum* are characteristic of deposition in cold conditions (Cameron and Holmes, 1999). Pebbly, shelly, sand was collected in the nearby well 206/08-2. Farther to the north-east, wells 206/05-1 and 207/01-1 recovered pebbly, sandy deposits from the outer shelf, whereas well 208/27-1 proved about 110 m of interbedded mud and pebbly, shelly, sand with thin limestone. On the West Shetland Slope, well 214/27-1 tested a mud-dominated section, whereas well 208/17-1 revealed a basal sandy section overlain by mud (STRATAGEM Partners, 2002).

The sediments above the Glacial Unconformity have been tested in many BGS boreholes along the length of the West Shetland margin, and are characterised by a variable assemblage of diamicton, gravel, sand and mud (Figures 107, 108 and Table 7). Many of the stratigraphical units preserved on the outer shelf, such as the MacDonald, Wyville Thomson Ridge, Mackay, Rona, Murray, Skerry and Otter Bank sequences, and the Ferder Formation (Figure 109), are dominated by diamicton, sand and gravel. The diamicton is commonly dark grey-brown, firm to hard and very poorly sorted, varying from crudely stratified to massive, and clast to matrix-supported, to massive and matrix-supported (Stoker, 1988; Stoker et al., 1993; Davison, 2005). The mud content is generally between 60 and 75%, and sand is between 25 and 40%. The variation between clast-poor and clast-rich units is reflected in the range in pebble content from 1 to 45%. Clast composition is highly variable and includes grey quartzite, gneiss, amphibolite, red sandstone and siltstone and dolomitic limestone. All of these lithologies are typical of north-west Scotland and the adjacent shelf. Recovered clasts are up to 8 cm maximum dimension with a mode of about 1 to 2 cm, which vary from angular and faceted to rounded and smooth. Shell debris is rare to common, and often considerably reworked. Lamination in the stratified diamicton is a result of alternating sandy and muddy layers, 3 to 10 mm thick. Pebbles locally distort the lamination (Davison, 2005).

These diamicton-dominated sequences were probably constructed by a variety of ice-contact and ice-marginal processes active during the periodic growth and decay of the British Ice Sheet (Stoker and Holmes, 1991). The Otter Bank sequence includes numerous mounded accumulations that form prominent sea-bed ridges that range from 5 to 50 m high and 1.5 to 8 km wide (Figure 115), and, as a system of moraines, can be traced for up to 200 km along the outer West Shetland Shelf (Davison, 2005). These ridges are interpreted as submarine end-moraines or morainal banks (Stoker et al., 1993; Long et al., 2004; Davison, 2005; Stoker et al., 2006). An ice-contact fan, 50 to 150 m thick, up to 10 km wide and at least 25 km long has also been reported in association with the MacDonald sequence at the junction of the Hebrides–West Shetland Shelf and Wyville Thomson Ridge (Stoker and Holmes, 1991). These ice-contact bedforms are interpreted to mark former positions of the British Ice Sheet, which reached the edge of the West Shetland Shelf (see below). In Figure 115a, smaller moraines, such as m1 to m3, may represent short-term halts in the retreat of the ice sheet, whereas the larger moraines (M1 to M3) mark a relatively longer halt (stillstand).

Discrete, ponded accumulations of acoustically layered strata, belonging to the upper part of the Otter Bank and the Stormy Bank sequence, locally overlie the moraines (Figure 115b), and were deposited at points of stillstand during ice-sheet decay. In older, buried sequences, such as the Murray sequence and the Ferder Formation, any diagnostic landform morphology has been eroded prior to or during the deposition of the overlying younger sequences, resulting in their sheet-like geometry. Processes of lodgement, ice-marginal mass-flow, iceberg rafting and iceberg scouring have been invoked for the origin of these diamicton-dominated sequences, in association with mass-flow sands, and glacifluvial sands and gravels (Stoker, 1988; Stoker and Holmes, 1991; Davison, 2005). The common occurrence of a restricted, low-diversity, yet indigenous microflora and microfauna, including dinoflagellate cysts such as *Bitectatodinium tepikiense* and *Protoperidinium* (round brown) cysts, and benthic foraminifera such as *Elphidium clavatum, Cassidulina reniforme, Cibicides lobatulus* and the left-coiling planktonic foraminifera *Neogloboquadrina pachyderma*, supports an ice-marginal setting.

On the West Shetland Slope, the bulk of the prograding shelf-margin succession above the Glacial Unconformity is assigned to the Morrison (unit 2) sequence, although the late to postglacial MacAulay sequence is locally differentiated in the uppermost part of the succession (Figure 109). At the edge of the West Shetland Shelf, the diamicton-dominated shelf sequences interdigitate with, or pass laterally into, the slope apron, which developed during phases of maximum shelf-edge glaciation (Stoker, 1995). During these phases, vast amounts of glacial sediment were delivered directly to the Rona (Figures 113 and 114) and Foula wedges.

BGS borehole 99/03 penetrated a 56.3 m section of the Rona Wedge preserved on the lower West Shetland Slope (Figures 95 and 112c), and recovered massive diamictons up to 17 m thick from the acoustically structureless packages of hummocky mass-flow deposits (Davison and Stoker, 2002). The diamictons are predominantly muddy, consisting of 60 to 75% mud, 25 to 35% sand and 1 to 5% gravel. The latter are matrix supported, and are up to 8 cm diameter. Shelf-derived arctic foraminifera are also present. The mass-flow deposits are separated by thinner (<10 m), acoustically layered packages that commonly represent the higher-amplitude clinoforms separating the mass-flow packages, and which consist of sandy mud with scattered pebbles, up to 11 cm diameter, and subordinate sand and gravel layers that locally form lag deposits.

These sediments contain a mix of cold and temperate, shelf and deep-water microfossils, which implies intervals of reduced sediment supply to the slope, such as in an interstadial or interglacial phase (Stoker, 1995). A number of adjacent short sediment cores (<6 m) revealed a more variable sandy and muddy diamicton facies (Stoker, et al., 1991; Paul et al., 1993). A 10.5 m thick section of debris-flow diamictons was proved in borehole 85/01 from the Foula Wedge, on the mid West Shetland Slope (Figures 107 and 116), overlain by

3.5 m of glacimarine and marine mud of the MacAulay sequence (Stoker, 1995). The latter represents the uppermost clinoform unit on the Foula Wedge. The occurrence of both onlapping and draped reflection patterns is indicative of bottom-current activity combined with hemipelagic rain out of sediment during the deposition of the MacAulay sequence.

Between 60°30' and 61°N and 3° and 4°W, the surface of the West Shetland Slope is cut by a number of linear gullies that extend from the mid to the lower slope, where they connect with a series of basin-floor fans (Figure 116). The latter form the uppermost part of the Morrison (unit 2) sequence, locally referred to as the Morrison Y subsequence (BGS, 1991b). The gullies are about 25 km long, 50 to 250 m wide, and between 5 and 30 m deep (Bulat and Long, 2001, 2005; Masson, 2001). They are straight-sided and subparallel along most of their lengths, though there is some indication of merging of gullies in the midslope region. They can be followed upslope to about 450 m water depth, where they are traced into the area of debris-flow deposits associated with the Foula Wedge. The gullies appear to cut some of the debris-flow deposits on the midslope, which imply that they either postdate, or are the last increment in, the generation of the uppermost part of the Foula Wedge.

The basin-floor fans are triangular-shaped with a base length of about 15 km (Bulat and Long, 2005). Individual fans have a relief of about 20 m, though in areas of overlap this can increase to 50 m. The association of gullies and fans has been interpreted to reflect an origin due to high-energy, downslope flow (Bulat and Long, 2005; Davison, 2005). The fan deposits are overlain by sediments of the late to postglacial MacAulay sequence, though it is unclear whether these represent a subsequent drape or contemporary overbank deposits. Moreover, the reason for the localisation of these features on this part of the slope remains uncertain. Further consideration of their origin and implications for ice-sheet reconstruction is presented below.

Faroe–Shetland Channel

Basinal sediments of the FSN-1 megasequence are best developed in the north-east part of the Faroe–Shetland Channel, where the succession exceeds 200 m in thickness in water depths greater than 1.2 km, and migrates up the West Shetland Slope to a water depth of about 600 m (BGS 1991a and d; Figures 107, 111 and 117). The absence of a comparable development on the East Faroe Slope has resulted in an asymmetric geometry to this part of the FSN-1 megasequence, which has been termed the West Shetland Drift (Figure 107) by Knutz and Cartwright (2003, 2004). The thickness of the basinal succession may be, in part, due to the interbedding of debris-flow deposits derived from the West Shetland and East Faroe margins (BGS 1991a, b; Knutz and Cartwright, 2003, 2004). At the south-west end of the Channel, where the Rona Wedge dominates the slope apron, the basinal deposits are locally absent in the area of the Judd Deeps.

The entire basinal succession has previously been assigned to the Faroe–Shetland Channel sequence in the UK sector (Stoker et al., 1993), and to the equivalent Unit 1C in the Faroese sector (Faroes GEM Network, 2001a, b, c and d; Austin, 2004; Figure 109). Unit 1C has been further divided into three sub-units, $1C_I$ to $1C_{III}$. Unit 1C incorporated the undifferentiated basinal equivalent of units FPC-D.2 and FPC-D.3 of Andersen et al. (2000) (Table 6) established on the East Faroe margin.

Faroes GEM Network (2001a, b, c and d) correlated the entire stratigraphical range of Unit 1C to the Late Pleistocene and Holocene, though no reliable age dating exists other than for sub-unit $1C_I$, for which Atomic Mass Spectroscopy (AMS) dates indicate a Late Pleistocene to Holocene age (see also Kuijpers et al., 1998). Austin (2004) subsequently refined this scheme and assigned sub-unit $1C_{III}$ an undefined Pleistocene age, albeit above the Glacial Unconformity, which implies a Mid Pleistocene age. Both of these age models are incompatible with the Plio-Pleistocene assignment of the Faroe–Shetland Channel sequence (Stoker et al., 1993), which is confirmed by well 214/04-1 from which the presence of the foraminifera *Sigmoilopsis schlumbergeri* and *Neogloboquadrina atlantica* are regarded as benthic and planktonic index species, respectively, for the Early Pliocene (Davies and Cartwright, 2002; Knutz and Cartwright, 2003; Figures 108 and 111). Indeed, Middle Pleistocene sediments (Eemian, marine isotope stage 5: MIS 5) have been proved from as shallow as 5 m below the sea bed in BGS vibrocore 61-04/37, in the central part of the Faroe–Shetland Channel (Akhurst et al., 2002; Figure 107). On the basis of these data, it seems probable that the stratigraphical range of Unit 1C, especially sub-unit $1C_{III}$, will extend beyond the Mid/Late Pleistocene into the Pliocene to Early Pleistocene (see Figure 109), though a detailed basinal stratigraphy remains to be developed.

The basin-floor succession is acoustically layered and commonly displays subparallel, flat-lying, continuous reflections that display evidence of downlap and onlap onto the intra-Neogene Unconformity (locally composite with the Top Palaeogene and intra-Miocene unconformities), internal convergence over highs on the channel floor, and low-angle truncation of reflections within the basin (Stoker et al., 1998; Akhurst et al., 2002; Howe et al., 2002; Knutz and Cartwright, 2003, 2004). This seismic reflection configuration has been interpreted to be the result of bottom-current activity, and forms the basinal component of the West Shetland Drift (Knutz and Cartwright, 2003, 2004). At the south-west end of the Faroe–Shetland Channel, where the bottom currents are probably strongest (Masson et al., 2004), erosion of the sea bed is evident where the acoustically layered basinal sediments pinchout onto the exhumed upper surface of interbedded glacigenic debris-flow deposits and the exposed bedrock associated with the Judd Deeps (Stoker et al., 1991, 1998). On the margin of the Judd Deeps, BGS borehole 99/06 (Figure 112c) penetrated a 22 m thick section of Pleistocene to Holocene sediments plastered against the wall of the deeps, displaying onlap onto the flank, and downlap onto the floor of the scour. The sediments consist of 10.5 m of soft, dark greenish grey to greyish green, sandy mud with scattered pebbles, overlying about 3 m of brown to brown-grey, medium to coarse-grained sand with thin muddy interbeds, resting on a basal gravelly interval, about 9 m thick, overlying Eocene bedrock.

By way of contrast, Damuth and Olson (2001) interpret the floor of the Faroe–Shetland Channel to be underlain by thin conformable sediments of glacimarine and hemipelagic origin, with less common turbidites and debris-flows. The West Shetland and Faroes shelf-margins have undoubtedly been significant sources of sediment,

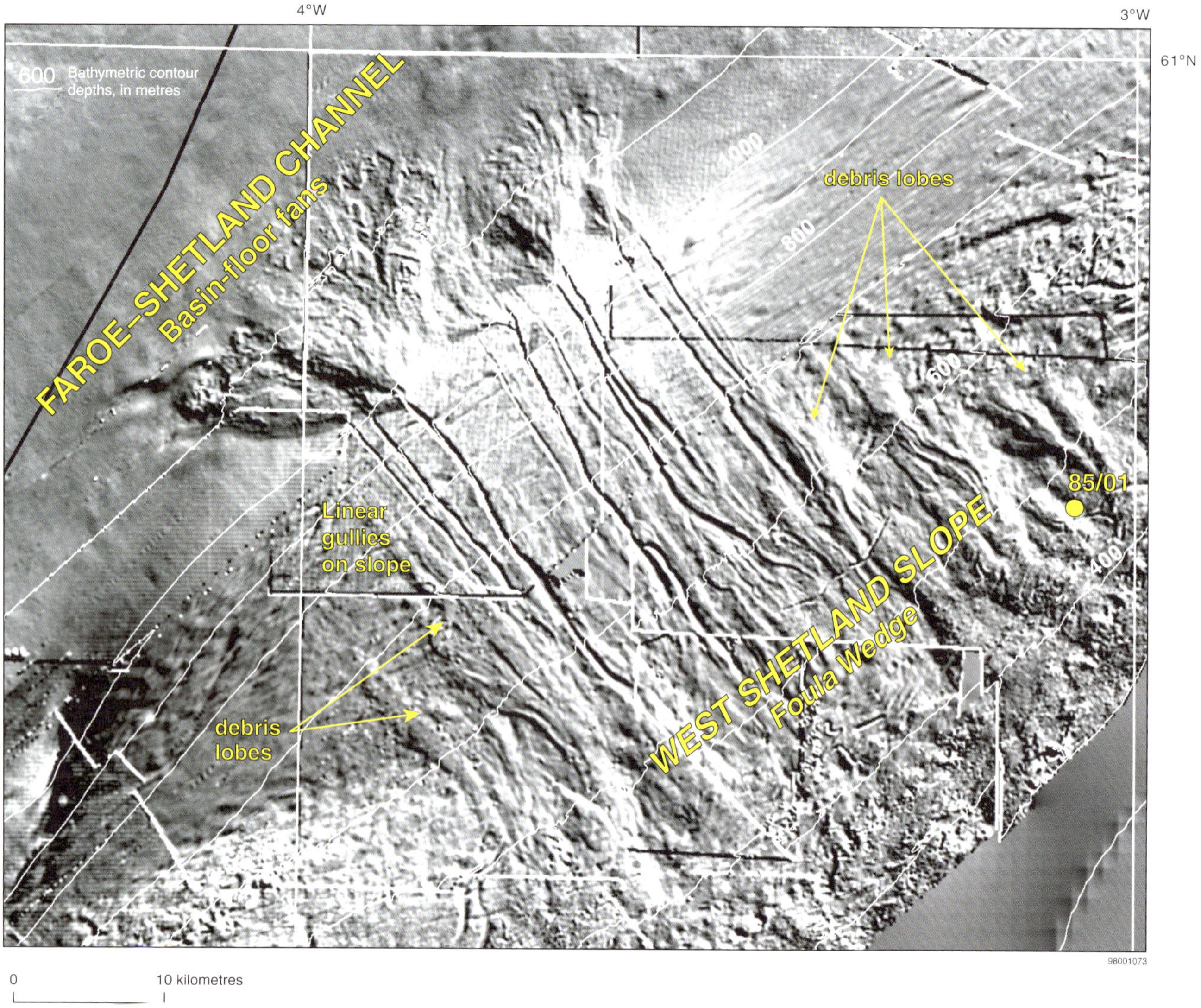

Figure 116 Sea-bed image (illuminated from the north-east) showing debris lobes associated with the Foula Wedge, location of BGS borehole 85/01, and the linear gullies that terminate in basin-floor fans at the base of the West Shetland Slope (modified from Bulat and Long, 2005). For location of image, see Figure 107. Sea-bed image derived from Figure 138.

with mass-flow (including turbidity currents) and hemipelagic fluxes important in the transfer of sediment into the Channel (Stoker, 1995; Kuijpers et al., 2001; Stoker et al., 2005a and b). However, detailed sedimentological evidence from numerous BGS short sediment cores indicates that strong contour current erosion and redistribution of sediment has occurred along the channel floor. This evidence includes; i) a range of sediment facies typical of contourites including continuous bioturbation with very rare primary sediment structures, grain size cyclicity, winnowing of fines and sorting of fine-grained sand, mixed composition and reworking of microfossil assemblages; ii) a clear cyclic arrangement of facies and grain size in vertical section showing both complete and partial contourite sequences, together with clearly erosive boundaries; iii) lateral correlation of facies cycles over at least 50 km along the Faroe–Shetland Channel (Stoker et al., 1998; Akhurst et al., 2002). The sedimentological evidence for bottom-current activity is consistent with the large variety of present day bedforms that has been documented from both the Faroe–Shetland and Faroe Bank channels, including scours, furrows, comet marks, barchan dunes and sand sheets (Masson et al., 2004). Large-scale subsurface sediment ripples, up to 1 km in wavelength and 11 to 12 m in amplitude, have also been imaged within the basinal section using 3D seismic data (Austin, 2004).

Towards the north-east margin of the Faroe–Shetland Channel, large sediment waves are observed on seismic reflection profiles and on sea-bed images derived from 3D data (Bulat and Long, 2001; Knutz and Cartwright, 2003, 2004; Long et al., 2004; Figure 117). These sediment waves form part of the West Shetland Drift development on the mid to lowerslope, between about 700 m and 1 km water depth, which forms a major, elongate-mounded, sediment drift (Figure 107). In Figure 117, it is clear that the present-day sea-bed waves form part of a succession of upslope-migrating sediment waves that have developed throughout the accumulation of the Plio-Pleistocene FSN-1 megasequence. The crests of the sea-bed features are generally spaced between 500 m and 1 km, and can be traced for up to 10 to 15 km. The heights of the waves range from 30 to 40 m. Interbedded debris-flow packages represent periods of enhanced mass movement on the West Shetland margin, linked either to shelf-margin glaciation or to slope failure, including the Mid Pleistocene Miller and Palaeo-Afen slides (see Chapter 11). Farther to the north-east, the West Shetland Drift becomes intercalated within, and overwhelmed by, the North Sea Fan (Nygård et al., 2005; Figure 107). On the western flank of the basin, onlap and contourite mound development is restricted to the base of the slope east of the Faroe Islands (Austin, 2004; Figure 111).

Faroe margin

The east and west Faroe margins are characterised by the development of prograding sediment wedges, whereas the north Faroe margin has been subjected to extensive mass failure. The East Faroe and West Faroe wedges represent two large Plio-Pleistocene depocentres, both exceeding 600 ms TWTT (Dahlgren et al., 2005; Figures 107 and 114). The East Faroe Wedge is the larger of the two depocentres, and during its development, the shelfbreak has shifted up to 75 km seaward since the Early Pliocene. The West Faroe Wedge has resulted in an advance of the shelfbreak between 5 and 10 km. On seismic profiles, both wedges display packages of prograding clinoforms that exhibit a chaotic and structureless internal reflection pattern characteristic of debris-flow deposits.

The construction of the East Faroe Wedge can be divided into two phases separated by a shelf-wide unconformity, reflector CN-050 of Andersen et al. (2000) (Table 6), which truncates the bulk of the prograding clinoforms in the wedge (Figure 114a). This unconformity was termed the Late Glacial Unconformity by Faroes GEM Network (2001a, b, c and d), and separated their units' 1A and 1B. However, by comparison with the West Shetland margin, Andersen et al. (2002), Stoker et al. (2002, 2005a), and Nielsen et al. (2007) correlated this boundary to the early Mid Pleistocene Glacial Unconformity. This correlation and terminology has been adopted in this report, and the stratigraphical ranges of units' 1A and 1B have been modified accordingly (Figure 109). This subdivision matches that originally proposed by Andersen et al. (2000), who recognised units' FSP-D.2 and FSP-D.3 separated by the intra-Pleistocene reflector, CN-050.

On the East Faroe Shelf, the Glacial Unconformity is overlain by an aggradational succession of stacked sheet-like sequences (Nielsen et al. 1979). These include mounded accumulations that locally form prominent ridges on the sea bed, including at the shelfbreak, up to 50 m high and 10 km wide (Nielsen et al., 2007, Figure 114a). These may represent moraines comparable with those developed on the West Shetland margin, though their age remains largely unknown. Several, discrete, transverse glacial troughs are also developed on the shelf, including the East Marginal Trough, the Sandoy Trough and the Skeivi Trough (Figure 118), which extend from the Faroe Islands to the shelfbreak (Waagstein and Rasmussen, 1975). Dredge samples from the south-east Faroe shelf recovered erratics of local (Faroese) derivation (Waagstein and Rasmussen, 1975), consistent with a glacial origin for these sequences.

Towards the shelfbreak, the Glacial Unconformity most probably becomes conformable with the base of one of the outermost prograding packages, though the trace of the boundary into the slope apron remains uncertain. The most recent sedimentation on the East Faroe Wedge is linked to latest Pleistocene to Holocene slope instability on the central and south-western parts of the slope apron known as the Sandoy and Suðuroy fans (Figure 107); the former has developed in front (downslope) of the Sandoy Trough on the adjacent shelf (Kuijpers et al., 2001; Andersen et al., 2002; Austin, 2004). Short sediment cores (<5 m) from the Sandoy Fan proved the occurrence of discrete beds of mass-flow, muddy diamicton, with gravel-sized clasts of mostly basaltic origin. The diamicton dates from the last glacial (late Weichselian) maximum, between about 17 000 to 20 000 years ago, and the Pleistocene–Holocene boundary (Kuijpers et al., 2001). The youngest instability event on the adjacent Suðuroy Fan is dated to c.27 000 ^{14}C BP, and its deposits have subsequently been buried beneath a contourite veneer (Nielsen et al., 2007). Elsewhere on the East Faroe Slope, the discovery of the GEM raft (see Figure 140) is further indication of slope instability, though no large-scale failures have so far been identified (Evans et al., 2005).

The construction of the West Faroe Wedge, also known as the Skeivi Bank Wedge (Andersen et al., 2002), is less well known. The wedge is located in front of the Skeivi Trough (Figure 118). Inspection of Figure 114c suggests that the prograding clinoforms of the wedge are truncated at or near to the present day sea bed, at the shelf edge, though a cover of diamicton including morainal banks is reported to overlie them farther landward towards the Faroe Islands (Nielsen et al., 2007). Dredge samples recovered glacial erratics of Oligocene sandstone on the outer shelf (Waagstein and Heilmann-Clausen, 1995). On the adjacent slope, the West Faroe Wedge interdigitates with upslope-migrating sediment-drift deposits, which occur at the base of the wedge as well as being interbedded with the debris-flow deposits. The latter pass laterally into the basinal succession of the Faroe Bank Channel. This sedimentary architecture is similar to that observed on the Sula Sgeir Fan, on the Hebridean margin south of the Wyville Thomson Ridge (Stoker et al., 2005a; Dahlgren et al., 2005). The upper part of the West Faroe Wedge has been termed the Skeivi Fan, which was subjected to mass-flow activity during the late Weichselian (Nielsen et al., 2007).

On the North Faroe margin, the Pliocene to Holocene succession has been extensively modified by mass failure. On the North Faroe Slope, the sediments are thickest (>200 ms TWTT) and least disturbed in the midslope region, between about 700 m and 1.5 km water depth, where the base of the succession is marked by the intra-Neogene Unconformity (Nielsen and van Weering, 1998; STRATAGEM Partners, 2002). On this part of the slope, acoustically chaotic deposits above 800 m pass downslope into acoustically layered strata that locally display mounded sediment drift geometry, though acoustically chaotic slump deposits occur at the base of this section, infilling the irregular basal unconformity. The upper slope deposits are interpreted to have been reworked by grounded icebergs during glacial intervals (Nielsen and van Weering, 1988). DSDP site 336, on the northern flank of the Iceland–Faroe Ridge, sampled the layered slope deposits at a water depth of about 830 m, and proved 168.5 m of grey mud, in part interstratified with thin graded sands, with common scattered pebbles (Talwani et al., 1976; Figure 104). The basal part of the section is dated as late Early Pliocene on the basis of the

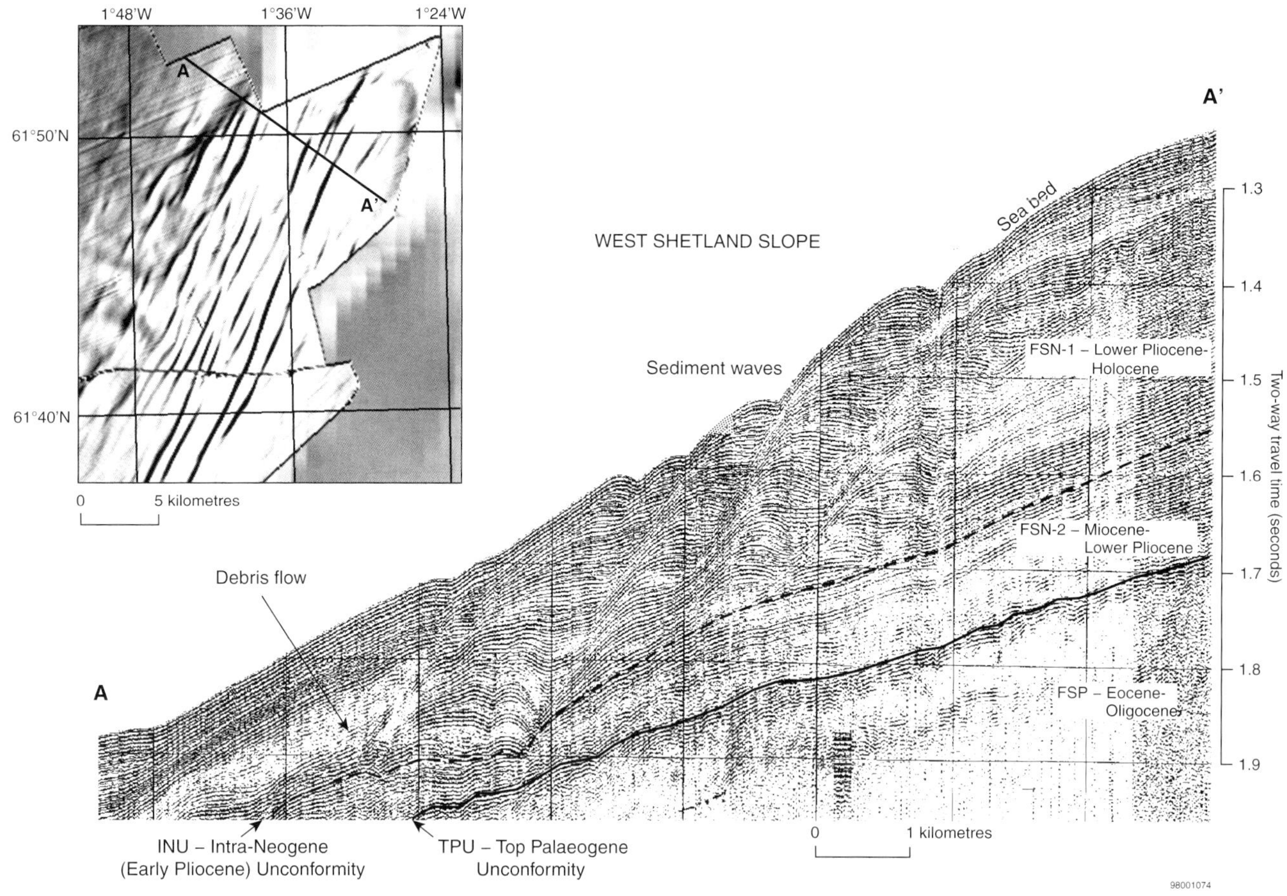

Figure 117 BGS sparker profile A–A' showing the development of Lower Pliocene to Holocene sediment waves on the West Shetland Slope, developed between 1 and 1.4 km water depth, and locally interbedded with debris-flow deposits. Inset shows sea-bed image (illuminated from the east) highlighting the sea-bed expression of the sediment waves, the axes of which trend subparallel to the strike of the slope (modified after Bulat and Long, 2001; Long et al., 2004). For location of sea-bed image, see Figure 107. Sea-bed image derived from Figure 138.

Distaphanus speculum silicoflagellate Zone (NN14/15–19) (Martini and Müller, 1976), and the *Thalassiosira kryophila* diatom Zone (dated at 3 to 4 Ma) (Schrader and Fenner, 1976). The boundary between the Pliocene and the Pleistocene is placed at about 25 m, based on the recognition of the diatom *Thalassiosira oestrupii* Zone (Schrader and Fenner, 1976). It is envisaged that sedimentation was controlled by a combination of glacimarine (ice-rafted detritus), hemipelagic and contouritic processes, though the occurrence of slump deposits at the base of the succession is indicative of slope instability early in its late Neogene development (Nielsen and van Weering, 1998; Taylor et al., 2000). This has been linked to Late Miocene or Pliocene uplift (Andersen et al., 2000).

Below 2 km water depth, the effects of more recent mass-movement processes are expressed by extensive headwalls and erosive scars as well as shallow retrogressive slumps that have dominated the development of the slope succession (van Weering and Nielsen, 1998; Taylor et al., 2000, 2003a; Kuijpers et al., 2001). The residual sea-floor topography has created a channel system that has focused the deposition of distal turbidite sediments into the Norwegian Basin, beyond 2.9 km (Taylor et al., 2003b). Taylor et al. (2000) suggested that the mass-wasting occurred predominantly during glacial intervals. This is consistent with AMS ^{14}C dates derived for distal turbidites in short sediment

Figure 118 *(opposite)* Schematic palaeogeographical reconstruction showing the inferred extent of the Late Devensian (MIS 2) ice sheet on the West Shetland margin, its potential late Weichselian correlative limit on the Faroe margin, and the preserved, relict, geomorphological features tentatively correlated with an Early Devensian (MIS 4) glaciation off north-west Britain (see text for details). UK margin reconstruction based on information derived from Stoker and Holmes, (1991), Stoker et al. (1993), Ballantyne et al. (1998), Davison, (2005) and Stoker and Bradwell (2005); Faroes reconstruction based on Waagstein and Rasmussen (1975), Nielsen et al., (1979), Jørgensen and Rasmussen (1986), Humlum et al. (1996) and Nielsen et al. (2007); outline of North Sea Fan based on Nygård et al. (2005). It should be noted that since the compilation of this figure, a more detailed reconstruction of the glacial geomorphology of the West Shetland Shelf has been published by Bradwell et al. (2008), with the south-east to west-directed ice-sheet flow lines sourced from a confluent British-Fennoscandian ice sheet; the zone of confluence is thought to be located between the Orkney and Shetland islands (Bradwell et al., 2008; figure 10).

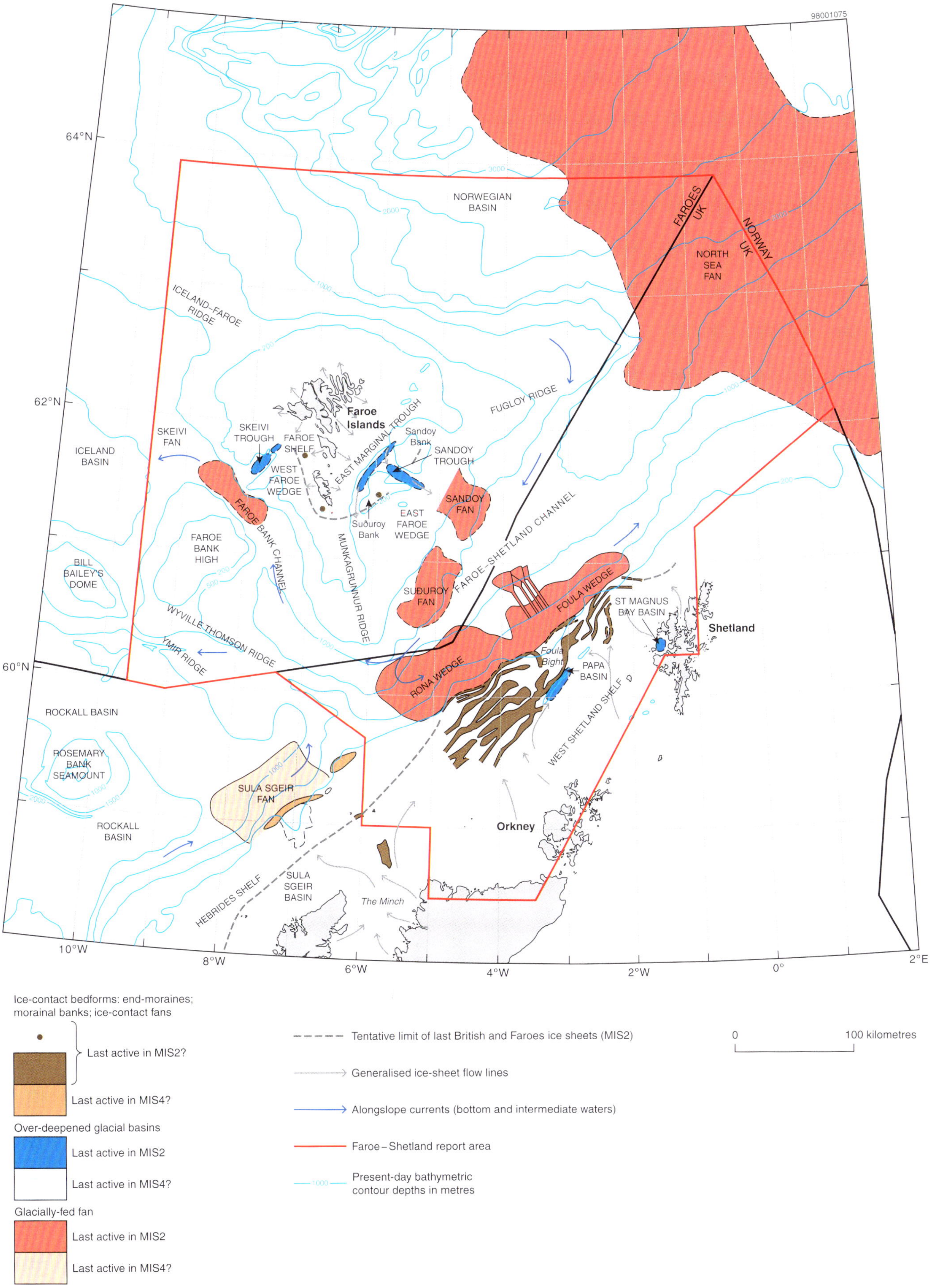
98001075
64°N
62°N
60°N
10°W
8°W
6°W
4°W
2°W
0°
2°E
NORWEGIAN BASIN
FAROES UK
NORWAY UK
NORTH SEA FAN
ICELAND–FAROE RIDGE
FUGLOY RIDGE
Faroe Islands
EAST MARGINAL TROUGH
Sandoy Bank
SANDOY TROUGH
SKEIVI FAN
SKEIVI TROUGH
FAROE SHELF
WEST FAROE WEDGE
ICELAND BASIN
SANDOY FAN
Suðuroy Bank
EAST FAROE WEDGE
FAROE BANK CHANNEL
FAROE BANK HIGH
MUNKAGRUNNUR RIDGE
FAROE–SHETLAND CHANNEL
BILL BAILEY'S DOME
SUÐUROY FAN
FOULA WEDGE
ST MAGNUS BAY BASIN
Shetland
WYVILLE THOMSON RIDGE
YMIR RIDGE
Foula Bight
PAPA BASIN
RONA WEDGE
WEST SHETLAND SHELF
ROCKALL BASIN
ROSEMARY BANK SEAMOUNT
SULA SGEIR FAN
ROCKALL BASIN
Orkney
HEBRIDES SHELF
SULA SGEIR BASIN
The Minch
Ice-contact bedforms: end-moraines; morainal banks; ice-contact fans
Last active in MIS2?
Last active in MIS4?
Over-deepened glacial basins
Last active in MIS2
Last active in MIS4?
Glacially-fed fan
Last active in MIS2
Last active in MIS4?
Tentative limit of last British and Faroes ice sheets (MIS2)
Generalised ice-sheet flow lines
Alongslope currents (bottom and intermediate waters)
Faroe–Shetland report area
Present-day bathymetric contour depths in metres
0
100 kilometres

cores taken within the slumped region, which indicate that mass-flow events correlate to the last glacial maximum, shortly after 23 000 years ago, and at the Pleistocene–Holocene boundary (Kuijpers et al., 2001).

Faroe Bank Channel and adjacent banks

In the Faroe Bank Channel, the base of the FSN-1 megasequence is strongly delineated by the angular, erosional, intra-Neogene Unconformity (Figures 110 and 114c). The basinal succession is generally less than 100 ms TWTT thick in the south-eastern part of the channel, but thickens in the area between the Faroe Bank and the Faroe Shelf, especially in association with the West Faroe Wedge (Figure 107). Mounded sediment drifts have been described adjacent to the northern slope of the Wyville Thomson Ridge (Masson et al., 2004) and the southern slope of the Munkagrunnur Ridge (Stoker et al., 1998; Howe et al., 2002). However, away from the margins of the channel, the sediments display an acoustically layered configuration that consists predominantly of subparallel, continuous, flat-lying reflections interpreted in this area, and in the adjacent Rockall Trough and Faroe–Shetland Channel, to represent sheeted sediment-drift deposits (Stoker et al., 1998; Akhurst et al., 2002; Howe et al., 2002; Masson et al., 2002). Scours, furrows, comet marks, barchan dunes and sand sheets indicate the continuation of bottom-current at the present day as the prime process of erosion and resedimentation within the Faroe Bank Channel (Masson et al., 2004).

The complex bathymetry south-west of the Faroe Shelf has juxtaposed a number of deep-water pathways and shallower banks and ridges (Figure 107). This has resulted in a comparable style of Early Pliocene to Holocene sedimentation and erosion in the deep-water channels between the Wyville Thomson and Ymir ridges, and between the Faroe, Bill Bailey and Lousy banks (Boldreel et al., 1995, 1998; Vanneste et al., 1995; Kuijpers et al. 1998). This pattern continues west of the outlet of the Faroe Bank Channel (Kenyon et al., 2003), and along the southern flank of the Iceland–Faroe Ridge (Bowles and Jahn, 1983; Dorn and Werner, 1993), though the latter is complicated by southerly overflow of bottom water across the ridge, directly from the Norwegian Basin, and the continuation of northerly flow along the edge of the Rockall Plateau.

Dredge samples from the top of the Faroe, Bill Bailey and Lousy banks reveal a cover of largely basaltic gravel overlying basalt, interpreted to be of local (to each bank) derivation, and possibly representing submerged shoreline deposits (Waagstein, 1988). The top of Faroe Bank was also tested by BGS borehole 99/07 (Figure 107) which confirmed a 0.3 m thick veneer of Quaternary sandy gravel on basalt. At the south-east end of the Wyville Thomson Ridge, borehole 99/04 proved an undifferentiated Quaternary sequence of about 5 m of soft, dark greenish grey, sandy mud with scattered basaltic pebbles, overlying 10.2 m of very coarse-grained sand and gravel, including basalt, siltstone, sandstone and gneiss, together with abundant shell debris, itself overlying weathered basalt. Farther to the south-east, a comparable sand and gravel deposit, 33.7 m thick, was recovered in borehole 99/05 in an area where mounded deposits of diamicton have been previously reported (Stoker, 1988).

North Sea Fan

The North Sea Fan is a large Pliocene to Holocene prograding sediment wedge that is mainly developed on the south-west Norwegian continental margin, extending about 500 km from the shallow (200 to 500 m) mouth of the Norwegian Channel into the deeper Norwegian Basin where it terminates at about 3.5 km depth (Dahlgren et al., 2005; Figures 107 and 118). The fan extends into the report area to the north of Shetland, where the preserved succession locally exceeds 1.2 s TWTT, and terminates against the lower slope of the North Shetland and north-east Faroe margins (Figure 107). In the Norwegian sector, it is reported that up to 1.7 km of sediment have been deposited on the North Sea Fan in the last 2.5 Ma (Nygård et al., 2005).

The growth of the North Sea Fan has been attributed to enhanced erosion due to the combined effects of Late Neogene uplift of the Fennoscandian mainland and initiation of glaciation (Rise et al., 2005). Construction of the North Sea Fan has been dominated by the accumulation of debris-flow deposits, especially during glacial intervals, with subordinate deposition of hemipelagic and contourite sediments (King et al., 1996; Nygård et al., 2005). The architecture of the fan has been periodically modified by episodes of mass wasting and large-scale sliding (Bryn et al., 2005; Dahlgren et al., 2005; Evans et al., 2005).

Offshore Norway, the North Sea Fan is included within the Naust Formation, the Norwegian equivalent of the FSN-1 megasequence (Stoker et al., 2005a). On the northern part of the West Shetland Shelf, the sediments that comprise the western part of the fan were previously assigned to a number of seismic-stratigraphical units, including the Shackleton, Mariner and Ferder formations, and the Undifferentiated Upper Pleistocene sequence (Stoker et al., 1993; BGS, 1989b; Figure 109). However, some revision of this scheme is now necessary in the light of recent stratigraphical studies focused on the subdivision of the Naust Formation (e.g. Evans et al., 2005; Rise et al., 2005).

Glacial history and ice-sheet reconstruction in the Faroe–Shetland region

Glaciation curves for south-west Norway and north-west Britain indicate that these continental margins have been subjected to repeated glaciation during the last 0.5 Ma (Stoker et al., 1993, 1994; Holmes, 1997; Sejrup et al., 2005; Figure 119). The very large scale of the North Sea Fan (Figure 118), which is envisaged to have accumulated up to 32 000 km^3 of sediment since MIS 12, is an indication that this prograding sediment wedge has been a major outlet for the drainage of the Fennoscandian ice sheet throughout this interval (Nygård et al., 2005). By comparison, the Sula Sgeir, Rona and Foula wedges on the Hebrides–West Shetland margins are orders of magnitude smaller (up to 330 km^3) (Stoker and Bradwell, 2005; Figure 118), but represent equally distinct glacially fed depocentres linked to the drainage of the north-west sector of the British Ice Sheet. On the Faroe Shelf, it has been established that the East Faroe and West Faroe wedges were important depocentres during Pleistocene glaciation (Nielsen et al., 2005; Sejrup et al., 2005). This is consistent with the suggestion that the Faroe Islands have been covered several times by an ice cap (Humlum et al., 1996), and is testament to their intense erosion.

In the West Shetland region, a system of mid to outer shelf moraines has been mapped as part of the Otter Bank sequence, and has been attributed to the late Devensian (MIS 2) expansion of the British Ice Sheet out to the edge of the West Shetland Shelf (Stoker and

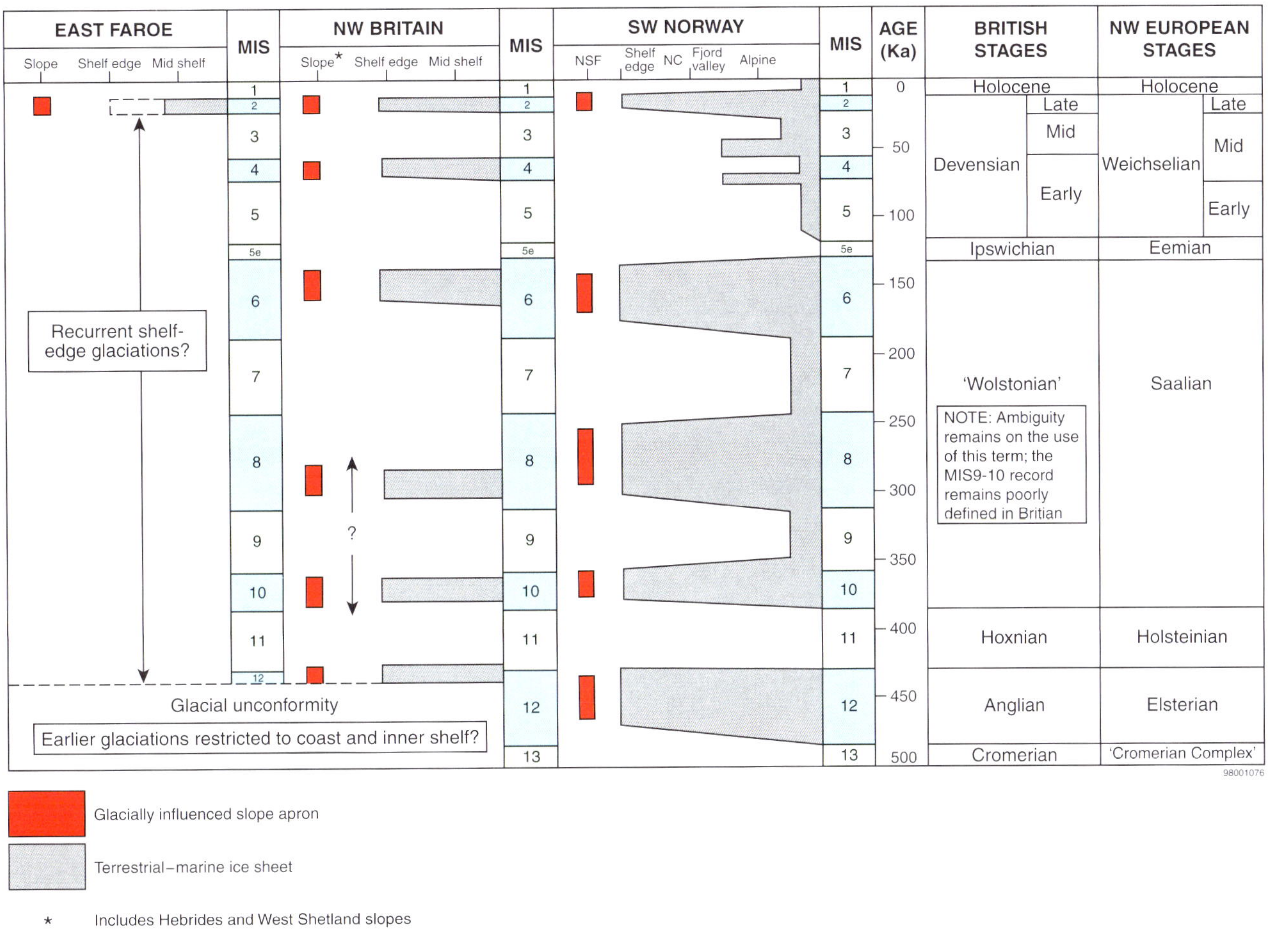

Figure 119 Glaciation curves for south-west Norway, north-west Britain and East Faroe, for the last 0.5 Ma, based on information derived from Waagstein and Rasmussen, (1975), Stoker et al., (1994), Holmes, (1997) and Sejrup et al., (2005). British and north-west European chronostratigraphical stages are based on Bowen, (1999) and Gradstein et al., (2004); the ages of marine isotope stage boundaries are based on Martinsen et al., (1987) and Williams et al., (1988). Abbreviations: NSF=North Sea Fan; NC=Norwegian Channel.

Holmes, 1991; Davison, 2005; Figure 119 for timescale). This contrasts with the northern Hebrides Shelf to the south-west, where the late Devensian ice limit is probably located in the mid-shelf region, landward of a presumed earlier Devensian (MIS 4) or older ice limit associated with the Sula Sgeir Fan (Stoker and Bradwell, 2005). The latter depocentre was repeatedly fed by The Minch palaeo-ice stream, which drained north-west Scotland and Lewis (Figure 118). A number of comparable pathways have been envisaged for the West Shetland Shelf sourced from Northern Scotland, and the Orkney and Shetland Islands (Stoker, 1995; Davison, 2005). Late Devensian till deposits in Caithness and Orkney include an upper unit of shelly till, which also contains clasts of Mesozoic sediment derived from the Moray Firth (Jardine and Peacock, 1973; Sutherland, 1984). Tills on Foula contain boulders derived from the eastern side of Shetland (Flinn, 1977b, 1978). This suggests that some of the sediment supplied to the West Shetland margin was derived from the North Sea region (Hall and Whittington, 1989). In a recent reconstruction of the last British Ice Sheet, it is evident that this ice sheet was confluent with the Fennoscandian ice sheet, with a zone of confluence located between Orkney and Shetland (Graham et al., 2007; Bradwell et al., 2008). On this basis, the prograding Rona and Foula wedges represent trough-mouth fans deposited at the shelf edge, at glacial maximum time, as the outlet drainage points of palaeo-ice streams that flowed north-westward across the North Sea and onto the West Shetland Shelf. The subsequent development of a local ice dome over Shetland, with an ice divide located along the axis of the island (Mykura, 1976; Sutherland, 1991), is now believed to represent a separate stage during deglaciation when the British and Fennoscandian ice sheets had become decoupled and separated by a marine incursion in the northern North Sea (Bradwell et al., 2008).

Substantial decay of the late Devensian ice sheet was underway by about 15 200 ^{14}C years BP (Boulton et al., 2002). The Otter Bank moraines represent the pattern of ice recession, which probably began in the area of the Foula Bight, an embayment on the outer shelf linked to the glacially overdeepened marginal trough of the Papa Basin (Davison, 2005; Figure 118). It may be no coincidence that linear gullies and basin floor fans (Figure 116) are located immediately downslope from the Foula Bight and Papa Basin, and may reflect high-energy meltwater discharges related to decay of this part of the ice margin. On the shelf, the association between the end-moraines and ponded stillstand deposits (Figure 115b) indicates an oscillatory retreat of the ice margin; locally halting for a time, and occasionally undergoing a slight advance albeit within an overall period of retreat (Stoker and Holmes, 1991).

A minimum age for its termination on Shetland is provided by radiocarbon dates from the organic sediments overlying the till, the oldest date being 12 090 ± 900 ^{14}C years BP (Hoppe, 1974). The final remnants of the ice sheet may have consisted of corrie glaciers and ice patches that were restricted to Shetland associated with the Loch Lomond stadial (between 12 500 and 11 500 ^{14}C years BP) (Mykura, 1976; Sutherland, 1991; Boulton et al., 2002). Farther south, the Loch Lomond stadial may have been responsible for the regrowth of small valley glaciers in north-west Scotland and northern Hoy (Orkney), which deposited end moraines (Sissons, 1977; Sutherland, 1991).

Field observations from the Faroe Islands indicate that an ice cap covered the terrain up to at least 700 m (above sea level), and thus is likely to have extended far beyond the present coastline (Jørgensen and Rasmussen, 1986). On the Faroe Shelf, a limit to the maximum extent of Pleistocene glaciation has been tentatively placed between 300 and 400 m water depth on the basis of analysis of glacial erratics and dropstones (Waagstein and Rasmussen, 1975; Nielsen et al., 1979; Figure 118). This glacial limit would fully incorporate the glacially overdeepened East Marginal, Sandoy and Skeivi troughs, which represent drainage pathways leading away from the Faroe Islands (Waagstein and Rasmussen, 1975). The ice drainage system was probably fed by individual ice caps developed on the islands of Sandoy and Suðuroy, whereas the larger conglomeration of northern islands contained a larger ice dome that had a north-east-trending ice divide (Jørgensen and Rasmussen, 1986). The direction of ice movements based on glacial striae and *roches moutonnées* are shown on Figure 118, and display a radial pattern of ice dispersal that is not inconsistent with the offshore pathways.

The preservation of moraines on the Faroe Shelf (Nielsen et al., 2007), including indications of outer shelf morainic deposits (Figures 114a and 118), indicates episodic coalescence of the individual ice caps into an expanded offshore ice sheet. However, the offshore moraine system remains poorly mapped to date, and the limit of the last, late Weichselian (MIS 2) glaciation remains one of conjecture.

Although sedimentation occurred on the Sandoy, Suðuroy and Skeivi fans in the late Weichselian, this has largely been attributed to mass-wasting triggered by changes in sea level and oceanographical conditions, and not to the direct effects of shelf-edge glaciation (Kuijpers et al., 2001; Nielsen et al., 2007). Instead, Nielsen et al. (2007) suggest that the late Weichselian ice sheet did not extend much beyond the present day 200 m water depth contour, on the basis that most of the moraines that they observed were located between 100 and 200 m water depth. Whereas the glacially eroded troughs may have acted as conduits for meltwater plumes for dispersal onto the slope apron, their proposed reconstruction implies an ice-free outer shelf during the late Weichselian.

During the ensuing deglaciation, several moraine systems associated with valley and cirque glaciers were deposited on the Faroe Islands, and a tentative Younger Dryas (Loch Lomond stadial equivalent) age has been suggested for the youngest glacial stage (Humlum et al., 1996). The Faroe Islands became ice-free at about 10 000 ^{14}C years BP (Jóhansen, 1975, 1985), and many relict late Weichselian features are probably covered by a thick layer of Holocene peat (Rasmussen, 1982).

9 Cenozoic (igneous)

Simon Passey[1] and Ken Hitchen[2]
[1] *Jarðfeingi (Faroese Earth and Energy Directorate)*
[2] *British Geological Survey*

The Cenozoic igneous rocks of the Faroe–Shetland region form part of the North Atlantic Igneous Province (NAIP), which extends over a pre-rift distance of approximately 2000 km from Baffin Island, Canada to Lundy Island in the Bristol Channel (Saunders et al., 1997). The arrival, beneath Greenland, of a spatially constrained melting anomaly during the Early Paleocene (Danian) resulted in the igneous activity that produced the NAIP (e.g. White and Mackenzie, 1989; Saunders et al., 1997; Ritchie et al., 1999a). Whether this melting anomaly is the result of a mantle plume, *sensu stricto* (i.e. the proto-Icelandic plume), or by some other process is currently of great debate and is beyond the scope of this report (e.g. Foulger, 2002; Lundin and Doré, 2005). During the Early Paleocene, possibly as early as 62 Ma, magma exploited weak spots in the thinned and heavily rifted lithosphere of the north-west European region (Figure 8). This resulted in surface volcanism, the emplacement of numerous volcanic centres (with associated intrusions) and shallow sills in the following principal areas: Baffin Island and West Greenland, East Greenland, Faroe Islands, western Scotland, and Antrim (Northern Ireland) (Saunders et al., 1997). This pre-rift igneous activity (or Phase 1) was followed by a hiatus in the volcanism which is represented in the report area by, for example, the Prestfjall Formation (formerly the Coal-bearing Formation) on the Faroe Islands (Ellis et al., 2002).

Igneous activity resumed with the initiation of sea-floor spreading between Greenland and north-west Europe at about 56 to 54 Ma, during C24r, in latest Paleocene to Early Eocene (Ypresian) times. Syn-rift volcanism (or Phase 2) associated with the opening of the north-east Atlantic Ocean is principally restricted to the East Greenland, Hatton–Rockall and Faroe–Shetland regions (Saunders et al., 1997; Larsen et al., 1999; Ritchie et al., 1999a; Jolley and Bell, 2002a; Figure 90). However, the Balder Formation, consisting of reworked and altered basaltic tuff deposits, is widespread throughout the North Atlantic region and is believed to be synchronous with an increase in volcanic activity at the site of sea-floor spreading, possibly due to a marine incursion resulting in large scale phreatomagmatic eruptions (Knox et al., 1997; Ritchie et al., 1999a; Jolley and Bell, 2002a). Subsequently, igneous activity was restricted to the north-east-trending rifts (Mid Atlantic and Aegir ridges) located to the west and north-west of the report area (Saunders et al., 1997). The Aegir Ridge, sited to the north of the Faroe Islands (Figure 91), was active from the commencement of sea-floor spreading to about 25 Ma ago (Late Oligocene, Chattian) (Jung and Vogt, 1997), and continues today along the Mid Atlantic Ridge, notably in Iceland (Saunders et al., 1997).

The temporal relationships between the various igneous lithologies within the report area have been difficult to determine due to the lack of suitable material and through the inaccuracies of previously used dating techniques (e.g. K–Ar isotopic dating). However, with increased interest in exploration in the Faroe–Shetland region from the oil and gas industry, and significant amounts of new well and seismic data becoming available, the establishment of a chronology for the order and emplacement duration of the various igneous structures and events has become possible. Geochemical, biostratigraphical and palaeomagnetic analytical techniques, supported by improved radiometric dating (e.g. Ar–Ar) have been applied to igneous and sedimentary core and cuttings sample material, derived from commercial wells, in order to enhance correlation between the various igneous rock units throughout the report area. The characterisation of igneous rocks, through major, trace, rare earth element and isotopic data, not only establishes their geochemical evolution, but has also enabled direct correlation, or differentiation, between similar lithologies in separate lava fields (Larsen et al., 1999). Biostratigraphy of associated sedimentary units has helped to constrain these correlations, as has, to a lesser extent, palaeomagnetism, which shows whether the magmatic rocks crystallised when the Earth's magnetic field was in the normal or reversed polarity state. Previous attempts have been made to establish an all-encompassing timescale for the NAIP, commonly using the timescales of Berggren et al. (1995) and Cande and Kent (1995). However, all ages mentioned in the text correspond to the revised timescale of Gradstein et al. (2004) (Figure 6).

The igneous rocks within the report area consist of widespread lava fields (flood and low shield basalts), volcaniclastic units (sedimentary as well as pyroclastic), localised volcanic mounds, intrusive complexes and eight volcanic centres (Figure 120). Except for a relatively small (approximately 1400 km^2) onshore outcrop on the Faroe Islands, the igneous rocks are located offshore, extending from the Møre Basin in the north-east to the Hatton–Rockall High located outside the report area to the south-west. Some of the offshore igneous rocks crop out at the sea bed but the majority are buried by younger sediments, particularly in the Faroe–Shetland Basin.

EXTRUSIVE ROCKS

The main expression of extrusive volcanic activity in the report area is the presence of large volumes of basaltic lava (several kilometres thick in places) which were erupted either as 'flood basalts' from fissure systems (e.g. Beinisvørð Formation of the Faroe Islands Basalt Group) or radially from point sourced, volcanic centres (e.g. Erlend and West Erlend volcanic centres) (Figures 120 and 121). However, two other styles of volcanic activity are also represented within the report area. These are volcaniclastic rocks produced from air-fall tuffs and their subsequently reworked deposits and small, conical volcanic mounds which appear to be associated with the emplacement of shallow intrusions.

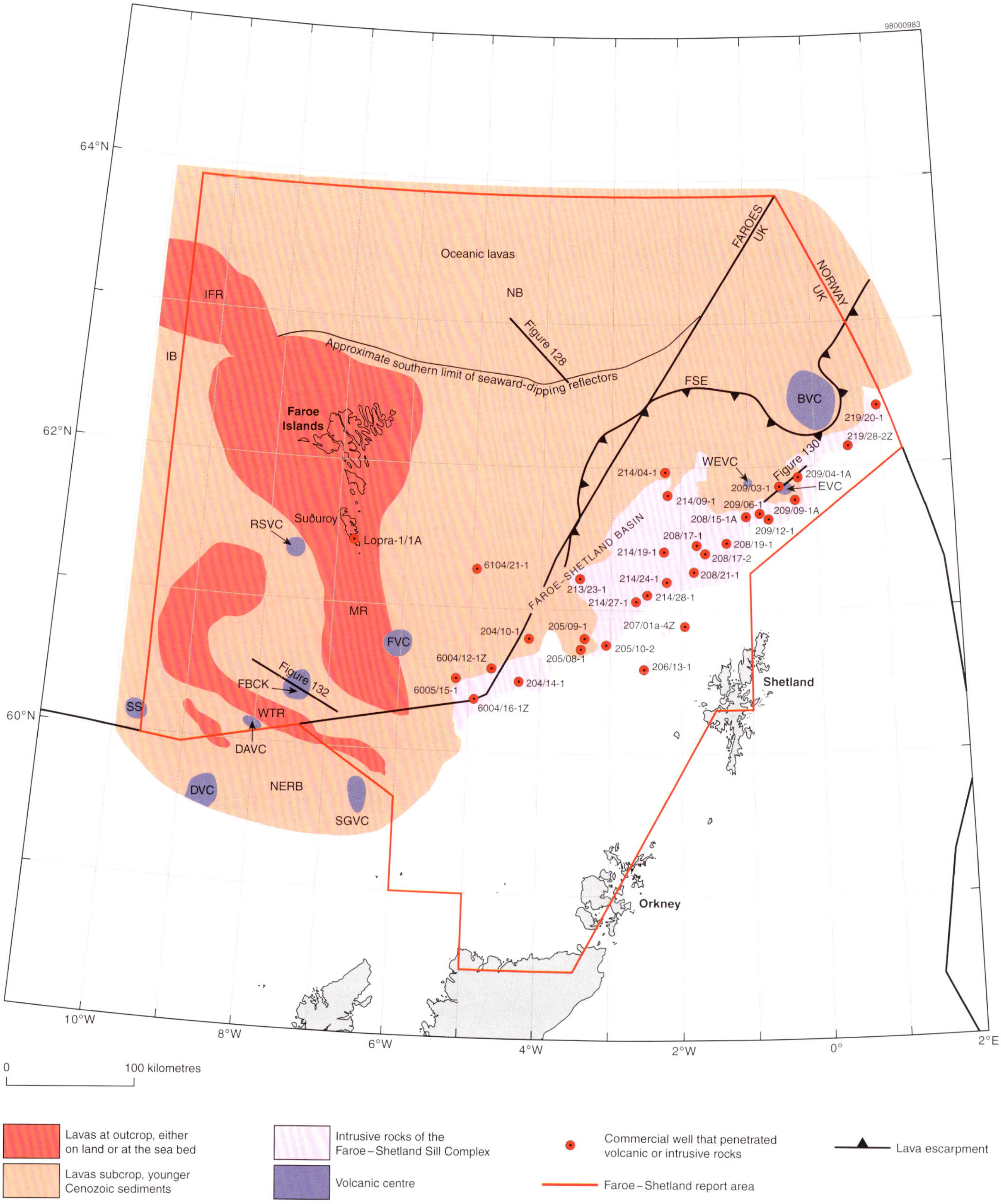

Figure 120 Distribution of Palaeogene igneous rocks within the Faroe–Shetland report area (modified from Ritchie et al., 1999a (and references therein); Ellis et al., 2002; Keser Neish, 2003 and Smallwood and Kirk, 2005). Wells shown are those which penetrated lavas or sills. Abbreviations: see Figures 7 and 24.

Lava fields (Faroe Islands Basalt Group)

Lithology

The Faroe Islands Basalt Group (FIBG) (Figures 122 and 123) is a predominantly subaerial basalt lava field that covers an estimated area of 120 000 km^2. The FIBG covers much of the western extent of the report area and merges with the North Rockall Trough–Hebrides Lavas Group to the south-west of the Wyville Thomson Ridge (Ritchie and Hitchen, 1996). The majority of the FIBG occurs offshore, although it does crop out over an area of 1400 km^2 on the Faroe Islands

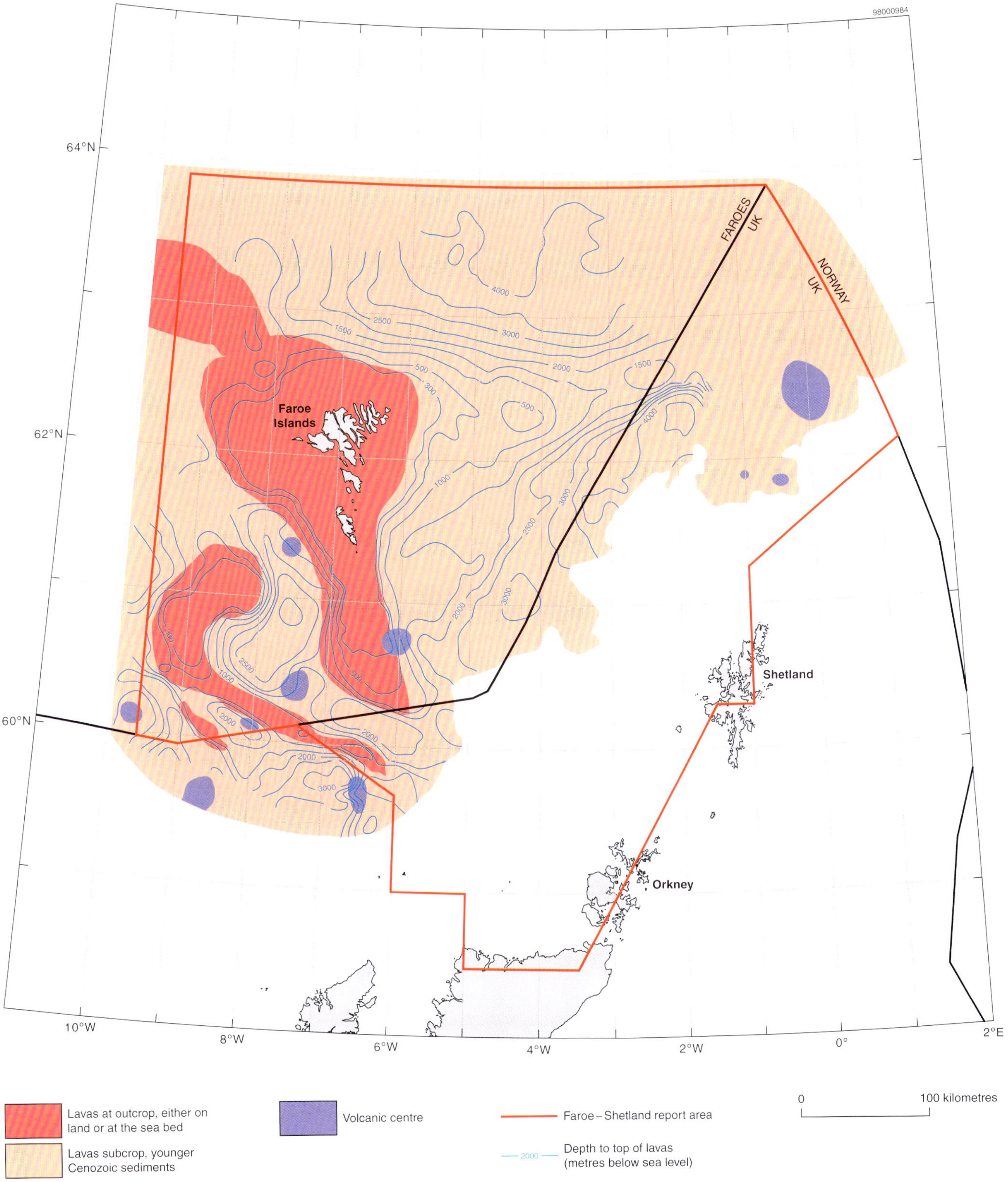

Figure 121 Depth (metres below sea level) to the top of the Palaeogene lavas within the Faroe–Shetland report area (courtesy of Jarðfeingi). See Figure 120 for names of volcanic centres.

where it has a stratigraphical thickness approaching 6.6 km and has been subdivided into seven formations (Figures 122 and 123) based on lithological characteristics (Passey and Jolley, 2009). Approximately 3.4 km of FIBG lavas were proven by the onshore well Lopra-1/1A (Figures 120 and 124) which was intended to drill below the base of the volcanic sequence and determine the underlying substratum. Unfortunately, this was not accomplished, but it has been suggested through geochemistry (Gariépy et al., 1983; Hald and Waagstein, 1983; Holm et al., 2001) and geophysics (Casten, 1973; Bott et al., 1974; Richardson et al., 1998; Richardson et al., 1999) that the FIBG is underlain by continental crust, most likely Precambrian

Rasmussen et al., (1956)	Rasmussen and Noe-Nygaard (1970)	Waagstein (1988) Faroe Basalt Plateau	Ellis et al., (2002) Faroe Lava Group	Passey and Jolley (2009) Faroe Islands Basalt Group
Upper Basalt Sequence	Upper Basalt Series	Upper Basalt Formation	Upper Lava Formation	Enni Formation
				Sneis Formation
	Middle Basalt Series	Middle Basalt Formation	Middle Lava Formation	Malinstindur Formation
Tuff-Agglomerate Sequence	Tuff-Agglomerate Zone	Tuff-Agglomerate Zone	Volcaniclastic Sandstone Sequence	Hvannhagi Formation
Coal-bearing Sequence	Coal-bearing Sequence	Coal-bearing Formation	Coal-bearing Formation	Prestfjall Formation
Lower Basalt Sequence	Lower Basalt Series	Lower Basalt Formation	Lower Lava Formation	Beinisvørð Formation
			Lopra Sequence	Lopra Formation

98000985

Figure 122 Proposed lithostratigraphical terminologies for the Faroe Islands volcanic rocks. The scheme according to Passey and Jolley (2009) is adopted for this report.

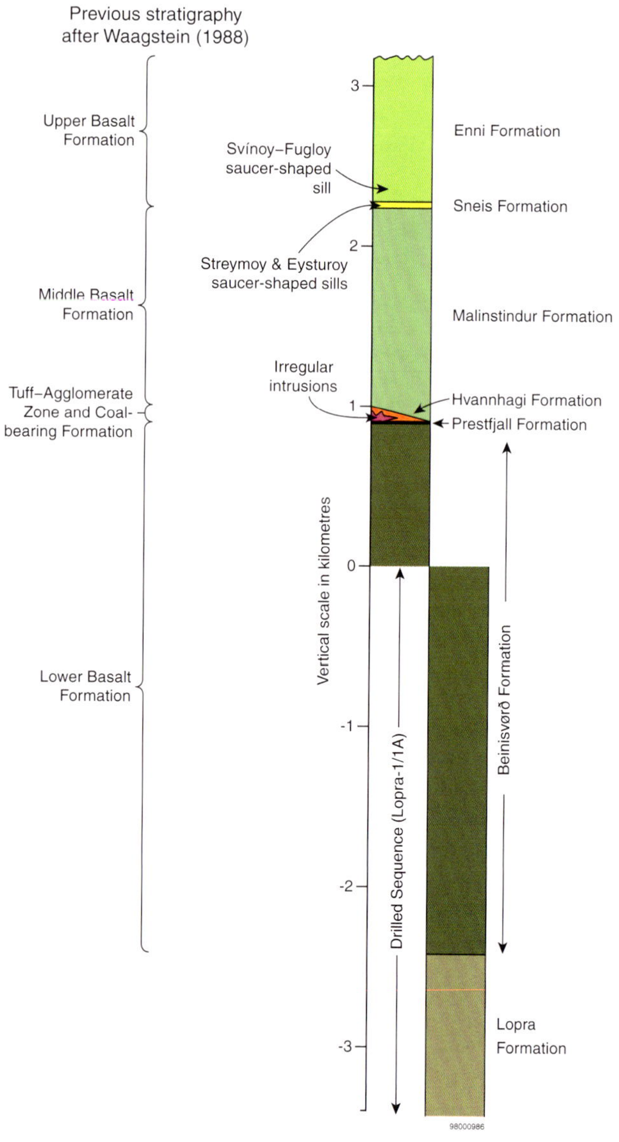

Figure 123 *(left)* Representative lithostratigraphy and thickness of the Faroe Islands Basalt Group (modified from Passey and Jolley, 2009).

crystalline basement, with or without an overlying section of Mesozoic or Palaeozoic rocks.

The Lopra Formation, at the base of the FIBG (Figures 122, 123 and 124), has a minimum thickness of 1 km and is composed of a sequence of volcaniclastic rocks intruded by sills (Ellis et al., 2002). The sequence most likely represents a prograding hyaloclastite deltaic succession, which formed when subaerial lava flows entered a basin, were quenched, fragmented and produced a sequence of angular glassy fragments. The palynoflora assemblage within the volcaniclastic rocks implies the hyaloclastites were emplaced in an estuarine to shallow marine environment with water depths up to 200 m (Ellis et al., 2002). As the hyaloclastite delta prograded into the basin, so the associated shoreline migrated seawards and consequently, the hyaloclastite sequence is overlain by geochemically affinitive subaerial lava flows, represented on the Faroe Islands by the Beinisvørð Formation.

Figure 124 *(opposite)* Biostratigraphical and lithofacies correlation between the onshore Faroe Islands Basalt Group and wells that penetrated lavas offshore. The UK offshore lithostratigraphy is an idealised section composed of the maximum thicknesses observed for each lithostratigraphical unit (after Knox et al., 1997). Timescale is based on 'A Geologic Time Scale 2004' by F M Gradstein, J G Ogg, A G Smith, et al. (2004) and the International Stratigraphical Chart, 2006 (ICS) with additions and D W Jolley (personal communication, 2005). Abbreviations: F=Formation and SM=Sandstone Member. For location of wells, see Figure 120.

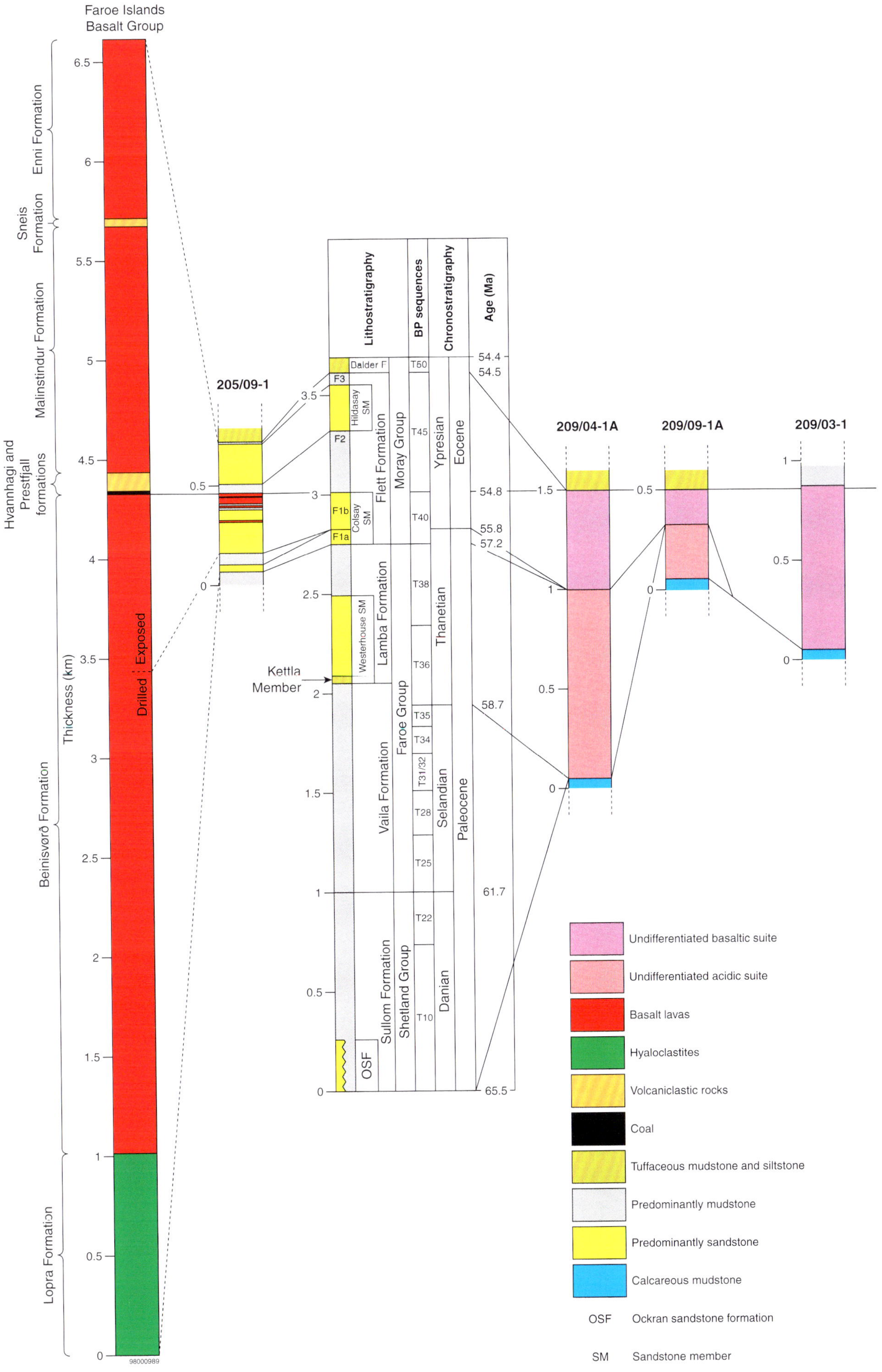
Faroe Islands Basalt Group
Enni Formation
Sneis Formation
Malinstindur Formation
Hvannhagi and Prestfjall formations
Beinisvørð Formation
Lopra Formation
Thickness (km)
Drilled
Exposed
205/09-1
Kettla Member
Lithostratigraphy
BP sequences
Chronostratigraphy
Age (Ma)
Dalder F
F3
F2
F1b
F1a
Hildasay SM
Colsay SM
Westerhouse SM
Flett Formation
Lamba Formation
Vaila Formation
Sullom Formation
OSF
Moray Group
Faroe Group
Shetland Group
T50
T45
T40
T38
T36
T35
T34
T31/32
T28
T25
T22
T10
Ypresian
Thanetian
Selandian
Danian
Eocene
Paleocene
54.4
54.5
54.8
55.8
57.2
58.7
61.7
65.5
209/04-1A
209/09-1A
209/03-1
Undifferentiated basaltic suite
Undifferentiated acidic suite
Basalt lavas
Hyaloclastites
Volcaniclastic rocks
Coal
Tuffaceous mudstone and siltstone
Predominantly mudstone
Predominantly sandstone
Calcareous mudstone
OSF Ockran sandstone formation
SM Sandstone member
98000989

Lava-fed Gilbert-type deltas have been identified offshore in the Faroe–Shetland Basin (Smythe, 1983; Smythe et al., 1983). The most notable is the Faroe–Shetland Escarpment (Figures 24, 120 and 125), which represents the leading edge of a hyaloclastite delta, as identified on seismic sections by prograding foresets (Kiørboe, 1999; Ritchie et al., 1999a). An approximately 450 m thick hyaloclastite sequence has also been penetrated by well 214/04-1 (50 km south-east of the Faroe–Shetland Escarpment) (Figures 120 and 125). This is stratigraphically older than the hyaloclastite delta that occurs at the Faroe–Shetland Escarpment, and could imply that others exist within the basin but cannot be resolved on seismic data (Passey, 2004). The hyaloclastite sequence in well 214/04-1 is overlain by about 52 m of subaerial basalt lava flows, which has been subdivided into at least four individual flows based on gamma-ray spikes which are suggested to represent interlava lithologies (Passey, 2004). Whole-rock geochemistry on the volcanic sequence in well 214/04-1 suggests there is an affinity with the basalt lava flows of the Beinisvørð Formation on the Faroe Islands and consequently, along with the Faroe–Shetland Escarpment hyaloclastite delta, the sequence is interpreted to represent the offshore continuation of the Beinisvørð Formation (Passey, 2004). It should be noted that the hyaloclastite sequences encountered offshore may not be stratigraphically equivalent to the hyaloclastites of the Lopra Formation, but may represent younger, lateral facies variations of the subaerial lava flows of the Beinisvørð Formation on the Faroe Islands.

The Beinisvørð Formation (formerly the Lower Basalt Formation) on the Faroe Islands has a stratigraphical thickness of approximately 3.3 km of which about 900 m is exposed (Figures 122, 123, 124 and 126). The formation is characterised by subaerial tholeiitic basalt lava flows with a tabular-classic facies architecture (Rasmussen and Noe-Nygaard, 1970; Waagstein, 1988; Ellis et al., 2002; Passey, 2004; Passey and Bell, 2007). The lava flows are commonly massive with weathered flow tops and have a typical average thickness of about 20 m, although flows up to 70 m thick also occur (Rasmussen and Noe-Nygaard, 1970; Waagstein, 1988; Ellis et al., 2002; Passey, 2004; Passey and Bell, 2007). The Beinisvørð Formation, comprising mainly aphyric basalts, shows little compositional variation over its entire thickness (Waagstein, 1988). The lava flows are far-reaching and were erupted from fissures most likely situated to the west of the Faroe Islands (Waagstein, 1988). Towards the top of the formation, interlava volcaniclastic lithologies become more abundant, due to a decrease in eruption frequency, thus allowing the development of swamp, lake and fluvial environments (Ellis et al., 2002; Passey, 2004). The interlava deposits, which comprise coal, sandstone, mudstone and palaeosol horizons, have been biostratigraphically correlated to the interlava lithologies within a 70 m thick sequence of subaerial basalt lava flows encountered in well 205/09-1, approximately 200 km east-south-east of the Faroe Islands (Figure 120; Ellis et al., 2002). However, Ellis et al. (2002) suggested that the lava flows in well 205/09-1 may have been locally sourced, rather than being contiguous with the main body of Beinisvørð Formation lava flows to the north-west. The lobate subcrop pattern of the lavas in this vicinity supports this view. Based on the biostratigraphy, the lava flows of the exposed section of the Beinisvørð Formation correspond to the sedimentary rocks of the Flett Formation unit 1b (F1b) in the Faroe–Shetland Basin (Figure 124), equivalent to the upper part of the T40 sequence on the BP scheme (Ebdon et al., 1995; Ellis et al., 2002; Jolley et al., 2002).

Erosion down to the second highest lava flow of the Beinisvørð Formation followed a cessation in the volcanism and led to the deposition of the coal-bearing formation (Rasmussen and Noe-Nygaard, 1970; Ellis et al., 2002), now called the Prestfjall Formation (Passey and Jolley, 2009; Figures 122, 123, 124 and 126). The Prestfjall Formation has a maximum thickness of about 15 m and consists of two coal seams, claystone and sandstone deposited in small lake, peat swamp and overbank floodplain environments (Rasmussen and Noe-Nygaard, 1970; Ellis et al., 2002; Passey, 2004). The palynoflora assemblage of the Prestfjall Formation is comparable to that in the sedimentary rocks that occur at the base of the Flett Formation unit 2 (F2) in the Faroe–Shetland Basin, equivalent to the base of the T45 sequence (Ellis et al. 2002; Jolley et al. 2002a; Figure 124).

The resumption in volcanism on the Faroe Islands is marked by the Hvannhagi Formation (formerly the tuff-agglomerate zone) (Figures 122, 123 and 124), which is at least 50 m thick and has been disrupted by numerous irregular doleritic sills (Passey, 2004; Passey an Jolley, 2009). The Hvannhagi Formation principally consists of two cyclic sections, each characterised by a basal sequence of olivine-phyric airfall tuffs that were subsequently reworked and incorporated into mass-flow (conglomerates and sandstones) and floodplain (sandstones and mudstones) deposits (Passey, 2004).

This phase of pyroclastic activity was followed by the effusion of basalt lava flows of the 1.2 to 1.3 km thick Malinstindur Formation (formerly the Middle Basalt Formation) (Rasmussen and Noe-Nygaard, 1970; Waagstein, 1988; Ellis et al., 2002; Passey, 2004) (Figures 122, 123, 124 and 126). The lava flows of the Malinstindur Formation have a compound-braided facies architecture and are composed almost exclusively of pahoehoe lava, which was transported through ubiquitous lava tubes (Rasmussen and Noe-Nygaard, 1970; Ellis et al., 2002; Passey, 2004; Passey and Bell, 2007). The lava flows have a typical average thickness of 20 m and are composed of thinner flow units up to 4 m in thickness (Rasmussen and Noe-Nygaard, 1970; Ellis et al., 2002; Passey, 2004; Passey and Bell, 2007). The Malinstindur Formation exhibits an evolution up sequence from olivine-phyric and aphyric basalts to plagioclase-phyric basalts, both of which are relatively high in titanium (Waagstein, 1988). However, at a stratigraphical height approximately 800 m above the base of the Malinstindur Formation, a >100 m thick suite of aphyric and olivine-phyric low-Ti basalts occur, which not only show a depletion in titanium but also in phosphorous, potassium and other incompatible elements making them geochemically similar to Mid Ocean Ridge Basalts (MORB) (Waagstein, 1988).

The Malinstindur Formation lava flows were locally erupted from point-sourced low shields, which had diameters in the order of 15 km and slopes of 0.5° (Noe-Nygaard, 1968). Through the morphologies of the lava flows and their calculated emplacement durations, Passey and Bell (2007) suggest that the lava flows of the Malinstindur Formation are laterally restricted to the Faroe Islands and possibly do not extend offshore for any significant distance. This does not imply that morphologically similar lava fields do not exist offshore, but if they do, they are more likely to be restricted to point sourced volcanoes (Passey and Bell, 2007).

The top of the Malinstindur Formation is marked by an erosional surface that is reddened due to weathering (Rasmussen and Noe-Nygaard, 1970; Passey and Jolley, 2009). The overlying Sneis Formation (Figures 122, 123 and 124) is approximately 20 m thick and comprises a sequence of volcaniclastic lithologies (Passey and Jolley, 2009). Principally, the sequence is characterised by a basal reddened volcaniclastic sandstone containing abundant woody material overlain by beds of greyish volcaniclastic conglomerate believed to have been deposited by mass-flow processes (Passey and Jolley, 2009). The conglomerate locally fines upwards into sandstone and there is a general fining of the entire sequence to the south across the Faroe Islands (Passey and Jolley, 2009).

The Sneis Formation is overlain by the Enni Formation, which represents a sequence of interfingering compound-braided lava flows, comparable to those of the Malinstindur Formation, and tabular-classic lava flows with average thicknesses in the region of 10 m (Passey and Jolley, 2009). The Enni Formation (formerly the Upper Basalt Formation) has a thickness in the region of 900 m (Waagstein, 1988; Passey and Jolley, 2009; Figures 122, 123 and 124). It has been estimated from zeolite studies that only a few hundred metres have been eroded from the top of the formation (Waagstein et al., 2002; Jørgensen, 2006). Geochemically, the formation is characterised by having a mixed population of plagioclase-phyric, high-Ti basalts and aphyric to olivine-phyric, low-Ti basalts (Waagstein, 1988). These low-Ti basalts are MORB-like, similar to the basalts that occur nearly two-thirds up in the Malinstindur Formation (Waagstein, 1988). The basalt flows are commonly intercalated with sediments which contain palynoflora assemblages and the whole sequence is suggestive of a former low-lying lava field characterised by ephemeral lakes, fluvial channels, estuaries and shallow marine embayments (Ellis et al., 2002).

The offshore distribution of the three main lava formations of the FIBG (Beinisvørð, Malinstindur and Enni formations) has been of much debate since Smythe (1983) first suggested that the Beinisvørð Formation extended the furthest into the Faroe–Shetland Basin, with the Malinstindur and Enni formations progressively 'stepping backwards' from the Faroe–Shetland Escarpment towards the Faroe Islands (Figure 125). Subsequently, Ritchie et al. (1999a) suggested that the Beinisvørð Formation did not extend much further into the Faroe–Shetland Basin beyond the Faroe–Shetland Escarpment and was progressively overlapped by the younger Malinstindur and Enni formations. This was implied, in part, by the then inferred Enni Formation age of the lava flows in well 205/09-1, which have subsequently been biostratigraphically correlated to the Beinisvørð Formation (Ellis et al., 2002). However, Ellis et al. (2002), as mentioned above, suggested that the lava flows in well 205/09-1 are locally restricted and that the main body of the Beinisvørð Formation extended only as far as a major north-east-trending magnetically reversed polarity feature, about 100 km to the south-east of the Faroe Islands i.e. the 'Annika Anomaly' (Figure 5). Ellis et al. (2002) also concurred with Ritchie et al. (1999a) in that the Beinisvørð Formation was progressively overlapped by the Malinstindur and Enni formations. However, only Beinisvørð Formation equivalent lavas, based on biostratigraphy and geochemistry, have been encountered in wells (e.g. 205/09-1, 214/04-1, 214/09-1) furthest to the east and south-east (i.e. about 200 to 250 km) from the Faroe Islands (Ellis et al., 2002; Passey 2004). In addition, Kiørboe (1999), based on seismic facies analysis, suggested that the Malinstindur Formation is locally restricted to the Faroe Islands, which is supported by field observations on the Faroe Islands (Passey and Bell, 2007).

The lateral extent of the Enni Formation is problematic as it is composed of a mixture of laterally restricted compound-braided flows and laterally extensive tabular-classic flows (Passey and Bell, 2007). The tabular-classic flows of the Enni Formation will, most likely, have a similar seismic signature, consisting of parallel-bedded reflectors, as the Beinisvørð Formation making it difficult to distinguish between the two formations. However, the average flow thickness of about 10 m for the tabular-classic flows of the Enni Formation is less than the 20 to 25 m thicknesses expected for the flows of the Beinisvørð Formation (Rasmussen and Noe-Nygaard, 1970; Waagstein, 1988; Passey and Bell, 2007). This tentatively suggests that the tabular-classic flows of the Enni Formation have not travelled as far as those of the Beinisvørð Formation, but have, most likely, travelled further than the locally restricted compound-braided flows of the Malinstindur Formation. The premise that the offshore sections of the FIBG comprise three layer-cake formations is far too simplistic and, most likely, consists of interdigitating lava flows of different facies architectures (compound-braided verses tabular-classic) sourced from disparate point-sourced and fissure vents, as exemplified by the Enni Formation on the Faroe Islands (Passey and Bell, 2007). This is demonstrated offshore by the lava flows of Erlend and West Erlend which have a compound-braided facies architecture but are time equivalent to the tabular-classic lava flows of the Beinisvørð Formation (Figure 124). Until wells are drilled through the FIBG on the eastern side of the Faroe Platform, the offshore distribution of the three lava formations will remain poorly constrained.

Age data

The Beinisvørð Formation has been geochemically correlated to the Nansen Fjord Formation of East Greenland and has been suggested to represent a pre-break-up succession covering an estimated area of 70 000 km^2 and having an estimated volume of 120 000 km^3 (Larsen et al., 1999). Radiometric dating of the drilled section of the Beinisvørð Formation (approximately the upper 2.2 km of Lopra-1/1A) yields K–Ar ages between 57.9 ± 0.8 and 58.9 ± 1.3 Ma and Ar–Ar ages between 60.0 ± 2.1 and 63.1 ± 1.8 Ma (Waagstein et al., 2002; Table 8). This is presently at odds with biostratigraphical data which suggest that the Lopra Formation, and therefore, the overlying Beinisvørð Formation, postdates the Lamba Formation (i.e. postdates the T38 sequence) (Ellis et al., 2002), which according to the timescale of Gradstein et al. (2004) implies that the Lopra and Beinisvørð formations were emplaced after 57.2 Ma (D W Jolley, pers comm. 2005).

To account for this apparent discrepancy between the biostratigraphical and radiometric dating for the drilled section of the Beinisvørð Formation, this would require

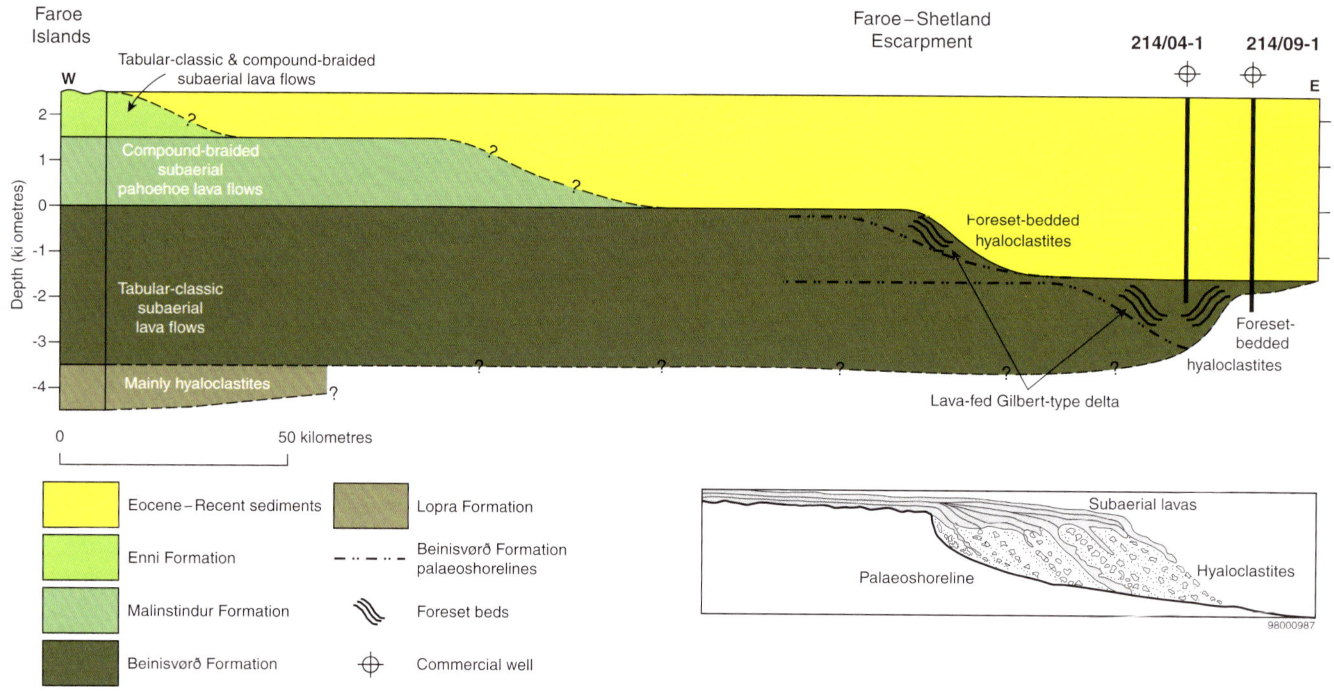

Figure 125 Schematic cross-section to illustrate the offshore relationships between the four principal lava formations of the Faroe Island Basalt Group (courtesy of S Passey). For location of wells, see Figure 120.

either of the following: (1) the interlava lithologies have been biostratigraphically miscorrelated to either the radiometrically dated rocks or the PETM which is globally dated to 55.8 ± 0.2 Ma (Paleocene–Eocene boundary) on the timescale of Gradstein et al. (2004); or (2) the radiometrically determined ages are unreliable. The latter is more likely, as Waagstein et al. (2002) suggested that the Ar–Ar ages, obtained for the drilled section of the Beinisvørð Formation, were too old and should be ignored because of possible ^{39}Ar recoil loss or its relocation during irradiation. It should be noted, however, that the Urbjerget Formation of East Greenland, which is believed to be stratigraphically equivalent to the Nansen Fjord Formation and in turn the Beinisvørð Formation (Hansen et al., 2002), has basalt lava flows with Ar–Ar ages between 61.0 ± 1.1 and 61.6 ± 1.3 Ma (Hansen et al., 2002), comparable to the Ar–Ar ages obtained for the drilled section of the Beinisvørð Formation (Waagstein et al., 2002). However, the exact relationship between the Urbjerget and Beinisvørð formations requires further geochemical and biostratigraphical investigation. Thus, until the discrepancy between the radiometric and biostratigraphical ages is resolved, the age of the drilled section of the Beinisvørð Formation will remain somewhat problematic (Jolley et al., 2002a).

The exposed section of the Beinisvørð Formation and the sedimentary rocks (Vandfalsdalen and 'upper' Ryberg formations) underlying the Nansen Fjord Formation of East Greenland, have been correlated to the F1b unit of the Flett Formation (i.e. the upper part of the T40 sequence) and have palynomorph assemblages that are consistent with postdating the PETM (Jolley et al., 2002a; Jolley and Whitham, 2004). The PETM is marked by a pronounced negative carbon isotope excursion that is associated with major turnovers in deep-sea benthic foraminifera, calcareous nannoplankton and mammals, allowing for global correlations in a wide variety of environments (Gradstein et al., 2004). The assignment of the exposed section of the Beinisvørð Formation and the Nansen Fjord Formation as postdating the PETM implies that any isotopic ages obtained for these sections should yield ages younger than 55.8 Ma. Waagstein et al. (2002) obtained K–Ar ages between 56.5 ± 1.3 and 57.1 ± 2.5 Ma and Ar–Ar ages of 55.0 ± 1.1 and 57.1 ± 1.6 Ma for basalt lava flows from the exposed section of the Beinisvørð Formation (Table 8). The biostratigraphically determined age range (54.8 to 55.8 Ma) for the exposed section of the Beinisvørð Formation (Jolley et al., 2002a; Jolley and Whitham, 2004) falls within the margin of error for these radiometrically dated samples (Figure 127).

The timing of the cessation of the volcanism that resulted in the emplacement of the Beinisvørð Formation is constrained by the coal-bearing Prestfjall Formation on the Faroe Islands. The Prestfjall Formation has a palynofloral assemblage that correlates to the base of the F2 unit of the Flett Formation (i.e. the base of the T45 sequence) (Ellis et al., 2002; Jolley et al., 2002a; Figure 124), which on the timescale of Gradstein et al. (2004), corresponds to 54.8 Ma (D W Jolley, pers comm. 2005). This differs from Lamers and Carmichael (1999) where the T40–T45 boundary, which according to Ebdon et al. (1995) is associated with the extinction of the dinoflagellate cyst *Apectodinium augustum*, corresponds to the Paleocene–Eocene boundary. It can, however, be clearly seen from the timescale of Gradstein et al. (2004) that the extinction of *Apectodinium augustum* postdates the Paleocene–Eocene boundary by about one million years (i.e. at 54.8 Ma) and consequently, that the Paleocene–Eocene boundary occurs within the T40 sequence as suggested by Jolley et al. (2002a) and Jolley and Whitham (2004).

The Prestfjall Formation was followed by a resumption in volcanism, represented by the syn-break-up succes-

Figure 126 General view of the Hæddin cliff section consisting of the Beinisvørð, Prestfjall and Malinstindur formations of the Faroe Islands Basalt Group, about 2.5 km north-west of Soyðistangi, Suðuroy, Faroe Islands (Figure 120). In the centre of the photograph the cliffs are about 470 m high. The Beinisvørð Formation is about 260 m thick and is characterised by having a tabular-classic facies architecture composed of eight simple lava flows, which are highlighted by their reddened flow tops and/or overlying volcaniclastic lithologies. The eight flows have an average thickness of about 35 m, with the thickest flow (number four) approximately 55 m. Overlying the Beinisvørð Formation is the poorly exposed 15 m thick Prestfjall Formation, consisting of coal and claystone. The Prestfjall Formation is overlain by a 195 m thick sequence of the Malinstindur Formation, which is composed of compound lava flows defining the compound-braided facies architecture. The flow boundaries for these compound flows are much more difficult to identify due, in part, to the overlapping and anastomosing nature of the constituent flow lobes.

sion, consisting of the Hvannhagi Formation through to the Enni Formation on the Faroe Islands and the Milne Land Formation in East Greenland (Larsen et al., 1999). This geochemically correlated syn-break-up succession covers an estimated area of 220 000 km^2 and has an estimated volume of 250 000 km^3 (Larsen et al., 1999).

The syn-break-up succession has not yielded enough characteristic age-diagnostic dinoflagellate cysts for the exposed sequence on the Faroe Islands to be accurately dated, but it is constrained by its stratigraphical relationship to the Prestfjall Formation (base of the T45 sequence) and the Balder Formation (T50 sequence) (Ellis et al., 2002; Jolley et al., 2002a; Jolley and Whitham, 2004). The Balder Formation overlies the FIBG offshore and therefore restricts the emplacement of the syn-break-up succession of the Faroe Islands to the T45 sequence (i.e. the F2 unit of the Flett Formation) (Figure 124) which, on the current timescale, implies emplacement between 54.5 and 54.8 Ma (D W Jolley, pers comm. 2005). Hansen et al. (2002) obtained Ar–Ar ages between 54.9 ± 1.6 and 56.7 ± 1.7 Ma for basalt lava flows of the Milne Land Formation. This is comparable to Ar–Ar ages (54.5 ± 4.1 to 57.8 ± 4.5 Ma) obtained for a single basalt lava flow from the Plateau Basalts, which includes the Milne Land Formation (Hansen et al., 1995). The biostratigraphical age suggested by Jolley et al. (2002a) and Jolley and Whitham (2004) for the syn-break-up succession falls within the margin of error for the samples dated by Hansen et al. (1995) and with one of the samples dated by Hansen et al. (2002).

The exposed section of the Beinisvørð Formation has a normal-reversed-normal-reversed polarity and the Malinstindur through to the Enni formations have a reversed polarity (Tarling and Gale, 1968; Tarling, 1970). The palaeomagnetic results for the drilled section of the Beinisvørð Formation is more tentative and is based on five to eight samples taken from each of five cores from depths of 338, 862, 1218, 1923 and 2177 m, respectively. These five cores gave the following results: normal, reversed, mixed, reversed, and normal (Schoenharting and Abrahamsen, 1984). However, Schoenharting and Abrahamsen (1984) postulated that the normal and

Table 8 Radiometric age dates for igneous rocks within the Faroe–Shetland report area. Note that some previously published ages, not included here, are regarded as unreliable. For location of wells, see Figure 120.

Occurrence	Location	Method	Age (Ma)		References	Comments
Sill	206/13-1	K-Ar whole rock	48.7±0.5	Ypresian	Hitchen and Ritchie (1993)	
		K-Ar whole rock	53.0±0.5	Ypresian	Hitchen and Ritchie (1993)	
Sill	207/01a-4Z	K-Ar whole rock	52.4±1.5	Ypresian	Ages previously unpublished but see Ritchie et al. (1999a) Figure 3	
		K-Ar whole rock	52.5±1.5	Ypresian	Ages previously unpublished but see Ritchie et al. (1999a) Figure 3	
Sill	208/15-1A	K-Ar whole rock	52.9±1.0	Ypresian	Hitchen and Ritchie (1993)	
		K-Ar whole rock	53.6±1.0	Ypresian	Hitchen and Ritchie (1993)	
Sill	209/03-1	K-Ar whole rock	54.3±2	Ypresian	Hitchen and Ritchie (1987)	From sill below Erlend lavas
		K-Ar whole rock	54.8±2	Ypresian	Hitchen and Ritchie (1987)	From sill below Erlend lavas
Sill	209/12-1	K-Ar whole rock	54.1±0.7	Ypresian	Hitchen and Ritchie (1993)	
		K-Ar whole rock	55.3±0.7	Ypresian	Hitchen and Ritchie (1993)	
Sill	219/28-2Z	K-Ar biotite	79.1±2.4	Campanian	Fitch et al. (1988)	From the lower sill and average of 3 analyses
		K-Ar whole rock	81.0±4.1	Campanian	Fitch et al. (1988)	From the lower sill and average of 3 analyses
Sill	6004/16-1Z	Ar-Ar whole rock	56±0.4	Ypresian	Smallwood (pers comm)	
Tuff	219/28-2Z	K-Ar whole rock	51.7±0.8	Ypresian	Fitch et al. (1988)	Average of 3 analyses and should be regarded as a minimum age
Basalt lava flows	Exposed section of the Beinisvørð Formation	Ar-Ar whole rock	55.0±1.1	Ypresian	Waagstein et al. (2002)	
		K-Ar whole rock	56.5±1.3	Thanetian	Waagstein et al. (2002)	
		Ar-Ar whole rock	57.1±1.6	Thanetian	Waagstein et al. (2002)	
		K-Ar whole rock	57.1±1.7	Thanetian	Waagstein et al. (2002)	
		K-Ar whole rock	57.1±2.5	Thanetian	Waagstein et al. (2002)	
Basalt lava flows	Drilled section (Lopra-1/1A) of the Beinisvørð Formation	K-Ar whole rock	57.9±0.8	Thanetian	Waagstein et al. (2002)	
		K-Ar whole rock	58.3±0.9	Thanetian	Waagstein et al. (2002)	
		K-Ar whole rock	58.9±1.3	Selandian	Waagstein et al. (2002)	
		Ar-Ar whole rock	60.0±2.1	Selandian	Waagstein et al. (2002)	Age regarded as too old by Waagstein et al. (2002)
		Ar-Ar whole rock	60.5±1.0	Selandian	Waagstein et al. (2002)	Age regarded as too old by Waagstein et al. (2002)
		Ar-Ar whole rock	62.4±1.8	Danian	Waagstein et al. (2002)	Age regarded as too old by Waagstein et al. (2002)
		Ar-Ar whole rock	62.8±1.4	Danian	Waagstein et al. (2002)	Age regarded as too old by Waagstein et al. (2002)
		Ar-Ar whole rock	63.1±1.8	Danian	Waagstein et al. (2002)	Age regarded as too old by Waagstein et al. (2002)

98000988

mixed polarity samples are the result of overprinting by later low temperature oxidation of less stable samples. Consequently, they suggested that the upper 2.2 km thick section from Lopra-1/1A of the Beinisvørð Formation was emplaced during a reversed magnetic period. Using the favoured isotopic ages of Waagstein et al. (2002) and the timescale of Gradstein et al. (2004), it suggests that the Beinisvørð Formation was erupted during magnetochrons C26r through to C24r and the Malinstindur through to the Enni formations were emplaced during C24r (Figure 127). This conforms with Waagstein (1988) and Waagstein et al. (2002) who proposed the emplacement of the FIBG during C26r to C24r (59 to 55 Ma). However, the biostratigraphical data imply that the entire exposed sequence of the FIBG postdates the PETM and must therefore have been emplaced during C24r (Jolley et al., 2002a). This is because the PETM has been globally tied to C24r (Norris and Röhl, 1999) and consequently, the two normal polarity zones recognised within the exposed section of the Beinisvørð Formation would, potentially, represent two cryptochron excursions (Jolley and Bell, 2002a).

Seaward dipping reflector sequences (SDRS)

The transition zone between continental and oceanic crust, on passive volcanic margins, is commonly identified on seismic sections by the presence of seaward dipping reflector sequences (SDRS) (e.g. Mutter et al., 1982; Smythe, 1983; Parson and the ODP Leg 104 Scientific Party, 1988). SDRS are believed to represent lavas that were emplaced as flows into rapidly subsiding rift basins under subaerial or shallow marine conditions (Mutter et al., 1982; Larsen et al., 1998; Larsen and Saunders, 1998; Planke et al., 2000). The subsidence is assumed to occur synchronously with sea-floor spreading and is believed to cause the pronounced dip of the reflectors seen on seismic sections (Mutter et al., 1982; Larsen et al., 1998; Larsen and Saunders, 1998; Planke et al., 2000). SDRS are clearly observed on seismic profiles to the north-east of the Faroe Islands where continental crust passes into oceanic crust of the Norway Basin (Smythe, 1983; Parson and the ODP Leg 104 Scientific Party, 1988; White et al., 2005; Figure 128). They have also been observed to the south-west of the Faroe Islands (Roberts et al., 1984; Parson and the ODP Leg 104 Scientific Party, 1988; Barton and White, 1997). The SDRS to the north-east of the Faroe Islands commonly consist of an inner and outer package (Figure 128). SDRS have not been recognised from seismic profiles to the west and north-west of the Faroe Islands, including the area between the Faroe Platform and the Faroe–Iceland Ridge (Richardson et al., 1998).

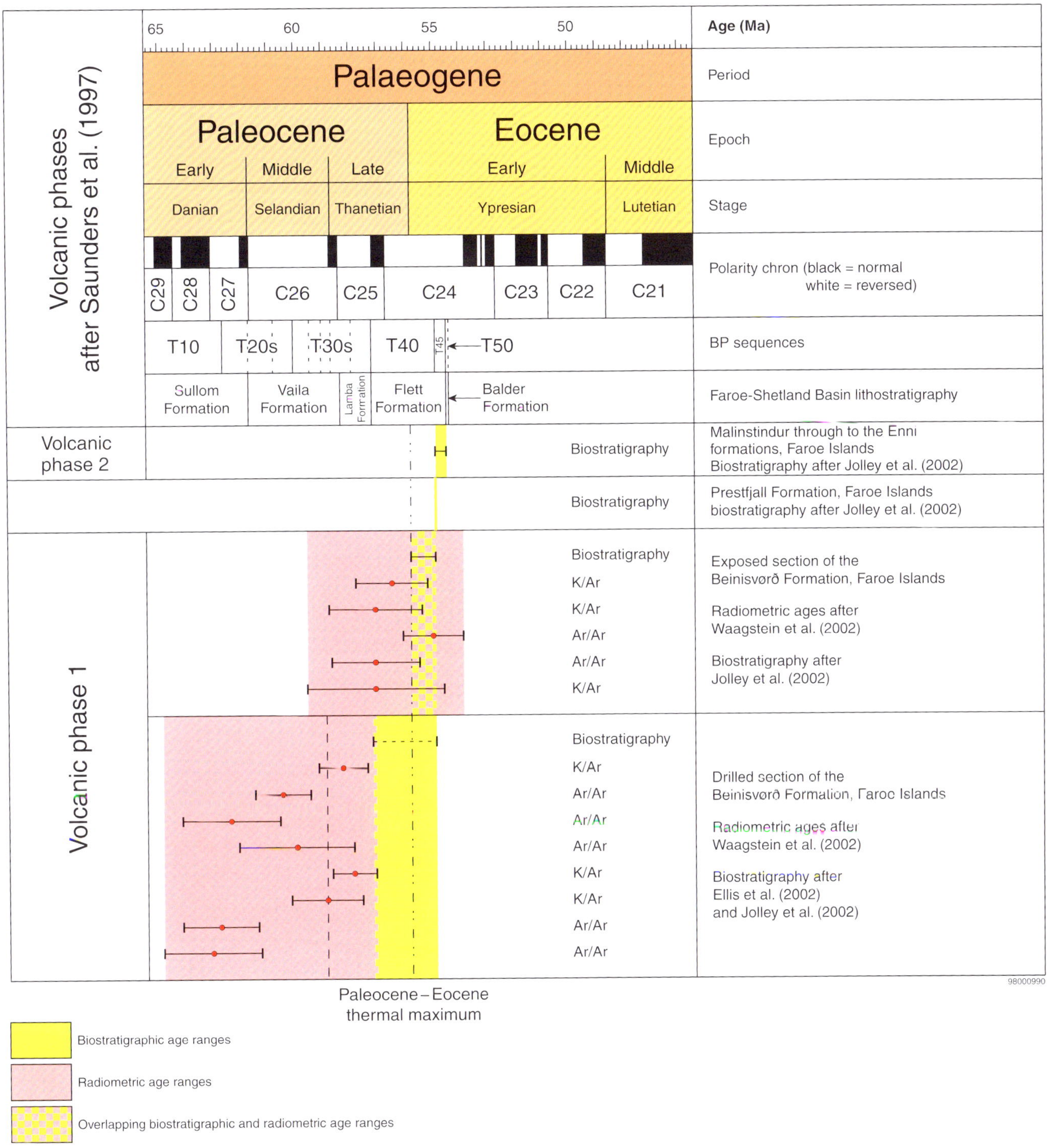

Figure 127 Comparison of the biostratigraphical and radiometric age data obtained for the Faroe Islands Basalt Group. The five radiometric ages to the left of the dashed line for the drilled section of the Beinisvørð Formation are considered unreliable by Waagstein et al., (2002).

Volcaniclastic rocks

Two principal volcaniclastic intervals have been recognised offshore west of Shetland, the Kettla Member and the Balder Formation (Knox et al., 1997; Figures 124 and 129). These typically consist of airfall tuffs interbedded mainly with siltstones and mudstones. The tuffaceous material may have been redistributed from the original airfall deposits either by debris-flows or due to basin-floor reworking.

Kettla Member

The older of the two volcaniclastic intervals has been formally defined as the Kettla Member and forms the lowermost subdivision of the Lamba Formation (Figure 124). It occurs at the same stratigraphical level as the Balmoral Tuffite (Andrew Tuff or Glamis Member) in the Outer Moray Firth and Central North Sea (Mudge and Copestake, 1992; Knox et al., 1997; Mudge and Bujak, 2001). It was deposited in the Early Thanetian

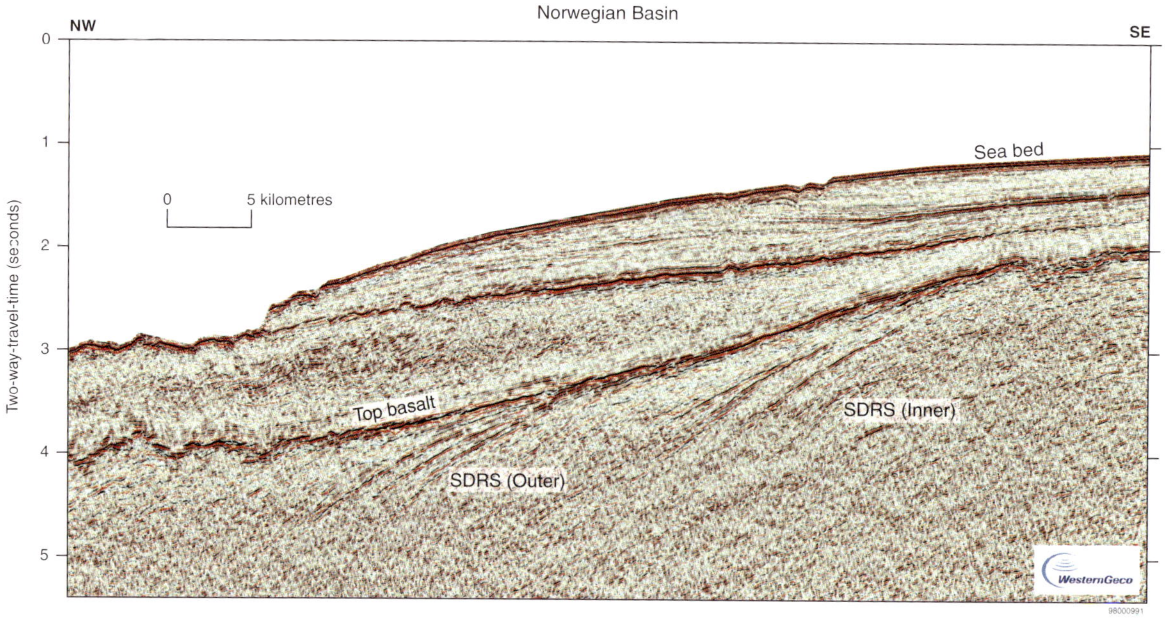

Figure 128 Seismic profile illustrating seaward dipping reflector sequences (SDRS) thought to represent extrusive lava flows emplaced at the time of continental break-up. The seaward dip is due to subsidence after emplacement. The SDRS occur in two packages, inner (to the south-east) and outer (to the north-west). For location of profile, see Figure 120. Seismic data courtesy of WesternGeco.

during T36 sequence time (Lamers and Carmichael, 1999) and is time equivalent to the pyroclastic activity phase 1 of Knox and Morton (1988). Lithologically it has been described as a dark grey-green tuffaceous siltstone grading to a silty tuffite (Knox et al., 1997) and consisting of degraded volcanic ash with volcanogenic lithoclasts (Jolley et al., 2005). It has been encountered, for example, in wells 204/19-1, -2, 206/02-1A, 208/19-1 and 214/27-2 but appears to have a restricted geographical extent and is best developed in the Flett Sub-basin (Figure 2). Although, the precise source area for the volcanogenic material of the Kettla Member is unknown, palynological and heavy mineral studies on core material from well 205/09-1 suggest that it is more likely to be sourced from west of the Faroe–Shetland Basin, rather than from the area of the Hebrides or Orkney–Shetland High (Jolley et al., 2005). This is supported by a westerly thickening of the member and it has been suggested that it represents debris-flow deposits (Knox et al., 1997).

The well log signature for the tuffaceous Kettla Member displays characteristic low gamma-ray and high velocity, density, and resistivity signatures (Knox et al., 1997; Figure 129). In wells 205/09-1, 204/19-1 and -2, it forms a characteristic double peak on the gamma-ray well log (Sørensen, 2003). The Kettla Member causes an easily identifiable reflector on seismic data, especially where it is thickest in the Flett Sub-basin. It is less easy to identify where it is thinner or reworked in the Judd Sub-basin (Sørensen, 2003).

From a depth of 22.7 m in BGS borehole 82/12 (Figure 2), Morton et al. (1988b) described a 1 cm thick tuff within a claystone sequence. The shards of volcanic glass are consistently high in TiO_2, tholeiitic and olivine and hypersthene-normative. The claystone was originally dated as late Danian (Early Paleocene) on the basis of dinoflagellate cyst data. However, Morton et al. (1988b) subsequently amended this age on the basis of lithological correlations to nearby commercial wells and further microfossil analyses. The tuff in borehole 82/12 is now correlated with a tuffaceous claystone at a depth of 344 m in nearby well 204/30-1 (Figure 2). This corresponds to the pyroclastic activity phase 1 of Knox and Morton (1988) and hence the Upper Paleocene Kettla Member of Knox et al. (1997). The original Danian dinoflagellate cysts occur with other Jurassic and Cretaceous species and are all considered to be reworked (Morton et al., 1988b).

Balder Formation

The younger volcaniclastic interval has been formally defined as the Balder Formation (Knox et al., 1997), being directly equivalent to the Balder Formation in the North Sea Basin and taking its name from the Balder oilfield in Norway (Deegan and Scull, 1977). It is generally better developed than the Kettla Member, typically being between 50 and 150 m thick, with its greatest thicknesses occurring in the sand-rich marginal sections adjacent to local clastic sediment sources (Knox et al., 1997). The formation is informally subdivided into a lower tuff-rich unit (B1) and an upper relatively tuff-poor unit (B2) (Knox et al., 1997). The Balder Formation is Ypresian (Early Eocene) in age and was deposited between 54.5 and 54.4 Ma during T50 sequence time (Figure 124). It is equivalent to the pyroclastic activity phase 2 of Knox and Morton (1988). Lithologically, the Balder Formation is characterised by grey, variably silty and carbonaceous mudstone with abundant layers of greenish-grey tuff, some of which may have been reworked (Knox et al., 1997). Towards the south-west of its wide geographical extent, the formation contains a higher sandstone content and still farther to the south-west it becomes a tuffaceous lignitic mudstone (Knox et al., 1997). The tuffaceous material of the Balder Formation is generally basaltic in composition, although, in the North Sea

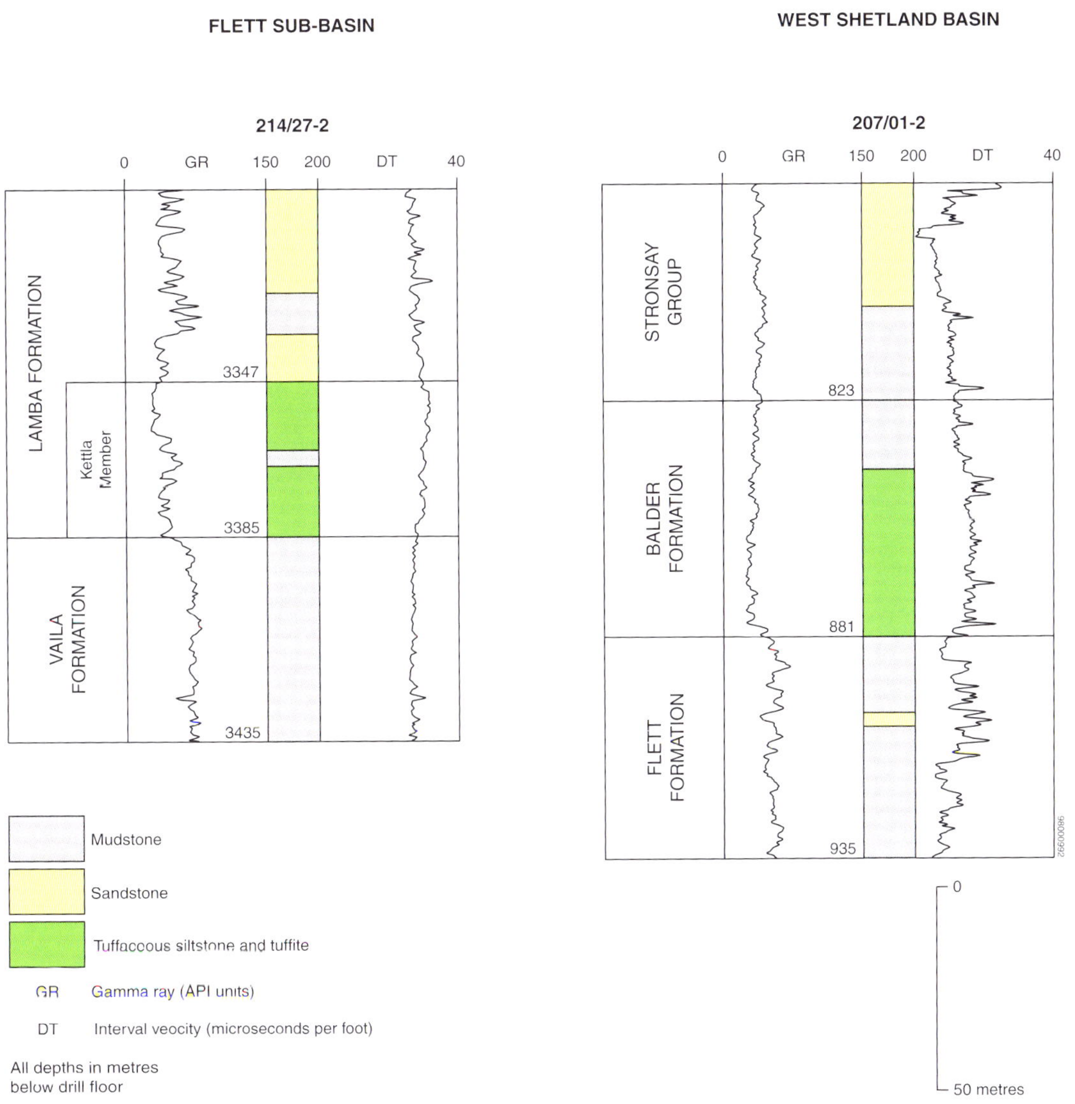

Figure 129 Examples of well logs through the Kettla Member (early Thanetian) and Balder Formation (early Ypresian). Both units have high tuffaceous content typically resulting in a lower gamma-ray well log response but with higher interval velocities. For location of wells, see Figure 2.

more intermediate and peralkaline compositions have been recorded. Volumetrically, the source for the bulk of the tuffs is usually considered to be north and west of the report area, in the 'Faroe–Greenland Province', although, some input from the British Hebridean volcanic province cannot be excluded (Knox and Morton, 1988; Morton and Knox, 1990).

As with the Kettla Member, the Balder Formation has lower gamma-ray and higher velocity, density, and resistivity signatures producing characteristic 'bell' or 'barrel' shaped gamma-ray and sonic log responses, relative to the underlying and overlying mudstones (Figure 129). The Balder Formation forms a prominent, distinctive, nondiachronous, geographically extensive seismic horizon, which can be used to determine regional structure at this level (Hitchen and Ritchie, 1987). It is capped by coal, tuff or highly porous sandstone, all of which return a distinctive high amplitude seismic reflection, although, the polarity of the event varies laterally with changing lithology (Smallwood and Gill, 2002).

In well 219/28-1 (Figure 2), a 3 m thick basic tuff is overlain by 85 m of tuffaceous claystone and siltstone, most of which can probably be attributed to the Balder Formation (Fitch et al., 1988). Above this interval, the lithology becomes increasingly less silty and the tuffaceous content decreases. In nearby well 219/28-2Z (Figure 120), the Balder Formation is 104 m thick and comprises numerous primary basic airfall tuffs intercalated with tuffaceous claystone and siltstone. It is clear that much of the tuffaceous material in the fine-grained sediments above the principal tuff layers is reworked. The main ash units are basaltic in character, consisting of vitric and lithic fragments. Crystal fragments, mainly pyroxene and plagioclase feldspar, are rare. Much of the glassy material is altered to palagonite and chlorite. A K–Ar whole rock age (average of three analyses) of 51.7 ± 0.8 Ma (Ypresian) obtained from the tuff in well 219/28-2Z (Table 8) must be regarded as a minimum age (Fitch et al., 1988).

The BGS borehole 85/02 was drilled on the south-east end of the Wyville Thomson Ridge. Beneath 23.4 m of Quaternary diamict, the borehole penetrated 4.95 m of dark grey to dark greenish grey, massive to poorly bedded, shelly, muddy, fine-grained tuffaceous sandstone containing abundant basaltic fragments, quartz, heavy minerals (hornblende, pyroxene and zircon) and a clay mineralogy composed entirely of smectite (Stoker et al., 1988). The absence of graded bedding, and the high degree of intermixing with terrigenous clastic material, indicates that the basaltic debris is reworked rather than deposited in situ (Stoker et al., 1988). The age of the sandstone can be estimated from two sources. Firstly, it contains abundant examples of the foraminifer *Nummulites rockallensis*, which has been dated as Early Eocene (Morton et al., 1983). Secondly, it overlies regional basalts that have been K–Ar dated at 57.3 ± 5.2 to 62.3 ± 4.7 Ma in the nearby BGS borehole 85/07 (Stoker et al., 1988) and thus, the sandstone is likely to be latest Paleocene to earliest Eocene in age. Although these ages suggest the tuffaceous sandstones of borehole 85/02 maybe equivalent to the Balder Formation, no direct correlation has been made.

Volcanic mounds

Recent high-quality 3D seismic surveys have imaged numerous buried mounds in the Faroe–Shetland Basin, both to the south-west and north-east of the Erlend and West Erlend volcanic centres (Bell and Butcher, 2002; Davies et al., 2002). The mounds are approximately conical in shape and vary from 30 to 300 m in height and from 500 m to 1.7 km in diameter (Bell and Butcher, 2002; Davies et al., 2002). Some are so close together that their outer margins almost coalesce (Bell and Butcher, 2002; Davies et al., 2002). Internal dips are perhaps up to 15° at the centre, but only 1° at the margins (Bell and Butcher, 2002; Davies et al., 2002). The mounds have a higher density and seismic interval velocity than the surrounding claystone and appear to rest on a continuous flat surface, which has been dated at 54.9 Ma (Bell and Butcher, 2002; Davies et al., 2002). An overlying seismic reflector, which onlaps the mounds, has been dated at 54.6 Ma (Bell and Butcher, 2002; Davies et al., 2002).

There is a striking spatial correlation between these mounds and the sills of the Faroe–Shetland Sill Complex (FSSC). All the mounds appear to occur directly above the terminations of sills and commonly there is a 'seismic chimney' linking the two features. This can be interpreted either as a dyke or a zone of disturbed, mobilised sediment (Bell and Butcher, 2002; Davies et al., 2002). The mounds are constructional in origin and thought to comprise either; (1) pillowed lava and/or hyaloclastite volcanic breccias or (2) explosion breccias of volcanic/sedimentary material caused by steam-driven expulsion of pore water (Bell and Butcher, 2002; Davies et al., 2002). Whichever interpretation is correct, there is a direct genetic link between the intrusion of the sills, at a relatively shallow level, and the formation of the mounds on the contemporary sea floor. As a result, the age of the mounds can be used to date the age of the sills of the FSSC.

INTRUSIVE ROCKS

Intrusive rocks within the report area take the form of the widespread FSSC and the spatially constrained intrusive components of eight recognised volcanic centres.

Faroe–Shetland Sill Complex

The FSSC is an elongate, north-east-orientated belt of sills, and associated dykes, which intrudes the sedimentary succession of the Faroe–Shetland Basin (e.g. Hitchen and Ritchie, 1987). The extent of the complex is shown in Figure 120 and its area has been estimated at a minimum of 22 500 km^2. However, this area is likely to be much greater as the complex almost certainly extends beneath the subcrop of the FIBG (as speculatively postulated by Smallwood and Maresh (2002, fig. 1)) and possibly as far north-west as beneath the Faroe Islands (Ritchie et al., 1999a), where the basaltic lava flows of the Malinstindur through to the Enni formations are known to be cut by dykes (Hald and Waagstein, 1991). These dykes possess similar geochemistry to the sills of the Faroe–Shetland Basin (Gibb and Kanaris-Sotiriou, 1988).

The FSSC should be viewed as part of a much larger intrusive belt which extends beyond the report area from the southern Rockall Basin (west of Ireland) to the Norwegian margin. Most of the sills in this large region were probably intruded approximately synchronously. The Wyville Thomson Ridge, although primarily a Miocene compressional structure, has been used as an arbitrary boundary to separate the FSSC from sills within the Rockall Basin to the south.

In most cases the host rocks for the intrusions of the FSSC appear to be mainly Late Cretaceous in age, although some late stage ?Thanetian sills intrude Paleocene sedimentary rocks, especially in the area of the Flett Sub-basin (Lamers and Carmichael, 1999).

Recognition of sills from seismic data

On conventional industry 2D seismic data acquired across the Faroe–Shetland Basin, sills are imaged as short, high-amplitude reflectors, which may be either parallel to, or cut across, normal sedimentary interfaces (e.g. Figures 24, 30, 33 and 37). Sills may also appear to possess a saucer-shaped geometry with up-turned terminations (e.g. Figure 31), comparable in geometry to the Streymoy and Eysturoy sills that intruded the Malinstindur, Sneis and Enni formations on the Faroe Islands. Sills commonly occur at several different geological levels and always degrade the seismic response from the deeper geology. Their size is commonly in the order of a few kilometres across, but individual sill 'leaves' may be linked to form composite transgressive structures (best seen on 3D seismic data) of the order of 10 km or more across. Other published examples of seismic data imaging sills of the FSSC are shown by Gibb and Kanaris-Sotiriou (1988), Naylor et al. (1999), Bell and Butcher (2002), Smallwood and Maresh (2002), Hansen et al. (2004) and Trude (2004).

Recognition of sills from well data

Sills of the FSSC generally intrude the Upper Cretaceous claystone and siltstone rocks of the Faroe–Shetland Basin. They exhibit a range of thicknesses from just a few metres to nearly 100 m (see Gibb et al. (1986) for examples of sills in well 219/20-1; Figure 120). Individual wells may intersect either isolated sills (e.g. well 219/28-2Z) or whole 'families' of sills. The contrast in lithologies between the igneous sill and the host sedimentary rock usually causes a marked response on well logs. Typically, the sills cause an increase in the resistivity, density and sonic velocity log values and a decrease in gamma-ray values compared to the surrounding host rock. On well logs, sills may be distinguished from extrusive rocks by their higher interval velocity and the presence of a thermal aureole. Sills have a high interval sonic velocity commonly over 5 km/s (and sometimes in excess of 6 km/s), whereas extrusive igneous rocks are typically in the range of 3.5–4.5 km/s. This is due to extrusive units commonly being composed of multiple flows with irregular weathered tops and thin sedimentary interbeds, all of which slow the passage of sound through the interval. Extrusive lava flows do not 'bake' the underlying sedimentary rocks to the same extent as sills and cannot thermally affect the overlying sedimentary units. Examples of typical log responses are shown by Stoker et al. (1993), Bell and Butcher (2002) and Smallwood and Maresh (2002). The regional thermal impact of the FSSC within the Faroe–Shetland Basin appears very variable. In well 214/28-1 (Figure 120), elevated vitrinite reflectance (R_0) values, determined from well cuttings, have been recorded over a 2 km vertical interval and attributed to contact thermal metamorphism and associated hot fluids (Holmes et al., 1999). However, elsewhere in the same basin, apatite fission track analyses and more R_0 values do not yield evidence for elevated regional palaeo-heat flow (Holmes et al., 1999).

Occurrence of sills in the Faroe–Shetland Basin

Most of the sills appear to be intruded into the axial region of the Faroe–Shetland Basin where the crystalline crust has been extended and thinned to the greatest extent and may be only 10 km thick compared to about 30 km thick beneath the Shetland Islands (Smallwood and Maresh, 2002). It might be assumed that the melt required for the intrusions was generated by upper crustal extension at about the time of the Paleocene to Eocene transition. However, as there is only limited evidence for extension at this time, the melt must have been derived either as a result of elevated temperatures below the lithosphere, or else would have been channelled laterally from the west, where continental break-up was approaching completion. The former explanation is preferred by Smallwood and White (2002).

On a qualitative basis, sills appear to be most common where the Faroe–Shetland Basin is most segmented (Smallwood and Maresh, 2002) by north-west-trending transfer zones (Rumph et al., 1993; Figure 7), although no volumetric assessment has been undertaken. Furthermore, sills appear to be intruded at shallower levels over existing basement highs and at the basin margin. Forceful intrusion of large volumes of igneous material at relatively shallow depth can inflate the overlying sedimentary pile. This observation may also help to date the age of intrusion. On this basis, Gibb and Kanaris-Sotiriou (1988) suggest that 'at least the youngest of the sills' in the vicinity of well 219/20-1 (Figure 120) are post-Paleocene in age. This is supported by the sills of the Faroe Islands that have intruded the syn-break-up succession of basalt lava flows, which are Ypresian in age. In part of UK Quadrant 205, the volume of material intruded appears to have inflated the contemporary sea floor and controlled the deposition of potential hydrocarbon-bearing reservoir sandstones (Smallwood and Maresh, 2002).

With the increasing availability of 3D seismic data sets in the Faroe–Shetland Basin, it has not only become possible to map the size, shape and overall geometry of the sills, but also draw some conclusions about their relative ages, relationships and evolutionary development. Sills within the basin have a variety of shapes, though they are commonly saucer-shaped and several kilometres in diameter. They are commonly multileaved and possess segments which are both concordant and discordant to the bedding. The vertical relief may extend to several hundred metres (Hansen et al., 2004). Detailed mapping of the Corona Sill has revealed ridges on its upper surface taken to be the result of a viscous magma being intruded into waterlogged sediments at a shallow level beneath the contemporary sea bed (Trude, 2004).

Internal structure of the sills

By analogy with exposed thick sills onshore, it is highly likely that the sills of the FSSC are internally structured, both physically and geochemically (differentiated). According to Bell and Butcher (2002), the uppermost and lowermost portions of the sills, adjacent to the host rock, may have slightly reduced interval velocity and density values due to the presence of gas cavities (typically with amygdale mineral infills), together with closely spaced fractures, which formed during and subsequent to crystallisation and cooling. Conversely, Smallwood and Maresh (2002) state that some velocity and density log values have a weak tendency to decrease from the margins towards the centre of the sills and attribute these changes to variations in phenocrysts, vesicles and flow fabrics.

The internal mineralogy and petrology of the sills can vary markedly even if the geochemical composition is relatively uniform. A 9 m length of core from the 'number 8' sill in well 219/20-1 is composed of olivine-dolerite and is described by Gibb et al. (1986). The grain size generally increases from 'chilled' or fine-grained dolerite at the top, to a very coarse-grained, or pegmatitic, facies at the base, where plagioclase feldspar crystals up to 1 cm in length occur sub-ophitically intergrown with augites over 2 cm across. This coarse-grained facies probably formed late in the crystallisation process. A 22.5 m length of core from the 'number 18' sill penetrated by the same well, but at a lower stratigraphical level, also increases in grain size downwards, but to a lesser degree and over a much longer interval. The lower half of this core comprises alternations of fine-grained and pegmatitic dolerite (Gibb et al., 1986).

Geochemistry of the sills

Most of the sills of the FSSC are tholeiitic olivine-dolerites. The main exceptions are those associated with the Erlend Volcanic Centre where the lithologies of the main acidic suite and some of the sills intruding the underlying Upper Cretaceous shales are described as felsite, rhyolite, microgranite, obsidian, pitchstone and dacite. Within some basic sills fractional crystallisation has also resulted in more acidic lithologies being preserved (see below).

Detailed geochemical analyses, based on cored intervals, have been published for sills penetrated in four wells located in the northern part of the FSSC i.e. 208/21-1, 209/03-1, 219/20-1 and 219/28-2Z (Figure 120). Sills cored in well 208/21-1 and 219/20-1 are tholeiitic olivine-dolerites. Despite being relatively low in SiO_2 they plot in the oceanic field of Pearce et al. (1975), in the ocean-floor basalts field on the Ti–Zr–Y and Ti–Zr plots of Pearce and Cann (1973) and in the N-type (normal) MORB field on the Nb–Zr–Y plot of Meschede (1986) (Gibb et al., 1986; Gibb and Kanaris-Sotiriou, 1988). However, on the basis of Ti:V ratios, it has been suggested by Gibb and Kanaris-Sotiriou (1988) that the cored sills are actually T-type (transitional) MORB (between N-type and E-type (plume-enriched)). The sills differ geochemically from the aphyric and plagioclase-phyric tholeiite basalt lava flows of the FIBG, but have a striking similarity to the low-Ti olivine tholeiites that occur two-thirds of the way up the Malinstindur Formation and throughout the Sneis and Enni formations on the Faroe Islands (Gibb et al., 1986).

The extent of internal differentiation within the sills is emphasised by an 8 m long core from near the top of a 110 m thick sill in well 208/21-1. Most of the core is doleritic but the interval from 2.5 to 6 m below the top (i.e. 8 to 11.5 m below the top of the actual sill) is described as a plagiogranite and is characterised by micrographic intergrowths of quartz and plagioclase feldspar (Kanaris-Sotiriou and Gibb, 1989). Compared to the adjacent dolerite, the SiO_2 content is much greater (exceeding 61 wt.% whereas the dolerites are typically 46.5–47.2 wt.%) but the TiO_2, CaO and K_2O contents are lower. The transition from tholeiitic olivine-dolerite to plagiogranite within the sill is strongly suggestive that oceanic plagiogranite can be produced from MORB-type magmas by fractional crystallisation at relatively low pressures (Kanaris-Sotiriou and Gibb, 1989).

Two sills, 'upper' and 'lower', were penetrated by well 219/28-2Z within Upper Cretaceous claystone. They are 34 and 46 m thick, respectively, and the central parts of both sills are microgabbroic. Both sills have extensive thermal aureoles atypical for intrusions of this thickness, which suggests that they may have acted as flow conduits for large volumes of hot magma. Although both were probably derived from tholeiitic magmas, the upper sill is relatively low in K_2O and appears sub-alkaline, whereas the lower sill has a low $Zr:P_2O_5$ ratio and plots as an alkali basalt (Fitch et al., 1988).

Well 209/03-1 drilled through the basalt lava flows of the Erlend Volcanic Centre into underlying Santonian to Maastrichtian claystones. These are intruded by numerous basalt, dolerite and quartz-andesite sills (information from composite well log). The well terminated in a sill, which was subsampled at four separate depths. The samples are described as dacitic glasses, in various stages of devitrification and contain up to 10 wt.% cordierite or pinitic mica pseudomorphs after cordierite. They are all quartz-corundum-hypersthene-normative and are therefore, peraluminous. All the dacite samples contain carbon (up to 0.61 wt.%) in the form of graphite (Kanaris-Sotiriou et al., 1993; Kanaris-Sotiriou, 1997). The distinctive geochemistry of the dacites suggests that they may have been generated by the original magmatic source being contaminated with upper crustal argillaceous sedimentary rocks through the process of anatexis (partial melting). A similar process has been invoked for the origin of other peraluminous dacites drilled in well 163/6-1A in the Rockall Basin (Morton et al., 1988a) and in ODP borehole 642 on the Vøring Plateau (Viereck et al., 1988).

Age of the sills

Various authors have tried to date the sills of the FSSC using radiometric methods (e.g. Hitchen and Ritchie, 1987; Fitch et al., 1988; Hitchen and Ritchie, 1993; Ritchie and Hitchen, 1996; Ritchie et al., 1999a). The ages deemed most reliable are given in Table 8. Other authors have used regional correlations and the age and attitude of the host rocks as indicators (e.g. Duindam and van Hoorn, 1987; Gibb and Kanaris-Sotiriou, 1988) or the similarity in geochemistry between sills and other igneous rocks of known age.

With the exception of the age for the lower sill in well 219/28-2Z, and ignoring the error bars quoted in Table 8, the reliable radiometric ages cluster between 55 and 52 Ma (Ypresian). This very broadly agrees with a biostratigraphically derived age of 54.9 to 54.6 Ma obtained for conical igneous mounds erupted on the contemporary sea floor and sourced from underlying sills and dykes of the FSSC (Davies et al., 2002). It is also compatible with the observation that the lava flows of the Malinstindur and Enni formations (emplaced between 54.8 and 54.5 Ma) are cut by sills of the FSSC.

The 80 Ma (Campanian) average age for the lower sill in well 219/28-2Z is problematic. However, other Cretaceous ages have been reported for volcanic events in the Rockall Basin to the south. These are 83 Ma for Helen's Reef, part of the Rockall complex (Pankhurst, 1982), Maastrichtian or older for the Anton Dohrn Seamount (Jones et al., 1974), also a 70 Ma age determination for an early volcanic episode connected with the Anton Dohrn Seamount (O'Connor et al., 2000) and a 'Late Cretaceous' age for basalts from the Rosemary Bank Seamount (Morton et al., 1995).

Volcanic centres

Presently, eight volcanic centres have been identified within the report area (Brendan, Drekaeyga, Erlend and West Erlend, Faroe Bank Channel Knoll, Frænir, Regin Smiður and Sigmundur) (Figure 120). Most consist of an intrusive core and an overlying volcanic sequence, suggesting they are the sites of former (conical) volcanoes. All the centres apart from Drekaeyga are associated with a large, positive, circular free-air and isostatic gravity anomaly (Figures 3 and 4). The existence of two previously interpreted volcanic centres, Judd (Hitchen and Ritchie, 1987) and Westray (Rumph et al., 1993), remains debatable. The positive gravity anomalies, which were originally interpreted to define these centres, are now thought to be caused by faulted metamorphic basement blocks of the Westray and Corona highs.

It has been suggested that the centres formed along either east-north-east or north-west volcanic trends (Ritchie et al., 1999b; Archer et al., 2005). The east-north-east trend may represent incipient break-up lineaments that the igneous centres exploited. Alternatively the centres may have utilised fracture zones that formed orthogonally (i.e. north-west-trending) to the general break-up trend for the north-east Atlantic region (Archer et al., 2005).

Brendan

The Brendan Volcanic Centre is situated between the Møre and the Faroe–Shetland basins, approximately 200 km north of the Shetland Islands and 70 km north-north-east of the Erlend Centre (Figure 120). The centre was first recognised from potential field data, forming a pronounced near-circular 80 mGal free-air gravity anomaly, which was modelled as a body that increased in diameter downwards from 10 km at a depth of 5 km to 50 km at a depth of just over 10 km. This structure was named the Brendan Seamount and dated as 'at least as old as Turonian' based on the age of the

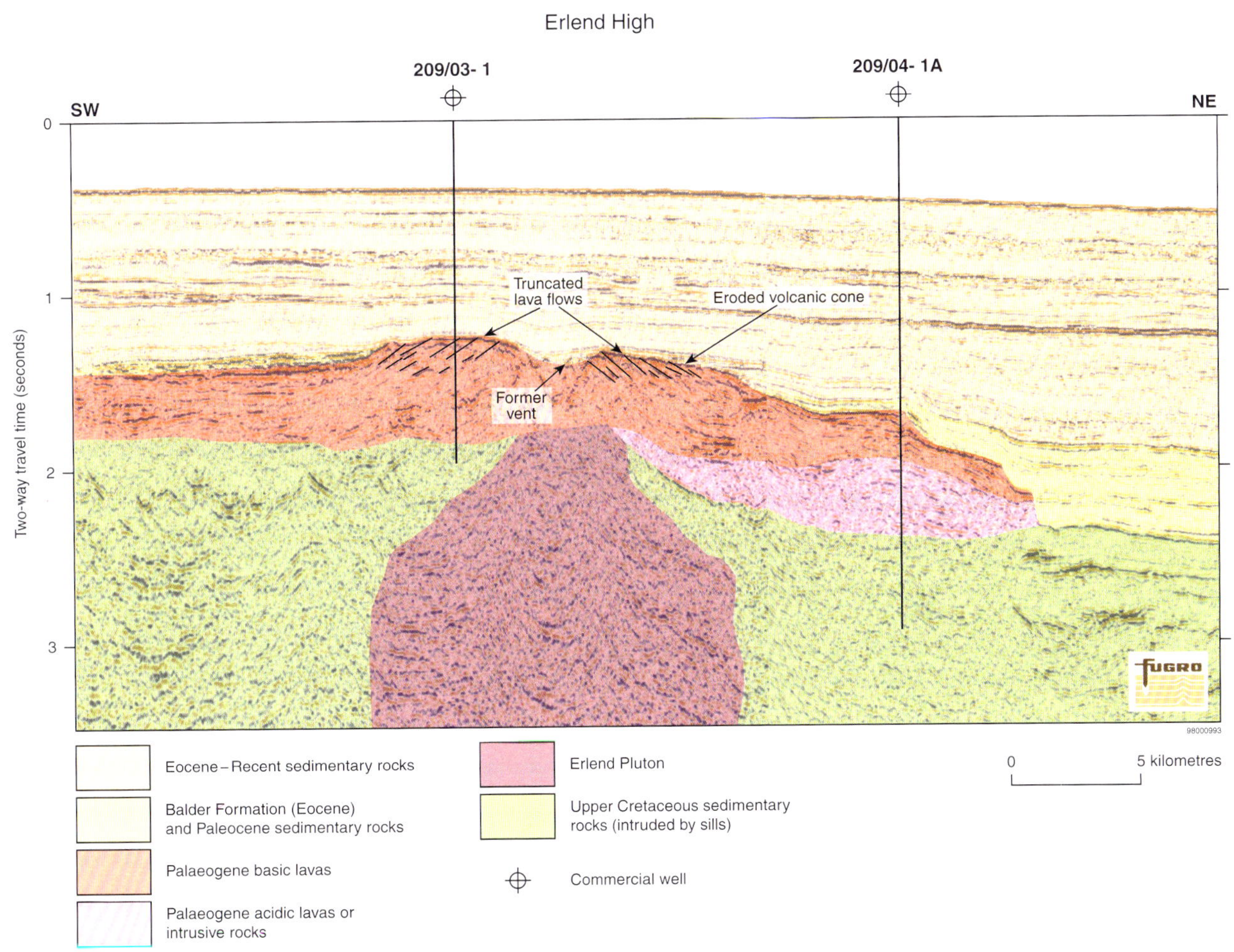

Figure 130 Merged seismic profile across the Erlend Volcanic Centre. The deep structural level of the Erlend pluton is derived from gravity modelling. For location of profile, see Figure 120. Seismic data courtesy of Fugro Multi Client Services.

sediments stacked up against the flanks of the seamount (Smythe et al., 1983). Above the seamount, the subcrop of the FIBG lava flows forms a southward-protruding lobe, which is also a structural high. This is the Brendan Dome of Smythe et al. (1983) who suggest that the spatial coincidence of the seamount and the dome of lava flows can be explained by 'Early Eocene' lavas flowing over a subaerial promontory created by differential compaction of the sediments around the seamount. This theory requires an age difference of 'at least 20 million years' between formation of the seamount and the extrusion of the lavas that define the dome (Smythe et al., 1983). In reality the age gap between the end of the Turonian and the end of the Paleocene is nearer 34 million years. An alternative explanation, given by Hodges and Evans (1999) is that the seamount is Late Paleocene in age, and that the lavas of the dome were emplaced during the Eocene. As the Brendan Volcanic Centre as a whole has not been drilled by any deep boreholes, the nature and age of the structure remains speculative.

Drekaeyga

The Drekaeyga Volcanic Centre is located in the Auðhumla Basin, a synclinal fold between the Wyville Thomson and Ymir ridges, and is completely buried by younger Cenozoic sediments (Figure 120). At the level of the top Palaeogene lavas, the centre appears to form a caldera-like crater (Keser Neish, 2005, fig. 6) which has an area of approximately 225 km^2. The lavas that define the structure are folded. Unlike the other basic volcanic centres in the report area, the observed gravity field does not contain the strong positive feature indicative of an underlying pluton. However, when the gravity effect of the Auðhumla Basin is accounted for by modelling, there is evidence to suggest that a dense pluton is present (Smith et al., 2009). The origin of the present configuration of the Drekaeyga structure remains problematic. It may represent an eroded former volcano, but it has also been described as an intrusion (Keser Neish, 2005) or an intrusive centre (Keser Neish and Ziska, 2005).

Erlend and West Erlend

The Erlend and West Erlend volcanic centres are situated on the Erlend High adjacent to the north-east Faroe–Shetland Basin, approximately 120 km north of the Shetland Islands (Figures 120 and 130). The existence of a plutonic centre here was originally postulated by Chalmers and Western (1979) using potential field data. It was named 'Erland' by J A Chalmers, in Ridd (1983). This centre was confirmed by Gatliff et al. (1984), who changed the spelling to Erlend, and who also recognised a second, but smaller centre, approximately 26 km to the west, which was subsequently named as West Erlend. The centres consist of plutonic bodies overlain by associated igneous lithologies.

Chalmers and Western (1979) produced two gravity models to explain the observed gravity values across the Erlend Centre. Both were similar with a postulated vertical cylinder of dense rock, with a diameter of 14 km, extending from about 2 km below sea level down to 25 or 30 km. The model of Gatliff et al. (1984) proposed a cylinder 14 km in diameter extending down to only 15 km below sea level. The depth to the base of the cylinder is the least well-constrained parameter. Differences between the models can be explained by varying assumptions made for density contrasts between the cylinder and the host rock at various depths. A small volcanic plug may be present at the West Erlend centre and Gatliff et al. (1984) used gravity modelling to illustrate possible alternative geometries for this body. Either a cylinder with a base at 15 km and a diameter of 5 km or one with a base at 8 km and a diameter of 7 km were shown to give an acceptable fit to the observed data (Gatliff et al., 1984).

The Erlend centre has been drilled by three exploration wells i.e. 209/03-1, 209/04-1A, and 209/09-1A, all of which proved a sequence of igneous rocks (Figure 131). In wells 209/04-1A and 209/09-1A a suite of acidic rocks intercalated with sedimentary units 948 m and 274 m thick, respectively, were encountered. The acidic suite consists of the following lithologies: felsite, rhyolite, microgranite, obsidian, pitchstone and dacite (information from composite well logs; Ridd, 1983; Mitchell and Euwe, 1988; Jolley and Bell, 2002b). Some authors have regarded the acidic suite to be extrusive in origin based partly on the presence of peperite (the product of explosive interaction between wet sediments and lava or magma) and the fine grain size of the acidic rocks (Ridd, 1983; Hitchen and Ritchie, 1987; Mitchell and Euwe, 1988). However, the in situ palynomorphs recovered from the intercalated sedimentary units have a variable, commonly high Thermal Alteration Index (TAI) which Jolley and Bell (2002b) suggest was generated by increased heatflow from intrusive bodies rather than volcanics. Thus, Jolley and Bell (2002b) interpreted the acidic suite to be intrusive in origin, which they suggest is supported by the presence of similar acidic rocks interpreted to be sills intruding the underlying Campanian sedimentary rocks beneath well 209/03-1.

The acidic suite is overlain by a basaltic suite, the Erlend lava field, which is 823 m, 499 m and 175 m thick in wells 209/03-1, 209/04-1A and 209/09-1A, respectively (Figure 131). These volcanic sections are dominated by hyaloclastite and basaltic lava interbedded with siltstone and sandstone (Jolley and Bell, 2002b). The basalt lava flows are subaerial in character and olivine and plagioclase porphyritic varieties occur (Kanaris-Sotiriou et al., 1993; Jolley and Bell, 2002b). These lava flows are olivine-normative tholeiitic basalts with a N- to T- type MORB character (Kanaris-Sotiriou et al., 1993).

The basaltic lava flows around the Erlend volcano (Figure 130) form a southward-extending bilobate subcrop pattern which merges with the lava flows of the FIBG (Figure 120). The Erlend lava field has a separate origin from the lava flows of the FIBG, although they maybe of a similar age. The West Erlend Volcanic Centre is more deeply eroded than Erlend and it has been postulated that up to 1 km of basaltic lava flows have been eroded off the top of the West Erlend volcano (Gatliff et al., 1984). No central vent or radially disposed lava flows have been imaged on seismic data for the West Erlend Centre. The lava flows erupted from the Erlend and West Erlend centres do not crop out at the sea bed and are covered by younger Eocene to Recent sediments (Figure 130).

On seismic profiles, the Erlend volcano exhibits dipping seismic reflectors, which represent the original depositional attitude of the lava flows (Figure 130). These are radially disposed around a central vent which has a diameter of 2 km and a depth of 300 to 400 m (Gatliff et al., 1984). The top of the original volcanic cone has been planed off by erosion. In places, the distal limits of the flows, especially on the northern and western flank of the volcano, terminate as depositional escarpments such as the Erlend Escarpment (Figure 26), suggested to represent the possible locations of former shorelines (Gatliff et al., 1984).

The problem of dating the Erlend Volcanic Centre has been addressed by a combination of seismic mapping, radiometric dating and biostratigraphy. The igneous rocks encountered in all three of the Erlend wells overlie mudstones of Late Cretaceous age (Campanian to possibly Maastrichtian) (Jolley and Bell, 2002b). Mitchell and Euwe (1988) obtained a K–Ar model formation age of between 58 ± 3 and 55 ± 4 Ma (mainly Thanetian) for the acidic suite as a whole, which is significantly different to the 64.8 ± 0.8 Ma Ar–Ar age (close to the Maastrichtian–Paleocene boundary), obtained from a rhyolite by Hitchen and Ritchie (1987) at the base of the acidic suite. However, this latter age should be regarded with some caution as Thanetian (T36 toT38 sequences) palynomorphs have been recovered from above and below the sampled rhyolite in well 209/04-1A (Jolley and Bell, 2002b). Also, Ridd (1983) reported the occurrence of foraminifera with a Late Cretaceous (Campanian) age from sedimentary interbeds within the acidic suite from well 209/09-1A, which is corroborated by the palynomorph assemblages reported by Jolley and Bell (2002b) for the same well. Thus, if the dated microfossils are all in situ, and the acidic suite is extrusive, then it was emplaced in two phases, the first being during the Late Cretaceous and the second during the Thanetian. This represents an unlikely minimum time gap of nearly 12 million years (according to the timescale of Gradstein et al., 2004).

Seismic mapping of the area shows that the overlying Erlend basaltic lava field is onlapped by the Balder Formation (Gatliff et al., 1984; Stoker et al., 1993), which therefore, constrains the younger age limit of the lavas to 54.5 Ma. Biostratigraphical data from below, within and above the lava flows suggest they have a palynomorph assemblage that occurs in the Flett Formation unit 1b of the Faroe–Shetland Basin (i.e. upper T40 sequence) and which postdates the PETM (Jolley and Bell 2002b; D W Jolley pers comm. 2005), dated to 55.8 Ma on the timescale of Gradstein et al. (2004). The Erlend lava field also biostratigraphically correlates to the exposed section of the Beinisvørð Formation of the FIBG on the Faroe Islands (Figure 124), currently the only section of the Beinisvørð Formation that has been analysed palynologically (Jolley and Bell, 2002b; D W Jolley pers comm. 2005). Therefore, the Erlend lava field is suggested to have been emplaced during the Ypresian (Early Eocene) between 55.8 and 54.8 Ma (Jolley and Bell 2002b, D W Jolley pers comm. 2005). If so, the interpretation of Jolley and Bell (2002b) that the acidic suite is intrusive suggests that the intrusive phase was

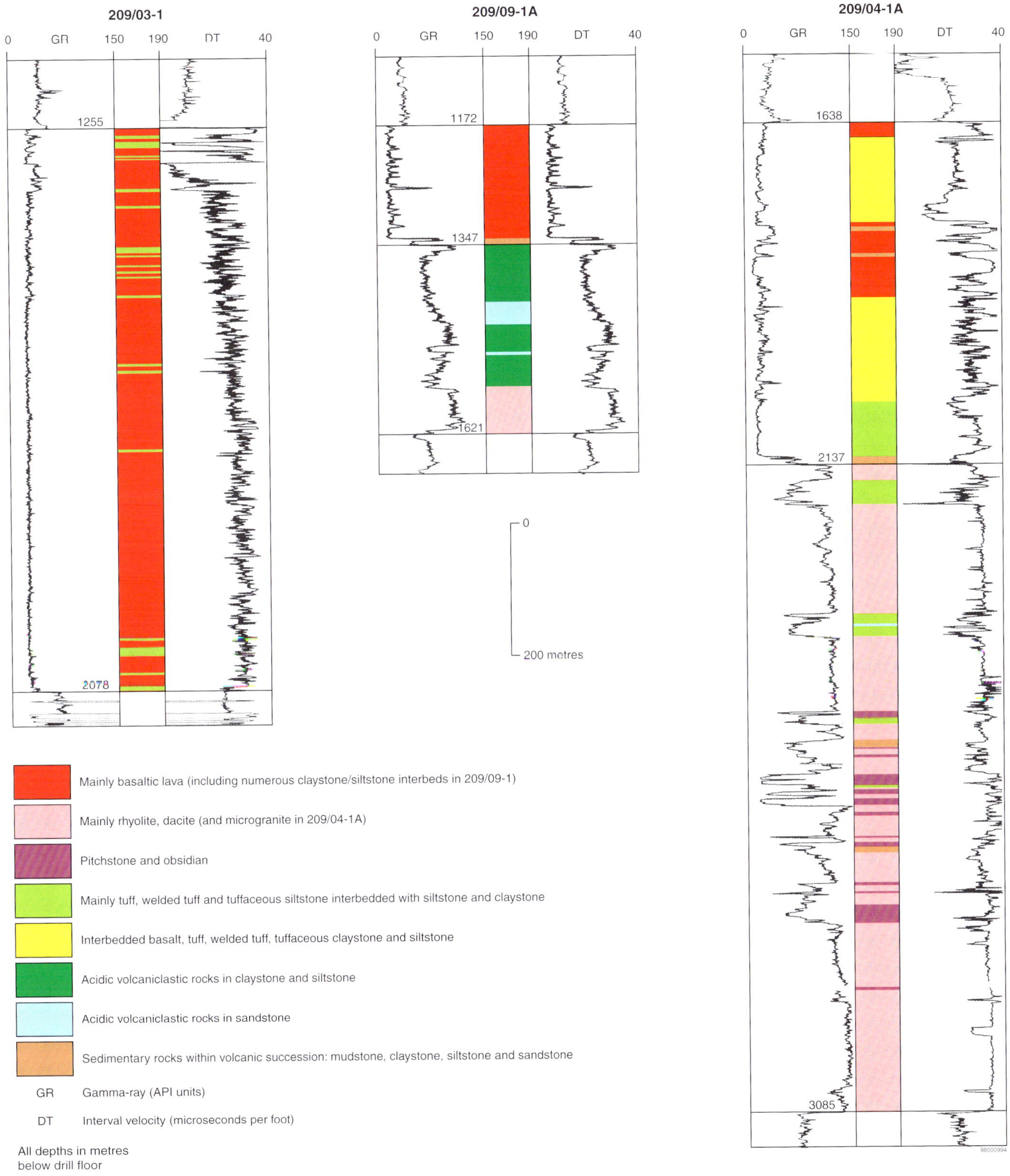

Figure 131 Well logs of wells 209/03-1, 209/09-1A and 209/04-1A drilled on the Erlend High that penetrated igneous rocks of the Erlend Volcanic Centre. A wide variety of igneous and interbedded sedimentary lithologies were proved. For location of wells, see Figure 120.

broadly coincident with the extrusive phase, being emplaced during the Late Paleocene or Early Eocene. This is supported by the K–Ar whole rock ages of 54.3 ± 2 and 54.8 ± 2 Ma (Early Eocene) obtained for a sill intruding the Campanian sedimentary section from well 209/03-1 (Hitchen and Ritchie, 1987).

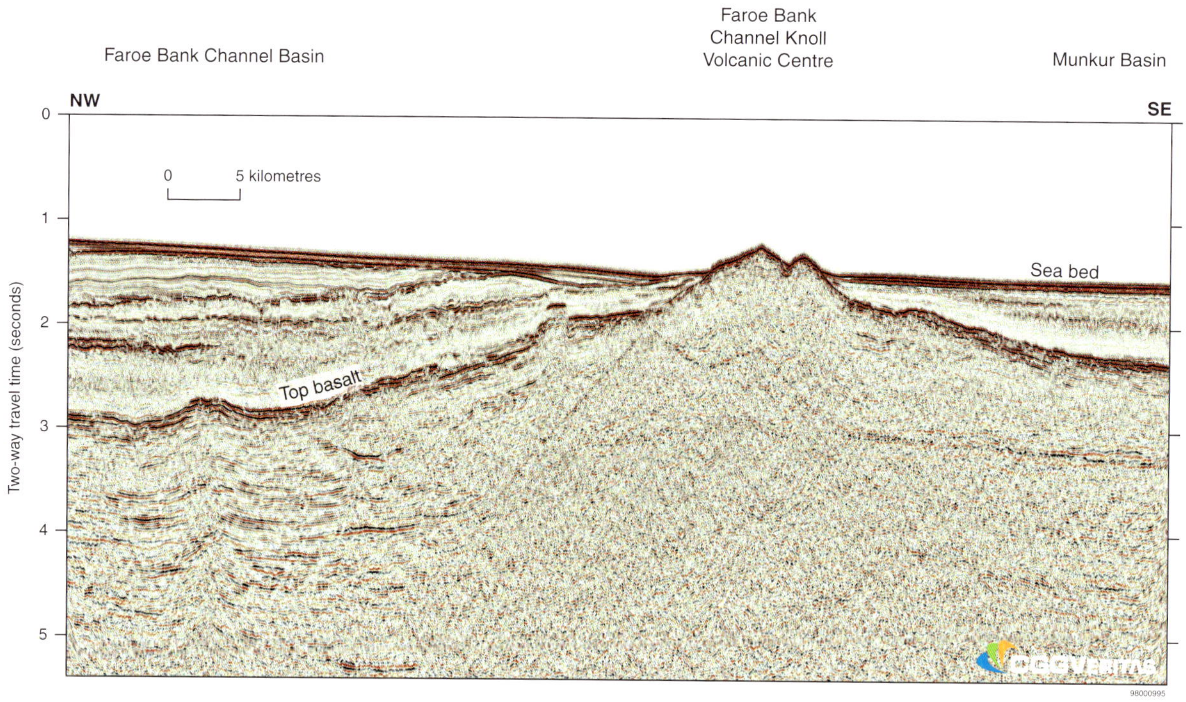

Figure 132 Seismic profile across the Faroe Bank Channel Knoll. For location of profile, see Figure 120. Seismic data courtesy of CGG Veritas.

Faroe Bank Channel Knoll

The Faroe Bank Channel Knoll, so named because it is situated at the southern entrance to the Faroe Bank Channel Basin (Keser Neish, 2003; Keser Neish and Ziska, 2005; Figure 132), was originally identified by Roberts et al. (1983) and was initially named as the Faroe Channel Knoll. The centre occurs at the boundary of the Munkur and Faroe Bank Channel basins and is bounded to the north-east and south-west by the Munkagrunnur and Wyville Thomson ridges (Figure 120), respectively. The Faroe Bank Channel Knoll is roughly circular in shape and exhibits a high-amplitude magnetic anomaly response (Figure 5). On seismic sections, the Faroe Bank Channel Knoll is observed to consist of basalt lava flows that crop out at the sea bed but which become buried by younger sediments away from the centre (Keser Neish and Ziska, 2005; Figure 132). It is also seen on seismic data that the basalt lava flows of the centre onlap older flows of the Wyville Thomson Ridge to the south-west (Keser Neish and Ziska, 2005).

Frænir

The Frænir Volcanic Centre was first recognised and named by Keser Neish (2003) and is located 80 km to the south-south-east of the Faroe Islands (Figure 120). It is clearly resolved by gravity data, particularly the isostatic gravity anomaly (Figure 4). Frænir is situated along a north-north-west fault trend but seismic imaging is restricted by overlying basalt lava flows. However, the basalt lava flows above the centre are disrupted, possibly implying the presence of a volcanic vent (Keser Neish, 2003).

Regin Smiður

The Regin Smiður Volcanic Centre is located 50 km to the south-west of the Faroe Islands (Figure 120) and is recognised from potential field data. The centre was initially named as West Suduroy by Ritchie et al. (1999a) but was subsequently described and renamed as the Regin Smiður Volcanic Centre by Keser Neish (2003). The centre is characterised by having a large, circular positive free-air gravity anomaly and a strong magnetic response (Figures 3 and 5). Keser Neish and Ziska (2005) suggested from seismic data across Regin Smiður that basalt lava flows originating from it are onlapped by younger basalt lava flows to the south-east, therefore tentatively suggesting the presence of a volcanic vent.

Sigmundur

Sigmundur was initially recognised as 'High B' by Roberts et al. (1983) and was subsequently renamed as the Sigmundur Seamount by Andersen (1988) and then the Sigmundur Complex by Ritchie et al. (1997) (Figure 120). The Sigmundur Seamount has a circular intrusive centre with a free-air gravity anomaly up to 60 mGal that was modelled by Ritchie et al. (1997) as a large mafic pluton with an asymmetrical shape. The pluton has a sloping top and base at depths of approximately 2.5 to 3 km and 25 to 25.5 km, respectively. The diameter of the intrusion increases downwards from approximately 3.5 to 7 km at the top to 14.5 to 21 km at the base (Ritchie et al., 1997).

From seismic data it is seen that the intrusive centre is overlain by basalt lava flows (Ritchie et al., 1997). The margins of the volcanic expression of the centre are represented by scarp-like features that define a roughly circular feature. At the margins of the volcanic centre, outward-dipping reflectors are observed within the basalt sequence which may reflect the original attitude of the lava flows. The radial nature of the volcanic sequence suggests that Sigmundur represents the eroded remnant of a former conically shaped volcano where the lava flows were extruded radially from a central vent (Ritchie et al., 1997). The majority of the volcanic sequence is buried by younger Eocene to Recent sediments but at the centre of the complex the lava flows crop out at the sea bed.

10 Sea-bed geology and environment

Alan Stevenson[1], Heather Stewart[1] and Heri Ziska[2]

[1] *British Geological Survey*

[2] *Jarðfeingi (Faroese Earth and Energy Directorate)*

Present day sea-bed sediment distributions and processes reflect the interaction between the relict Pleistocene sea floor and Holocene sediment redistribution, the latter mainly driven by bottom currents. Medium to large-scale mobile sedimentary bedforms provide reliable indications of the areas where persistent bottom currents transport sediments along the sea floor whereas smaller scale bedforms may record only the last significant event to affect the area in which they occur. Interpretation of the geological features and sediment distribution observed on the sea bed therefore requires an understanding of the modern hydrodynamic regime and the sources of sediment.

BATHYMETRY

The bathymetry of the sea floor, shown in Figure 1, can exert a considerable influence on hydrodynamic regime. The distribution of sea-bed sediments and mobile bedforms is described in relation to the three main physiographical types that occur in the report area; shelf, slope and channel/basin areas. The shelf is defined as areas in water depths generally less than 300 m, the slope between 300 and 900 m water depth and the channel/basin region includes areas where water depths exceed 900 m.

The main bathymetric features within the report area include the West Shetland Shelf, Faroe Shelf, Faroe–Shetland Channel, the Faroe Bank Channel and a series of banks to the west of that channel including the Wyville Thomson Ridge and Faroe Bank (Figure 1). To the south-west is the Rockall Trough, and to the north-west the Faroe Bank Channel runs into the Iceland Basin. The Norwegian Basin is located to the north of the Faroe Shelf and Faroe–Shetland Channel.

Shelf areas

West Shetland Shelf

The West Shetland Shelf is a north-easterly continuation of the Hebrides Shelf (Figure 1). The sea floor from the mainland of northern Scotland to Shetland comprises a broad, relatively flat platform, with waters ranging in depth from 70 to 120 m. The sea bed in the outer area of the West Shetland Shelf has low relief and is underlain by sequences of sheet-like, mounded or ridged glacigenic deposits that form broad, low-amplitude rises and depressions. These range in length from several hundreds of metres to upwards of 20 km, generally with a relief of between 20 and 50 m. Submerged moraine complexes form distinctive ridges running parallel to the shelfbreak to the west of Shetland (Stoker and Holmes, 1991). The major change of gradient between the shelf and slope, the shelfbreak, varies in water depth between 120 and 250 m (Stoker et al., 1993).

Faroe Shelf

Between Iceland and the Faroe Islands, the Iceland–Faroe Ridge forms a broad ridge (Figure 1) with minimum water depths along the crest of 300 to 500 m, generally deepening in an eastward direction. The south-eastern part of this ridge joins the Faroes Shelf, a platform with the Faroe Islands at its centre. The eastern continuation of the Faroes Shelf, the Fugloy Ridge, separates the Norwegian Sea from the Faroe–Shetland Channel. To the south-east, the Munkagrunnur Ridge forms a prominent spur to the Faroe Shelf. Water depths increase away from the Faroe Islands and the shelfbreak occurs most commonly between 300 to 500 m.

Large areas of the sea bed on the outer shelves and upper slopes have an irregular surface due to cross-cutting iceberg ploughmark furrows (see Chapter 11). These furrows typically occur in water depths ranging from 140 to 510 m both west of Shetland and to the north and east of the Faroes. However, there is evidence of features similar to iceberg ploughmarks in water depths of greater than 700 m on the Wyville Thomson Ridge (Masson et al., 2004).

Iceland–Faroe Ridge and Wyville Thomson Ridge

The Iceland–Faroe Ridge and Wyville Thomson Ridge form components of the Greenland–Scotland Ridge that extends from East Greenland to Scotland (Figure 92). Below water depths of 840 m, the Greenland–Scotland Ridge forms a continuous barrier between the North Atlantic and the Norwegian Sea (Hansen and Østerhus, 2000), which connects to the Arctic Ocean through the Fram Strait between Greenland and Svalbard. North of the Greenland–Scotland Ridge, the sea floor generally deepens north-eastwards, with maximum water depth in the report area of 3 km in the Norwegian Sea.

The Wyville Thomson Ridge forms the south-west flank of the Faroe Bank Channel (Figure 1). Rising to between 350 and 600 m water depth, the ridge forms a narrow, generally north-westerly trending topographical barrier between the Faroe–Shetland Channel and the Rockall Trough. The ridge joins the West Shetland Shelf at its south-eastern limit and at the northern end joins the Faroe Bank, a large, flat-topped bank, much of which is less than 100 m below the sea surface. The shape of the Faroe–Shetland Channel, which narrows to the south-west, and the orientation of the Wyville Thomson Ridge have a significant effect on the strength of present day bottom currents by funneling the currents, then deflecting most of the water mass through 90 degrees into the Faroe Bank Channel. Slope gradients on the north-east flank of the ridge are steep and can be as much as 8°.

Slope areas

West Shetland Slope

The slope area west of Shetland is relatively narrow compared to the shelf. Slope angles are typically less than 1°, increasing to between 1.5° to 2° midslope, before decreasing towards the base of slope. The topography of the upper and middle slopes can be attributed mainly to the volume of sediment transported from the shelves to form the underlying prograding Plio-Pleistocene wedges, whereas the lower slope reflects the angularity of underlying Cenozoic erosion surfaces. A characteristic convex profile towards the base of slope may be partly the result of bottom-current erosion.

Slope instability, deposition and erosion by alongslope currents and iceberg scouring also locally shape the sea floor. Small, scarp-like features and depressions related to both erosion and slope failure are widespread, as are gullies and channels. The upper slopes, to a maximum water depth of over 500 m, generally show small-scale irregularities due to iceberg scouring whereas the middle and lower slopes range from smooth to undulating and hummocky, caused by mass-flow deposits and contourite drifts.

Faroes Slope

The slope areas around the Faroes Shelf are typically steep along the north-eastern side of the Faroe Bank Channel as well as on the flanks of the Fugloy Ridge (where the slope gradient at 62°30'N is about 7°). A significant feature of the Faroe Shelf is a slight embayment to the south-east of the Faroes that is clearly defined by the 200 m bathymetric contour (Figure 1).

Channel/basin areas

The relatively broad, deep Faroe–Shetland Channel and the narrower Faroe Bank Channel form a continuous deep-water channel separating the Faroes Shelf from the West Shetland Shelf to the east, and the Munkagrunnur Ridge from the Wyville Thomson Ridge and Faroe Bank to the south. The Faroe–Shetland Channel narrows and shallows southward from about 190 km wide and 2 km water depth at 63°N to a sill of approximately 1 km water depth and 90 km width at 60° 25' N. Farther to the south-west it gradually deepens again reaching 1.2 km water depth at 60°10'N, 6°00'W. At this point the deep-water channel turns abruptly to the north-west into the Faroe Bank Channel, the eastern part of which is a relatively broad, flat-floored basin with water depths of 1.1 to 1.2 km in the channel floor. Farther north-west, the channel narrows and shallows forming a sill of approximately 840 m depth between the Faroes Shelf and Faroe Bank at 8°20'W. To the west of this sill, the Faroe Bank Channel increases rapidly in depth into the Iceland Basin.

In a small area along the axis of the southern Faroe–Shetland Channel (60°20'N 5°30'W and 60°40'N 4°20'W), an area of complex sea-floor topography identifies the Judd Deeps (Figure 112b and c; see chapters 8 and 11). Slopes frequently exceed 5° and are locally greater than 25°.

OCEANOGRAPHICAL REGIME

Throughout the Holocene, the composition and distribution of sea-bed sediments have been influenced by a combination of tidal, wave-induced, wind-driven and oceanic currents. All four processes have operated on the shelves, whereas oceanic currents are the predominant influence on the slopes and within the channels (Stoker et al., 1993). The interaction between the surface and deep-water masses influences the current patterns and the distribution of sea-bed sediments in the area. Turrell et al. (1999) have described the complex interplay of five major water masses in the Faroe–Shetland Channel (Figure 133) which are known to generate strong, persistent bottom currents capable of eroding and transporting sediments with grain sizes up to and including gravel (Masson et al., 2004). It is possible that the present day hydrographical regime has existed continuously for about the last 14 000 years (Rasmussen et al., 1996), although some of the currents, such as the Faroe Shetland Channel Bottom Water (FSCBW) (Figure 133), may have influenced sediment build up at the shelf margins since the Late Eocene and Oligocene (Miller and Tucholke, 1983).

Published data on deep-water or bottom current flow velocities in the southern Faroe–Shetland Channel consist of a few relatively short-term measurements, which are highly variable. Saunders (1990) gives current velocities of 0.1 to 0.2 m/s towards the south-west. Stoker et al. (1998) give bottom current estimates of between 0.13 to 0.22 m/s, with a peak velocity of 0.33 m/s. Masson (2001) gives typical velocities of 0.1 to 0.2 m/s, with a peak velocity of 0.6 m/s. All sources indicate that the mean velocity of bottom currents is slower than that of the surface water (0.3 to 0.6 m/s, Saunders, 1990; Masson, 2001). Some of the strongest and most persistent bottom currents occur where deep-water flow is funnelled through narrow topographical passages. The shape of the Faroe–Shetland Channel and orientation of the Wyville Thomson Ridge have a significant effect on the strength of present day bottom currents within the report area, by funnelling the FSCBW into the Faroe Bank Channel (Bulat and Long, 2001). Direct measurements show that this flow constriction generates bottom current velocities that can significantly exceed 1.0 ms^{-1} (Hansen and Østerhus, 2000).

Oceanographical studies in the Faroe Bank Channel have mainly concentrated on the sill region at its north-western end (Masson et al., 2004). Here, a sharp boundary, typically lying between 500 and 600 m water depth, occurs between the upper, warm North Atlantic water and the cold overflow water. The 200 to 300 m thick cold layer consists primarily of FSCBW and the contribution of the intermediate water masses is much reduced, probably because of mixing and recirculation in the south-west Faroe–Shetland Channel (Hansen and Østerhus, 2000). Studies of bottom water velocity in the sill region show a marked asymmetry across the channel with the highest velocity (>1.0 m/s) occurring about 100 m above the sea floor on the southern side of the channel adjacent to the Faroe Bank (Hansen and Østerhus, 2000). However, Saunders (1990) has shown that the high-velocity core then switches to the north side of the channel further downstream over the sill. Hansen et al. (2001) have given some evidence for a reduction in flow velocities through the Faroe Bank Channel since 1950.

SEA-BED SEDIMENTS

In general, the sea-bed sediments in the Faroe–Shetland report area are sand or gravel-rich on the shelf areas and more fine-grained on the slopes and channel floors. Previous studies in the region indicate that the sea bed is covered by a patchy veneer of unconsolidated sediments consisting of an admixture of terrigenous and biogenic carbonate material (BGS 1:250 000 sea-bed sediment map series; Faroes GEM Network 2001a; b; c; d; Stoker et al., 1993; Masson et al., 2004). The terrigenous com-

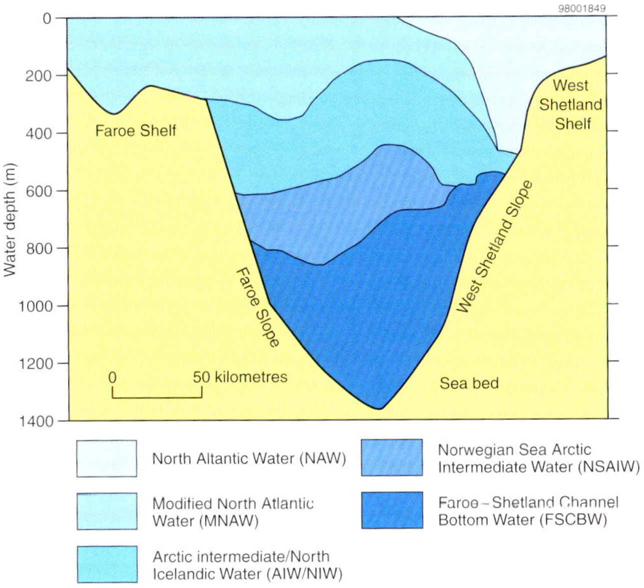

Figure 133 Water masses in the Faroe–Shetland Channel (modified from Turrell et al., 1999).

ponents are mainly derived from underlying Pleistocene deposits and bedrock, such that the boundary between the sea-bed sediments and underlying glacial deposits is not easily discernible. During the Holocene, most of the area has been starved of terrestrial sediment input. Masson et al. (2004), for example, suggest that postglacial sedimentation in the Faroe Bank and Faroe–Shetland channels is almost entirely controlled by bottom currents. Postglacial sedimentation rates in these areas have been typically very low (<1 cm per thousand years, Masson, 2001). In contrast, the biogenic carbonate component has been accumulating throughout the Holocene (Stoker et al., 1993).

The distribution of sea-bed sediments in the area is shown in Figure 134. The classification is based mainly on the GEM sea-bed association approach, which is an interpretation of sea-bed facies from the seismic character of the sea-bed reflector on high-resolution seismic and a limited number (approximately 100) of core samples (Faroes GEM Network, 2001a). As a result of the GEM approach, the area west of Shetland is broadly generalised, however a detailed description of the sea-bed sediment types in this area can be found in Stoker et al. (1993).

The sea-bed sediments can be categorised under seven groups, as shown on Figure 134. These are principally areas of sand and gravel with significant shell contribution; diamicton either exposed at sea bed or lying beneath a veneer of finer-grained sediment; soft sediment disturbed by slope failure; hard sea bed/rock outcrop; gravel, cobbles and boulders exposed at sea bed or lying beneath a thin veneer of sand; fine-grained sediment, mainly mud; deep-water, fine to medium-grained sand and mud with gravel patches. The distribution of these sediment associations is summarised below according to the three main physiographical features in the area.

Surficial sediments on shelf areas

West Shetland Shelf

On the West Shetland Shelf, there is great variability in sediment grain size, however, in general, sand and gravel-rich sediments are predominant not only across the shelf, but on the adjacent slopes beyond the shelf-break, where they can occur down to water depths between about 900 m and 1 km. Muddy sediments are rare on the shelf (Figure 134), except in sea lochs and other sheltered sites in coastal areas and bathymetric depressions on the inner shelf. They generally occur in areas which have been previously infilled with late Pleistocene to early Holocene mud, where reduced currents have helped to maintain the deposition of fine-grained sediments throughout the Holocene.

In many areas, especially on topographical highs, the sediments are coarse-grained lag deposits that are either exposed at sea bed or are overlain by a thin cover of finer-grained mobile sediment. Gravel lags are formed by the winnowing of fine-grained sediments by bottom currents, leaving the coarser material at the sea bed. These gravel and bedrock outcrops form a favourable environment for a prolific and diverse calcareous fauna. The faunal remains make a significant contribution to the gravel fraction in the region, especially on the inner shelf in areas such as between Orkney and Shetland, a major high-latitude centre of carbonate production where sands can be 100% biogenic debris (Farrow et al., 1984). The lithic portion of the gravel fraction ranges in size from granules up to boulders, and depending on location consists of varying mixtures of metamorphic, sedimentary and igneous rocks.

The carbonate component of the gravel consists mainly of bivalve and echinoid fragments, serpulid tubes, barnacle plates and bryozoans. The composition of the carbonate fraction varies with location; near Shetland, barnacles and attached serpulids are dominant whereas west of Orkney, bivalves predominate (Wilson, 1979). Patches of cold-water coral occur in the vicinity of the shelfbreak (Figure 135). Wilson (1979, 1982) reported that radiocarbon dating indicates that the carbonate material has been accumulating throughout the Holocene, with a mean age range of between 8000 to 3000 years BP. Sediments on the outer shelf and deeper sea floor have low carbonate content.

Sand-grade sediments have been reworked, sorted and transported by bottom currents throughout the Holocene. On the shelf, the sand is transported away from the elevated parts of the sea bed to be redeposited on the surrounding, lower lying sea bed where current strengths are reduced. This is most readily apparent on the middle and inner shelf where abundant, homogeneous sand-rich sediments form sheet-like deposits up to 2 m thick. Like gravelly sediments, sands on the shelf area between Orkney and Shetland are carbonate-rich, typically containing between 25 to 50% biogenic material. On the current-swept outer shelf, sand commonly occurs as mobile, thin longitudinal patches and streaks overlying coarser-grained lag and Plio-Pleistocene deposits.

Areas with iceberg ploughmarks are characterised by irregular gravel ridges raised above the general level of the sandy sea floor. Individual ploughmarks typically consist of raised ridges separated by a shallow depression generally between 20 to 80 m in width (see Chapter 11). Sand also partially infills many ploughmarks and other local hollows and depressions.

Areas of rock outcrop are located to the west of Shetland and Orkney and to the north of Cape Wrath. These typically occur in areas of strong tidal currents that remove the mobile sediment fraction to expose a rugged

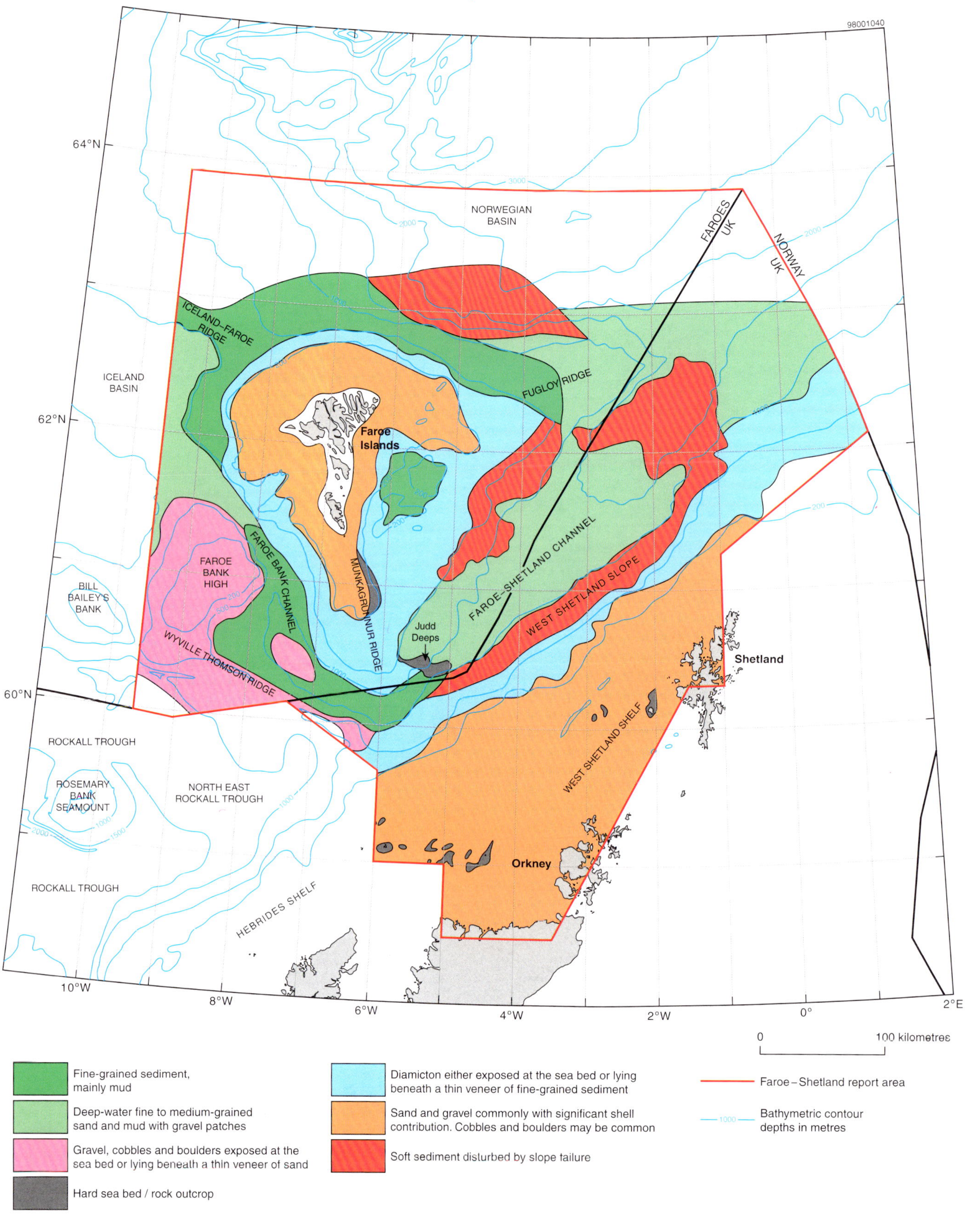

Figure 134 An interpretation of the nature of sea-bed sediments within the Faroe–Shetland report area based on the sea-bed sediment associations of Faroes GEM Network (2001a, b, c and d). Samples within the Faroes national sector were provided by BIOFAR (Nørrevang et al., 1994) and the Faroes GEM Network (2001a). Interpretation within the UK sector is very broadly based on existing BGS Sea-bed sediment 1:250 000 series of map sheets and Masson et al., (2004).

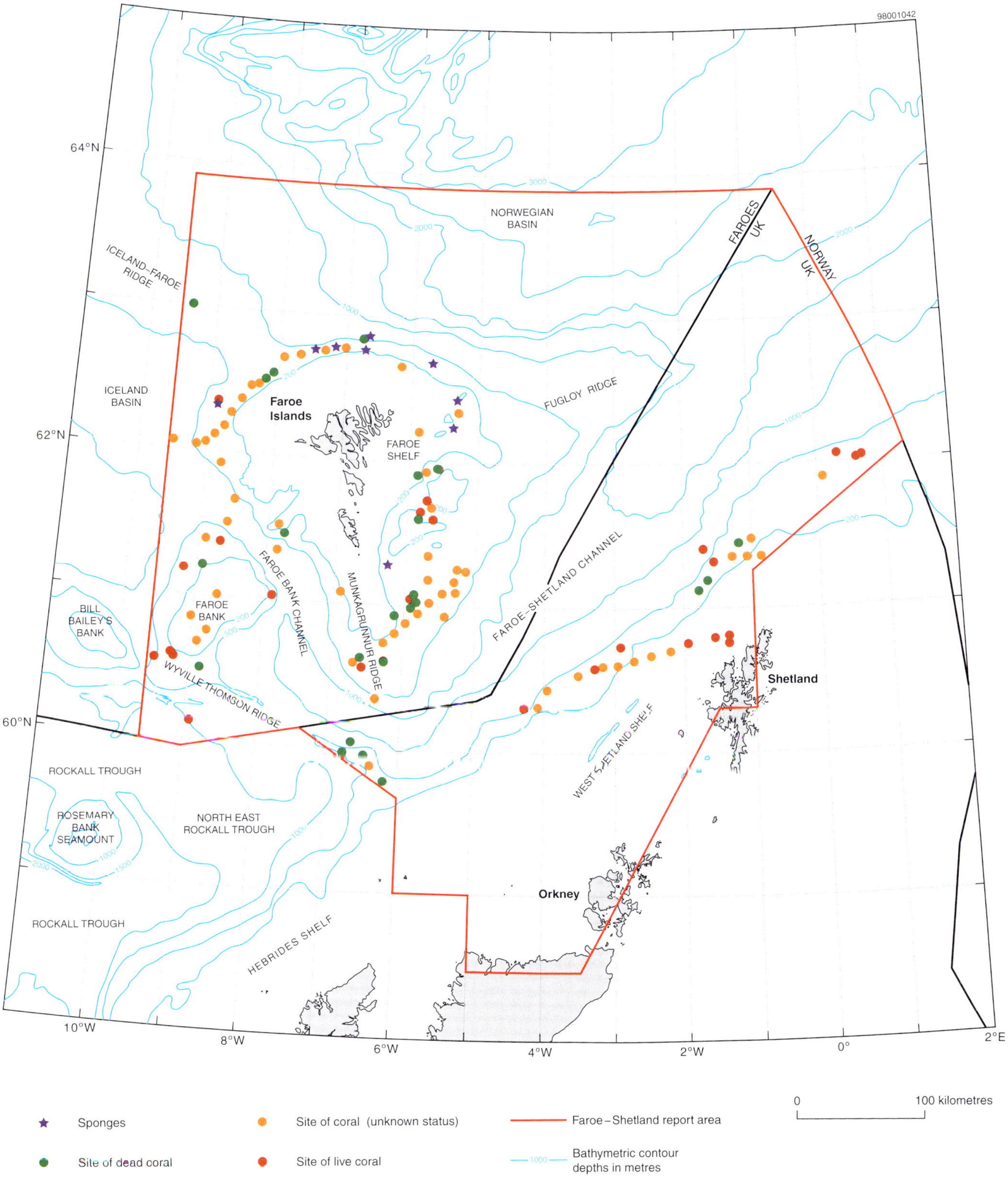

Figure 135 Sample locations of coral and sponges (modified from Wilson, 1979; Shelton, 1980; Zibrowius, 1980; Chesher, 1987; Frederiksen et al., 1992; Jensen and Frederiksen, 1992; Nørrevang et al., 1994; Bett, 1997, 1999; Bell and Smith, 1999 and Roberts et al., 2003).

rock platform comprising primarily Devonian, Permo-Triassic and Precambrian.

Faroe Shelf

In water depths less than 200 m in the inner parts of the Faroe Shelf, the sea floor is predominantly formed by Palaeogene basalt (Figure 120; Waagstein, 1988). The basalt occurs close to, or in places at the sea bed in the vicinity of the Faroe Islands and along the Fugloy and Munkagrunnur ridges. A thin layer of surficial, unconsolidated sediment comprising sand, gravel and cobbles up to a few metres thick occurs more generally and has accumulated within depressions or hollows or at the surface of morainic deposits. Boulders may also be common in this area (Figure 134; Faroes GEM Network, 2001a).

Around the Faroe Islands, extensive diamicton deposits occur at the sea bed, or are covered by a thin layer of sand that is typically less than 0.3 m thick and may be mobile (Faroes GEM Network, 2001a). The sand is poorly sorted and varies between coarse and fine-grained material consisting of foraminifera and shell fragments. Along the eastern margin of the northern Faroe Shelf, this sand layer is typically less than 0.15 m thick; the grain size also becomes more variable, ranging from very fine to locally coarse-grained. On the shelf to the east and south-east of the Faroe Islands there is a local increase in the thickness of superficial sediment cover to generally greater than 0.3 m. The sediments comprise mainly silty sand. The thickness of sea-bed sediments gradually increases to the west from approximately 0.3 metres at the shelfbreak to between 1 and 1.5 m towards the southern Faroe Islands. The boundary with the basalt outcrop at the sea bed in the vicinity of the islands is poorly defined by available data, but it may be abrupt. Clay clasts are common and pebbles and gravels, or even boulders, may be present. Local reworking and remoulding of these sediments has occurred due to the effects of iceberg grounding. Photographs from the southern part of the Faroes Shelf show mainly coarse gravel and rock outcrop at the sea bed (Masson et al., 2004).

Wyville Thomson Ridge and Faroe Bank

Photographs, samples and sidescan-sonar images from the crest and upper slopes of the Wyville Thomson Ridge show that the sea-bed sediments in the area mainly comprise coarse-grained gravel ranging in size from granules to boulders, formed by the reworking of glacial sediments by bottom currents (Figure 134; Masson et al., 2004). Sea-floor photographs show the gravel deposits extend to water depths >800 m. Although there are few samples from the Faroe Bank, seismic facies suggest a thin veneer of sediment lies at the sea bed.

Surficial sediments in slope areas

West Shetland Slope

Sea-bed sediments on the slope regions in the report area generally consist of soft, silty clay (Figure 134) with varying amounts of sand, clasts and shell fragments. On the West Shetland Slope, muddy sediments are widespread with a progressive increase in the mud content with increasing water depth. In many areas, the thin veneer of muddy sediments is indistinguishable from the underlying glacial deposits from which they are derived.

Masson et al. (2004) described samples and photographs form the midslope area of the West Shetland Shelf near the Wyville Thomson Ridge that are characterised by a thin (0.05 to 0.2 m thick) surficial sand and gravel layer of Holocene age. These sediments were termed 'gravel lag contourites' by Stoker et al. (1998) and are considered to represent the result of reworking and winnowing of Pleistocene glacial sediments by strong Holocene bottom currents. Some carbonate material may also be introduced into the slope sediments by biological production (Light and Wilson, 1998). In areas that have been studied in detail, there is no evidence of active downslope sediment transport by gravity-flow processes at present (Masson et al., 2004).

Between 500 m and about 800 m to 1 km water depth, much of the sediment underlying the sea floor on the West Shetland Slope has been formed by downslope processes such as debris-flows associated with sediment deposition at a glaciated shelf margin (Bulat and Long, 2001). This part of the slope is also disturbed by slope failures, such as the AFEN and Miller slides (see Figure 140). The sea-bed sediments in this area are typically soft, fine-grained sandy mud or mud, which is commonly indistinguishable from the underlying sediments.

Faroe Slope and flanks of the Faroe Bank and Wyville Thomson Ridge

Sea-bed sediments in this area are predominantly sands and silty sands which become finer grained towards the base of slope, where the sea bed veneer is mainly very fine to fine-grained slightly silty sands (Faroes GEM Network, 2001a). However, photographs from the flanks of the Wyville Thomson Ridge show gravel extending to >800 m water depth (Masson et al., 2004). As observed on the West Shetland Slope, there is evidence of ice-rafted material and reworking by the ploughing action of icebergs and sediment slides on the slopes, possibly down to water depths greater than 700 m. Significant modification and erosion of those sediments by bottom currents has led to the generation of gravel lag deposits, with clasts ranging in size from pebbles to boulders, and associated sandy bedforms (Masson et al., 2004).

On the western slope of the Faroe–Shetland Channel, sediments that have been disturbed by slope failure or other mass-wasting processes consist of a relatively thin veneer of fine to medium-grained sand overlying interbedded sandy silty clay debris-flow units and silty to silty sandy clay layers containing sand lenses and dropstones (Faroes GEM Network, 2001a).

Surficial sediments in channel/basin areas

Sea-bed sediments in the axial parts of the Faroe Bank and Faroe–Shetland channels show evidence of the influence of significant deep-water currents. The dominant material in the axial part of the channels is a discontinuous veneer of very fine to medium-grained silty sand (Figure 134) with occasional gravel lag deposits (Faroes GEM Network, 2001a). Photographs from the base of the northern flank of the Wyville Thomson Ridge confirm a sandy or slightly muddy sand sea floor with some gravel, whereas in the Faroe Bank Channel, sidescan sonar images and core samples suggest that much of the channel floor west of 5°40'W is characterised by areas of mainly fine-grained sand or muddy sand deposition (Masson et al., 2004). Sub-bottom profiles from the channel area west of 5°40'W show a low-relief hummocky sea floor underlain by parallel-bedded sediments interpreted as sediment drifts and contourite sheets (Masson et al., 2002). The sediments are usually well-sorted with an absence of glacial dropstones, which characterise most other sediments at the sea bed in this area, suggesting that they are of postglacial age. Sidescan sonar records from the northern edge of the basin floor suggest that gravel lag deposits extend on to the lower slope of the Faroe Shelf.

In the area around the Judd Deeps (Figure 134), seismic data suggest that Eocene–Oligocene sediments crop out at the sea bed. A cover of coarse gravel is

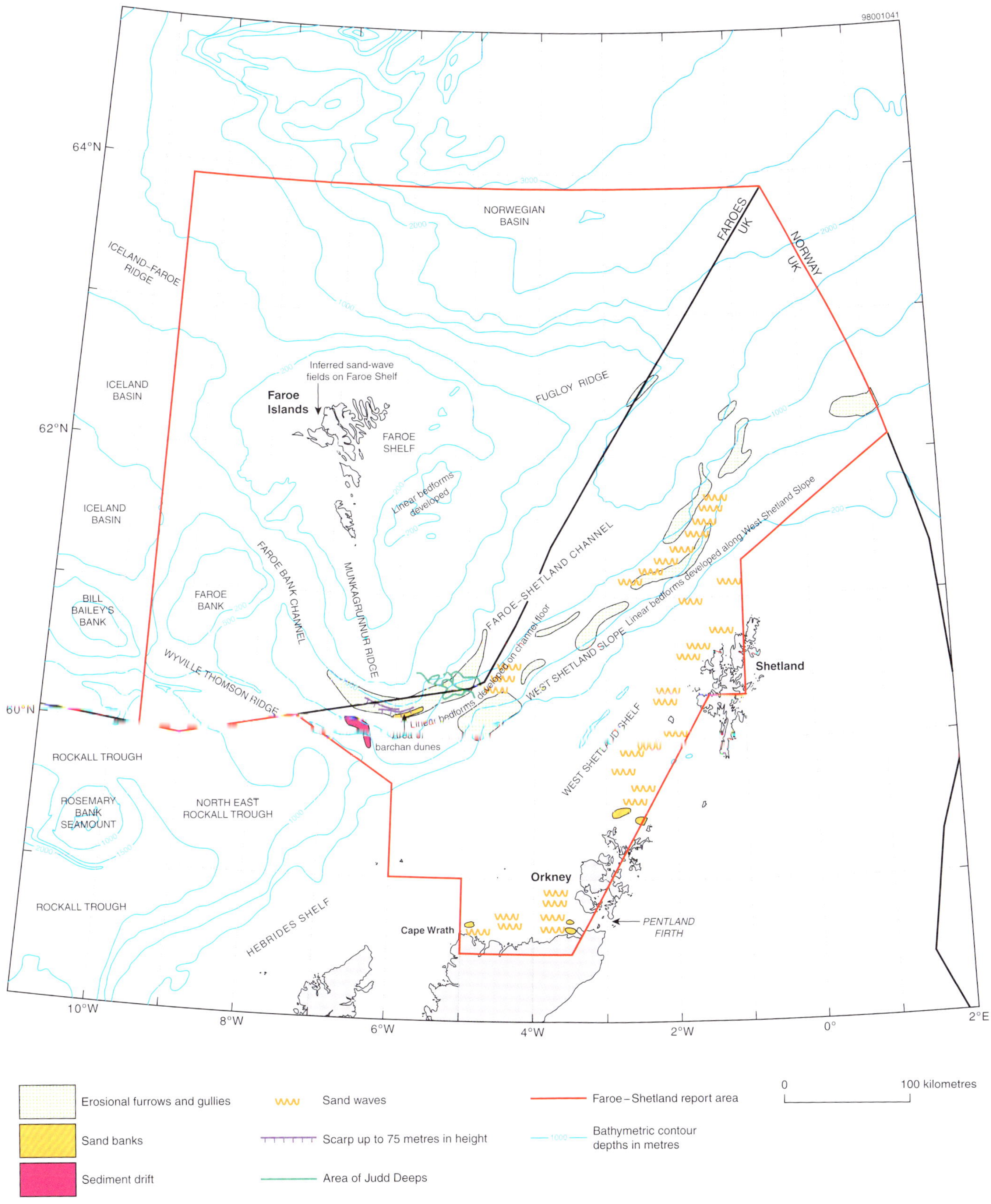

Figure 136 Sedimentary bedforms and erosional features in the Faroe–Shetland report area (modified from Bulat and Long, 2001; Masson et al., 2003; Stoker et al., 1993 (and references therein) and Wynn et al., 2002). For extent of iceberg ploughmarks, see Figure 139.

present in places on the exposed rock (Faroes GEM Network, 2001a).

At the mouth of the Faroe–Shetland Channel and the Norwegian Basin, and further north, the sea floor typically comprises a veneer of mud below about 1 km water depth. Glacigenic debris-flows of the North Sea Fan underlie the thin sediment veneer over much of the area. The presence of mud diapirs, termed the Pilot Whale Mud Diapirs (see Figure 147), introduce localised variability at the north-east end of the Faroe–

Shetland Channel. The sea bed in the area of the diapirs comprises competent mud blocks with gravel and cobbles located between the blocks (Bett, 2001).

BEDFORMS AND EROSIONAL FEATURES

A large variety of sedimentary bedforms occur in the area, including sand waves, streaks and ribbons and barchan dunes (Figure 136). Erosional bedforms such as scours, furrows and comet marks are also seen in more limited areas. The bedforms indicate areas where the sea bed interacts with bottom currents that erode, transport and redeposit the sea-floor sediments.

Shelf areas

West Shetland Shelf

Sand streaks, sand ribbons and longitudinal sand patches are widespread on the shelf where they range from several metres up to several hundred metres in width and can be up to a few kilometres in length. The streaks and ribbons are of negligible thickness, and usually occur where there are relatively strong bottom currents or where supplies of mobile sand are limited. The sand patches overlie a gravel substrate and can be up to 1m thick, and appear to be common on the middle and inner shelf, where sand is more abundant (Stoker et al., 1993). The long axes of linear bedforms are generally aligned along-shelf, parallel to the tidal currents.

Sand waves are relatively common on the middle and inner shelf, both as isolated features and in groups (Figure 136). In many cases they are asymmetric in profile and therefore provide an indication of net sediment transport direction. Sand waves are particularly well developed in areas of carbonate-rich sediments such as between Orkney and Shetland, where sand-wave fields have amplitudes in the range of 3 to 8 m (Flinn, 1973; Allen, 1983). Sand waves in the vicinity of Orkney and Cape Wrath (Figure 136) occur in association with sandbanks that range in height from about 10 to 30 m above the surrounding sea bed (Allen, 1983). Large sandbanks to the north of Orkney are up to 10 km long, 0.5 km wide and 30 m high (Farrow et al., 1984; Figure 136). Sand waves with heights of up to 20 m, and wavelengths of 200 m, are developed on the slopes of these banks; their orientation may indicate a clockwise movement of sediment around the banks.

The combined evidence of bedform orientation and bottom currents measurement indicates that the main sediment transport paths are probably aligned along the shelf or parallel to the land. There is no evidence for the transfer of sediment from the outer shelf to the slope (Stoker et al., 1993). Provenance studies indicate that material is being transported northwards along the shelf, and eastwards towards the North Sea, a pattern of dispersal very similar to the water circulation. South of Orkney, a bed-load parting zone occurs in the Pentland Firth where sediments are being transported both to the east and west (Kenyon and Stride, 1970).

Slope areas

West Shetland Slope

Furrows are common on the West Shetland Slope between 500 m and 1.2 km water depth (Figure 136). In addition, broader, alongslope oriented depressions, with the appearance of shallow (<10 m deep) channels are observed on the slope between 400 and 500 m water depth; they are up to 800 m wide and 20 km long. Masson (2001) attributed these to alongslope sediment transport by bottom currents. Sonar data from the West Shetland Slope show that sand ribbons are a common feature in water depths down to 500 m (Kenyon, 1986).

Pockmarks are shallow depressions on the sea bed that are relatively common in areas of muddy sediments on shelf areas (see Chapter 11); they are also thought to occur on the West Shetland Slope (Hovland and Judd, 1998) although Stoker et al. (1993) suggest that most of these features are probably a result of iceberg ploughmarking or slope instability.

An extensive network of elongate mounds running subparallel to the strike of the slope is observed between the 900 m and 1.4 km isobaths in the area west of Shetland. The network is restricted to the slope area but appears to cover it completely. The mounds are approximately 1 km in width and at least 10 km in length and could cover an area approximately 60 km in length and between 5 km and 15 km in width, narrowing southwards (Bulat and Long, 2001). These have been interpreted as longitudinal sediment waves generated by contour currents (Kenyon, 1987; Damuth and Olsen, 1993).

Faroes Slope

Bedforms and erosional features on the lower slope at 500 m to 1 km water depth show evidence of strong bidirectional currents (both south-west and north-east), interpreted as indicating that some features have preserved the signature of both the glacial and postglacial (Holocene) bottom current flow. Furrows are common in a small area to the north-east of the Faroe Shelf at about 1 km water depth (Figure 136). The south-east slope of the Faroe Shelf also has a small area of bedforms including erosional scarps and narrow lineations interpreted as furrows (Masson et al., 2004). The furrows occur mainly below 800 m water depth extending on to the channel floor to 1.2 km (see below).

An erosional channel extends for some distance almost parallel to the 1 km bathymetric contour at the confluence of the Faroe–Shetland and Faroe Bank channels (Figure 136). It has a depth below the surrounding sea bed up to about 85 to 90 m. To the east and west, the erosional channel is less well developed and fades out. The channel appears to partially follow a now inactive and locally buried slide scar, which initially was a conduit for, and subsequently was modified by, the locally erosional contour currents. The erosional character may be caused by the slope forming an obstacle for the southerly flow of NSDW along the margin of the Norwegian Basin. The southerly flowing water mass is subsequently diverged and deflected to the east, along and over the Fugloy Ridge (Masson et al., 2004).

Channel/basin areas

The most common bedforms in the channel areas are contourite sands, comet marks, barchan dunes and furrows (Figure 136). All are assumed to be active at the present day indicating the persistent strong bottom-current regime (Masson et al., 2004). However, the most dramatic erosional features on the floor of the channel areas are the Judd Deeps (Figures 112 and 136). These scarps are up to 200 m high and are believed to be largely relict features, initiated during a period of unusually strong bottom water flow in Miocene times (Smallwood, 2004), although the continuing bottom current flow in the area is responsible for keeping the scours from being completely infilled by sediment.

Several south-facing scarps, some up to 75 m high, cut the southern flank of the Faroe Shelf (Figure 136). These occur mainly at water depths between 1 and 1.2 km, within an area characterised by gravel and rock outcrop at sea bed. They are interpreted as being erosional in origin, most likely associated with the strong bottom current activity in the area (Masson et al., 2004).

Masson et al. (2004) identified a narrow band of furrows between 800 m and 1.2 km water depth on the southern flank of the Faroe Shelf extending at least 90 km westward from the Judd Deeps. The furrows form a 10 to 15 km wide band that curves around the axis of the Faroe Bank and Faroe–Shetland channels (Figure 136). These were interpreted as erosional furrows similar to those observed on the tidally swept gravel sea floor on the shelf, an interpretation supported by sea-floor photographs in the furrowed area, which show a predominantly gravel substrate.

Barchan dunes occur in a narrow, contour-parallel band on the floor of the Faroe–Shetland Channel at approximately 1.1 to 1.2 km water depths (Figure 136), immediately to the south-east of the Munkagrunnur Ridge. The band extends for about 25 km, the largest barchans occurring immediately south of the zone of erosional furrows where sidescan images show a few isolated dunes up to 120 m across. The orientation of barchan dunes on the floor of the southern Faroe–Shetland Channel indicate long-term flow to the south-west (Wynn et al., 2002).

Strongly elongate areas of erosion, marked by a lag deposit of relatively coarse-grained sediment lining a depression are associated with many of the boulders observed on the sea floor at about 1.1 km water depth on the northern flank of the Faroe–Shetland Channel. The examples observed by Masson et al. (2004) are up to about 1 m wide and several metres long and were interpreted as comet marks, erosional bedforms formed by increased bottom stress as currents flow around an obstacle.

A number of mounded sediment accumulations were observed on seismic profile data from the Faroe Bank Channel and Faroe–Shetland Channel (Masson et al., 2004) and interpreted as sediment drifts. In the Faroe Bank Channel, the largest drift can be traced for at least 20 km (Figure 136), parallel to the contours and separated from the northern slope of the Wyville Thomson Ridge by a distinct moat. It has the typical form of an elongate drift, formed where a bottom current is constrained against a steep slope. A similar drift appears to occur at the base of the slope adjacent to the south-western edge of the Faroe Shelf. Sea-floor photographs show little evidence of bedforms other than fine-scale lineations that appear to result from accumulations of fine-grained detritus in the lee of sea floor irregularities. Masson et al. (2004) concluded that these observations are consistent with sediment drift accumulation in a weak bottom current environment.

In a small area of the Faroe Bank Channel and in a linear patch of sediment extending southward from east of the Faroe Islands to the Faroe–Shetland Channel, an area of fine-grained Holocene sands are thought to occur, corresponding to sediment waves. Elongate bands interpreted as longitudinal bedforms such as sand ribbons or sand patches have been observed by Kuijpers et al. (2002) and Masson et al. (2004) in areas characterised by strong bottom currents in the channel floors.

Much of the sea floor in the north-eastern part of the Faroe–Shetland Channel where it joins the Norwegian Sea is underlain by the glacigenic debris-flow sediments of the North Sea Fan (Figure 118). Downslope lineations on the fan mark the edge of individual debris-flows. Mud diapirs, marked by areas of rough, elevated topography rising 100 m or more above the general level of the sea floor occur in a limited area between 62°30'N–63°N and 1°–2°W. A north-west-trending scarp, up to 100 m high crosses the northern part of the area between 63°10'–63°40'N, marking the sidewall of the Tampen Slide (see Figure 140). A number or relatively small channels and gullies, a few tens of metres deep, extend upslope away from the Tampen Slide sidewall. These appear to postdate the last episode of fan debris-flows and are unlikely to be active today (Masson et al., 2003).

ENVIRONMENTAL DATA

Although the ecology and environment of the Faroe–Shetland Channel area is generally well known, there is only a limited amount of detailed information on the biology (Bett, 2001). The major controls on the distribution of benthic fauna in the area are likely to be hydrography/bathymetry and sea-bed sediment type.

Biological data

Until the last part of the 19th Century, most of the surveys of the Faroese benthic fauna focussed on molluscs (Bett et al., 2001). The earliest observations of invertebrate species (mostly molluscs and crustaceans) were collected by the Faroese naturalist Jens Christian Svabo in 1781–1782, but were not published until 1959 (Svabo, 1959). Landt (1800) published a list of about 60 marine invertebrate species, about half of them molluscs, the rest being sponges, cnidarians, polychaetes, crustaceans, bryozoans and echinoderms. A detailed account of collections of molluscs from this period is given in Mörch (1868).

During the 1860s, the recognition that the deep seas were the habitat for a rich, undescribed fauna, led to a number of sounding, dredging and hydrographical surveys, which were followed by numerous surveys during the next 100 years. The surveys of the 1860s, during which the Wyville Thomson Ridge was discovered and mapped, led to a new explanation of the hydrography of the area and the description of the fauna, much of which was included in the series of volumes produced by the Challenger Expedition. A comprehensive review of the surveys during the 20th Century is given in Bett et al. (2001).

The distribution of fauna around the Faroe Islands in waters shallower than 100 m were studied as part of a series of investigations known as the 'The Zoology of the Faroes' (Spärck, 1928). The Internordic BIOFAR Programme was created in 1987 to investigate the Faroese marine bottom fauna between 100 m and 1 km water depth (Nørrevang et al., 1994). These were augmented by the results of BIOFAR-2, run from 1995 to

1998 in shallow waters (see www.biofar.fo). There are 168 BIOFAR sample sites in the area between 60–62°N and 3–6°W, of which 48 are from water depths greater than 500 m. The samples were recovered by six different methods and photographs were also taken. Information on sea-bed nature, temperature and salinity exists for all 48 stations; 21 stations have reported between 1 and 58 animal species, the 10 most common of which are *Bythocaris payeri; Alvania wyvillethompsoni; Admete viridula; Astarte acuticostata; Haploops setosa; Haploops tubicola; Imidronea atlantica; Ophiactis abyssicola; Pseudophyrapus anomalus* and *Glycera lapidum*. Bett et al. (2001) compared the work done on the west of Shetland area in water depths in excess of 500 m with the work done on the Faroese side of the Faroe–Shetland Channel and found that, of the frequently occurring species, only two are the same (*Glycera lapidum* and *Haploops setosa*). According to Bett et al. (2001), one of the major factors controlling the benthic communities is the variation in sediment type.

Coral banks

The coral *Lophelia pertusa* is known to have existed within the Faroe–Shetland area for a long time (Bett, 2000; Figure 137) and is a cold-water variety that can survive in temperatures as low as 4°C (Bett, 2000). Long et al. (1999) report that all known *Lophelia* finds in the Faroe Bank Channel are within water depths of between 100 and 450 m; most findings are between 225 and 300 m. This correlates well with the BIOFAR findings. The coral banks are known to be large, with lengths of several hundred metres and heights of a few metres above sea bed.

Sponges

Bruntse et al. (1999) described the distribution of the larger sponges to be found in Faroese waters (Figure 135). Scientific trawling for these sponges has shown a relatively large number of redfish associated with them, and it is possible that the sponges are an important nursery and/or feeding area for these fish (Klitgaard, 1997). The general water depth range of the species of sponge reported during the BIOFAR Programme seems to be about 230 to 950 m, however

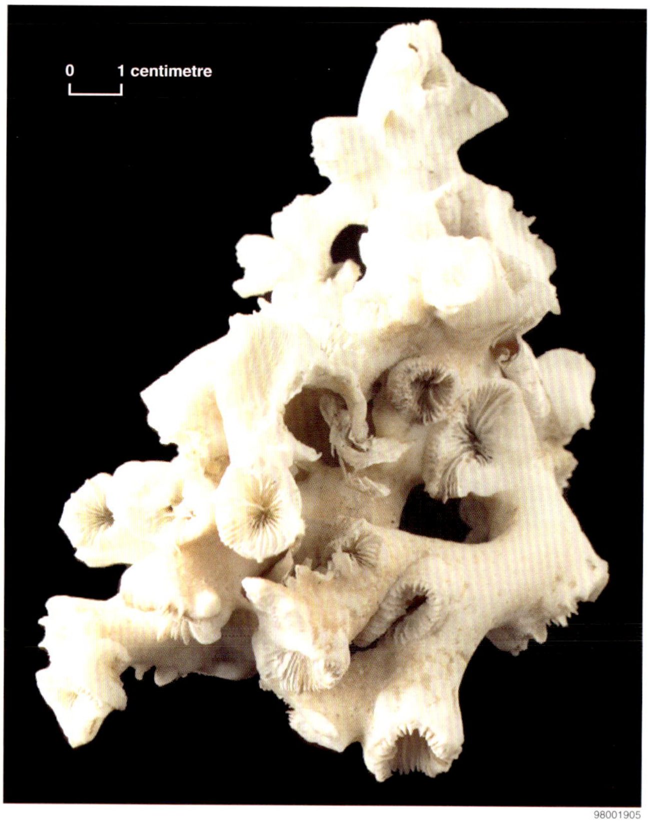

Figure 137 An example of the cold water coral *Lophelia pertusa* (modified from Long, 2006).

Bett et al. (2001) state that there is good reason to believe that other sponges exist in the deeper parts of the Faroe–Shetland Channel. Figure 135 shows sponge occurrences to the north and south-east of the Faroe Islands. However sponge spicules are a common occurrence on the Fugloy Ridge, to the south-east of the Faroe Islands and the middle and lower slopes of the Faroe–Shetland Channel (BGS 1988d and 1990a) indicating that sponges may have a much larger distribution than previously suggested.

11 Geohazards

David Long[1], Heri Ziska[2] and Roger Musson[1]
[1] *British Geological Survey*
[2] *Jarðfeingi (Faroese Earth and Energy Directorate)*

Geohazards are geological features or events that involve a degree of risk of detriment to life, property or the environment. For the purpose of this chapter they are primarily located at or close to the sea bed and would be evaluated by anyone proposing to use the sea bed for any activity. The details presented here result from a wide range of studies, both regional and site specific. The distribution of data is not even but is focused primarily in the Faroe–Shetland Channel, due to oil and gas industry interest. A range of industry-funded surveys have been undertaken and their integration allows detailed images of the sea floor to be produced for geohazard recognition and analysis (Figure 138). For the purpose of discussion, geohazards within the report area are presented in terms of shelf, slope or basinal settings and also in relation to seismicity.

SHELF

Carbonate sand waves

The sea-floor sediments on the shelf west of Shetland and Orkney and possibly around the Faroes are carbonate rich (Light and Wilson, 1998; Allen, 1983). They commonly occur as migrating sand waves and therefore are a hazard to constructions on the sea bed through burial and scour. The bedforms migrate in a clockwise direction on the UK shelf moving from the Minch and west of the Hebrides across the shelf and exiting into the North Sea, mainly through the Fair Isle Channel (Figure 1). The rate of migration is poorly understood. The largest and potentially most active sediment bodies are located at sites of high velocity, near sea-bed currents, such as in the Pentland Firth.

Morainic deposits

During the glacial episodes, ice sheets flowed across the shelf and locally extended beyond the shelf break. They modified the landscape and deposited glacial material. In some instances the glacial material is very hard close to sea bed and therefore moraines and drumlins can create difficult foundation conditions. The variable and often coarse-grained lithology of these mounds, together with their uneven surface and occasional buried boulders, can cause problems for drilling and pile installation, and make estimates of shear strength profiles very difficult.

The extent of moraines can be mapped from 3D sea-bed imaging. They are most notable on the outer shelf, where prominent ridges have been mapped parallel to the shelf break (Figure 115). Just north-east of the Wyville Thomson Ridge, there is an area with five similarly sized ridges, regularly spaced across 40 km of the West Shetland Shelf. The ridges are up to 50 m high, 8 km wide and 60 km in length and they have been mapped as part of the Otter Bank sequence (BGS 1990b, d, 1991b; Figure 109). The north-western edge of each ridge interdigitates with acoustically well-layered deposits revealing that the oldest ridge occurs to the north-west and the bathymetric pattern may reflect a series of halts in the retreat of an ice sheet. Similar ridges have been shown in greater detail on the sea-bed images derived from 3D seismic data from the vicinity of the Clair Field (Bulat and Long, 2001). The distribution of individual moraines has not been mapped in detail around the Faroes but morainic deposits are extensive on the surrounding shelf.

Commonly, small basins infilled with very soft sediments overlying morainic material are observed between morainic ridges on the outer shelf, or between drumlins seen closer in to Shetland (Stoker and Holmes, 1991). Geotechnical testing of cores taken through the soft clays infilling basins between some Otter Bank ridges west of Shetland locally reveal a softened, possibly weathered, surface to the underlying diamicton below (Long and Moore, 1980), often identifiable by a colour change. This zone may be up to a metre thick, with shear strengths only 50% of that of the 'unweathered' diamicton below. Recovered cores yield major variations in shear strengths, from firm to hard, and some extremely hard diamictons occur with undrained shear strengths in excess of 500 KiloPascals (kPa) within a few metres of the sea bed.

Shallow gas

In basinal areas that contain very soft to soft, late glacial to Holocene mud close to the coasts of Shetland, Orkney and Faroes, small areas of acoustic blanking can be observed on seismic data. This blanking is considered to be caused by the presence of shallow methane gas. For example, in the St Magnus Bay Basin located close to the west coast of Shetland (Figure 7), acoustically well-layered sediments extend to at least 50 m below sea bed, with layers of blanking occurring within 10 m of the sea bed. Here, the gas is probably biogenic, derived from organic matter deposited with the host sediments rather than being derived from petrogenic sources at depth. Compared with the North Sea and other continental shelves, the West Shetland and Faroese shelves show few occurrences of shallow gas blanking. This may reflect absence of source material but as there are few occurrences of thick fine muds with sufficiently low permeability to provide a cap, much of the gas generated may have leaked soon after formation thereby preventing the development of accumulations. The sediments have very low shear strengths, for example less than 10 kPa at 6 m depth in Loch Eribol on the north Scottish coast (Figure 43) or 35 m depth in St Magnus Bay.

In basins such as the St Magnus Bay Basin, some of the gas has leaked to leave pockmarks on the sea floor. Pockmarks are shallow depressions on the sea bed, and are relatively common in areas of muddy sediments on continental shelves. They are typically 50 to 100 m in diameter and 1 to 2 m deep, though they can be bigger

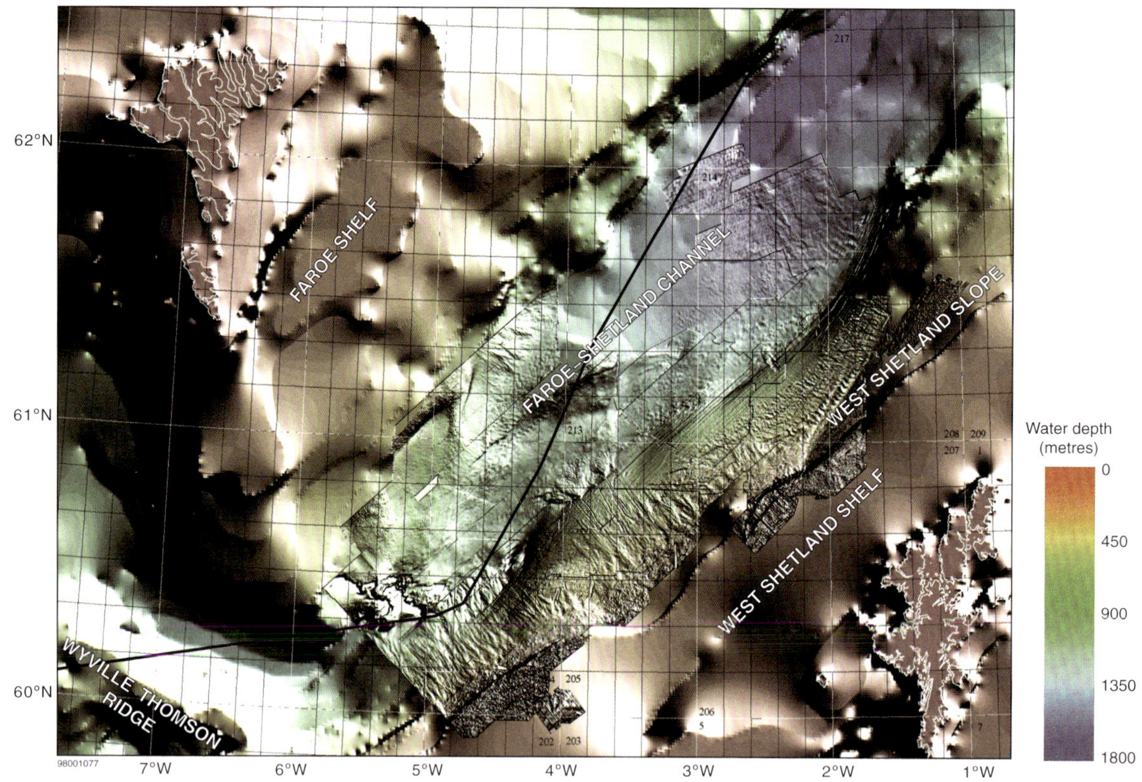

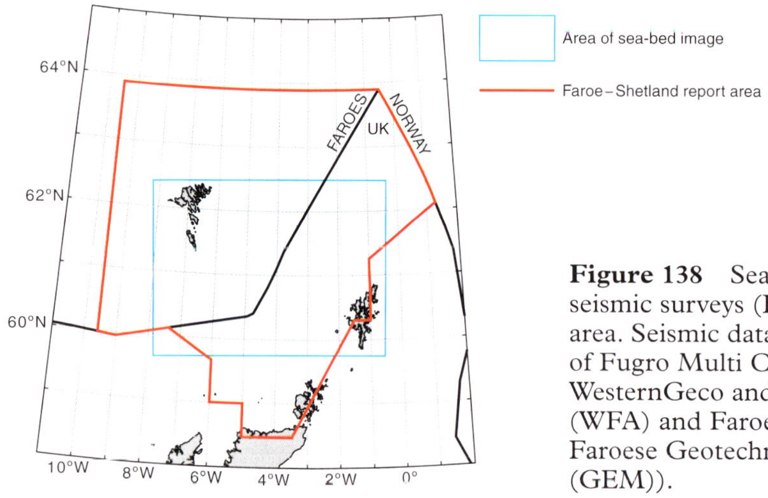

Figure 138 Sea-bed image derived from first returns of 3D seismic surveys (Long et al., 2004) within the Faroe–Shetland area. Seismic data used in the production of the image courtesy of Fugro Multi Client Services, PGS Geophysical, CGG Veritas, WesternGeco and members of the Western Frontiers Association (WFA) and Faroes Oil Industry Group (FOIB) (formerly the Faroese Geotechnical, Environmental and Meteorological group (GEM)).

or smaller depending on the flow rate of the seeping gas and the grain size of the surficial deposits. They occur widely in the North Sea, most notably in the Witch Ground Basin and Norwegian Trench, but appear to be very rare on the West Shetland Shelf. Most pockmarks are thought to be the result of a single gas-eruptive event. However, when gas continues to leak, its anaerobic consumption can be developed by chemosynthetic communities leading to the formation of carbonate crusts at or just below the sea bed. Crusts like this have been recovered from St Magnus Bay Basin and elsewhere around Shetland (Hovland and Judd, 1988; Sabine, 1970).

There is also evidence for relict pockmarks in a small basin approximately 55 km north-east of Rona (approximately 59°30'N 5°W), where a few very small pockmarks occur in association with gas-charged sediments of the Stormy Bank sequence (BGS, 1990d). This area of pockmarks differs from those typical of the North Sea, in that the soft, silty sediment in which the pockmarks are formed appears to be covered by a thin layer of sand and gravel. This suggests that they are relict features subsequently covered by Holocene sand.

Rock platform

Exposed rock creates constraints to pipeline or cable routing, as these structures cannot easily be buried, and can also create problems for anchoring and 'spudding-in' of commercial wells. Pipelines require to be covered with sediments to protect them from trawling activity. Sharp changes in topography also create spanning and flexing points for pipelines and cables.

A large rock platform extends to 120 m water depths around Orkney and Shetland, comprising mainly rocks of Precambrian, Devonian and Permo-Triassic age (Figures 43, 48 and 54). These strata provide a smooth surface and although there is some differential erosion along fracture patterns into gullies that are usually filled with gravel, mobile carbonate sand that commonly migrates across the rock platform. Similar bed rock is exposed west and south of Orkney, the latter area scoured by the strong currents through the Pentland Firth and crossed by coarse-grained sediment waves (Figure 136). In places, the smooth bedrock of the Devonian succession is interrupted by interbedded intrusive and extrusive igneous rocks which locally cause the development of more rugged sea-bed topography.

Near the west coast of Shetland, offshore extensions to Devonian granites in the Walls Peninsula area can be anticipated (BGS, 1984; Figure 29). Onshore, these granites show extreme variability in weathering profiles (with localised deep pockets of granitic sand) that may continue offshore. Crystalline Precambrian basement crops out at the sea bed between Foula to the Ve Skerries west of Shetland, forming the north-east-trending Foula High (Figures 7 and 29). There are extensive areas of exposed rock at the sea bed around the Faroes that are very uneven reflecting the variable composition and fracture pattern of the outcropping Palaeogene basalts. Similar areas of rock outcrop occur on Faroe Bank and on the Munkagrunnur and Wyville Thomson ridges (Figure 35).

Coastal processes

Following the last deglaciation, the landmasses of Shetland, Orkney and the Faroes were subject to isostatic rebound (Lambeck, 1995). However, uplift did not keep pace with the glacioeustatic rise in sea level, as the former ice thicknesses were limited, thereby creating a drowned coastal landscape with numerous inlets and virtually no coastal sedimentation or beach building. This is confirmed by the finding of terrestrial archaeological features at and below sea level (Mather and Smith, 1974). Over a longer time period, peat deposited 6000 years ago in Shetland is now 9 m below sea level indicating the scale of coastal submergence (Hoppe, 1965; Birnie et al., 1993). This soft submerged peat contributes to considerable variation in the physical properties of the surficial deposits and is a potential source of shallow gas in coastal waters.

Although the bedrock of the land within the report area is predominantly hard, there is ample evidence that the coastline is changing. Cliffs around the Faroes and Shetland in particular are often steep, with abundant jointing and planes of weakness. Consequently, there are regular collapses of cliffs into the sea. Recent studies of block falls suggest an average retreat of 1.3–6 mm per year of a cliff face in Shetland (Hall et al., 2008). Although the fallen rubble temporarily changes the sea floor topography, the erosive action of the waves is sufficient to remove debris from the cliff falls within a few years.

SLOPE

Iceberg ploughmarks

The outer shelf and upper slope west of Shetland and east and north of the Faroe Islands are characterised by abundant iceberg ploughmark furrows that are often flanked by ridges (Figure 139). These landforms are also likely to occur at similar water depths to the west of the Faroes and around Faroe Bank. In cross-section, the furrows average around 1 to 3 m in depth, exceptionally up to 10 m, and 20 to 80 m in width, extraordinarily up to 300 m. They generally occur in water depth ranging from 140 to 510 m but exceptionally, up to 700 m (J Trueman, pers comm.). The furrows normally consist of a channel with a flat or U-shaped floor, descending a few metres below the surrounding sea floor and often flanked by low ridges rising slightly above (<1.5 m) the surrounding level. In profile, the furrows may be symmetrical or asymmetrical, occasionally with a stepped flank indicating slumping along the sidewall. Their distribution is not uniform, being most abundant in water depths between 200 and 250 m and showing a decrease in density both upslope and downslope. In plan, the furrows are up to 5.5 km in length and form a cross-cutting network (Belderson et al., 1973). Their orientations appear chaotic at the shelf break and slope, but become increasingly orientated alongslope in deeper water. The ploughmarks usually follow an approximately straight or slightly curved course, although locally this may be sinuous or contorted. Where ploughmarks locally exceed 100 m in width, these could be sites where the icebergs have lodged in the sea floor for a period of time forming meltout pits. Buried late glacial ploughmarks have been reported in several areas, particularly in sea bed depressions such as the Foula Bight (Figure 139) where late to postglacial sedimentation, up to 25 m thick, has occurred (Masson, 2001).

The ploughmarks are of Late Pleistocene or early Holocene age and formed during periods of glaciation and deglaciation that repeatedly affected the area. Those at the sea bed were probably formed during the last deglaciation, although a few rogue icebergs have been reported stationary (and therefore possibly grounded) in these waters during the Little Ice Age (approximately 400 to 250 years ago) (Lamb, 1977). The ridges on either side of a ploughmark are known as berms and are believed to have formed initially as horizontally displaced sediments were forced out of the channel by the passage of the iceberg. As the keel of the iceberg moved past, stress release caused high-angle normal faulting along the ridges, creating a blocky topography (Woodworth-Lynas and Guigne, 1990). Consequent remoulding of the upper slope sediments by iceberg scouring may not significantly modify the shear strength profile but may disturb the lithology of layered sediments and alter permeabilities. The berms have subsequently been subjected to winnowing, leaving a gravel lag on the sea bed. The ploughmark troughs usually contain sand or mud winnowed from the berms or carbonate sand that has been swept along the shelf edge. The latter lithology predominates at the top of the infill, with the bulk of the underlying sediment comprised of very soft, fine, sandy mud.

When ploughmarks associated with pre-Weichselian glaciations are buried by a thick sequence of glaciomarine sediments, they can become associated with another geohazard. If coarse-grained sediments swept off the shelf during interglacial periods infill them they create a reservoir and potential trap for the accumulation of shallow gas. This occurrence has proved hazardous in offshore Norway. Currently, no such deep horizons of iceberg ploughmarks are known from the Faroe–Shetland report area.

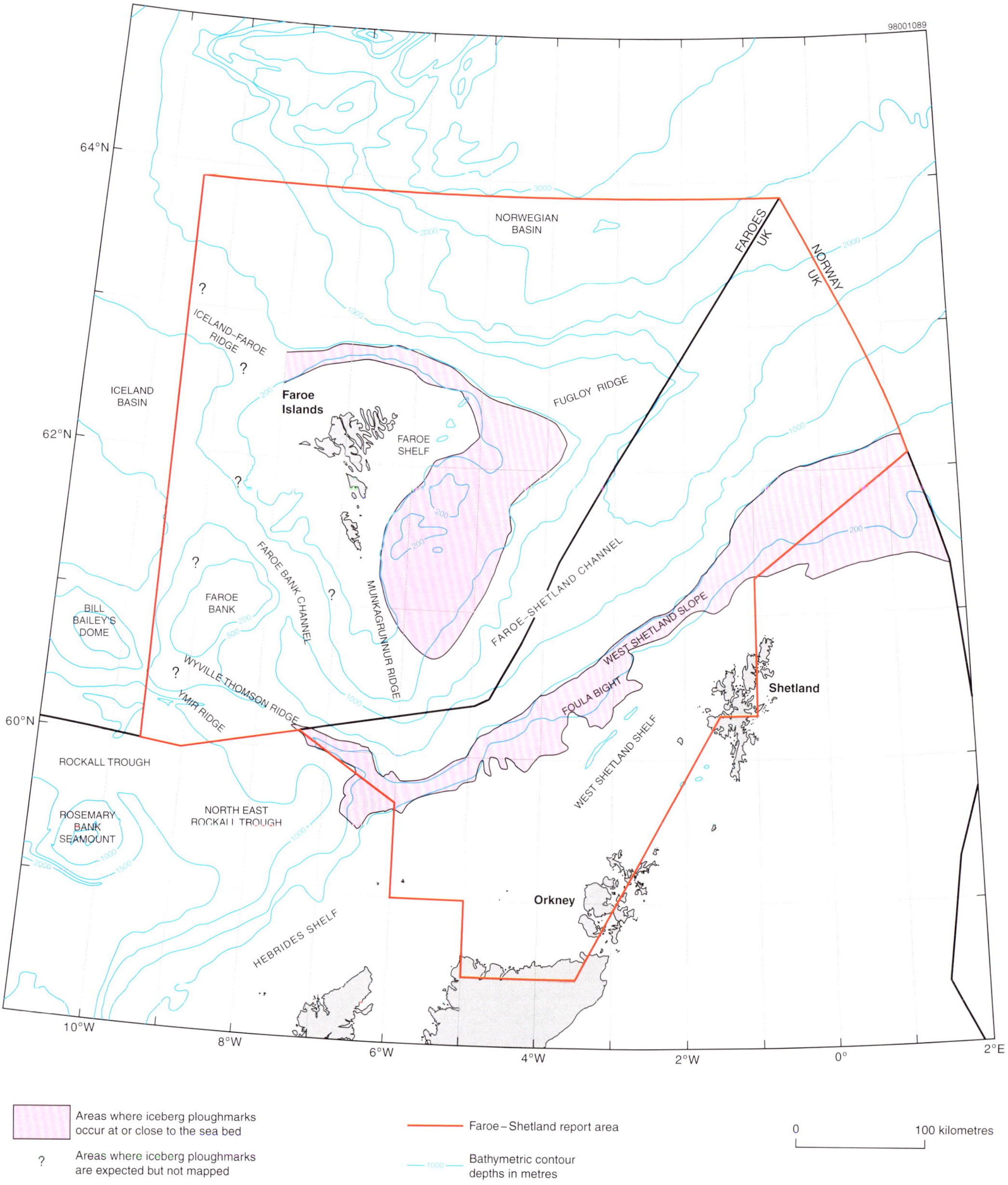

Figure 139 Distribution of iceberg ploughmarks within the Faroe–Shetland report area (courtesy of WFA).

Contourite deposits/sediment waves

On the lower slope on both sides of the Faroe–Shetland Channel, there is evidence for alongslope movement of the sea-bed sediment (Stoker et al., 1998). This is most marked between 800 m and 1 km water depth, where the associated sediments are the thickest and best sorted. These deposits thin to the south-west, reflecting the increasing strength of current preventing deposition. They also form sediment waves that migrate upslope (Figure 117). Although the speed of sediment movement is not known, there is the potential for burial and/or scour of sea-bed installations in this area. Contouritic sedimentation is stronger during deglaciation and interglacial periods, with reduced currents and the intermixing with downslope sedimentation and iceberg rain out occurring during glacial periods. The well-sorted fine sand layers of buried interglacial deposits play an important role in slope instability.

Unconsolidated lowstand sands

Early site investigation studies on the West Shetland Slope identified a sedimentary unit of sandy lithology termed the 'Taylor (Sand) Formation' that posed difficulties for drillers. It was reported that these loose sands caused stability problems in the upper sections of some drillholes. Subsequently, units of various ages and depths have been referred to as the Taylor Formation. This has been identified as a lowstand unit in a seismo-stratigraphical unit within the southern part of Quadrant 204 (Figure 1) and tentatively suggested that it may have been penetrated by a shallow borehole that recovered, pale olive-grey, fine to coarse-grained, massive sands with scattered shell debris and sporadic bands of weakly cemented calcareous sand. The unit lies at the top of the FSN-2 megasequence (Miocene to earliest Pliocene age) (Figure 94), and is thought to be a debris-flow deposit comprising highstand glauconitic sands eroded from the West Shetland Shelf (Stoker, 1999). The debris-flow may be up to 120 m thick and similar units may occur elsewhere along the margin.

Submarine slides

The slopes of the Faroe–Shetland Channel, like other glaciated margins, show evidence of downslope movement associated with sediment deposition e.g. debris-flows (Figure 116). Similar to offshore mid Norway, the slopes also show evidence for submarine slides (Figure 140). These slides are sites of significant deposition, often with >500 m of Quaternary deposits. With the exception of the North Sea Fan (Figure 140), the West Shetland Slope typically contains less than 200 m of Quaternary deposits. At the southern end of the Faroe–Shetland Channel, debris-flows extend down the slope to its floor and were deposited by ice sheets that extended beyond the shelfbreak. North of about 61°N, debris-flow deposits terminate about half way down the slope and are associated with occasional turbidite flows. Further north, the debris deposits are even more bathymetrically restricted except on the North Sea Fan where they form a major part of the Plio-Pleistocene succession (King et al., 1996). However, these deposits have been subject to mass movement following deposition and slides have been reported in the Faroe–Shetland Channel north of 61°N and on the north Faroese slope. The location of slides may reflect the distribution of seismic surveys, with small buried slides the most difficult to detect. Contourites are considered to provide the failure surface for slide development and the distribution of these deposits a major factor in slide occurrence. Seismic activity has been suggested as a plausible triggering mechanism for major slope failure on the Norwegian continental margin (Bugge, 1983). For example, a seismic trigger was considered the most likely cause for the development of the Storegga Slide offshore mid Norway (Solheim et al., 2005). However, short-time seismicity monitoring of the Faroe–Shetland Channel area suggests a low frequency of seismic events (see below).

Evaluation of ground motion amplification specifically for the west of Shetland area has been considered by Jackson et al. (2004), who suggest that the Factor of Safety (the ratio of the breaking stress of a structure to the estimated stress in ordinary use) to slope failure for many sections, although apparently high, is reduced to a value of near 1, and therefore vulnerable to failure by the presence of thin cohesionless layers within a thick succession of unconsolidated sediment. Based on the geological model established for the Storegga Slide (Solheim et al., 2005) and results from some of the Faroe–Shetland Channel slides (Wilson et al., 2004), it is suggested that layers of very well-sorted contourite deposits laid down during deglaciation and interglacial stages provide a slide plane that potentially can be mobilised by ground acceleration leading to slope failure.

Miller Slide

The Miller Slide (Long and Bone, 1990) occurs on the western edge of the North Sea Fan (Figure 140). It is a composite feature comprising a series of debris-flows, up to 95 km in length that extends from the upper slope to the basin floor. The headwall of the slide is at least 35 km wide, with a height locally exceeding 180 m. A total of 5700 km^2 of the sea bed was affected by the slope failure and the volume of displaced sediment has been estimated at about 800 km^3. However, the slide headwall has only a minor surface expression (approximately 15 m) due to the near complete infill of the evacuation hollow, leaving just the top of the scarp exposed. The debris-flows were blocked to the north-west by the Fugloy Ridge, and are thickest in the north where they bank against a zone of diapirs or mud mounds. The age of the slide has been estimated from the sediment it cuts, including equivalents of the Middle to Upper Pleistocene Mariner Formation (Figure 109), while contourites that postdated the slide have been correlated with hemipelagic sediments associated with Oxygen Isotope Stage 7, to give an age of about 200 000 years (Long et al., 2003).

Tampen Slide

The Tampen Slide (Figure 140) forms one of several very large slides identified on the mid Norwegian margin. It was first identified by King et al. (1996) and was originally suggested to predate the Mid Pleistocene onset of outer shelf glaciation (Evans et al., 1996). Recent studies, however, have correlated the slide with late Oxygen Isotope Stage 6 (end Saalian) (Nygård et al., 2005). The area affected by this slide includes the north-eastern edge of the report area where a 100 m high buried cliff has been left along the sidewall beneath the North Sea Fan. This sidewall shows a stepped geometry suggestive of multistage failure. North of 63° 10'N, there is a pronounced sea-bed scarp up to 80 m high that follows a similar orientation to the buried sidewall scarp. However, the relationship between these two features remains unclear. The Tampen Slide scar is presently confined to the North Sea Fan, although it is apparent from its truncation by the Holocene Storegga Slide that it was originally more widely developed offshore mid Norway and may have covered an area even larger than that of the Holocene Storegga Slide (Evans et al., 2005).

Afen Slide

The Afen Slide was initially identified on sonar data (Masson, 2001; UKOOA, 2000). It has also been imaged from 3D seismic data (Long et al., 1998; Figures 140 and 141). The slide is elongate downslope and covers an area of approximately 45 km^2. The thickness of displaced sediments is 10 to 20 m and the transport involved about 0.4 km^3 of sediment. The 3D and high-

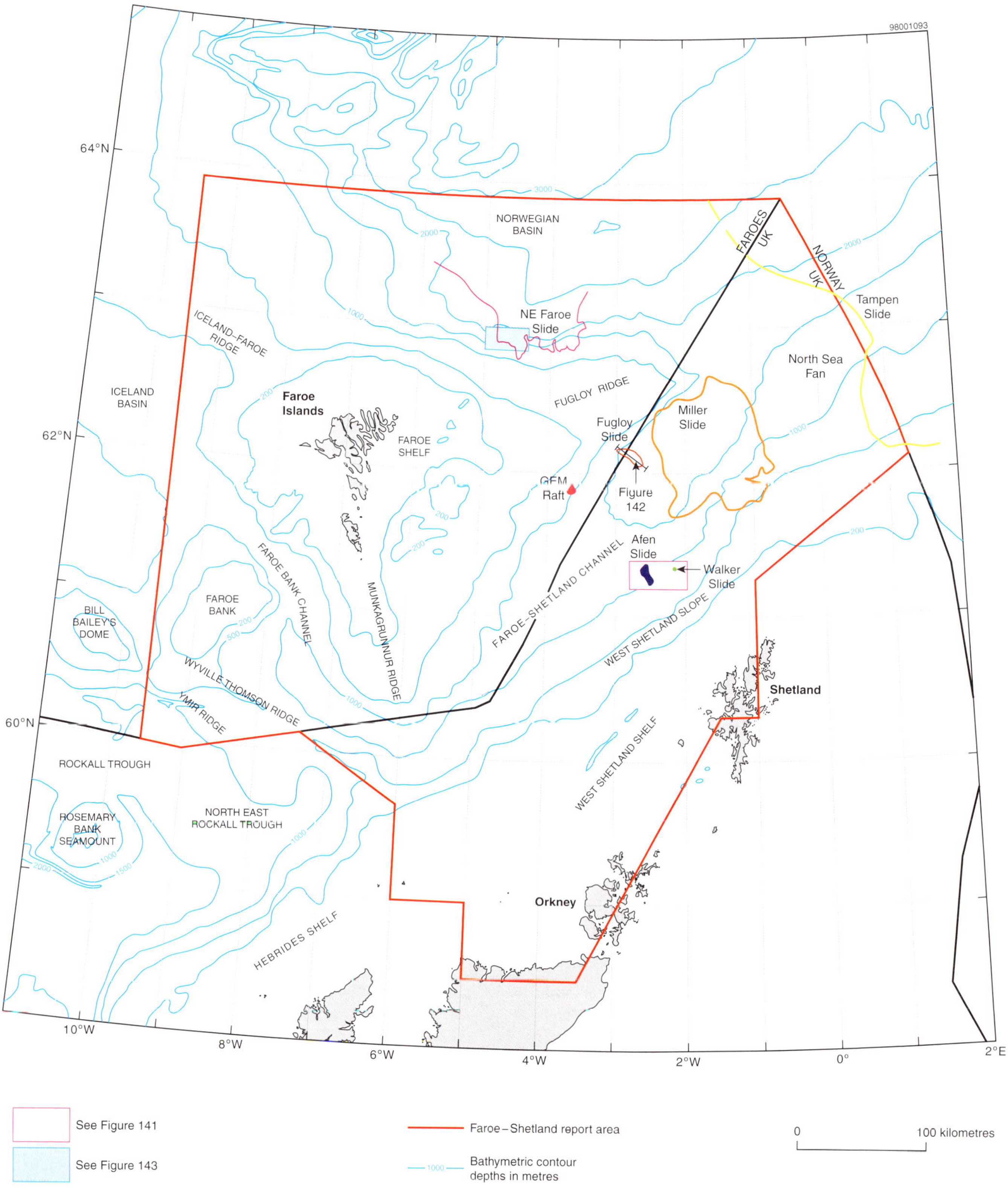

Figure 140 Location of submarine slides within the Faroe–Shetland report area (courtesy of WFA).

resolution 2D seismic data show that the slide as determined by its topographical expression is more extensive than that initially observed on the sidescan data, and is a multistage event involving failure along reflectors interpreted as contouritic horizons (Wilson et al., 2004). Failure probably took place as a series of debris-flows, with only the final and smallest stage occurring as block failure. Recent dating of sediments located at the base of this slide yielded a very young age of less than 2880 years, indicating that the development of the Afen Slide failure occurred in the very recent past.

High-resolution seismic data across the Afen Slide have revealed the presence of a similar-sized slide buried about 50 m below the sea-bed slide (Long et al., 2003). This slide has its headwall in the same water depth as the Afen Slide. Based on the low sedimenta-

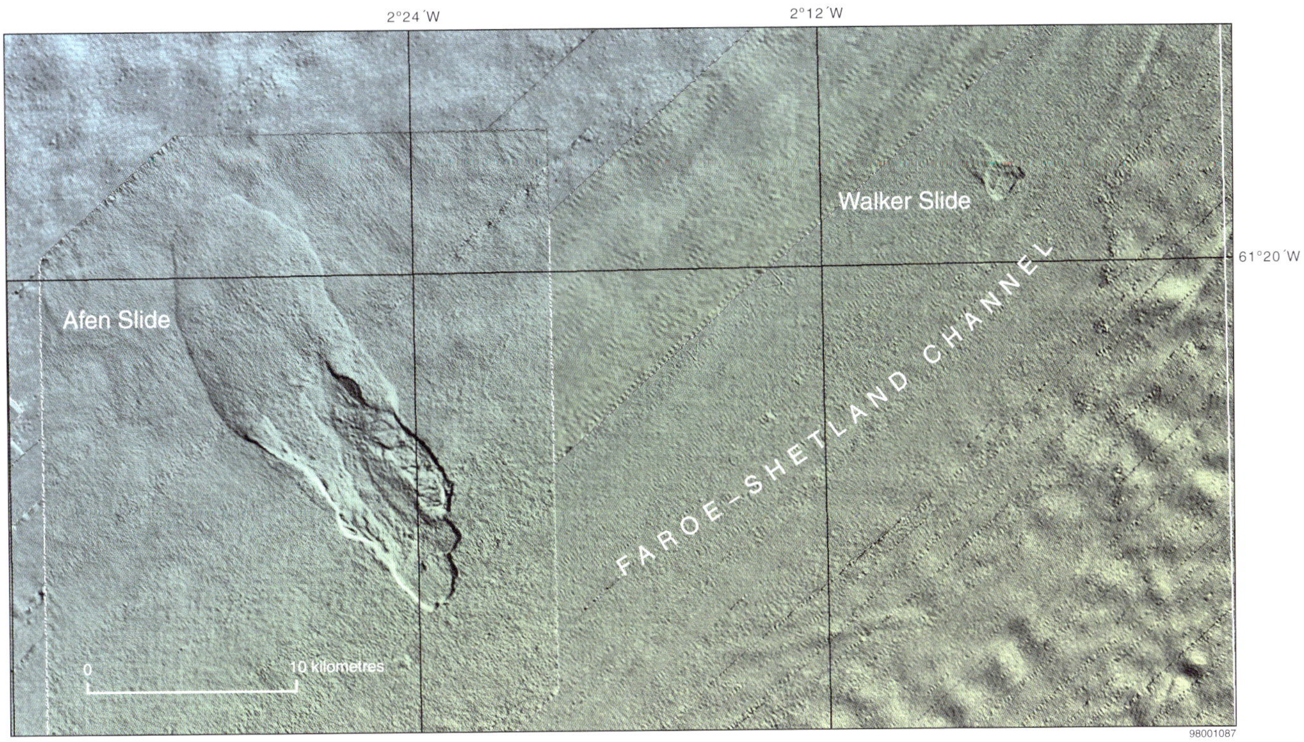

Figure 141 Sea-bed image showing Afen and Walker slides (derived from Figure 138). For location of image, see Figure 140 (courtesy of WFA).

tion rate in the area, the age of this palaeo-Afen slide is estimated to be several 100 000 years old. Both the Afen and palaeo-Afen slide failures are located with headwalls on the slope where the best developed contourite sediments occur, and where 80 cm of such sediments have been deposited at the foot of the headwall since the Afen Slide event. Physical property analysis of very well-sorted contourite deposits suggests that ground motion amplification within the shallow sediments could cause liquefaction and slope failure (Jackson et al., 2004).

Walker Slide

The Walker Slide is a small landslide approximately 1.5 km² located less than 20 km north-east of the Afen Slide (Figure 140 and 141). The displaced sediments associated with the slide are only about 4 m in thickness and therefore the approximate volume of transported sediment is 0.002 km³. Like the Afen Slide, the Walker Slide is located in 850 m water depth where contourite sediments are thickest. A slide mechanism comparable to that proposed for the Afen Slide is considered most likely. This feature is only poorly imaged on deep-towed sidescan images, indicating that the acoustic properties of the sea-bed sediments are little changed. Its subdued appearance on the 3D sea-bed image compared with the Afen Slide may indicate that the feature predates the latter's late Holocene age.

Gem Raft

Regional studies of the Faroese Slope on the north-east flank of the Faroe–Shetland Channel on behalf of GEM revealed a small slide termed the GEM Raft (Figure 140). The raft covers an area of approximately 7 km², and consists of upper and lower parts separated by an approximately 15 km wide plateau. The lower part of the raft shows a variety of mass-failure features including detached blocks, pressure ridges, slumps and debris flows, both surficial and buried, indicating repeated slope failure. This lower part is covered by a sediment layer that has been sampled. There is no clear indication that the failure processes are still active here and a Late Pleistocene age is considered likely. The upper part of the raft is the smaller, and appears on side-scan sonar records to be a simple slab approximately 21 km² that has moved only a short distance (<1 km) downslope with some rotation. This part of the raft is possibly younger than the lower failure, for there is no indication from acoustic records or core samples that the latest slump deposits are covered by sediment. AMS ^{14}C measurements confirm that mass-flow activity post-dates the last glacial maximum. The GEM raft occurs on the opposite flank of the Faroe–Shetland Channel from the Afen Slide (Figure 140) and is associated with similar contourite deposits. Consequently, it may have experienced the same failure mechanism.

Fugloy Slide

The Fugloy Slide occurs just north of 62°N on the southern flank of the Fugloy Ridge (note that the alternative term FOIB Slide has also been used) (Figures 140 and 142). This slide extends approximately 25 km south-eastwards and crosses the Faroes/UK median line. The back scarp of the slide reaches 70 m in height and the associated debris lobe is 50 m thick, infilling hollows above an erosive base (Figure 142). This slide is probably just one of many on the south-eastern flank of the Fugloy Ridge, as this area displays a 'blocky' sea-bed topography, with small scarps and a covering of undisturbed sediments. Jacobs and Masson (in UKOOA, 2000) reported evidence of slope failure at 62° 42'N 02° 00'W on the lower slopes, with a series of

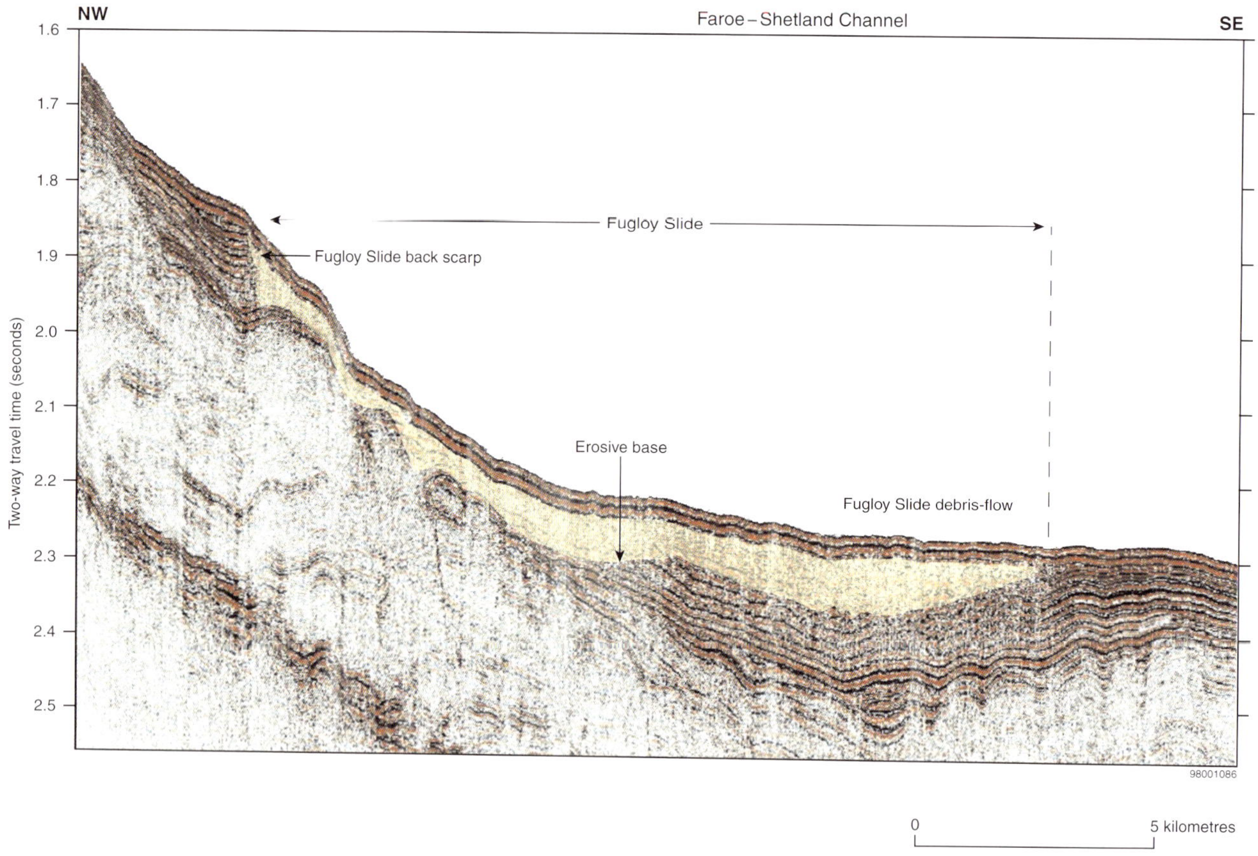

Figure 142 BGS seismic profile across the Fugloy Slide. The distribution of the slide is represented by the yellow shaded area. For location of slide and seismic profile, see Figure 140.

headwall scarps from several hundred metres to 30 km long and usually 10 to 30 m (but exceptionally up to 50 m) in height.

NORTH-EAST FAROE SLIDE

An extensive area of sliding along the north-east Faroes margin (Figure 140) has been described by Nielsen and van Weering (1998) and van Weering et al. (1998). Most of the slope leading into the Norwegian Basin shows evidence of large-scale failure, which Taylor et al. (2000) found to be of a scale similar to that of the Holocene Storegga Slide on the Norwegian margin. The main failures of the North-east Faroe Slide occur on the middle and lower slope and there is a pronounced surface expression with a headwall up to 300 m high (Figure 143) and a slide almost 400 km long. Associated detached blocks, slumped material and runouts extend far into the Norwegian Basin. A buried part of the slide has been recognised on the more gently sloping upper slope. This upper part of the slide has been largely infilled with contourite deposits, but remains detectable at the sea bed around the headwall where the infilling process is incomplete.

The age of the slide complex is not well constrained, but based on seismic stratigraphical analysis, Nielsen and van Weering (1998) suggested an Early Pliocene initiation. Van Weering et al. (1998) found evidence that this part of the slope has been prone to repeated failure since that time, with the last stage of mass-flow activity dated near the Pleistocene–Holocene boundary, close to 10 000 ^{14}C years ago (Kuijpers et al., 2001). Differential basin-margin steepening, the onset of NSDW current, or initiation of the Northern hemisphere glaciation are all suggested as possible triggering mechanisms of the slide complex (Nielsen and van Weering, 1998), whilst the youngest failure event has been related to maximum glaciation and glacioeustatic lowstands (Kuijpers et al., 2001). Taylor et al. (2000) suggest that failures during interglacials resulted from synsedimentary faulting.

Side-scan sonar surveys of the lower slope, at about 2.3 km water depth and below the main area of sliding, show the presence of a large number of downslope trending tracks on a low slope gradient, locally criss-crossing, and occasionally with a markedly irregular pattern (Nielsen and Kuijpers, 2004; Figure 144). At the termination of the tracks, outrunner blocks of sediment were observed, up to 18 m high, and with a maximum length of 70 m. Some of the blocks were found 25 km from the main headwall. The sub-bottom profiles in these areas indicate that most of the tracks have been filled with acoustically transparent sediment comparable to the Holocene hemipelagic surface unit, and thus the block sliding may predate the Holocene.

BASIN

Strong bottom currents

The basin floor of the Faroe–Shetland Channel is gently inclined to the north-east, deepening from about 1 to 1.6 km before opening into the Norwegian Basin (Figure 1). Currents are generally gentle but they increase in strength where the Faroe–Shetland Channel narrows (see Chapter 10). Where bottom currents are strong, they

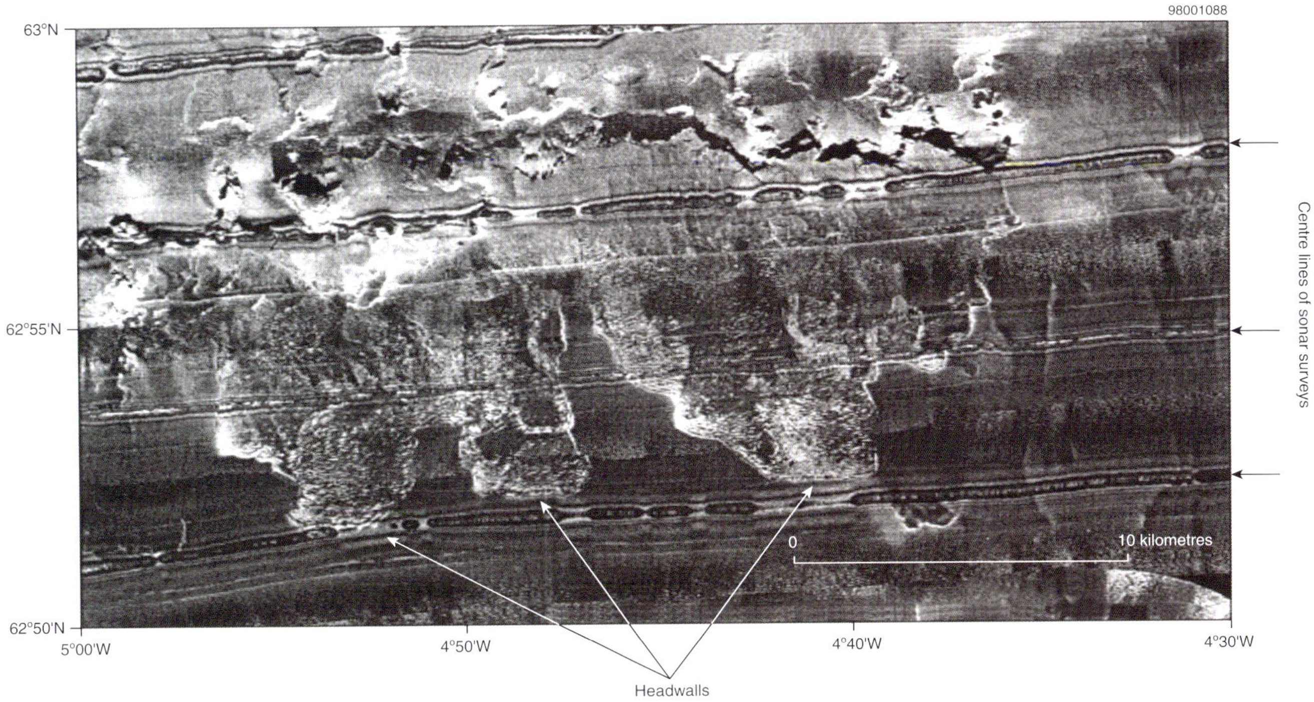

Figure 143 Sidescan image of North-east Faroes Slide (slightly modified from Kuijpers et al., 2001, by kind permission of Springer Science and Business Media; data courtesy of Geological Survey of Denmark and Greenland (GEUS)). For location of image, see Figure 140.

have modified the sea floor lithology and bedforms within the channel and could impact on foundation conditions. Where the floor of the Faroe–Shetland Channel narrows to less than 20 km (between about 60° 20' N 5° 30' W and 60° 40' N 4° 20' W) towards the south-west part of the report area, the enclosed Judd Deeps occur, with slopes frequently exceeding 5° and locally greater than 25° (Figure 112c; see Chapters 8 and 10). The Deeps previously extended over a wider area than that presently exposed on the sea floor with some buried scarps. Periods of erosion have occurred since the Miocene (Smallwood, 2004) and are associated with deep-water flow through the Faroe–Shetland Channel from the north-east. This flow is abruptly deflected to the north-west through the Faroe Bank Channel by the Wyville Thomson Ridge growth fold. The cliff faces and upstream shoulders of the Judd Deeps reveal either exposed Eocene rocks (Figure 112b) or have a cover of coarse gravels that could pose difficulties for the anchoring or spudding-in of drillings rigs. Although the Judd Deeps are presumed to be have been created by strong currents falling over the eroded 'cliff face', current measurements within the features indicates low velocities within the deepest part of the depression when compared with those around the rims (W Turrell pers comm.). Additional supporting evidence for the existence of strong currents is derived from the presence of barchan dunes within the Faroe Bank Channel (Wynn et al., 2002) and Faroe–Shetland Channel (Masson et al., 2004; Figure 136) and large scours systems buried beneath a sediment drape to the north-east of the Judd Deeps (BGS, 1990c).

Methane hydrates

Methane hydrates are a type of clathrate in which a 3D framework of water molecules is stabilised by molecules of methane held in the centre of molecular cages. Methane hydrates are only stable at low temperatures and/or high pressures. Assuming a gas source is present, the vertical and lateral extent of any particular accumulation is dependent on the interaction between the controlling factors of bottom water temperature, pressure and geothermal gradient. The base of a hydrate may be recognised on seismic profiles by a BSR reflector that tends to mimic the sea-bed topography and cross-cuts stratigraphical reflectors. However, BSRs only occur where there is free gas trapped beneath the hydrate. It should be noted that hydrate can occur within sediments without any BSR being evident on seismic profiles.

Accumulations of hydrate may cover hundreds of square kilometres, and be several hundred metres in thickness. Hydrates could become an important resource in themselves, or may act as a cap rock for underlying gaseous hydrocarbons. However, they are a potential hazard as disturbance of a hydrate layer (such as by drilling induced heating) may cause the sudden release of large volumes of methane, both from melting the hydrate as well as any free gas trapped beneath the hydrate level. Similarly, free gas released from deep below the sea bed could rise and form hydrates as it enters the hydrate stability zone at or near the sea bed, resulting in the freezing of drilling equipment.

Hydrate formation is possible within the bottom water temperatures and pressures found below 400 m water depth in the Faroe–Shetland Channel (Long et al., 2005; Figure 145). Hydrates within the Faroe–Shetland area have been reported to form around sea-bed installations, where drilling activity created a pathway for deep gas to rise close to the sea bed (HSE pers comm. 1995). However, there is no definitive evidence for naturally occurring gas hydrates within the Faroe–Shetland Channel

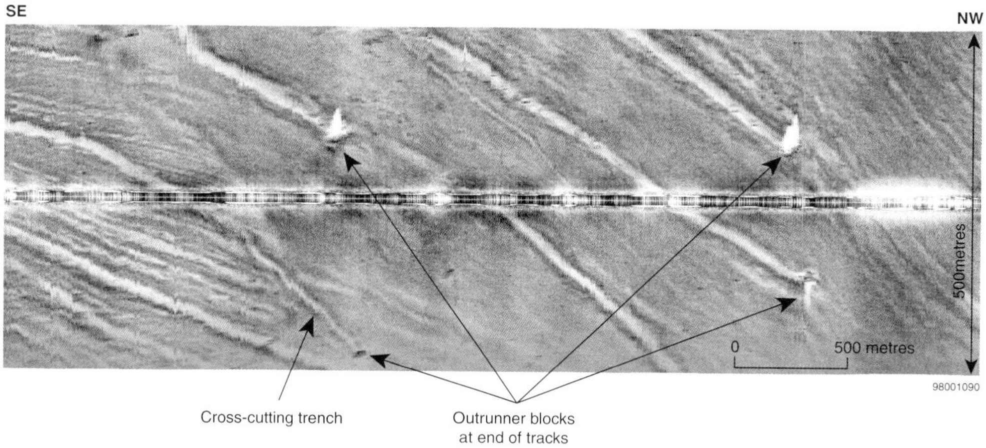

Figure 144 Deep-tow sidescan sonar record showing outrunner blocks and tracks (slightly modified from Nielsen and Kuijpers, 2004, Copyright: Geological Survey of Denmark and Greenland, (GEUS); data courtesy of GEUS)).

area (Long et al., 2005). A cross-cutting seismic reflector observed within the Oligocene to Miocene succession of the Faroe–Shetland Channel north of 61°N (Figures 26 and 111) was once considered a possible BSR (Long et al., 2005), but is now known to relate to silica diagenesis, the Opal A–Opal C/T transition formed in Miocene times (Davies and Cartwright, 2002). This silica-rich porcellanite could provide a minor obstruction to drilling.

Polygonal faulting

Images of the sea bed derived from first returns of 3D seismic data within the Faroe–Shetland Channel show parts of the floor display a polygonal pattern (Figure 146). This pattern is not seen on other sea-bed surveying techniques and may indicate that it is developed over the upper few metres of strata close of the sea bed, rather than at the sea bed itself. High-resolution seismic profiles across these areas show numerous vertical growth faults that coincide with the boundaries of the polygonal patterns noted on the sea-bed images. This suggests that the polygonal cells form part of a polygonal fault system that originates at depth. This faulting is considered to have been initiated in Eocene times and relates to smectite diagenesis (Davis et al., 1999). From analysis of deep-tow boomer records, sea-bed mounds (approximately 1 to 2 m high) occur vertically above fault tips associated with the development of polygonal features observed on the sea-bed images. This suggests that fluids migrate up the faults and transport sediments which are deposited at the sea bed to form small mounds.

Diapirism and mud mounds

Mud mounds at the sea bed have been described at the north-east entrance of the Faroe–Shetland Channel (e.g. Haflidason et al., 1996, Ritchie et al., 2003; Figures 24 and 147). High-resolution and commercial seismic, sidescan sonar and video surveys indicate that these features are most likely a manifestation of mud diapirs originating from subsurface sediment mobilisation. This sediment mobilisation may be due to the excessive load of dense, glacigenic sediments of the North Sea Fan deposited rapidly on top of Miocene low-density diatomaceous ooze, or may reflect a response to neotectonic stresses. These diapiric and mud mound structures show several stages of maturity: (1) an initial phase where the diapirs do not pierce the sea bed (2) an early extrusive phase with mounds on the sea bed up to 50 m high and (3) a mature phase with mounds up to 100 m high and having a smoother appearance. The buried features cover an area of about 3000 km^2 with an area of about 350 km^2 of outcropping diapirs (Figure 147). They are similar to, and may be the southernmost occurrence of a series of diapirs extending down the Norwegian margin (Hjelstuen et al., 1997). Preliminary results from AMS ^{14}C dating of sediment cores collected from the mud mounds suggest an episode of major activation of the diapirs around the time of the Last Glacial Maximum (Nielsen and Kuijpers, 2004).

Globally, submarine mud volcanoes, or diapirs, can range in size between 0.5 and 800 m in height (Guliyev and Feyzullayea, 2003). Two main mechanisms are considered to lead to the formation of mud diapirs and mounds, i.e. high sedimentation rates and/or lateral tectonic compression. Both mechanisms can result in overpressure of a mobile sediment layer.

SEISMICITY

Seismicity is not a sea-bed hazard as such, rather the consequences of seismicity causes hazards, most notably slope failure. Any assessment of the risk due to seismicity needs to involve an evaluation of the historical record. This record shows that the Faroe–Shetland region is an area of low seismicity (Musson, 1998; Musson et al., 2001). The number of recorded earthquakes in the report area is 14 (Baptie, 2005), and the largest is only 3.1 ML (Richter's local magnitude) in size (Figure 148). In the Faroes, the only known earthquakes are a short sequence of events recorded on Suðuroy in April–May 1967, of which the largest was 2.2 ML (Musson et al., 2001).

There is presently good seismic network coverage of the Faroe–Shetland region, with stations located in the Outer Hebrides, northern Scotland, Shetlands, and the Faroes (Figure 149). This coverage results from an expansion of the UK network in 1995 and the installation of a Faroese network in 1999 (Walker et al., 2003), supported by the oil and gas exploration industry. As a result, an earthquake ≥ 3ML can be detected anywhere in the report area, and for most of the area the threshold is 2.5 ML, as seen in Figure 149 (Walker et al., 2003).

The distribution of seismicity within the Faroe–Shetland report area is mainly confined to the south and east (Figure 148). This is partly due to limits on detection, but it should be noted that in the few years of its

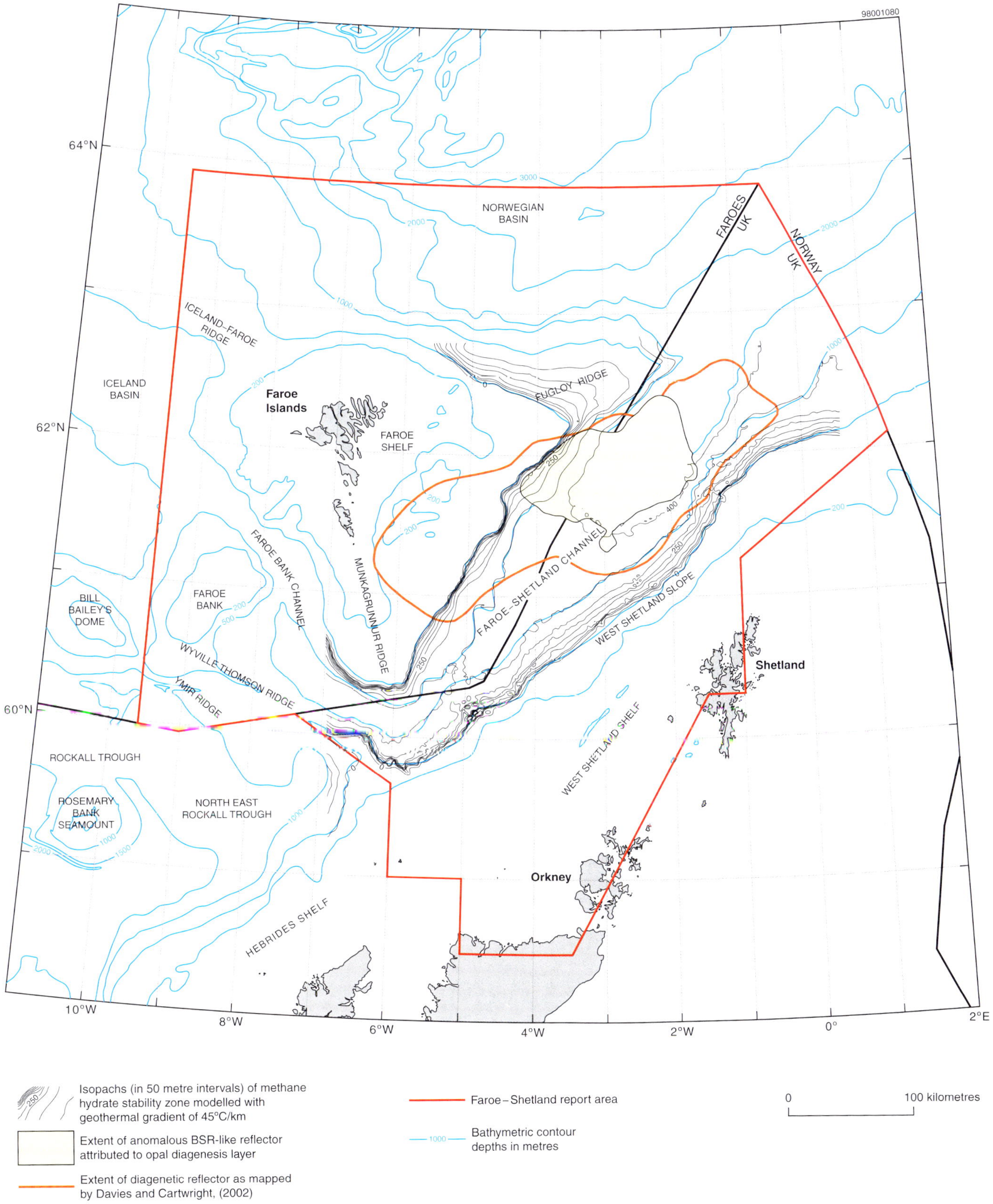

Figure 145 Location of a methane hydrate stability zone with potential isopachs and the extent of opal diagenesis layer (modified from unpublished maps, courtesy of WFA).

operation to date, the Faroese network has recorded no local events whatsoever. A few events in the north-east of the report area appear to be associated with the more active Møre Basin and North Viking Graben area (Figure 148). Estimates of focal depth for earthquakes in the region give a mean depth of 12 km, but the depth estimates have a large associated uncertainty. The uncertainty in the earthquake locations means that any attempt to match up epicentres with known geological features should be treated with caution.

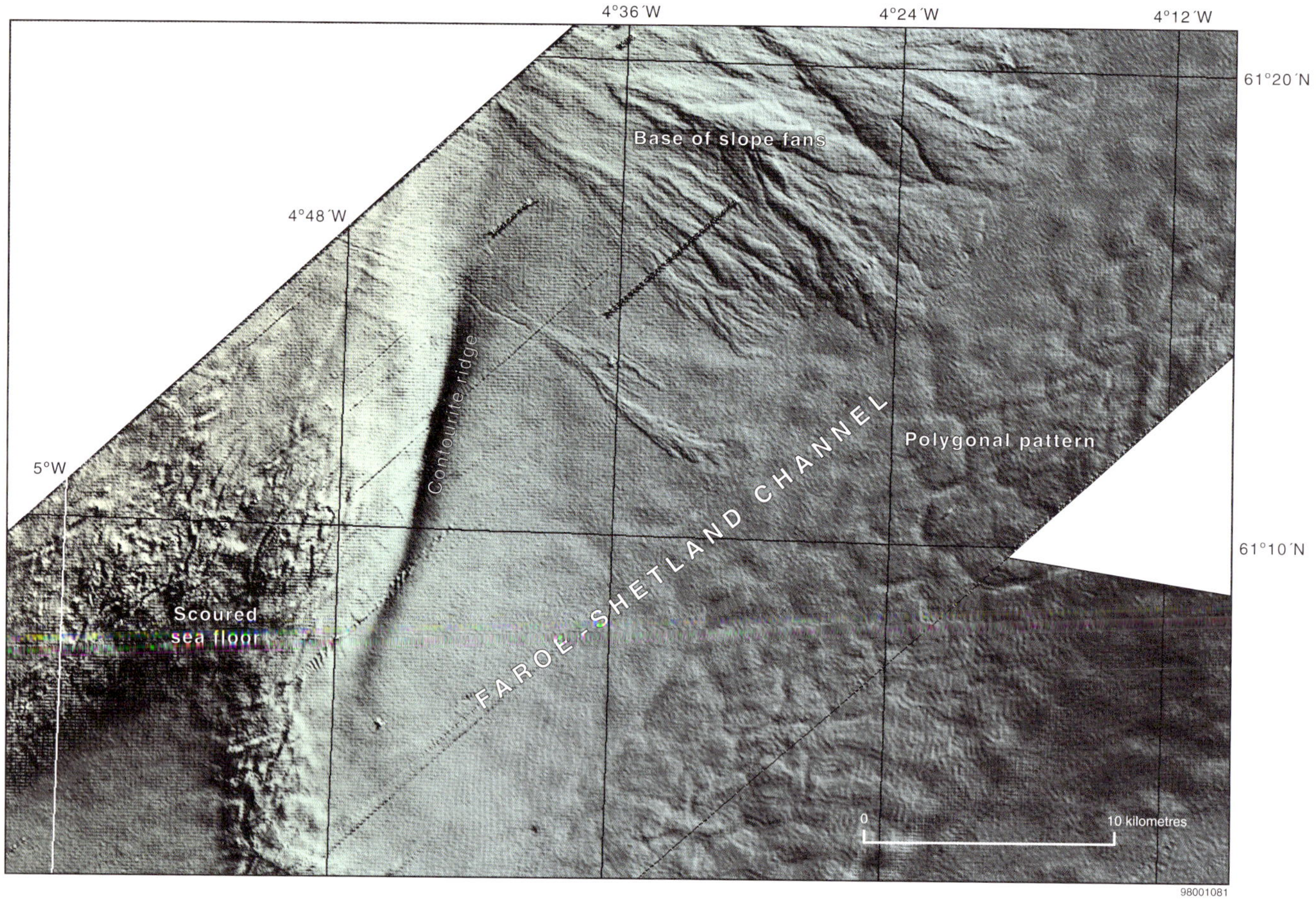

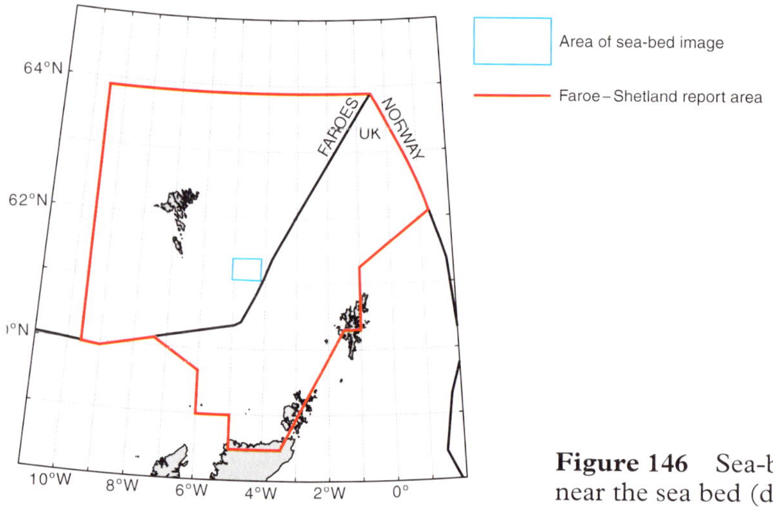

Figure 146 Sea-bed image showing polygonal patterns near the sea bed (derived from Figure 138).

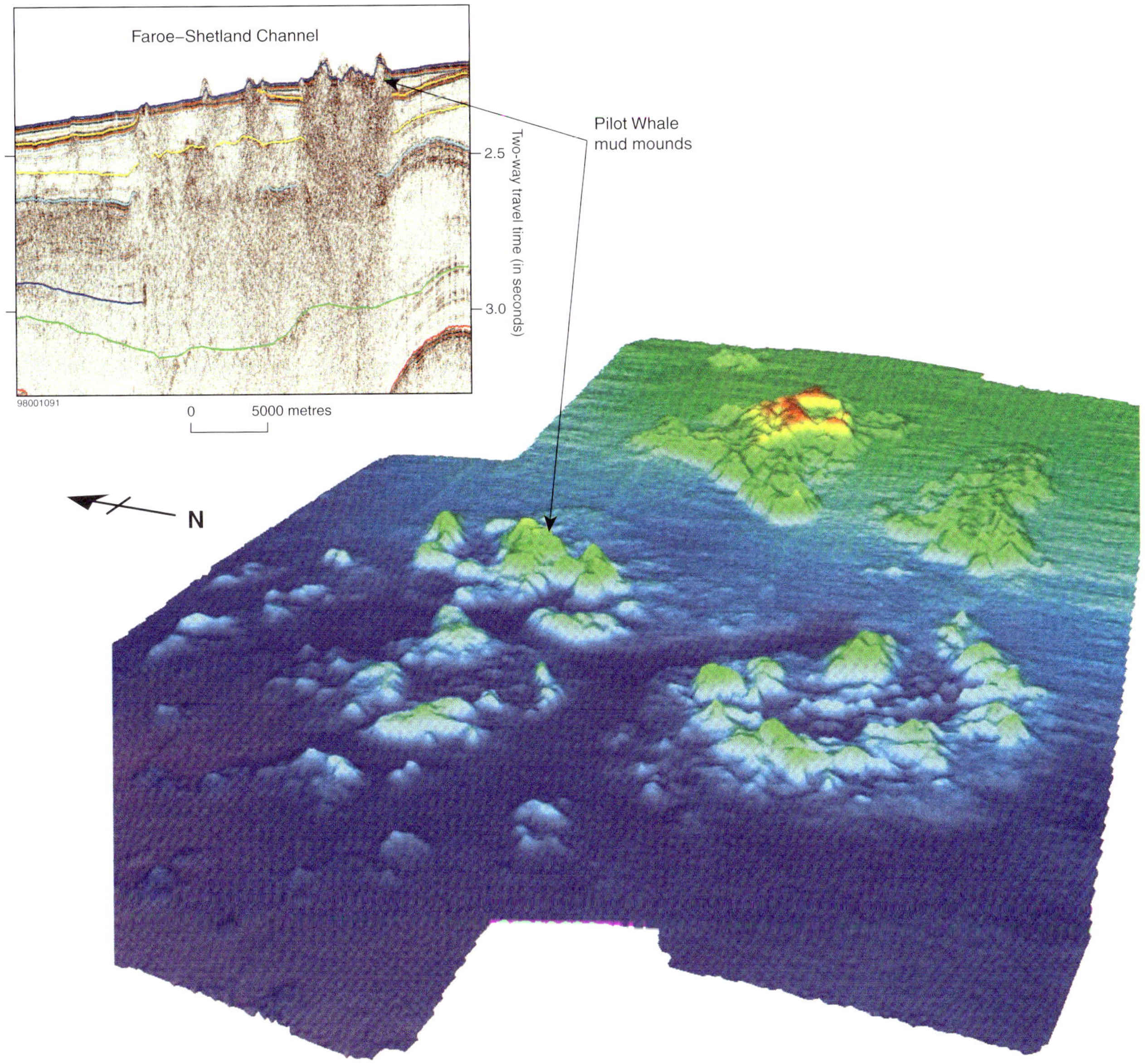

Figure 147 Examples of Pilot Whale mud mounds at the north-east end of the Faroe–Shetland Channel displayed on a BGS seismic profile and on a sea-bed bathymetric image (reproduced courtesy of DECC (formerly DBERR), from SEA4 report available from www.Offshore-sea.org.uk).

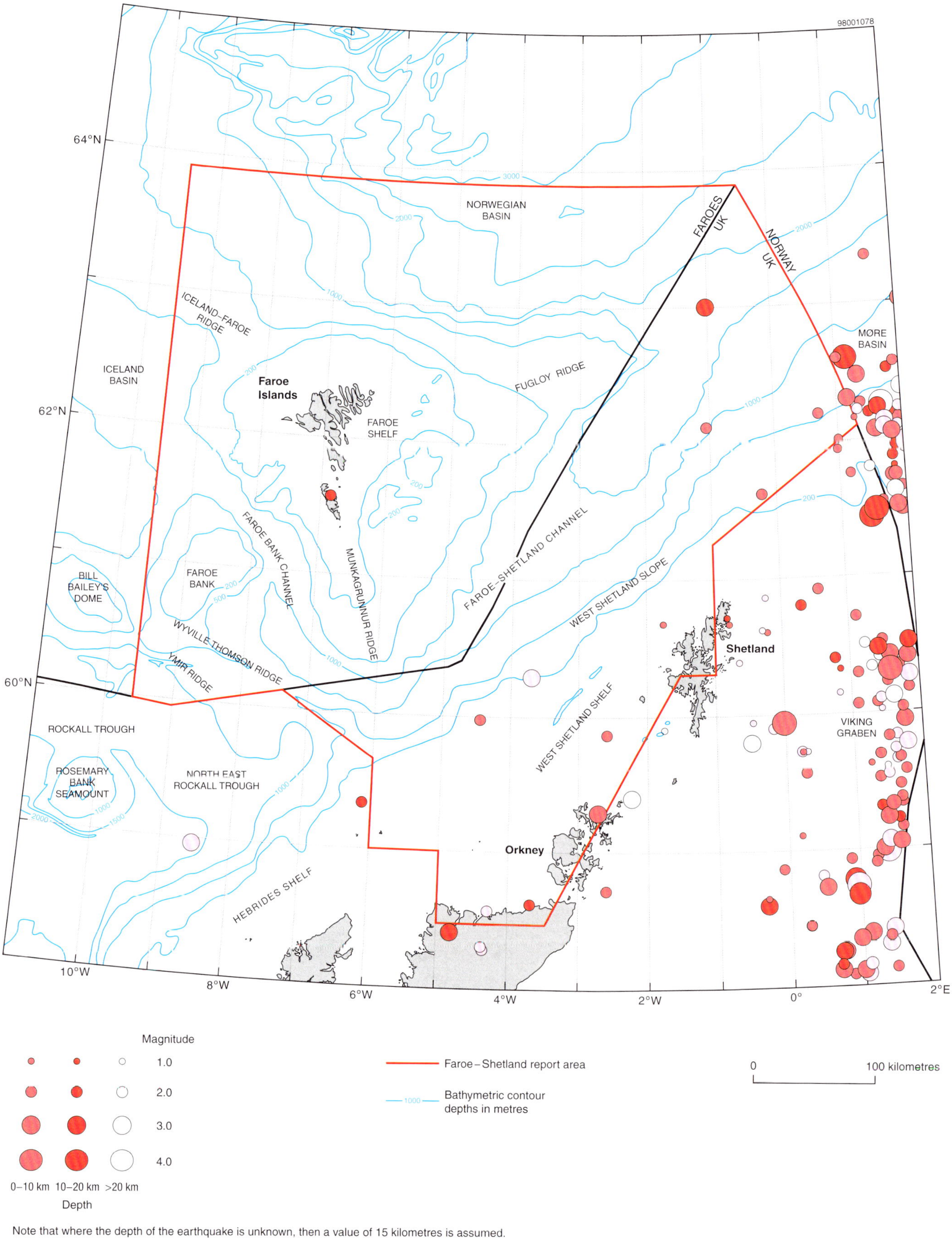

Figure 148 Location of historical and detected seismicity events of the Faroe–Shetland region and surrounding area (the UK earthquake database, with the addition of the 1967 Faroes events).

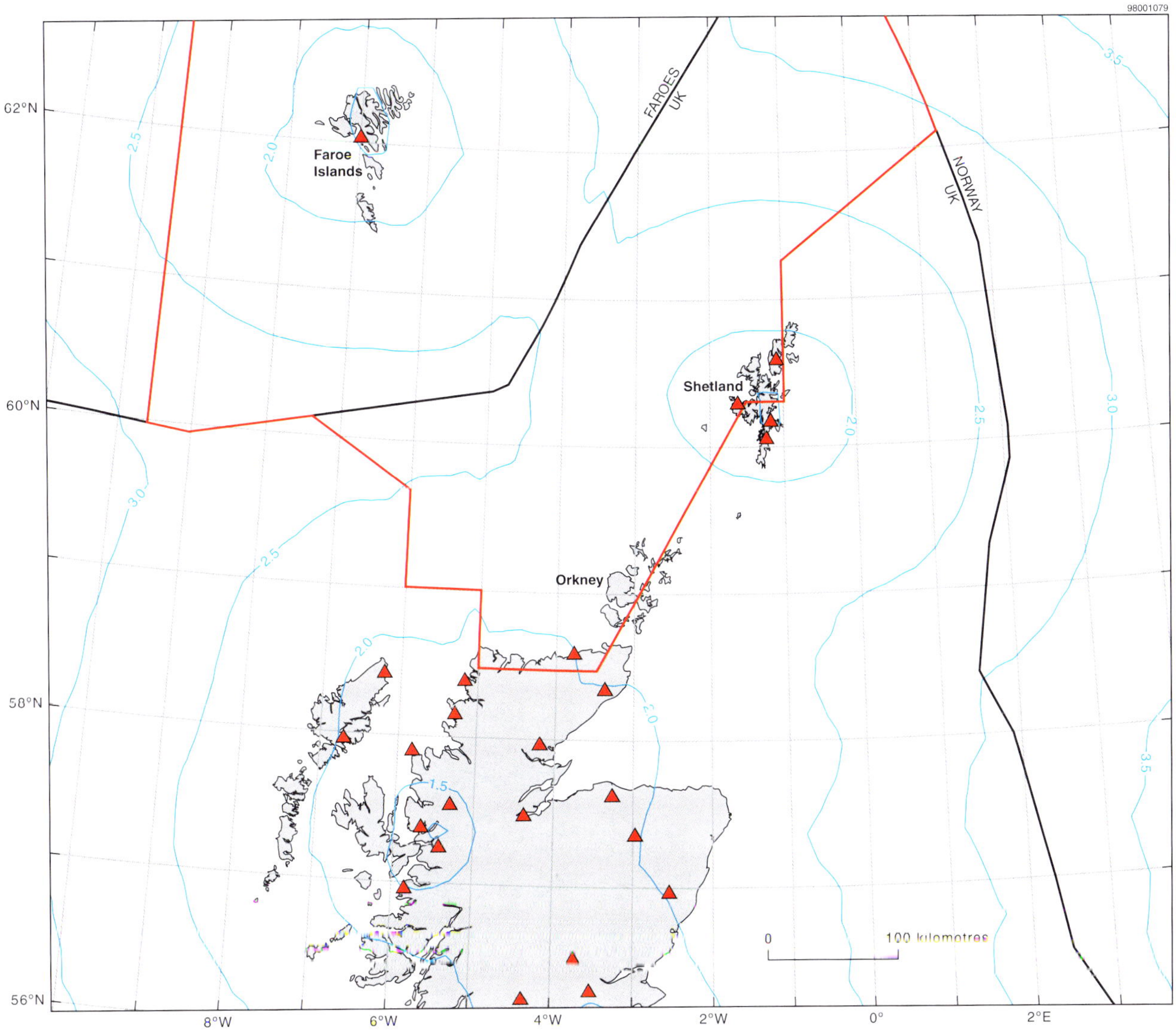

Figure 149 Present locations of seismic stations in the UK and the Faroese networks. The contours represent the magnitude detection threshold from these networks alone. Detection capability east of Shetland is better than this in practice because of the availability of data from Norwegian stations (modified from Walker et al., 2003).

12 Petroleum geology

Martyn Quinn[1], Thomas Varming[2] and Jana Ólavsdottir[2]
[1] *British Geological Survey*
[2] *Jarðfeingi (Faroese Earth and Energy Directorate)*

A proven petroleum system exists within the Faroe–Shetland report area, with a Middle and Upper Jurassic source rock, reservoirs ranging in age and composition from fractured Archaean Lewisian gneiss to Eocene fan sands, laterally continuous and effective seals, and a history of burial and uplift that has matured the source rocks and created structural, stratigraphical and combination traps. However, uncertainties remain regarding aspects of the petroleum system. For example, the full distribution of the Jurassic source rock is poorly understood due to limited seismic resolution and a lack of calibration by well penetrations (see Chapter 6). Individual reservoirs may vary in quality because of changes in their depositional environments and subsequent diagenesis. Some seals are laterally extensive and very effective, even causing over-pressuring in some areas, whereas others are thin and/or breached by faults. A significant number of oil and gas discoveries have been confirmed so far, three of which have been developed i.e. Foinaven, Schiehallion/Loyal development and Clair (Figure 150). Several others could have the potential to become producing fields when additional discoveries and/or economic conditions justify the costs involved in extraction of hydrocarbons.

Approximately 200 exploration and appraisal wells have been drilled in the Faroe–Shetland report area over the past 30 years in water depths that vary between less than 200 to 1500 m or so (Figure 1). Results from these wells have yielded a vast amount of information from the subsurface and helped to calibrate the thousands of line-kilometres of 2D and 3D seismic reflection data that have also been acquired.

EXPLORATION HISTORY

The first exploration well 206/12-1 drilled within the Faroe–Shetland report area was spudded on the 11th July 1972 on the Rona High (Figures 2 and 7). This well terminated within basement after encountering hydrocarbon shows in fractured Upper Cretaceous chalk and Archaean Lewisian gneiss. Early exploration activity occurred in relatively shallow waters, focusing on the drilling of tilted fault blocks in the North Rona and West Shetland basins where results were disappointing, and on the Rona High where oil and gas shows were recorded within several stratigraphical intervals. Exploration also took place in the shallow water areas close to the south-east fringes of the Faroe–Shetland Basin. For example, wells 205/22-1A and 206/05-1 were drilled within the Judd Sub-basin in 1974 and the Foula Sub-basin in 1976, respectively, with both recording oil shows in the Mesozoic. This early phase of exploration led to the discovery in 1977 of the Clair and Victory fields on the Rona High, within oil-charged Devono-Carboniferous and gas-charged Lower Cretaceous reservoirs, respectively (Figures 150 and 151). In addition, a significant gas discovery was made within the Lower to Upper Cretaceous Commodore Formation sands in well 206/11-1, drilled within the Foula Sub-basin. However, the high cost of developing an infrastructure, compounded by potential production problems associated with the fractured nature of the Clair reservoir and the relatively small size of the Victory gas discovery, resulted in no early exploitation of these discoveries. Between 1979 and 1980, the Erlend High was drilled for the first time, but all five wells drilled within this very short time span were dry (Figures 2 and 7).

In the early 1980s, exploration 'stepped out' to the north-west into the deeper waters of the Faroe–Shetland Basin. For example in 1980, well 206/02-1A (water depth of 611 m) was the first to be drilled on the Flett High (Figures 2 and 7), where it penetrated more than 3 km of Eocene and Paleocene sediments and recorded gas shows in Paleocene sandstone. In 1981, the Judd High was drilled by well 204/28-1 (water depth of 373 m) and proved 49 m of oil-bearing Upper Jurassic Rona Member sandstone and conglomerate resting on fractured Lewisian basement that also contained oil. However, well testing resulted in disappointing flow rates.

In 1982, the East Solan Basin was first drilled by well 205/26a-2 (Figures 2, 7, and 39) which tested a Lower Cretaceous succession with poor permeabilities. In the same year, the Papa Basin was drilled by well 205/27a-1, but this was dry. In 1984, the adjacent West Orkney Basin was tested by well 202/19-1 which penetrated nearly 3 km of Permo-Triassic sediments subcropping the Quaternary but with no hydrocarbon indications.

As deep-water technology developed, targets within the Faroe–Shetland Basin became accessible and comprised not only tilted fault blocks, but also stratigraphical traps within the extremely thick, sand-rich Paleocene succession. In 1984 for example, well 214/30-1 drilled the Laxford prospect (water depth of 548 m), and tested gas from Paleocene sandstone in the Foula Sub-basin (Figure 150) whereas in the Møre Basin, well 219/28-2Z flowed gas from Lower Cretaceous limestone (water depth of 517 m) (Figures 2 and 7). In 1985, well 214/27-1 drilled the Torridon prospect in the Flett Sub-basin and flowed gas from the Paleocene succession (Loizou, 2005). However, all these gas discoveries were relatively small and uneconomic at that time. Well 206/01-2, the Laggan gas discovery drilled in 1986 on the Flett High (water depth of 623 m), arguably represents the only significant Paleocene discovery during this phase of exploration (Figures 150 and 151; Loizou, 2005).

In 1987, well 204/23-1 drilled within the Judd Sub-basin flowed gas condensate from fractured Lewisian basement (Figures 2 and 7). In 1989, well 205/09-1 drilled within the Flett Sub-basin encountered slightly

Figure 150 *(opposite)* Location of producing fields, selected significant hydrocarbon discoveries and conceptualised extents of play fairways within the Faroe–Shetland region. Abbreviations: see Figure 7: Present day structural elements (grey) derived from Figure 7.

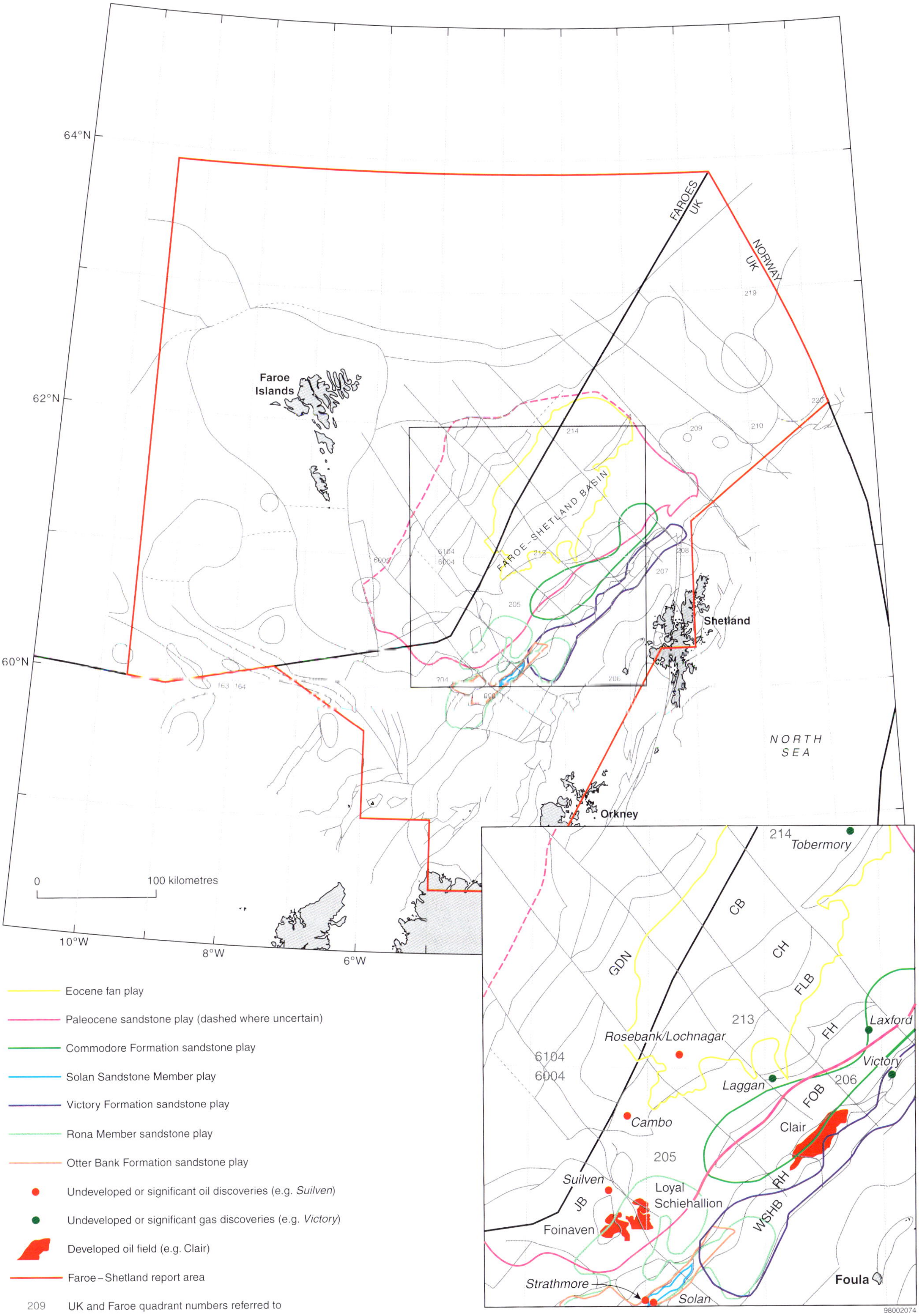

Eocene fan play

Paleocene sandstone play (dashed where uncertain)

Commodore Formation sandstone play

Solan Sandstone Member play

Victory Formation sandstone play

Rona Member sandstone play

Otter Bank Formation sandstone play

Undeveloped or significant oil discoveries (e.g. *Suilven*)

Undeveloped or significant gas discoveries (e.g. *Victory*)

Developed oil field (e.g. Clair)

Faroe–Shetland report area

209 UK and Faroe quadrant numbers referred to

overpressured Paleocene sandstones with very good porosity and permeability characteristics within the T30s succession (Figure 94) at depths of up to 4 km below sea bed. This well was significant, stimulating further exploration within the area and in the subsequent development of the Paleocene play, particularly within the Flett Sub-basin (Smallwood and Kirk, 2005).

Within the East Solan Basin, the Triassic Strathmore and the Upper Jurassic Solan discoveries were made in 1990 and 1991, respectively (Figures 150 and 151). Prior to these discoveries, and despite the presence of oil shows recorded in several wells, there was a concern that insufficient hydrocarbons had been generated to form commercially viable accumulations in these so called 'back basins' that are separated from the Faroe–Shetland Basin to the north (Herries et al., 1999). The Strathmore discovery well 205/26a-3 (Figure 2) was drilled primarily to test Middle and Upper Jurassic sandstone reservoirs. Although these targets were absent, an oil column was discovered within a Triassic red-bed succession. It was an appraisal of the Strathmore discovery well 205/26a-4 that proved oil within Upper Jurassic intra-Kimmeridge Clay Formation sandstone, which was subsequently ascribed to the Solan Sandstone Member (Herries et al., 1999; see Chapter 6).

Advances in seismic technology in the early 1990s enabled the significance to be recognised of a discovery that would eventually become the Foinaven Field (Figures 150 and 151). Well 204/24-1A was drilled in 1990 to test a Mesozoic target in the Judd Sub-basin (Figures 2 and 7) and encountered 11 m of oil and gas-bearing Paleocene sands. Reprocessing of 2D seismic data and analysis of seismic amplitude anomalies that had helped to image this secondary Paleocene target resulted in the drilling of well 204/24a-2 in 1992. This well proved 48 m of oil-bearing sands associated with amplitude anomalies. These encouraging results led to the acquisition of a 3D seismic survey and a successful appraisal of the Foinaven Field (Cooper et al., 1999). This approach, along with the application of modelling techniques imported from the Gulf of Mexico (Lamers and Carmichael 1999), led to the discovery of the Schiehallion and Loyal fields and other significant discoveries within Paleocene sands (Figures 150 and 151; Loizou, 2005).

Throughout the 1990s, appraisal and development of the Clair, Foinaven, Schiehallion and their satellite fields continued. Regional exploration also progressed with varying success, with wells drilled in the West Solan and North Rona basins and on the Judd, Flett, Foula and Corona sub-basins and associated highs (Figure 7). The successful application of seismic attribute extraction to aid in the recognition of 'direct hydrocarbon indicators' (DHIs) helped with appraisal of Paleocene discoveries in the Judd Sub-basin. This approach was utilised in the adjacent Flett Sub-basin as a key discriminatory factor in prospect evaluation but led to several disappointing well tests (Smallwood and Kirk, 2005). For example in 1997, wells 205/10-4 and -5A (Figure 2) were sited on prospects identified from relatively stronger seismic amplitudes, but these were later shown to have been caused by igneous extrusive and intrusive rocks. In the same year, well 205/14-3, was located on a mapped pinchout of Paleocene sands (T35 and T36 sequences) on the Flett High that was associated with a seismic amplitude anomaly (Smallwood and Kirk, 2005). However, the well was dry, with no reservoir development and post-well analysis indicating that the amplitude anomaly was due to positive tuning between reflections defining top and base of the sequence (Smallwood and Kirk, 2005). Other potential problems associated with the identification of prospects on the basis of seismic amplitude anomalies include adjustments during the seismic processing stage, resulting in manufactured amplitude anomalies or seismic artefacts. Finally, lateral variation of shale anisotropy can result in an attractive 'seismic amplitude variation with offset' (AVO) response (Smallwood and Kirk, 2005). However, there were some exploration successes too. For example in 1996, well 204/19-8 was drilled within the Judd Sub-basin and encountered oil in Paleocene sandstone. This was to become the Suilven discovery, and was later appraised by well 204/14-1 (Figures 2, 7, 150 and 151; Loizou, 2005). On the Rona High in 1998, well 205/23-2 drilled the Bombardier prospect and tested gas from Upper Jurassic sandstone (possibly equivalent to the Rona Member, see Chapter 6), revitalising the Upper Jurassic sandstone play.

In 1998, the first wells were drilled on the Corona High in the centre of the Faroe-Shetland Basin. Well 214/17-1 (Figures 2 and 7), drilled on the south-east flank of the Corona High, was designed to test an Eocene deep marine fan sandstone prospect (Strachan Fan) but was unsuccessful. However, well 213/23-1 reached total depth in Lewisian basement after discovering gas and oil shows in Upper Cretaceous, Triassic and Carboniferous strata. In the Corona Sub-basin, well 214/04-1 drilled the Tobermory prospect in 1999, flowing gas from a Middle Eocene sandstone that was deposited in a deep marine fan setting (Figures 103, 150, and 151; Loizou, 2005).

In 2001, well 6005/15-1 was the first of three wells to be drilled in Faroese waters, testing a Paleocene target located on the Sjúrður Ridge that was dry with only hydrocarbon traces (Figures 2 and 7). Wells 6004/12-1Z and 6004/16-1Z were both drilled in the Faroese part of the Judd Sub-basin. Well 6004/12-1Z, which included a sidetrack to test a deeper target in the Paleocene, confirmed the presence of oil and gas. Finally, on deepening, to a terminal depth of 4247.5 m (below sea level), well 6004/16-1Z reported a gross hydrocarbon column of approximately 170 m (Smallwood and Kirk, 2005; Cawley et al., 2005) in a low permeability Paleocene T10 reservoir with an observed oil–water contact at approximately 4199 m (T Varming, Jarðfeingi, pers comm.). The reservoir consists of deep marine turbidites, deposited in proximal to mid fan lobe setting.

In the UK sector, wells 214/09-1 (2000), 204/10-1 (2002), 204/10-2 (2004), 213/27-1Z (2004) and 213/27-2 (2007) have been drilled on the Corona High (Figures 2 and 7). Field appraisal and development wells continue to be drilled on Foinaven, Schiehallion, Laggan and Clair, and further exploratory wells have been drilled in the Judd Sub-basin and on the Rona, Judd and Møre Marginal highs, for example 219/21-1, the Ben Nevis prospect on the latter. In the Faroese part of the Judd Sub-basin, well 6004/17-1 was drilled in 2003 to a depth of 3823 m (below sea level) but registered no oil or gas shows (see Smallwood and Kirk, 2005). In 2006, well 6104/21-1 (Figure 2) was drilled on the basalt covered East Faroe High in a water depth of 478 m and reached a terminal depth of 4201 m. However, only traces of gas were encountered (see www. jardfeingi.fo).

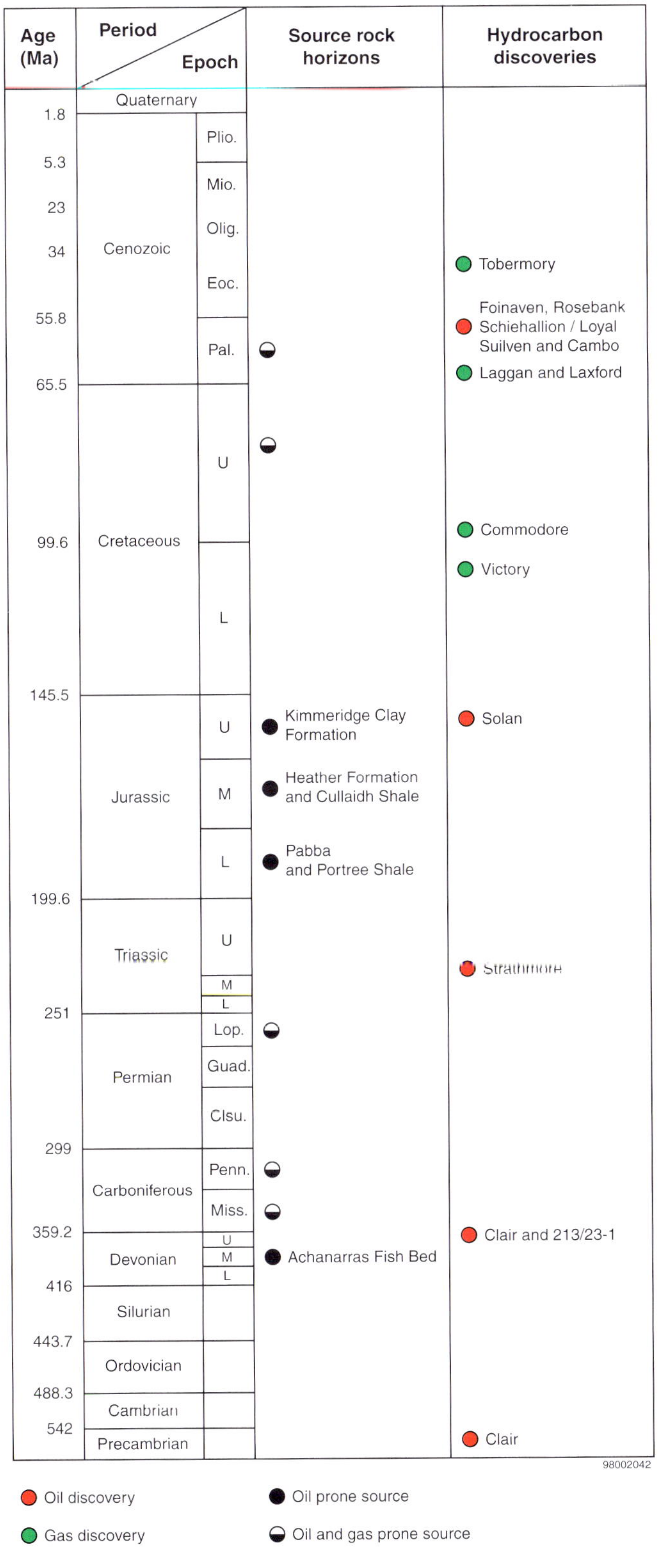

Figure 151 Stratigraphical distribution of known source rocks (including their potential for oil and gas generation) and significant hydrocarbon discoveries within the Faroe–Shetland region (modified from Scotchman and Dore, 1995; Scotchman et al., 1998; Scotchman and Carr, 2005 and Ziska and Andersen, 2005). Timescale is based on 'A Geologic Time Scale 2004' by Gradstein F M, Ogg, J G, Smith A G (2004) and the International Stratigraphic Chart, 2006 (ICS) with additions.

SOURCE ROCKS AND MIGRATION

Nine potential petroleum source rock successions have been identified within the north-east Atlantic Margin (Figure 151; Scotchman and Doré, 1995; Scotchman et al., 1998). To date, regional assessments show that only Middle Devonian (Peters et al., 1989; Bailey et al., 1990), Lower Jurassic (Scotchman and Thomas, 1995) and Middle and Upper Jurassic (Bailey et al., 1987) source rocks have been correlated to oil recovered from commercial wells. The bulk of the oils in the Faroe–Shetland area, for instance in the Clair, Foinaven, Schiehallion and Strathmore fields, have been geochemically correlated to the Late Jurassic to earliest Cretaceous, Kimmeridge Clay Formation (Scotchman et al., 1998; Herries et al., 1999). Unfortunately, there are relatively few well penetrations of source rock facies within the Upper, and particularly the Middle Jurassic successions within the Faroe–Shetland region (Figures 59 and 60).

Upper Jurassic to Early Cretaceous (Kimmeridge Clay Formation)

The organic-rich Kimmeridge Clay Formation is the principal source for hydrocarbons in the UK northern and central North Sea basins and is also regarded as the main source rock within the Faroe–Shetland area (e.g. Scotchman et al., 1998; Holmes et al., 1999; Figure 152). In total, some 37 exploration wells in the report area have penetrated the Kimmeridge Clay Formation (Figures 59 and 60), which mainly comprises organic-rich mudstone that was deposited in fully marine, outer shelf to bathyal slope conditions. The formation is commonly characterised by a high gamma-ray well log signature due to its uranium content (Figure 64). Uranium can be adsorbed onto organic matter and its association with organic-rich mudstones generally results in a high gamma-ray response. The Kimmeridge Clay Formation is interpreted to be relatively widespread throughout the graben areas, deposited during a series of marine transgressions that progressively overstepped contemporaneous basement highs (Herries et al., 1999; Holmes et al., 1999; Cawley et al., 2005). It typically reaches 10's of metres in thickness, with maximum recorded thicknesses of 295.5 m and 236.8 m encountered in wells 206/05-1 (Foula Sub-basin) and 204/15-2 (Westray High), respectively (Figure 60b). Towards the north-west part of the Faroe–Shetland Basin, the full extent of the Kimmeridge Clay Formation remains uncertain because of limited well penetrations and poor seismic resolution (Figures 65 and 152). However, its presence is supported by results from analyses of oil shows from the Lopra-1 borehole (Laier et al., 1997) located on the island of Suðuroy in the Faroes (Figure 120).

The organic matter of the Kimmeridge Clay Formation is mainly marine algal in origin although some studies have pointed to a minor contribution from land-derived organic material (Bertrand et al., 1990; Ramanampisoa and Disnar, 1994). Source rock quality is variable, with typical values of the order of 6 weight percent (wt. %) Total Organic Carbon content (TOC) and typical Hydrogen Indices (HI) of 350 milligrams of hydrocarbon per gram of TOC (mgHC/gTOC) (oil-prone) (Iliffe et al., 1999). However, TOC values of more than 10 wt. %, HI up to 600 mgHC/gTOC (very oil-prone), and petroleum yields up to 50 kg/ton are not

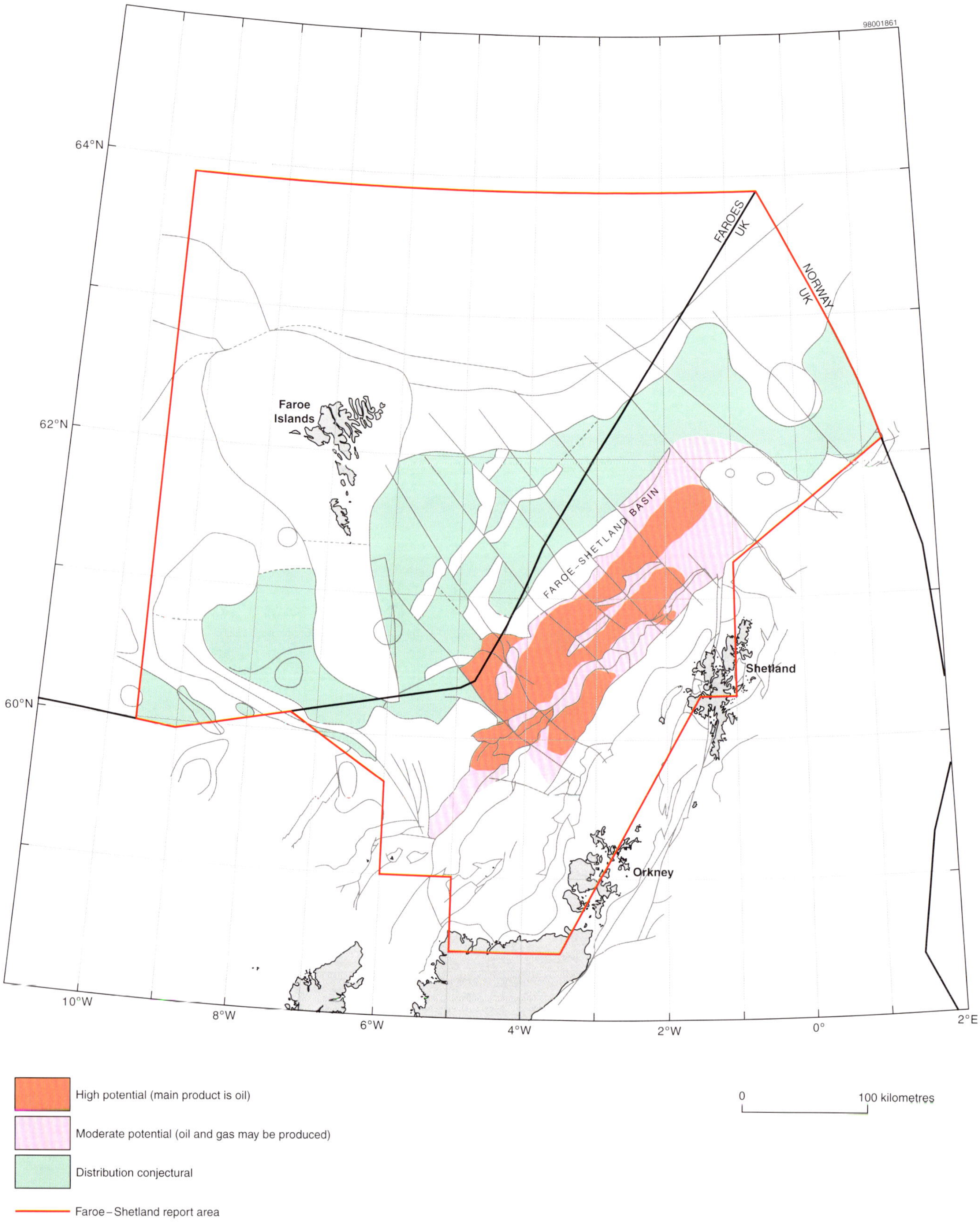

Figure 152 Distribution of Upper Jurassic to earliest Lower Cretaceous Kimmeridge Clay Formation source rocks and their estimated original generative potential (modified from Holmes et al., 1999; Cawley et al., 2005 and DTI, 2006). Present day structural elements (grey) derived from Figure 7.

uncommon from thermally immature Kimmeridge Clay Formation mudstone (Scotchman et al., 1998; Holmes et al., 1999; Iliffe et al., 1999; Cawley et al., 2005). The Kimmeridge Clay Formation shows considerable facies variation with kerogen types ranging from Type II (mainly derived from marine algae and with a high hydrocarbon-generative capacity), including the high sulphur content Type IIS found in the Solan basins

(Figure 7) and Type II-III and Type III (composed of terrestrial organic material, mainly cellulose and lignin with much lower hydrocarbon-generative capacities) (Scotchman et al., 1998; Holmes et al., 1999; Isaksen et al., 2000). The elevated sulphur concentrations seen in the Type IIS kerogen might be the result of a more carbonate-rich lithology, probably reflecting shallower water but in a more oxygen-depleted environment (Scotchman et al., 1998; Holmes et al., 1999; Cawley et al., 2005). In most depositional models for the Kimmeridge Clay Formation (see Oschmann, 1988; 1991), anoxia of the palaeowater column plays an important role in preserving the organic material, but the presence of benthic macrofauna (up to 10%) in mudstones containing very high amounts of organic matter (Wignall, 1990) has been interpreted to indicate recurring or seasonal periods of anoxia rather than continuous bottom water anoxia (Tyson, 1996). However, diagenetic products derived from green sulphur bacteria indicates that periods of oxic conditions may have been relatively short-lived and that sedimentation predominantly took place in a euxinic water column extending into the photic zone (Koopmans et al., 1996; van Kaam-Peters et al., 1997).

In terms of geophysical well log response, there is generally a very good correlation between organic content and gamma-ray values (Lüning and Kolonic, 2003). For instance, in well 205/26a-3 within the East Solan Basin (Figures 2, 7, 39 and 68), which drilled approximately 64 m of the Kimmeridge Clay Formation, gamma-ray well log values increase to more than 150 API (Scotchman et al., 1998). The high gamma-ray mudstone corresponds to high values in TOC content (maximum values of 15.4 wt.%), and a maximum HI of 500 mgHC/gTOC and is a very oil-prone unit.

Middle to Upper Jurassic (Heather Formation and older Middle Jurassic)

Bathonian to Oxfordian mudstone of the Heather Formation locally underlies the Kimmeridge Clay Formation in the Faroe–Shetland region (Figure 60). Within the Foula Sub-basin and on the Judd High, the Heather Formation has been penetrated in wells 206/05-1 (Figure 63) and 204/22-1, respectively. Pyrolysis of rock samples from the Heather Formation in well 206/05-1 show a rich source rock with TOC values of 6.3 and 7.3 wt. % and a HI of 73 and 149 mgHC/gTOC (gas-prone). In well 204/22-1, pyrolysis data available for the Heather Formation show a much leaner source rock with TOC values of 1.02 to 1.98 wt. % and a HI of 59 to 118 mgHC/gTOC (gas-prone) (Scotchman et al., 1998). South of the report area, within the West Lewis Basin (Figure 7), Middle Jurassic source rocks sampled in BGS Boreholes 88/01 and 90/02 show TOC higher than 4 wt. % (Isaksen et al., 2000).

The abundance of *Botryococcus* (fresh to brackish water algae) throughout the Bathonian interval in well 204/22-1 is consistent with deposition in a marginal marine setting, an environment proposed by Isaksen et al. (2000) for the Middle Jurassic source rock in the West Lewis Basin. A pre-Heather Formation source rock has only been encountered in well 204/22-1 and includes a rich oil-prone unit of lacustrine/lagoonal origin overlain by a poorer and less oil-prone succession deposited in a marine environment. In the more oil-prone lower section, TOC values up to 5 wt. % have been recorded with a HI of 400 to 800 mgHC/gTOC.

It is of interest to note that organic compound 28, 30 Bisnorhopane is a biomarker found only in the Kimmeridge Clay Formation and never in Middle Jurassic rocks. Analyses of oils from discoveries in the Faroe–Shetland region show a wide range of relative concentrations of this biomarker. This suggests that oils sourced from Middle Jurassic sediments, may have contributed to the hydrocarbon charge in reservoirs and that these source rocks may be more widespread than current drilling results indicate (Cawley et al., 2005).

Other potential source rocks

It is possible that Cenozoic coal and Late Cretaceous mudstone may be a source for hydrocarbons within the Faroe–Shetland report area (Figure 151). On the Faroe Islands, waxy bitumen associated with Paleocene basalts has been shown to be partly derived from a deltaic source rock of Late Cretaceous to Cenozoic age. This type of source rock has also been related to oils trapped in lavas on West Greenland (Laier et al., 1997; Scotchman et al., 1998). South of the report area, good quality, rich oil-prone source rocks of Lower Jurassic age, equivalent to the Portree and Pabba shale of Skye, occur in the Slyne Basin on the eastern flank of the Rockall Basin (Scotchman and Thomas, 1995). Within the report area, Iliffe et al. (1999) noted that a Lower Jurassic succession, that includes a 262.3 m thick continuous succession of mudstone penetrated in well 202/03a-3 (Figure 62), has source rock potential. Coal and shale of Carboniferous age such as those encountered in well 213/23-1 (Figure 52) on the Corona High could also be potential sources of hydrocarbons in the report area. The Middle Devonian contains potential source rock intervals of lacustrine-facies (e.g. Achanarras Fish Bed and equivalents) and is present on the northern coast of Caithness and on the islands of Orkney and Shetland (Figure 49). The presence of the Achanarras Fish Bed, or equivalents, suggest these occurrences formed part of the same basin or were closely connected as sub-basins (Marshall et al., 1985). Analysis of the Achanarras Fish Bed at Holburn Head near Thurso on the northern coast of mainland Scotland indicates that it is a rich oil-prone source rock with TOC values of 5.2 and 7.2 wt. % recorded from two samples (Scotchman et al., 1998).

Thermal maturity

The present day thermal maturity of a source rock can be measured in various ways, such as by vitrinite reflectance, thermal alteration index (TAI), T_{max} (the temperature at which the maximum rate of pyrolysis occurs), apatite fission track analysis and by epimer-based molecular parameters such as C_{27}–C_{29} non-rearranged steranes and C_{31}–C_{33} hopanes. Generally, vitrinite reflectance is the technique most commonly used within the oil and gas industry. These measurements, along with analyses of source rock quality, help to constrain source rock maturation models that attempt to quantify the timing, amount and type of hydrocarbons generated from a source 'kitchen'. Maturity models have been based on either time-temperature indexes (TTI) produced by calibrating the maturity, constrained from vitrinite reflectance and other maturity indicators, with time and temperature in the basin (Waples, 1980; Goff, 1983) or, increasingly, on kinetically based models based on the rates at which chemical reactions of maturation proceed (Middleton, 1982; Burnham and Sweeney, 1989; Sweeney and Burnham, 1990). In conventional maturity modelling based on kinetics, temperature is the dominant control on the generation of petroleum and the estimation of palaeogeothermal gradient and heat flow provides a major input to maturity modelling. In the Faroe–Shetland area, the effects of rifting and volcanism would generally be expected to result in higher heat flows. However, studies using both vitrinite reflectance and apatite fission track analysis (e.g. Green et al., 1999; Iliffe et al., 1999; Parnell et al., 1999) show no direct evidence for significant elevated values. Thus, despite Cretaceous to Early Paleocene rifting and Palaeogene volcanism (e.g. Figure 8), many workers have assumed a relatively constant heat flow through time (e.g. see Dean et al., 1999; Iliffe et al., 1999; Holmes et al., 1999).

Predictions for the onset of oil generation from Upper Jurassic source rocks in the report area vary, mainly as a result of model location and assumptions made regarding burial and thermal histories. For example, Holmes et al. (1999) suggested that local early oil generation began in the Judd Sub-basin during the mid Cretaceous, with generation peaking by the end of the Cretaceous, and gas being generated during the Paleocene. Iliffe et al. (1999), by modelling maturation in a pseudowell located in the centre of the Corona Sub-basin, indicated that hydrocarbon generation principally took place during the Paleocene coinciding with the development of major overpressuring. Lamers and Carmichael (1999) reported that the onset of oil generation in the Judd Sub-basin was during the Late Cretaceous, while Jowitt et al. (1999) modelled the majority of hydrocarbons within the report area as having been generated by end Late Cretaceous times. It should be noted that these authors assumed little or no variation in heat flow through time.

Maturity modelling of the Upper Jurassic source rocks suggests that most of the report area is either postmature for oil or is gas mature, with only a small number of highs being within the oil window around the eastern and south-western fringes of the Faroe–Shetland Basin (Figure 153; Scotchman and Carr, 2005). In parts of the Judd Sub-basin, modelling of deeply buried source rocks predicts gas generation, yet to date, only oil has been discovered (Scotchman et al., 1998; Iliffe et al., 1999; Scotchman and Carr, 2005). In order to explain the lack of gas accumulations, workers have invoked the so-called 'motel' (Doré et al., 1997; Lamers and Carmichael, 1999) and 'whoopee cushion' holding mechanism models (Iliffe et al., 1999). These models suggested temporary residence of the generated oil within Mesozoic reservoirs followed by remigration into the Paleocene. Lower Cretaceous sandstone units are thought to have acted as conduits for hydrocarbon migration towards the crestal areas of structural highs, to form palaeo-accumulations within the Mesozoic reservoirs. These models also suggested that rapid subsidence during the Early Palaeogene caused an increase in overburden pressure in basinal areas, which was transferred updip, resulting in hydrofracturing within the overlying Upper Cretaceous mudstone succession. This allowed vertical migration into the Paleocene T10–T30 sequences with ensuing lateral migration to present day traps (Doré et al., 1997; Lamers and Carmichael, 1999; Iliffe et al., 1999).

More recently, laboratory studies have shown that high water pressures appear to inhibit organic matter breakdown, retard the aromatisation reactions and shift the maximum soluble extract to higher temperatures, resulting in retardation of hydrocarbon generation (Price and Wegner, 1992; Hatcher et al., 1994; Landis et al., 1994). This led Carr (1999) to propose a maturation model incorporating pressure dependence. Application of this model in the Faroe–Shetland Basin dispenses with the need for transient trapping of hydrocarbons and allows heat flow to vary in response to rifting and volcanism (Carr and Scotchman, 2003; Scotchman and Carr, 2005). In this model, rapid Paleocene subsidence induced overpressure which inhibited further thermal maturation, and consequently, Jurassic source rocks are thought to be producing oils at present day (Scotchman and Carr, 2005).

Oil families, mixing of source rocks and charging

Analyses of oils from the Clair, Foinaven and Schiehallion/Loyal fields (Scotchman et al., 1998), the undeveloped Strathmore Field (Figures 150 and 151), and from discoveries 204/28-1 and 204/22-1 on the Judd High and Sub-basin respectively (Figures 2 and 7), show them all to be heavy oils (9°–28° API), with sulphur content ranging from 0.42% in Clair to 7.6% in the former well.

Scotchman et al. (1998) recognised two main oil families in the Faroe–Shetland area, namely the 'Clair' Family (mixed Heather and Kimmeridge Clay Formation-sourced biodegraded and undegraded oils) and the 'northern North Sea' Family (oils from rich, marine, highly anoxic Kimmeridge Clay Formation). Additionally, Scotchman et al. (1998) also recognised the 'Lopra' Family, consisting of oils sourced from Upper Cretaceous to Cenozoic deltaic rocks and in the south-eastern margin of the report area, 'Devonian' Family oils sourced from Middle Devonian lacustrine rocks.

Fluid inclusion studies (e.g. Parnell et al., 1999) and analyses of oils (Scotchman et al., 1998) show that the Faroe–Shetland region was subjected to several episodes of hydrocarbon expulsion from the deep basinal source rocks and migration towards the adjacent structural highs. Evidence of mixing of different oils from different source rocks and source rock facies has been established for the fields and discoveries within the report area (Rooney et al., 1998; Scotchman et al., 1998). For example, isotope analysis of oils from across the Clair Field suggests that there were two pulses of oil charge to the field. The oil is a fairly uniform mixture of a severely biodegraded early accumulation and a non-biodegraded later oil pulse, with the latter less volumetrically important (Scotchman et al., 1998). Compound specific isotope analysis (CSIA) of components of the Clair Field oils show differences that led Rooney et al. (1998) to conclude that the two charges of oil to Clair had different compositions and that the early pulse was sourced from Kimmeridge Clay Formation Type II–IIS facies and the later pulse from a more oxic, Type II/III source similar to the Heather Formation.

Gas chromatography and isotope analysis of the oils from the Foinaven and Schiehallion/Loyal fields also show that the reservoirs are charged with a mixture of two phases of oil migration, both of which have been biodegraded, with the first pulse being more biodegraded than the second pulse (Scotchman et al., 1998). In con-

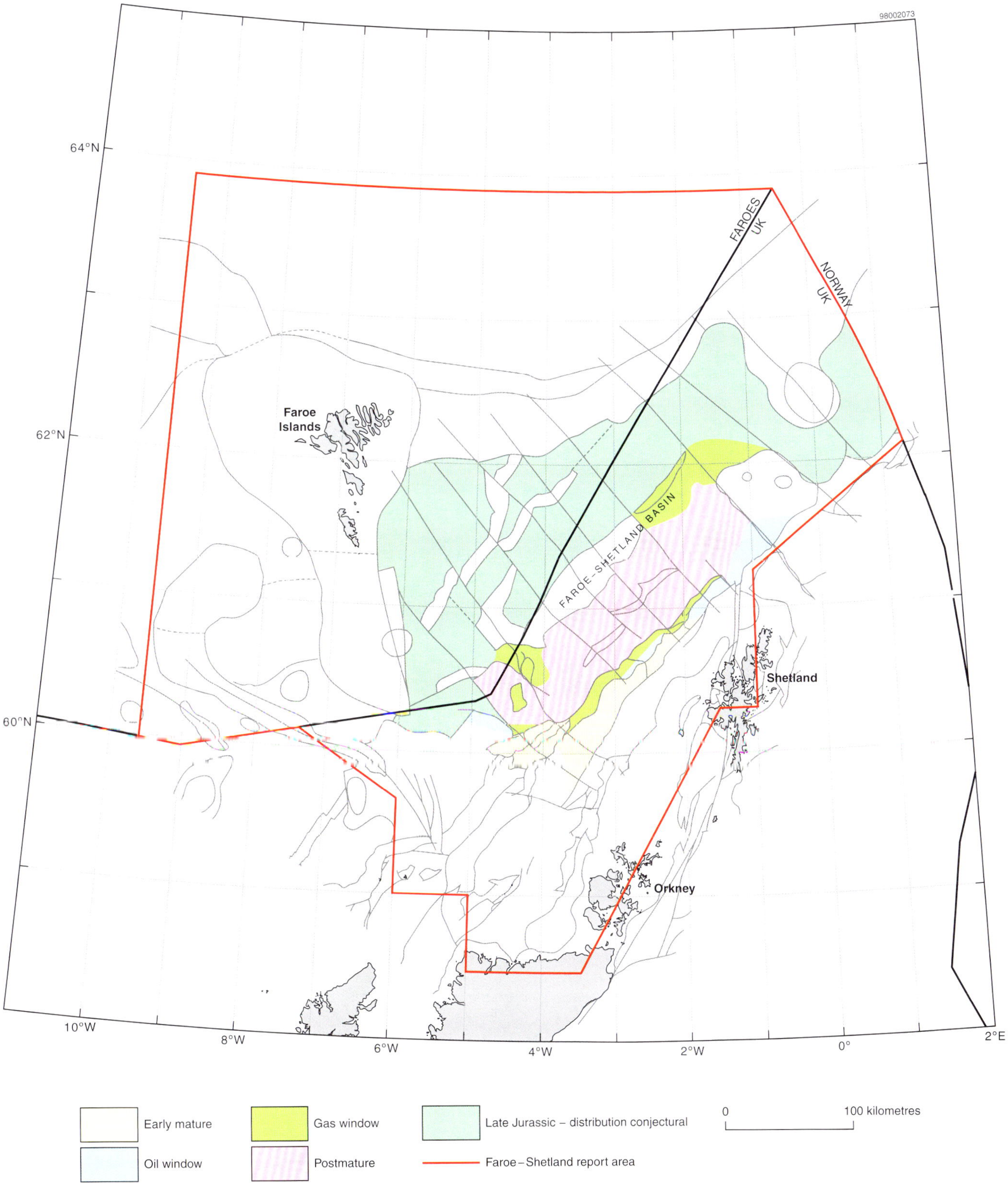

Figure 153 Inferred present day maturity of the Upper Jurassic based on a constant heat flow model (modified from Scotchman and Carr, 2005). Present day structural elements (grey) derived from Figure 7.

trast to the Clair Field, the second pulse of oil generation and migration to the Foinaven Field had a greater volumetric contribution (Scotchman et al., 1998). From isotopic analysis of the oil, the first pulse is shown to be nearly identical to the first pulse in Clair, suggesting a similar Kimmeridge Clay Formation source facies. However, the second pulse into the reservoir shows geochemical similarities with the Middle Jurassic, pre-Heather Formation lacustrine mudstone in 204/22-1 (Scotchman et al., 1998). The observation by Scotchman et al. (1998) that the first pulse of oil in both the Clair Field and Foinaven Field has been biodegraded suggests that this event was of regional importance and was probably a result of meteoric water flushing. They suggested that this regional flushing could have been caused either by the development of semiregional convection cells related to the intense igneous activity during the Paleocene, or as a result of the late Palaeogene compressional events (e.g. Doré and Lundin, 1996). They also suggested that the second late stage biodegradation observed in the Foinaven Field may be related to meteoric water flushing during the last ice age (Damuth and Olson, 1993).

Oil from the Strathmore Field is possibly derived from a different Kimmeridge Clay Formation source facies to both the Clair Field and Foinaven Field. The source consists of a Type IIS kerogen formed within a very anoxic, calcareous depositional environment (Scotchman et al., 1998; Holmes et al., 1999). Oil samples from well 205/26a-3 (Figure 39) display a relatively high sulphur content of 2.9% and no evidence of biodegradation. These oils are considered to be derived from a local source and were generated at low thermal stress with an early level of maturity resulting in heavy 22° API oil. The Strathmore and adjacent Solan oil accumulations are thought to be the result of two charging episodes. An initial Late Oligocene charge may have protected higher parts of the reservoir from permeability damaging cementation. Hydrocarbon generation may have temporarily ceased during Oligocene to Miocene inversion, but commenced again towards the end of the Miocene (Herries et al., 1999).

In the undeveloped Victory Field at the north-east end of the Rona High, oils reservoired in the Lower Cretaceous succession in wells 207/01-3 and 207/01a-5 (Figures 2 and 7) share a common Upper Jurassic source. Gas chromatography and mass spectrometry results show that a high level of biodegradation characterises all the analysed samples (Goodchild et al., 1999). According to Goodchild et al. (1999), a significant contribution from a pre-Kimmeridge Clay Formation (i.e. Heather Formation and/or older Middle Jurassic) source is also indicated by an absence of the 28, 30 Bisnorhopane biomarker in some of the tested oils. Maturity related biomarkers indicate that charging occurred in multiple stages, although the hydrocarbons were derived from source rocks of similar maturity with the first pulse being the most volumetrically important. The charging is thought to have occurred during Late Cretaceous times, and possibly contemporaneously with the first migration of hydrocarbons into the Clair Field (Goodchild et al., 1999). This first charge is thought to have been largely lost during a regional Late Cretaceous to early Cenozoic inversion episode which resulted in local uplift and faulting on the Rona Ridge and breaching of faults in the Victory Field. Any remaining oils were biodegraded due to fresh water influx. As subsidence resumed during the Late Paleocene, a second more limited charge of oil, prone to biodegradation may have occurred, synchronous with the first phase of gas migration into the reservoir. From Eocene times, the field received mainly gas, but a nonbiodegraded component of the residual oil in the Victory Field is thought to be a late stage, local recharge derived from the western flank of the Rona High (Goodchild et al., 1999).

Within the Faroe–Shetland Basin, results from 28, 30 Bisnorhopane biomarker studies relating to the Faroese wells (Figure 2) also suggest that mixing of hydrocarbons from the Kimmeridge Clay Formation and a Middle Jurassic source has taken place. This implies that the same source rock system is present throughout the whole of the Judd Sub-basin (Cawley et al., 2005).

HYDROCARBON EXPLORATION PLAYS

A petroleum exploration 'play' consists of a source, migration route, reservoir and seal that together, have the potential to generate, attract and hold a hydrocarbon accumulation. As demonstrated from a number of producing fields and significant discoveries in the Faroe–Shetland area, proven hydrocarbon exploration plays range in age from Precambrian to Eocene and are described below.

Fractured Lewisian basement play

Well 206/12-1 encountered oil shows in fractured Lewisian gneiss basement on the Rona High that also forms a proven reservoir in the Clair Field (Figure 151). Towards the north-western edge of the field, well 206/07-1 (Figure 2) flowed 963 barrels of oil per day (BOPD) through fractured gneiss. Studies of the Clair Field intimate that the frequency of fractures and their orientations are important factors when considering optimisation of oil extraction from fractured basement. On the north coast of Scotland and the Orkney Islands, onshore analogue studies of Precambrian Moinian basement and overlying Devonian Old Red Sandstone show a spatial variation in the frequency of fractures mapped in the former. In areas where there are a high number of fractures, they commonly pass up into the overlying Devono-Carboniferous sediments (Coney et al., 1993). Fracture corridors of this type, together with the larger fault zones, usually form the best producing areas for hydrocarbon extraction in the Clair Field (Coney et al., 1993). Although the most common fracture orientation within the Field is north-east-trending, the north-west-trending fractures tend to be open whereas the former tend to be cemented. This is also the case in north-west-trending basement fractures in the Victory Field (Goodchild et al., 1999).

Other examples of fractured Lewisian reservoir rocks are found in the Judd Sub-basin and High where well 204/22-1 recorded weak oil shows and 204/23-1 (Figures 2 and 7) recorded a condensate flowrate of 117 barrels per day and 1.25 million standard cubic feet of gas per day (MMSCFD). On the Judd High, well 204/28-1 encountered oil in fractured basement rocks that produced 42 barrels at an average flowrate of 153 BOPD. Potentially, this play may occur on any culmination of Lewisian basement within the report area.

Clair Group sandstone play

In the UK sector of the report area, the Devono-Carboniferous Clair Group play is restricted to parts of the Rona and Corona highs where Devonian and Carboniferous sediments have been preserved resting unconformably on Lewisian basement (Figures 31, 52 and 154). The play consists of a Devonian and Carboniferous reservoir top-sealed by Mesozoic argillaceous rocks. Typically, trapping is facilitated by three-way dip closure, with updip pinchout onto the Rona High and truncation by a south-east-dipping fault on the Corona High (Figure 154). Fractured Lewisian basement can form a significant component of the reservoir in this play.

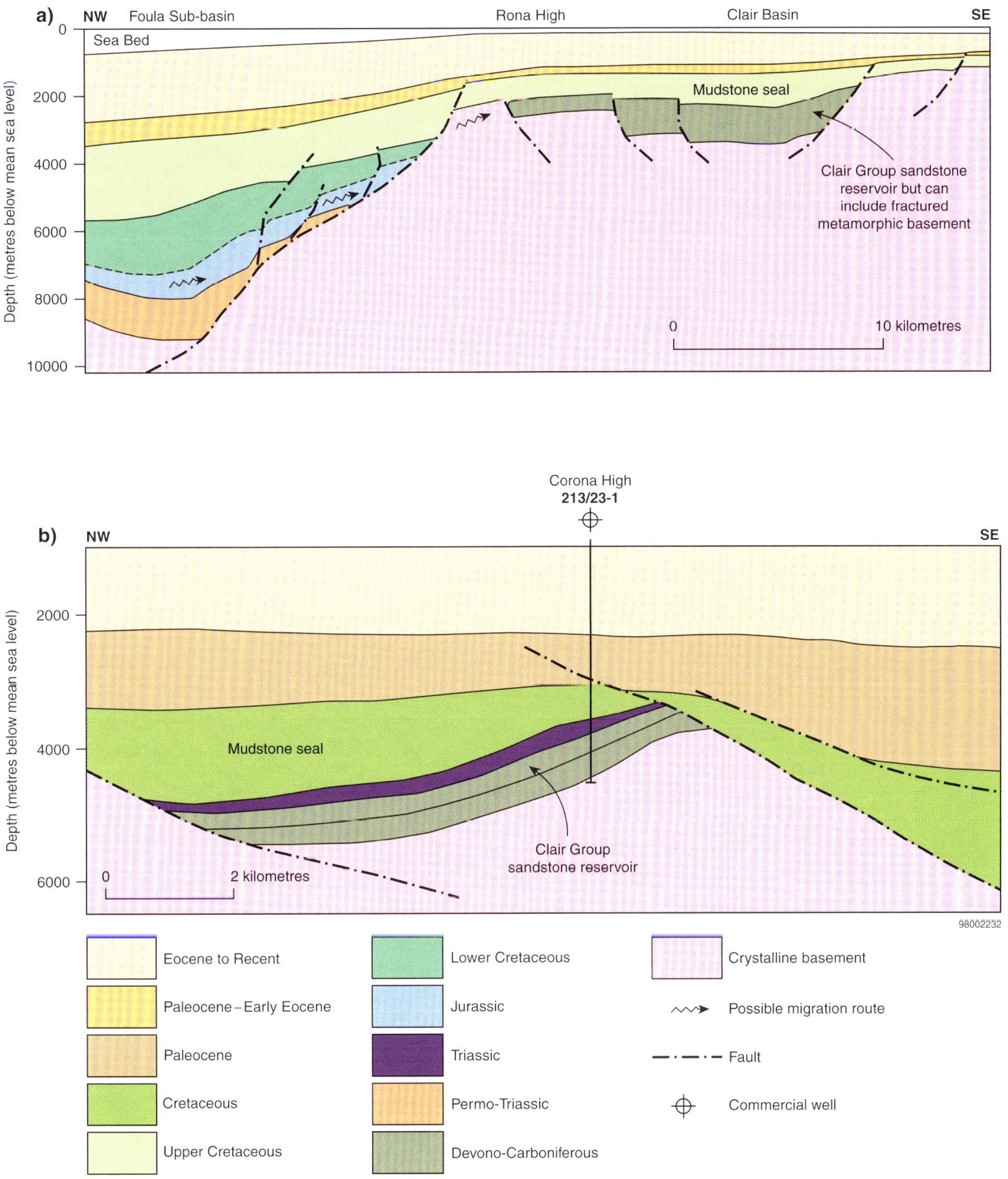

Figure 154 Schematic representation of the Clair Group sandstone play within a) the Rona High and Clair Basin b) the Corona High (see text for details). Note that vertical and horizontal scales are approximate. For location of structural features and well, see Figures 2 and 7.

The Clair Field is a successful example of this play and is located on the Rona High (Figures 48 and 150). Here, the Devono-Carboniferous succession has been divided into the Lower and Upper Clair groups, separated by an unconformity (Figure 50; Allen and Mange-Rajetzky, 1992). The Lower Clair Group, of Middle Devonian age, consists of six reservoir units derived from high and low energy fluviatile, aeolian and lacustrine sedimentary environments with porosities varying from 12–15% and permeabilities from 30 millidarcies (mD) to 360 mD. The Upper Clair Group, of Carboniferous (youngest Viséan) age, consists of four reservoir units deposited in fluviatile to marine sedimentary environments. Porosities range from 12–15% with permeabilities generally less than 10 mD. In the Clair Field, oil extraction rates are primarily dependant upon the distribution of fractures (Barr et al., 2007). Charging of the reservoirs is considered to have been facilitated from a Jurassic source to the west within the Foula Sub-basin (Figure 154a).

On the Corona High, well 213/23-1 drilled on a tilted fault block, proved oil-bearing Devono-Carboniferous sediments resting on Lewisian basement (Figure 154b). An updip fault was shown to be sealing despite having moved during early Eocene times. Core analysis of the Carboniferous succession indicated a low to moderate reservoir quality with a mean porosity of 7.5% (0–13% range) that is mainly secondary due to potassium feldspar leaching, but with low pore space connectivity. Charging of the reservoirs on the Corona High is expected to have been from a Jurassic source located downdip to the west within the Corona Sub-basin.

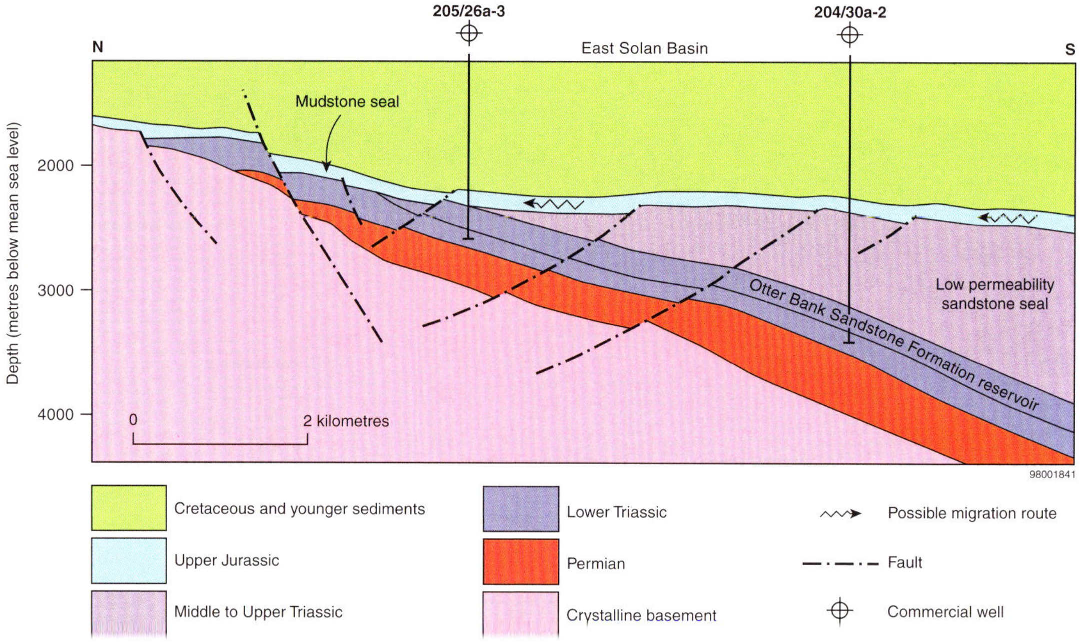

Figure 155 Schematic representation of the Otter Bank Sandstone Formation play within the East Solan Basin (see text for details) (modified from Herries et al., 1999). Note that vertical and horizontal scales are approximate. For location of structural feature and wells, see Figures 2 and 7.

Otter Bank Sandstone Formation play

The Strathmore discovery within the East Solan Basin is the only successful example of the Lower Triassic Otter Bank Sandstone Formation play in the report area (Figures 150 and 155). However, the Otter Bank Sandstone Formation reservoir has been drilled 70 km to the south in the West Orkney Basin (Figure 57). The main factor controlling the extent of this play is the presence and maturity of a hydrocarbon source within the West, South and East Solan basins in particular, which currently lies at a shallower depth than the main source kitchens that supplied other discoveries in the Faroe–Shetland Basin. In the Solan basins, the Upper Jurassic Kimmeridge Clay Formation mudstone (Figure 39) is the source for the undeveloped Strathmore and Solan oilfields. Crucially, the mudstone contains a sulphur-rich Type IIS kerogen that has a lower thermal threshold than the more common Type II kerogen present in the main part of the Faroe–Shetland Basin to the west, and therefore reaches maturity despite the shallower burial depths found in the Solan basins. This enables generation and migration of hydrocarbons in these less deeply buried marginal basins. A migration route for the oil, sourced from the younger Upper Jurassic Kimmeridge Clay, into the Lower Triassic reservoir is not straightforward and would require juxtaposition of the reservoir either against a carrier bed such as the intra-Upper Jurassic Solan Sandstone Member or directly against basal Kimmeridge Clay Formation mudstone (Herries et al., 1999).

In the Strathmore discovery, the trap consists of three-way dip closure with truncation of the reservoir by an angular unconformity that juxtaposes the Early Triassic Otter Bank Sandstone Formation reservoir with the overlying Upper Jurassic Kimmeridge Clay Formation mudstone. Elsewhere, the reservoir is sealed by overlying Middle to Upper Triassic low permeability Foula Sandstone Formation (Figure 155).

The Otter Bank Sandstone Formation reservoir is of Early Triassic age (Figure 56), the lower part of which consists of braided sandy, often pebbly, fluvial channel deposits while the upper part consists of aeolian sandy sabkha deposits becoming interbedded with fluvial deposits towards the top of the formation and into the lower part of the overlying Foula Formation (Herries et al., 1999). Reservoir quality depends upon grain size and also primary depositional relationships such as whether beds are horizontally layered or cross-stratified, as this will affect grain packing. Facies also exert an influence on reservoir quality. The best quality reservoirs are seen in cross-stratified aeolian and fluvial layers, with the latter's quality decreasing with an increase in the presence of intra-formational clasts. The sandy sabkha deposits correlatable over the Strathmore Field have the lowest reservoir quality and form permeability barriers that partition the reservoir. These primary depositional influences on reservoir quality have been modified by the presence of diagenetic chlorite, calcite and kaolinite minerals. Grain-rimming chlorite has had a positive effect on porosity by preventing the development of blocky pore-filling calcite minerals whilst pore-filling kaolinite is more common in the overlying Foula Formation. Mean porosity of the Otter Bank Sandstone Formation typically varies from 14–17.7% with effective permeabilities varying from 5–185 mD (Herries et al., 1999).

Rona Member sandstone play

The Rona Member has been proved mainly within the North Rona, Solan and West Shetland basins (Figures 60, 65 and 150). The Rona Member is Kimmeridgian to Tithonian in age (Verstralen et al., 1995; Ritchie et al., 1996) and rests on Lewisian basement, Middle Triassic Foula Formation or Lower Jurassic rocks (Figure 156a). On the basis of core description, Verstralen et al. (1995) divided the succession into seven main facies (A–G), grouped into two main facies associations, namely a basal coarse-grained facies and an overlying fine-grained facies (see Chapter 6). The main lithologies are conglomerate, pebbly sandstone, very fine to very coarse-grained sandstone, siltstone and mudstone. The Rona Member has been interpreted to have been deposited in an overall transgressive regime with the conglomerate, which includes a palaeosol horizon, deposited in the

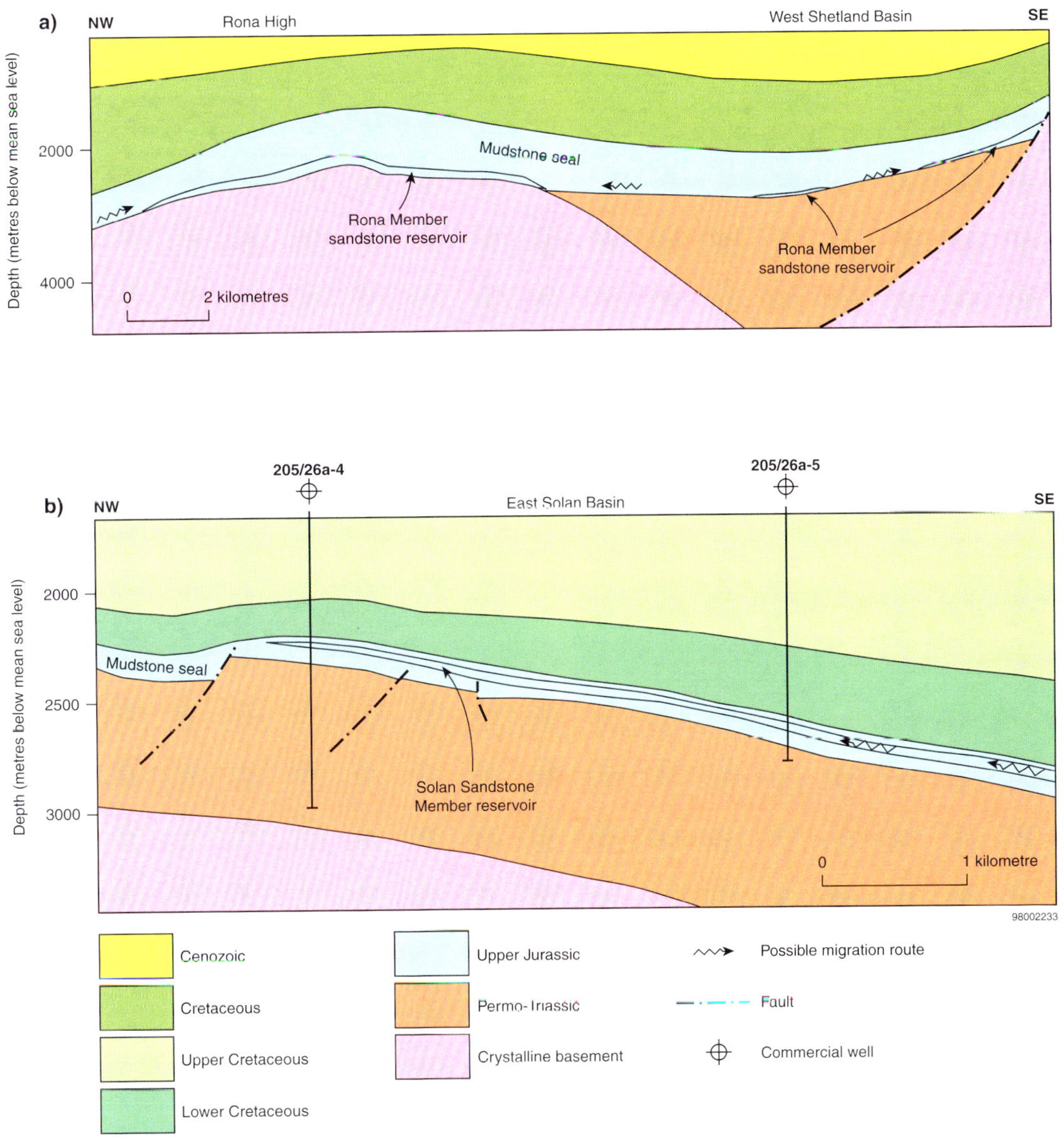

Figure 156 Schematic representation of a) the Rona Member sandstone play within the West Shetland Basin b) the Solan Sandstone Member play within the East Solan Basin (see text for details) (modified from Herries et al., 1999). Note that vertical and horizontal scales are approximate. For location of structural features and wells, see Figures 2 and 7.

subaerial part of a fan delta with overlying lithologies deposited in beach, shoreface and marine shelf environments (Verstralen et al., 1995). There is a wide variation in the thickness of the Rona Member (4.6 to 105.5 m proved in well penetrations) due to deposition on an irregular topography; the thickest sections that include the coarser-grained basal facies association were deposited in topographical lows on a weathered Lewisian surface (Verstralen et al., 1995). In terms of porosity and permeability characteristics, core analysis from well 205/22-1A (Figure 2) indicated porosity generally less than 10% and variable permeability usually less than 10 mD.

The Rona Member is generally overlain by Kimmeridge Clay Formation mudstone which has the potential to act as an effective seal, and downdip, as a hydrocarbon source. However, Herries et al. (1999) noted that the lack of success of this play may be due to the restricted extent of the Rona Member sandstone reservoir that would result in a lack of an extensive migration pathway from the deeper source areas to the updip sandstone reservoir.

Well 205/23-2, located on the western flank of the Rona High (Figure 2), penetrated 34.1 m of Upper Jurassic mudstone, lignite and sandstone. The sandstone is gas-bearing with oil shows and is 17.1 m thick, coarse to very coarse-grained and locally conglomeratic and rests on Triassic sediments. The succession is currently undivided but the possibility exists that it may prove to form part of the Rona Member.

Solan Sandstone Member play

The proven extents of this play, defined by the presence of the Solan Sandstone Member reservoir, are restricted to the East Solan Basin (Figures 60a and 150; Herries et al., 1999). The play elements consist of a north-east-dipping Solan sandstone reservoir pinching out updip to the south-west adjacent to the north-east-trending Rona High, top and laterally sealed by mudstone of the Upper Jurassic Kimmeridge Clay Formation (Figure 156b). The reservoir is charged via the downdip Solan Sandstone Member from adjacent mature Kimmeridge Clay Formation mudstone. The play was revealed by well 205/26a-4 (Figure 156b), an appraisal for the Triassic Strathmore discovery, that drilled a 82 m thick Upper Jurassic succession that included a 15.8 m oil-bearing sandstone that flowed at 2080 BOPD. The stratigraphical relationships between the Solan and Rona successions and the Kimmeridge Clay Formation (Herries et al., 1999) are discussed more fully in Chapter 6.

The Solan Sandstone Member is derived from an amalgamation of a series of high-density gravity flows deposited in a marine shelf environment (Herries et al., 1999). Although grain size of the Solan Sandstone Member increases downdip to the north-east within the East Solan Basin, reservoir quality decreases, mainly due to kaolinite cementation that has reduced permeability significantly. This cementation is thought to be the result of high water saturation below a palaeo oil-water contact prior to later charging of the reservoir. However, in well 205/26a-4 positioned in an updip location (Figure 156b), reservoir quality is excellent with an average porosity of 25.7% and permeability of 185 mD. Porosity and permeability decrease downdip and in wells 205/26a-5Z and 205/26a-5, average permeability is 63 mD and 13 mD, respectively (Herries et al., 1999).

OTHER POTENTIAL JURASSIC SANDSTONE PLAYS

Well 206/05-1 proved two Jurassic medium to coarse-grained units not recorded elsewhere in the Faroe–Shetland Basin but which have the potential to form the reservoir and carrier beds in plays similar to the Solan and Rona plays described above (Ritchie et al., 1996). The Fair Sandstone Member forms the stratigraphically oldest part of the Heather Formation, has a wide age range i.e. late Pliensbachian to latest Oxfordian, and consists of 216 m of poorly sorted, fine to coarse-grained sandstone (Figures 60b and 63). The Ridge Conglomerate Member occurs within the Upper Jurassic Kimmeridge Clay Formation and consists of a 64 m thick interval of mudstone grading to sandstone and conglomerate (Figures 60b and 63).

Victory Formation sandstone play

The Victory Formation sandstone play comprises a Lower Cretaceous syn-rift succession of sandstone and conglomerate, located on the footwall of the north-east-trending Rona High (Figure 157a). The succession dips south-eastwards into the north-east part of the West Shetland Basin and forms a proven play that may extend south-westwards along the Rona High and the West Shetland Basin (Figures 72a and 150). The extents of the play are defined by the presence of the potential reservoir, which is of variable quality (Goodchild et al., 1999). The Victory gas Field is a successful example of this play, forming a three-way dip-closure with updip sealing provided by a north-east-trending, north-west-dipping fault and a top seal of thick Upper Cretaceous mudstone (Figure 157a; Goodchild et al., 1999).

The Victory Formation rests either directly on Lewisian basement or locally, Devono-Carboniferous sediments. Goodchild et al. (1999) divided the Victory Formation into two seismic packages. The lower package, that maintains a relatively constant thickness across the footwall of the structure, has broadly parallel reflections and no obvious onlap surfaces. The upper package has divergent and internally disconformable reflections thickening to the south-east into the West Shetland Basin half-graben (see Victory Field below). In the Victory Field, the upper package occurs downdip of the gas accumulation (Figure 157a). The lower package, which is thought to be an early syn-rift sequence, forms the reservoir to this play, and has been divided into four correlatable informal units (V1–V4) (Figure 70; Goodchild et al., 1999). The stratigraphically oldest unit (V1) consists of conglomerate, interbedded with clean well sorted sandstone that thickens and becomes more dominant towards the top of the unit. Unit V2 comprises sandstone interbedded with coarser pebbly sandstone. Units V1 and V2 are thought to have been deposited as a fan delta that built out south-eastwards from the Rona High into the West Shetland Basin (Goodchild et al., 1999). The conglomerate in unit V1 has poor reservoir quality (3.9–8.7% porosity, 0.3–5.6 mD permeability) and although good thin (<0.75 m) reservoir quality sandstone beds (11.8–27.5% porosity, 213–1280 mD permeability) are present, the overall quality of V1 is poor. Unit V2 has very good porosity and permeability characteristics within the pebble-rich sandstones (13–24.7% porosity, 4.4–4800 mD permeability) where minimal cementation has taken place. However, the interbedded massive sandstone beds have relatively poorer porosity and permeability characteristics due to quartz overgrowths and local carbonate cements (5.8–19.5% porosity, 17.8–759 mD permeability). Unit V3 consists of fine-grained feldspathic sandstone (21.8–35.5% porosity, 54–1640 mD permeability) with disrupted argillaceous laminae and is overlain by massive and coarser sandstone (10.7–11.7% porosity, 219–299 mD permeability) of unit V4. Units V3 and V4 are interpreted to have been deposited in a shallow marine shoreface setting and have excellent reservoir characteristics despite local diagenesis.

There is a general risk that, as is thought to have happened in the Victory Field, any charging by oil during Late Cretaceous times may have been lost through leakage and biodegradation following uplift on the Rona High and reactivation of major faults controlling the basement horst.

Commodore Formation sandstone play

The likely extents of this play are defined by the presence of the potential reservoir, the Commodore Formation sandstone (Figures 70, 72b and 150) and the structural elements that may provide updip closure.

Well 206/03-1 drilled on the south-east flank of the Foula Sub-basin penetrated a succession of more than 300 m of sandstone of variable porosity and permeability (Figure 157b). Ritchie et al. (1996) recognised these and other sandstone-rich penetrations in the Foula Sub-basin as being of similar age and origin and assigned them to the Commodore Formation of ?early Albian to Cenomanian age. The sandstones form part of a series of gravity flow sediments deposited in a sandy submarine fan environment during a highstand that drowned the Rona High (Figure 157b). The Rona High may have provided a break in slope for the sands being shed from the uplifted hinterland to the south-east, and hence a focus for their deposition.

Interpretation of the Commodore Formation sandstone in well 206/03-1 by log evaluation and Routine Formation Test (RFT) pressure data, enabled the succession to be informally sub-divided into three units. Specifically, a 70 m thick middle unit comprising blocky sandstone with porosities ranging from 20–25% sandwiched above and below by thinly bedded sandstone of poorer reservoir quality. The basal unit is approximately 100 m thick and tightly cemented while the uppermost unit has porosities between 5–10% and includes a limestone bed. Taken together, the sandstone units form a

Figure 157 Schematic representation of a) the Victory Formation sandstone play within the West Shetland Basin (modified from Goodchild et al., 1999) b) the Commodore Formation sandstone play within the Foula Sub-basin (see text for details) (slightly modified reproduction from Grant et al., 1999 by permission of The Geological Society, London). Note that vertical and horizontal scales are approximate. For location of structural features and wells, see Figures 2 and 7.

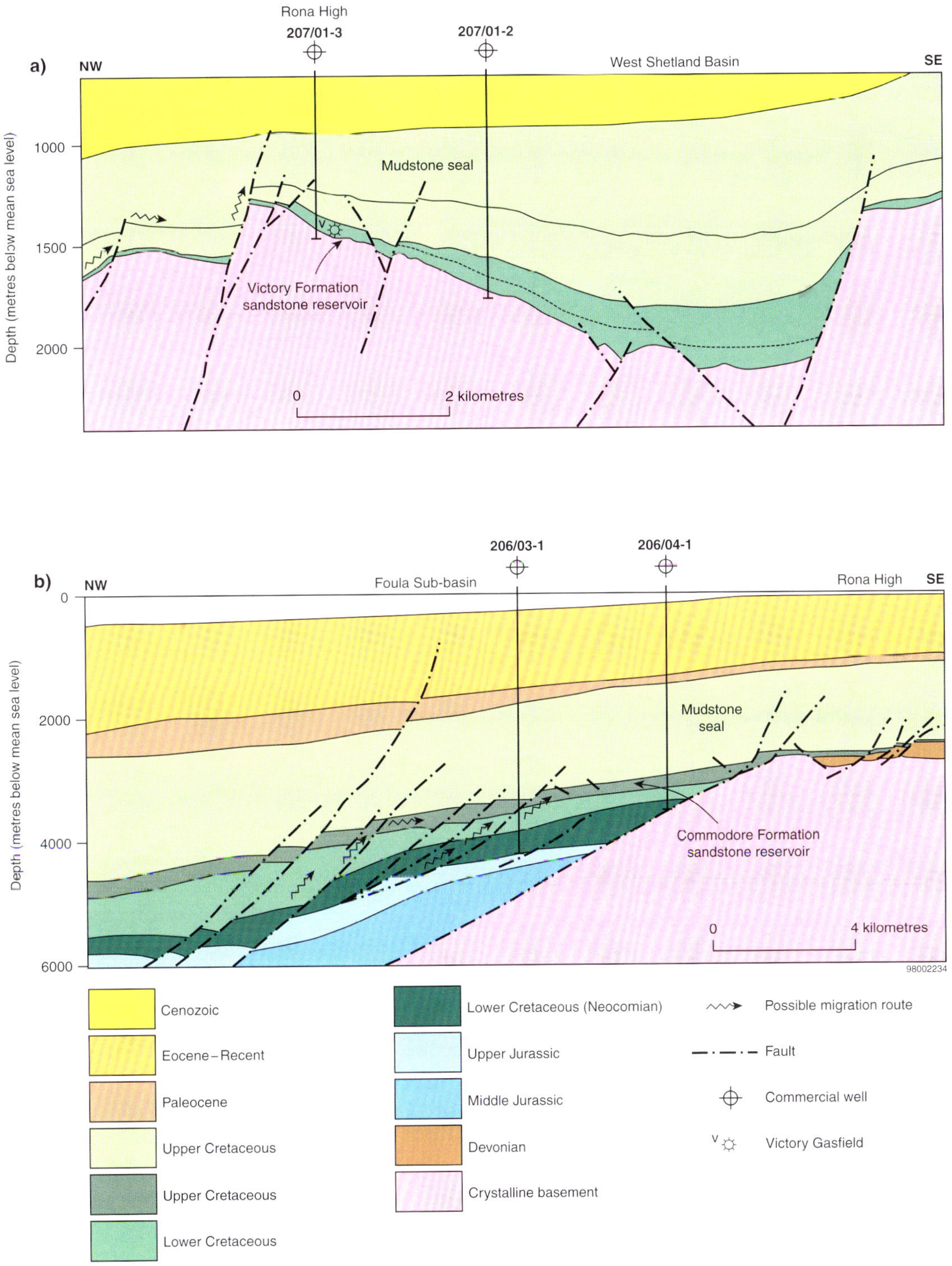

distinctive seismic package consisting of high amplitude sub-parallel reflections which, when mapped, suggest that it is restricted to the Foula Sub-basin (Grant et al., 1999). Upper Cretaceous mudstone provides top and lateral sealing to the reservoir sandstone while updip closure relies either on a fault sealing upward migration or the reservoir becoming mud prone (Figure 157b). Charging of the reservoir is expected to be from the underlying Jurassic source via lateral and vertical migration paths.

The Foula Sub-basin has a strong regional dip towards the north-west with sedimentary successions including the Commodore Formation sandstone reservoir, dipping steeply away from the Rona High resulting in only rare occurrences of four-way dip closures. However, well 206/11-1 (Figure 2) drilled a four-way dip closure that tested 1.65 MMSCFD of gas from Commodore Formation sandstone of Albian age. The more common down-faulted three-way dip closure of the Lower to Upper Cretaceous play was tested by well 206/04-1. This was drilled updip from well 206/03-1 and penetrated a reservoir interval more than 480 m thick. Although shows were encountered during drilling, including isolated gas-bearing sand interpreted from petrophysical logs and oil patches and bitumen staining from core cut within the reservoir section, no hydrocarbon column was found. It was speculated that updip leakage, via sandstone juxtaposed at the south-west-trending fault bounding the prospect, was the reason for the lack of success (Grant et al., 1999).

Deep-water Paleocene sandstone play

The deep-water Paleocene play has a potentially large areal extent and is located within the Judd and Flett sub-basins, on the Westray and Flett highs and in adjacent areas (Figure 150). However, its occurrence over large areas of the north-west part of the Faroe–Shetland Basin is largely speculative. Traps are formed from a combination of structural and stratigraphical components where the Paleocene sandstone reservoir either pinches or shales out updip (e.g. Figure 158). The sands are interbedded with commonly laterally continuous mudstone that provides an effective top seal when not breached by faults.

Several regional stratigraphical schemes have been proposed for the Paleocene succession (e.g. Lamers and Carmichael, 1999; Jowitt et al., 1999; Figure 94) but arguably, the T-sequence scheme of BP is now the most commonly used and is based on the correlation of maximum flooding surfaces that bound the T-sequences (Mitchell et al., 1993; Ebdon et al., 1995; Lamers and Carmichael, 1999). The depositional histories of the Judd and Flett sub-basins are broadly similar. For example, regression resulted in thick sands being developed within the T22–25 and T31–35 sequences in the Judd and the Flett sub-basins beneath the Kettla and Andrew tuffs, respectively (see Chapter 9). However, the older T10 sequence is mudstone-prone in the Flett Sub-basin but in contrast, significant sands are present in the T10 sequence within the Judd Sub-basin. The Judd Sub-basin is mudstone-rich following deposition of the Kettla Tuff whereas in the Flett Sub-basin, significant developments of sands occur in the T38 sequence (post-Andrew Tuff). Within the Faroese part of the Judd Sub-basin the well results generally exhibit a more sand prone Paleocene interval than its UK counterpart (Smallwood, 2005a and b; Woodfin et al., 2005).

According to Lamers and Carmichael (1999), during the Early Paleocene, a series of sand-rich channel complexes, derived from the south-east, consisting of T10, T22–25 and T31–35 sequences were deposited in the Judd Sub-basin in a base of slope submarine fan environment. In Late Paleocene to Early Eocene times, widespread regression took place with associated fluvio-deltaic deposition, commencing as aggradational but becoming in the Late Paleocene, progradational. By T50 times, deposition of the Balder Tuff in a non-marine paralic environment had taken place (Knox et al., 1997). In the Flett Sub-basin, mainly deep marine environments prevailed with sedimentation influenced by instability of the mudstone-prone T10 sequence that resulted in the development of small basins, enhanced by differential loading that filled with T20s and T30s sandstones. The T20s sandstones are generally poorly sorted and argillaceous, whereas the T30s sandstones are clean with a blocky gamma-ray well log response.

The main control on Paleocene reservoir porosity in the Judd and Flett sub-basins is depth of burial. In the Judd Sub-basin, T10 and T20s sandstones are mainly at depths greater than 2000 m below the sea bed and have porosity values in the region of 12% and permeabilities less than 100 mD (Lamers and Carmichael, 1999). In the Flett Sub-basin, T20s sandstones that have been drilled are generally at greater depths with consequently poorer porosity and permeability characteristics i.e. less than 12% and 1 mD. In both sub-basins the T30s sandstones at less than 2000 m below sea bed have porosities of over 20% and permeabilities of several hundred mD. In the Flett Sub-basin, T30s sandstones at greater depths have generally poorer reservoir characteristics. However, some more deeply buried T30s sandstones (T31–T35) exhibit excellent porosity values of 25% and permeabilities of several 100 mD at depths greater than 3000 m below the sea bed. These sands are moderately clean, possibly due to reworking prior to final transport and deposition. The preservation of porosity occurred as a result of early precipitation of chlorite on the sandstone grains preventing later quartz overgrowth. The chlorite formed either as a clay coat prior to burial or as a product of in situ early diagenesis and potentially, is sourced from abundant tuffaceous material deposited with the sandstones (Sullivan et al., 1999; Lamers and Carmichael, 1999).

In the Judd Sub-basin, the top seal to all the main hydrocarbon accumulations is a T35 prograding lowstand wedge that rests on top of the hydrostatically pressured T30s sandstones (Figure 158b). In contrast, in the Flett Sub-basin, T30s sandstones underlying the Andrew Tuff and associated mudstones (T35–T36 sequence) are significantly overpressured (Figure 158a). In well 205/09-1 within the south-west part of the Flett Sub-basin (Figure 2), overpressures of 350 pounds per square inch (psi) have been recorded. To the north-east in well 208/19-1, the overpressure is 550 psi, while over most of the Flett Sub-basin overpressures of 600 psi are recorded (Lamers and Carmichael, 1999). The amount of overpressure does not vary significantly over the Flett Sub-basin area suggesting significant connectivity within the key target sandstones. Much higher overpressures of 1000 to 1400 psi are recorded in the T20 sequence below the widespread T28 mudstone interval.

Eocene basin-floor fan play

At least three interdigitating basin-floor fans form an Eocene fan complex up to 760 m thick that occurs within the central part of the Faroe–Shetland Basin, lying above the Corona High, Guðrun and Corona sub-basins and western edge of the Flett Sub-basin (Figures 103 and 150; Davies et al., 2004).

Five wells have drilled this fan complex and confirm the presence of a dominantly sandy Middle Eocene deposit sealed both laterally and vertically by Eocene to Oligocene mudstone (Figure 159). In three of these wells, 214/04-1, 214/17-1 and 214/26-1, the Eocene was the primary target (Figure 103). Wells 214/17-1 and 214/26-1 were unsuccessful with no hydrocarbon indications, despite the presence of a good reservoir, but well 214/04-1, the Tobermory discovery, successfully tested this play proving 28 m of gas-bearing Eocene sandstone. The hydrocarbon accumulation was imaged on seismic data as a 'flat spot' (DHI) (DTI, 2007b). The well proved a gross reservoir thickness of 178 m with a gross pay of 34 m and with porosities averaging 35%. The internal seismic character of the Eocene fan varies from high amplitude continuous reflectors to a lower amplitude chaotic character. It is uncertain whether this chaotic facies represents primary massive unstructured deposits or a secondary effect caused by dewatering and possible remobilisation of the sediments. This variation in seismic character, suggesting different modes of deposition and possible later disruption to parts of the Eocene fan, may indicate a significant variation in reservoir quality (DTI, 2007b).

Figure 158 Schematic representation of a) the Paleocene play in the Flett Sub-basin b) the Paleocene play within the Judd Sub-basin (see text for details) (slightly modified reproduction from Lamers and Carmichael, 1999 by permission of The Geological Society, London). Note that vertical and horizontal scales are approximate. For location of structural features and wells, see Figures 2 and 7.

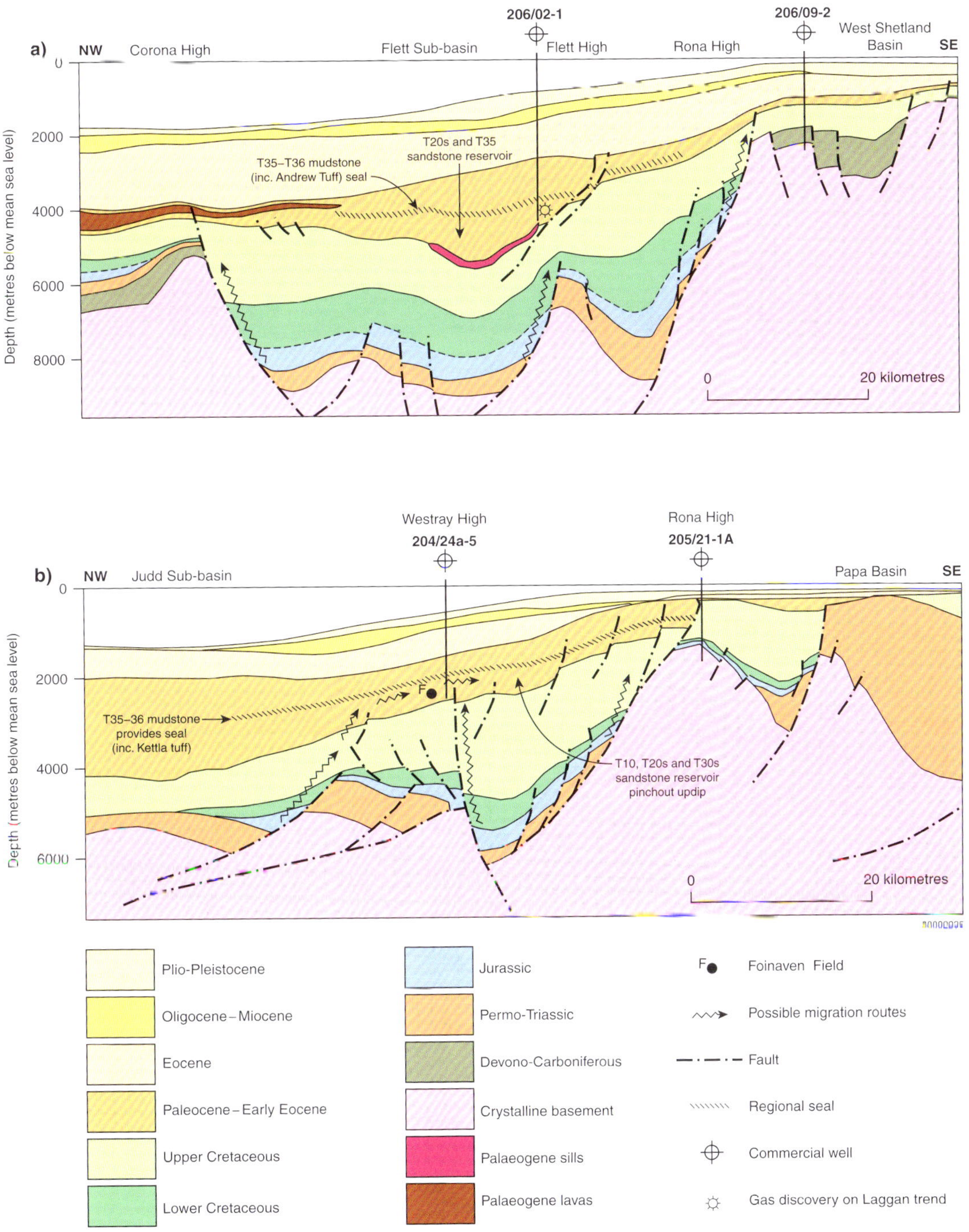

The most likely trapping mechanism in the Eocene play is four-way dip closure, especially along axial crestal ridges. Structural inversion in Cenozoic times has contributed to the formation of the structural closures. For instance, well 214/04-1, drilled on a north-east-trending anticline (Figure 33) was described by Ritchie et al. (2003) as being the result of mainly Miocene compression (see also Davies et al., 2004).

It has been suggested that migration of hydrocarbons in the Tobermory discovery was facilitated by basement faults and also that the edge of the lavas acted as a focus for this migration (e.g. DTI 2007b; Figure 120). Davies et al. (2004), suggested that a late Neogene phase of mainly gas migration, triggered by Middle and Late Miocene compression, charged the structures. Any oil already present in these structures may have been lost through seal leakage or updip spillage, with the result that these Eocene anticlinal traps are more likely to contain gas than oil (Davies et al., 2004). The gas is thought to have migrated from the east, from a deeply buried source within the Flett Sub-basin. Well 213/23-1 tested a four-way dip-closed structure within the Eocene but no hydrocarbons were found at this level. This is thought in retrospect to be due to either a lack of a viable migration route or timing of the charge to trap. Well 213/23-1 lies outwith the area of Neogene compression defined by Davies et al. (2004) and therefore may not have benefited from the reduction in the effectiveness of the underlying seal.

PRODUCING OIL AND GAS FIELDS

There are three producing oil fields in the Faroe–Shetland area, namely the Foinaven Field, the Schiehallion/Loyal development and the Clair Field (Figures 150

Figure 159 Schematic representation of the Eocene fan sandstone play on the Corona High (see text for details) (modified after DTI 2006). Note that vertical and horizontal scales are approximate. For location of the Corona High, see Figure 7.

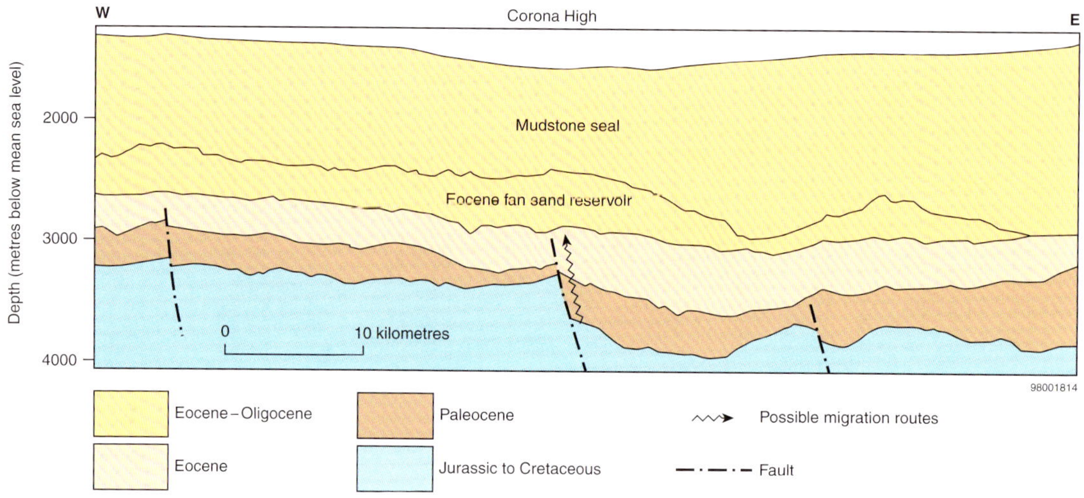

and 151; Coney et al., 1993; Cooper et al., 1999; Leach et al., 1999; DTI, 2007a; 2007). Collectively (to the end of 2005), they have produced 73.6 million tonnes or 539.5 million barrels (MMBBL) of oil (DTI, 2007a).

The Foinaven Field

The Foinaven Field is located in blocks 204/19a, 20a, 24a and 25b overlying the Westray High close to the south-western end of the Faroe–Shetland Basin (Figures 150 and 160a). The accumulation was first drilled in 1990 by well 204/24-1A (Figure 2) which proved 11 m of oil and gas-bearing T30 Paleocene sandstone. However, although the proven presence of hydrocarbons in the Cenozoic succession was very important, the thin sands and lack of mapped closure on a 2D seismic dataset delayed appraisal until 1992 when the drilling of 204/24a-2 encountered 48 m of net oil-bearing sands that flowed 3800 BOPD. In both wells the oil-bearing sands coincide with seismic amplitude anomalies. The results of the first two wells and their sidetracks encouraged the acquisition of a 3D seismic survey and an appraisal programme that was followed by development approval in 1994 (Figure 160; Cooper et al., 1999).

Hydrocarbons have accumulated in a combined structural/stratigraphical trap consisting of a faulted anticline dip-closed to the west and partially to the east, but here with an element of stratigraphical pinchout (Cooper et al., 1999). The Foinaven Field is fault bounded to the south, whereas to the north it is either fault and/or stratigraphically closed (Figure 160). The hydrocarbon accumulation consists of a fully gas saturated oil leg of between 24–27° API with a mixed biogenic–thermogenic gas cap (Carruth, 2003). The Foinaven Field is thought to have been sourced from the south-west, with oil accumulating in the reservoirs from top to bottom by a fill and spill mechanism, resulting in multiple oil–water contacts within the field. The northernmost area of the field has the youngest and most mature oil (Carruth, 2003). The presence of multiple gas caps within the field is thought to be due to the oil being subjected to varying degrees of biodegradation, related to the amount of exposure to formation waters, one of the results of this being a reduction of the oil capacity to hold gas in solution. Biodegradation results in heavier oil in place with generally the shallower and younger reservoirs being less degraded than deeper older reservoirs (Carruth, 2003).

The Foinaven Field has been divided into five areas or 'Panels' (0, 1, 2, 3, 4) whose boundaries are mainly defined by a series of west-north-west-trending faults and by changes in lithology (Figures 160; Cooper et al., 1999). It has been recognised that some of the west-north-west-trending faults that define the boundaries of the Panels may not be sealing, resulting in pressure communication and fluid movement between them. The Foinaven Field has also been divided vertically into three reservoir layers comprising stacked slope and basin floor submarine fan sands, separated by mudstone units (Carruth, 2003). The mudstone is relatively continuous, but may be thin or absent in some areas which could result in communication between the different reservoir layers. The three reservoir layers are defined within the T-sequence stratigraphical scheme as the T31, T32 and T34 sand sequences that are separated by laterally continuous T32 and T34 mudstone, respectively. A reservoir within the T35 sequence provides an additional pool of hydrocarbons that is not in pressure communication with the other T30 sandstone reservoirs (Carruth, 2003).

The mudstone units that separate the different reservoir intervals may affect hydraulic communication between the reservoir sands (Cooper et al., 1999). In the south of the field, the T32 mudstone cannot be resolved seismically and may be absent, allowing the possibility of communication between the T31 and T32 reservoirs. The T34 mudstone, lying between the T32 and T34 reservoirs has been mapped over Panels 1 to 4 but is interpreted to be eroded over Panel 0 in the north of the Field (Figure 160a). Top seal is provided by mudstone that overlies the T34 sand. Each of the T30s reservoir layers consist of a series of submarine fan sands that differ in their petrography reflecting their distinct source areas (Cooper et al., 1999). Reservoir quality improves in the stratigraphically younger strata (T31 to T34) as the amount of ductile rock content (illite, smectite and mica) decreases. The reservoir is generally poorly cemented. Porosity ranges from 20–30% and permeability between 500–2000 mD. Each Panel has been developed separately in order to best exploit differences in reservoir characteristics, disposition of sand and intervening mudstone within the T31, T32 and T34 sand sequences and allow for the presence of a regional aquifer

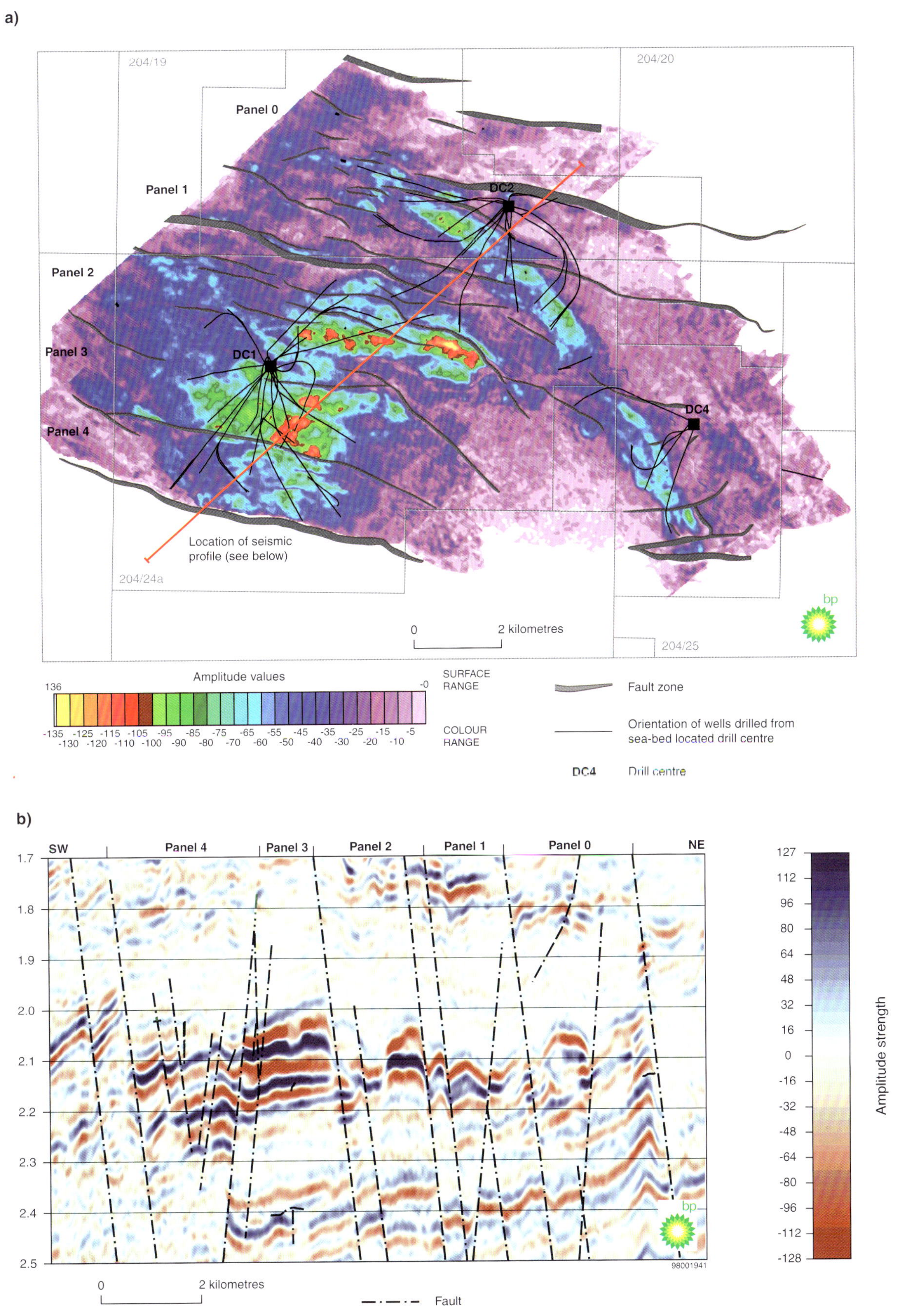

Figure 160 a) Seismic amplitude extraction map (sum of negative amplitudes) over the Foinaven Field b) north-east-trending seismic profile across the Foinaven Field (see text for details) (courtesy of BP). For general location of Foinaven Field, see Figure 150.

and gas caps (Cooper et al., 1999; Carruth, 2003). The majority of wells within the Foinaven Field are drilled from two drill centres (DC) i.e. DC1 and DC2 (Figure 160a), at high angles to vertical, to maximise exposure to the reservoir sands by targeting specific channels within each of the T30 reservoir layers. Water injectors are located where aquifer connection is thought to be poor and unable to provide sufficient pressure, or where a large gas cap requires controlling.

The Foinaven Field is estimated to contain 1070 MMBBL of oil and 440 BCF of gas in place with reserves, based on an average recovery factor of 25%, of 250 MMBBL of oil. This reserves estimate is based on well and 3D seismic information from the first development phase that covered part of the field area, where recovery factors vary between 12.5–50%, and extrapolated to the whole field (Carruth, 2003). Oil production rates at start-up were 40 thousand barrels of oil per day (MBOPD) rising to an average rate of 80 MBOPD. In 2005 the Foinaven Field produced nearly 22 MMBBL of oil (DTI, 2007a).

The Schiehallion and Loyal fields

These fields comprise the Schiehallion development and are located in blocks 204/20a, 25a, 25b and 205/16a and 21b (Figures 150 and 161a; Leach et al., 1999). The Schiehallion development comprises oil accumulations reservoired in Paleocene T30s sandstone. The fields are cut by a series of east-west-trending normal faults that compartmentalise the reservoirs as well as contributing to the overall trap (Figure 161; Leach et al., 1999). The Schiehallion Field was discovered in 1993 by well 204/20-1 and its sidetrack that encountered oil trapped within Paleocene T31 and T34 sandstones (Figure 161a). To the north, the Loyal Field discovery well 204/20-3 drilled in 1994, encountered oil within Paleocene T35 sandstones.

Trapping in the Schiehallion development has both structural and stratigraphical components. A north-westerly 2 to 3° dip over the development area results in dip closure of the reservoirs to the north and west. Updip to the south, the Schiehallion and Loyal fields are both sealed by east-west-trending normal faults that juxtapose the reservoirs against mudstones (Leach et al., 1999). To the east, reservoirs pinchout onto an intrabasinal high. Top and bottom seals are provided by Mid Paleocene Vaila Formation T35 and T28 mudstone respectively (Figure 94). High-resolution 3D seismic data, designed to focus on the reservoir, has enabled three seismic facies to be recognised, namely channelised, parallel and chaotic, and this, integrated with core and well log information, has been used in formulating the development strategy. The channelised seismic facies, interpreted as scours on a submarine slope that have subsequently been back-filled with sandstone dominated material, are thought to form the main part of the reservoir at all stratigraphical levels. Seismic attribute mapping delineates hydrocarbon-bearing successions and images the channelised nature of the reservoirs (Leach et al., 1999; Figure 161a).

The main reservoir sandstones are generally grain-supported with minor ductile and authigenic (most commonly calcite) components. Tightly cemented calcite intervals have been described from core but are thought to be laterally restricted. Reservoir quality is excellent with average porosity and permeability values for the T30 reservoirs in Schiehallion and Loyal Fields being 27–29% and 800–1600 mD (horizontal air permeability), respectively (Leach et al., 1999).

The Schiehallion Field has been divided into four segments whose boundaries are defined by the cross-cutting east-west-trending faults (Figure 161). Fault throw tends to decrease towards the east where displacement has not been great enough to create a barrier to migration, whereas to the west, greater displacement may have created a barrier to flow (Leach et al., 1999). Observations following an extended well test (EWT) suggest connectivity between segments is limited, at least within a production timescale, however, over a longer geological timescale the faults are not thought to form a barrier to fluid flow (Leach et al., 1999). In the Schiehallion Field, the reservoir is situated within sandstones of the T31 sequence, which are the most important within the development, and also the T34 sequence (Figure 161b; Leach et al., 1999). The connectivity between mapped channel complexes within the T30 reservoirs, crucial when planning the number of producing wells required to best exploit the field, was tested by drilling a horizontal appraisal well (204/20-5). An EWT carried out on this well, that drilled 1697 m of T31 reservoir sandstone, demonstrated good connectivity between these different channel complexes (Leach et al., 1999).

The Loyal Field is divided into a 'Core' area consisting of oil reservoired in three Paleocene T35 sand bodies with an oil–water contact at 2397 m TVDSS, and a separate segment to the south (Leach et al., 1999). This is located on an east–west-trending fault block where oil is reservoired in T34 sandstones with an inferred (from reservoir pressure data) oil–water contact at 2160 m TVDSS (Leach et al., 1999).

Oil from the Schiehallion and Loyal fields ranges in specific gravity between 22–28° API and is partially biodegraded, with the degree of degradation increasing with depth, probably linked with its proximity to the oil–water contact. The overall Schiehallion development has mean recoverable reserves of 425 MMBBL of oil (Schiehallion has 340 MMBBL and Loyal has 85 MMBBL) (Leach et al., 1999). In 2005, the Schiehallion and Loyal fields produced just over 25 and 3.5 MMBBL of oil respectively (DTI, 2007a).

The Clair Field

The Clair Field covers an area of approximately 200 square km and is located mainly in blocks 206/07, 08, 09, 12 and 13 (Figures 150 and 162a). The field may be broadly divided into two segments, the 'Ridge' representing the Lewisian-cored, north-east-trending Rona High and to the south-east, the 'Rollover' representing a terrace comprising a thick Devono-Carboniferous red bed succession (Coney et al., 1993; Figure 162b). Later studies further divide the Rollover into the 'Core', 'Graben' and 'Horst' segments (Barr et al., 2007; Figure 162). The field was discovered in 1977 by BP well 206/08-1A (Figure 2), drilled to test an anticlinal trap prognosed to contain Jurassic/Cretaceous sands. However, the well proved a 700 m succession of Carboniferous to Devonian clastic rocks of which the upper 568 m was oil-bearing beneath thick Upper Cretaceous mudstone (Coney et al., 1993). The well flowed at an average rate of 1502 BOPD. The field has undergone a long period of appraisal mainly due to problems

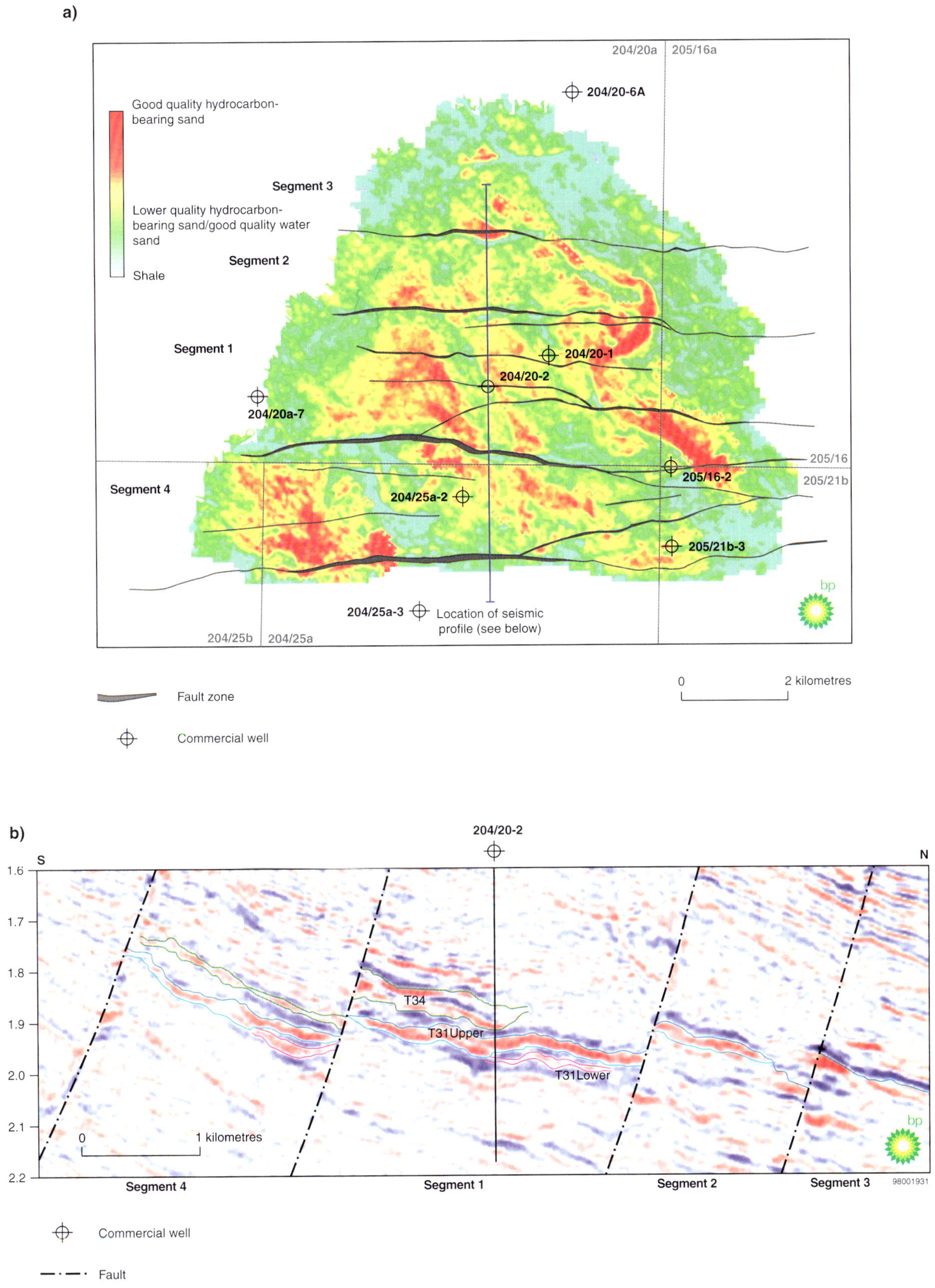

Figure 161 a) Seismic amplitude extraction map (sum of negative amplitudes) over the Schiehallion Field b) north-trending seismic profile across the Schiehallion Field (see text for details) (courtesy of BP). For general location of Schiehallion Field and wells, see Figures 2 and 150.

associated with extraction of its heavy oil accumulation from a low average porosity reservoir with highly variable matrix and fracture permeability. Appraisal drilling began with well 206/07-1 (Figure 2) that tested 963 BOPD from fractured Lewisian basement. Further drilling by nine wells before the end of 1980 delineated the field and identified a hydrocarbon column of over 800 m, two separate gas caps and oil in place of several

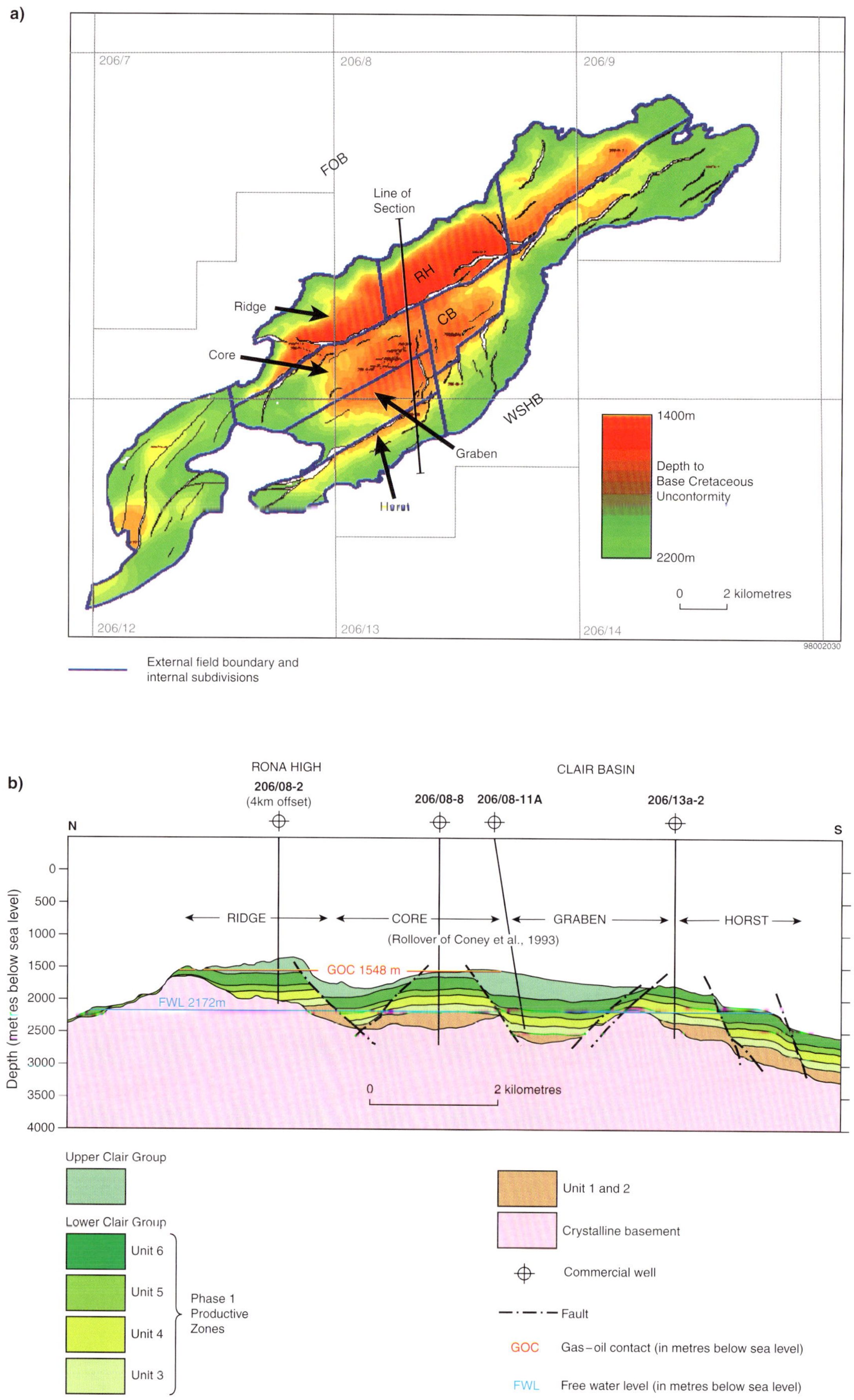

Figure 162 a) Depth to Base Cretaceous Unconformity over the Clair Field b) geological cross-section across the Clair Field (see text for details) (slightly modified from Barr et al., 2007 by permission of the Geological Society of London, courtesy of BP). For abbreviations and general location of Clair Field and wells, see Figures 2, 7 and 150.

billion barrels (Coney et al., 1993). However, all drill stem tests (DSTs) carried out in these wells failed to reach the flows measured in the discovery well 206/08-1A. The appraisal drilling highlighted the presence of a reservoir whose quality depended upon the distribution and permeability of a pervasive system of fractures. Technical problems during appraisal drilling hampered attempts to stimulate flow and in addition, seismic imaging beneath the Cretaceous has proved difficult (Coney et al., 1993).

By the end of the 1980s, a collaborative approach by the Clair Field partners led to a joint programme of work which included acquisition of a 3D survey over the central part of the discovery and the initiation of a series of engineering and sub-surface studies aimed at resolving some of the problems that beset the field (Coney et al., 1993). Studies were initiated in two different parts of the field. In the Ridge area, studies focused on the fractures identified within the Lewisian basement while in the Rollover (Coney et al., 1993; Figure 162), an understanding of appraisal test results in relation to the geology of the Devono-Carboniferous red beds was attempted.

Ridge area

Analysis of core taken from the first appraisal well 206/07-1 on the Ridge section of the field (Figure 162) showed that the fractures within the Lewisian basement were subvertical, suggesting that the best way of optimising extraction of the oil was through the drilling of a horizontal well (Coney et al., 1993). Studies of fracture systems affecting Devonian Old Red Sandstone resting on Moinian metamorphic rocks on the north coast of Scotland and Orkney led to the identification of areas of higher concentration of fractures, fracture corridors and fault zones that continued, albeit with decreasing frequency, up into the overlying Devonian strata. Potentially, these zones should have better permeability and could be exploited in the attempt to extract economic amounts of oil from the field. Coney et al. (1993) identified two main fault trends in the basement, a north-north-east-trend truncating the Rona High and a north-east- to east-north-east-trend running along the high. In 1992, horizontal well 206/07a-2 was sited to drill fracture zones predicted by the Ridge study. However, the well encountered few open fractures as most were filled with calcite. Although the well did not reach the modelled flow rates predicted in the study, it did prove the existence of fracture corridors and confirmed that oil could drain from the Devono-Carboniferous red beds (Coney et al., 1993).

Rollover area

Studies on the Rollover part of the field (Figure 162) focused on the Devono-Carboniferous continental red bed succession. Allen and Mange-Rajetzky (1992) developed a lithostratigraphical scheme for the Clair Field sediments dividing the fluviatile, aeolian and lacustrine facies associations into 10 units (I to X) (Figure 50). Reservoir quality in the field is best defined by net reservoir ratio and permeability, with the net reservoir classified as having porosity greater than 9% and V_{SHALE}, the volume of shale or mud present in sandstone, of less than 35% (Coney et al., 1993). Three reservoir units (III, V, VI) were identified as having better net reservoir ratios and permeability than surrounding units. Lithofacies modelling, fracture and fault analysis were carried out and data from this used as inputs to a series of simulation models to attempt to predict flow rates from the Devono-Carboniferous succession in the Rollover sector of the field (Coney et al., 1993). The simulation models suggested that a vertical well, located optimally and with hydraulic fracturing, would improve significantly on previous flow rates in this part of the field. A horizontal well could also improve on earlier flow rates, but only if it encountered a natural open fracture system, none of which had yet been positively identified. A 3D seismic survey acquired during this time improved greatly on imaging of faults and seismic packages within the reservoir and aided in the positioning of the vertical well in the Rollover area of the field (Coney et al., 1993). In 1991, well 206/08-8 was drilled with encouraging results (Figure 162b). Four DSTs were run with one of these achieving a rate of 3100 BOPD, the highest so far since the discovery well and without stimulation and any drop in rate or pressure throughout a six day test (Coney et al., 1993). Similar rates were obtained on another interval after fracturing. Results showed that production is dependant on a combination of matrix quality and the presence of open fractures. Although the dominant trend of fractures is north-east, open fractures are most common in those with north-north-west and west-north-west trends (Figure 53). The north-north-west trend matches a Late Cretaceous fault trend mapped on 3D seismic data whose acquisition formed part of this early 1990s study (Coney et al., 1993).

In the first year of production in 2005, and as part of a phased development, the Clair Field produced just over 5 MMBBL of oil. This development phase focuses on units III-VI of the Lower Clair Group (Figure 50; Barr et al., 2007) in the Core, Graben and Horst parts of the Field, that includes the Rollover of earlier studies (Figure 162b; Coney et al., 1993) and is estimated to contain 30% of the oil in place (Barr et al., 2007). Stratigraphically, only the Lower Clair Group is included in this phase of the development, as a higher concentration of smectite in the Upper Clair Group degrades reservoir quality at this level (Barr et al., 2007).

SIGNIFICANT UNDEVELOPED DISCOVERIES

Within the Faroe–Shetland report area, Cambo, Laggan, Laxford, Rosebank/Lochnagar, Solan, Strathmore, Suilven, Tobermory and Victory represent some of the more important discoveries (Figures 150 and 151). In addition, Alligin, Arkle and Cuillin are significant discoveries in the Judd Sub-basin adjacent to the Foinaven Field (see Loizou, 2005; Loizou et al., 2006). Of all these undeveloped discoveries, only Cambo, Laggan, Laxford, Rosebank/Lochnagar, Solan, Strathmore and Victory have been described in varying detail below, as there is little or no publicly available data regarding the others. It should be noted that significant additional information regarding the Victory and Rosebank/Lochnagar, Laggan and Cambo discoveries/fields has been provided by Chevron, Total and Hess respectively.

Cambo

Towards the south-west end of the Corona High, the Cambo discovery well 204/10-1 (Figures 2 and 150) encountered oil and gas within a reservoir consisting of T45 Hildasay Member deltaic sands (Flett Formation) occurring in the post-basalt section. There is still considerable uncertainty as to the volume of this discovery and further appraisal is planned.

Laggan

The Laggan Field was discovered in 1986 by well 206/01-2, and is a Paleocene gas discovery in water depths of 600 m (Figures 2, 150 and 163a). The field is located towards the south-western end of the north-east-trending Flett High, dipping north-westwards into the Flett Sub-basin. Trapping is facilitated by a combination of structural and stratigraphical elements, with updip truncation in the hanging wall of an arcuate, but broadly north-east-trending major fault and a combination of lateral reservoir pinchout and faulting along the strike of the field (Figure 163a and b). Overlying mudstone forms the top seal to the accumulation (Figure 163c). The crest of the structure is at 3500 m below sea level and the gas–water contact (GWC) is at 3909 m below mean sea level.

The Laggan Field reservoir lies within the Late Paleocene T35 sand interval and has a gross thickness of less than 70 m that includes two field-wide interbedded silty mudstone units. The mudstone units divide the reservoir into three sand intervals (A, B and C) resulting in a net reservoir thickness of up to 50 m (Figure 163c). The reservoir is poorly resolved on seismic data (Figure 163b) but consists of a sand-rich turbidite lobe facies, with its architecture controlled by offset stacking (a sequential process whereby the reservoir fairway builds outwards as its adjacent depocentre is filled and becomes the positive topography with a new depocentre forming next to it) and normal faults. These gas-charged sands may be seen as high amplitude reflections on seismic data (Figure 163b). Reservoir properties are good due in part to the presence of chlorite, that tends to prevent formation of pore filling cements (average porosities of 22–27%, permeabilities of 30–300 mD). The reservoir sands have a high net to gross ratio with a field-wide average of approximately 95%, excluding the two ubiquitous interbedded silty mudstone units (Figure 163c). The occurrence of calcite cemented zones, which are of limited lateral extent, make poorer reservoir intervals. Charging of the reservoir is likely to have taken place from a deeply buried Upper Jurassic source within the Flett Sub-basin.

To date, a total of four exploration and appraisal wells have been drilled in the field area (Figure 163a). Two DSTs performed in well 206/01-2 recorded a maximum rate of 25 MMSCFD of gas and 500 BBLS/D of condensate. A DST carried out on well 206/01a-4AZ flowed at a maximum rate of 37.8 MMSCFD of gas and 800 BBLS/D of condensate from the 'B' sands (Figure 163c).

Laxford

The Laxford prospect located in the Foula Sub-basin, was drilled by well 214/30-1 (Figures 2 and 150) and tested gas from a T34 Paleocene sandstone (Loizou et al., 2006).

Rosebank/Lochnagar

The Rosebank/Lochnagar discovery well 213/27-1Z, drilled on the crest of a large anticlinal complex on the Corona High, encountered oil and gas accumulations in Paleocene sediments approximately 100 m thick within a several hundred metre thick volcanic sequence of Paleocene age, and a pre-Cretaceous succession (Loizou, 2005; Helland-Hansen, 2006; 2009; Figures 2 and 150). Total net pay is 52 m and oil gravity ranges between 27°–36° API (Loizou, 2005; Helland-Hansen, 2006).

The Rosebank discovery comprises three generally north-east-trending anticlinal closures entitled, South, Main and North (Helland-Hansen, 2006; 2009). Rosebank Main was tested by well 213/27-1Z and encountered at least three reservoir horizons with separate oil and gas accumulations within the Colsay Sandstone Member of the Flett Formation (Figure 94). These reservoir units are interpreted to have been formed within shallow marine to delta top depositional environments and are considered to represent the distal equivalents of the north-west-prograding 'Flett delta'. The sandstone units are interbedded with north-westerly derived lavas, volcaniclastics and possibly hyaloclastites and intruding sills. There are still some uncertainties regarding the Rosebank discovery including hydrocarbon contacts, the lateral extent of the reservoir sands, their connectivity, permeability and vertical net to gross ratios (Helland-Hansen, 2006; 2009), that were appraised through a 3 well, 1 sidetrack and 1 DST program in 2006–2007. Further exploration wells of the North and South structure are in planning.

Strathmore and Solan

The Strathmore Field straddles blocks 204/30 and 205/26a within the East Solan Basin while the Solan Field is restricted to Block 205/26a (Figure 150). They comprise the partly overlapping Lower Triassic Strathmore and Upper Jurassic Solan discoveries, respectively. The fields are separated from the Faroe–Shetland Basin to the north by the Rona and Judd highs and from the Papa and West Shetland basins to the east by the Otter Bank Fault (Figures 7 and 150).

The Triassic reservoir of the Strathmore Field was discovered in 1990 by well 205/26a-3 (Figure 39) that was drilled on a structural closure mapped at base Cretaceous level with a Middle Jurassic sandstone target. However, the Middle Jurassic was absent and instead Upper Jurassic Kimmeridge Clay Formation mudstone was proven, resting unconformably on Triassic sandstone that contained a 204 m oil column (Herries et al., 1999). Appraisal wells 204/30a-2 and 205/26a-4 drilled in 1991 (Figure 2) indicated that the oil was reservoired in the south-easterly-dipping, three-way dip-closed Lower Triassic Otter Bank Sandstone (Figures 56 and 155), that was sealed by the overlying impermeable Foula Formation sandstone. The Upper Jurassic Kimmeridge Clay Formation mudstone provides the top seal where the Otter Bank Sandstone Formation reservoir subcrops the Jurassic unconformity. The reservoir is bottom-sealed by the Lower Triassic Otter Bank Shale Formation.

Appraisal well 205/26a-4 also became the discovery well of the Solan Field, proving 15 m of oil-filled Upper Jurassic Solan Sandstone Member totally enclosed within Kimmeridge Clay Formation mudstone that flowed at 2080 BOPD (Herries et al., 1999). Downdip, well 205/26a-5, an appraisal to the Solan discovery, identified two separate sand bodies with a combined thick-

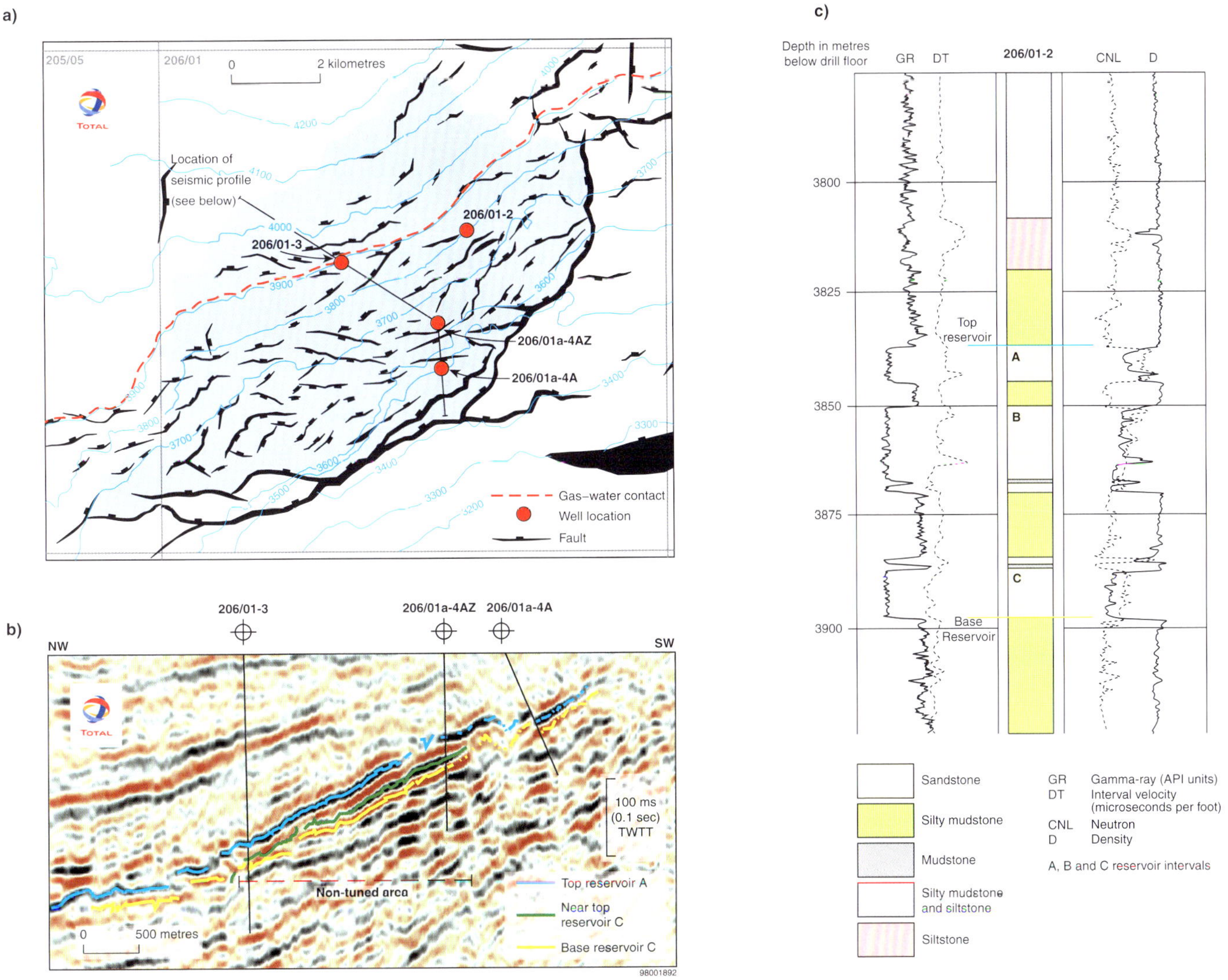

Figure 163 a) Depth to Top Laggan 'A' sandstone reservoir. Area shaded blue represents an amplitude anomaly which may be partly related to the limit of the reservoir sands b) seismic profile across the Laggan Field in the Foula Sub-basin c) lithology and well log response from Laggan Field discovery well 206/01-2 (courtesy of Total). For location of Laggan Field and wells, see Figures 2 and 150.

ness of 26 m. However, the well was below the OWC of the main Solan Field and a sidetrack was drilled in an updip direction to further test the oil leg. Well 205/26a-5Z (Figure 2) also proved two oil-bearing sandstones that are in pressure communication with the 205/26a-4 discovery, but flowed at a considerably lower rate (204 BOPD) than the discovery well (Herries et al., 1999). Trapping in the Solan Field is facilitated by the updip south-westerly pinchout of the upper and lower sandstone bodies that are totally enclosed and sealed by the Kimmeridge Clay Formation mudstone (Figure 156b).

The Otter Bank Sandstone Formation reservoir is divided into three sedimentary units that reflect its primary depositional properties; overall it exhibits a change from a fluvially dominated to a more arid environment with increasing frequency of interbedded aeolian and sabkha deposits (Herries et al., 1999). The Solan Sandstone Member basin-floor fan reservoir consists of two units, the lower unit being thinner and less extensive (Figure 68).

The precise location of the oil–water contact in the fields is difficult to define. In the Strathmore reservoir, recognition of oil–water boundaries is equivocal due to the presence of permeability barriers and the difficulty of interpreting formation pressure data, made harder by the compositional heterogeneity of the oil (Herries et al., 1999). In well 205/26a-4, the oil–water contact in the Lower Triassic reservoir is recognised at 2682 m below drill floor. However, in well 205/26a-3, a downward increase in asphaltene culminates in a tar mat whose base occurs higher at 2640 m. Herries et al. (1999) interpret the lack of oil beneath the tar mat in this well as being due to the presence of a very low permeability interval and with supporting RFT data, place the OWC at the deeper position.

In the Upper Jurassic Solan Sandstone Member reservoir only one, deeper, OWC has been penetrated at 2750 m below drill floor and this is thought to be a separate accumulation. Pressure data suggests an OWC at 2682 m, the same as that found in the Triassic reservoir (Herries et al., 1999). However, although data suggests a common OWC between accumulations this may be coincidence and not be a true reflection of oil distribution in the two fields (Herries et al., 1999).

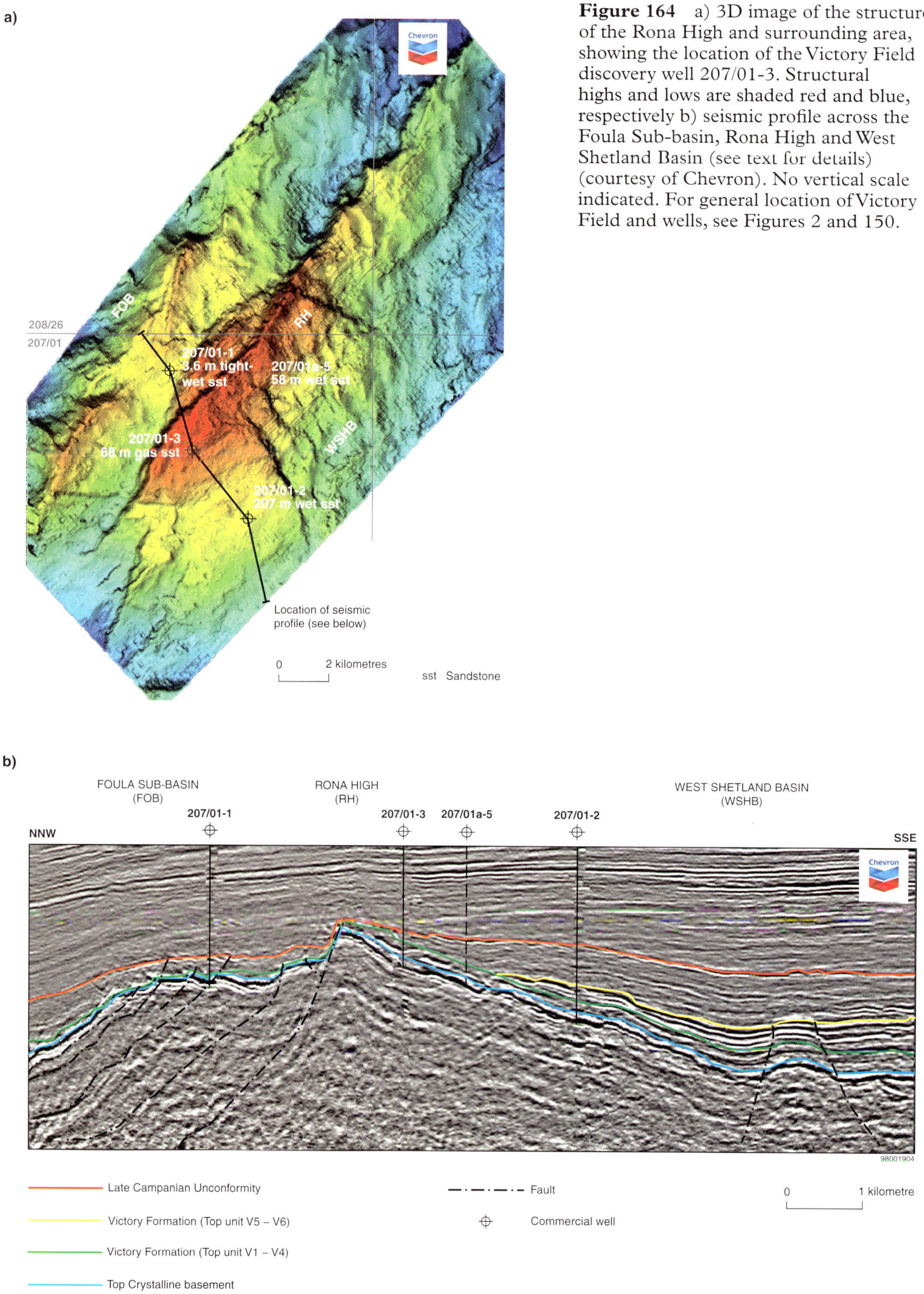

Figure 164 a) 3D image of the structure of the Rona High and surrounding area, showing the location of the Victory Field discovery well 207/01-3. Structural highs and lows are shaded red and blue, respectively b) seismic profile across the Foula Sub-basin, Rona High and West Shetland Basin (see text for details) (courtesy of Chevron). No vertical scale indicated. For general location of Victory Field and wells, see Figures 2 and 150.

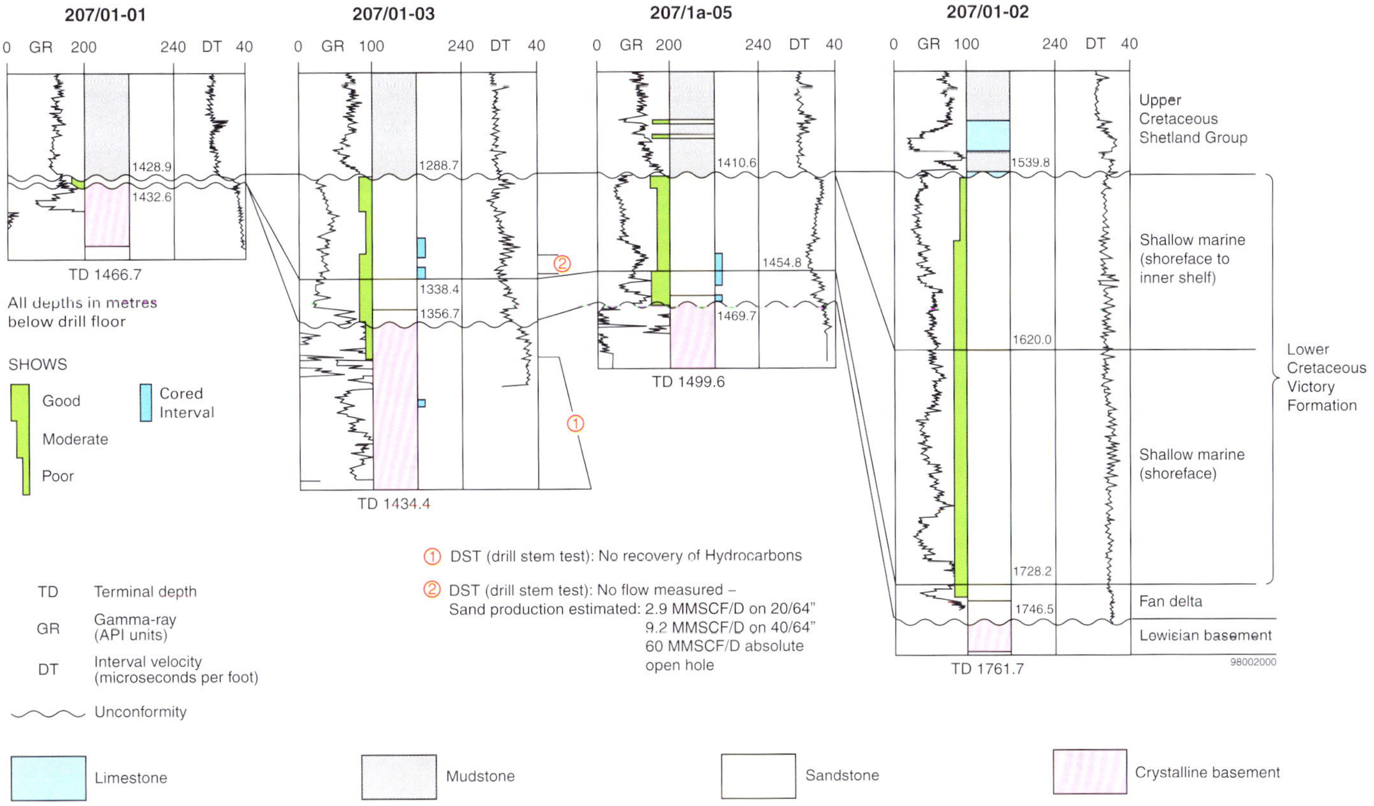

Figure 165 Well correlation across the Victory Field (courtesy of Chevron). For location of wells, see Figures 2 and 164.

Oil composition varies considerably in both the Solan and Strathmore fields. The oil becomes heavier downdip, varying between 26.7° API at 2423 m and 23.7° API at 2606 m below drill floor. It is thought that two phases of hydrocarbon charging best explains the heterogeneous nature of the Solan and Strathmore oils with a lighter, more mature, later charge displacing the early heavier oil. Oil in place for the fields amounts to 278 x 10^6 BBLS (78 x 10^6 BBLS for Solan and 200 x 10^6 BBLS for Strathmore) but recoverable reserves of only 59 x 10^6 BBLS (23 x 10^6 BBLS for Solan and 36 x 10^6 BBLS for Strathmore), reflecting low permeabilites for the Lower Triassic Strathmore Field reservoir in particular (Herries et al., 1999).

Victory

The Victory Field is situated toward the north-eastern end of the north-east-trending Rona High in Block 207/01a and was discovered in 1977 by well 207/01-3, the third of three back-to-back exploration wells (Figures 150 and 164). There are two dominant fault trends mapped within the field, a north-east trend (Figure 164a) that is offset by a north-west trend (Goodchild et al., 1999). The latter trend is consistent with the orientation of the transfer faults mapped by Rumph et al. (1993) (Figure 7). The first exploration well 207/01-1 drilled by Texaco, was primarily focused on a Jurassic target, but encountered only 3.7 m of tight Lower Cretaceous sand with oil shows, directly overlying Lewisian basement (Figures 164 and 165). The second well, 207/01-2 was drilled downdip on the structure and encountered 207 m of reservoir quality Lower Cretaceous sandstone with poor to moderate oil shows (Goodchild et al., 1999). The discovery well 207/01-3, drilled closer to the crest of the structure, encountered dry gas in 68 m of moderate to good reservoir quality Lower Cretaceous sandstone. In 1990, well 207/01a-4Z unsuccessfully tested a separate structure adjacent to the Victory Field. In 1996, following acquisition of a 3D seismic dataset, well 207/01a-5 was drilled on a possible oil leg below the proven gas column but the reservoir sands were water-bearing with only residual oil shows (Figures 164 and 165).

In the Victory Field, gas is trapped in a south-easterly-dipping tilted fault block that is dip-closed to the north-east, south-east and south-west, bounded updip by a north-east-trending normal fault and top-sealed by Upper Cretaceous mudstone (Figure 164b). The structure is not filled to its spill point, structural closure is mapped at 1600 m subsea and the GWC is estimated at between 1350 to 1359 m subsea. The fault trends observed at the regional scale are also seen in fractures identified in core and log data. Fractures in the Lewisian basement mainly show a north-west trend and tend to be open, whereas the north-east-trending fractures tend to be cemented by quartz. The reservoir occurs within the Victory Formation and has been divided into four units (V1 to V4) (Figure 164b) that vary in reservoir quality depending on the mineralogy of the sandstone and depositional environment (Goodchild et al., 1999).

Hydrocarbons reservoired in the Victory Field are thought to represent several charges of oil and gas and, at the present day, consist of both non-moveable oil and free gas (Goodchild et al., 1999). Analysis of the oil that occurs both in inclusions within quartz overgrowths and as staining seen in both cores and cuttings suggests a significant early charge of oil, the remnants of which are now heavily biodegraded, and a late charge consisting of live oil preferentially trapped in more porous zones of the reservoir. The initial Late Cretaceous charge of oil has been largely lost through leakage and biodegradation during subsequent uplift and influx of fresh water (Goodchild et al., 1999). The gas in the Victory Field is likely to have a thermogenic origin, with methane levels ranging from 84% up to 98%, and may represent a late stage (Cenozoic) charging episode (Goodchild et al., 1999). Goodchild et al. (1999) estimated gas-in-place of between 250 to 350 x 10^9 standard cubic feet (SCF).

References

ABRAHAM, D, and RITCHIE, J D. 1991. The Darwin Complex, a Tertiary igneous centre in the Northern Rockall Trough. *Scottish Journal of Geology*, Vol. 27, 113–126.

ABRANTES, F, BAAS, J, HAFLIDASON, H, RASMUSSEN, T L, KLITGAARD, D, LONCARIC, N, and GASPAR, L. 1998. Sediment fluxes along the northeastern European Margin: inferring hydrological changes between 20 and 8 kyr. *Marine Geology*, Vol. 152, 7–23.

ABREU, V S, and ANDERSON, J B. 1998. Glacial eustacy during the Cenozoic: sequence stratigraphic implications. *American Association of Petroleum Geologists Bulletin*, Vol. 82, 1385–1400.

AKHURST, M C, STOW, D A V, and STOKER, M S. 2002. Late Quaternary glacigenic contourite, debris flow and turbidite process interaction in the Faroe–Shetland Channel, NW European Continental Margin. 73–84 *in* Deep-Water Contourite Systems: Modern Drifts and Ancient Series, Seismic and Sedimentary Characteristics. STOW, D A V, PUDSEY, C J, HOWE, J A, FAUGÈRES, J-C, and VIANA, A R (editors). *Geological Society of London Memoir*, No. 22.

ALLEN, N H. 1983. Recent temperate carbonate deposits on the continental shelf north and west of Scotland: distribution, sedimentology and reserves. Unpublished PhD thesis, University of Strathclyde.

ALLEN, P A, and MANGE-RAJETZKY, M A. 1992. Devonian–Carboniferous sedimentary evolution of the Clair area, offshore north-western UK: impact of changing provenance. *Marine and Petroleum Geology*, Vol. 9, 29–52.

ANDERSEN, M S. 1988. Late Cretaceous and early Tertiary extension and volcanism around the Faroe Islands. 115–122 *in* Early Tertiary volcanism and the opening of the NE Atlantic. MORTON, A C, and PARSON, L M (editors). *Geological Society of London Special Publication*, No. 39.

ANDERSEN, M S, and BOLDREEL, L O. 1995. Effect of Eocene–Miocene compression structures on bottom-water currents in the Faroe–Rockall area. 141–143 *in* The tectonics, sedimentation and palaeoceanography of the North Atlantic region. SCRUTTON, R A, STOKER, M S, SHIMMIELD, G B, and TUDHOPE, A W (editors). *Geological Society of London Special Publication*, No. 90.

ANDERSEN, M S, NIELSEN, T, SØRENSEN, A B, BOLDREEL, L O, and KUIJPERS, A. 2000. Cenozoic sediment distribution and tectonic movements in the Faroe region. *Global and Planetary Changes*, No. 24, 239–259.

ANDERSEN, M S, SØRENSEN, A B, BOLDREEL, L O, and NIELSEN, T. 2002. Cenozoic evolution of the Faroe Platform: comparing denudation and deposition. 291–311 *in* Exhumation of the North Atlantic Margin: timing, mechanisms and implications for petroleum exploration. DORÉ, A G, CARTWRIGHT, J A, STOKER, M S, TURNER, J P, and WHITE, N. (editors). *Geological Society of London Special Publication*, No. 196.

ANDERSON, T B, PARNELL, J, and RUFFELL, A H. 1995. Influence of basement on the geometry of Permo-Triassic basins in the north-west British Isles. 103–122 *in* Permian and Triassic rifting in Northwest Europe. BOLDY, S A R (editor). *Geological Society of London Special Publication*, No. 91.

ANDRESEN, C S, BJÖRCK, S , RUNDGREN, M, CONLEY, D J, and JESSEN, C. 2006. Rapid Holocene climate changes in the North Atlantic: evidence from lake sediments from the Faroe Islands. *Boreas*, Vol. 35, 23–34.

ANDREWS, I J. 1985. The deep structure of the Moine Thrust, south-west of Shetland. *Scottish Journal of Geology*, Vol. 21, 213–217.

ARCHER, S G, BERGMAN, S C, ILIFFE, J, MURPHY, C M, and THORNTON, M. 2005. Palaeogene igneous rocks reveal new insights into the geodynamic evolution and petroleum potential of the NE Rockall Trough. *Basin Research*, Vol. 17, 171–201.

ASTIN, T R. 1982. The Devonian geology of the Walls peninsula, Shetland. Unpublished PhD thesis, University of Cambridge.

AUBRY, M-P, BERGGREN, W A, VAN COUVERING, J, MCGOWRAN, B, PILLANS, B, and HILGEN, F. 2005. Quaternary: status, rank, definition, survival. *Episodes*, Vol. 28, No. 2, 118–120

AUSTIN, B. 2004. Integrated use of 3D seismic in field development, engineering and drilling: examples from the shallow section. 279–296 *in* 3D seismic technology: application to the exploration of sedimentary basins. DAVIES, R M, CARTWRIGHT, J A, STEWART, S A, LAPPIN, M, and UNDERHILL, J R (editors). *Geological Society of London Memoir*, No. 29.

BABA, S. 1999. Evolution of the Lewisian Complex in South Harris, north-west Scotland, and its relationship to the North Atlantic craton in the Palaeoproterozoic (2.0 Ga). *Journal of Geosciences, Osaka University*, Vol. 42, 115–125.

BADHAM, J P N. 1982. Strike-slip orogens — an explanation for the Hercynides. *Journal of the Geological Society of London*, Vol. 139, 493–504.

BAILEY, N J L, WALKO, P, and SAUER, M J. 1987. Geochemistry and source rock potential of the west of Shetlands. 711–721 in *Petroleum geology of north west Europe, proceedings of the 3rd conference*. BROOKS, J, and GLENNIE, K W (editors). (London: Graham and Trotman.)

BAILEY, N J L, BURWOOD, R, and HARRIMAN, G E. 1990. Application of pyrolysate carbon isotope and biomarker technology to organofacies definition and oil correlation problems in North Sea basins. *Organic Geochemistry*, Vol. 16, 1157–1172.

BALLANTYNE, C K, MCCARROLL, D, NESJE, A, DAHL, S O, STONE, J O, and FIFIELD, L K. 1998. High-resolution reconstruction of the last ice sheet in NW Scotland. *Terra Nova*, Vol. 10, 63–67.

Bamford, D, Nunn, K, Prodehl, C, and Jacob, B. 1978. LISPB-IV. Crustal structure of northern Britain. *Geophysical Journal of the Royal Astronomical Society*, Vol. 54, 43–60.

Baptie, B. (editor). 2005. UK earthquake monitoring 2004/2005. *British Geological Survey Internal Report*, No. IR/05/088.

Barr, D, Savoray, K E, Fowler, S R, Arman, K, and McGarrity, J P. 2007. Pre-development fracture modelling in the Clair field, west of Shetland. 199–219 *in* Fractured Reservoirs. Lonergan, L, Jolly, R J H, Sanderson, D J, and Rawnsley, K (editors). *Geological Society of London Special Publication*, No. 270.

Barton, A J, and White, R S. 1997. Volcanism on the Rockall continental margin. *Journal of the Geological Society of London*, Vol. 154, 531–536.

Barton, P J. 1992. LISPB revisited: a new look under the Caledonides of northern Britain. *Geophysical Journal International*, Vol. 110, 371–391.

Belderson, R H, Kenyon, N H, and Wilson, J B. 1973. Iceberg plough marks in the northeast Atlantic. *Palaeogeography, Palaeoclimatology, Palaeoecology*, Vol. 13, 215–224.

Bell, B R, and Butcher, H. 2002. On the emplacement of sill complexes: evidence from the Faroe–Shetland Basin. 307–329 *in* The North Atlantic Igneous Province: stratigraphy, tectonic, volcanic and magmatic processes. Jolley, D W, and Bell, B R (editors). *Geological Society of London Special Publication*, No. 197.

Bell, N, and Smith, J. 1999. Coral growing on North Sea oil rigs. *Nature*, Vol. 402, 601.

Berggren, W A, and Schnitker, D. 1983. Cenozoic marine environments in the North Atlantic and Norwegian–Greenland Sea. 495–548 in *Structure and Development of the Greenland–Scotland Ridge: new methods and concepts*. Bott, M H P , Saxov, S, Talwani, M, and Thiede, J (editors). (New York: Plenum Press.)

Berggren, W A, Kent, D V, Swisher, III, C C, and Aubry, M-P. 1995a. A revised Cenozoic geochronology and chronostratigraphy. 129–212 *in* Geochronology, time scales and stratigraphic correlation: framework for an historical geology. Berggren, W A, Kent, D V, Aubry, M-P, and Hardenbol, J (editors). *Society for Economic Palaeontologists and Mineralogists Special Publication*, No. 54.

Berggren, W A, Hilgen, F J, Langereis, C G, Kent, D V, Obradovich, J D, Raffi, I, Raymo, M E, and Shackleton, N J. 1995b. Late Neogene chronology: new perspectives in high-resolution stratigraphy. *Geological Society of America Bulletin*, Vol.107, 1272–1287.

Berndt, C, Planke, S, Alvestad, E, Tsikalas, F, and Rasmussen, R. 2001. Seismic volcanostratigraphy of the Norwegian Margin: constraints on tectono-magmatic break-up processes. *Journal of the Geological Society of London*, Vol. 158, 413–426.

Bertrand, P, Lallier-Vergès, E, Martinez, L, Pradier, B, Tremblay, P, Huc, A, Jouhannel, R, and Tricart, J P. 1990. Examples of spatial relationships between organic matter and mineral groundmass in the microstructure of the organic-rich Dorset Formation rocks, Great Britain. *Organic Geochemistry*, Vol. 16, 661–675.

Bett, B J. 1997. RRS *Charles Darwin* cruise 101C Leg 2, 14 Jul–20 Aug 1996. Atlantic Margin Environmental Survey: seabed survey of the shelf edge and slope west of Shetland. *Southampton Oceanography Centre Cruise Report*, No. 7.

Bett, B J. 1999. RRS *Charles Darwin* cruise 112C Leg 2, 19 May–24 Jun 1998. Atlantic Margin Environmental Survey: seabed survey of deep-water areas (17th round Tranches) to the north and west of Scotland. *Southampton Oceanography Centre Cruise Report*, No. 25.

Bett, B J. 2000. Benthic ecology of the Faroe–Shetland Channel. Section 4.3.1 *in Atlantic Margin Environmental Surveys of the Seafloor 1996 and 1998*. UKOOA (United Kingdom Offshore Operators Association) [CD-ROM]. ISBN 09538399-0-7.

Bett, B J. 2001. UK Atlantic Margin Environmental Survey: introduction and overview of bathyl benthic ecology. *Continental Shelf Research*, Vol. 21, 917–956.

Bett, B J, Pearson, T H, and Sørenson, J. 2001. Expert review of seabed fauna and chemistry. (report to GEM). (Kaldbak Marine Biological Laboratory.)

Birnie, J F, Gordon, J E, Bennett, K J, and Hall, A M. 1993. *The Quaternary of Shetland: field guide*. (Quaternary Research Association.) ISBN X780863223.

Blackbourn, G A. 1987. Sedimentary environments and stratigraphy of the late Devonian–Carboniferous Clair Basin, west of Shetland. 75–91 in *European Dinantian Environments*. Miller, J, Adams, A E, and Wright, V P (editors). (London: John Wiley and Sons.)

Blanc, P-L, Rabussier, D, Vernaud-Grazzini, C, and Duplessy, J-C. 1980. North Atlantic water formed by the later middle Miocene. *Nature*, Vol. 283, 553–555.

Blystad, P, Brekke, H, Faerseth, R B, Larsen, B T, Skogseid, J, and Tørrudbakken, B. 1995. *Structural elements of the Norwegian continental shelf. Part II: The Norwegian Sea Region*. NPD-Bulletin, No. 8. (Stavanger: Norwegian Petroleum Directorate.)

Boldreel, L O, and Andersen, M S. 1993. Late Pliocene to Miocene compression in the Faroes–Rockall area. 1026–1034 in *Petroleum geology of northwest Europe, proceedings of the 4th conference*. Parker, J R (editor). (London: The Geological Society.)

Boldreel, L O, and Andersen, M S. 1995. The relationship between the distribution of Tertiary sediments, tectonic processes and deep-water circulation around the Faroe Islands. 145–158 *in* The tectonics, sedimentation and palaeoceanography of the North Atlantic region. Scrutton, R A, Shimmield, G B, Stoker, M S, and Tudhope, A W (editors). *Geological Society of London Special Publication*, No. 90.

Boldreel, L O, and Andersen, M S. 1998. Tertiary compressional structures on the Faroe–Rockall Plateau in relation to north-east Atlantic ridge-push and Alpine foreland stresses. *Tectonophysics*, Vol. 300, 13–28.

Boldreel, L O, Andersen, M S, and Kuijpers, A. 1998. Neogene seismic facies and deep-water gateways in the Faroe Bank area, NE Atlantic. *Marine Geology*, Vol. 152, 129–140.

Bonhoff, M, and Makris, J. 2004. Crustal structure of the south-eastern Iceland–Faroe Ridge (IFR) from wide aperture seismic data. *Journal of Geodynamics*, Vol. 37, 233–252.

Booth, J, Swiecicki, T, and Wilcockson, P. 1993. The tectono-stratigraphy of the Solan Basin, west of Shetland. 987–998 in *Petroleum geology of northwest Europe, proceedings of the 4th conference*. Parker, J R (editor). (London: The Geological Society.)

Borg, S G, De Paolo, D J, and Smith, B M. 1990. Isotopic structure and tectonics of the central Transantartctic Mountains. *Journal of Geophysical Research*, Vol. 95, 6647–6667.

Bott, M H P. 1984. Deep structure and origin of the Faroe–Shetland Channel. 341–347 in *Petroleum Geology of the North European Margin*. Spencer, A M (editor). (London: Graham and Trotman.)

Bott, M H P, and Gunnarsson, K. 1980. Crustal structure of the Iceland–Faroe Ridge. *Journal of Geophysics*, Vol. 47, 221–227.

Bott, M H P, and Smith, P J. 1984. Crustal structure of the Faroe–Shetland Channel. *Geophysical Journal of the Royal Astronomical Society*, Vol. 76, 383–398.

Bott, M H P, and Stacey, A P. 1967. Geophysical evidence on the origin of the Faroe Bank Channel — II. A gravity and magnetic profile. *Deep-Sea Research*, Vol. 14, 7–11.

Bott, M H P, and Watts, A B. 1971. Deep structure of the continental margin adjacent to the British Isles. 89–109 *in* Geology of the East Atlantic Continental Margin, 2. Delaney, F M (editor). *Report of the Institute of Geological Sciences*, No. 70/14.

Bott, M H P, Sunderland, J, Smith, P J, Casten, U, and Saxov, S. 1974. Evidence for continental crust beneath the Faroe Islands. *Nature*, Vol. 248, 202–204.

Bott, M H P, Nielsen, P H, and Sunderland, J. 1976. Converted P-waves originating at the continental margin between the Iceland–Faroe Ridge and the Faroe Block. *Geophysical Journal of the Royal Astronomical Society*, Vol. 44, 229–238.

Boulton, G S, Peacock, J D, and Sutherland, D G. 2002. The Quaternary. 409–430 in *The Geology of Scotland*. Fourth edition. Trewin, N H (editor). (London: The Geological Society.)

Bowen, D Q. 1999. On the correlation and classification of Quaternary deposits and land–sea correlations. 1–9 *in* A revised correlation of quaternary deposits in the British Isles. Bowen, D Q (editor). *Geological Society of London Special Report*, No. 23.

Bowles, F A, and Jahn, W H. 1983. Geological/geophysical observations and inferred bottom-current flow: south flank Iceland–Faroe Ridge. *Marine Geology*, Vol. 52, 159–185.

Bradley, D C. 1982. Subsidence in late Palaeozoic basins in the northern Appalachians. *Tectonics*, Vol. 1, 107–123.

Bradwell, T, Stoker, M S, Golledge, N, Wilson, C, Merritt, J, Long, D, Everest, J, Hestvik, O B, Stevenson, A G, Hubbard, A, Finlayson, A, and Mathers, H. 2008. The northern sector of the last British Ice Sheet: maximum extent and demise. *Earth Science Reviews*, Vol. 88, 207–226.

Brekke, H. 2000. The tectonic evolution of the Norwegian Sea continental margin with emphasis on the Vøring and Møre basins. 327–378 *in* Dynamics of the Norwegian margin. Nøttvedt, A (editor). *Geological Society of London Special Publication*, No. 167.

Brekke, H, Dahlgren, S, Nyland, B, and Magnus, C. 1999. The prospectivity of the Vøring and Møre basins on the Norwegian Sea continental margin. 261–274 in *Petroleum geology of northwest Europe, proceedings of the 5th conference*. Fleet, A J, and Boldy, S A R (editors). (London: The Geological Society.)

Brewer, J A, and Smythe, D K. 1984. MOIST and the continuity of crustal reflector geometry along the Caledonian–Appalachian orogen. *Journal of the Geological Society of London*, Vol. 121, 105–120.

Brewer, J A, Matthews, D H, Warner, M R, Hall, J, Smythe, D K, and Whittington, R. 1983. BIRPS deep seismic reflection studies of the British Caledonides. *Nature*, Vol. 305, 206–210.

British Geological Survey. 1984. Shetland. Sheet 60°N–02°W. Solid geology. 1:250 000. (Southampton: Ordnance Survey for the British Geological Survey.)

British Geological Survey. 1985. Orkney. Sheet 59°N–04°W. Solid geology. 1:250 000. (Southampton: Ordnance Survey for the British Geological Survey.)

British Geological Survey. 1986a. Rona. Sheet 59°N–06°W. Solid geology. 1:250 000. (Southampton: Ordnance Survey for the British Geological Survey.)

British Geological Survey. 1986b. Rona. Sheet 59°N–06°W. Sea-bed sediments. 1:250 000. (Southampton: Ordnance Survey for the British Geological Survey.)

British Geological Survey. 1987. Cormorant. Sheet 61°N–00°W. Solid geology. 1:250 000. (Southampton: Ordnance Survey for the British Geological Survey.)

British Geological Survey. 1988a. Fair Isle. Sheet 59°N–02°W. Solid geology. 1:250 000. (Southampton: Ordnance Survey for the British Geological Survey.)

British Geological Survey. 1988b. Foula. Sheet 60°N–04°W. Solid geology. 1:250 000. (Southampton: Ordnance Survey for the British Geological Survey.)

British Geological Survey. 1988c. Sula Sgeir. Sheet 59°N–08°W. Solid geology. 1:250 000. (Southampton: Ordnance Survey for the British Geological Survey.)

British Geological Survey. 1988d. Sula Sgeir. Sheet 59°N–08°W. Sea-bed sediments. 1:250 000. (Southampton: Ordnance Survey for the British Geological Survey.)

British Geological Survey. 1989a. Miller Sheet. Sheet 61°N–02°W. Solid geology. 1:250 000. (Southampton: Ordnance Survey for the British Geological Survey.)

British Geological Survey. 1989b. Sutherland. Sheet 58°N–06°W. Solid geology. 1:250 000.

(Southampton: Ordnance Survey for the British Geological Survey.)

British Geological Survey. 1990a. Foula. Sheet 60°N–04°W. Sea-bed sediments. 1:250 000. (Southampton: Ordnance Survey for the British Geological Survey.)

British Geological Survey. 1990b. Judd. Sheet 60°N–06°W. Quaternary. 1:250 000. (Southampton: Ordnance Survey for the British Geological Survey.)

British Geological Survey. 1990c. Judd. Sheet 60°N–06°W. Sea-bed sediments. 1:250 000. (Southampton: Ordnance Survey for the British Geological Survey.)

British Geological Survey. 1990d. Rona. Sheet 59°N–06°W. Quaternary. 1:250 000. (Southampton: Ordnance Survey for the British Geological Survey.)

British Geological Survey. 1991a. Flett. Sheet 61°N–04°W. Quaternary. 1:250 000. (Southampton: Ordnance Survey for the British Geological Survey.)

British Geological Survey. 1991b. Foula. Sheet 60°N–04°W. Quaternary. 1:250 000. (Southampton: Ordnance Survey for the British Geological Survey.)

British Geological Survey. 1991c. Judd. Sheet 60°N–06°W. Solid geology. 1:250 000. (Southampton: Ordnance Survey for the British Geological Survey.)

British Geological Survey. 1991d. Miller Sheet. Sheet 61°N–02°W. Quaternary. 1:250 000. (Southampton: Ordnance Survey for the British Geological Survey.)

British Geological Survey. 1992. Peach. Sheet 56°N–10°W. Solid geology. 1:250 000. (Southampton: Ordnance Survey for the British Geological Survey.)

British Geological Survey. 1996. Tectonic Map of Britain, Ireland and adjacent areas. Pharaoh, T C, Morris, J H, Long, C B, and Ryan, P D (compilers). 1:500 000. (Keyworth, Nottingham: British Geological Survey.)

British Geological Survey. 2007. Northern Rockall Basin. Solid geology. 1:500 000. (Keyworth, Nottingham: British Geological Survey.)

British Geological Survey and Petroleum Affairs Division (Ireland). 2002. Central Rockall Basin. Solid geology. 1:500 000. (Keyworth, Nottingham: British Geological Survey.)

Brodie, J, and White, N. 1995. The link between sedimentary basin inversion and igneous underplating. 21–38 *in* Basin Inversion. Buchanen, J G, and Buchanen, P G (editors). *Geological Society of London Special Publication*, No. 88.

Brooks, J R V , Stoker, S J, and Cameron, T D J. 2001. Hydrocarbon exploration opportunities in the twenty-first century in the United Kingdom. 167–199 *in* Petroleum provinces of the twenty-first century. Downey, M W, Threet, J C, and Morgan, W A (editors). *American Association of Petroleum Geologists Memoir,* No. 74.

Bruntse, G, Lein, T E. and Nielsen, R (editors). 1999. Marine benthic algae and invertebrate communities from the shallow waters of the Faroe Islands. The Faroe Islands: Kaldbak Marine Biological Laboratory. ISBN 99918-3-065-0.

Bryn, P, Berg, K, Forsberg, C F, Solheim, A, and Kvalstad, T J. 2005. Explaining the Storegga Slide. *Marine and Petroleum Geology,* Vol. 22, 11–19.

Buchan, K L, Mertanen, S, Park, R G, Pesonen, L J, Elming, S-A, Abrahamsen, N, and Bylund, G. 2000. The drift of Laurentia and Baltica in the Proterozoic: a comparison based on key palaeomagnetic poles. *Tectonophysics,* Vol. 319, 167–198.

Bugge, T. 1983. Submarine slides on the Norwegian continental margin, with special reference on the Storegga Slide. *Institutt for kontinentalsokkelundersokelser Publication,* No. 110. Trondheim.

Bulat, J, and Long, D. 2001. Images of the seabed in the Faroe–Shetland Channel from commercial 3D seismic data. *Marine Geophysical Researches,* Vol. 22, 345–367.

Bulat, J, and Long, D. 2005. Images of debris fans and other deep-sea sediments on the seabed of the Faroe–Shetland Channel based on 3D seismic data. *Scottish Journal of Geology,* **41**, 81–86.

Burnham, A K, and Sweeney, J J. 1989. A chemical kinetic model of vitrinite maturation and reflectance. *Geochimica et Cosmochimica Acta,* Vol. 53, 2649–2657.

Butler, R W H, and Coward, M P. 1984. Geological constraints, structural evolution and deep geology of the north-west Scottish Caledonides. *Tectonics,* Vol. 3, 347–365.

Cameron, T D J. 1993a. 5. Carboniferous and Devonian of the Southern North Sea. *In Lithostratigraphic nomenclature of the UK North Sea.* Knox, R W O'B, and Cordey, W G (editors). (Keyworth, Nottingham: British Geological Survey.)

Cameron, T D J. 1993b. 4. Triassic, Permian and Pre-Permian of the Central and Northern North Sea. *In Lithostratigraphic nomenclature of the UK North Sea.* Knox, R W O'B, and Cordey, W G (editors). (Keyworth, Nottingham: British Geological Survey.)

Cameron, T D J, and Holmes, R. 1999. The Continental Shelf. 125–139 *in* A revised correlation of quaternary deposits in the British Isles. Bowen, D Q (editor). *Geological Society of London Special Report,* No. 23.

Cande, S C, and Kent, D V. 1995. Revised calibration of the geomagnetic polarity timescale for the Late Cretaceous and Cenozoic. *Journal of Geophysical Research,* Vol. 100 (B4), 6093–6095.

Carr, A D. 1999. A vitrinite reflectance kinetic model incorporating overpressure retardation. *Marine and Petroleum Geology,* Vol. 16, 355–377.

Carr, A D, and Scotchman, I C. 2003. Thermal history modelling in the southern Faroe–Shetland Basin. *Petroleum Geoscience,* Vol. 9, 333–345.

Carruth, A G. 2003. The Foinaven Field, Blocks 204/19 and 204/24a, UK North Sea. 121–130 *in* United Kingdom Oil and Gas Fields, Commemorative Millennium Volume. Gluyas, J G, and Hichens, H M (editors). *Geological Society of London Memoir,* No. 20.

Cashion, W W. 1975. The geology of the West Shetland Basin. 216.1–216.7 in *Proceedings of the Offshore Europe Conference.* Woodland, A (editor). (Kingston-upon-Thames: Spearhead Publications.)

CASTEN, U. 1973. The crust beneath the Faroe Islands. *Nature Physical Science*, Vol. 241, 83–84.

CAWLEY, S, MATHIESON, H, and STALKER, G. 2005. An updated view of the Faroes Shetland Petroleum System. 109–130 *in* Faroe Islands Exploration Conference: Proceedings of the 1st Conference. ZISKA, H, VARMING, T, and BLOCH, D (editors). *Annales Societatis Scientarum Færoensis, Supplementum*, Vol. 43.

CERAMICOLA, S, STOKER, M S, PRAEG, D, SHANNON, P M, DE SANTIS, L, HOULT, R, HJELSTUEN, B O, LABERG, J S, and MATHIESEN, A. 2005. Anomalous post-rift subsidence from 2D flexural backstripping of Cenozoic megasequences along the NW European margin. *Marine and Petroleum Geology*, Vol. 22, 1045–1067.

CHADWICK, R A, and EVANS, D J. 1995. The timing and direction of Permo-Triassic extension in southern Britain. 161–192 *in* Permian and Triassic rifting in Northwest Europe. BOLDY, S A R (editor). *Geological Society of London Special Publication*, No. 91.

CHADWICK, R A, and PHARAOH, T C. 1998. The seismic reflection Moho beneath the United Kingdom and adjacent areas. *Tectonophysics*, Vol. 299, 255–279.

CHADWICK, R A, JACKSON, D I, BARNES, R P, KIMBELL, G S, JOHNSON, H, CHIVERRELL, R C, THOMAS, G S P, JONES, N S, RILEY, N J, PICKETT, E A, YOUNG, B, HOLLIDAY, D W, BALL, D F, MOLYNEUX, S G, LONG, D, POWER, G M, and ROBERTS, D H. 2001. Geology of the Isle of Man and its offshore area. *British Geological Survey Research Report*, RR/01/06.

CHALMERS, J A. 1997. The continental margin off southern Greenland: along-strike transition from an amagmatic to a volcanic margin. *Journal of the Geological Society of London*, Vol. 154, 571–576.

CHALMERS, J A, and WESTERN, P G. 1979. A Tertiary igneous centre north of the Shetland Islands. *Scottish Journal of Geology*, Vol. 15, 333–341.

CHAMBERS, L, DARBYSHIRE, F, NOBLE, S, and RITCHIE, D. 2005. NW UK continental margin; chronology and isotope geochemistry. *British Geological Survey Commissioned Report*, CR/05/095.

CHAPMAN, T J, BROOKS, T M, CORCORAN, D V, DUNCAN, L A, and DANCER, P N. 1999. The structural evolution of the Erris Trough, offshore north-west Ireland, implications for hydrocarbon generation. 455–469 in *Petroleum geology of northwest Europe, proceedings of the 5th conference*. FLEET, A J, and BOLDY, S A R (editors). (London: The Geological Society.)

CHESHER, J A. 1987. Cruise report of Magnus Heinason. 17 September–14 October, NE Atlantic. *British Geological Survey Report*, No. 87/43.

CHESHER, J A, SMYTHE, D K, and BISHOP, P. 1983. The geology of the Minches, Inner Sound and Sound of Raasay. *Report of the Institute of Geological Sciences*, No. 83/6.

CLIFT, P D. 1999. The thermal impact of Paleocene underplating in the Faroe–Shetland–Rockall region. 585–593 in *Petroleum geology of northwest Europe, proceedings of the 5th conference*. FLEET, A J, and BOLDY, S A R (editors). (London: The Geological Society.)

CLIMAP MEMBERS. 1981. Seasonal reconstruction of the Earth's surface at the Last Glacial Maximum. Geological Society of America, Map and Chart Series MC-36. Boulder, Colorado.

CLOETINGH, S, GRADSTEIN, F M, KOOI, H, GRANT, A C, and KAMINSKI, M. 1990. Plate reorganization: a cause of rapid late Neogene subsidence and sedimentation around the North Atlantic. *Journal of the Geological Society of London*, Vol. 147, 495–506.

COCKCROFT, D N. 1987. The Quaternary sediments of the Shetland Platform and adjacent continental margin. Unpublished PhD thesis, University of Keele.

COFFIELD, D Q. 1992. Evidence for wrenching during mid Permian extension in East Greenland. 89–98 *in Arctic Geology and Petroleum Potential*. VORREN, O T, et al. (editors). *Norwegian Petroleum Directorate Special Publication*, No. 2. (Amsterdam: Elsevier.)

COFFIN, M F, and ELDHOLM, O. 1992. Volcanism and continental break-up: a global compilation of a large igneous provinces. 17–30 *in* Magmatism and the causes of continental break-up. STOREY, B C, ALABASTER, T, and PANKHURST, R J (editors). *Geological Society of London Special Publication*, No. 68.

COLE, J E, and PEACHEY, J. 1999. Evidence for the pre-Cretaceous rifting in the Rockall Trough: an analysis using quantitative plate tectonic modelling. 359–370 in *Petroleum geology of northwest Europe, proceedings of the 5th conference*. FLEET, A J, and BOLDY, S A R (editors). (London: The Geological Society.)

COLLINS, A G, and DONOVAN, N R. 1977. The age of two Old Red Sandstone sequences in southern Caithness. *Scottish Journal of Geology*, Vol. 13, 53–57.

CONEY, D, FYFE, T B, RETAIL, P, and SMITH, P J. 1993. Clair appraisal: the benefits of a co-operative approach. 1409–1420 in *Petroleum geology of northwest Europe, proceedings of the 4th conference*. PARKER, J R (editor). (London: The Geological Society.)

COOPER, M M, EVANS, A C, LYNCH, D J, NEVILLE, G, and NEWLEY, T. 1999. The Foinaven Field: managing reservoir development uncertainty prior to start-up. 675–682 in *Petroleum geology of northwest Europe, proceedings of the 5th conference*. FLEET, A J, and BOLDY, S A R (editors). (London: The Geological Society.)

COPESTAKE, P, SIMS, A P, CRITTENDEN, S, HAMAR, G P, INESON, J R, ROSE, P T, and TRINGHAM, M E. 2003. Lower Cretaceous. 191–211 in *Millennium Atlas: petroleum geology of the central and northern North Sea*. EVANS, D, GRAHAM, C G, ARMOUR, A, and BATHHURST, P (editors). (London: The Geological Society.)

CORFU, F, HEAMAN, L M, and ROGERS, G. 1994. Polymetamorphic evolution of the Lewisian complex, NW Scotland, as recorded by U-Pb isotopic compositions of zircon, titanite and rutile. *Contributions to Mineralogy and Petrology*, Vol. 117, 215–228.

CORFU, F, CRANE, A, MOSER, D, and ROGERS, G. 1998. U-Pb zircon systematics at Gruinard Bay, north-west Scotland: implications for the early orogenic evolution of the Lewisian complex. *Contributions to Mineralogy and Petrology*, Vol. 133, 329–345.

COWARD, M P. 1993. The effects of Late Caledonian and Variscan continental escape tectonics on basement structure, Palaeozoic basin kinematics and subsequent Mesozoic basin development in NW Europe. 1095–1108 in *Petroleum geology of northwest Europe, proceedings of the 4th conference*. PARKER, J R (editor). (London: The Geological Society.)

COWARD, M P, and ENFIELD, M A. 1987. The structure of the West Orkney and adjacent basins. 687–696 in *Petroleum geology of northwest Europe, proceedings of the 3rd conference*. BROOKS, J, and GLENNIE, K W (editors). (London: Graham and Trotman.)

COWARD, M P, ENFIELD, M A, and FISHER, M W. 1989. Devonian basins of northern Scotland: extension and inversion related to Late Caledonian–Variscan tectonics. 275–308 *in* Inversion tectonics. COOPER, M A, and WILLIAMS, G D (editors). *The Geological Society of London Special Publication*, No. 44.

COWARD, M P, DEWEY, J F, HEMPTON, M, and HOLROYD, J. 2003. Tectonic evolution. 17–33 in *Millennium Atlas: petroleum geology of the central and northern North Sea*. EVANS, D, GRAHAM, C G, ARMOUR, A, and BATHHURST, P (editors). (London: The Geological Society.)

CRITTENDEN, S, COLE, J M, and KIRK, M J. 1997. The distribution of Aptian Sandstones in the Central and Northern North Sea (UK Sectors): A Lowstand Systems Tract 'Play'. Part 1. Stratigraphy, biostratigraphic age determination and genesis of the sandstones. *Journal of Petroleum Geology*, Vol. 20, 3–25.

DAHLGREN, K I T, VORREN, T O, STOKER, M S, NIELSEN, T, NYGÅRD, A, and SEJRUP, H P. 2005. Late Cenozoic prograding wedges on the NW European continental margin: their formation and relationship to tectonics and climate. *Marine and Petroleum Geology*, Vol. 22, 1089–1110.

DALLAND, A, WORSLEY, D, and OFSTAD, K. 1988. A lithostratigraphic scheme for the Mesozoic and Cenozoic succession offshore mid and northern Norway. *NPD-Bulletin*, No. 4. (Stavanger: Norwegian Petroleum Directorate.)

DAMUTH, J E, and OLSON, H C. 1993. Preliminary observations of Neogene–Quaternary depositional processes in the Faroe–Shetland Channel revealed by high-resolution seismic facies analysis. 1035–1045 in *Petroleum geology of northwest Europe, proceedings of the 4th conference*. PARKER, J R (editor). (London: The Geological Society.)

DAMUTH, J E, and OLSON, H C. 2001. Neogene Quaternary contourite and related deposition on the West Shetland Slope and Faroe–Shetland Channel revealed by high-resolution seismic studies. *Marine Geophysical Researches*, Vol. 22, 369–399.

DANCER, P N, ALGAR, S T, and WILSON, I R. 1999. Structural evolution of the Slyne Trough. 445–453 in *Petroleum geology of northwest Europe, proceedings of the 5th conference*. FLEET, A J, and BOLDY, S A R (editors). (London: The Geological Society.)

DAVIES, R, and CARTWRIGHT, J. 2002. A fossilized Opal A to Opal C/T transformation on the north-east Atlantic margin: support for a significantly elevated palaeogeothermal gradient during the Neogene? *Basin Research*, Vol. 14, 467–486.

DAVIES, R, CARTWRIGHT, J, and RANA, J. 1999. Giant hummocks in deep-water marine sediments: Evidence for large-scale differential compaction and density inversion during early burial. *Geology*, Vol. 27, 907–910.

DAVIES, R, CARTWRIGHT, J, PIKE, J, and LINE, C. 2001. Early Oligocene initiation of North Atlantic deep-water formation. *Nature*, Vol. 410, 917–920.

DAVIES, R, BELL, B R, CARTWRIGHT, J A, and SHOULDERS, S. 2002. Three-dimensional seismic imaging of Paleogene dyke-fed submarine volcanoes from the north-east Atlantic margin. *Geology*, Vol. 30, 223–226.

DAVIES, R, CLOKE, I, CARTWRIGHT, J, ROBINSON, A, and FERRERO, C. 2004. Post-breakup compression of a passive margin and its impact on hydrocarbon prospectivity: An example from the Tertiary of the Faroe–Shetland Basin, United Kingdom. *American Association of Petroleum Geologists Bulletin*, No. 88, 1–20.

DAVISON, S. 2005. Reconstructing the Last Pleistocene (Late Devensian) Glaciation on the Continental Margin of Northwest Britain. Unpublished PhD thesis, University of Edinburgh.

DAVISON, S, and STOKER, M S. 2002. Late Pleistocene glacially influenced deep-marine sedimentation off NW Britain: implications for the rock record. 129–147 *in* Glacier-influenced sedimentation on high-latitude continental margins. DOWDESWELL, J A, and Ó' COFAIGH, C (editors). *Geological Society of London Special Publication*, No. 203.

DEAN, K, MCLAUCHLAN, K, and CHAMBERS, A. 1999. Rifting and the development of the Faroe–Shetland Basin. 533–544 in *Petroleum geology of northwest Europe, proceedings of the 5th conference*. FLEET, A J, and BOLDY, S A R (editors). (London: The Geological Society.)

DEEGAN, C E, and SCULL, B J. 1977. A Standard Lithostratigraphic Nomenclature for the Central and Northern North Sea. *Institute of Geological Sciences Report*, No. 77/25.

DEFANT, M J, and DRUMMOND, M S. 1990. Derivation of some modern arc magmas by melting of young subducted lithosphere. *Nature*, Vol. 347, 662–665.

DE PAOLA, N, HOLDSWORTH, R E, MCCAFFREY, K J W, and BARCHI, M R. 2005. Partitioned transtension: an alternative to basin inversion models. *Journal of Structural Geology*, Vol. 27, 607–625.

DE PAOLO, D J, LINN, A M, AND SCHUBERT, G. 1991. The continental crustal age distribution: Methods of determining mantle separation ages from Sm-Nd isotopic data and application to the south-western United States. *Journal of Geophysical Research*, Vol. 96, 2071–2088.

DEPARTMENT OF TRADE AND INDUSTRY (DTI). 2006. https://www.og.berr.gov.uk/UKpromote/geoscientific/UKCS_prospectivity_2007.pdf

DEPARTMENT OF TRADE and INDUSTRY (DTI). 2007a. https://www.og.berr.gov.uk/information/bb_updates/appendices/Appendix9.htm.

DEPARTMENT OF TRADE and INDUSTRY (DTI). 2007b. https://www.og.berr.gov.uk/UKpromote/geoscientific/Eocene_play_WoS.pdf

DEWEY, J F, and STRACHAN, R A. 2003. Changing Silurian–Devonian relative plate motion in the Caledonides: sinistral transpression to sinistral transtension. *Journal of the Geological Society, London*, Vol. 160, 219–229.

DICKIN, A P. 1992. Evidence for an Early Proterozoic crustal province in the North Atlantic region. *Journal of the Geological Society of London*, Vol. 149, 483–486.

DOBINSON, A. 1970. 1. The development of a Marine Seismic Recording System, 2. A Magnetic Survey of the Faroe Bank. Unpublished PhD thesis, University of Durham.

DORÉ, A G, and LUNDIN, E R. 1996. Cenozoic compressional structure of the NE Atlantic margin: nature, origin and potential significance for hydrocarbon exploration. *Petroleum Geoscience*, Vol. 2, 299–311.

DORÉ, A G, LUNDIN, E R, FICHLER, C, and OLSEN, O. 1997. Patterns of basement structure and reactivation along the NE Atlantic margin. *Journal of the Geological Society of London*, Vol. 154, 85–92.

DORÉ, A G, LUNDIN, E R, JENSEN, L N, BIRKELAND, O, ELIASSEN, P E, and FICHLER, C. 1999. Principal tectonic events in the evolution of the northwest European Atlantic margin. 41–61 in *Petroleum geology of northwest Europe, proceedings of the 5th conference*. FLEET, A J, and BOLDY, S A R (editors). (London: The Geological Society.)

DORN, W U, and WERNER, F. 1993. The contour-current flow along the southern Iceland–Faroe Ridge as documented by its bedforms and asymmetrical channel fillings. *Sedimentary Geology*, Vol. 82, 47–59.

DUINDAM, P, and VAN HOORN, B. 1987. Structural evolution of the West Shetland continental margin. 765–773 in *Petroleum geology of northwest Europe, proceedings of the 3rd conference*. BROOKS, J, and GLENNIE, K W (editors). (London: Graham and Trotman.)

EARLE, M M, JANKOWSKI, E J, and VANN, I R. 1989. Structural evolution of the Faroe–Shetland Channel and northern Rockall Trough. 461–489 *in* Extensional tectonics and stratigraphy of the North Atlantic margins. TANKARD, A J, and BALKWILL, H R (editors). *American Association of Petroleum Geologists Memoir*, No. 46. Tulsa, Oklahoma, USA.

EBDON, C C, GRANGER, P J, JOHNSON, H D, and EVANS, A M. 1995. Early Tertiary evolution and sequence stratigraphy of the Faroe–Shetland Basin: implications for hydrocarbon prospectivity. 51–69 *in* Tectonics, sedimentation and palaeoceanography of the North Atlantic region. SCRUTTON, R A, STOKER, M S, SHIMMIELD, G B, and TUDHOPE, A W (editors). *Geological Society of London Special Publication*, No. 90.

EIRÍKSSON, J. 1981. Lithostratigraphy of the upper Tjörnes sequence, North Iceland: The Breidavík Group. *Acta Naturalia Islandica*, Vol. 29, 1–37.

ELLETT, D J, and ROBERTS, D G. 1973. The overflow of Norwegian Sea deep water across the Wyville–Thomson Ridge. *Deep Sea Research*, Vol. 20, 819.

ELLIS, D, BELL, B R, JOLLEY, D W, and O'CALLAGHAN, M. 2002. The stratigraphy, environment of eruption and age of the Faroes Lava Group, NE Atlantic Ocean. 253–269 *in* The North Atlantic Igneous Province: Stratigraphy, Tectonic, Volcanic and Magmatic Processes. JOLLEY, D W, and BELL, B R (editors). *Geological Society of London Special Publication*, No. 197.

ENFIELD, M A, and COWARD, M P. 1987. The Structure of the West Orkney Basin, northern Scotland. *Journal of the Geological Society of London*, Vol. 144, 871–884.

ENGLAND, R J, MCBRIDE, J H, and HOBBS, R W. 2005. The role of Mesozoic rifting in the opening of the NE Atlantic: evidence from deep seismic profiling across the Faroe–Shetland Trough. *Journal of the Geological Society of London*, Vol. 162, 661–673.

EVANS, D, KING, E L, KENYON, N H, BRETT, C, and WALLIS, D. 1996. Evidence for long-term instability in the Storegga Slide region off western Norway. *Marine Geology*, Vol. 130, 281–292.

EVANS, D, MORTON, A C, WILSON, S, JOLLEY, D, and BARREIRO, B A. 1997. Palaeoenvironmental significance of marine and terrestrial Tertiary sediments on the NW Scottish Shelf in BGS borehole 77/7. *Scottish Journal of Geology*, Vol. 33, 31–42.

EVANS, D, HARRISON, Z, SHANNON, P M, LABERG, J S, NIELSEN, T, AYERS, S, HOLMES, R, HOULT, R J, LINDBERG, B, HAFLIDASON, H, LONG, D, KUIJPERS, A, ANDERSEN, E S, and BRYN, P. 2005. Palaeoslides and other mass failures of Pliocene to Pleistocene age along the Atlantic continental margin of NW Europe. *Marine and Petroleum Geology*, Vol. 22, 1131–1148.

EVANS, D J. 1997. Estimates of eroded overburden and the Permian–Quaternary subsidence history of the area west of Orkney. *Scottish Journal of Geology*, Vol. 33, 169–182.

FÆRSETH, R B, and LIEN, T. 2002. Cretaceous evolution of the Norwegian Sea — a period characterised by tectonic quiescence. *Marine and Petroleum Geology*, Vol. 19, 1005–1027.

FALEIDE, J I, KYRKJEBØ, R, KJENNERUD, T, GABRIELSEN, R H, JORDT, H, FANAVOLL, S, and BJERKE, M D. 2002. Tectonic impact on sedimentary processes during Cenozoic evolution of the northern North Sea and surrounding areas. 235–269 *in* Exhumation of the North Atlantic margin: timing, mechanisms and implications for petroleum exploration. DORÉ, A G, CARTWRIGHT, J A, STOKER, M S, TURNER, J P, and WHITE, N (editors). *Geological Society of London Special Publication*, No. 196.

FAROES GEM (Faroe Oil Industry Group formerly Geotech Environmental and Meterological Group). 2001a. Seabed and Shallow Geology of the Faroes 1st Round Licence Area, Vol. 1. Project 23-811.

FAROES GEM (Faroe Oil Industry Group formerly Geotech Environmental and Meterological Group). 2001b. Seabed and Shallow Geology of the Faroes 1st Round Licence Area, Vol. 2. Project 23-811.

FAROES GEM (Faroe Oil Industry Group formerly Geotech Environmental and Meterological Group). 2001c. Seabed and Shallow Geology of the Faroes 1st Round Licence Area, Vol. 3. Project 23-811.

FAROES GEM (Faroe Oil Industry Group formerly Geotech Environmental and Meterological Group). 2001d. Seabed and Shallow Geology of the Faroes 1st Round Licence Area, Vol. 4. Project 23-811.

FARROW, G E, ALLEN, N H, and AKPAN, E T. 1984. Bioclastic carbonate sedimentation on a high-latitude, tide-dominated shelf: Northeast Orkney Islands, Scotland. *Journal of Sedimentology and Petrology*, Vol. 54, 373–393.

FITCH, F J, HEARD, G L, and MILLER, J A. 1988. Basaltic magmatism of late Cretaceous and Palaeogene age recorded in wells NNE of the Shetlands. 253–262 *in* Early Tertiary Volcanism and the Opening of the NE Atlantic. Morton, A C, and Parson, L M (editors). *Geological Society of London Special Publication*, No. 39.

Flack, C A, and Warner, M. 1990. Three dimensional mapping of seismic reflections from the crust and upper mantle, north-west of Scotland. *Tectonophysics*, Vol. 173, 469–482.

Fleet, A J, and Boldy, S A R (editors). 1999. *Petroleum geology of northwest Europe, proceedings of the 5th conference.* (London: The Geological Society.)

Fliedner, M M, and White, R S. 2003. Depth imaging of basalt flows in the Faroe–Shetland Basin. *Geophysical Journal International*, Vol. 152, 353–371.

Flinn, D. 1956. On the deformation of the Funzie conglomerate, Fetlar, Shetland. *Journal of Geology*, Vol. 64, 480–505.

Flinn, D. 1961. Continuation of the Great Glen Fault beyond the Moray Firth. *Nature*, Vol. 191, 581–591.

Flinn, D. 1973. The topography of the sea floor around Orkney and Shetland. *Journal of the Geological Society of London*, Vol. 129, 39–58.

Flinn, D. 1977a. Transcurrent faults and associated cataclasis in Shetland. *Journal of the Geological Society of London*, Vol. 133, 231–248.

Flinn, D. 1977b. The erosion history of Shetland: a review. *Proceedings of the Geologists' Association*, Vol. 88, 129–146.

Flinn, D. 1978. The most recent glaciation of the Orkney–Shetland Channel and adjacent areas. *Scottish Journal of Geology*, Vol. 14, 109–123.

Flinn, D. 1988. The Moine rocks of Scotland. 74–85 in *Later Proterozoic stratigraphy of the northern Atlantic regions*. Winchester, J A (editor). (Glasgow: Blackie.)

Flinn, D. 1992. The history of the Walls Boundary fault: the northward continuation of the Great Glen Fault from Scotland. *Journal of the Geological Society of London*, Vol. 149, 721–726.

Flinn, D, Miller, J A, Evans, A L, and Pringle, I R. 1968. On the age of the sediments and contemporaneous volcanic rocks of western Shetland. *Scottish Journal of Geology*, Vol. 4, 10–19.

Flinn, D, Roberts, J L, and Treagus, J E. 1972. A revision of the stratigraphic succession of the East Mainland of Shetland. *Scottish Journal of Geology*, Vol. 8, 335–343.

Flinn, D, Frank, P L, Brook, M, and Pringle, I R. 1979. Basement–cover relations in Shetland. 115–119 *in* The Caledonides of the British Isles reviewed. Harris, A L, Holland, C H, and Leake, B E (editors). *Geological Society of London Special Publication*, No. 8.

Foulger, G R. 2002. Plumes, or plate tectonic processes? *Astronomy and Geophysics*, Vol. 43(6), 19–23.

Foulger, G R, and Anderson, D L. 2005. A cool model for the Iceland hotspot. *Journal of Volcanology and Geothermal Research*, Vol. 41, 1–22.

Fowler, S R, White, R S, Spence, G D, and Westbrook, GK. 1989. The Hatton Bank continental margin — II. Deep structure from two-ship expanding spread seismic profiles. *Geophysical Journal*, Vol. 96, 295–309.

Francis, E H. 1982. Emplacement mechanisms of late Carboniferous tholeiitic sills in Northern Britain. *Journal of the Geological Society of London*, Vol. 139, 1–120.

Frederiksen, R, Jensen, A, and Westerberg, H. 1992. The distribution of the scleractinian coral *Lophelia pertusa* around the Faroe Islands and the relation to internal tidal mixing. *Sarsia*, Vol. 77, 157–171.

Friend, C R L, and Kinny, P D. 2001. A reappraisal of the Lewisian Gneiss Complex: geochronological evidence for tectonic assembly of gneiss terranes in the Proterozoic. *Contributions to Mineralogy and Petrology*, Vol. 142, 198–218.

Friend, P F, Williams, B P F , Ford, N, and Williams, E A. 2000. Kinematics of the Old Red Sandstone basins. 29–60 *in* New perspectives on the Old Red Sandstone. Friend, P D, and Williams, B P J (editors). *Geological Society of London Special Publication*, No. 180.

Fyfe, J A, Long, D, and Evans, D. 1993. *The geology of the Malin–Hebrides sea area*. (London: HMSO for the British Geological Survey.)

Gabrielsen, R H, Færseth, R, Hamar, G, and Rønnevik, H C. 1984. Nomenclature of the main structural features on the Norwegian Continental Shelf north of the 62nd parallel. 41–60 in *Petroleum Geology of the North European Margin*. Spencer, A M, et al. (editors). *Norwegian Petroleum Society.* (London: Graham and Trotman.)

Gabrielsen, R H, Odinsen, T, and Grunnaleite, I. 1999. Structuring of the Northern Viking Graben and the Møre Basin; the influence and role of basement structural grain and the particular role of the Møre–Trøndelag Fault Complex. *Marine and Petroleum Geology*, Vol. 16, 443–465.

Gallagher, J W, and Dromgoole, P. 2005. Seeing Below the Basalt, Offshore Faroes, EAGE 67th Conference and Exhibition, Madrid, Spain, 13–16 June 2005 (extended abstract).

Galloway, W E. 2001. Seismic expressions of deep-shelf depositional and erosional morphologies, Miocene Utsira formation, North Sea Basin. *Marine Geophysical Researches*, Vol. 22, 309–321.

Gariépy, C, Ludden, J, and Brooks, C. 1983. Isotopic and trace element constraints on the genesis of the Faroe lava pile. *Earth and Planetary Science Letters*, Vol. 63, 257–272.

Gatliff, R W, Hitchen, K, Ritchie, J D, and Smythe, D K. 1984. Internal structure of the Erlend Tertiary volcanic complex, north of Shetland, revealed by seismic reflection. *Journal of the Geological Society of London*, Vol. 141, 555–562.

Gibb, F G F, and Kanaris-Sotiriou, R. 1988. The geochemistry and origin of the Faroe–Shetland sill complex. 241–252 *in* Early Tertiary volcanism and the opening of the NE Atlantic. Morton, A C, and Parson, L M (editors). *Geological Society of London Special Publication*, No. 39.

Gibb, F G F, Kanaris-Sotiriou, R, and Neves, R. 1986. A new Tertiary sill complex of mid ocean ridge basalt type NNE of the Shetland Islands: a preliminary report. *Transactions of the Royal Society of Edinburgh: Earth Sciences*, Vol. 77, 223–230.

GIBBARD, P L, SMITH, A G, ZALASIEWICZ, J A, BARRY, T L, CANTRILL, D, COE, A L, COPE, J W C, GALE, A S, GREGORY, F J, POWELL, J H, RAWSON, P F, STONE, P, and WATERS, C N. 2005. What status for the Quaternary? *Boreas*, Vol. 34, 1–6.

GJELBERG, J, ENOKSEN, T, KJÆRNES, P, MANGERUD, G, MARTINSEN, O J, ROE, E, and VÅGNES, E. 2001. The Maastrichtian and Danian depositional setting, along the eastern margin of the Møre Basin (mid Norwegian Shelf): implications for reservoir development of the Ormen Lange Field. 421–440 *in* Sedimentary environments offshore Norway — Palaeozoic to Recent. MARTINSEN, O J, and DREYER, T (editors). *Norwegian Petroleum Directorate Special Publication*, No. 10.

GJELBERG, J, MARTINSEN, O J, CHARNOCK, M, MØLLER, N, and ANTONSEN, P. 2005. The reservoir development of the Late Maastrichtian–Early Paleocene Ormen Lange gas field, More Basin, mid Norwegian Shelf. 1165–1184 in *Petroleum geology: northwest Europe and global perspectives, proceedings of the 6th conference*. DORÉ, A G, and VINING, B (editors). (London: The Geological Society.)

GLENNIE, K W. 2002. Permian and Triassic. 301–322 in *The Geology of Scotland*. Fourth edition. TREWIN, N H (editor). (London: The Geological Society.)

GLENNIE, K, HIGHAM, J, and STEMMERIK, L. 2003. Permian. 91–103 in *Millennium Atlas: petroleum geology of the central and northern North Sea*. EVANS, D, GRAHAM, C G, ARMOUR, A, and BATHHURST, P (editors). (London: The Geological Society.)

GOFF, J C. 1983. Hydrocarbon generation and migration from Jurassic source rocks in the East Shetland Basin and Viking Graben of the Northern North Sea. *Journal of the Geological Society of London*, Vol. 140, 445–474.

GOODCHILD, M W, HENRY, K L, HINKLEY, R J, and IMBUS, S W. 1999. The Victory gas field, west of Shetland. 713–724 in *Petroleum geology of northwest Europe, proceedings of the 5th conference*. FLEET, A J, and BOLDY, S A R (editors). (London: The Geological Society.)

GOODENOUGH, K M, EVANS, J A, and KRABBENDAM, M. 2006. Constraining the maximum age of movements in the Moine Thrust Belt: dating the Canisp Porphyry. *Scottish Journal of Geology*, Vol. 42, 77–81.

GRADSTEIN, F M, and BERGGREN, W A. 1981. Flysch-type agglutinating foraminifera and the Maastrichtian to Palaeogene history of the Labrador and North seas. *Marine Micropalaeontology*, Vol. 6, 211–268.

GRADSTEIN, F M, OGG, J G, SMITH, A G, and 37 OTHERS. 2004. *A Geologic Time Scale 2004*. (Cambridge: Cambridge University Press.) (with modifications).

GRAHAM, A G C, LONERGAN, L, and STOKER, M S. 2007. Evidence for Late Pleistocene ice stream activity in the Witch Ground Basin, Central North Sea, from 3D seismic reflection data. *Quaternary Science Reviews*, Vol. 26, 627–643.

GRANT, N, BOUMA, A, and MCINTYRE, A. 1999. The Turonian play in the Faroe–Shetland Basin. 661–673 in *Petroleum geology of northwest Europe, proceedings of the 5th conference*. FLEET, A J, and BOLDY, S A R (editors). (London: The Geological Society.)

GREEN, P F, DUDDY, I R, HEGARTY, K A, and BRAY, R J. 1999. Early Tertiary heat flow along the UK Atlantic margin and adjacent areas. 348–357 in *Petroleum geology of northwest Europe, proceedings of the 5th conference*. FLEET, A J, and BOLDY, S A R (editors). (London: The Geological Society.)

GULIYEV, I S, and FEYZULLAYEV, A A. 2003. *All about Mud Volcanoes*. (Azerbaijan: Academy of Sciences Geology Institute.)

HAFLIDASON, H, KING, E L, BRETT, C P, STEVENSON, A G, WALLIS, D G, CAMPBELL, N, SEJRUP, H P, and WAAGE, B. 1996. Marine geological/geophysical cruise report on the North Sea margin. Upper North Sea Fan, Miller Slide and Faroe–Shetland Channel. R/V Håkon Mosby cruise, HM110-96. University of Bergen.

HALD, N, and WAAGSTEIN, R. 1983. Silicic basalts from the Faroe Islands: evidence of crustal contamination. 343–349 in *Structure and Development of the Greenland–Scotland Ridge: New Methods and Concepts*. BOTT, M H P, SAXOV, S, TALWANI, M, and THIEDE, J (editors). (New York: Plenum Press.)

HALD, N, and WAAGSTEIN, R. 1991. The dykes and sills of the Early Tertiary Faroe Island basalt plateau. *Transactions of the Royal Society of Edinburgh: Earth Sciences*, Vol. 82, 373–388.

HALL, A M, and WHITTINGTON, G. 1989. Late Devensian glaciation of southern Caithness. *Scottish Journal of Geology*, Vol. 25, 307–324.

HALL, A M, HANSOM, J D, and JARVIS, J. 2008. Patterns and rates of erosion produced by high energy wave processes on hard rock headlands: The Grind of the Navir, Shetland, Scotland. *Marine Geology*, Vol. 248, 28–46.

HAMAR, G P, and HJELLE, K. 1984. Tectonic framework of the Møre Basin and the northern North Sea. 349–358 in *Petroleum Geology of the North European Margin*. SPENCER, A M, et al. (editors). Norwegian Petroleum Society. (London: Graham and Trotman.)

HANISCH, J. 1984. The Cretaceous opening of the northeast Atlantic. *Tectonophysics*, Vol. 101, 1–23.

HANSEN, B, and ØSTERHUS, S. 2000. North Atlantic–Nordic Seas exchanges. *Progress in Oceanography*, Vol. 45, 109–208.

HANSEN, B, TURRELL, W R, and ØSTERHUS, S. 2001. Decreasing overflow from the Nordic seas into the Atlantic Ocean through the Faroe Bank Channel since 1950. *Nature*, Vol. 411, 927–930.

HANSEN, D M, CARTWRIGHT, J A, and THOMAS, D. 2004. 3D seismic analysis of the geometry of igneous sills and sill junction relationships. 199–208 *in* 3D seismic technology: application to the exploration of sedimentary basins. DAVIES, R J, CARTWRIGHT, J A, STEWART, S A, LAPPIN, M, and UNDERHILL, J R (editors). *Geological Society of London Memoir*, No. 29.

HANSEN, H, REX, D C, GUISE, P G, and BROOKS, C K. 1995. $^{40}Ar/^{39}Ar$ ages on early Tertiary basalts from the Scoresby Sund area, East Greenland. *Newsletters on Stratigraphy*, Vol. 32, 103–116.

Hansen, H, Pedersen, A K, Duncan, R A, Bird, D K, Brooks, C K, Fawcett, J J, Gittens, J, Gorton, M, and O'Day, P. 2002. Volcanic stratigraphy of the southern Prinsen af Wales Bjerge region, East Greenland. 183–218 *in* The North Atlantic Igneous Province: stratigraphy, tectonic, volcanic and magmatic processes. Jolley, D W, and Bell, B R (editors). *Geological Society of London Special Publication*, No. 197.

Haq, B U, Hardenbol, J, and Vail, P R. 1987. Chronology of fluctuating sea levels since the Triassic. *Science*, Vol. 235, 1156–1167.

Harland, W B, Armstrong, R L, Cox, A V, Craig, L E, Smith, A G, and Smith, D G. 1990. *A Geologic Time Scale 1989*. (Cambridge: Cambridge University Press.)

Harker, S D. 2002. Cretaceous. 351–360 in *The Geology of Scotland*. Fourth edition. Trewin, N H (editor). (London: The Geological Society.)

Harris, A L, Bradbury, H J, Johnson, H D, and Smith, R A. 1978. Ensialic basin sedimentation: the Dalradian Group. 115–138 in *Crustal evolution in north-western Britain and adjacent regions*. Bowes, D R, and Leake, B E (editors). (Liverpool: Seal House Press.)

Haszeldine, R S. 1984. Carboniferous North Atlantic palaeogeography: stratigraphic evidence for rifting, not megashear or subduction. *Geological Magazine*, Vol. 121, 443–463.

Haszeldine, R S, and Russell, M J. 1987. The Late Carboniferous northern Atlantic Ocean: implications for hydrocarbon exploration from Britain to the Arctic. 1163–1175 in *Petroleum geology of north west Europe, proceedings of the 3rd conference*. Brooks, J, and Glennie, K W (editors). (London: Graham and Trotman.)

Haszeldine, R S, Ritchie, J D, and Hitchen, K. 1987. Seismic and well evidence for the early development of the Faroe–Shetland Basin. *Scottish Journal of Geology*, Vol. 23, 283–300.

Hatcher, P G, Wenzel, K A, and Cody, G D. 1994. Coalification reactions of vitrinite derived from coalified wood: transformations to rank of bituminous coal. 112–135 *in* Vitrinite reflectance as a maturity parameter: applications and limitations. Mukhopadhyay, P K, and Dow, W G (editors). *American Chemical Society, ACS Symposium Series*, No. 570.

Haug, G H, and Tiedemann, R. 1998. Effect of the formation of the Isthmus of Panama on Atlantic Ocean thermohaline circulation. *Nature*, Vol. **393**, 673–676.

Helland-Hansen, D. 2006. Rosebank — Challenges to development from a subsurface perspective. *Faroe Islands Exploration Conference 2006*, 12–14 September, 2006, Tórshavn, Faroe Islands (abstract).

Helland-Hansen, D. 2009. Rosebank — Challenges to development from a subsurface perspective. 241–245 *in* Faroe Islands Exploration Conference: Proceedings of the 2nd Conference. Varming, T, and Ziska, H (editors). *Annales Societatis Scientiarum Færoensis, Supplementum*, Vol. 50.

Herries, R, Poddubuik, R, and Wilcockson, P. 1999. Solan, Strathmore and the back basin play, West of Shetland. 693–712 in *Petroleum geology of northwest Europe, proceedings of the 5th conference*. Fleet, A J, and Boldy, S A R (editors). (London: The Geological Society.)

Himsworth, E M. 1973. Marine geophysical studies between northwest Scotland and the Faroes Plateau. University of Durham, unpublished PhD thesis.

Hinz, K, Dostman, H J, and Hanisch, J. 1984. Structural elements of the Norwegian Sea continental margin. *Geologisches Jahrbuch*, A 75, 193–211.

Hitchen, K, and Ritchie, J D. 1987. Geological review of the West Shetland area. 737–749 in *Petroleum geology of northwest Europe, proceedings of the 3rd conference*. Brooks, J, and Glennie, K W (editors). (London: Graham and Trotman.)

Hitchen, K, and Ritchie, J D. 1993. New K-Ar ages, and a provisional chronology, for the offshore part of the British Tertiary Igneous Province. *Scottish Journal of Geology*, Vol. 29, 73–85.

Hitchen, K, and Stoker, M S. 1993. Mesozoic rocks from the Hebrides Shelf and implications for hydrocarbon prospectivity in the northern Rockall Trough. *Marine and Petroleum Geology*, Vol. 10, 246–254.

Hitchen, K, Jolley, D W, Harland, R, and Jones, J M. 1995a. Regional implications of micropalaeontological and palaeo-environmental investigations of the early Eocene interval in BGS borehole 90/3, offshore NW Scotland. *Scottish Journal of Geology*, Vol. 31, 117–124.

Hitchen, K, Stoker, M S, Evans, D, and Beddoe-Stephens, B. 1995b. Permo-Triassic sedimentary and volcanic rocks in basins to the north and west of Scotland. 87–102 *in* Permian and Triassic rifting in Northwest Europe. Boldy, S A R (editor). *Geological Society of London Special Publication*, No. 91.

Hjelstuen, B O, Eldholm, O, and Skogseid, J. 1997. Vøring Plateau diapir fields and their structural and depositional settings. *Marine Geology*, Vol. 144, 33–57.

Hjelstuen B O, Eldholm, O, and Skogseid, J. 1999. Cenozoic evolution of the northern Vøring margin. *Geological Society of America Bulletin*, Vol. 111, 1792–1807.

Hodges, S, and Evans, R. 1999. The Other Millennium Dome. 1999 Offshore Europe Conference, Society of Petroleum Engineers, SPE 56895.

Holdsworth, R E, Harris, A L, and Roberts, A M. 1987. The stratigraphy, structure and regional significance of the Moine rocks of Mull, Argyllshire, W Scotland. *Geological Journal*, Vol. 22, 83–107.

Holdsworth, R E, Strachan, R A, and Harris, A L. 1994. Precambrian rocks in northern Scotland east of the Moine Thrust: the Moine Supergroup. 23–32 *in* A revised correlation of Precambrian rocks in the British Isles. Gibbons, W, and Harris, A L (editors). *Geological Society of London Special Report*, No. 22.

Hollingsworth, R. 2002. *BP Faroes Licence 004 Status Report*. Offshore Faroes Conference, Tórshavn (from Smallwood, 2005a and b).

HOLM, P M, HALD, N, and WAAGSTEIN, R. 2001. Geochemical and Pb-Sr-Nd isotopic evidence for separate hot depleted and Iceland plume mantle sources for the Paleogene basalts of the Faroe Islands. *Chemical Geology*, Vol. 178, 95–125.

HOLMES, A J, GRIFFITH, C E, and SCOTCHMAN, I C. 1999. The Jurassic petroleum system West of Britain Atlantic margin — an integration of tectonics, geochemistry and basin modelling. 1351–1365 in *Petroleum geology of northwest Europe, proceedings of the 5th conference*. FLEET, A J, and BOLDY, S A R (editors). (London: The Geological Society.)

HOLMES, R. 1997. Quaternary stratigraphy: the offshore record. 72–94 *in* GORDON, J E (editor). *Reflections on the Ice Age in Scotland.* (Glasgow: Scottish Natural Heritage.)

HOPPE, G. 1965. Submarine peat in the Shetland Islands. *Geografiska Annaler Stockholm*, Vol. 47A, 195–203.

HOPPE, G. 1974. The glacial history of the Shetland Islands. *Institute of British Geographers Special Publication*, No. 7.

HOVLAND, M, and JUDD, A G. 1988. *Sea-bed pock-marks and seepages: impact on geology, biology and marine environment.* (London: Graham and Trotman.)

HOWE, J A, STOKER, M S, STOW, D A V, and AKHURST, M C. 2002. Sediment drifts and contourite sedimentation in the north-eastern Rockall Trough and Faroe–Shetland Channel, North Atlantic Ocean. 65–72 *in* Deep-water contourite systems: modern drifts and ancient series, seismic and sedimentary characteristics. STOW, D A V , PUDSEY, C J, HOWE, J A, FAUGÈRES, J-C, and VIANA, A R (editors). *Geological Society of London Memoir*, No. 22.

HUDSON, J D, and TREWIN, H. 2002. Jurassic. 323–350 in *The Geology of Scotland.* Fourth edition. TREWIN, N (editor). (London: The Geological Society.)

HUGHES, S, BARTON, P J, and HARRISON, D. 1997. Characterising the mid Faroe Ridge using seismic velocity measurements. *Journal of Geophysical Research*, Vol. 102 (B4), 7837–7847.

HUGHES, S, BARTON, P J, and HARRISON, D. 1998. Exploration in the Shetland–Faroe Basin using densely spaced arrays of ocean-bottom seismometers. *Geophysics*, Vol. **63**, 490–501.

HUMLUM, O, CHRISTIANSEN, H H, SVENSSON, H, and MORTENSEN, L E. 1996. Moraine systems in the Faroe Islands: glaciological and climatological implications. *Danish Journal of Geography*, Vol. 96, 21–31.

HUUSE, M. 2002. Cenozoic uplift and denudation of southern Norway: insights from the North Sea Basin. 209–233 *in* Exhumation of the North Atlantic margin: timing, mechanisms and implications for petroleum exploration. DORÉ, A G, CARTWRIGHT, J A, STOKER, M S, TURNER, J P, and WHITE, N (editors). *Geological Society of London Special Publication*, Vol. 196.

ILIFFE, J E, ROBERTSON, A G, WARD, G H F , WYNN, C, PEAD, S D M, and CAMERON, N. 1999. The importance of fluid pressures and migration to the hydrocarbon prospectivity of the Faroe–Shetland White Zone. 601–611 in *Petroleum geology of northwest Europe, proceedings of the 5th conference.* FLEET, A J, and BOLDY, S A R (editors). (London: The Geological Society.)

INSTITUTE OF GEOLOGICAL SCIENCES. 1982. Caithness. Sheet 54°N–04°W. Solid geology. 1:250 000. (Southampton: Ordnance Survey for the Institute of Geological Sciences.)

ISAKSEN, G H, WILKINSON, D R, and HITCHEN, K. 2000. Geochemistry of organic-rich Cretaceous and Jurassic mudstones in the West Lewis and West Flannan basins, offshore north-west Scotland: Implications for source rock presence in the north-east Rockall Trough. *Marine and Petroleum Geology*, Vol. 17, 27–42.

JACKSON, P D, GUNN, D A, and LONG, D. 2004. Predicting variability in the stability of slope sediments due to earthquake ground motion in the AFEN area of the western UK continental shelf. *Marine Geology*, Vol. 213, 363–378.

JACQUES, J M, and REAVY, R J. 1994. Caledonian plutonism and major lineaments in the SW Scottish Highlands. *Journal of the Geological Society of London*, Vol. 151, 955–969.

JANSEN, E, and RAYMO, M E. 1996. 1. Leg 162: New Frontiers on Past Climates. 5–29 *in* JANSEN, E, RAYMO, M E, BLUM, P, *et al.* (editors). *Proceedings of the Ocean Drilling Program, Initial Reports*, Vol. 162.

JAPSEN, P, and CHALMERS, J A. 2000. Neogene uplift and tectonics around the North Atlantic: overview. *Global and Planetary Changes*, Vol. 24, 165–173.

JARDINE, W G, and PEACOCK, J D. 1973. Scotland. 53–59 *in* A correlation of Quaternary deposits in the British Isles. MITCHELL, G F, PENNY, L F, SHOTTON, F W, and WEST, R G (editors). *Geological Society of London Special Report*, No. 4.

JENSEN, A, and FREDERIKSEN, R. 1992. The fauna associated with the bank-forming deep-water coral *Lophelia pertusa* (Scleractinaria) on the Faroe Shelf. *Sarsia*, Vol. 77, 53–69.

JÓHANSEN, J. 1975. Pollen diagrams from the Shetland and Faroe Islands. *New Phytologist*, Vol. 75, 369–387.

JÓHANSEN, J. 1985. Pollen diagrams from the Shetland and Faroe Islands. *Annales Societatis Scientarium Fœroensis Supplementum*, Vol. 11.

JOHNS, C R, and ANDREWS, I J. 1985. The petroleum geology of the Unst Basin. *Marine and Petroleum Geology*, Vol. 2, 361–372.

JOHNSON, G S, SMITH, D I, and HARRIS, A L. 1969. The Moinian Assemblage of Scotland. 159–180 *in* North Atlantic Geology and Continental Drift. KAY, M (editor). *American Association of Petroleum Geologists Memoir*, No. 12. Tulsa, Oklahoma, USA.

JOHNSON, H, and LOTT, G K. 1993. 2. Cretaceous of the Central and Northern North Sea. In *Lithostratigraphic nomenclature of the UK North Sea.* KNOX, R W O'B, and CORDEY, W G (editors). (Keyworth, Nottingham: British Geological Survey.)

JOHNSON, H, RICHARDS, P C, LONG, D, and GRAHAM, C C. 1993. *United Kingdom offshore regional report: the geology of the northern North Sea.* (London: HMSO for the British Geological Survey.)

Johnson, H, Leslie, A B, Wilson, C K , Andrews, I J, and Cooper, R M. 2005a. Middle Jurassic, Upper Jurassic and Lower Cretaceous of the UK Central and Northern North Sea. *British Geological Survey Research Report*, No. RR/03/001.

Johnson, H, Ritchie, J D, Hitchen, K, McInroy, D B, and Kimbell, G S. 2005b. Aspects of the Cenozoic deformational history of the north-east Faroe–Shetland Basin, Wyville–Thomson Ridge and Hatton Bank areas. 993–1007 in *Petroleum geology: northwest Europe and global perspectives, proceedings of the 6th conference*. Doré, A G, and Vining, B (editors). (London: The Geological Society.)

Jolley, D W, and Bell, B R. 2002a. The evolution of the North Atlantic Igneous Province and the opening of the NE Atlantic rift. 1–13 *in* The North Atlantic Igneous Province: stratigraphy, tectonic, volcanic and magmatic processes. Jolley, D W, and Bell, B R (editors). *Geological Society of London Special Publication*, No. 197.

Jolley, D W, and Bell, B R. 2002b. Genesis and age of the Erlend volcano, NE Atlantic Margin. 95–109 *in* The North Atlantic Igneous Province: stratigraphy, tectonic, volcanic and magmatic processes. Jolley, D W, and Bell, B R (editors). *Geological Society of London Special Publication*, No. 197.

Jolley, D W, and Whitham, A G. 2004. A stratigraphicaland palaeoenvironmental analysis of the sub-basaltic Palaeogene sediments of East Greenland. *Petroleum Geoscience*, Vol. 10, 53–60.

Jolley, D W, Clarke, B, and Kelley, S. 2002. Paleogene time scale miscalibration: evidence from the dating of the North Atlantic igneous province. *Geology*, Vol. 30, 7–10.

Jolley, D W, Morton, A, and Prince, I. 2005. Volcanogenic impact on phytogeography and sediment dispersal patterns in the NE Atlantic. 969–975 1184 in *Petroleum geology: northwest Europe and global perspectives, proceedings of the 6th conference*. Doré, A G, and Vining, B (editors). (London: The Geological Society.)

Jones, E J W. 1981. Seismic refraction shooting on the continental margin west of the Outer Hebrides, north-west Scotland. *Journal of Geophysical Research*, Vol. 86, 11553–11574.

Jones, E J W, Ramsay, A T S, Preston, N J, and Smith, A C S. 1974. A Cretaceous guyot in the Rockall Trough. *Nature*, Vol. 251, 129–131.

Jones, R W, and Milton, N J. 1994. Sequence development during uplift: Palaeogene stratigraphy and relative sea-level history of the outer Moray Firth, UK North Sea. *Marine and Petroleum Geology*, Vol. 11, 157–165.

Jones, S M, White, N, Clarke, B J, Rowley, E, and Gallagher, K. 2002. Present and past influence of the Iceland Plume on sedimentation. 13 25 *in* Exhumation of the North Atlantic Margin: timing, mechanisms and implications for petroleum exploration. Doré, A G, Cartwright, J A, Stoker, M S, Turner, J P, and White, N (editors). *Geological Society of London Special Publication*, Vol. 196.

Jongepier, K, Rui, J C, and Grue, K. 1996. Triassic to early Cretaceous stratigraphic and structural development of the northeastern Møre Basin margin, off mid Norway. *Norsk Geologisk Tidsskrift*, Vol. 76, 199–214.

Jordt, H, Thyberg, B I, and Nøttvedt, A. 2000. Cenozoic evolution of the central and northern North Sea with focus on differential vertical movements of the basin floor and surrounding clastic source areas. 219–243 *in* Dynamics of the Norwegian Margin. Nøttvedt, A (editor). *Geological Society of London Special Publication*, Vol. 167.

Jørgensen, G, and Rasmussen, J. 1986. Glacial striae, roches moutonnées and ice movements in the Faroe Islands. *Geological Survey of Denmark series C*, No. 7.

Jørgensen, O. 2006. The regional distribution of zeolites in the basalts of the Faroe Islands and the significance of zeolites as palaeotemperature indicators. 123–156 *in* Scientific Results from the Deepened Lopra-1 Borehole, Faroe Islands. Chalmers, J A, and Waagstein, R (editors). *Geological Survey of Denmark and Greenland Bulletin*, No. 9.

Jowitt, R, Hindle, D, Jones, D, and Rose, P. 1999. Petroleum systems analysis of the Paleocene play in the West of Shetland area. 1367–1382 in *Petroleum geology of northwest Europe, proceedings of the 5th conference*. Fleet, A J, and Boldy, S A R (editors). (London: The Geological Society.)

Jung, W-Y, and Vogt, P R. 1997. A gravity and magnetic anomaly study of the extinct Aegir Ridge, Norwegian Sea. *Journal of Geophysical Research*, Vol. 102 (B3), 5065–5089.

Kalsbeek, F, Nutnam, A P, and Taylor, P N. 1993. Palaeoproterozic basement province in the Caledonian fold belt of north-east Greenland. *Precambrian Research*, Vol. 63, 163–178.

Kaminski, M A, and Austin, W E N. 1999. Oligocene deep-water agglutinated foraminifers at site 985, Norwegian basin, Southern Norwegian Sea. 169–177 *in* Jansen, E, Raymo, M E, Blum, P, et al. (editors). *Proceedings of the Ocean Drilling Program, Initial Reports*, Vol. 162. Ocean Drilling Program, College Station, Texas, USA.

Kanaris-Sotiriou, R. 1997. Graphite-bearing peraluminous dacites from the Erlend volcanic complex, Faroe–Shetland Basin, North Atlantic. *Mineralogical Magazine*, Vol. 61, 175–184.

Kanaris-Sotiriou, R, and Gibb, F G F. 1989. Plagiogranitic differentiates in MORB-type sills of the Faroe–Shetland Basin. *Journal of the Geological Society of London*, Vol. 146, 607–610.

Kanaris-Sotiriou, R, Morton, A C, and Taylor, P N. 1993. Palaeogene peraluminous magmatism, crustal melting and continental break-up: the Erlend complex, Faroe–Shetland Basin, NE Atlantic. *Journal of the Geological Society of London*, Vol. 150, 903–914.

Kenyon, N H. 1986. Evidence from bedforms for a strong poleward current along the upper continental slope of northwest Europe. *Marine Geology*, Vol. 72, 187–198.

Kenyon, N H. 1987. Mass-wasting features on the continental slopes of northwest Europe. *Marine Geology*, Vol. 74, 57–77.

Kenyon, N H, and Stride, A H. 1970. The tide-swept continental shelf sediments between the Shetland Isles and France. *Sedimentology*, Vol. 14, 159–173.

KENYON, N H, AKHMETZHANOV, A, NIELSEN, T, KUIJPERS, A, IVANOV, M, and SHASHKIN, P. 2003. Sandy contourites and pathways of the Norwegian Sea overflow water, west of the Faroe Bank Channel. 153–157 in *European margin sediment dynamics: side-scan sonar and seismic images*. MIENERT, J, and WEAVER, P (editors). (Berlin: Springer.)

KERSHAW, S. 2000. *Oceanography: An Earth Science Perspective*. (Cheltenham: Stanley Thornes.)

KESER NEISH, J. 2003. *The Faroese region: a standard structural nomenclature system*. (Tórshavn, Faroe Islands: Faroese Geological Survey.)

KESER NEISH, J. 2005. Faroese area: structural interpretation of seismic data in a basalt environment. 131–145 *in* Faroe Islands Exploration Conference: Proceedings of the 1st Conference ZISKA, H, VARMING, T, and BLOCH, D (editors). *Annales Societatis Scientarum Færoensis Supplementum*, Vol. 43.

KESER NEISH, J, and ZISKA, H. 2005. Structure of the Faroe–Bank Channel. 873–885 in *Petroleum geology: northwest Europe and global perspectives, proceedings of the 6th conference*. DORÉ, A G, and VINING, B (editors). (London: The Geological Society.)

KIMBELL, G S, GATLIFF, R W, RITCHIE, J D, WALKER, A S D, and WILLIAMSON, J P. 2004. Regional three-dimensional gravity modelling of the NE Atlantic margin. *Basin Research*, Vol. 16, 259–278.

KIMBELL, G S, RITCHIE, J D, JOHNSON, H, and GATLIFF, R W. 2005. Controls on the structure and evolution of the NE Atlantic margin revealed by regional potential field imaging and 3D modelling. 933–945 in *Petroleum geology: northwest Europe and global perspectives, proceedings of the 6th conference*. DORÉ, A G, and VINING, B (editors). (London: The Geological Society.)

KING, E L, SEJRUP, H P, HAFLIDASON, H , ELVERHOI, A, and AARSETH, I. 1996. Quaternary seismic stratigraphy of the North Sea Fan: glacially fed gravity flow aprons, hemipelagic sediments, and large submarine slides. *Marine Geology*, Vol. 130, 293–315.

KINNY, P D, FRIEND, C R L, and LOVE, G J. 2005. Proposal for a terrane-based nomenclature for the Lewisian Gneiss Complex of NW Scotland. *Journal of the Geological Society of London*, Vol. 162, 175–186.

KIØRBOE, L. 1999. Stratigraphic relationships of the Lower Tertiary of the Faroe Basalt Plateau and the Faroe–Shetland Basin. 559–572 in *Petroleum geology of northwest Europe, proceedings of the 5th conference*. FLEET, A J, and BOLDY, S A R (editors). (London: The Geological Society.)

KIRTON, S R, and HITCHEN, K. 1987. Timing and style of crustal extension N of the Scottish mainland. 501–510 *in* Continental extensional tectonics. COWARD, M P, DEWEY, J F, and HANCOCK, P L (editors). *Geological Society of London Special Publication*, No. 28.

KLEMPERER, S, and HOBBS, R. 1991. *The BIRPS Atlas. Deep seismic reflection profiles around the British Isles*. (Cambridge: Cambridge University Press.)

KLINGELHÖFER, F, EDWARDS, R A , HOBBS, R W, and ENGLAND, R W. 2005. Crustal structure of the NE Rockall Trough from wide-angle seismic data modelling. *Journal of Geophysical Research*, Vol. 110 (B11105), 1–25.

KLITGAARD, A B, TENDAL, O S, and WESTERBERG, H. 1997. Mass occurrences of large sponges (*Porifera*) in Faroe Island (NE Atlantic) shelf and slope areas: characteristics, distribution and possible causes. 129–142 *in* The responses of marine organisms to their environment. HAWKINS, L E, and HUTCHINS, S (editors). Proceedings of the 30th European Marine Biological Symposium, Southampton, UK.

KNOTT, S D, BURCHELL, M T, JOLLEY, E J, and FRASER, A J. 1993. Mesozoic to Cenozoic plate reconstructions of the North Atlantic and hydrocarbon plays of the Atlantic margins. 953–974 in *Petroleum geology of northwest Europe, proceedings of the 4th conference*. PARKER, J R (editor). (London: The Geological Society.)

KNOX, R W O'B. 2002. Tertiary sedimentation. 361–370 in *The Geology of Scotland*. Fourth edition. TREWIN, N H (editor). (London: The Geological Society.)

KNOX, R W O'B, and MORTON, A C. 1988. The record of early Tertiary N Atlantic volcanism in sediments of the North Sea Basin. 407–419 *in* Early Tertiary volcanism and the opening of the NE Atlantic. MORTON, A C, and PARSON, L M (editors). *Geological Society of London Special Publication*, No. 39.

KNOX, R W O'B, HOLLOWAY, S, KIRBY, G A, and BAILEY, H E. 1997. *Stratigraphic nomenclature of the UK North West Margin. 2. Early Palaeogene lithostratigraphy and sequence stratigraphy*. (Keyworth, Nottingham: British Geological Survey.)

KNUDSEN, T-L. 2000. The provenance of Devonian sandstones from Shetland: a Sm-Nd and trace element study. *Scottish Journal of Geology*, Vol. 36, 61–72.

KNUTZ, P K, and CARTWRIGHT, J. 2003. Seismic stratigraphy of the West Shetland Drift: implications for late Neogene paleocirculation in the Faroe–Shetland gateway. *Paleoceanography*, Vol. 18, 4.

KNUTZ, P K, and CARTWRIGHT, J. 2004. 3D anatomy of late Neogene contourite drifts and associated mass flows in the Faroe–Shetland Basin. 63–71 *in* 3D seismic technology: application to the exploration of sedimentary basins. DAVIES, R M, CARTWRIGHT, J A, STEWART, S A, LAPPIN, M, and UNDERHILL, J R (editors). *Geological Society of London Memoir*, No. 29.

KOOPMANS, M P, KÖSTER, J, VAN KAAM-PETERS, H M E, KENIG, F, SCHOUTEN, S, HARTGERS, W A, DE LEEUW, J W, and SINNINGHE DAMSTÉ, J S. 1996. Diagenetic and catagenetic products of isorenieratene: molecular indicators for photic zone anoxia. *Geochimica et Cosmochimica Acta*, Vol. 60, 4467–4496.

KRABBENDAM, M, PRAVE, T, and CHEER, D. 2008. A fluvial origin for the Neoproterozoic Morar Group, NW Scotland; implications for Torridon–Morar Group correlation and the Grenville Orogen foreland basin. *Journal of the Geological Society of London*, Vol. 165, 379–394.

KUIJPERS, A, ANDERSEN, M S, KENYON, N H, KUNZENDORF, H, and VAN WEERING, T C E. 1998. Quaternary sedimentation around Bill Bailey Bank, north-eastern Atlantic. *Marine Geology*, Vol. 152, 101–127.

Kuijpers, A, Nielsen, T, Akhmetzhanov, A, de Haas, H, Kenyon, N H, and van Weering, T C E. 2001. Late Quaternary slope instability on the Faroe margin: mass flow features and timing of events. *Geo-Marine Letters*, Vol. 20, 149–159.

Kuijpers, A, Hansen, B, Hühnerbach, V, Larsen, B, Nielsen, T, and Werner, F. 2002. Norwegian Sea overflow through the Faroe–Shetland gateway as documented by its bedforms. *Marine Geology*, Vol. 188, 147–164.

Kusznir, N. 2005. The state of stress within rifted continental margin lithosphere: geodynamic constraints. Conference proceedings on compressional deformation within passive margins: nature causes and effects, October 2005. (London: The Geological Society.) (abstract).

Laberg, J S, Stoker, M S, Dahlgren, K I T, de Haas, H, Haflidason, H, Hjelstuen, B O, Nielsen, T, Shannon, P, Vorren, T O, van Weering, T C E, and Ceramicola, S. 2005. Cenozoic alongslope processes and sedimentation on the NW European Atlantic margin. *Marine and Petroleum Geology*, Vol. 22, 1069–1088.

Laier, T, Nytoft, H P, Jørgensen, O, and Isaksen, G H. 1997. Hydrocarbon traces in the Tertiary basalts of the Faroe Islands. *Marine and Petroleum Geology*, Vol. 14, 257–266.

Lamb, H H. 1977. *Climate: Past, Present and Future, Volume 2. Climatic History and the Future.* (London: Methuen.)

Lambeck, K. 1995. Late Devensian and Holocene shorelines of the British Isles and North Sea from models of glacio-isostatic rebound. *Journal of the Geological Society of London*, Vol. 152, 437–448.

Lamers, E, and Carmichael, S M M. 1999. The Paleocene deepwater sandstone play West of Shetland. 645–659 in *Petroleum geology of northwest Europe, proceedings of the 5th conference.* Fleet, A J, and Boldy, S A R (editors). (London: The Geological Society.)

Landis, P, Michels, R, and Elie, M. 1994. Are time and temperature the only constraints to the simulation of organic matter maturation? *Organic Geochemistry*, Vol. 22, 617–630.

Landt, J. 1800. Forsøg til en beskrivelse over Færøerne. Kiøbenhavn. 479pp.

Larsen, H C, and Marcussen, C. 1992. Sill-intrusion, flood basalt emplacement and deep crustal structure of the Scoresby Sund region, East Greenland. 365–386 *in* Magmatism and the causes of continental break-up. Storey, B C, Alabaster, T, and Pankhurst, R J (editors). *Geological Society of London Special Publication*, No. 68.

Larsen, H C, and Saunders, A D. 1998. Tectonism and volcanism at the southeast Greenland rifted margin: a record of plume impact and later continental rupture. 503–533 *in* Saunders, A D, Larsen, H C, and Wise, S W, Jr (editors). *Proceedings of the Ocean Drilling Program, Scientific Results*, Vol. 152. Ocean Drilling Program, College Station, Texas, USA.

Larsen, H C, Dahl-Jensen, T, and Hopper, J R. 1998. Crustal structure along the Leg 152 drilling transect. 463–475 *in* Saunders, A D, Larsen, H C, and Wise, S W, Jr (editors). *Proceedings of the Ocean Drilling Program, Scientific Results*, Vol. 152. Ocean Drilling Program, College Station, Texas, USA.

Larsen, L M, Waagstein, R, Pedersen, A K, and Storey, M. 1999. Trans-Atlantic correlation of the Palaeogene volcanic successions in the Faroe Islands and East Greenland. *Journal of the Geological Society of London*, Vol. 156, 1081–1095.

Larsen, M, and Whitham, A G. 2005. Evidence for a major sediment input point into the Faroe–Shetland Basin from the Kangerlussuaq region of southern East Greenland. 913–922 in *Petroleum geology: northwest Europe and global perspectives, proceedings of the 6th conference.* Doré, A G, and Vining, B (editors). (London: The Geological Society.)

Larsen, M, Hamberg, L, Olaussen, S, Preuss, T, and Stremmrik, L. 1999. Sandstone wedges of the Cretaceous–Lower Tertiary Kangerlussuaq Basin, East Greenland–outcrop analogues to the offshore North Atlantic. 337–348 in *Petroleum geology of northwest Europe, proceedings of the 5th conference.* Fleet, A J, and Boldy, S A R (editors). (London: The Geological Society.)

Larsen, P-H, and Bengaard, H-J. 1991. Devonian basin initiation in East Greenland: a result of sinistral wrench faulting and Caledonian extensional collapse. *Journal of the Geological Society of London*, Vol. 148, 355–368.

Lassen, S, Jansen, E, Knudsen, K L, Kuijpers, A, Kristensen, M, and Christensen, K. 1999. North-east Atlantic surface circulation during the last 30,000 to 10 000 C-14 years. *Paleoceanography*, Vol. 14, 616–625.

Lassen, S, Kuijpers, A, Kunzendorf, H, Lindgren, H, Heinemeier, J, Jansen, E, and Knudsen, K L. 2001. Intermediate water signal leads surface water response during north-east Atlantic deglaciation. *Global and Planetary Change*, Vol. 32, 111–125.

Leach, H M, Herbert, N, Los, A, and Smith, R L. 1999. The Schiehallion development. 683–692 in *Petroleum geology of northwest Europe, proceedings of the 5th conference.* Fleet, A J, and Boldy, S A R (editors). (London: The Geological Society.)

Lear, C H, Rosenthal, Y, and Wright, J D. 2003. The closing of a seaway: ocean water masses and global climate change. *Earth and Planetary Sciences*, Vol. 210, 425–436.

Leeder, M R. 1988. Recent developments in Carboniferous geology: a critical review with implications for the British Isles and NW. Europe. *Proceedings of the Geologists' Association*, Vol. 99, 73–100.

Light, J M, and Wilson, J B. 1998. Cool-water carbonate deposition on the West Shetland Shelf: a modern distally steepened ramp. 73–105 *in* Carbonate ramps. Wright, V P, and Burchette, T P (editors). *Geological Society of London Special Publication,* No. 149.

Linnard, S, and Nelson, R. 2005. Effects of Tertiary Volcanism and later Events upon the Faroese Hydrocarbon system. 44–53 *in* Faroe Islands Exploration Conference: Proceedings of the 1st Conference. Ziska, H, Varming, T, and Bloch, D (editors). *Annales Societatis Scientarum Færoensis Supplementum,* No. 43.

LOIZOU, N. 2005. West of Shetland exploration unravelled — an indication of what the future may hold. *First Break*, Vol. 23, 53–59.

LOIZOU, N, ANDREWS, I J, STOKER, S J, and CAMERON, D. 2006. West of Shetland revisited: the search for stratigraphic traps. 225–245 *in* The deliberate search for the stratigraphic trap. ALLEN, M R, GOFFEY, G P, MORGAN, R K, and WALKER, I M (editors). *Geological Society of London Special Publication,* No. 254.

LONG, D. 2006. Cold-water corals. *Earthwise*, Vol. 23, 17. (Keyworth, Nottingham: British Geological Survey.)

LONG, D, and BONE, B D. 1990. Sediment instability on the continental slope of north-western Europe. Proceedings of Oceanology International, 1990.

LONG, D, and MOORE, I C. 1980. Geotechnical aspects of the 1980 MV Whitethorn sea-bed sampling cruises. *Institute of Geological Sciences, Engineering Geology Unit (Scotland) Report,* No. 80/4.

LONG, D, WALKER, P M, HILL, A W, and ARDUS, D A. 1998. The Western Frontiers Association – an example of joint industry-funded regional studies of the shallow geology in a frontier area. *Proceedings of the Offshore Site Investigation and Foundation Behaviour – 'new frontiers' Conference 22–24 September 1998.* ARDUS, D A, HOBBS, R, HORSNELL, M, JARDINE, R, LONG, D, and SOMMERVILLE, J (editors). (London: Society for Underwater Technology.)

LONG, D, ROBERTS, J M, and GILLESPIE, E J. 1999. Occurrences of *Lophelia pertusa* on the Atlantic margin. *British Geological Survey Technical Report*, No. WB/99/24.

LONG, D, STEVENSON, A G, WILSON, C K, and BULAT, J. 2003. Slope failures in the Faroe–Shetland Channel. 281–289 in *Submarine mass movements and their consequences*. LOCAT, J, and MIENERT, J (editors). Advances in Natural and Technological Hazards Research Series. (Dordrecht: Kluwer Academic Publishers.)

LONG, D, BULAT, J, and STOKER, M S. 2004. Sea-bed morphology of the Faroe–Shetland Channel derived from 3D seismic datasets. 53–61 *in* 3D seismic technology: application to the exploration of sedimentary basins. DAVIES, R J, CARTWRIGHT, J A, STEWART, S A, LAPPIN, M, and UNDERHILL, J R (editors). *Geological Society of London Memoir*, No. 29.

LONG, D, JACKSON, P D, LOVELL, M A, ROCHELLE, C A, FRANCIS, T J G, and SCHULTHEISS, P J. 2005. Methane hydrates: problems in unlocking their potential. 723–730 in *Petroleum geology: northwest Europe and global perspectives, proceedings of the 6th conference*. DORÉ, A G, and VINING, B (editors). (London: The Geological Society.)

LØSETH, H, and HENRIKSEN, S. 2005. A Middle to Late Miocene compression phase along the Norwegian passive margin. 845–859 in *Petroleum geology: northwest Europe and global perspectives, proceedings of the 6th conference*. DORÉ, A G, and VINING, B (editors). (London: The Geological Society.)

LOVE, G J, KINNY, P D, and FRIEND, C R L. 2004. Timing of magmatism and metamorphism in the Gruinard Bay area of the Lewisian Gneiss Complex: comparisons with the Assynt Terrane and implications for terrane accretion. *Contributions to Mineralogy and Petrology*, Vol. 46, 620–636.

LUNDIN, E. 2002. North Atlantic–Arctic: overview of sea-floor spreading and rifting history. 40–47 in *BATLAS – mid Norway plate reconstruction atlas with global and Atlantic perspectives*. EIDE, E A (co-ordinator). (Geological Survey of Norway.)

LUNDIN, E, and DORÉ A G. 1997. A tectonic model for the Norwegian passive margin with implications for the NE Atlantic: Early Cretaceous to break-up. *Journal of the Geological Society of London*, Vol. 154, 545–550.

LUNDIN, E, and DORÉ, A G. 2002. Mid Cenozoic post-breakup deformation in the 'passive' margins bordering the Norwegian–Greenland Sea. *Marine and Petroleum Geology*, Vol. 19, 79–93.

LUNDIN, E, and DORÉ, A G. 2005. NE Atlantic break-up: a re-examination of the Iceland mantle plume model and the Atlantic–Arctic linkage. 933–945 in *Petroleum geology: northwest Europe and global perspectives, proceedings of the 6th conference*. DORÉ, A G, and VINING, B (editors). (London: The Geological Society.)

LÜNING, S, and KOLONIC, S. 2003. Uranium spectral gamma-ray response as a proxy for organic richness in black shales: applicability and limitations. *Journal of Petroleum Geology*, Vol. 26(2), 153–174.

MACLENNAN, J, and LOVELL, B. 2002. Control of regional sea level by regional uplift and subsidence caused by magmatic underplating of Earth's crust. *Geology*, Vol. 30, 675–678.

MAKRIS, J, LANGE, K, SAVOSTIN, L, and SEDOV, V. 1995. A wide-angle reflection profile across the Iceland Faroe Ridge. 459–466 *in* The Petroleum Geology of Ireland's offshore basins. CROKER P F, and SHANNON P M (editors). *Geological Society of London Special Publication*, No. 93.

MARSHALL, J E A. 1988. Devonian miospores from Papa Stour, Shetland. *Transactions of the Royal Society of Edinburgh: Earth Sciences*, Vol. 79, 13–18.

MARSHALL, J E A. 1991. Palynology of the Stonehaven Group, Scotland: evidence for a Mid Silurian age and its geological implications. *Geological Magazine*, Vol. 128, 283–286.

MARSHALL, J E A. 2000. Devonian (Givetian) miospores from the Walls Group, Shetland. *in New Perspectives on the Old Red Sandstone*. FRIEND, P F, and WILLIAMS, B P J (editors). *Geological Society of London Special Publication*, No. 180, 473–483.

MARSHALL, J E A, and HEWETT, A J. 2003. Devonian. 65–81 in *Millennium atlas: petroleum geology of the central and northern North Sea*. EVANS, D, GRAHAM, C G, ARMOUR, A, and BATHHURST, P (editors). (London: The Geological Society.)

MARSHALL, J E A, BROWN, J F, and HINDMARSH, S. 1985. Hydrocarbon potential of the Devonian rocks of the Orcadian Basin. *Scottish Journal of Geology*, Vol. 21, 301–320.

MARSHALL, J E A, ROGERS, D A, and WHITELEY, M J. 1996. Devonian marine incursions into the Orcadian Basin, Scotland. *Journal of the Geological Society of London*, Vol. 153, 451–466.

MARTINI, E, and MÜLLER, C. 1976. 28. Eocene to Pleistocene silicoflagellates from the Norwegian–

Greenland Sea (DSDP Leg 38). 857–895 *in* TALWANI, M, UDINTSEV, G, et al. (editors). *Initial reports of the Deep Sea Drilling Project,* No. 38. US Government Printing Office, Washington.

MARTINSEN, D G, PISIAS, N G, HAYS, J D, IMBRIE, J, MOORE, T C, and SHACKLETON, N J. 1987. Age dating and the orbital theory of the ice ages; development of a high-resolution 0 to 300,000-year chronostratigraphy. *Quaternary Research,* Vol. 27, 1–29.

MARTINSEN, O J, BØEN, F, CHARNOCK, M A, MANGERUD, G, and NØTTVEDT, A. 1999. Cenozoic development of the Norwegian margin 60–64° N: sequences and sedimentary response to variable basin physiography and tectonic setting. 293–304 in *Petroleum geology of northwest Europe, proceedings of the 5th conference.* FLEET, A J, and BOLDY, S A R (editors). (London: The Geological Society.)

MARTINSEN, O J, LIEN, T, and JACKSON, C. 2005. Cretaceous and Palaeogene turbidite systems in the North Sea and Norwegian Sea basins: source, staging area and basin physiography controls on reservoir development. 1147–1164 in *Petroleum geology: northwest Europe and global perspectives, proceedings of the 6th conference.* DORÉ, A G, and VINING, B (editors). (London: The Geological Society.)

MASSON, D G. 2001. Sedimentary processes shaping the eastern slope of the Faroe–Shetland Channel. *Continental Shelf Research,* Vol. 21, 825–857.

MASSON, D G, LE BAS, T P, BETT, B J, HUEHNERBACH, V, JACOBS, C L, WYNN, R B, and DEPARTMENT OF TRADE AND INDUSTRY. 2003. Seafloor sediments and sedimentary processes on the outer continental shelf, continental slope and basin floof. Strategic Environmental Assessment: SEA 4. Consultation Document. London, Department of Trade and Industry.

MASSON, D G, HOWE, J A, and STOKER, M S. 2002. Bottom current sediment waves, sediment drifts and contourites in the northern Rockall Trough. *Marine Geology,* Vol. 192, 215–237.

MASSON, D G, WYNN, R B, and BETT, B J. 2004. Sedimentary environment of the Faroe–Shetland and Faroe Bank Channels, north-east Atlantic, and the use of bedforms as indicators of bottom current velocity in the deep ocean. *Sedimentology,* Vol. 51, 1207–1241.

MATHER, A S, and SMITH, J S. 1974. Beaches of Shetland. (Aberdeen: Department of Geography, University of Aberdeen.)

MCBRIDE, J H. 1994. Structure of a continental strike-slip fault from deep seismic reflection: Walls Boundary fault, northern British Caledonide. *Journal of Geophysical Research,* Vol. 99, 23985–24005.

MCBRIDE, J H, and ENGLAND, R W. 1994. Deep seismic structure of the Caledonian orogenic front west of Shetland. *Journal of the Geological Society of London,* Vol. 151, 9–16.

MCBRIDE, J H, and ENGLAND, R W. 1995. Problems resolving deep crustal reflectors beneath the Moine Thrust between Shetland and the Scottish Highlands. *Geological Magazine,* Vol. 132, 465–472.

MCBRIDE, J H, SNYDER, D B, TATE, M P, ENGLand, R W, and HOBBS, R W. 1995. Upper mantle reflector structure and origin beneath the Scottish Caledonides. *Tectonics,* Vol. 14, 1351–1367.

MCBRIDE, J H, WHITE, R S, SMALLWOOD, J R, and ENGLAND, R W. 2004. Must magmatic intrusion in the lower crust produce reflectivity? *Tectonophysics,* Vol. 388, 271–297.

MCCLAY, K R, NORTON, M G, CONEY, P, and DAVIS, G H. 1986. Collapse of the Caledonide Orogen and the Old Red Sandstone. *Nature,* Vol. 525, 147–149.

MCGEARY, S. 1989. Reflection seismic evidence for a Moho offset beneath the Walls Boundary strike-slip fault. *Journal of the Geological Society of London,* Vol. 146, 261–299.

MCINROY, D B, HITCHEN, K, and STOKER, M S. 2006. Potential Eocene and Oligocene stratigraphic traps of the Rockall Plateau, NE Atlantic Margin. 247–266 *in* The deliberate search for the stratigraphic trap. ALLEN, M R, GOFFEY, G P, MORGAN, R K, and WALKER, I M (editors). *Geological Society of London Special Publication,* No. 254.

MCKENZIE, D. 1978. Some remarks on the development of sedimentary basins. *Earth and Planetary Science Letters,* Vol. 40, 25–32.

MCKERROW, W S, MACNIOCAIL, C, and DEWEY, J F. 2000. The Caledonian Orogeny redefined. *Journal of the Geological Society of London,* Vol. 157, 1149–1157.

MCKIE, T, and GARDEN, I R. 1996. Hierarchical cycles in the non-marine Clair Group (Devonian), UKCS. 23–30 *in* High resolution sequence stratigraphy: innovations and applications. HOWELL, J A, and AITKEN, J F (editors). *Geological Society of London Special Publication,* No. 89.

MCLEAN, A C. 1978. Evolution of fault-controlled ensialic basins in north-western Britain. 325–346 *in* Crustal Evolution in North-western Britain and Adjacent Regions. BOWES, D R, and LEAKE, B E (editors). *Geological Society of London Special,* No. 10.

MCQUILLIN, R, DONATO, J A, and TULSTRUP, J. 1982. Development of basins in the Inner Moray Firth and the North Sea by crustal extension and dextral displacement of the Great Glen Fault. *Earth and Planetary Science Letters,* Vol. 60, 127–139.

MEADOWS, N S, MACCHI, L, CUBITT, J M, BAILEY, N J L, and JOHNSON, B. 1987. Sedimentary and reservoir potential in the West Shetland Basin, UK exploration area. 723–736 in *Petroleum geology of northwest Europe, proceedings of the 3rd conference.* BROOKS, J, and GLENNIE, K W (editors). (London: Graham and Trotman.)

MESCHEDE, M. 1986. A method of discriminating between different types of mid ocean ridge basalts and continental tholeiites with the Nb-Zr-Y diagram. *Chemical Geology,* Vol. 56, 207–218.

MIDDLETON, M F. 1982. Tectonic history from vitrinite reflectance. *Geophysical Journal of the Royal Astronomical Society,* Vol. 68, 121–138.

MILLER, J A, and FLINN, D. 1966. A survey of the age relations of Shetland rocks. *Geological Journal,* Vol. 5, 95–116.

MILLER, K G, and TUCHOLKE, B E. 1983. Development of Cenozoic abyssal circulation south of the Greenland–Scotland Ridge. 549–589 in *Structure and development of the Greenland–Scotland Ridge: new methods and concepts.* BOTT, M H P,

SAXOV, S, TALWANI, M, and THIEDE, J (editors). (New York: Plenum Press.)

MITCHELL, J G, and EUWE, M G. 1988. A model of single-stage concomitant potassium-argon exchange in acidic lavas from the Erlend Volcanic Complex, north of Shetland Islands. *Chemical Geology (Isotope Geoscience Section)*, No. 72, 95–109.

MITCHELL, S M, BEAMISH, G W, WOOD, M V, MALECEK, S J, ARMENTROUT, J A, DAMUTH, J E, and OLSON, H C. 1993. Palaeogene sequence stratigraphic framework of the Faroe Basin. 1011–1023 in *Petroleum geology of northwest Europe, proceedings of the 4th conference*. PARKER, J R (editor). (London: The Geological Society.)

MJELDE, R, DIGRANES, P, SHIMAMURA, H, SHIOBARA, H, KODAIRA, S, BREKKE, H, EGEBJERG, T, SØRENES, N, and THORBJØRNSEN, S. 1998. Crustal structure of the northern part of the Vøring Basin, mid Norway margin, from wide-angle seismic and gravity data. *Tectonophysics*, Vol. 293, 175–205.

MOORBATH, S, and PARK, R G. 1971. The Lewisian chronology of the southern region of the Scottish mainland. *Scottish Journal of Geology*, Vol. 8, 51–74.

MÖRCH, O A L. 1868. Faunula molluscorum insularum Færoënsium. Beretning om de hidtil fra Færøerne bekjendte bløddyr. *Videnskabelige Meddelelser fra Dansk naturhistorisk Forening*, 1–47.

MORGAN, J V, BARTON, P J, and WHITE, R S. 1989. The Hatton Bank continental margin — III. Structure from wide-angle OBS and multichannel seismic refraction profiles. *Geophysical Journal International*, Vol. 98, 367–384.

MORGAN, J V, HADWIN, M, WARNER, M, BARTON, P, and MORGAN, R P LL. 1994. The polarity of deep seismic reflections from the lithospheric mantle: evidence for a relic subduction zone. *Tectonophysics*, Vol. 232, 319–328.

MORGAN, R P LL, BARTON, P J, WARNER, M, MORGAN, J, PRICE, C, and JONES, K. 2000. Lithospheric structure north of Scotland – I. P-wave modelling, deep reflection profiles and gravity. *Geophysical Journal International*, Vol. 142, 716–736.

MORTON, A C. 1995. Provenance and correlation of Triassic, Jurassic and Cretaceous sandstones, UKCS blocks 205/26: heavy mineral constraints. Unpublished report for Amerada Hess (from Herries et al., 1999.)

MORTON, A C, and KNOX, R W O'B. 1990. Geochemistry of late Palaeocene and early Eocene tephras from the North Sea Basin. *Journal of the Geological Society of London*, Vol. 147, 425–437.

MORTON, A C, BACKMAN, J, and HARLAND, R. 1983. A reassessment of the stratigraphy of DSDP hole 117A, Rockall Plateau: implications for the Paleocene–Eocene boundary in NW Europe. *Newsletters on Stratigraphy*, Vol. 12, 102–111.

MORTON, A C, DIXON, J E, FITTON, J G, MACINTYRE, R M, SMYTHE, D K, and TAYLOR, P N. 1988a. Early Tertiary volcanic rocks in well 163/6-1A, Rockall Trough. 293–308 *in* Early Tertiary volcanism and the opening of the NE Atlantic. MORTON, A C, and PARSON, L M (editors). *Geological Society of London Special Publication,* No. 39.

MORTON, A C, EVANS, D, HARLand, R, KING, C and RITCHIE, J D. 1988b. Volcanic ash in a cored borehole W of the Shetland Islands: evidence for Selandian (late Paleocene) volcanism in the Faroes region. 263–269 *in* Early Tertiary volcanism and the opening of the NE Atlantic. MORTON, A C, and PARSON, L M (editors). *Geological Society of London Special Publication,* No. 39.

MORTON, A C, HITCHEN, K, RITCHIE, J D, HINE, N M, WHITEHOUSE, M, and CARTER, S G. 1995. Late Cretaceous basalts from the Rosemary Bank, northern Rockall Trough. *Journal of the Geological Society of London*, Vol. 152, 947–952.

MORTON, A C, BOYD, J D, and EWEN, D F. 2002a. Evolution of Paleocene sediment dispersal systems in the Foinaven Sub-basin, west of Shetland. 69–93 *in* The North Atlantic Igneous Province: stratigraphy, tectonic, volcanic and magmatic processes. JOLLEY, D W, and BELL, B R (editors). *Geological Society of London Special Publication*, No. 197.

MORTON, A C, KNOX, R W O'B, and HALLSWORTH, C. 2002b. Correlation of reservoir sandstones using quantitative heavy mineral analysis. *Petroleum Geoscience*, Vol. 8, 251–262.

MORTON, A C, SPICER, P J, and EWEN, D. 2003. Geosteering of high-angle wells using heavy-mineral analysis: the Clair Field, west of Shetland, UK. 249–260 *in* Horizontal wells: focus on the reservoir. CARR, T R, MASON, E P, and FEAZEL, C T (editors). *American Association of Petroleum Geologists, Methods in Exploration*, Vol. 14.

MORTON, N. 1987. Jurassic subsidence history in the Hebrides, NW Scotland. *Marine and Petroleum Geology*, Vol. 4, 226–242.

MORTON, N. 1989. Jurassic sequence stratigraphy in the Hebrides Basin. *Marine and Petroleum Geology*, Vol. 6, 243–260.

MOSAR, J, LEWIS, G, and TORSVIK, T H. 2002. North Atlantic sea-floor spreading rates: implications for the Tertiary development of inversion structures of the Norwegian–Greenland Sea. *Journal of the Geological Society of London*, Vol. 159, 503–515.

MUDGE, D C, and BUJAK, J P. 2001. Biostratigraphic evidence for evolving palaeoenvironments in the Lower Palaeogene of the Faroe–Shetland Basin. *Marine and Petroleum Geology*, Vol. 18, 577–590.

MUDGE, D C, and COPESTAKE, P. 1992. Revised Lower Palaeogene lithostratigraphy for the Outer Moray Firth, North Sea. *Marine and Petroleum Geology*, Vol. 9, 53–69.

MUDGE, D C, and RASHID, B. 1987. The geology of the Faroese region. 751–736 in *Petroleum geology of northwest Europe, proceedings of the 3rd conference*. BROOKS, J, and GLENNIE, K W (editors). (London: Graham and Trotman.)

MUSGROVE, F W, and MITCHENER, B. 1996. Analysis of the pre-Tertiary history of the Rockall Trough. *Petroleum Geoscience*, Vol.2, 353–360.

MUSSON, R M W. 1998. Historical seismicity of the Western Frontiers area. *British Geological Survey Technical Report*, No. WL/98/26.

MUSSON, R M W, MIKKELSEN, T, and ZISKA, H. 2001. Historical seismicity of the Faroe Islands. *Annali di Geofisica*, Vol. 44, 1031–1047.

MUTTER, J C, TALWANI, M, and D, STOFFA, P L. 1982. Origin of seaward-dipping reflectors in oceanic crust off the Norwegian margin by 'subaerial sea-floor spreading'. *Geology*, Vol. 10, 3–12.

MYHRE, A M, ELDHOLM, O, FALEIDE, J I, SKOGSEID, J, GUDLAUGSSON, S T, PLANKE, S, STUEVOLD, L, and VÅGNES, E. 1992. Norway–Svalbard continental margin: structural and stratigraphic styles. 157–185 in *Geological evolution of Atlantic continental margins*. POAG, C W, and DE GRACIANSKY, P C (editors). (New York: Van Nostrand Reinhold.)

MYKURA, W. 1976. *British regional geology: Orkney and Shetland*. (London: HMSO for the British Geological Survey.)

MYKURA, W. 1991. *The Old Red Sandstone*. 297–346 in *The Geology of Scotland*. Third edition. CRAIG, G Y (editor). (London: The Geological Society.)

MYKURA, W, and PHEMISTER, J. 1976. The Geology of Western Shetland. *Memoir of the Geological Survey of Great Britain*.

NADIN, P A, KUSZNIR, N J, and CHEADLE, M J. 1997. Early Tertiary plume uplift of the North Sea and Faroe–Shetland Basins. *Earth and Planetary Science Letters*, Vol. 148, 109–207.

NADIN, P A, HOUCHEN, M A, and KUZNIR, N J. 1999. Evidence for pre-Cretaceous rifting in the Rockall Trough: an analysis using quantitative 2D structural/stratigraphic modelling. 371–378 in *Petroleum geology of northwest Europe, proceedings of the 5th conference*. FLEET, A J, and BOLDY, S A R (editors). (London: The Geological Society.)

NAYLOR, D, and SHANNON P W. 2005. The structural framework of the Irish Atlantic Margin. 1009–1022 in *Petroleum geology: northwest Europe and global perspectives, proceedings of the 6th conference*. DORÉ, A G, and VINING, B (editors). (London: The Geological Society.)

NAYLOR, D, SHANNON, P, and MURPHY, N. 1999. Irish Rockall Basin region– a standard structural nomenclature system. *Petroleum Affairs Division. Special Publication*, 1/99.

NAYLOR, P H, BELL, B R, JOLLEY, D W, DURNALL, P, and FREDSTED, R. 1999. Palaeogene magmatism in the Faroe–Shetland Basin: influences on uplift history and sedimentation. 545–558 in *Petroleum geology of northwest Europe, proceedings of the 5th conference*. FLEET, A J, and BOLDY, S A R (editors). (London: The Geological Society.)

NELSON, P H H, and LAMY, J M. 1987. The More/West Shetland area: a review. 775–784 in *Petroleum geology of northwest Europe, proceedings of the 3rd conference*. BROOKS, J, and GLENNIE, K W (editors). (London: Graham and Trotman.)

NICHOLS, G J. 2005. Sedimentary evolution of the Lower Clair Group, Devonian, west of Shetland: climate and sediment supply controls on fluvial, aeolian and lacustrine deposition. 957–967 in *Petroleum geology: northwest Europe and global perspectives, proceedings of the 6th conference*. DORÉ, A G, and VINING, B (editors). (London: The Geological Society.)

NIELSEN, P H, WAAGSTEIN, R, RASMUSSEN, J, and LARSEN, B. 1979. Marine seismic investigation of the shelf around the Faroe Islands. *Fróðskaparrit*, Vol. 27, 102–112.

NIELSEN, T, and KUIJPERS, A. 2004. Geohazard studies offshore the Faroe islands: slope instability, bottom currents and sub-seabed sediment mobilisation. *Bulletin of the Geological Survey of Denmark and Greenland*, Vol. 4, 57–6.

NIELSEN, T, and VAN WEERING, T C E. 1998. Seismic stratigraphy and sedimentary processes at the Norwegian Sea margin northeast of the Faroe Islands. *Marine Geology*, Vol. 152, 141–157.

NIELSEN, T, DE SANTIS, L, DAHLGREN, K I T, KUIJPERS, A, LABERG, J S, NYGÅRD, A, PRAEG, D, and STOKER, M S. 2005. A comparison of the NW European glaciated margin with other glaciated margins. *Marine and Petroleum Geology*, Vol.22. 1149–1183.

NIELSEN, T, RASMUSSEN, T L, CERAMICOLA, S, and KUIJPERS, A. 2007. Quaternary sedimentation, margin architecture and ocean circulation variability around the Faroe Islands, North Atlantic. *Quaternary Science Reviews*, Vol. 26, 1016–1036.

NILSEN, T H. 1978a. Lower Tertiary laterite on the Iceland–Faroe Ridge and the Thulean land bridge. *Nature*, Vol. 274, 786–788.

NILSEN, T H. 1978b. Sedimentation in the North-east Atlantic Ocean and Norwegian Sea. 433–454 *in* Crustal evolution in northwestern Britain and adjacent regions. BOWES, D R, and LEAKE, B E (editors). *Geological Society of London Special Issue*, No. 10.

NILSEN, T H, and KERR, D R. 1978. Turbidites, red beds, sedimentary structures and trace fossils observed in DSDP leg 38 cores and the sedimentary history of the Norwegian–Greenland Sea. 259–288 *in Supplement to Volumes XXXVIII, XXXIX, XL and XLI of the Deep Sea Drilling Project*. US Government Printing Office, Washington.

NISBET, H C, and BOWES, D R. 1961. The geology of north Rona. *Transactions of the Geological Society of Glasgow*, Vol. 24, 169–189.

NOE-NYGAARD, A. 1968. On extrusion forms in plateau basalts. Shield volcanoes of 'scutulum' type. *Vísindafélag Íslendinga*, Anniversary Volume, 10–13.

NØRREVANG, A, BRATTEGARD, T, JOSEFSON, A B, SNELI, J-A, and TENDAL, O S. 1994. List of BIOFAR stations. *Sarsia*, Vol. 79, 165–180.

NORRIS, R D, and RÖHL, U. 1999. Carbon cycling and chronology of climate warming during the Palaeocene/Eocene transition. *Nature*, Vol. 401, 775–778.

NORTON, G M, MCCLAY, K R, and WAY, N A. 1987. Tectonic evolution of Devonian basins in northern Scotland and southern Norway. *Norsk Geologisk Tidsskrift*, Vol. 67, 323–338.

NUNNS, A G. 1983. Plate tectonic evolution of the Greenland–Scotland Ridge and surrounding regions. 91–108 in *Structure and development of the Greenland–Scotland Ridge: new methods and concepts*. BOTT, M H P , SAXOV, S, TALWANI, M, and THIEDE, J (editors). (New York: Plenum Press.)

NYGÅRD, A, SEJRUP, H P, HAFLIDASON, H, and BRYN, P. 2005. The glacial North Sea Fan, southern Norwegian Margin: architecture and evolution from the upper continental slope to the deep-sea basin. *Marine and Petroleum Geology*, Vol. 22, 71–84.

OAKMAN, C D, and PARTINGTON, M A. 1998. Cretaceous. 294–349 *in Petroleum geology of the North Sea, basic concepts and recent advances*. Fourth edition. GLENNIE, K W (editor). (Oxford: Blackwell Scientific Publication.)

O'CONNOR, J M, STOFFERS, P, WIJBRANS, J R, SHANNON, P M, and MORRISSEY, T. 2000. Evidence from episodic seamount volcanism for pulsing of the Iceland plume in the past 70 Myr. *Nature*, Vol. 408, 954–958.

ODIN, G S, and MORTON, A C. 1988. Authigenic green particles from marine environments. 213–264 *in* Diagenesis II. CHILINGARIAN, G V, and WOLF, K H. (editors). *Developments in Sedimentology*, Vol. 43.

OLAFSSON, I, SUNDVOR, E, ELDHOLM, O, and GRUE, K. 1992. Møre margin: crustal structure from analysis of expanded spread profiles. *Marine Geophysical Researches*, Vol. 14, 137–162.

OLIVER, G J H. 2002. Chronology and terrane assemblies, new and old controversies. 201–212 in *The Geology of Scotland*. Fourth edition. TREWIN, N H (editor). (London: The Geological Society.)

OSCHMANN, W. 1988. Kimmeridge Clay sedimentation — a new cyclic model. *Palaeogeography, Palaeoclimatology, Palaeoecology*, Vol. 65, 217–251.

OSCHMANN, W. 1991. Distribution, dynamics and palaeoecology of Kimmeridgian (Upper Jurassic) shelf anoxia in western Europe. 381–395 *in* Modern and ancient continental shelf anoxia. TYSON, R V, and PEARSON, T H (editors). *Geological Society of London Special Publication*, No. 58.

PANKHURST, R J. 1982. Geochronological tables for British igneous rocks. 575–581 in *Igneous rocks of the British Isles*. SUTHERLAND, D S (editor). (Chichester: John Wiley and Sons.)

PARK, R G. 1995. Palaeoproterozoic Laurentia–Baltica relationships: a view from the Lewisian. 211–224 *in* Early Precambrian Processes. COWARD, M P, and REIS, A C (editors). *Geological Society of London Special Publication*, No. 95.

PARK, R G. 2005. The Lewisian terrane model: a review. *Scottish Journal of Geology*, Vol. 41, 105–118.

PARK, R G, and TARNEY, J. 1987. The Lewisian complex: a typical Precambrian high-grade terrain? 13–25 *in* Evolution of the Lewisian and comparable Precambrian high-grade terrains. PARK, R G, and TARNEY, J (editors). *Geological Society of London Special Publication*, No. 27.

PARK, R G, CLIFF, R A, FETTES, D G, and STEWART, A D. 1994. Precambrian rocks in north-west Scotland west of the Moine Thrust: the Lewisian Complex and the Torridonian. 6–22 *in* A revised correlation of Precambrian rocks in the British Isles. GIBBONS, F C, and HARRIS, A L (editors). *Geological Society of London Special Report*, No. 22.

PARK, R G, STEWART, A D, and WRIGHT, D T. 2002. The Hebridean terrane. 45–80 in *The Geology of Scotland*. Fourth edition. TREWIN, N H (editor). (London: The Geological Society.)

PARNELL, J, CAREY, P W, GREEN, P, and DUNCAN, W. 1999. Hydrocarbon migration history, West of Shetland: integrated fluid inclusion and fission track studies. 613–625 in *Petroleum geology of northwest Europe, proceedings of the 5th conference*. FLEET, A J, and BOLDY, S A R (editors). (London: The Geological Society.)

PARSON, L M, and THE ODP LEG 104 SCIENTIFIC PARTY. 1988. Dipping reflector styles in the NE Atlantic Ocean. 57–68 *in* Early Tertiary volcanism and the opening of the NE Atlantic. MORTON, A C, and PARSON, L M (editors). *Geological Society of London Special Publication*, No. 39.

PASSEY, S R. 2004. The volcanic and sedimentary evolution of the Faroe Plateau Lava Group, Faroe Islands and the Faroe–Shetland Basin, NE Atlantic. Unpublished PhD thesis, University of Glasgow.

PASSEY, S R, and BELL, B R. 2007. Morphologies and emplacement mechanisms of the lava flows of the Faroe Islands Basalt Group, Faroe Islands, NE Atlantic Ocean. *Bulletin of Volcanology*, Vol. 70, 139–156.

PASSEY, S R, and JOLLEY, D W. 2009. A revised lithostratigraphic nomenclature for the Palaeogene Faroe Islands Basalt Group, NE Atlantic Ocean. *Earth and Environmental Science Transactions of the Royal Society of Edinburgh*, Vol. 99, 127–158.

PAUL, M A, TALBOT, L A, and STOKER, M S. 1993. Geotechnical properties of sediments from the continental slope northwest of the British Isles. 77–106 in *Offshore site investigation and foundation behaviour*. ARDUS, D A, CLARE, D, HILL, A, HOBBS, R, JARDINE, R J, and SQUIRE, J M (editors). (Netherlands: Kluwer Academic Publishers.)

PEARCE, J A, and CANN, J R. 1973. Tectonic setting of basic volcanic rocks determined using trace element analysis. *Earth and Planetary Science Letters*, Vol. 19, 290–300.

PEARCE, T H, GORMAN, B E, and BIRKETT, T C. 1975. The TiO_2-K_2O-P_2O_5 diagram: a method of discriminating between oceanic and non-oceanic basalts. *Earth and Planetary Science Letters*, Vol. 24, 419–426.

PETERS, K E, MOLDOWAN, J M, DRISCOLE, A R, and DERMAISON, G J. 1989. Origin of Beatrice oil by co-sourcing from Devonian and Middle Jurassic source rocks, Inner Moray Firth, UK. *American Association of Petroleum Geologists Bulletin*, Vol. 73, 454–471.

PHARAOH, T C. 1999. Palaeozoic terranes and their lithospheric boundaries within the Trans-European Suture Zone (TESZ); a review. 17–41 *in* Geophysical investigations of the Trans-European Suture zone. THYBO, H, PHARAOH, T, and GUTERCH, A (editors). *Special Issue of Tectonophysics*, Vol. 314.

PLANKE, S, SYMONDS, P A, ALVESTAD, E, and SKOGSEID, J. 2000. Seismic volcanostratigraphy of large-volume basaltic extrusive complexes on rifted margins. *Journal of Geophysical Research*, Vol. 105 (B8), 19335–19351.

PLANKE, S, RASMUSSEN, T, REY, S S, and MYKLEBUST, R. 2005. Seismic characteristics and distribution of volcanic intrusions and hydrothermal vent complexes in the Vøring and Møre basins. 833–844 in *Petroleum geology: northwest Europe and global perspectives, proceedings of the 6th conference*. DORÉ, A G, and VINING, B (editors). (London: The Geological Society.)

PRAEG, D, STOKER, M S, SHANNON, P M, CERAMICOLA, S, HJELSTUEN, B O, LABERG, J S, and MATHIESEN, A.

2005. Episodic Cenozoic tectonism and the development of the NW European 'passive' continental margin. *Marine and Petroleum Geology,* Vol. 22, 1007–1030.

PRICE, C, and MORGAN, J. 2000. Lithospheric structure north of Scotland – II. Poisson's ratios and waveform modelling. *Geophysical Journal International,* Vol. 142, 737–754.

PRICE, I, and RATTEY, R P. 1984. Cretaceous tectonics off mid Norway: implications for the Rockall and Faroe–Shetland toughs. *Journal of the Geological Society of London,* Vol. 141, 985–992.

PRICE, L C, and WEGNER, L M. 1992. The influence of pressure on petroleum generation and maturation as suggested by aqueous pyrolysis. *Organic Geochemistry,* Vol. 19, 141–159.

PRINGLE, I R. 1970. The structural geology of the North Roe area, Shetland. *Geological Journal,* Vol. 7, 147–170.

RAINBIRD, R H, HAMILTON, M A, and YOUNG, G M. 2001. Detrital zircon geochronology and provenance of the Torridonian, NW Scotland. *Journal of the Geological Society of London,* Vol. 158, 15–27.

RAMANAMPISOA, L, and DISNAR, J R. 1994. Primary control of paleoproduction of organic matter preservation and accumulation in the Kimmeridge rocks of Yorkshire (UK). *Organic Geochemistry,* Vol. 21, 1153–1167.

RASMUSSEN, J. 1982. Færøernes geologi. 7–28 in *Færøernes nature, Danmarks nature,* Part 12. NØRREVANG, A, and LUNDØ, J (editors). (Copenhagen: Politikens Forlag.)

RASMUSSEN, J, and NOE-NYGAARD, A. 1970. *Geology of the Faroe Islands (Pre-Quaternary).* Translated by HENDERSON, G. from original 1969 Danish text. Geological Survey of Denmark, Copenhagen, 1/25.

RASMUSSEN, J, NOE-NYGAARD, A, and TROELSEN, J-C. 1956. Faröe Islands (Føroyar, Færøerne, îles Faeröe). 13–17 *in* Congrés Géologique International & Commission De Stratigraphie (editors). *Lexique Stratigraphique International: Europe.* (Paris: Centre National De La Recherche Scientifique.)

RASMUSSEN, T L, THOMSEN, E, VAN WEERING, T C E, and LABEYRIE, L. 1996. Rapid changes in surface and deep water conditions at the Faroe margin during the last 58 000 years. *Paleoceanography,* Vol. 11, 757–771.

RASMUSSEN, T L, BÄCKSTRÖM, D, HEINEMEIER, J, KLITGAARD-KRISTENSEN, D, KNUTZ, P K, KUIJPERS, A, LASSEN, S, THOMSEN, E, TROELSTRA, S R, and VAN WEERING, T C E. 2002. The Faroe–Shetland Gateway: Late Quaternary water mass exchange between the Nordic seas and the north-eastern Atlantic. *Marine Geology,* Vol. 188, 165–192.

RATTEY, R P, and HAYWARD, A W. 1993. Sequence stratigraphy of a failed rift system; the Middle Jurassic to Early Cretaceous basin evolution of the Central and Northern North Sea. 215–249 in *Petroleum geology of northwest Europe, proceedings of the 4th conference.* PARKER, J R (editor). (London: The Geological Society.)

RAUM, T, MJELDE, R, BERGE, A M, PAULSEN, J T, DIGRANES, P, SHIMAMURA, H, SHIOBARA, H, KODAIRA, S, LARSEN, V B, FREDSTED, R, HARRISON, D J, and JOHNSON, M. 2005. Sub-basalt structures east of the Faroe Islands revealed from wide-angle seismic and gravity data. *Petroleum Geoscience,* Vol. 11, 291–308.

RAWSON, P F, and RILEY, L A. 1982. Latest Jurassic–Early Cretaceous events and the 'Late Cimmerian Unconformity' in the North Sea. *American Association of Petroleum Geologists Bulletin,* Vol. 66, 2628–2648.

READ, W A. 1988. Controls on Silesian sedimentation in the Midland Valley of Scotland. 222–241 in *Sedimentation in a synorogenic basin complex: The Upper Carboniferous of Northwest Europe.* BESLY, B M, and KELLING, G (editors). (Glasgow: Blackie.)

READ, W A. 1990. The interplay of sedimentation, volcanicity and tectonics in the Passage Group (Arnsbergian, E2 to Westphalian A) in the Midland Valley of Scotland. 143–152 in *The role of tectonics in Devonian and Carboniferous sedimentation in the British Isles.* ARTHURTON, R S, GUTTERIDGE, P, and NOLAN, S C (editors). (London: Yorkshire Geological Society.)

READ, W A, BROWNE, M A E, STEPHENSON, D, and UPTON, B G J. 2002. Carboniferous. 251–299 in *The Geology of Scotland.* Fourth edition. TREWIN, N H (editor). (London: The Geological Society.)

REDFERN, R. 2000. *Origins: the Evolution of Continents, Ocean and Life.* (London: Cassell and Co.)

RICHARDS, P C, LOTT, G K, JOHNSON, H, KNOX, R W O'B, and RIDING, J B. 1993. Jurassic of the Central and Northern North Sea. *In Lithostratigraphic Nomenclature of the UK North Sea.* KNOX, R W O'B and CORDEY, W G (editors). (Keyworth, Nottingham: British Geological Survey.)

RICHARDSON, K R, SMALLWOOD, J R, WHITE, R S, SNYDER, D B, and MAGUIRE, P K H. 1998. Crustal structure beneath the Faroe Islands and the Faroe–Iceland Ridge. *Tectonophysics,* Vol. 300, 159–180.

RICHARDSON, K R, WHITE, R S, ENGLAND, R W, and FRUEHN, J. 1999. Crustal structure east of the Faroe Islands: mapping sub-basalt sediments using wide-angle seismic data. *Petroleum Geoscience,* Vol. 5, 161–172.

RIDD, M F. 1981. Petroleum Geology West of Shetlands. 414–425 in *Petroleum geology of the Continental Shelf of Northwest Europe.* ILLING, L V, and HOBSON, G D (editors). (Heydon, London: Institute of Petroleum.)

RIDD, M F. 1983. Aspects of the Tertiary geology of the Faroe–Shetland Channel. 91–108 in *Structure and development of the Greenland–Scotland Ridge: new methods and concept.* BOTT, M H P, SAXOV, S, TALWANI, M, and THIEDE, J (editors). (New York: Plenum Press.)

RIDING, J B, and FENSOME, R A. 2002. A review of *Scriniodinium* Klement 1957, *Endoscrinium* (Klement 1960) Vozzhennikova 1967 and related dinoflagellate cyst taxa. *Palynology,* Vol. 26, 5–33.

RIPPON, J, READ, W A, and PARK, R G. 1996. The Ochil Fault and the Kincardine basin: key structures in the tectonic evolution of the Midland Valley of Scotland. *Journal of the Geological Society of London,* Vol. 153, 573–587.

RISE, L, OTTESEN, D, BERG, K, and LUNDIN, E. 2005. Large-scale development of the mid Norwegian margin during the last 3 million years. *Marine and Petroleum Geology*, Vol. 22, 33–44.

Ritchie, J D, and Darbyshire, D P F. 1984. Rb-Sr dates on Precambrian rocks from in and around the West Shetland Basin. *Scottish Journal of Geology*, Vol. 20, 31–36.

Ritchie, J D, and Hitchen, K. 1993. Discussion on the location and history of the Walls Boundary Fault. *Journal of the Geological Society of London*, Vol. 150, 1001–1008.

Ritchie, J D, and Hitchen, K. 1996. Early Paleogene offshore igneous activity to the northwest of the UK and its relationship to the North Atlantic Igneous Province. 63–78 *in* Correlation of the Early Paleogene in Northwest Europe. Knox, R W O'B, Corfield, R M, and Dunay, R E (editors). *Geological Society of London Special Publication*, No. 101.

Ritchie, J D, Hitchen, K, and Mitchell, J G. 1987. The offshore continuation of the Moine Thrust north of Shetland as deduced from basement isotopic ages. *Scottish Journal of Geology*, Vol. 23, 163–173.

Ritchie, J D, Gatliff, R W, and Riding, J B. 1996. *Stratigraphic nomenclature of the UK north west margin. 1. Pre-Tertiary Lithostratigraphy.* (Keyworth, Nottingham: British Geological Survey.)

Ritchie, J D, Hitchen, K, and Edwards, J W F. 1997. The Sigmundur Complex, a ?Tertiary igneous complex in the North Rockall Trough. *Scottish Journal of Geology*, Vol. 33, 97–103.

Ritchie, J D, Gatliff, R W, and Richards, P C. 1999a. Early Tertiary magmatism in the offshore NW UK margin and surrounds. 573–584 in *Petroleum geology of northwest Europe, proceedings of the 5th conference.* Fleet, A J, and Boldy, S A R (editors). (London: The Geological Society.)

Ritchie, J D, Johnson, H, and Gatliff, R W. 1999b. Palaeogene sill-complex architecture within the NE Atlantic Margin: as revealed from seismic reflection data. *The North Atlantic Igneous Province: Magmatic Controls on Sedimentation.* April 1999. (London: The Geological Society) (abstract.)

Ritchie, J D, Johnson, H, and Kimbell, G S. 2003. The nature and age of Cenozoic contractional dating within the NE Faroe–Shetland Basin. *Marine and Petroleum Geology*, Vol. 20, 399–409.

Ritchie, J D, Johnson, H, Quinn, M F, and Gatliff, R W. 2008. Cenozoic compressional deformation within the Faroe–Shetland Basin and adjacent areas. 121–136 *in* The nature and origin of compression in passive margins. Johnson, H, Doré, Holdsworth, R E, Gatliff, R W, Lundin, E R, and Ritchie, J D (editors). *Geological Society of London Special Publication*, No. 306.

Roberts, A W, White, R S, Lunnon, Z C, Christie, P A F, Spitzer, R, and the iSIMM Team. 2005. Imaging magmatic rocks on the Faroes Margin. 755–766 in *Petroleum geology: northwest Europe and global perspectives, proceedings of the 6th conference.* Doré, A G, and Vining, B (editors). (London: The Geological Society.)

Roberts, D G. 1975. Marine geology of the Rockall Plateau and Trough. *Philosophical Transactions of the Royal Society of London*, Vol. 278, 447–509.

Roberts, D G, Bott, M H P, and Uruski, C. 1983. Structure and origin of the Wyville–Thomson Ridge. 113–158 in *Structure and development of the Greenland–Scotland Ridge: new methods and concepts.* Bott, M H P, Saxov, S, Talwani, M, and Thiede, J (editors). (New York: Plenum Press.)

Roberts, D G, Schnitker, D, Baldauf, J G, Desprairies, A, Homrighausen, R, Huddlestun, P F, Kaltenback, A J, Krumsiek, K A O, Morton, A C, Murray, J W, Westberg-Smith, J, Zimmerman, H B, Keene, J B, and Backman, J (editors). 1984. *Initial Reports of the Deep Sea Drilling Project.* Ocean Drilling Program, College Station, Texas, USA, Vol. 81.

Roberts, D G, Ginzberg, A, Nunn, K, and McQuillin, R. 1988. The structure of the Rockall Trough from seismic refraction and wide-angle reflection measurements. *Nature*, Vol. 332, 632–635.

Roberts, D G, Thomson, M, Mitchener, B, Hossack, J, Carmichael, S, and Bjornseth, H-M. 1999. Palaeozoic to Tertiary rift and basin dynamics: mid Norway to the Bay of Biscay – a new context for hydrocarbon prospectivity in the deep water frontier. 7–40 in *Petroleum geology of northwest Europe, proceedings of the 5th conference.* Fleet, A J, and Boldy, S A R (editors). (London: The Geological Society.)

Roberts, J M, Long, D, Wilson, J B, Mortensen, P B, and Gage, J D. 2003. The cold-water coral *Lophelia pertusa* (Scleractinia) and enigmatic sea-bed mounds along the north-east Atlantic margin: are they related? *Marine Pollution Bulletin*, Vol. 46, 7–20.

Robinson, A M. 2004. Stratigraphic development and controls on the architecture of Eocene depositional systems in the Faroe–Shetland Basin. Unpublished PhD thesis, University of Cardiff.

Robinson, A M, Cartwright, J, Burgess, P M, and Davies, R J. 2004. Interactions between topography and channel development from 3D seismic analysis: an example from the Tertiary of the Flett Ridge, Faroe–Shetland Basin, UK. 73–82 *in* 3D seismic technology: application to the exploration of sedimentary basins. Davies, R J, Cartwright, J A, Stewart, S A, Lappin, M, and Underhill, J R (editors). *Geological Society of London Memoir*, No. 29.

Rogers, D A, Marshall, J E A, and Astin, T R. 1989. Devonian and later movements on the Great Glen fault system, Scotland. *Journal of the Geological Society of London*, Vol. 146, 369–372.

Rønnevik, H C, Bersager, E I, Moe, A, and Navrestad, Ø. 1975. The geology of the Norwegian Continental Shelf. 71–93 in *Petroleum and the Continental Shelf of north-west Europe.* Woodland, A W (editor). (Oxford: Blackwell.)

Rooney, M A, Vuletich, A K, and Griffith, C E. 1998. Compound-specific isotope analysis as a tool for characterizing mixed oils: an example from the West of Shetland area. *Organic Geochemistry*, Vol. 29, 241–254.

Ruffell, A, and Shelton, R. 1999. The control of sedimentary facies by climate during phases of crustal extension: examples from the Triassic of onshore and offshore England and Northern Ireland. *Journal of the Geological Society of London*, Vol. 156, 779–789.

Rumph, B, Reaves, C M, Orange, V G, and Robinson, D L. 1993. Structuring and transfer zones in the Faroe Basin in a regional tectonic context. 999–1009 in *Petroleum geology of northwest Europe, proceedings of the 4th conference*. PARKER, J R (editor). (London: The Geological Society.)

RUSSELL, M J. 1968. Structural controls of base metal mineralisation in Ireland in relation to continental drift. *Transactions of the Institute of Mining and Metallurgy*, Vol. B77, 117–128.

RUSSELL, M J. 1972. North-south geofractures in Scotland and Ireland. *Scottish Journal of Geology*, Vol. 8, 75–84.

RUSSELL, M J. 1987. The late Carboniferous, northern North Atlantic Ocean: Implications for hydrocarbon exploration from Britain to the Arctic. 737–749 in *Petroleum geology of northwest Europe, proceedings of the 3rd conference*. BROOKS, J, and GLENNIE, K W (editors). (London: Graham and Trotman.)

RUSSELL, M J, and SMYTHE, D K. 1978. Evidence for an early Permian oceanic rift in the North Atlantic. 973–979 in *Petrology and geochemistry of continental rifts*. NEUMANN, E R, and RAMBERG, I B (editors). (Dordrect: Reidal.)

SABINE, P A. 1970. A bowl-like object of aragonite-limestone from Shetland waters. *Proceedings of the Geologists' Association*, Vol. 81, 539–548.

SAUNDERS, A D, FITTON, J G, KERR, A C, NORRY, M J, and KENT, R W. 1997. 45–93 in *The North Atlantic Igneous Province*. MAHONEY, J J, and COFFIN, M F (editors). American Geophysical Union Geophysical Monograph, No. 100.

SAUNDERS, P M. 1990. Cold outflow from the Faroe Bank Channel. *Journal of Physical Oceanography*, Vol. 20, 29–43.

SCHMITZ, B, and PUJALTE, V. 2003. Sea-level, humidity, and land-erosion records across the initial Eocene thermal maximum from a continental-marine transect in northern Spain. *Geology*, Vol. 31, 689–692.

SCHOENHARTING, G, and ABRAHAMSEN, N. 1984. Magnetic investigations on cores from Lopra-1 drillhole, Faroe Islands. 109–114 *in* The Deep Drilling Project 1980–1981 in the Faroe Islands. BERTHELSEN, O, NOE-NYGAARD, A, and RASMUSSEN, J (editors). *Annales Societatis Scientiarum Færoensis, Supplementum*, No. 9.

SCHRADER, H-J, and FENNER, J. 1976. 30. Norwegian Sea Cenozoic diatom biostratigraphy and taxonomy; Part 1, Norwegian Sea Cenozoic diatom biostratigraphy. 921–962 *in* TALWANI, M, UDINTSEV, G, et al. (editors). *Initial Reports of the Deep Sea Drilling Project*, Vol. 38. US Government Printing Office, Washington.

SCHRADER, H-J, BJØRKLUND, K, MANUM, S, MARTINI, E, and HINTE, J. 1976. 33. Cenozoic biostratigraphy, physical stratigraphy and palaeoceanography in the Norwegian–Greenland Sea, DSDP Leg 38: palaeontological synthesis. 1197–1211 *in* TALWANI, M, UDINTSEV, G, *et al.* (editors). *Initial Reports of the Deep Sea Drilling Project*, Vol. 38. US Government Printing Office, Washington.

SCHRØDER, N F. 1971. Magnetic anomalies around the Faroe Islands: Fróðskaparrit. *Annales Societatis Scientarum Færoensis*, Vol. 19, 20–29.

SCOTCHMAN, I C, and CARR, A D. 2005. Modelling the effects of transient overpressure on the petroleum systems of the UK North East Atlantic Margin and northern North Sea areas: implications for the deep water South Atlantic and Gulf of Mexico. 1247–1265 in *Petroleum geology: northwest Europe and global perspectives, proceedings of the 6th conference*. DORÉ, A G, and VINING, B (editors). (London: The Geological Society.)

SCOTCHMAN, I C, and DORÉ, A G. 1995. A Regional Appraisal of Source Rocks North and West of Britain. *American Association of Petroleum Geologists Bulletin*, Vol. 79, 1247 (abstract).

SCOTCHMAN, I C, and THOMAS, J R W. 1995. Maturity and hydrocarbon generation in the Slyne Trough, Northwest Ireland. 385–411 *in* The Petroleum Geology of Ireland's Offshore Basins. CROKER, P F, and SHANNON, P M (editors). *Geological Society of London Special Publication*, Vol. 93.

SCOTCHMAN, I C, GRIFFITH, C E, HOLMES, A J, and JONES, D M. 1998. The Jurassic petroleum system north and west of Britain: a geochemical oil-source correlation study. *Organic Geochemistry*, Vol. 29, 671–700.

SCRUTTON, R A, and BENTLEY, P A D. 1988. Major Cretaceous volcanic province in southern Rockall Trough. *Earth and Planetary Science Letters*, Vol. 91, 198–204.

SEIDLER, L, STEEL, R J, STEMMERIK, L, and SURLYK, F. 2004. North Atlantic marine rifting in the Early Triassic: new evidence from East Greenland. *Journal of the Geological Society of London*, Vol. 161, 583–592.

SEJRUP, H P, HJELSTUEN, B O, DAHLGREN, K I T, HAFLIDASON, H, KUIJPERS, A, NYGÅRD, A, PRAEG, D, STOKER, M S, and VORREN, T O. 2005. Pleistocene glacial history of the NW European continental margin. *Marine and Petroleum Geology*, Vol. 22, 1111–1129.

SERANNE, M. 1992. Devonian extensional tectonics versus Carboniferous inversion in the northern Orcadian Basin. *Journal of the Geological Society of London*, Vol. 149, 27–37.

SHANMUGAN, G, BLOCH, R B, MITCHELL, S M, BEAMISH, G W J, HODGKINSON, R J, DAMUTH, J E, STRAUME, T, SYVERTSEN, S E, and SHIELDS, K E. 1995. Basin-floor fans in the North Sea: sequence stratigraphic models vs. sedimentary facies. *American Association of Petroleum Geologists Bulletin*, Vol. 79, 477–512.

SHANNON, P M, JACOB, A W B, O'REILLY, B M, HAUSER, F, READMAN, P W, and MAKRIS, J. 1999. Structural setting, geological development and basin modelling in the Rockall Trough. 421–432 in *Petroleum geology of northwest Europe, proceedings of the 5th conference*. FLEET, A J, and BOLDY, S A R (editors). (London: The Geological Society.)

SHELTON, G. 1980. *Lophelia pertusa* (L.). Electrical conduction and behavior in a deep-water coral. *Journal of the Marine Biological Association*, Vol. 60, 517–528.

SHIPBOARD SCIENTIFIC PARTY. 1976. Sites 336 and 352. *Initial Reports of the Deep Sea Drilling Program*, Vol. 38, 23–49. US Government Printing Office, Washington.

SISSONS, J B. 1977. The Loch Lomond readvance in the northern mainland of Scotland. 45–59 in *Studies in the Scottish late glacial environment*. GRAY, J M, and LOWE, J J (editors). (Oxford: Pergamon Press.)

SMALLWOOD, J R. 2004. Tertiary inversion in the Faroe–Shetland Channel and the development of major erosional scarps. 187–198 *in* 3D seismic technology: application to the exploration of sedimentary basins. DAVIES, R J, CARTWRIGHT, J A, STEWART, S A, LAPPIN, M, and UNDERHILL, J R (editors). *Geological Society of London Memoir*, No. 29.

SMALLWOOD, J R. 2005a. Quantity, Distribution and Provenance of Paleocene Sediments in the Faroe–Shetland area 82–95 *in* Faroe Island Exploration Conference: Proceedings of the 1st Conference. ZISKA, H, VARMING, T, and BLOCH, D (editors). *Annales Societatis Scientarum Færoensis Supplementum*, Vol. 43.

SMALLWOOD, J R. 2005b. Lithology Prediction from Velocity Data; Paleocene Sediments in the Faroe–Shetland Area. 71–81 *in* Faroe Island Exploration Conference: Proceedings of the 1st Conference. ZISKA, H, VARMING, T, and BLOCH, D (editors). *Annales Societatis Scientarum Færoensis Supplementum*, Vol. 43.

SMALLWOOD, J R, and GILL, C E. 2002. The rise and fall of the Faroe–Shetland Basin; evidence from seismic mapping of the Balder Formation. *Journal of the Geological Society of London*, Vol. 159, 627–630.

SMALLWOOD, J R, and KIRK, W. 2005. Exploration in the Faroe–Shetland Basin: disappointments and discoveries. 977–991 in *Petroleum geology: northwest Europe and global perspectives, proceedings of the 6th conference*. DORÉ, A G, and VINING, B (editors). (London: The Geological Society.)

SMALLWOOD, J R, and MARESH, J. 2002. The properties, morphology and distribution of igneous sills: modelling, borehole data and 3D seismic from the Faroe–Shetland area. 271–306 *in* The North Atlantic Igneous Province: Stratigraphy, Tectonic, Volcanic and Magmatic Processes. JOLLEY, D W, and BELL, B R (editors). *Geological Society of London Special Publication*, No. 197.

SMALLWOOD, J R, and WHITE, R S. 2002. Ridge–plume interaction in the North Atlantic and its influence on continental breakup and sea-floor spreading. 15–37 *in* The North Atlantic Igneous Province: Stratigraphy, Tectonic, Volcanic and Magmatic Processes. JOLLEY, D W, and BELL, B R (editors). *Geological Society of London Special Publication*, No. 197.

SMALLWOOD, J R, STAPLES, R K, RICHARDSON, K R, and WHITE, R S. 1999. Crust generated above the Iceland mantle plume: from continental rift to oceanic spreading centre. *Journal of Geophysical Research*, Vol. 104 (B10), 22885–22902.

SMALLWOOD, J R, TOWNS, M J, and WHITE, R S. 2001. The structure of the Faroe–Shetland Trough from integrated deep seismic and potential field modelling. *Journal of the Geological Society of London*, Vol. 158, 409–412.

SMALLWOOD, J R, PRESCOTT, D, and KIRK, W. 2004. Alternatives in Palaeocene exploration West of Shetland: a case study. *Scottish Journal of Geology*, Vol. 119, 131–143.

SMETHURST, M A. 2000. Land–offshore tectonic links in western Norway. *Journal of the Geological Society of London*, Vol. 157, 769–781.

SMITH, K, WHITLEY, P W, KIMBELL, G S, JOHNSON, H, and KUBALA, M. 2009. Enhancing the prospectivity of the Wyville Thomson Ridge. 286–302 *in* Faroe Islands Exploration Conference: Proceedings of the 2nd Conference. VARMING, T, and ZISKA, H (editors). *Annales Societatis Scientiarum Færoensis*, Vol. 50.

SMITH, P J, and BOTT, M H P. 1975. Structure of the crust beneath the Caledonian foreland and Caledonian belt of the north Scottish shelf region. *Geophysical Journal of the Royal Astronomical Society*, Vol. 40, 187–205.

SMITH, W H F, and SANDWELL, D T. 1997. Global sea-floor topography from satellite altimetry and ship depth soundings. *Science*, Vol. 277, 1956–1962.

SMYTHE, D K. 1983. Faroe–Shetland escarpment and continental margin north of the Faroes. 109–119 in *Structure and Development of the Greenland–Scotland Ridge: new methods and concepts*. BOTT, M H P, SAXOV, S, TALWANI, M, and THIEDE, J (editors). (New York: Plenum Press.)

SMYTHE, D K. 1989. Rockall Trough–Cretaceous or Late Palaeozoic? *Scottish Journal of Geology*, Vol. 25, 5–43.

SMYTHE, D K, DOBINSON, A, MCQUILLIN, R, BREWER, J A, MATTHEWS, D H, BLUNDELL, D J, and KELK, B. 1982. Deep structure of the Scottish Caledonides revealed by MOIST reflection profile. *Nature*, Vol. 299, 338–340.

SMYTHE, D K, CHALMERS, J A, SKUCE, A G, DOBINSON, A, and MOULD, A S. 1983. Early opening of the North Atlantic – 1. Structure and origin of the Faroe–Shetland escarpment. *Geophysical Journal of the Astronomical Society*, Vol. 72, 373–398.

SMYTHE, D K, RUSSELL, M J, and SKUCE, A G. 1995. Intra-continental rifting inferred from the major late Carboniferous quartz dolerite dyke swarm of NW Europe. *Scottish Journal of Geology*, Vol. 31, 151–162.

SNYDER, D B. 1990. The Moine thrust in the BIRPS data set. *Journal of the Geological Society of London*, Vol. 147, 81–86.

SNYDER, D B, and FLACK, C A. 1990. A Caledonian age for reflectors within the mantle lithosphere north and west of Scotland. *Tectonics*, Vol. 9, 903–922.

SNYDER, D B, and HOBBS, R. 1999. *The BIRPS Atlas II. A second decade of deep seismic reflection profiling*. (London: The Geological Society.) [CD-ROM].

SNYDER, D B, ENGLAND, R W, and MCBRIDE, J H. 1997. Linkage between mantle and crustal structures and its bearing on inherited structures in north-western Scotland. *Journal of the Geological Society of London*, Vol. 154, 79–83.

SOLHEIM, A, BRYN, P, SEJRUP, H P, MIENERT, J, and BERG, K. 2005. Ormen Lange — an integrated study for the safe development of a deep-water gas field within the Storegga Slide Complex, NE Atlantic continental margin; executive summary. *Marine and Petroleum Geology*, Vol. 22, 1–9.

SOPER, N J. 1988. Timing and geometry of collision, terrane accretion and sinistral strike-slip events in the British Caledonides. 481–492 *in* The Caledonian–Appalachian Orogen. HARRIS, A L, and FETTES, D J (editors). *Geological Society of London Special Publication*, No. 38.

Soper, N J, and Woodcock, N H. 2003. The lost Lower Old Red sandstone of England and Wales: a record of post-Iapetan flexure or Early Devonian transtension? *Geological Magazine*, Vol. 140, 627–647.

Soper, N J, Webb, B C, and Woodcock, N H. 1987. Late Caledonian (Acadian) transpression in north-west England: timing, geometry and geotectonic significance. *Proceedings of the Yorkshire Geological Society*, Vol. 46, 175–187.

Soper, N J, Strachan, R A, Holdsworth, R E, Gayer, R A, and Greiling, R O. 1992. Sinistral transpression and the Silurian closure of Iapetus. *Journal of the Geological Society of London*, Vol. 149, 871–880.

Soper, N J, Strachan, R A, and Harris, A L. 1998. Tectonostratigraphy of the Moine Supergroup; a synthesis. *Journal of the Geological Society of London*, Vol. 155, 13–24.

Sørensen, A B. 2003. Cenozoic basin development and stratigraphy of the Faroes area. *Petroleum Geoscience*, Vol. 9, 189–207.

Spärck, R. 1928. Introduction. *The Zoology of the Faroes I.* 1928–1942 *in* Jensen, A S, Lundbeck, W, Mortensen, T, and Spärck, R (editors). (Copenhagen: A F Høst and Søn.)

Spencer, A M, Birkeland, Ø, Knag, G Ø, and Fredstaed, R. 1999. Petroleum systems of the Atlantic margin of northwest Europe. 231–246 in *Petroleum geology of northwest Europe, proceedings of the 5th conference.* Fleet, A J, and Boldy, S A R (editors). (London: The Geological Society.)

Spitzer, R, White, R S, and the iSIMM Team. 2005. Advances in seismic imaging through basalts: a case study from the Faroe–Shetland Basin. *Petroleum Geoscience*, Vol. 11, 147–156.

Srivastava, S P, and Tapscott, C R. 1986. Plate kinematics of the North Atlantic. 379–404 in *The Geology of North America, Vol. M. The western North Atlantic Region.* Vogt, P R, and Tucholke, B E (editors). (Geological Society of America.)

Stein, A M. 1988. Basement controls upon development in the Caledonian foreland, NW Scotland. *Basin Research*, Vol. 1, 107–119.

Stein, A M. 1992. Basin development and petroleum potential in The Minches and Sea of Hebrides Basin. 17–20 *in* Basins on the Atlantic seaboard: petroleum geology, sedimentology and basin evolution. Parnell, J (editor). *Geological Society of London Special Publication*, No. 62.

Stein, A M, and Blundell, D J. 1990. Geological inheritance and crustal dynamics of the north-west Scottish continental shelf. *Tectonophysics*, Vol. 173, 455–467.

Stephenson, M H, Leng, M J, Michie, U, and Vane, C H. 2006. Palaeolimnolgy of Palaeozoic lakes, focussing on a single lake cycle in the Middle Devonian of the Orcadian Basin, Scotland. *Earth Science Reviews*, Vol. 75, 177–197.

Stewart, A D (editor). 2002. The later Proterozoic Torridonian rocks of Scotland: their sedimentology, geochemistry and origin. *Geological Society of London Memoir*, No. 24, 209–217.

Stoker, M S. 1988. Pleistocene ice-proximal glaciomarine sediments in boreholes from the Hebrides Shelf and Wyville–Thomson Ridge, NW UK Continental Shelf. *Scottish Journal of Geology*, Vol. 24, 249–262.

Stoker, M S. 1995. The influence of glacigenic sedimentation on slope–apron development on the continental margin off Northwest Britain. 159–177 *in* The tectonics, sedimentation and paleoceanography of the North Atlantic region. Scrutton, R A, Stoker, M S, Shimmield, G B, and Tudhope, A W (editors). *Geological Society of London Special Publication*, No. 90.

Stoker, M S. 1999. *Stratigraphical nomenclature of the UK north-west margin. 3. Mid to late Cenozoic stratigraphy.* (Keyworth, Nottingham: British Geological Survey.)

Stoker, M S. 2002. Late Neogene development of the UK Atlantic margin. 313–329 *in* Exhumation of the North Atlantic margin: timing, mechanisms and implications for petroleum exploration. Doré, A G, Cartwright, J A, Stoker, M S, Turner, J P, and White, N (editors). *Geological Society of London Special Publication*, No. 196.

Stoker, M S, and Bradwell, T. 2005. The Minch palaeo-ice stream, NW sector of the British–Irish Ice Sheet. *Journal of the Geological Society of London*, 162, 425–428.

Stoker, M S, and Holmes, R. 1991. Submarine end-moraines as indicators of Pleistocene ice limits off NW Britain. *Journal of the Geological Society of London*, Vol. 148, 431–434.

Stoker, M S, Morton, A C, Evans, D, Hughes, M J, Harland, R, and Graham, D K. 1988. Early Tertiary basalts and tuffaceous sandstones from the Hebrides Shelf and Wyville–Thomson Ridge, NE Atlantic. 271–282 *in* Early Tertiary volcanism and the opening of the NE Atlantic. Morton, A C, and Parson, L M (editors). *Geological Society of London Special Publication*, No. 39.

Stoker, M S, Harland, R, and Graham, D K. 1991. Glacially influenced basin plain sedimentation in the southern Faroe–Shetland Channel, north-west United Kingdom continental margin. *Marine Geology*, Vol. 100, 185–199.

Stoker, M S, Hitchen, K, and Graham, C C. 1993. *United Kingdom Offshore Regional Report: the Geology of the Hebrides and West Shetland shelves, and adjacent deep-water areas.* (London: HMSO for the British Geological Survey.)

Stoker, M S, Leslie, A B, Scott, W D, Briden, J C, Hine, N M, Harland, R, Wilkinson, I P, Evans, D, and Ardus, D A. 1994. A record of Late Cenozoic stratigraphy, sedimentation and climate change from the Hebrides slope, NE Atlantic Ocean. *Journal of the Geological Society of London*, No. 151, 235–249.

Stoker, M S, Akhurst, M A, Howe, J A, and Stow, D A V. 1998. Sediment drifts and contourites on the continental margin off north-west Britain. *Sedimentary Geology*, Vol. 115, 33–51.

Stoker, M S, Nielsen, T, van Weering, T C E, and Kuijpers, A. 2002. Towards an understanding of the Neogene tectonostratigraphic framework of the

NE Atlantic margin between Ireland and the Faroe Islands. *Marine Geology*, Vol. 188, 233–248.

STOKER, M S, LONG, D, and BULAT, J. 2003. A record of mid Cenozoic strong deep-water erosion in the Faroe–Shetland Channel. 145–148 in *European margin sediment dynamics: side-Scan sonar and seismic images.* MIENERT, J, and WEAVER, P (editors). (Berlin: Springer.)

STOKER, M S, PRAEG, D, HJELSTUEN, B O, LABERG, J S, NIELSEN, T, and SHANNON, P M. 2005a. Neogene stratigraphy and the sedimentary and oceanographic development of the NW European Atlantic margin. *Marine and Petroleum Geology*, Vol. 22, 977–1005.

STOKER, M S, PRAEG, D, SHANNON, P M, HJELSTUEN, B O, LABERG, J S, VAN WEERING, T C E, SEJRUP, H P, and EVANS, D. 2005b. Neogene evolution of the Atlantic continental margin of NW Europe (Lofoten Islands to SW Ireland): anything but passive. 1057–1076 in *Petroleum geology: northwest Europe and global perspectives, proceedings of the 6th conference.* DORÉ, A G, and VINING, B (editors). (London: The Geological Society.)

STOKER, M S, HOULT, R J, NIELSEN, T, HJELSTUEN, B O, LABERG, J S, SHANNON, P M, PRAEG, D, MATHIESEN, A, VAN WEERING, T C E, and MCDONNELL, A. 2005c. Sedimentary and oceanographic responses to early Neogene compression on the NW European margin. *Marine and Petroleum Geology*, Vol. 22, 1031–1044.

STOKER, M S, LONG, D, BULAT, J, and DAVISON, S. 2006. Seismic geomorphology and Pleistocene ice limits off NW Britain. 160–162 in *Glacier Science and Environmental Change.* Knight, P G (editor). (Blackwell Publishing).

STRACHAN, R A. 1985. The stratigraphy and structure of the Moine rocks of the Loch Eil area. *Scottish Journal of Geology*, Vol. 21, 9–22.

STRACHAN, R A, and HOLDSWORTH, R E. 1988. Basement–cover relationships and structure within the Moine rocks of central and south-east Sutherland. *Journal of the Geological Society of London*, Vol. 145, 23–36.

STRACHAN, R A, SMITH, M, HARRIS, A L, and FETTES, D J. 2002. Northern and Grampian Highlands. 81–147 in *The Geology of Scotland.* Fourth edition. TREWIN, N H (editor). (London: The Geological Society.)

STRATAGEM PARTNERS. 2002. *The Neogene stratigraphy of the glaciated European margin from Lofoten to Porcupine.* STOKER, M S. (compiler). A product of the EC-supported STRATAGEM Project. (Great Yarmouth: Svitzer.)

STRATAGEM PARTNERS. 2003. *Neogene evolution of the glaciated European margin from Lofoten to Porcupine.* STOKER, M S. (compiler). A product of the EC-supported STRATAGEM Project. (Great Yarmouth: Svitzer.)

STRIDE, A H, BELDERSEN, R H, CURRAY, J R, and MOORE, D G. 1967. Geophysical evidence on the origin of the Faroe Bank Channel — 1. Continuous reflection profiles. *Deep-Sea Research*, Vol. 14, 1–6.

SULLIVAN, M, COOMBES, T, IMBERT, P, and AHAMDACH-DEMARS, C. 1999. Reservoir quality and petrophysical evaluation of Paleocene sandstones in the West Shetland area. 627–633 in *Petroleum geology of northwest Europe, proceedings of the 5th conference.* FLEET, A J, and BOLDY, S A R (editors). (London: The Geological Society.)

SURLYK, F. 1990. Timing, style and sedimentary evolution of Late Palaeozoic–Mesozoic extensional basins of East Greenland. 107–125 *in* Tectonic events responsible for Britain's oil and gas reserves. HARDMAN, R F P, and BROOKS, J (editors). *Geological Society of London Special Publication*, No. 55.

SURLYK, F, DONS, T, CLAUSEN, C K, and HIGHAM, J. 2003. Upper Cretaceous. 213–233 in *Millennium Atlas: petroleum geology of the central and northern North Sea.* EVANS, D, GRAHAM, C G, ARMOUR, A, and BATHHURST, P (editors). (London: The Geological Society.)

SUTHERLAND, D G. 1984. The Quaternary deposits and landforms of Scotland and the neighbouring shelves: a review. *Quaternary Science Reviews*, Vol. 3, 157–254.

SUTHERLAND, D G. 1991. The glaciation of the Shetland and Orkney Islands. 121–127 in *Glacial deposits in Great Britain and Ireland.* EHLERS, J, GIBBARD, P L, and ROSE, J (editors). (Rotterdam: A A Balkema.)

SUTTON, J, and WATSON, J. 1951. The pre-Torridonian metamorphic history of the Loch Torridon and Scourie areas in the North-west Highlands and its bearing on the chronological classification of the Lewisian. *Quarterly Journal of the Geological Society of London*, Vol. 106, 241–307.

SVABO, J C. 1959. Indberetninger fra en reise i Færøe 1781 og 1782 — Selskabet til udgivelse af Færøske kildeskrifter og studier. Copenhagen: Genoptryk 1976 af Reitzels Boghandel A/S.

SWEENEY, J J, and BURNHAM, A K. 1990. Evolution of a simple model of vitrinite reflectance based on chemical kinetics. *American Association of Petroleum Geologists Bulletin*, Vol. 74, 1559–1570.

SWEETMAN, S. 1997. The integration of seismic interpretations with gravity and magnetic 2-D modelling — An example from offshore Faroes area. *The Leading Edge*, 15–20.

SWIECICKI, T, WILCOCKSON, P, CANHAM, A, WHELAN, G, and HOMAN, H. 1995. Dating, correlation and stratigraphy of the Triassic sediments in the West of Shetlands area. 57–86 *in* Permian and Triassic rifting in northwest Europe. BOLDY, S A R (editor). *Geological Society of London Special Publication*, No. 91.

TALWANI, M, UDINTSEV, G, ET AL. 1976. 2. Sites 336 and 352. 23–116 *in* TALWANI, M, UDINTSEV, G, et al. (editors). *Initial reports of the Deep Sea Drilling Project*, Vol. 38. US Government Printing Office, Washington.

TARLING, D H. 1970. Palaeomagnetic results from the Faroe Islands. 193–208 in *Palaeogeophysics.* RUNCORN, S K (editor). (London: Academic Press.)

TARLING, D H, and GALE, N H. 1968. Isotopic dating and palaeomagnetic polarity in the Faroe Islands. *Nature*, Vol. 218, 1043–1044.

TATE, M P, DODD, C D, and GRANT, N T. 1999. The north-east Rockall Basin and its significance in the evolution of the Rockall–Faroes/East Greenland rift system. 391–406 in *Petroleum geology of northwest Europe, proceedings of the 5th conference.* FLEET, A J, and BOLDY, S A R (editors). (London: The Geological Society.)

Taylor, J, Dowdeswell, J A, Kenyon, N H, Whittington, R J, van Weering, T C E, and Mienert, J. 2000. Morphology and Late Quaternary sedimentation on the north Faroes slope and abyssal plain, North Atlantic. *Marine Geology*, Vol. 168, 1–24.

Taylor, J, Dowdeswell, J A, Kenyon, N H, Whittington, R J, and Mienert, J. 2003a. Long-range side-scan sonar imagery of the north Faroes margin. 133–137 in *European margin sediment dynamics: side-scan sonar and seismic images*. Mienert, J, and Weaver, P (editors). (Berlin: Springer.)

Taylor, J, Dowdeswell, J A, Kenyon, N H, Nielsen, T, Kuijpers, A, van Weering, T C E, and Ivanov, M. 2003b. A major channel system on the north Faroes margin from side-scan sonar records. 139–143 in *European margin sediment dynamics: side-scan sonar and seismic images*. Mienert, J, and Weaver, P (editors). (Berlin: Springer.)

Thiede, J, and Eldholm, O. 1983. Speculations about the paleodepth of the Greenland–Scotland Ridge during late Mesozoic and Cenozoic times. 445–456 in *Structure and development of the Greenland–Scotland Ridge: new methods and concepts*. Bott, M H P, Saxov, S, Talwani, M, and Thiede, J (editors). (New York: Plenum Press.)

Thiede, J, and Myhre, A M. 1996. The palaeo-oceanographic history of the North Atlantic–Arctic gateways: synthesis of the Leg 151 drilling results. 645–658 *in* Thiede, J, Myhre, A M, Firth, J V, Johnson, G L, and Ruddiman, W F. (editors). *Proceedings of the Ocean Drilling Program, Scientific Results*, Vol. 151. Ocean Drilling Program, College Station, Texas.

Thirlwall, M F. 1979. The petrochemistry of the British Old Red Sandstone province. Unpublished PhD thesis, University of Edinburgh.

Thomson, K, Underhill, J R, Green, P F, Bray, R J, and Gibson, H J. 1999. Evidence from apatite fission track analysis for the post-Devonian burial and exhumation history of the northern highlands, Scotland. *Marine and Petroleum Geology*, Vol. 16, 27–39.

Thrane, K. 2002. Relationships between Archaean and Palaeoproterozoic crystalline basement complexes in the southern part of the East Greenland Caledonides: an ion microprobe study. *Precambrian Research*, Vol. 113, 19–42.

Torsvik, T H, et al. 1996. Continental breakup and collision in the Neoproterozoic and Palaeozoic of Baltica and Laurentia. *Earth Science Reviews*, Vol. 40, 229–258.

Trewin, N H, and Rollin, K E. 2002. Geological history and structure of Scotland. 45–80 in *The Geology of Scotland*. Fourth edition. Trewin, N H (editor). (London: The Geological Society.)

Trewin, N H, and Thirlwall, M F. 2002. Old Red Sandstone. 213–249 in *The Geology of Scotland*. Fourth edition. Trewin, N H (editor). (London: The Geological Society.)

Trude, K J. 2004. Kinematic indicators for shallow level igneous intrusions from 3D seismic data: evidence of flow direction and feeder location. 209–217 *in* 3D seismic technology: application to the exploration of sedimentary basins. Davies, R J, Cartwright, J A, Stewart, S A, Lappin, M, and Underhill, J R (editors). *Geological Society of London Memoir*, No. 29.

Turnbull, M J M, Whitehouse, M J, and Moorbath, S. 1996. New isotope age determination for the Torridonian. *Journal of the Geological Society of London*, Vol. 153, 955–964.

Turner, J D, and Scrutton, R A. 1993. Subsidence patterns in western margin basins: evidence from the Faroe–Shetland Basin. 975–983 in *Petroleum geology of northwest Europe, proceedings of the 4th conference*. Parker, J R (editor). (London: The Geological Society.)

Turrell, W R, Slesser, G, Adams, R D, Payne, R, and Gillibrand, P A. 1999. Decadal variability in the composition of Faroe–Shetland Channel bottom water. *Deep-Sea Research*, Vol. 46, 1–25.

Tyson, R V. 1996. Sequence stratigraphical interpretation of organic facies in marine siliciclastic systems: general principles and application to the onshore Kimmeridge Clay Formation, UK. 75–96 *in* Sequence stratigraphy in British geology. Hesselbo, S, and Parkinson, N (editors). *Geological Society of London Special Publication*, No. 103.

UKOOA (United Kingdom Offshore Operators Association). 2000. Atlantic Frontiers Environment Network surveys 1996 and 1998 Final Report [CD-ROM]. Available from Geotek Limited, Daventry, Northants.

Underhill, J R. 1993. Discussion on the location and history of the Walls Boundary fault and Moine Thrust north and south of Shetland. *Journal of the Geological Society of London*, Vol. 150, 1003–1008.

Underhill, J R, and Partington, M A. 1993. Jurassic thermal doming and deflation in the North Sea: implications of the sequence stratigraphic evidence. 337–345 in *Petroleum geology of northwest Europe, proceedings of the 4th conference*. Parker, J R (editor). (London: The Geological Society.)

Upton, B G J, Stephenson, D, Smedley, P M, Wallis, S M, and Fitton, J G. 2004. Carboniferous and Permian magmatism in Scotland. 195–218 *in* Permo-Carboniferous magmatism and rifting in Europe. Wilson, M, Neumann, E-R, Davies, G R, Timmerman, M J, Heeremans, M, and Larsen, B T (editors). *Geological Society of London Special Publication*, Vol. 223.

Vågnes, E, Gabrielsen, R H, and Haremo, P. 1998. Late Cretaceous–Cenozoic intraplate contractional deformation at the Norwegian continental shelf: timing, magnitude and regional implications. *Tectonophysics*, Vol. 300, 29–46.

van Kaam-Peters, H M E, Schouten, S, De Leeuw, J W, and Sinnghe Damsté, J S. 1997. A molecular and carbon isotope biochemical study of biomarkers and kerogen pyrolysates of the Kimmeridge Clay Facies: palaeoenvironmental implications. *Organic Geochemistry*, Vol. 27, 399–422.

van Weering, T C E, Nielsen, T, Kenyon, N H, Akentieva, K, and Kuijpers, A H. 1998. Sediments and sedimentation at the NE Faroe continental margin: contourites and large-scale sliding. *Marine Geology*, Vol. 152, 159–176.

VANNESTE, K, HENRIET, J-P, POSEWANG, J, and THIELEN, F. 1995. Seismic stratigraphy of the Bill Bailey and Lousy Bank area: implications for subsidence history. 125–139 *in* The tectonics, sedimentation and palaeoceanography of the North Atlantic region. SCRUTTON, R A, STOKER, M S, SHIMMIELD, G B, and TUDHOPE, A W (editors). *Geological Society of London Special Publication,* No. 90.

VARMING, T. 2009. Results from drilling of the 1st round of wells in the Faroese part of the Judd Basin. 346–363 *in* Faroe Island Exploration Conference: Proceedings of the 2nd Conference. VARMING, T, and ZISKA, H (editors). *Annales Societatis Scientarum Fœroensis, Supplementum,* Vol. 50.

VARMING, T, and ZISKA, H (editors). 2009. Faroe Island Exploration Conference: Proceedings of the 2nd Conference. *Annales Societatis Scientarum Fœroensis, Supplementum,* Vol. 50.

VERHOEF, J, ROEST, W R, MACNAB, R, ARKANI-HAMED, J, and THE PROJECT TEAM. 1996. Magnetic anomalies of the Arctic and North Atlantic Oceans and adjacent Land areas. *Geological Survey of Canada, Dartmouth NS, Open File Report,* No. 3125a.

VESTRALEN, I. 1996. Sedimentology and reservoir characteristics of the Upper Jurassic, West of Shetlands. Unpublished PhD thesis, University of Aberdeen.

VESTRALEN, I, and HURST, A. 1994. Sedimentology, reservoir characteristics and exploration potential of the Rona Sandstone (Upper Jurassic), west of Shetland, UKCS. *First Break,* Vol. **12**, 11–120.

VESTRALEN, I, HARTLEY, A J, and HURST, A. 1995. The sedimentology of the Rona Sandstone (Upper Jurassic), West of Shetlands, UK. 155–176 *in* Characterization of deep marine clastic systems. HARTLEY, A J, and PROSSER, D J (editors). *Geological Society of London Special Publication,* No. 94.

VIERECK, L G, TAYLOR, P N, PARSON, L M, MORTON, A C, HERTOGEN, J , GIBSON, I L, and THE ODP LEG 104 SCIENTIFIC PARTY. 1988. Origin of the Palaeogene Vøring Plateau volcanic sequence. 69–83 *in* Early Tertiary volcanism and the opening of the NE Atlantic. MORTON, A C, and PARSON, L M (editors). *Geological Society of London Special Publication,* No. 39.

WAAGSTEIN, R. 1988. Structure, composition and age of the Faroe basalt plateau. 225–238 *in* Early Tertiary Volcanism and the Opening of the NE Atlantic. MORTON, A C, and PARSON, L M (editors). *Geological Society of London Special Publication,* No. 39.

WAAGSTEIN, R, and HEILMANN-CLAUSEN, C. 1995. Petrography and biostratigraphy of Palaeogene volcaniclastic sediments dredged from the Faroes Shelf. 179–197 *in* The tectonics, sedimentation and palaeoceanography of the North Atlantic region. SCRUTTON, R A, STOKER, M S, SHIMMIELD, G B, and TUDHOPE, A W. (eds). *Geological Society of London Special Publication,* No. 90.

WAAGSTEIN, R, and RASMUSSEN, J. 1975. Glacial erratics from the sea floor south-east of the Faroe Islands and the limit of glaciation. *Fróðskaparrit,* Vol. 23. 101–119.

WAAGSTEIN, R, GUISE, P, and REX, D. 2002. K/Ar and ^{39}Ar/^{40}Ar whole-rock dating of zeolite facies metamorphosed flood basalts: the upper Paleocene basalts of the Faroe Islands, NE Atlantic. 219–252 *in* The North Atlantic igneous province: stratigraphy, tectonic, volcanic and magmatic processes. JOLLEY, D W, and BELL, B R (editors). *Geological Society of London Special Publication,* No.197.

WADDAMS, P, and CORDINGLEY, T. 1999. The regional geology and exploration potential of the NE Rockall Basin. 379–390 in *Petroleum geology of northwest Europe, proceedings of the 5th conference.* FLEET, A J, and BOLDY, S A R (editors). (London: The Geological Society.)

WALKER, A B, BAPTIE, B, and OTTEMÖLLER, L. 2003. UK earthquake monitoring 2002/2003. *British Geological Survey Internal Report,* No. IR/03/67.

WAPLES, D W. 1980. Time and temperature in petroleum formation: application of Lopatin's method to petroleum exploration. *American Association of Petroleum Geologists Bulletin,* Vol. 64, 916–926.

WARNER, M. 1987. Seismic reflections from the Moho – the effect of isostasy. *Geophysical Journal of the Royal Astronomical Society,* Vol. 88, 425–435.

WARNER, M, MORGAN, J, BARTON, P, MORGAN, P, PRICE, C, and JONES, K. 1996. Seismic reflections from the mantle represent relict subduction zone within the continental lithosphere. *Geology,* Vol. 24, 39–42.

WATSON, J V. 1984. The ending of the Caledonian orogeny in Scotland. *Journal of the Geological Society of London,* Vol. 141, 193–214.

WATTS, L M, HOLDSWORTH, R E, SLEIGHT, J A, STRACHAN, R A, and SMITH, A F. 2007. The movement history and fault rock evolution of a reactivated crustal-scale strike-slip fault: the Walls Boundary Fault Zone, Shetland. *Journal of the Geological Society of London,* Vol. 164, 1037–1058.

WHITE, N, and LOVELL, B. 1997. Measuring the pulse of a plume with the sedimentary record. *Nature,* Vol. 387, 888–891.

WHITE, R S. 1988. A hot-spot model for early Tertiary volcanism in the N Atlantic. 3–13 *in* Early Tertiary volcanism and the opening of the NE Atlantic. MORTON, A C, and PARSON, L M (editors). *Geological Society of London Special Publication,* No. 39.

WHITE, R S, and MCKENZIE, D. 1989. Magmatism at rift zones: The generation of volcanic continental margins and flood basalts. *Journal of Geophysical Research,* Vol. 94, 7685–7729.

WHITE, R S, FRUEHN, J, RICHARDSON, K R, CULLEN, E, KIRK, W, SMALLWOOD, J R, and LATKIEWICZ, C. 1999. Faroes Large aperture Research Experiment (FLARE): imaging through basalt. 1243–1252 in *Petroleum geology of northwest Europe, proceedings of the 5th conference.* FLEET, A J, and BOLDY, S A R (editors). (London: The Geological Society.)

WHITE, R S, SMALLWOOD, J R, FLIEDNER, M M, BOSLAUGH, B, MARESH, J, and FRUEHN, J. 2003. Imaging and regional distribution of basalt flows in the Faroe–Shetland Basin. *Geophysical Prospecting,* Vol. 51, 215–231.

WHITE, R S, SPITZER, R, CHRISTIE, P A F, ROBERTS, A, LUNNON, Z, MARESH, J, and ISIMM WORKING GROUP. 2005. Seismic imaging through basalt flows on the

Faroe Shelf. 44–53 *in* Faroe Islands Exploration Conference: Proceedings of the 1st Conference. Ziska, H, Varming, T, and Bloch, D (editors). *Annales Societatis Scientarum Faroensis Supplementum*, Vol. 43.

Whitehouse, M J. 1989. Pb-isotopic evidence for U-Th-Pb behaviour in a prograde amphibolite to granulite facies transition from the Lewisian Complex of northwest Scotland: implications for Pb-Pb dating. *Geochimica et Cosmochimica Acta*, Vol. 53, 717–724.

Wignall, P B. 1990. Benthic palaeoecology of the Late Jurassic Kimmeridge Clay of England. *Special papers in Palaeontology*, Vol. 43.

Williams, D F, Thunell, R C, Tappa, E, Rio, D, and Raffi, I. 1988. Chronology of the Pleistocene oxygen isotope record; 0–1.88 my BP. *Palaeogeography, Palaeoclimatology, Palaeoecology*, Vol. 64, 221–240.

Williams, G E. 2001. Neoproterozoic (Torridonian) alluvial fan succession, north-west Scotland and its tectonic setting and provenance. *Geological Magazine*, Vol. 138, 161–184.

Wilson, C K, Long, D, and Bulat, J. 2004. The morphology, processes and setting of the Afen Slide. *Marine Geology*, Vol. 213, 149–167.

Wilson, J B. 1979. Biogenic carbonate sediments on the Scottish continental shelf and on Rockall Bank. *Marine Geology*, Vol. 33, 85–93.

Wilson, J B. 1982. Shelly faunas. 126–171 *in Offshore tidal sands*. Stride, A H (editor). (London: Chapman and Hall.)

Wilson, M, Neumann, E-R, Davies, G R, Timmerman, M J, Heeremans, M, and Larsen, B T. 2004. Permo-Carboniferous magmatism and rifting in Europe: introduction. 1–10 *in* Permo-Carboniferous magmatism and rifting in Europe. Wilson, M, Neumann, E-R, Davies, G R, Timmerman, M J, Heeremans, M, and Larsen, B T (editors). *Geological Society of London Special Publication*, No. 223.

Winchester, J A. 1988. Later Proterozoic environments and tectonic evolution in the northern Atlantic lands. 253–270 in *Later Proterozoic stratigraphy of the Northern Atlantic regions*. Winchester, J A (editor). (Glasgow: Blackie.)

Woodfin, M, Seedhouse, J, Spadini, G, and Cardamone, M. 2005. Elastic impedance inversion to aid lithology prediction in the Palaeocene of the Judd Sub-Basin. 54–69 *in* Faroe Islands Exploration Conference: Proceedings of the 1st Conference. Ziska, H, Varming, T, and Bloch, D (editors). *Annales Societatis Scientarum Færoensis Supplementum*, Vol. 43.

Woodworth-Lynas, C M T, and Guigne, J Y. 1990. Iceberg scours in the geological record: examples from glacial Lake Agassiz. 217 233 *in* Glaciomarine environments: processes and sediments. Dowdeswell, J A, and Scourse, J D (editors). *Geological Society of London Special Publication*, No. 53.

Wynn, R B, Masson, D G, and Bett, B J. 2002. Hydrodynamic significance of variable ripple morphology across deep-water barchan dunes in the Faroe–Shetland Channel. *Marine Geology*, Vol. 192, 309–319.

Zachos, J, Pagani, M, Sloan, L, Thomas, E, and Billups, K. 2001. Trends, rhythms, and aberrations in global climate 65 Ma to present. *Science*, Vol. 292, 686–693.

Zanella, E, and Coward, M P. 2003. Structural framework. 44–59 in *Millennium atlas: petroleum geology of the central and northern North Sea*. Evans, D, Graham, C G, Armour, A, and Bathhurst, P (editors). (London: The Geological Society.)

Zanella, E, Coward, M P, and McGrandle, A. 2003. Crustal structure. 35–43 in *Millennium atlas: petroleum geology of the central and northern North Sea*. Evans, D, Graham, C G, Armour, A, and Bathhurst, P (editors). (London: The Geological Society.)

Zibrowius, H. 1980. Les scleractinaires de la Mediterranee et de l'Atlantique nord-oriental. *Memoires de la Institut Oceanographique, Monaco*, No. 11.

Ziegler, P A. 1982. *Geological atlas of western and central Europe*. (Mijdrecht, Netherlands: Drunkkerji Verweij B V.)

Ziegler, P A. 1988. Evolution of the Arctic–North Atlantic and the western Tethys. *American Association of Petroleum Geologists Memoir*, No. 43.

Ziegler, P A. 1990. *Geological atlas of western and central Europe*. (The Hague: Shell Internationale Maatschappij, B V.)

Ziolkowski, A, Hanssen, P, Gatliff, R W, Jakubowicz, H, Dobson, A, Hampson, G, Xiang-Yang Li, and Enru Liu. 2003. Use of low frequencies for sub-basalt imaging. *Geophysical Prospecting*, Vol. 51, 169–182.

Ziska, H, and Andersen, C. 2005. Exploration Opportunities in the Faroe Islands. 146–162 *in* Faroe Islands Exploration Conference: Proceedings of the 1st Conference. Ziska, H, Varming, T, and Bloch, D (editors). *Annales Societatis Scientarum Faroensis Supplementum*, Vol. 43.

Ziska, H, and Morgan, R. 2005. Can paleo-magnetism be used as a mapping tool in basalt provinces? EAGE 67th Conference and Exhibition, Madrid, Spain, 13–16 June 2005 (extended abstract).

Ziska, H, Varming, T, and Bloch, D. 2005 (editors). Faroe Islands Exploration Conference: Proceedings of the 1st Conference. *Annales Societatis Scientarum Faroensis Supplementum*, Vol. 43.

Ziska, H, and Varming, T. 2008. Palaeogene evolution of the Ymir and Wyville Thomson ridges, European North Atlantic margin. 121–136 *in* The nature and origin of compression in passive margins. Johnson, H, Doré, A G, Holdsworth, R E, Gatliff, R W, Lundin, E R, and Ritchie, J D (editors). *Geological Society of London Special Publication*, No. 306.

Index